GEOMORPHOLOGY AND GLOBAL TECTONICS

GEOMORPHOLOGY AND GLOBAL TECTONICS

Edited by
Michael A. Summerfield
Department of Geography, University of Edinburgh

John Wiley & Sons
Chichester • New York • Weinheim • Brisbane • Singapore • Toronto

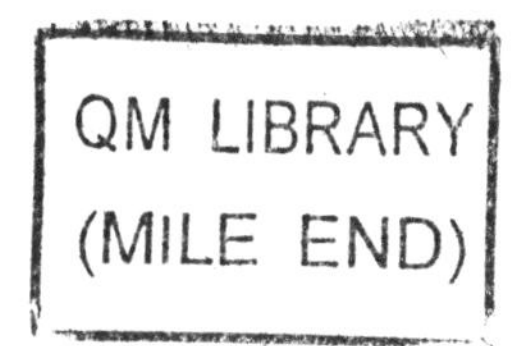

National 01243 779777
International (+44) 1243 779777
e-mail (for orders and customer service enquiries): cs-books@wiley.co.uk
Visit our Home Page on http://www.wiley.co.uk or http://www.wiley.com

Reprinted March 2000

Other Wiley Editorial Offices

John Wiley & Sons, Inc., 605 Third Avenue,
New York, NY 10158-0012, USA

WILEY-VCH Verlag GmbH, Pappelallee 3,
D-69469 Weinheim, Germany

Jacaranda Wiley Ltd, 33 Park Road, Milton,
Queensland 4064, Australia

John Wiley & Sons (Asia) Pte Ltd, 2 Clementi Loop #02-01,
Jin Xing Distripark, Singapore 129809

John Wiley & Sons (Canada) Ltd, 22 Worcester Road,
Rexdale, Ontario M9W 1L1, Canada

Library of Congress Cataloging-in-Publication Data

Geomorphology and global tectonics / edited by Michael A. Summerfield.
p. cm.
Includes bibliographical references and index.
ISBN 0-471-97193-6
1. Geomorphology. 2. Morphotectonics. I. Summerfield, M.A. (Michael A.)
GB401.5.G455 1999
551.41—dc21 98-43167
CIP

British Library Cataloguing in Publication Data

A catalogue record for this book is available from the British Library

ISBN 0-471-97193-6

Typeset in 9/11pt Times by Mayhew Typesetting, Rhayader, Powys
Printed and bound in Great Britain by Bookcraft (Bath) Ltd
This book is printed on acid-free paper responsibly manufactured from sustainable forestry, in which at least two trees are planted for each one used for paper production.

CONTENTS

LIST OF CONTRIBUTORS

Christopher Beaumont, Department of Oceanography, Dalhousie University, Halifax, Nova Scotia, B3H 4J1, Canada

Paul Bishop, Department of Geography and Topographic Science, University of Glasgow, Glasgow G12 8QQ, United Kingdom

Roderick W. Brown, Victorian Institute of Earth and Planetary Sciences, Department of Earth Sciences, University of Melbourne, Parkville, Victoria 3052, Australia

Eric J. Fielding, Mailstop 300-2333, Jet Propulsion Laboratory, 4800 Oak Grove Drive, Pasadena, California 91109, USA

L. Fleitout, Laboratoire de Géologie, Ecole Normale Supérieure, CNRS-URA, Paris, France

Kerry Gallagher, Department of Geology, Imperial College, London, Prince Consort Road, London SW7 2BP, United Kingdom

Thomas W. Gardner, Department of Geology, Trinity University, San Antonio, Texas 78212, USA

Andrew J.W. Gleadow, Victorian Institute of Earth and Planetary Sciences, Department of Earth Sciences, University of Melbourne, Parkville, Victoria 3052, Australia

Geoff Goldrick, Department of Geography and Environmental Sciences, Monash University, Clayton, Victoria 3168, Australia

Y. Gunnell, Laboratoire de Géographie Physique, Tour 54, 2e étage, 2, Place Jussieu, 75251 Paris, France

Niels Hovius, Department of Earth Sciences, University of Cambridge, Cambridge CB2 3EQ, United Kingdom

Lorcan Kennan, Department of Earth Sciences, University of Oxford, Parks Road, Oxford OX1 3PR, United Kingdom

Andrew Kerr, Department of Geography, University of Edinburgh, Drummond Street, Edinburgh EH8 9XP, United Kingdom

Henk Kooi, Department of Sedimentary Geology, Vrije Universiteit, De Boelelaan 1085, 1081 HV Amsterdam, The Netherlands

Jiun-Chuan Lin, Department of Geography, National Taiwan University, Taipei, 106 Taiwan

Larry Mayer, Department of Geology, Miami University, Oxford, Ohio 45056, USA

Hiroo Ohmori, Department of Geography, Graduate School of Science, University of Tokyo, Hong 7-3 2, Bunkyo-ku, Tokyo 113, Japan

Frank J. Pazzaglia, Department of Earth and Planetary Sciences, University of New Mexico, Northorp Hall, Albuquerque, New Mexico 87131 USA

David E. Sugden, Department of Geography, University of Edinburgh, Drummond Street, Edinburgh EH8 9XP, United Kingdom

Michael A. Summerfield, Department of Geography, University of Edinburgh, Drummond Street, Edinburgh EH8 9XP, United Kingdom

J. Mark Tippett, Telford Institute of Environmental Systems, University of Salford, Salford M5 4WT, United Kingdom

A.B. Watts, Department of Earth Sciences, University of Oxford, Parks Road, Oxford OX1 2PR, United Kingdom

Sean Willett, Department of Geological Sciences, University of Washington, Seattle, Washington 98195, USA

PREFACE

The past 10 years have witnessed a remarkable growth in interest in the relationships between global tectonics and the Earth's macroscale topographic features. This rejuvenation of research has arisen across the earth sciences, but particularly from geophysicists concerned with the mechanisms of orogenesis and continental break-up, from geologists whose interest lies in the large-scale controls of sediment supply to sedimentary basins, from geochronologists wanting to understand the factors influencing the cooling history of the shallow crust, and from those geomorphologists who have felt impelled to tackle long-standing problems of landscape evolution and rates of landscape change that had been largely ignored in the preceding decades. Many technical developments and new data sources have contributed to this change in emphasis, including the construction of coupled tectonic–surface process numerical models of large-scale landscape development, the creation of high-resolution digital elevation models based on both conventional cartographic sources and a range of remotely-sensed altimetric data, and the derivation of long-term denudational records, both from thermochronological techniques such as fission-track analysis, and from offshore sediment volume estimates based on seismic stratigraphy and borehole data.

This book attempts to present the state of current research on the interrelationships between global tectonics and macroscale landscape development across a wide range of topics and study areas. The idea for such a survey arose from the Working Group on Geomorphology and Global Tectonics that was initiated through the International Association of Geomorphologists at the Third International Geomorphology Conference held at McMaster University in Hamilton, Ontario, Canada, in 1993. It also follows on, in some respects, from the collection of papers on *Tectonics and Topography* published as special sections in the *Journal of Geophysical Research* in 1994 and based on the American Geophysical Union Chapman Conference convened by Dorothy Merritts and Mike Ellis in Snowbird, Utah, USA, in 1992.

In attempting to present an overview of innovative research within this theme, and to provide directions on potentially fruitful directions for future work, this book serves the role of the final report of the International Association of Geomorphologists Working Group. The members of the Working Group, a number of whom have contributed directly to this volume, were C. Beaumont, P. Bishop, R.W. Brown, A. Cinque, R. Dikau, D. Easterlow (Secretary), M. Fort, J.-C. Lin, D.L. Merritts, H. Ohmori, J.P. Peulvast, I. Stewart, M.A. Summerfield (Chair) and A.B. Watts. Dr Richard Dikau also chaired a sub-group on Geometric Global Relief Classification.

The 23 contributors to the book represent an international coverage across the sub-disciplines of geomorphology, geology and geophysics. They were invited to participate both because of their particular expertise and the innovative nature of their research, and in order to provide specialist insights across a wide range of morphotectonic environments around the world and from an international range of perspectives. Contributors were asked to provide an overview of their particular topic, to present key insights from their own work, and to suggest directions for future research. Inevitably, some differences of opinion are apparent between contributors, and the fact that these have been retained in the book rather than resolved reflects an editorial policy aimed at highlighting, rather than concealing, points of debate and reflecting our incomplete understanding of many issues. Differences also extend to the use of terminology, such as in the case of the term 'exhumation' which recently appears to have assumed a more general meaning among geophysicists and some geologists than its previously accepted much more specific definition. In these and other cases where the application of terminology differs somewhat between practitioners from different sub-disciplines, I have retained the usage of the individual contributors where the intended meaning is unambiguous.

The book is divided into four sections and aims to provide up-to-date surveys of key research questions, to report on important current work, and to highlight outstanding research issues. Following a brief introduction, four chapters address general issues involved in the application of digital elevation models to landscape analysis, the modelling of coupled tectonic–surface process systems, the use of thermochronological data, and the large-scale processes that contribute to the erosion of mountain belts. The third section focuses on the morphotectonics of interplate settings, with chapters on Taiwan, Japan, the Southern Alps of New Zealand, and the large complex orogens represented by the Andes, and the Himalayas and Tibetan Plateau. In the final section attention moves to intraplate settings, with discussions of oceanic islands and the passive margins of eastern Australia, eastern North America, western India, eastern South America and southwestern Africa, and the Transantarctic Mountains margin of Antarctica.

ACKNOWLEDGEMENTS

I am extremely grateful to those contributors who assisted in the refereeing of manuscripts, and I am also indebted to others who, although not contributing to the volume themselves, found the time to provide very valuable comments on individual chapters. I would like to thank Sally Wilkinson and the editorial and production staff at Wiley for their processing of the manuscript and final production of the book.

I would also like to express my appreciation to those organizations, institutions and scientific bodies who have supported my own research within the field of geomorphology and global tectonics over the past decade, specifically the Natural Environment Research Council, the Royal Society, London, the Royal Society of Edinburgh, the Carnegie Trust for the Universities of Scotland, the University of Edinburgh, the University of New South Wales, the Geological Survey of Namibia, Texaco Inc., and De Beers Mining Company Limited. I would also like to acknowledge the contribution to my research arising from collaboration with colleagues past and present in the Department of Geography at the University of Edinburgh, and fellow researchers from institutions both in the UK and elsewhere around the world.

Acknowledgements for individual contributors are to be found at the end of the relevant chapters, while acknowledgements to the source of copyright material are noted in the figure captions.

PART I

INTRODUCTION

1

Geomorphology and global tectonics: introduction

Michael A. Summerfield

1.1 Aim and Context

This volume provides an overview of current research into the relationships between geomorphology and global tectonics. More specifically, it focuses on interactions between macroscale, global tectonic phenomena and long-term, large-scale landscape development. No such survey could ever claim to be fully comprehensive, but it is intended that the broad range of issues considered here will at least enable the reader to appreciate the significance and potential of this rapidly growing field of research in the earth sciences.

The sudden awakening of interest over the past decade in the relationships between geomorphology and global tectonics among practitioners from a variety of sub-disciplines across the earth sciences is in some ways difficult to explain since it has not arisen directly from any obvious technical or conceptual advance. In many ways it might have been thought that the development of the plate tectonics model in the late 1960s and early 1970s would have had the same kind of immediate impact on geomorphology that was experienced by almost every other constituent of the earth sciences. This is especially so given the long-standing concern of geomorphology with the problem of landscape evolution, an issue that would have benefited substantially from the comprehensive global tectonic framework provided by the plate tectonics model.

It is now generally acknowledged that the primary reason why, in the event, the plate tectonics revolution had little initial impact on geomorphology was rooted in the growing disenchantment with the whole basis of the landscape evolution approach from the late 1950s, especially among British and American geomorphologists. Frustration with the lack of data on the mechanisms and rates of surface geomorphic processes, coupled with the perceived lack of rigour and testability of the 'classic' models of landscape development, precipitated a dramatic change in focus in geomorphology during this period. The new 'paradigm' was that of *surface* process geomorphology, which was almost universally, but misleadingly, represented simply as process geomorphology. This involved, in effect, a re-definition of the discipline which directed research towards short temporal and small spatial scales and very largely excluded endogenic processes and problems of long-term landscape change. This 'paradigm-shift' soon became engrained through textbooks (e.g. Ritter, 1978; Embleton and Thornes, 1979), and the exogenic perspective remains evident today both in textbook form (e.g. Ritter *et al.*, 1995) and in surveys of research in geomorphology (e.g. Rhoads and Thorn, 1996).

Given the overwhelming preponderance of geomorphologists concerned with small-scale phenomena within geomorphology over the past three decades, it is perhaps not surprising that, in general, little interest has been shown either in the conceptual advances provided by the plate tectonics model, or in the appearance of a range of new techniques and data sources relevant to the quantification of long-term denudation rates. An explanation of the present convergence of interests between geomorphologists and other earth scientists therefore lies elsewhere, and is arguably to be found in the fundamental change in emphasis that has been occurring in the earth sciences over the past decade. Tracing attempts over the past

Geomorphology and Global Tectonics. Edited by Michael A. Summerfield.

three centuries to understand the Earth, it is evident that the dominant research agenda have changed markedly during this time (Laudan, 1987; Oldroyd, 1996). From the grand-scale theories of systematizers such as Thomas Burnett and James Hutton in the 17th and 18th centuries, which had a theological understanding of the history of the Earth as their central concern, emerged the more familiarly 'scientific' geological research programme of the 19th century with its emphasis on questions of methodology (Rudwick, 1985a; Gould, 1987; Secord, 1997), the discovery of mineral resources, mapping and stratigraphic correlation, and the construction of a universal geological chronology (Rudwick, 1985b; Secord, 1986). The disciplinary infrastructure established by the beginning of the 20th century provided the foundation for the enormous subsequent growth in the application of experimental and measurement techniques directed to a more quantitative understanding of geological processes and earth history. These technical developments encouraged specialization among practitioners to the extent that significantly less emphasis was placed on broadly framed research questions about the Earth – the plate tectonics revolution notwithstanding; and from early this century such specialization contributed to the separation of the study of landforms from the geological mainstream.

By contrast, the past decade has seen a dramatic change in research priorities as the increasing 'globalization' of research agenda, driven in part by politically motivated concerns and funding priorities, has encouraged the development of a much more holistic and multidisciplinary approach to understanding the Earth. Supported by the advent of global-scale remote sensing and an increasing ability to model numerically complex phenomena at a regional or global scale, the balance has begun to shift dramatically towards integrative research linking the lithosphere – the traditional turf of the geologist – with the atmosphere, hydrosphere and biosphere. This change has been so profound that even the label 'geology', with its resonance of rock-based research, has begun to be replaced by 'earth sciences', or, even more explicitly, 'earth system science', in those institutions keen to proclaim a new, more expansive and integrative research agenda.

It is in this context of interrelating internal and surface processes that the present interest in linking geomorphology and global tectonics can best be understood. Such an interpretation is, I believe, exemplified in the following chapters, yet it is worth briefly recalling here some previous attempts to address the problem of the interaction of endogenic and exogenic processes. Such a backward glance provides both a salutory reminder that few apparently new fashions or insights are truly novel, and a historical background to the reports from the current research frontier that represent the *raison d'être* of this book. In briefly surveying these earlier contributions it is my intention to highlight some important developments, and to re-examine certain ideas that have entered the mythology of geomorphology and still entrap researchers from outside the discipline.

1.2 Early Ideas

Relationships between topography and endogenic processes provide a recurrent theme in discussions about the surface form of the Earth extending back to the 17th century and beyond. The development of these debates has recently been discussed by Oldroyd (1996) and there is no intention here to provide a comprehensive historical survey of these issues. It is relevant to note, however, that topographic features played a much more central role in understanding internal processes in these early studies given the lack of means of investigating structures below the Earth's surface. The signature of internal processes expressed as surface topography was also central to the early 19th century debate about the directionality or otherwise of earth history, and in Charles Lyell's pivotal analysis, originally presented in 1827 and developed and exemplified in his *Principles* (Lyell, 1830–1833), this was viewed in terms of a 'steady-state' model (Rudwick, 1985a). This stance has commonly been regarded as involving an empirical claim about earth history in which the Earth is held to have experienced substantial long-term cyclic shifts in the distribution of continents and oceans, in areas of uplift and subsidence, and in global climate – in short, Lyell considered that change is 'continuous but leads nowhere' (Gould, 1987, p. 123). Although this interpretation has been challenged, (with Lyell's public comments about Earth history being seen as 'almost entirely regulative' – that is, concerned with how geologists should carry out their investigations rather than providing a 'connected narrative history of the world' (Secord, 1997)), Lyell's notion of a 'balance' between areas of uplift and subsidence over the Earth's surface, whether substantive claim or procedural guideline, appears to have significantly influenced Charles Darwin during the early part of his scientific career in which he was keenly interested in the nature and causes of uplift, or, in his terms, the problem of 'continental elevation'.

Although long regarded as a brilliantly creative biologist and evolutionist, Darwin's primary interests during his formative years as a supernumerary on HMS *Beagle* (1832–1836) were in fact in geology, and his geological work continued long after the voyage had ended. During the *Beagle* voyage Darwin wrote substantially more geological than biological notes (Porter, 1985), and the only time Darwin referred to himself in print as a scientist it was as a geologist (Darwin, 1855). On returning to Britain at the completion of the voyage Darwin had two primary scientific aims – to complete his 'official' geological reports, and 'to develop a global synthesis . . . within which his extensive geological observations could be fitted' (Rhodes, 1991). It was the origin of topography and the internal mechanisms responsible for generating it which were clearly in Darwin's mind when he wrote 'I cannot avoid the conviction that some great law of nature remains to be discovered by geologists' (unpublished manuscript cited in Rhodes (1991)).

Darwin saw coral reefs as providing key insights into the behaviour of the Earth's crust, and in one of his earliest geological papers he set his developmental model of coral atoll formation in the context of Lyell's 'steady-state' conception of Earth history: "as continental elevations act over wide areas, so might we suppose continental subsidences would do, and in conformity to these views, that the Pacific and Indian seas could be divided into symmetrical areas of the two kinds; the one sinking, as deduced from the presence of encircling and barrier reefs, and lagoon islands, and the other rising, as known from uplifted shells and corals, and skirting reefs" (Darwin, 1838). The title of this paper clearly indicates that Darwin was using coral formations as a geophysical probe – that is, as a means of identifying areas of uplift and subsidence and of providing (literally) deeper insights into the internal mechanisms governing vertical motions of those parts of the crust on which they were located (see Chapter 16). In doing this Darwin "went further with Lyell than almost any other geologist of the time, when he interpreted the phenomena of oceanic islands not merely in terms of gradual depression or elevation, but also in terms of the *steady-state* movement of crustal masses." (Rudwick, 1985b, p. 189).

In identifying alternating bands of great uniformity of subsidence and elevation, Darwin also noted the correlation with volcanism, observing that "points of eruption all fall on areas of elevation". He developed this idea in a paper presented in 1838 on volcanism in South America, asserting that "no theory of the cause of volcanos which is not applicable to continental elevations can be considered as well-grounded" (Darwin, 1840). Considering what mechanism might link these different phenomena, he argued that if we "observe the continuity of the great chain of the Andes . . . we shall be deeply impressed with the grandeur of the one motive power, which, causing the elevation of the continent, has produced, as secondary effects, mountain-chains and volcanos" (Darwin, 1840). This notion required "that this large portion of the earth's crust floats . . . on a sea of molten rock". Noting evidence for surface uplift in other areas in South America, and elsewhere around the world, Darwin concluded that "we are urged to include the entire globe in the foregoing hypothesis". In an almost prescient conclusion to his paper exemplifying his characteristically measured speculation, the "furthest generalization" he would allow himself was "that the configuration of the fluid surface of the earth's nucleus is subject to some change, – its cause completely unknown, – its action slow, intermittent, but irresistible". Although Darwin was not the first to emphasize the links between continental uplift, earthquakes and volcanoes, the global scale of Darwin's concept was a distinctive feature of this seminal paper (Rhodes, 1991).

1.3 Davisian Dogma

The cycle of erosion model of Davis was a natural extension to the realm of landforms of the developmental and evolutionary ideas that had been permeating scientific thinking since the mid-19th century, and to which Darwin was such an influential contributor (although the sometimes inferred direct lineage from Darwin's evolutionary thinking to the Davisian cycle is debatable). Davis's enormous published output, coupled with the persuasive clarity of his prose and the elegance of his approach to understanding landscape change through time, more or less ensured that his model would play an important role in discussions about the interplay of tectonics and landscape development. That this remains the case is exemplified by references to Davisian concepts in modern modelling studies of landscape development, and by explicit attempts to reconcile modelled landscape sequences with Davis's ideas (Kooi and Beaumont, 1996).

The apparent simplicity and very ubiquity of the Davisian cycle of erosion has, to some extent, promoted a complacent attitude to understanding what Davis actually proposed, and the role he envisaged for his model – a potential confusion that Davis did not effectively rectify. The essence of the model is clear; the cycle of erosion provided a framework for geomorphology based on the role of structure and the qualitative description and understanding of denudational processes, but with a particular emphasis on "those aspects of landforms which are susceptible to progressive, sequential and irreversible change through time" (Chorley, 1965a).

A major source of confusion that has emerged from the Davisian cyclic model has been the relative timing and rate of crustal uplift that he envisaged. To use his own words: "The elementary presentation of the ideal cycle usually postulates a rapid uplift of a land mass, followed by a prolonged stillstand . . . [The] uplift may be of any kind and rate, but the simplest is one of uniform amount and rapid completion; hence plains and plateaus have an early place in a systematic classification of land forms . . . In my own treatment of the problem *the postulate of rapid uplift is largely a matter of convenience* [my italics], in order to gain ready entrance to the consideration of sequential processes and of the successive stages of development, – young, mature, and old, – in terms of which it is afterwards so easy to describe typical examples of land forms. Instead of rapid uplift, gradual uplift may be postulated with equal fairness to the scheme." (Davis, 1905). Gradual uplift, however, "requires the consideration of erosion during uplift" so it is therefore "preferable to speak of rapid uplift in the first presentation of the problem, and afterwards to modify this elementary and temporary view by a nearer approach to the probable truth" (Davis, 1905). In defending this simplification Davis noted that "uplift must usually be much faster than the downwear of general sub-aërial erosion, however nearly it may be equaled by the corrasion of large rivers. The original postulate of rapid uplift therefore requires only a moderate amount of modification to bring it into accord with most of the land forms that we have to consider." (Davis, 1905).

Although clearly acknowledged by Davis to be a simplifying assumption, the notion of crustal uplift rates greatly outpacing rates of denudation in the initial stage of the cycle was gradually transformed into a dogma of landscape evolution, in spite of the rather obvious difficulties of transforming a peneplain graded to sea level to a lofty orogen without a significant and rapid denudational response. Empirical data were even produced in apparent support of the idea, with Schumm (1963), for instance, arguing that his conclusion that modern rates of orogenic uplift are "about 8 times greater than the average maximum rate of denudation" provided some support for Davis's cyclic notions.

As Chorley (1965a) noted over three decades ago, Davis revised many aspects of his ideas on the nature of landscape change in the last phase of his career – for instance, by acknowledging the problem of applying the simple version of the cycle to areas of active orogeny, and by modifying his views on the early stage of the cycle and on peneplanation. But it is the views expressed in his essays written prior to 1905, together with the writings of his most influential proselytizers, notably Douglas Wilson Johnson, that have been most important in establishing the modern understanding of the Davisian vision of landscape development (Chorley, 1965a). Moreover, in spite of Davis's proclaimed attempts to present the cycle "in a more advanced manner, as opened by upheaval and erosion acting together and completed by the continued action of erosion after upheaval ceases" (Davis, 1923), it is very largely the highly abstract and simplified version that has provided the basis for discussion as to the validity and relevance of the cycle of erosion as a basis for understanding long-term landscape development.

Although Davis argued that initial presentation of the cyclic model in its simplified form was a necessary prerequisite to more advanced applications including coeval uplift and denudation, Chorley (1965a) has convincingly argued that in excluding "the possible effects of climatic change or of progressive movements of base level from his scheme, Davis was not . . . doing so to facilitate and simplify his *explanation* but to make the cyclic scheme *possible at all*." He continues: "One can only imagine what would have remained of the cycle if Davis had permitted the possible effects of continuous movements of base level to have been superimposed upon those associated with the progressive subaerial degradation of the landmass". This is "just what Walther Penck attempted to do, and is the reason why his model is much more confused and unsatisfactory than the cycle." (Chorley, 1965a). Penck envisaged an infinite number of developmental possibilities in the landscape as a result of different combinations of rates of uplift and denudation (Penck, 1953). He drew parallels with Davis's stages of landscape development, but saw these as different landscape-forming regimes, not as stages in an inevitable developmental sequence.

1.4 Erosion Surfaces, Denudation Chronology and Tectonics

The construction of landscape histories inherent in Davis's approach increasingly assumed the dominant role in geomorphology in the first half of the 20th century, at least in Britain, the United States and France. Such denudation chronology had as its primary aim deciphering the sequential development of erosional forms in relation to base-level changes. Indeed, Davis's cycle of erosion meshed so readily with the aims of denudation chronology that the two approaches effectively merged (Chorley, 1965a). This approach was further strengthened by the implications of Eduard Suess's eustatic theory. This claimed that, outside orogenic belts, continental transgressions and regressions recorded from widely separated localities across the globe indicated a high degree of synchroneity in oscillations of sea level and hence implied the stability of the continents through geological time (Suess, 1906, pp. 537–538). This model had an immediate and profound impact on stratigraphic studies in the first three decades of the 20th century, but it also had major implications for denudation chronology for "if one accepts the view that the geometry of landscape features is fashioned under the important control of base-level . . . then the eustatic theory provided a tempting key to a chronological correlation of landforms on a world scale" (Chorley, 1963). But problems in attempting to develop a world-wide correlation of Pliocene and Quaternary eustatic sea levels were already becoming apparent by the 1930s, and this led to the gradual acknowledgement of the importance of local isostatic, epeirogenic and orogenic controls (Chorley, 1963).

The culmination of the application of Suess's eustatic theory to landscape interpretation was Baulig's *Le Plateau Central de la France* (1928). Baulig recognized evidence of erosion surfaces at more or less uniform levels to which extrapolated river profiles also seemed to be graded (Chorley, 1963). By 1935, Baulig, while maintaining the importance of synchronous eustatic movements in the late Tertiary, recognized the importance of the local influence of tectonic movements. Even while the eustatic theory was on the wane elsewhere, in Britain it provided the framework on which regional denudation chronologies were constructed. For example, Wooldridge and Linton (1955) linked their paradigmatic synthesis of the denudation chronology of southeast England with a purported eustatic sea-level fall of ~200 m in the Pliocene.

Perhaps the strongest pronouncement as to the importance of denudation chronology and its potential role in deciphering the later stages of earth history was provided in S.W. Wooldridge's mid-century evaluation of geomorphology: "It is not too much to claim that, in its contribution to earth-history, geomorphology seems within grasping distance of a great unifying generalization. Just at the point where the stratigraphic record fails or becomes incomplete, an alternative principle of inter-regional correlation is offered in the fact that old sea-levels have engraved their mark on the margins of the lands and that the same levels are recognizable inland as 'terraces' or 'platforms'. . . Here, in one of its major fields of advance, geomorphology coverges on geophysics" (Wooldridge, 1951). These claims are not, however, merely of historical interest since a number of studies have appeared in the last decade that have cited the results of these traditional regional denudation chronologies and used them in support of particular geophysical interpretations of the interplay between tectonics and landscape development. The question of the validity of the procedures adopted in the construction of these denudation chronologies is, therefore, still of relevance, even though criticisms of the methodology of denudation chronology are long-standing.

The fundamental postulates (or 'articles of faith') of denudation chronology can be summarized as: "areas of low slope are probably erosional surfaces related to appropriate baselevels", "topographic flat means still-stand; higher is older and lower is younger; uplift is generally discontinuous" (Chorley, 1965a,b). The majority of geomorphologists today would have little difficulty in agreeing with the assessment of traditional denudation chronology as relying on "highly ambiguous evidence" and being "more a product of the means of analysis rather than a physical reality" (Chorley, 1965a). Indeed, "for most regions attempts at a denudation chronology commonly end with the presentation of a cleverly integrated body of ambiguous circumstantial evidence the interpretation of which is strongly coloured by the previous findings of similar workers in other areas" (Chorley, 1965b). Borrowing a metaphor from Carl Sauer, Chorley (1965a) concluded that "many studies of denudation chronology [looked] like the products of men set out to 'bag their own decoys'". He also underlined the fact that denudation chronologies are highly dependent on stratigraphic evidence in the landscape for chronological control: "One of the major limitations on the study of denudation chronology is that it cannot exist satisfactorily on a purely morphological plane. Many of the most impressive studies of this type rely heavily on the known origin and date of associated *deposits* (commonly terrace deposits or those resting on erosion surfaces), and it is characteristic that the higher the elevation or the longer the uninterrupted period of erosional history the more difficult it is to produce an unambiguous denudation chronology for the area" (Chorley, 1965b).

The difficulties in interpreting landscape components subject to prolonged erosion in the context of a denudation chronology is particularly evident in the case of accordant summits. These were seen by Davis and his disciples as representing raised peneplains and thus were thought to provide the means by which tectonic displacements of landscapes relative to base level could be established. Initially the raised peneplain model was limited to explaining constant summit elevations across plateau-like terrain, but it was soon extended to accordant summits in Alpine topography (Beckinsale and Chorley, 1991). The fact that the raised peneplain interpretation of accordant summits persists in the literature has important implications for the present-day attempts to use morphological evidence to elucidate tectonic processes. This is because there are other explanations of accordant summits, especially where represented by sharp-crested drainage divides, which allow no direct inferences to be made about the amount of vertical tectonic displacement involved during landscape development.

These alternative explanations were put forward soon after Davis had presented his raised peneplain model, and ironically the most telling was provided by Nathaniel Southgate Shaler, one of Davis's own teachers at Harvard. Shaler wrote his paper on the spacing of rivers in Kentucky in direct reaction to Davis's interpretation of accordant summits as remnants of uplifted surfaces originally graded to sea level (Shaler, 1899). He observed how the rivers of Kentucky exhibited highly regular spacing, and noted that this phenomenon had previously been reported by Albrecht Penck for the Alps and the North American Cordillera. He went on to argue that regularly spaced streams could well explain "the origin of the coincidences in mountain crests, which is so generally held to indicate the existence of ancient baselevels of erosion which have been lifted to a height above the level of the sea and then dissected by rivers". He argued instead for "the emergence of an integrated stream network pattern, whose regularly spaced components would show a tendency to be separated by divides with accordant summit levels". The clear implication of this simple geometric argument, of valley spacing and valley-side slope together determining valley depth and therefore drainage divide elevation, was that the broadly accordant summit levels in areas such as the Appalachians and the Alps could have been formed through the normal processes of fluvial erosion and in the absence of an uplifted peneplain acting as a reference surface. Davis responded by acknowledging Shaler's interpretation as 'interesting', but only accepted it as a possible alternative to accordant summits "initiated by dissection after peneplanation" (Davis, 1901).

Albrecht Penck had previously proposed a different explanation for accordant summits in which the height of mountain ranges was limited by the very high denudation rates characteristic of high altitudes which exceeded maximum crustal uplift rates (Beckinsale and Chorley, 1991). However, in his key paper 'Die Gipfelflur der Alpen', he suggested, under the influence of his son Walther, a similar explanation to Shaler, arguing that ridges are sharpened continuously as valley deepening proceeds in Alpine terrains. It is, therefore, pointless to consider these sharpened ridges as forms inherited from an uplifted peneplain; on the contrary, they are the inevitable products of the progressive erosion of an uplifted mountain range (Penck, 1919).

1.5 Current Issues

The enormous growth in topographic, stratigraphic and geochronologic data that has occurred over the past two decades, coupled with the ability to numerically model complex landscape systems that could previously only be conceptualized qualitatively, has provided new opportunities to address long-standing questions of landscape evolution. Many of these opportunities are explored in the following chapters, but some fundamental issues and difficulties remain which will provide challenges for the future. Before briefly listing these problems it is important to emphasize that arguably the greatest impediment to understanding the relationships between tectonics and landscape development appears to have been overcome. This is the lack of interchange between geomorphologists and structural geologists, geochronologists and geophysicists that characterized much previous research. The sea-change that has occurred over the past decade or so in this respect is exemplified by Peter Koons' observation on the strategies traditionally adopted to understand the evolution of mountain belts along convergent plate boundaries: "In retrospect, it is difficult to understand how models of collision zones could be constructed without reference to the inevitable topography but we somehow managed" (Koons, 1995). The fundamental nature of this turn is further exemplified by a comment on the role of surface processes in exposing rocks from deep within mountain belts: "Savor the irony should those orogens most alluring to hard-rock geologists owe their metamorphic muscles to the drumbeat of tiny raindrops" (Hoffman and Grotzinger, 1993).

An important adjustment that has to be made when moving from an analysis of the impact of tectonics on landforms at the local scale to the role of tectonics on landscape development at the regional and continental scale is the change between relative and absolute elevation changes. When analysing the role of mountain-front faulting on alluvial fan development, for instance, we only need be concerned with relative changes in base level and the absolute elevation of the system above sea level is not relevant. But in dealing with large landscape units, such as whole mountain belts, there is a much greater problem since we have to concern ourselves with changes in elevation with reference to sea level (more strictly the geoid) if the role of tectonics in creating topography is to be addressed. Unfortunately, changes in elevation through time are extremely difficult to constrain, especially over the time spans relevant to the development of macroscale landforms such as orogens or passive margins, and in sub-aerial environments dominated by denudation where marine deposits which could provide a palaeo-sea-level datum are unlikely to survive. Methods of inferring palaeoelevation, including palaeobotanical techniques (Gregory and Chase, 1994; Forest *et al.*, 1995), the use of vesicle size distributions in basalt flows to infer atmospheric pressure at the time of eruption (Sahagian and Maus, 1994) and the variation in cosmogenic isotope production rate with altitude (Brook *et al.*, 1996), have been suggested, but as general tools they are problematic.

The difficulties inherent in inferring past changes in topography have not prevented implicit claims in the literature that thermochronologic data can be used to quantify tectonically induced surface uplift. It is now acknowledged that such data can provide direct information only on denudation, and any inferences about associated topographic changes must rest on additional assumptions or data (Summerfield and Brown, 1998). This limitation restricts the testing of numerical coupled tectonic–surface process models aiming to replicate long-term landscape development. We can obtain reasonable empirical estimates of spatial and temporal patterns of denudation and calculate the resulting isostatic response to denudational unloading, but unless the initial elevation of a landscape before the phase of denudation was initiated is known, it is not possible to quantify the amount of active tectonic uplift that has occurred. This invariably leads to assumptions being made about the initial elevation of the landsurface (e.g. Gilchrist and Summerfield, 1990).

Modelling long-term landscape development not only requires the incorporation of tectonic mechanisms and the response of the lithosphere to changes in load, but also raises the question of how surface processes are to be treated at such extended spatial and temporal scales. There certainly seems little chance of (or little point in) modelling each specific surface process, since empirical verification of individual mechanisms, such as can be achieved in present-day instrumented catchments, is impossible over the temporal and spatial scales involved. This is a well-recognized problem (Church, 1980, 1996), and the strategies adopted so far of representing denudational processes in some aggregate form seem the only way to proceed (Chapter 3). Success in integrating some process understanding into large-scale tectonic–surface process numerical models is most likely through concentrating attention on key elements in the landscape system. The propagation of base-level changes through knickpoint retreat, and the broader question of bedrock channel incision, are obvious landscape components on which to focus since they provide the most direct and pervasive link between tectonics and the response of the denudational system (Howard *et al.*, 1994; Summerfield, 1996). It is unfortunate, indeed, that we know so little about bedrock channel behaviour, and that investigating bedrock channel erosion poses such major technical and practical difficulties.

1.6 Plan of the Book

Following this introductory chapter, the remainder of the book is divided into three sections. The first of these (Part II) considers general issues concerning the relationships between global tectonics and landscape development and focuses on four main issues. Chapter 2 outlines the development of digital elevation models (DEMs) and examines their use in identifying landscape deformation as a result of tectonic processes in a number of case studies up to the scale of continental rifting. In Chapter 3 the potential for using DEMs in the numerical modelling of the interaction of tectonic and surface processes is explored both in convergent plate boundary and divergent plate boundary (passive margin) settings. Emphasis here is placed on examining the relationships between tectonics and landscape development, and the feedback effects of surface processes on fundamental tectonic mechanisms. Quantifying rates of denudation at the spatial and temporal scale of first-order tectonic processes is a key requirement for the evaluation of numerical surface process–

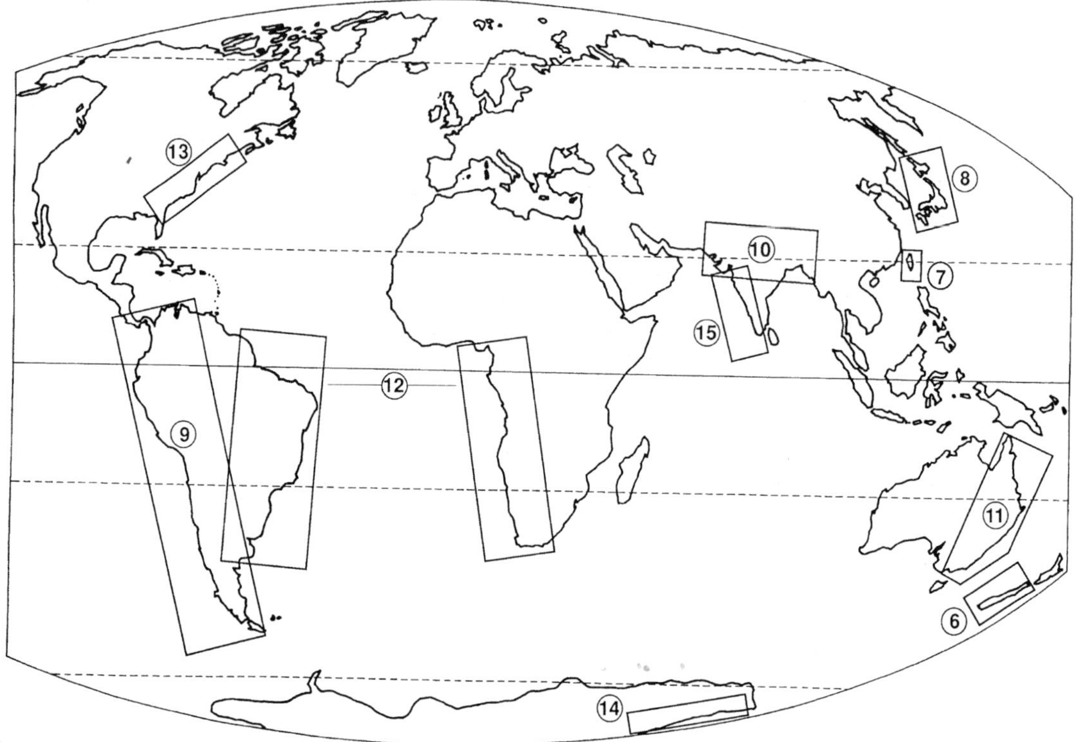

Figure 1.1 *Location of regional case studies identified by chapter number. Chapter 16 considers oceanic islands across the world's oceans*

tectonic models, and Chapter 4 explores the use of fission-track thermochronology to produce such data by recording the thermal history of the shallow crust. Some important issues concerning the interaction between tectonics and specific surface processes and landscape components, such as landsliding and longitudinal river profile development, are addressed in Chapter 5 in the context of tectonically active orogenic settings.

The remainder of the book explores current research on the relationships between large-scale tectonics and landscape development through a series of regionally based case studies. While not providing a comprehensive representation of morphotectonic environments, they do provide detailed assessments of several key areas across the globe (Figure 1.1). Part III examines interplate morphotectonic settings characterized, for the most part, by high rates of tectonic uplift and denudation, and a short landscape 'memory'. Chapter 6 considers the highly dynamic setting of the Southern Alps of New Zealand, for which a large body of data on rates of denudation and crustal displacement has been accumulated over the past decade or so with which to explore exogenic–endogenic interactions. In Chapters 7 and 8 Taiwan and Japan represent further examples of relatively simple but active orogens which provide the opportunity of understanding the relationship between topography and spatial and temporal variations in rates of crustal uplift and denudation. In Chapter 9 attention turns to the much larger and more complex Andean orogen. Reconstructing its morphological evolution and the changing interplay between tectonics and surface processes over a period of tens of millions of years poses a major challenge. This theme is continued in Chapter 10, which considers the morphotectonic development of the world's largest mountain range – the Himalaya–Tibetan Plateau orogen. The focus here is on explaining the conditions which have given rise to the remarkable elevation and extent of the Tibetan Plateau.

In the final section (Part IV) we turn to intraplate morphotectonic settings where the interplay between tectonics and landscape development have to be viewed over a period of 100 Ma or more. Chapter 11 examines the morphotectonic evolution of the East Australian margin, where the presence of Cenozoic lavas provides very useful constraints on landscape history not available in most other analogous morphotectonic settings. In Chapter 12 the tectonic and denudational history of the originally conjugate South Atlantic margins of Africa and South America are explored largely through the application of apatite fission-track thermochronology. In Chapter 13 we move from new continental margins formed through the break-up of Gondwana to the eastern margin of North America where stratigraphic data, both offshore and onshore, provide valuable constraints on the morphological evolution of one of Davis's classic field areas. In Chapter 14 the focus moves to the rifted margin of the Transantarctic Mountains, where a persistent frigid, hyper-arid environment has allowed the preservation of landscape elements dating back several million years. Chapter 15 examines the western margin of India through combining estimates of long-term denudation from fission-track thermochronology with modelling of the tectonic evolution of the margin. Finally, Chapter 16 explores the interplay of endogenic and exogenic processes affecting the evolution of the 'natural laboratories' represented by oceanic islands.

References

Beckinsale, R.P. and Chorley, R.J. 1991. *The History of the Study of Landforms*, Vol. 3, Routledge, London.

Brook, E.J., Brown, E.T., Kurz, M.D., Ackert, R.P., Raisbeck, G.M. and Yiou, F. 1996. Constraints on age, erosion, and uplift of Neogene glacial deposits in the Transantarctic Mountains determined from in situ cosmogenic ^{10}Be and ^{26}Al. *Geology*, **23**, 1063–1066.

Chorley, R.J. 1963. Diastrophic background to twentieth-century geomorphological thought. *Geol. Soc. Amer. Bull.*, **74**, 953–970.

Chorley, R.J. 1965a. A re-evaluation of the geomorphic system of W.M. Davis. In: R.J. Chorley and P. Haggett (eds) *Frontiers in Geographical Teaching*, Methuen, London, pp. 21–38.

Chorley, R.J. 1965b. The application of quantitative methods to geomorphology. In: R.J. Chorley and P. Haggett (eds) *Frontiers in Geographical Teaching*, Methuen, London, pp. 147–163.

Church, M. 1980. Records of recent geomorphological events. In: R.A. Cullingford, D.A. Davidson and J. Lewin (eds) *Timescales in Geomorphology*, Wiley, Chichester, pp. 13–29.

Church, M. 1996. Space, time and the mountain – how do we order what we see? In: B.L. Rhoads and C.E. Thorn (eds) *The Scientific Nature of Geomorphology*, Wiley, Chichester, pp. 147–170.

Darwin, C. 1838. On certain areas of elevation and subsidence in the Pacific and Indian Oceans, as deduced from the study of coral formations. *Proc. Geol. Soc. Lond.*, **2**, 552–554.

Darwin, C. 1840. On the connexion of certain volcanic phenomena in South America; and on the formation of mountain chains and volcanos, as the effect of the same power by which continents are elevated. *Trans. Geol. Soc. Lond.*, 2nd series, part 3, 5, 601–631.

Darwin, C. 1855. Does sea-water kill seeds? *Gardeners' Chron. Agric. Gaz.*, no. 21, 26 May 1855, 356–357.

Davis, W.M. 1901. Peneplains of central France and Brittany. *Bull. Geol. Soc. Amer.*, **12**, 481–483.

Davis, W.M. 1905. Complications of the geographical cycle. *Report of the Eighth International Geographical Congress, Washington, 1904*, 150–163.

Davis, W.M. 1923. The cycle of erosion and the summit levels of the Alps. *J. Geol.*, **31**, 1–41.

Embleton, C. and Thornes, J. 1979. *Process in Geomorphology*, Arnold, London.

Forest, C.E., Molnar, P. and Emanuel, K.A. 1995. Palaeoaltimetry from energy conservation principles. *Nature*, **374**, 347–350.

Gilchrist, A.R. and Summerfield, M.A. 1990. Differential denudation and flexural isostasy in formation of rifted-margin upwarps. *Nature*, **346**, 739–742.

Gould, S.J. 1987. *Time's Arrow, Time's Cycle*, Harvard University Press, Cambridge, Massachusetts.

Gregory, K.M. and Chase, C.G. 1994. Tectonic and climatic significance of a late Eocene low-relief, high-level geomorphic surface, Colorado. *J. Geophys. Res.*, **99**, 20 141–20 160.

Hoffman, P.F. and Grotzinger, J.P. 1993. Orographic precipitation, erosional unloading, and tectonic style. *Geology*, **21**, 195–198.

Howard, A.D., Dietrich, W.E. and Seidl, M.A. 1994. Modeling fluvial erosion on regional to continental scales. *J. Geophys. Res.*, **99**, 13 971–13 986.

Kooi, H. and Beaumont, C. 1996. Large-scale geomorphology: Classical concepts reconciled and integrated with contemporary ideas via a surface processes model. *J. Geophys. Res.*, **101**, 3361–3386.

Koons, P.O. 1995. Modeling the topographic evolution of collisional belts. *Ann. Rev. Earth Planet. Sci.*, **23**, 375–408.

Laudan, R. 1987. *From Mineralogy to Geology*, University of Chicago Press, Chicago.

Lyell, C. 1830–1833. *Principles of Geology*, 3 vols, Murray, London.

Oldroyd, D.R. 1996. *Thinking About the Earth*, Athlone, London.

Penck, A. 1919. Die Gipfelflur der Alpen. *Sitzungberichte der Preussischen Akademie der Wissenschaften zu Berlin*, **17**, 256–268.

Penck, W. 1953. *Morphological Analysis of Land Forms*, Macmillan, London.

Porter, D.M. 1985. The *Beagle* collector and his collections. In: D. Kohn (ed.) *The Darwinian Heritage*, Princeton University Press, Princeton, pp. 973–1019.

Rhoads, B.L. and Thorn, C.E. (eds) 1996. *The Scientific Nature of Geomorphology*, Wiley, Chichester.

Rhodes, F.H.T. 1991. Darwin's search for a theory of the earth: symmetry, simplicity and speculation. *Brit. J. Hist. Sci.*, **24**, 193–229.

Ritter, D.F. 1978. *Process Geomorphology*, Wm.C. Brown, Dubuque.

Ritter, D.F., Kochel, R.C. and Miller, J.R. 1995. *Process Geomorphology* (3rd edn), Wm.C. Brown, Dubuque.

Rudwick, M.J.S. 1985a. *The Meaning of Fossils* (2nd edn), University of Chicago Press, Chicago.

Rudwick, M.J.S. 1985b. *The Great Devonian Controversy*, University of Chicago Press, Chicago.

Sahagian, D.L. and Maus, J.E. 1994. Basalt vesicularity as a measure of atmospheric pressure and palaeoelevation. *Nature*, **372**, 449–451.

Schumm, S.A. 1963. The disparity between present rates of denudation and orogeny. *US Geol. Surv. Prof. Pap.*, **454-H**.

Secord, J.A. 1986. *Controversy in Victorian Geology*, Princeton University Press, Princeton.

Secord, J.A. 1997. Introduction in Charles Lyell, *Principles of Geology* (ed. J.A. Secord), Penguin, London.

Shaler, N.S. 1899. Spacing of rivers with reference to hypothesis of baseleveling. *Bull. Geol. Soc. Amer.*, **10**, 262–276.

Suess, E. 1906. *The Face of the Earth (Das Antlitz der Erde)*, vol. II, Clarendon Press, Oxford.

Summerfield, M.A. 1996. Understanding landscape development: the evolving interface between geomorphology and other earth sciences. *Area*, **28**, 211–220.

Summerfield, M.A. and Brown, R.W. 1998. Geomorphic factors in the interpretation of fission-track data. In: P. van den Haute and F. de Corte (eds) *Advances in Fission-Track Geochronology*, Kluwer, Dordrecht, pp. 269–284.

Wooldridge, S.W. 1951. The progress of geomorphology. In: G. Taylor (ed.) *Geography in the Twentieth Century*, Methuen, London, pp. 165–177.

Wooldridge, S.W. and Linton, D.L. 1955. *Structure, Surface and Drainage in South-East England*, Philip, London.

PART II

GENERAL MODELS AND EMPIRICAL APPROACHES

2

Application of digital elevation models to macroscale tectonic geomorphology

Larry Mayer

2.1 Introduction

Tectonics drive geodynamic processes that, over time, directly shape topography to form tectonic landscapes. The effects of tectonics on topography occur over a large range of spatial scales or wavelengths. Superimposed on tectonic landscapes are the effects of those processes related to factors such as climate, lithology and vegetation. Tectonic geomorphology, which includes the study of how tectonics shape the landscape, and at which scales, benefits from the ability to visualize topography at continental scales (see for instance, Chapter 10). Digital elevation models (DEMs) permit visualization of topography and a more detailed search into topographic data structure than is possible directly from conventional topographic contour maps. When combined with other sources of information, such as stratigraphic and tectonic mapping, topographic information derived from DEMs can provide a rapid and effective way to map active tectonic structures.

One purpose of using digital elevation data in a study of tectonic geomorphology is the straightforward identification of surface deformation. Here the obvious operative assumption is that tectonic strain has deformed the topographic surface. Surface deformation may, or may not, be associated with earthquakes. Inferences regarding whether or not topographic deformation is aseismic or coseismic are not based on the identification of strain itself.

Visualization of topographic features can be facilitated by image processing of digital topographic data. Image processing of digital elevation data may even permit a precise visualization of features from real terrain that would otherwise be difficult or impossible to see using traditional topographic contour maps. A particular strength of digital imaging is the ability to manipulate the display of topography in such a way as to emphasize one element of the landscape, such as streams, valleys, or interfluves.

Surface deformation may occur at many different spatial scales. Short wavelength features, such as fault scarps, distribute strain over a very short distance measured perpendicular to strike. Uplifts or depressions may be characterized by vertical strain distributed over very long distances. As the distance over which strain is distributed increases, the likelihood of it being directly observed decreases unless there is substantial amplitude. The scale at which observations are made from digital elevation images can, to some extent, be manipulated, thus allowing the observer to examine different topographic wavelengths independently. Finally, digital topography enables hypotheses concerning tectonic geomorphology to be formulated so that they can be tested in a reproducible fashion.

This chapter provides some background into the development of DEMs for tectonic geomorphology, briefly describes DEMs, and summarizes several applications of DEMs to tectonic problems. No attempt is made to provide a comprehensive discussion of all possible applications. The challenge to the geomorphologist using DEMs is the ability to recognize a deformed landscape despite its presence in a dynamic environment, one where geomorphic features are being generated or modified, that is overprinted by non-tectonic processes.

Geomorphology and Global Tectonics. Edited by Michael A. Summerfield.

2.2 Digital Elevation Models

A digital elevation model (DEM) is a digital representation of the Earth's topographic surface. Commonly these representations of topography are in the form of a two-dimensional array, or grid, of regularly spaced elevation points; however, irregularly spaced elevations (triangulated irregular networks or TINs) are also used. DEMs are constructed using a variety of techniques including digitizing topographic contours from existing topographic maps and converting the vector data into regularly spaced gridded data (US Geological Survey, 1978), photogrammetric methods using aerial photographs, and digital photogrammetry using stereo satellite images such as those from SPOT, or from stereo aerial photographs. Recent technologies for acquiring topographic data (Topographic Science Working Group, 1988) which are used in generating DEMs are based on remote sensing methods including laser altimetry, and synthetic aperture radar (SAR) such as obtained by the recent radar satellite RADARSAT.

DEMs are commonly displayed in a format which looks like a photograph, a satellite image, or a radar image in appearance (Plate 2.1). These images are often displayed using eight bits, or (2^8) 256 levels of grey, but pseudocolour is also used. A function converts the elevation range to one bounded by 256 levels. Most often, this function is referred to as the *look up table*. The look up table (or LUT), the mapping between elevation and grey level (or elevation and colour in the case of a *colour lookup table*, or CLUT), can be linear, or it can be any arbitrary mapping. Digital topographic data have an inherent granularity which is a result of the digitizing and gridding methods. Granularity is the coarseness of the grid spacing in the two-dimensional array used to display the elevations. The coarser the grid spacing, the larger the ground distance represented by a single elevation point. Thus, each data point in an elevation array represents an elevation for a block distance which can be displayed using a single pixel on a computer monitor. For example, in the case of the one-degree DEMs produced by the United States Defense Mapping Agency (DMA), this block distance is about 90 m, or 0.36 mm at map scale on a 1:250 000 US Geological Survey series topographic map (US Geological Survey, 1978). This can be displayed on a computer monitor as a brightness value or colour as a single screen pixel. Accuracy for those DEMs made by contouring existing topographic maps is determined by comparison of the two products. In other cases, DEM accuracy may be determined by matching control points. Digital topographic data at 30-arcsecond resolution are now available for the Earth (see appendix for Digital Chart of the World).

A distinct advantage of DEM-based images of topography over traditional topographic contour maps exists at the regional scale. DEMs contain more high-resolution information than can be portrayed using contour lines (see, for instance, Plate 10.1). The digital data can be easily manipulated for visual, statistical, or other analyses.

2.3 Topographic Analysis and DEMs

Topographic analysis, the study of topography from maps, digital data, or field observation, is an important element of tectonic geomorphology. Concepts which provide a framework for studying erosion surfaces, peneplains (Davis, 1899), accordant summits (Penck, 1919, in Pannekoek, 1967), and escarpment evolution (King, 1951), all rely on the identification of physical topographic features. Erosion surfaces, which may reflect periods of tectonic quiescence presumably separated temporally by periods of tectonic activity, required delineation on maps in order to describe their extent. Classical techniques, such as summit level maps (Pannekoek, 1967), subenvelope maps (Stearns, 1967; Hack, 1974), relief maps (Smith, 1935), available relief maps (Dury, 1951), and hypsometric analysis (Strahler, 1952; Pike and Wilson, 1971), provide tools to identify topographic features which may be related to tectonic processes. Relationships between topography and tectonics have long provided a rich area for research (Ahnert, 1970; Hack, 1973; Bull and McFadden, 1977; Keller, 1986). Analyses of DEMs are an extension of these techniques into the digital modelling domain, all designed to use topographic data to help visualize or analyse the landscape in the context of tectonic processes.

Implicit in the use of DEMs for tectonic geomorphology is the assumption that the form of the landscape contains information about tectonics. The precise nature of such information is a field of active research (cf. Ollier, 1981; Summerfield, 1985; Ellis and Merritts, 1994), and includes order reflected in spatial frequency, spatial correlation, correlation with tectonic forms, or inference. Importantly, DEMs or topographic data alone are rarely sufficient sources of information from which to derive detailed tectonic histories that include age information. For illustration, consider the Harrisburg West 1-degree DEM (Plate 2.2). This region is the classical area from which elaborate hypotheses regarding peneplains, some by

W.M. Davis, were drawn. Bascom (1921) suggested that five peneplains are represented in the topography. Best known are the Schooley peneplain which was postulated on the basis of accordant summits in the Valley and Ridge province (Plate 2.2), and the Harrisburg peneplain on the lowest surface before the Fall Line. Digital topographic data permit the effective evaluation of accordant summits and frequency of elevation categories, which may reflect erosional surfaces, but such analyses would suffer the same drawbacks as the original ones, except for the time required to delineate the topographic features.

Simple two-dimensional features within the topography that can be represented by a contour line are easily extracted by thresholding. Thresholding is used to separate an image into two objects – objects of interest and of background – on the basis of elevation, grey level, or value. All parts of the image, whose values are above or below some threshold value, are considered an object of interest where the user selects an appropriate threshold value. When thresholding is used, objects of interest are displayed in black, and background is white on the computer screen. Converting this displayed image to binary sets the values of all thresholded pixels to black and all background pixels to white. Thresholding can be used to identify, for example, elevations above some base level, or all regions lying at the same elevation. It is also useful for identifying the shapes of planforms that can be approximately expressed as a contour line.

Topography, and thus DEMs, reflect many spatial frequencies. Modifying DEMs to enhance a particular frequency is essentially an image processing technique. Consider a DEM to be an array of elevations in x rows and y columns. We can focus our attention on any spatial size within this array and modify the raw information, or complete spectrum, to enhance or suppress a particular frequency. For example, a local average is a method to smooth topography. The amount of smoothing depends on the size of the sample box over which the average is computed. For example, the local average, l, at point x, y is computed as the arithmetic mean of its neighbours.

$$l(x,y) = \frac{1}{8}\sum_{i=-1}^{1}\sum_{j=-1}^{1} f(x+i,y+j) \tag{1}$$

Implicit in this arithmetic mean is that each point is weighted equally, in this case by the coefficient one. A filter is an array of coefficients that is designed to enhance or suppress a particular aspect of the data. The arithmetic smoothing filter is simply,

$$\begin{bmatrix} 1 & 1 & 1 \\ 1 & 1 & 1 \\ 1 & 1 & 1 \end{bmatrix}$$

This smoothes the topography and thus suppresses the high-frequency information within the data. Alternatively, to enhance the high-frequency information in the topography and suppress the low-frequency, one can subtract the smoothed data from the original and the remainder represents the high-frequency information content. Commonly, directionality of the data is important, especially in enhancing linear features with specific azimuths. For example, a directional filter which enhances a linear feature of a given orientation can be designed,

$$\begin{bmatrix} 0 & -1 & -1 \\ 1 & 0 & -1 \\ 1 & 1 & 0 \end{bmatrix}$$

The early use of DEMs provided a way to visualize topography by portraying shaded views. Displaying DEMs as shaded-relief, or hill-shaded, images (Yoeli, 1967), is a common means of enhancing the high-frequency information in topographic data such as those reflecting geological structures and drainage patterns. Shaded-relief images of digital topography simulate the illumination and shadowing found under low sun angle conditions, but with the position of the illumination under the user's control. Batson *et al.* (1975) made the first synoptic shaded-relief images for parts of the western United States (at 1:500 000 scale), and Arvidson *et al.* (1982) published the first image of the conterminous United States. These were followed by small-scale, shaded-relief maps of Australia (Moore and Simpson, 1982) and South Africa (Lamb *et al.*, 1987), the first shaded-relief map of the Earth (Heirtzler, 1985), and a map of Sweden (Elvhage and Lidmar-Bergström, 1987). Among the latest synoptic images are those for the southwestern United States at 1:1 000 000 and 1:2 000 000 scale (Edwards and Batson, 1990).

Combined with GIS techniques, the use of simple box filters and box functions permits a wide range of analytical goals to be achieved. For example, simple methods can be used to estimate slope, relief, variance (or smoothness), aspect, hypsometric integral (cf. Lifton and Chase, 1992, and below), potential surface flow, and position in the landscape. Each one of these factors can also be combined with another derivative factor, slope and aspect for example, in order to determine patterns, delineate relationships, or classify landforms. The availability of digital topographic data, combined with rapidly evolving computational environments, has spurred the use of DEMs as a tool for neotectonic investigation.

2.4 Use of DEMs for Tectonic Studies

The early use of DEMs for neotectonic investigations included specific site studies. For example, Schowengerdt and Glass (1983) were interested in evaluating earthquake hazards associated with active faults in south-central Washington state, USA, just behind the Cascade volcanic arc. They used hill shading on DMA 1-degree DEMs to highlight faults reflected in the topography. Oronati *et al.* (1992) compared structural features obtained from the analysis of hill-shaded 230 m DEMs of Italy with those on geological maps and determined that even these coarse DEMs provided very valuable information on tectonic elements. Chorowicz *et al.* (1994) used 250 m resolution DEMs to map the structural elements at the Maras triple junction in Turkey. They were able, through hill shading of the DEM, to delineate the major structures accommodating the differential convergence of Africa and Arabia relative to Eurasia. Their analysis (Chorowicz *et al.*, 1994) suggests that the Dead Sea fault, which takes up the left-lateral motion between Africa and Arabia, may have been initiated in the north during the collision of Anatolia and the north-western corner of the continental crust of Africa/Arabia. Examining the DEM for the region (Plate 2.3), one can threshold the data to show the patterns of continental-scale highs and lows. The East African rift system, the Gulf of Aden, and the plate-scale uplift and tilting of the Arabian plate are highlighted. Even the simple display of the DEM provides a way of visualizing the scale of the processes operating on the landscape and should convince geomorphologists interested in understanding landscapes at the plate scale of the need to incorporate geodynamic and tectonic perspectives. Plate 2.4 illustrates the topographic segmentation of the East African rift system which reflects structural segmentation (Bosworth, 1986). The topography influences the drainage directions and thus directly modulates the sediment flux characteristics of the adjacent basins (Summerfield, 1985), but it also reflects the major influence of flexural uplift and isostasy (Weissel and Karner, 1989; Gilchrist and Summerfield, 1991; Wdowinski and Axen, 1992).

In addition to identification of tectonic structures by hill shading, or other enhancement techniques, workers have also examined broader geomorphic features. Deffontaines *et al.* (1994) employed several methods in studying the active tectonics of the Taiwan orogen. In addition to hill shading to determine the important structural elements of Taiwan's active fault systems, they computed summit-level and base-level maps which place envelopes on the upper and lower parts of the landscape. Using these envelope maps, they were able to estimate erosional volumes (cf. below). The effects of tectonic deformation on drainage may also be addressed using DEMs. Deffontaines and Chorowicz (1991) point out the usefulness of automated drainage extraction from DEMs and the role drainage anomalies play in the interpretation of the Zaire Basin and reactivation of older geological structures there. Similarly, Colleau and Lenôtre (1991) examined the relationship among planar surfaces, valley and river slopes, drainage patterns, and structure to determine whether a correlation existed between geomorphic and levelling data in the Bonnevaux–Chambaran area, France.

Lowry and Smith (1994) used tandem digital topographic data and Bouguer gravity anomalies to estimate the flexural rigidity in the Basin and Range, Colorado Plateau and Rocky Mountain topographic provinces. Consistent with earlier studies, they found the Basin and Range showed the lowest flexural rigidity, with a mean of 9×10^{21} Nm (for an elastic lithosphere = 10 km), while the Rocky Mountains showed the highest flexural rigidity of 3×10^{23} Nm (for an elastic lithosphere = 33 km).

To systematize the search for tectonic imprints on topography, Chase (1992) suggested the use of variograms which provide estimates of the spatial autocorrelation in topographic data. By using variograms, Chase opined that it might be possible to distinguish climatic from tectonic landscapes. Lifton and Chase (1992) evaluated such possible tectonic signatures on the topography of the San Gabriel Mountains through a combination of erosional modelling, morphometry (using the hypsometric integral) and fractal analysis. They estimated the fractal dimension by calculating the slope of the variogram which was based on 15-minute DEMs for the region.

Order (the opposite of randomness) in topography may also consist of symmetry. Gephart (1994) used a DEM from Isacks (1988) to investigate the relationship between subduction and non-collisional mountain building. He examined the relationship between symmetry elements of the data, defined as the difference between the elevations on either side of a symmetry plane. By minimizing this difference he was able to find an optimum model and compare that symmetry plane to the subduction pole of rotation. Further, he compared each symmetry model by calculating its resultant hypsometry.

Modelling of landscapes plays an increasingly important role in the use of DEMs for tectonic geomorphology. Modelling may provide a method of testing geomorphic hypotheses; for example, what effect would faulting of a specified magnitude have on a landscape represented by a DEM? Chase (1992) developed a computer landsculpting model based on cellular automata. The model uses simple rules to determine whether erosion occurs, how much occurs, and over what spatial scale. In this manner, Chase (1992) was able to consider two classes of processes: smoothing processes, which followed a diffusion model, and roughening processes. The cellular automata model, which Chase renamed the 'precipiton' model, uses DEMs as input. The simulations based on the precipiton model suggest that tectonic processes act mostly to increase the amplitude of topography whereas climatic processes act to smooth it. The roughness of topography, as measured by the slope of its variogram, is independent of its amplitude or relief at a 1 km grid spacing (and presumably at all spacings). Gregory and Chase (1994) used DEMs, palaeoclimatic data, and erosional modelling to constrain the evolution of the high-elevation, but low-relief, late Eocene erosion surface in the southern Rocky Mountains (Epis and Chapin, 1975) and partition the resultant topography into climatic and tectonic signals. They concluded that the dominant signal is a climatic one, and this is evident from the low fractal dimension indicating a smoothed topography.

2.5 DEMs for Macroscale Tectonic Geomorphology

DEMs can provide a context for studying tectonic processes. Mass flux is to some extent directly modulated by the landscape, and therefore examining the landscape using DEMs can provide a convenient framework to construct hypotheses involving mass balances. As erosion removes rock that may lie over a detachment fault, the normal stresses on the fault change dynamically with the landscape. Moreover, as erosion proceeds some form of isostatic response may modify the landscape. In addition to direct measurements, DEMs allow for sophisticated hypothesis testing (see, for example, Chapter 3). For example, modelled topography can be quantitatively compared to actual topography to evaluate the goodness of fit of the model. As we press for better models of topographic evolution in active tectonic settings, DEMs will become the way to determine whether such models are realistic in their results.

Below, I provide some examples of DEMs in the context of tectonic geomorphology. The selection of examples is based on my own field experiences and presents a spectrum of tectonic settings, from convergent margins and volcanic arcs to divergent plate settings and intraplate environments. Other applications of DEMs are to be found throughout the book, especially in Chapter 10.

2.5.1 GRAND CANYON EROSION

The Grand Canyon, which in places is more than 2 km deep, cuts across several plateaus within the Colorado Plateau topographic province. The timing of the formation of the Grand Canyon is constrained by K–Ar dating of lava flows which suggest that it may have started to develop about 6 Ma BP and reached a level near its present one about 3.5. Ma BP (Damon *et al.*, 1978; Luchitta and Young, 1986; Luchitta, 1990). However, at Lava Falls, flows adjacent to Vulcans Throne suggest that the Colorado River had cut to its present level by about 1 Ma BP (Damon *et al.*, 1978). Plate 2.5 shows the 1-degree DEM for the Grand Canyon east quadrangle displayed as a simple greyscale image, with elevation being simply mapped to grey level on a scale of 0 to 256, or 8-bits. The canyon that cuts north into the plateau is Kanab Creek, and this appears to demarcate two sections of the Grand Canyon that are morphologically distinct. The western Grand Canyon is cut by tributaries with well-developed side canyons, but to the east the tributary canyons are less developed. The Kaibab upwarp seems to be related to this difference in morphology, which is perhaps an inherited feature of the ancestral Colorado River network. DEMs are very useful for identifying these gross topographic differences.

The DEM also clearly shows the Kaibab Plateau and its margins which seem to be morphologically young. Because DEMs show the stream elevations, to some approximate extent, it is possible to evaluate the responses of streams to tectonic structures or rock control. For example, even at this coarse resolution, one can see that the streams crossing the Kaibab from east to west have steeper sections.

A useful procedure provided by DEMs is the reconstruction of the pre-Grand Canyon topography for the purpose of visualizing the pre-canyon landscape (Mayer, 1982, unpublished map), and also to estimate the volume of material eroded by canyon cutting. Prior to the use of DEMs, such calculations required extensive planimetry or physical models. For example, an erosion estimate was made by filling the canyons on a three-dimensional plastic model of topography with sand and then scaling the model and weighing the sand to estimate the volume of material. DEMs provide a way to make volumetric estimates in a manner analogous to numerical integration.

2.5.2 VOLCANIC ERUPTIONS: MOUNT ST HELENS

Prior to a magnitude 4.2 earthquake in March 1980, Mount St Helens was dormant for 123 years. On 27 March, Mount St Helens began to expel ash and steam, the first significant eruption in the conterminous United States since that of Lassen Peak, California from 1914 to 1917. From March until May, magma inflated the volcano and the north flank of the volcano bulged measurably. On 18 May 1980, a magnitude 5.1 earthquake apparently triggered the catastrophic collapse of the north flank of the volcano. On the 1-degree DEM for the Hoqium East quadrangle, which shows the post-eruption topography, the collapsed north flank is visible. Hill shading of the DEM highlights not only the volcanic edifice, but also the tectonic structures in the region (Plate 2.6). Pre- and post-eruption comparisons of DEMs provide a way to estimate quantitatively volumetric changes caused by the volcanic eruptions (Plate 2.7).

The collapse of the north flank of Mount St Helens resulted in a very large debris avalanche (Tilling *et al.*, 1990). Moving at speeds in excess of 240 km h^{-1}, the debris avalanche consisted of about 2.5 km^3 of unconsolidated rock debris covering some 60 km^2 of the North Fork Toutle River valley (Brantley and Glicken, 1986). Debris levees as high as 30 m were formed along the margins of the deposit, and individual ridges as high as 70 m are found. Tilling *et al.* (1990) report that at one location, about 6.4 km north of the summit, the avalanche front still had sufficient momentum to flow over a ridge more than 345 m high. After the eruption, the highest point on the summit was about 400 m lower than before (Tilling *et al.*, 1990). There are 7.5-minute DEMs for Mount St Helens which show both pre- and post-eruption topography. Subtracting one from the other shows the difference in the surfaces (Plate 2.7). As the resolution of topographic data increases, it will be possible to calculate before and after differences for a wide variety of tectonic and geomorphic processes.

2.5.3 INTRAPLATE SEISMICITY: NEW MADRID, MISSOURI, USA

The New Madrid earthquake, which consisted of three main shocks centred on the region between Marked Tree, Arkansas and New Madrid, Missouri, occurred between 16 December 1811 and 7 February 1812 (Nuttli, 1982) in a region of relatively muted topographic expression within the lower Mississippi valley (Plate 2.8). The shocks were felt over a very large area, approximately 5 000 000 km^2, and the damage area extended 600–700 km from the epicentre (Nuttli, 1982). The total seismic moment for the three shocks, estimated to be 6×10^{20} J (Herrmann *et al.*, 1978), is larger than the seismic moment estimate for the 1906 San Francisco earthquake of 4×10^{20} J (Thatcher, 1975). Since then, 20 more damaging earthquakes have occurred in this region which has become known as the New Madrid seismic zone (Nuttli, 1982).

Based on intensity data, Nuttli (1979) has shown that 488 earthquakes exceeding m_b=3.0 (excluding the New Madrid aftershocks) have occurred since the New Madrid earthquake within a $4° \times 4°$ region centred on 37°N, 89°30'W. The geographical distribution of these earthquakes given by Stauder (1982) indicates several distinct spatial clusters, one of which is located along the eastern edge of the Missouri bootheel. St Louis University established a microearthquake network in 1974 which has recorded or catalogued thousands of earthquakes. The geographical pattern of a portion of these earthquakes described by Stauder (1982) included a dense region of seismicity from 36°36'N, 89°36'W, just west of New Madrid, extending southeast across the Mississippi River to the town of Ridgely, Tennessee. From Ridgely extending to the southwest into Arkansas, Stauder described a 120-km-long linear seismic trend. A parallel line of seismicity extends 40 km from an area just to the southwest of New Madrid trending to the northeast (Stauder, 1982).

From composite focal plane solutions for a dense network of seismic stations operated between March and May 1978, it appeared that the two long parallel seismic trends were defining right-slip faults, and that the zone of seismicity in between these trends represented a left-step in this system experiencing northeast–southwest-directed compressive stress (O'Connell *et al.*, 1982). Surface features spatially associated with faults inferred from focal plane mechanisms or mapped from seismic reflection profiling (Zoback *et*

	A	B	C	D	E	F	G	H	I	J	K	L	M	N	O	P	Q
1	0	0	0	0	0	0	-1	-1	-1	-1	-1	0	0	0	0	0	0
2	0	0	0	0	1	1	-1	-1	-1	-1	-1	1	1	0	0	0	0
3	0	0	1	1	1	2	-3	-3	-3	-3	-3	2	1	1	1	0	0
4	0	0	1	1	2	3	-3	-3	-3	-3	-3	3	2	1	1	0	0
5	0	1	1	2	3	3	-3	-2	-3	-2	-3	3	3	2	1	1	0
6	0	1	2	3	3	3	0	2	4	2	0	3	3	3	2	1	0
7	1	1	3	3	3	0	4	10	12	10	4	0	3	3	3	1	1
8	1	1	3	3	2	2	10	18	21	18	10	2	2	3	3	1	1
9	1	1	3	3	3	4	12	21	24	21	12	4	3	3	3	1	1
10	1	1	3	3	2	2	10	18	21	18	10	2	2	3	3	1	1
11	1	1	3	3	3	0	4	10	12	10	4	0	3	3	3	1	1
12	0	1	2	3	3	3	0	2	4	2	0	3	3	3	2	1	0
13	0	1	1	2	3	3	-3	-2	-3	-2	-3	3	3	2	1	1	0
14	0	0	1	1	2	3	-3	-3	-3	-3	-3	3	2	1	1	0	0
15	0	0	1	1	1	2	-3	-3	-3	-3	-3	2	1	1	1	0	0
16	0	0	0	0	1	1	-1	-1	-1	-1	-1	1	1	0	0	0	0
17	0	0	0	0	0	0	-1	-1	-1	-1	-1	0	0	0	0	0	0

Figure 2.1 *Box filter referred to as a 'Hat filter' in text. This moving filter enhances the difference between interfluves and valleys. See text for discussion*

al., 1980) were further related to surface topographic features by O'Connell *et al.* (1982). The Ridgely and Cottonwood Grove faults define Ridgely ridge in a manner consistent with the apparent slip on the faults and the topography of the ridge. In addition, the Reelfoot and an unnamed subparallel fault appear to define the Tiptonville dome in a similar manner (see Figure 5 of O'Connell *et al.*, 1982).

A regional perspective on the topography of this region is afforded by applying a 17×17 (pixels) Hat filter (Figure 2.1) to the raw digital elevation data and producing a type of 'ridge map'. Following filtering, a threshold was applied so that the image was a binary map of high patterns in grey and low patterns in white. The final image was made by merging colour-coded topography with the ridge map. The result is shown in Plate 2.9, where the interfluves and ridges are patterned and elevations are colour coded, with cooler colours representing higher elevations. Crowleys Ridge is identifiable and provides an intuitive way to evaluate the scale at which the filter operates. Several features identified on the basis of the resulting topographic textures appear significant because they show changes in pattern. Note that the river appears to flow completely within a grey zone immediately to the west of the bluffs.

Topographic deformation resulting from the New Madrid earthquakes centred on the Lake County uplift (Plate 2.10). The geometry of the Lake County uplift as generally cited in the literature is based upon the investigation of Russ (Russ, 1982). In order to define the uplift, Russ used an idealized representation of the slope of the Mississippi River meander belt before modification. By dividing the area between the 270 and 300 foot (82 m and 91 m) contours into six equally spaced contour bands separated by five contour lines, the construction of this surface completely determines the resulting pattern of presumed uplift. The difference between the postulated pre-uplift surface and the modern post-uplift surface was used to construct isobase lines that describe the pattern of uplift. The resulting uplift pattern shows two strong uplifts, one at Tiptonville Dome and the other at Ridgely Ridge.

Density slicing, like thresholding, allows objects to be partitioned on the basis of grey level or elevation. All parts of the DEM whose values are between a minimum and maximum threshold criterion are highlighted. Density slicing is a simple but useful tool in geomorphic image processing used to identify those areas that fall within a specified vertical range above base level. For example, we can identify all areas that are within a specified elevation range above the Mississippi River, and highlight topographic features that could not result from overbank flood deposition. In rougher topography, density slicing permits the rapid identification of elevations that are accordant. These could represent erosion surfaces or some other correlative feature. Plate 2.11 shows a density slice, in red, through the topography of the Dyersburg 1-degree DEM. Reelfoot Lake, which is bounded by the Reelfoot fault, shows up as a depression in the density slice. A northerly trending slice boundary characterizes the New Madrid area, perhaps indicating a north–south structure which would cut the expected east–west trend of the valley contours. Thus, although we have not quantified a specific model as Russ (1982) did, we are following a similar logical track, assuming that the north–south Mississippi River should flow across east–west-trending contour lines, using the DEM as the basis of visualizing the results of our model. Also note the anomalously high area to the south of the New Madrid loop, along the Mississippi River. By rendering the elevation data one can achieve a more photorealistic view of the topography and perhaps better visualize the process (Plate 2.12). Essentially we are slicing through the topography with a surface which we believe represents the maximum range of elevations that could result from non-tectonic process.

2.5.4 CONTINENTAL RIFTING: BAJA CALIFORNIA

Lister *et al.* (1986a) have suggested that passive margins should reflect the asymmetry of the detachment structures and result in upper-plate margins and lower-plate margins. Upper-plate margins are those from above the detachment fault whereas lower-plate margins comprise crust from below the detachment. Significantly, upper- and lower-plate margins should differ in their pre-rift structure and subsidence histories. Axen (1995) refers to these settings as hangingwall and footwall margin segments.

Moustafa (cited in Bosworth, 1986) recognized the asymmetry of the Gulf of Suez and the interpretation of seismic stratigraphic data suggests that symmetrically opposed detachments are coevally active in the early phases of rifting (Bosworth, 1986). Accommodation zones, transfer fault systems separating half-grabens with opposing detachment vergence, are complex structural zones dominated by oblique wrenching (Bosworth, 1986). Accommodation zones cross the rift trend at an oblique angle but may also be roughly perpendicular to the rift (Lister *et al.*, 1986b). To some extent a more significant relation is between the accommodation zone and the rift-angle, the angle between the rift trend and the extension direction, produced in oblique rifts. For instance, if this angle is about 30° then strike-slip, oblique-slip, and normal faults may all develop (Withjack and Jamison, 1986).

Gawthorpe and Hurst (1993) distinguished 'interbasin' transfer zones, which are a kilometre to tens of kilometres in size, and 'intrabasin' transfer zones which are commonly less than a kilometre in size. Interbasin transfer zones accommodate changes between half-grabens in contrast to intrabasin transfer zones which accommodate changes in structure between fault segments. They define several geometries of interbasin transfer zones; antithetic interbasin ridges, antithetic interference zones, simple transfer faulting, and synthetic relay ramps. All of these structures strongly affect the topographic surface while they are active. Transfer zone structures, which have surface expression during their development, also affect hydrocarbon migration. Gawthorpe and Hurst (1993) point out that the largest oil fields in the Gulf of Suez rift are associated with such structures.

Topographic expression of upper- versus lower-plate margins, and development of topography above, or associated with, accommodation zones may provide a useful means of distinguishing between intrabasin transfer zones within an extensional allochthon, and those separating detachments with opposite polarity. Axen (1995) uses structural tilts as well as topography to infer the positions of accommodation zones separating detachment fault vergence reversals along the entire length of the Baja Peninsula. The use of topography is of particular interest for this study. Currently, Paul Umhoefer, Rebecca Dorsey and myself are

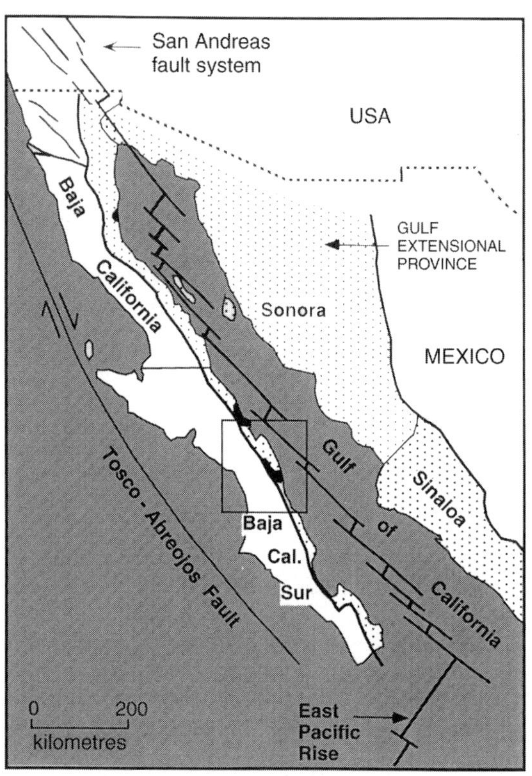

Figure 2.2 Location map showing Baja California, the Gulf Extensional Province, and the Tosco–Abreojos fault. The Main Gulf Escarpment generally defines the western boundary of the Gulf Extensional Province. Box shows location of Figure 2.3. Figure from Dorsey et al. *(1995)*

attempting to document these complex structural features in the region around Loreto, Baja California Sur (Figures 2.2 and 2.3). We believe that DEMs can be used to help locate the positions of these accommodation zones.

Baja California was part of the Mexican mainland while a volcanic arc was active along western North America during Oligocene times (Sawlan, 1991). Arc rocks in the La Paz area of Baja Sur are as young as about 12 Ma BP (Hausback, 1984) during which time the plate configuration rapidly changed, and the Riviera triple junction appeared at the southern end of the Gulf of California (Stock and Hodges, 1989), essentially turning off arc volcanism in Baja. During, and subsequent to, extinction of the volcanic arc, ENE-directed extension and possible strike-slip faulting began in what would become the Gulf of California. This early deformation is referred to as the 'proto-Gulf' stage (Moore and Buffington, 1968;

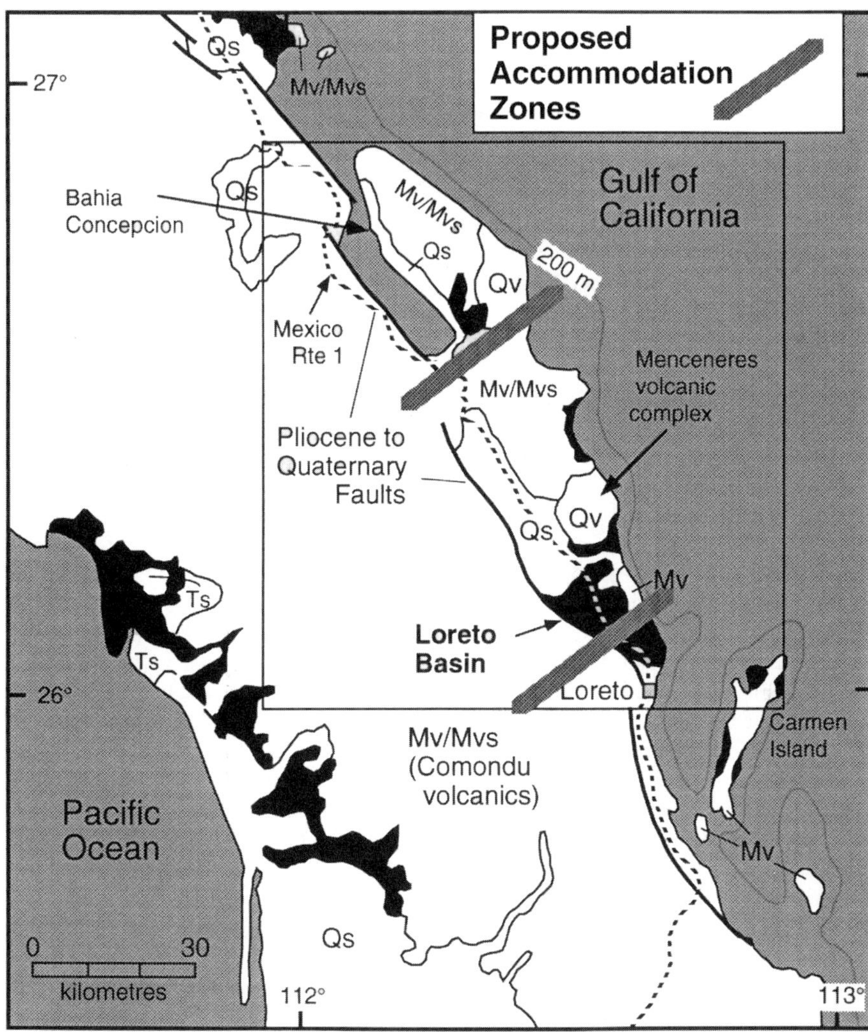

Figure 2.3 *Location map showing location of place names referred to in text, Pliocene sediments, and proposed accommodation zones (after Umhoefer* et al. *(1994). The inset box shows the approximate extent of the Landsat MSS image and DEM in Plate 2.14*

Hausback, 1984; Stock and Hodges, 1989) and describes the period between about 12 and 5 Ma BP. The ENE-directed extension is consistent with plate models which indicate oblique motion between the Pacific and North American plates between 5.5 and 10.6 Ma BP (Stock and Hodges, 1989). Following the proto-Gulf stage, north-trending pull-apart basins formed as transform faults developed east of Baja, in the Gulf of California, while the Tosco–Abreojos fault actively accommodated dextral slip on the west side of Baja. The final and current stage, which began about 3.5 Ma BP (Stock and Hodges, 1989), is characterized by the attachment of the Baja microplate to the Pacific plate and by transform faulting which is strongly oblique to the trend of the rift. Because extension in the Gulf of California was due first to orthogonal, and then to oblique, rifting, its structural characteristics and the role of accommodation zones are especially intriguing.

Axen places an accommodation zone at the southern end of the Bahia Concepcion characterized by a disrupted structural grain and terminating a footwall segment he names the Loreto segment. The Loreto segment is in part structurally defined by the north–south-trending Loreto fault (McLean, 1988). Work by Umhoefer *et al.* (1994) indicates that the Loreto basin is bounded by a listric east down fault forming a west-tilted half-graben which is consistent with Axen's footwall segment hypothesis.

On the Landsat MSS image of the region (Plate 2.13), one can recognize the tectonic controls on the coastline and the faulted nature of the block bounding the Bahia Concepcion on the northeast. The southern boundary of the Bahia Concepcion may represent a structure which terminates faulting to the north. Some of this faulting, located on the northeast side of the Bahia Concepcion, may be as young as Quaternary, on the basis of the linearity of the fault scarp. To the south, however, the topography again flattens into broad alluvial surfaces characteristic of a basin. In the centre of this alluvial basin, but near the coast, is the largely Pliocene (Zanchi, 1994) Mencenares volcanic complex which is cut by north–south and northwest–southeast-trending faults. The alluvial area is then terminated on the south by the Loreto fault where its orientation changes to a southeast–northwest trend (Umhoefer *et al.*, 1994). Despite the higher quality of the Landsat image, the DEM for the same area (Plate 2.14) displays the gross topographic features more clearly. The DEM, which is displayed using an arbritrary CLUT, shows higher topography in yellows, brown and white, and lower elevations in greens and pale blues. Because the CLUT directly maps to elevation, it is easier to manipulate the CLUT to enhance specific topographic features than to enhance other types of imagery such as Landsat.

The pronounced east-facing escarpment on the DEM (Plate 2.14) is the Main Gulf Escarpment (MGE) which essentially runs the length of the Baja Peninsula (Figure 2.2). On the DEM several features of the MGE are visible. First, the MGE is stepped in places. The L-shaped step along the MGE in the northwestern part of the DEM spatially corresponds to the dog-leg in the coastline of the Bahia Concepcion. Following the MGE to the south, we see that a saddle, largely erosional, in the MGE occurs at the southern termination of the Bahia Concepcion. At the southern boundary of the Bahia Concepcion is a transverse ridge trending east–northeast that segments the low topography into the valley in the south from the Bahia Concepcion to the north. This positive topography is where Gastil and Fenby (1991) indicate a disruption of the structural grain and where Axen (1995) places an accommodation zone. As we follow the valley southward on the proposed Loreto footwall segment, the most dramatic feature, other than the MGE which becomes higher and more linear, is the topography of the young Mencenares volcano. Three-dimensional rendering of a portion of the digital elevation data shown in Plate 2.14 illustrates the appearance of the Mencenares volcanic complex in the middle of the basin (Plate 2.15). Continuing southward we encounter another east–northeast topographic ridge. This feature is partly on Miocene Comandu arc volcanics, but also in part on much younger Pliocene rocks of the uplifted Loreto basin (Dorsey *et al.*, 1994; Umhoefer *et al.*, 1994, 1995).

At the coast, the ridge is represented by a bedrock high with Miocene Comandu volcanics and also some upper Pliocene limestones of the Loreto basin (Figure 2.2, Plate 2.15). These limestones crop out about 300 m above sea level, which, to the first approximation, represents post-Loreto basin tectonic uplift (Dorsey *et al.*, 1994). However, west of the coastal high, the ridge represents the modern drainage divide separating the south-flowing washes from north-flowing ones. This drainage divide was formerly part of the Loreto basin and therefore represents a post-Pliocene tectonic feature. Dorsey *et al.* (1994) have speculated that structures affecting the topography altered the sediment supply for the Loreto basin. Umhoefer *et al.* (1995) suggest that this zone represents an accommodation zone separating the Loreto footwall segment from the Nopolo segment to the south based on structural changes across the zone, and Dorsey *et al.* (1994) confirm that this zone appears to be an actively deforming feature, possibly associated with the late Pleistocene uplift of the coastal marine terraces (Figure 2.2). Unlike the proposed accommodation zone to the north of the Loreto footwall segment which corresponds to a low in the MGE, this proposed zone corresponds to a high in the MGE (Plate 2.16) and the topography over the zone is demonstrably young and possibly active. Thus the DEM serves to identify features very effectively; other data, however, are required to document fully the significance of the topographic features observed.

2.6 Future Research Directions

The use of DEMs in tectonic geomorphology is certainly still developing; however, the availability of DEMs today, and the prospect for worldwide topographic coverage acquired by NASA, will initiate a burgeoning interest in DEM applications. The most promising developments, in my opinion, will be those that integrate information on structure, tectonics, sedimentation and geodynamics with geomorphology. Mechanically, DEM use will migrate to GIS approaches and further integrate GPS technologies to accommodate the multilayered requirements of interdisciplinary research on active tectonic systems. The use of DEMs for macroscale tectonic geomorphology is not, by and large, limited in application by present technological constraints. Rather the widespread and effective use of DEMs is waiting for a group of inquisitive and innovative geomorphologists creatively to solve scientific problems in active tectonics.

Acknowledgements

Many thanks to Becky Dorsey and Paul Umhoefer who shared their ideas and figures on Baja, and continue to provide a stimulating environment for the geological study of margin segmentation. The work in Baja California Sur, Mexico, was supported by the National Science Foundation. The work in the New Madrid seismic zone was supported by the US Geological Survey through the NEHRP. Field work in the Cascades of Washington, Oregon and California was supported by the Ohio Board of Regents. Their support is gratefully appreciated.

References

Ahnert, F. 1970. Functional relationships between denudation, relief, and uplift in large mid-latitude drainage basins. *Amer. J. Sci.*, **268**, 243–263.

Arvidson, R.E., Guiness, E.A., Strebeck, J.W., Davies, G.F. and Schulz, K.J. 1982. Image processing applied to gravity and topography data covering the continental U.S. *EOS, Trans. Amer. Geophys. Un.*, **63**, 261–265.

Axen, G. 1995. Extensional segmentation of the Main Gulf Escarpment, Mexico and United States. *Geology*, **23**, 515–518.

Bascom, F. 1921. Cycles of erosion in the Piedmont province of Pennsylvania. *J. Geol.*, **29**, 540–559.

Batson, R.M., Edwards, K. and Eliason, E,M. 1975. Computer-generated shaded relief images. *U.S. Geol. Surv. J. Res.*, **3**, 401–408.

Bosworth, W. 1986. Comment on detachment faulting and the evolution of passive continental margins. *Geology*, **14**, 890–891.

Brantley, S. and Glicken, H. 1986. Volcanic debris avalanches. *Earthquakes and Volcanoes*, **18**, 195–197.

Bull, W.B. and McFadden, L.D. 1977. Tectonic geomorphology north and south of the Garlock fault, California. In: D.O. Doehring (ed.) *Geomorphology in Arid Regions*, SUNY, Binghamton, pp. 115–138.

Chase, C.G. 1992. Fluvial landsculpting and the fractal dimension of topography. *Geomorphology*, **5**, 39–57.

Chorowicz, J., Luxey, P., Lyberis, N., Carvalho, J., Parrot, J.-F., Yürür, T. and Gündogdu, N. 1994. The Mara Triple Junction (southern Turkey) based on digital elevation model and satellite imagery interpretation. *J. Geophys. Res.*, **99**, 20 225–20 242.

Colleau, A. and Lenôtre, N. 1991. A new digital method for analysis of neotectonics applied to the Bonnevaux–Chambaran area, France. *Tectonophysics*, **194**, 295–305.

Damon, P.E., Shafiquallah, M. and Scarborough, R.B. 1978. Revised chronology for critical stages in the evolution of the lower Colorado River. *Geol. Soc. Amer. Absts with Progs*, **10**, 101–102.

Davis, W.M. 1899. The geographical cycle. *Geogr. J.*, **14**, 481–504.

Deffontaines, B. and Chorowicz, J. 1991. Principles of drainage basin analysis from multisource data – Application to the structural analysis of the Zaire Basin. *Tectonophysics*, **194**, 237–263.

Deffontaines, B., Lee, J.-C., Angelier, J., Carvalho, J. and Rudant, J.-P. 1994. New geomorphic data on the active Taiwan orogen: A multisource approach. *J. Geophys. Res.*, **99**, 20 243–20 266.

Dorsey, R., Mayer, L. and Umhoefer, P. 1994. Interplay between structural and geomorphic processes in the Pliocene–Quaternary Loreto basin, Baja California Sur, Mexico. *EOS, Trans. Amer. Geophys. Un.*, **75**, H520-08.

Dorsey, R.J., Umhoefer, P.J. and Renne, P.R. 1995. Rapid subsidence and stacked Gilbert-type fan deltas, Pliocene Loreto basin, Baja California Sur, Mexico. *Sed. Geol.*, **98**, 181–204.

Dury, G.H. 1951. Quantitative measurement of available relief and depth of dissection. *Geol. Mag.*, **88**, 339–343.

Edwards, K. and Batson, R.M. 1990. *Experimental Digital Shaded-Relief Map of Southwestern United States*. U.S. Geological Survey Miscellaneous Investigations Series Map 1-1850, scale 1:2 000 000.

Ellis, M. and Merritts, D. (eds) 1994. *Tectonics and Topography*. American Geophysical Union, Washington DC.

Elvhage, C. and Lidmar-Bergström, K. 1987. Some working hypotheses on the geomorphology of Sweden in the light of a new relief map. *Geogr. Ann.*, **69**, 343–358.

Epis, R.C. and Chapin, C.E. 1975. Geomorphic and tectonic implications of the post-Laramide, late Eocene erosion surface in the Southern Rocky Mountains. In: B.F. Curtis (ed.) *Cenozoic History of the Southern Rocky Mountains, Geol. Soc. Amer. Mem.*, **144**, 45–74.

Gastil, R.G. and Fenby, S.S. 1991. Geologic–tectonic map of the Gulf of California and surrounding areas. In: J.P. Dauphin and B.R.T. Simoneit (eds) *The Gulf and*

Peninsular Provinces of the Californias, Amer. Ass. Petrol. Geol. Mem., **47**, 79–83.

Gawthorpe, R.L. and Hurst, J.M. 1993. Transfer zones in extensional basins: their structural style and influence on drainage development and stratigraphy. *J. Geol. Soc. Lond.*, **150**, 1137–1152.

Gephart, J.W. 1994. Topography and subduction geometry in the central Andes. Clues to the mechanics of a noncollisional orogen. *J. Geophys. Res.*, **99**, 12 279–12 288.

Gilchrist, A.R. and Summerfield, M.A. 1991. Denudation, isostasy and landscape evolution. *Earth Surf. Processes Ldfms*, **16**, 555–562.

Gregory, K.M. and Chase, C.G. 1994. Tectonic and climatic significance of a late Eocene low-relief, high level geomorphic surface, Colorado. *J. Geophys. Res.*, **99**, 20 141–20 160.

Hack, J.T. 1973. Stream-profile analysis and stream-gradient index. *U.S. Geol. Surv. J. Res.*, **1**, 421–429.

Hack, J.T. 1974. Drainage adjustment in the Appalachians. In: M. Morisawa (ed.) *Fluvial Geomorphology*, SUNY, Binghamton, pp. 51–69.

Harrington, H.J., Simpson, C.J. and Moore, R.F. 1982. Analysis of continental structures using a digital terrain model (DTM) of Australia. *BMR J. Austr. Geol. Geophys.*, **7**, 68–72.

Hausback, B.P. 1984. Cenozoic volcanic and tectonic evolution of Baja California, Mexico. In: A. Frizzell (ed.) *Geology of the Baja California Peninsula, Bakersfield, California*, Pacific Section, Society of Economic Paleontologists and Mineralogists, pp. 219–236.

Heirtzler, J.R. 1985. *Relief of the Surface of the Earth, Computer Generated Shaded Relief* (1:39 000 000). US National Oceanic and Atmospheric Administration, National Geophysical Data Center.

Herrmann, R.B., Cheng, S.H. and Nuttli, O.W. 1978. Archeoseismology applied to the New Madrid earthquakes of 1811–12. *Seismol. Soc. Amer. Bull.*, **68**, 1751–1759.

Horn, B.K.P. 1981. Hill shading and the reflectance map. *Proc. IEEE*, **69**, 14–47.

Isacks, B.L. 1988. Uplift of the central Andean plateau and bending of the Bolivian orocline. *J. Geophys. Res.*, **93**, 3211–3231.

Johnson, D. 1931. Stream sculpture on the Atlantic slope. Columbia University Press, New York, 142 pp.

Keller, E.A. 1986. Investigation of active tectonics: use of surficial earth processes. In: R.E. Wallace (ed.) *Active Tectonics*, National Academy Press, Washington, pp. 136–147.

King, L.C. 1951. The study of the world's plainlands: a new approach in geomorphology. *Q. J. Geol. Soc. Lond.*, **106**, 101–127.

Lamb, A.D., Malan, O.G. and Merry, C.L. 1987. Application of image processing techniques to digital elevation models of southern Africa. *S. Afr. J. Sci.*, **83**, 43–47.

Lifton, N.A. and Chase, C.G. 1992. Tectonic, climatic and lithologic influence on landscape fractal dimension and hypsometry: implication for landscape evolution in the San Gabriel Mountains, California. *Geomorphology*, **5**, 77–114.

Lister, G.S., Etheridge, M.A. and Symonds, P.A. 1986a. Detachment faulting and the evolution of passive continental margins. *Geology*, **14**, 246–250.

Lister, G.S., Etheridge, M.A. and Symonds, P.A. 1986b. Reply on "Detachment faulting and the evolution of passive continental margins", *Geology*, **14**, 891–892.

Lowry, A.R. and Smith, R.B. 1994. Flexural rigidity of the Basin and Range–Colorado Plateau–Rocky Mountain transition from coherence analysis of gravity and topography. *J. Geophys, Res.*, **99**, 20 123–20 140.

Luchitta, I. 1990. History of the Grand Canyon and Colorado River in Arizona. In: J. Jenny and S.J. Reynolds (eds) *Geological Evolution of Arizona*, Arizona Geological Society, Tucson.

Luchitta, I. and Young, R.A. 1986. Structure and geomorphic character of the western Colorado Plateau in the Grand Canyon–Lake Mead region. In: J.D. Nations, C.M. Conway and G.A. Swann (eds) *Geology of Central and Northern Arizona*, Geological Society of America Rocky Mountain Section Field Trip Guidebook, pp. 159–176.

McLean, H. 1988. *Reconnaissance Geologic Map of the Loreto and Part of the San Javier Quadrangles, Baja California Sur, Mexico (1:50 000)*. US Geological Survey Map.

Moore, D.G. and Buffington, E.C. 1968. Transform faulting and growth of the Gulf of California since the late Pliocene. *Science*, **161**, 1238–1241.

Moore, R.F. and Simpson, C.J. 1982. Computer manipulation of a digital terrain model (DTM) of Australia. *BMR J. Austr. Geol. Geophys.*, **7**, 63–67.

Moustafa, A.M. 1976. Block Faulting in the Gulf of Suez, 5th Egyptian General Petroleum Organization Exploration Seminar, Cairo, 19.

Nuttli, O.W. 1979. The Seismicity of the Central United States, Geology in the Siting of Nuclear Power Plants. Reviews in Engineering Geology, Geological Society of America, 67–93.

Nuttli, O.W. 1982. Damaging earthquakes in the central Mississippi Valley. *U.S. Geol. Surv. Prof. Pap.*, **1236-B**, 15–20.

O'Connell, D.R., Bufe, C.G. and Zoback, M.D. 1982. Microearthquakes and faulting in the area of New Madrid, Missouri–Reelfoot Lake, Tennessee. *U.S. Geol. Surv. Prof. Pap.*, **1236-D**, 31–38.

Ollier, C.D. 1981. *Tectonics and Landforms*. Longman, London.

Oronati, G., Poscolieri, M., Ventura, R., Chiarini, V. and Crucilla, U. 1992. The digital elevation model of Italy for geomorphology and structural geology *Catena*, **19**, 147–178.

Pannekoek, A.J. 1967. Generalized contour maps, summit level maps, and streamline surface maps as geomorphological tools. *Z. Geomorph.*, **11**, 169–182.

Penck, A. 1919. Die Gipfelflur der Alpen. *Sitzungberichte der Preussischen Akademie der Wissenschaften zu Berlin*, **17**, 256–268.

Pike, R.J. and Wilson, S.E. 1971. Elevation relief ratio, hypsometric integral and geomorphic area–altitude analysis. *Geol. Soc. Amer. Bull.*, **82**, 1079–1084.

Russ, D.P. 1982. Style and significance of surface deformation in the vicinity of New Madrid, Missouri. *U.S. Geol. Surv. Prof. Pap.*, **1236**, 95–114.

Sawlan, M.G. 1991. Magmatic evolution of the Gulf of California rift. In: J.P. Dauphin and B.R.T. Simoneit (eds) *The Gulf of California Peninsular Province of the Californias*, American Association of Petroleum Geologists, Tulsa.

Schowengerdt, R.A. and Glass, C.E. 1983. Digitally processed topographic data for regional tectonic evaluations. *Geol. Soc. Amer. Bull.*, **94**, 549–556.

Simpson, D.W. and Anders, M.H. 1992. Tectonics and topography of the western United States – an application of digital mapping. *GSA Today*, **2**, 117–118, 121–122.

Smith, G.-H. 1935. The relative relief of Ohio. *Geogr. Rev.*, **25**, 272–284.

Stauder, W. 1982. Present-day seismicity and identification of active faults in the New Madrid seismic zone. *U.S. Geol. Surv. Prof. Pap.*, **1236-C**, 21–30.

Stearns, R.G. 1967. Warping of the Western highland rim peneplane in Tennessee by ground water sapping. *Geol. Soc. Amer. Bull.*, **78**, 1111–1124.

Stock, J.M. and Hodges, K.V. 1989. Pre-Pliocene extension around the Gulf of California and the transfer of Baja California to the Pacific plate. *Tectonics*, **8**, 99–115.

Strahler, A.N. 1952. Hypsometric (area–altitude) analysis of erosional topography. *Bull. Geol. Soc. Amer.*, **63**, 1117–1142.

Summerfield, M.A. 1985. Plate tectonics and landscape development. In: M. Morisawa and J.T. Hack (eds) *Tectonic Geomorphology*, Allen and Unwin, Boston, pp. 27–51.

Thatcher, W. 1975. Strain accumulation and release mechanism of the 1906 San Francisco earthquake. *J. Geophys. Res.*, **80**, 4862–4880.

Thelin, G.P. and Pike, R.J. 1991. *Landforms of the United States – A Digital Shaded Relief Portrayal.* US Geological Survey.

Tilling, R.I., Topinka, L. and Swanson, D.A. 1990. *Eruptions of Mount St Helens. Past, Present and Future.* US Geological Survey, US Government Printing Office, Washington, DC., 56 pp.

Topographic Science Working Group. 1988. *Topographic Working Group Report to the Land Processes Branch, Earth Science and Applications Division, NASA, Headquarters, Houston, Lunar and Planetary Institute*, 64 pp.

Umhoefer, P.J., Dorsey, R.J. and Mayer, L. 1995. Tectonic segmentation of the Gulf Extensional Province, Loreto region, Baja California Sur, Mexico. *Geol. Soc. Amer. Absts with Progs*, **27**, A189.

Umhoefer, P.J., Dorsey, R.J. and Renne, P. 1994. Tectonics of the Pliocene Loreto basin, Baja California Sur, Mexico, and the evolution of the Gulf of California. *Geology*, **26**, 649–652.

US Geological Survey. 1978. *Digital Terrain Tapes Users Guide*. National Cartographic Information Center, 12 pp.

Wdowinski, S. and Axen, G.J. 1992. Isostatic rebound due to tectonic denudation: a viscous flow model of a layered lithosphere. *Tectonics*, **11**, 303–315.

Weissel, J.K. and Karner, G.D. 1989. Flexural uplift of rift flanks due to mechanical unloading of the lithosphere during extension. *J. Geophys. Res.*, **94**, 13 919–13 950.

Withjack, M.O. and Jamison, W.R. 1986. Deformation produced by oblique rifting. *Tectonophysics*, **126**, 99–124.

Yoeli, P. 1965. Analytical hill shading. *Surv. Mapping*, **25**, 573–579.

Yoeli, P. 1967. The mechanization of analytical hill shading. *Cartogr. J.*, **4**, 82–88.

Zanchi, A. 1994. The opening of the Gulf of California near Loreto, Baja California, Mexico, from basin and range extension to transtensional tectonics. *J. Struct. Geol.*, **16**, 1619–1639.

Zoback, M.D., Hamilton, R.M., Crone, A.J., Russ, D.P., KcKeown, F.A. and Brockman, S.R. 1980. Recurrent intraplate tectonism in the New Madrid seismic zone. *Science*, **209**, 971–976.

Appendix: Some sources of global elevation data

Digital Chart of the World was digitized from 1:1 000 000 and 1:2 000 000 scale base maps and provides 30-arcsecond data. UNEP/GRID Sioux Falls: http://grid2.cr.usgs.gov/dem/sources.html

Also check: http://edcwww.cr.usgs.gov/landdaac/30asdcwdem/30asdcwdem.html

Defense Mapping Agency provides a listing of existing digital products such as elevations and bathymetric data including the 1-degree digital elevation models. One-arcsecond data (about 30 m) cells with about 50 m horizontal accuracy, 30 m vertical.

http://164.214.2.53/publications/guides/dtf/dtf.html

Digital shaded relief map of the United States, MAP 1-2206, by Gail P. Thelin and Richard J. Pike, US Department of the Interior, US Geological Survey, 1991, can be found at: http://www.usgs.gov/reports/misc/Misc._Investigations_Series_Maps_(I_Series)/I_2206/I_2206.html. The data in tiff format can be retrieved from:

ftp://wrgis.wr.usgs.gov/pub/topo/usa/i2206/usa_shade.tif.tar

More information on digital topography can be found at: http://www.geo.ed.ac.uk/home/ded.html

3

Coupled tectonic–surface process models with applications to rifted margins and collisional orogens

Christopher Beaumont, Henk Kooi and Sean Willett

3.1 Introduction

Plate tectonics successfully explains the primary topographic features of the Earth's surface, the division between continents and oceans, and the location of mountain belts and sedimentary basins at plate boundaries. Central to this explanation are concepts concerning divergent, transcurrent and convergent plate boundaries which provide the basic framework for physical models of the processes that form the plates. Now that the plate tectonics paradigm has matured, attention has been refocused on mesoscale problems. Among these problems are the more detailed tectonic and geomorphic evolution of plate boundaries, and the feedback effects of surface processes, such as denudation and sedimentation, on the primary tectonic processes active at these boundaries. In this chapter, we discuss aspects of progress in our understanding of surface processes and tectonics derived from models of these processes acting at convergent and divergent plate boundaries.

Passive, or Atlantic-type, continental margins and active rift valleys are now recognized to arise from the same processes of extension, rifting and ocean basin formation that stem from the breaking apart of tectonic plates and the divergent motions of the lithospheric fragments. Although the primary result of this divergence is subsidence and the formation of basins, the morphology of the onland portion of passive margins is distinctive. Features, such as the extensive escarpments ringing southern Africa and portions of South America, India, Australia and the Red Sea, remain problematic and not well explained by the basic tenets of plate tectonics (Ollier, 1985). They need to be addressed in the context of surface processes and tectonics. In contrast, convergent plate boundaries are characterized by both the flexural subsidence of the lithosphere at oceanic trenches and in foreland basins, and by extensive regions of surface and tectonic uplift which develop mountain and/or plateau topography. Dramatic high-elevation topography is typical of convergent continent–continent boundaries (collision zones) both in small (e.g., Alps) and large (Himalayan–Tibetan) orogens. Ocean–continent plate boundaries can either have equally dramatic topography (e.g. Andes) or be more subdued as in the Coast Ranges of the Pacific Northwest region of North America.

When viewed from an 'earth systems' perspective, we now recognize that crustal tectonic processes, although fundamentally generated by mantle convection acting on lithosphere, are also strongly linked to the atmosphere–hydrosphere system through denudational surface processes (Figure 3.1). The atmospheric system in turn depends on tectonic processes through climatic feedback with mountain topography. Other links exist between these systems, such as the dependence of climate on ocean circulation which, in turn, depends on the global distribution of the continents and, therefore, ultimately on mantle convection patterns in space and time (Figure 3.1). At the largest scale even mantle convection patterns evolve with the configuration of lithospheric plates and respond to the creation and fragmentation of supercontinents.

Geomorphology and Global Tectonics. Edited by Michael A. Summerfield.

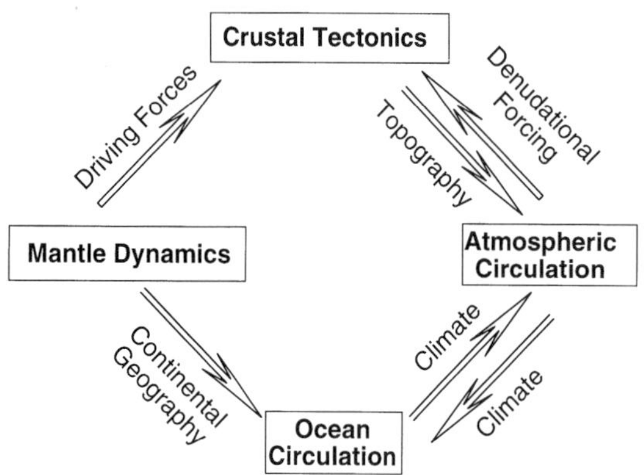

Figure 3.1 *Interactions and feedback mechanisms between earth systems responsible for the tectonic formation of topography and development of macroscale landforms*

Conceptually, a picture of a complex system of interdependency and feedback is emerging in which individual systems cannot be considered in isolation. To progress beyond concepts it is necessary to develop a quantitative understanding of the component processes so that the interactions can also be quantified. A suitable starting point is a brief review of the kinematic and mechanical models proposed for convergent and divergent plate boundaries because these models form the basis for the tectonic component of coupled surface–tectonic processes. We also indicate how surface processes enter the problem.

3.1.1 CONVERGENT PLATE BOUNDARIES

For convergent settings two classes of continuum mechanical models have been used extensively. The two classes are two-dimensional specialized forms of the fully three-dimensional problem which, until recently, proved computationally intractable except for simplified cases.

The first type is the thin-sheet models of Bird and Piper (1980), England and McKenzie (1982), Vilotte *et al.* (1982), England *et al.* (1985), England and Houseman (1986), Sonder *et al.* (1986), and subsequent work by these authors and others. In these models, lithospheric deformation is simulated in planform by assuming that the entire crust and lithospheric mantle deform together by pure shear shortening and thickening. This approach provides predictions of planform patterns of deformation, crustal thickening, isostatic balance and surface elevation, but it does not allow for vertical variations in deformation within the model lithosphere, except in special cases such as multiple, coupled sheets (Bird, 1989). The thin-sheet modelling parallels the plate description of tectonics. The perspective is that plates deform in planform and that the deformation is much the same throughout the lithosphere.

Although the thin-sheet assumption of no simple shear strain on horizontal planes may be a reasonable approximation for large regions of relatively uniform deformation (Tibet, for example), it is not likely to be valid for smaller size convergent orogens (such as the Alps and the continental collision zone of New Zealand) or regions bounding large plateaus (such as the Himalayan orogen). In these orogens, crustal deformation clearly shows the importance of near-horizontal detachments within the crust and the underthrusting of one lithospheric plate beneath another (Elliott, 1976a, b; Suppe, 1980; Price, 1981; Davis *et al.*, 1983; Mattauer, 1986; Cook and Varsek, 1994). This interpretation involving different levels of the crust deforming in dissimilar ways is also a common theme in older descriptive tectonic models (e.g. Argand, 1924), where the focus was on vertical cross-sections through the Earth, not planform structure. Observations concerning the deformation of crustal layers above decoupling horizons has led to the second class of two-dimensional models, the vertical section plane-strain models. The simplifying assumption here is that all motion is confined to the plane of the vertical section.

The plane-strain critical wedge models of Chapple (1978), Davis *et al.* (1983), Stockmal (1983) and Dahlen (1984, 1990) have proved to be particularly useful because they make predictions concerning deformation of brittle, near-surface material above a frictional detachment. Critical wedge models are based on the concept that convergence results in the formation of a crustal wedge of deforming material, analogous to a pile of sand in front of an advancing bulldozer blade. The surface slope of an ideal, uniform taper plastic (frictional) wedge will steepen until it reaches the minimum angle at which gravitational stresses balance the basal traction while keeping the material within the wedge everywhere at its yield stress.

While the critical wedge models were applied initially to fold-and-thrust belts on the periphery of major mountain belts or to accretionary wedges at ocean–continent boundaries, it was subsequently realized that entire orogenic belts could be modelled similarly through the use of numerical models which allowed for more general boundary conditions, stress distributions and material properties. For example, the reversal in the sense of basal traction beneath the crest of the mountain belt implicit in mantle subduction models can be included, as described by Willett *et al.* (1993). In this model, convergence between rigid mantle plates is achieved by underthrusting of one plate beneath the other, a process analogous to oceanic plate subduction (Figure 3.2a). The overlying crust is shortened, thickened and deformed by the opposing basal shear stresses applied by the moving mantle.

The importance of erosion or, more generally, mass redistribution by earth surface processes, to these models of convergent orogenic belts was recognized early (e.g. Elliott, 1976b; England and Richardson, 1977), although in most cases interest focused on rock exhumation and its role in determining metamorphic conditions and the explanation of pressure–temperature–time observations (e.g. England and Thompson, 1984). The importance of erosion to the mechanics of convergence was emphasized by Adams (1980) in his qualitative description of the New Zealand Southern Alps (see Chapter 6), and by Suppe (1980) in his critical wedge model of Taiwan (see Chapter 7). Since then, Davis *et al.* (1983), Dahlen and Barr (1989), Koons (1989), England and Molnar (1990), Sinclair *et al.* (1991), Beaumont *et al.* (1992) and Willett *et al.* (1993), have all amplified our understanding of the interaction between surface and internal processes, to the point where Hoffman and Grotzinger (1993) could write: "Savor the irony should those orogens most alluring to hard-rock geologists owe their metamorphic muscles to the drumbeat of tiny raindrops" (p. 198).

These observational and modelling studies consistently show that processes of crustal deformation cannot be understood independently from their interactions with surface processes which redistribute material through denudation of uplifted regions and deposition in basins, thereby changing the gravitational forces. Critical wedge or orogen models are particularly sensitive to this mass redistribution as we will demonstrate later. In fact, denudational forcing is potentially as important as the mantle forcing that is generally acknowledged to drive convergence and orogenesis (Figure 3.2a), as is testified by a range of observations from collisional orogens (Figure 3.2b). Modified deformation patterns are reflected in modified topography, which in turn modifies regional and sometimes global climate. Finally, a 'modified climate' may, on geological time scales, dictate the style of deformation within an orogen by varying the spatial patterns of denudation, thus defining a complex, coupled system with multiple feedback mechanisms (Isacks, 1992).

3.1.2 DIVERGENT PLATE BOUNDARIES

The main focus of research concerning divergent plate boundaries and passive margins was understandably the syn- and post-rift subsidence, the thermal history and the evolution of the sedimentary basins (e.g. see Allen and Allen (1990) for review). Kinematic models for the rifting process focused on lithospheric stretching by pure shear (McKenzie, 1978) and simple shear (Wernicke, 1985), or hybrid combinations of those mechanisms (Lister *et al.*, 1991) (Figure 3.2c). Subsequent mechanical models were designed to investigate the syn- and post-rift thermal and mechanical evolution of margins. These plane-strain vertical section models predict the extension and necking styles of lithospheres with complex stratified rheologies subjected to stress boundary conditions (e.g. Buck, 1986; Braun and Beaumont, 1989; Dunbar and Sawyer, 1989; Chèry *et al.* 1992; Kooi and Cloetingh, 1992; Bassi *et al.* 1993).

Although sedimentation is itself a surface process of mass redistribution, the geomorphic problem that has captured the most attention is the uplift and denudation of rift flanks. Geodynamic explanations for the tectonic flank uplift include differential stretching between crust and mantle (Royden and Keen, 1980), lateral heat flow (Cochran, 1983), dynamical effects due to secondary mantle convection (Buck, 1986), magmatic underplating (White and McKenzie, 1989), and flexural unloading of deep lithospheric necking. The last mechanism is favoured (Braun and Beaumont, 1989) when flank uplifts at rifted margins persist for 100 Ma or more – that is, longer than the time required for the lithosphere to cool, contract and subside from a hot syn-rift state. The margin flank uplift affects the morphology and general elevation of passive margins far inland (see Chapters 11, 12 and 13), and may also initiate the formation of the large erosional escarpments observed on some margins.

The correlation of erosional 'great escarpments' with rifted margins has been recognized and described (Ollier, 1985), but quantification of the strong coupling between tectonics and denudation has received less attention than similar processes in convergent orogens. There are several reasons for this difference. Uplifted rift flanks are the undeformed passively uplifted external parts of extension zones (Figure 3.2c). They are not regions of strong tectonic activity and there is probably a stronger coupling between isostasy and surface processes by onshore denudation flexurally coupled to offshore deposition (Gilchrist and Summerfield, 1990)

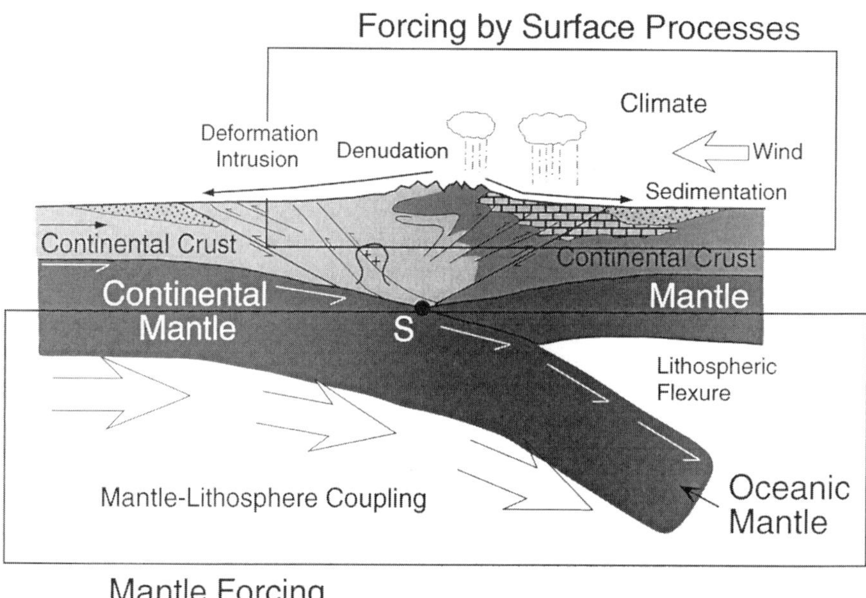

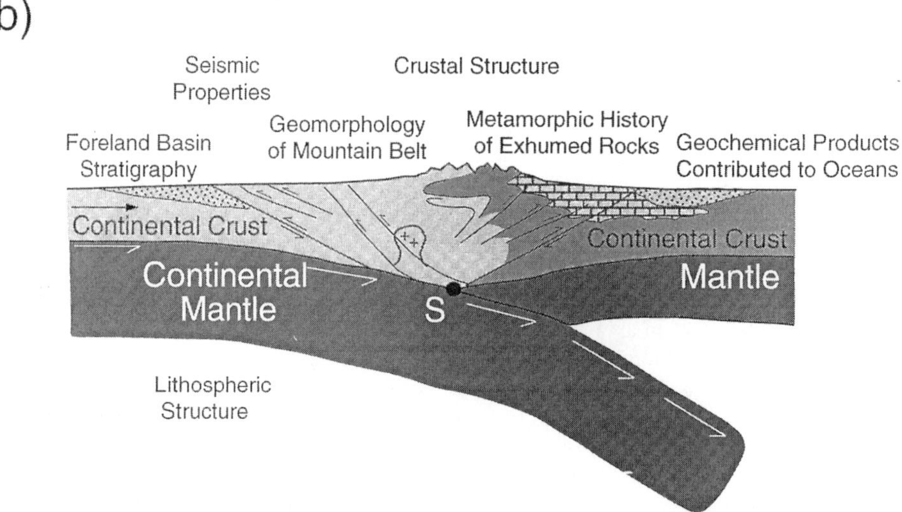

Figure 3.2 *(a) Primary processes affecting topography, morphology and structure of small- to medium-size convergent orogens. (b) Types of observations available from convergent orogens*

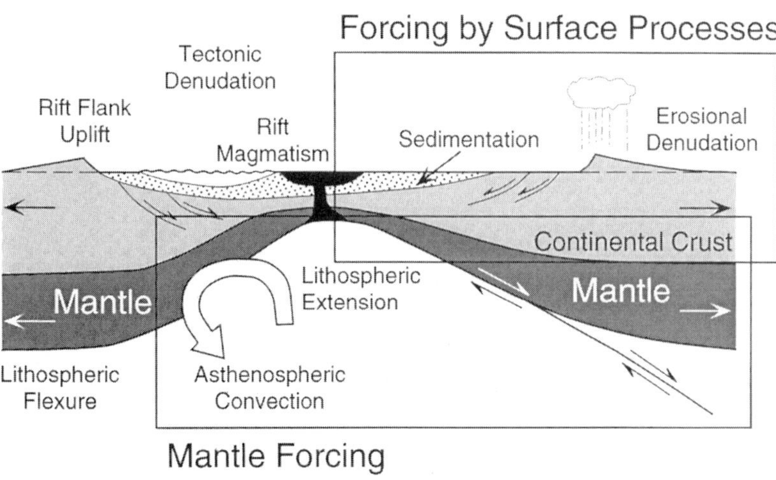

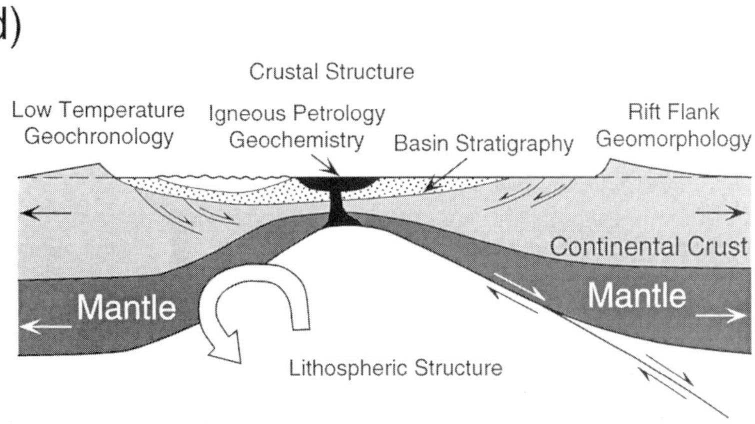

Figure 3.2 *(continued) (c) Processes affecting topography, morphology and structure of divergent or rift margins. (d) Types of observations available from divergent margins*

(see Chapter 13). Furthermore, the state of stress in uplifted rift flanks is generally not at frictional yield. Consequently, denudational redistribution of mass does not have the immediate gravitational feedback to deformation. However, within the zone of active lithospheric extension, surface processes and tectonics are certainly linked through tectonic controls on sediment deposition by extensional faulting. The idea that sediment deposition, in particular, should link to extensional faulting can be understood as a general application of extensional critical wedge mechanics (Xiao *et al.*, 1991). Sedimentation modifies both the surface slope of the wedge and the basal slope through isostasy and, therefore, influences the attainment of critical conditions. To our knowledge this aspect, which is a generalization of the slope failure–surface process interaction, has not been analysed in a quantitative manner. This coupled problem is the converse of the interactions that occur between crustal tectonics and surface processes in convergent settings, which we consider later.

3.1.3 STATING THE PROBLEM

The problem of interest in the present context concerns the mechanisms and strengths of the couplings and feedbacks between tectonics and surface processes as evidenced by deformation, the development of topography, growth of relief and the transport of material by surface processes. Our focus is not on the details of either component process, particularly at small spatial scales or short time scales; rather, it is on the interactions and circumstances under which the components form one dynamical system.

The approach used in this review is to consider only models of the component processes that are quantitative and have the potential to be coupled to investigate the dynamics of the surface processes–tectonics system. In Section 3.2 we present an overview of current surface process models (SPMs) with the potential for coupling to tectonic models (TMs). An assessment is then given of the status of coupled models. Our overview includes coupled models where the coupling is unidirectional; that is, either the TM or SPM is prescribed and drives the other model component, with no feedback. Sections 3.3, 3.4 and 3.5 are mainly drawn from our own research and are arranged to illustrate several facets of the coupled behaviour of SPMs and TMs. The focus of Section 3.3 is the denudational response and topographic development predicted by an SPM acting on simple kinematic uplift functions. In Section 3.4 conclusions drawn from experiments with the same and similar SPMs applied to passive margins are discussed. The post-tectonic evolution of the uplifted flanks of margins provides an example of tectonic and surface processes where it can be argued that the tectonics occur on shorter time scales than the denudation and, consequently, the system is only weakly coupled. By contrast, tectonics and denudation are often strongly coupled in convergent orogens, the subject of Section 3.5. In the last section (Section 3.6), we offer some thoughts on the limitations of current models and possible directions for improvements.

3.2 Surface Process Models and Coupling to Tectonic Models

The current state of research is well represented by the special *Journal of Geophysical Research* issues on tectonics and topography edited by Ellis and Merritts (1994) where coupled geodynamic–surface process models are used to better understand: (1) rift margins (Gilchrist *et al.*, 1994; Kooi and Beaumont, 1994; Tucker and Slingerland, 1994); (2) oblique collisional margins (Koons, 1994); (3) compressional zones (Zhou and Stüwe, 1994); and (4) continental plateaus (Gregory and Chase, 1994; Masek *et al.*, 1994; Montgomery, 1994). Koons (1995) has recently reviewed his own and other work in the context of modelling of the topographic evolution of collisional belts. *Basin Research* (vol. 8, no. 3, 1996, pp. 199–369) includes additional articles with both an observational and modelling perspective.

Instead of an attempt to review the development of SPMs and coupled TM-SPMs, we tabulate (Table 3.1) some of the research which we believe contributed to progress in these fields. Depending on the application, many of the techniques proposed may still be useful. For example, at the largest of scales and where a kinematic approach suffices, models that couple diffusion of mass from uplifted regions to neighbouring sedimentary basins, and flexurally compensate the redistributed mass (e.g. Flemings and Jordan, 1989; Sinclair *et al.*, 1991) reproduce the first-order behaviour. Such models are, however, limited because they lack dynamical coupling between the tectonics and surface processes and because the scale dependence of both processes cannot be investigated.

The models (Table 3.1) can be classified according to the degree to which the model calculates or predicts component processes based on basic principles, and the degree to which the coupling includes the interaction and feedback processes anticipated for natural systems.

The minimum requirement for an SPM that will be coupled to a TM is that the SPM predicts lateral mass fluxes among the various components of the model landscape. Correspondingly, a TM must calculate the solid earth mass fluxes in response to the surface processes and as a consequence of tectonic forces or velocity fields. When integrated spatially through time, the coupled system predicts the topographic and landscape evolution and the deformation within the crust or lithosphere. As an example, the minimal coupled model (not included in Table 3.1) could be as simple as a vertical one-dimensional column through

Table 3.1 *Surface process models (SPMs) and coupled models (TM-SPMs)*

Models	Comments
SPMs	
Single-process landscape models	
Stephenson (1984); Stephenson and Lambeck (1985)	2-D, erosion rate proportional to elevation but varies with wavelength of topography
Moretti and Turcotte (1985)	1-D diffusion of homogeneous substrate
Dahlen and Suppe (1988)	1-D, steady, erosion of homogeneous substrate proportional to elevation
Koons (1989)	2-D diffusion and *ad hoc* fluvial transport of homogeneous substrate
Stark (1994)	2-D cluster growth algorithm
Zhou and Stüwe (1994)	1-D, steady, total erosion proportional to elevation
Shaw and Mooers (1988)	2-D, heuristic algorithm where erosion is proportional to slope
Multi-process landscape models	
Carson and Kirkby (1972)	Summary including review of process–response models at hillslope scale
Ahnert (1976, 1987a, b)	2-D hillslope model (without channels), simulating bedrock weathering, splash, wash, and viscous or plastic flows
Armstrong (1976)	Diffusive soil creep, exponential weathering with depth, fluvial erosion
Kirkby (1986)	2-D slope and evolution model
Musgrave *et al.* (1989)	2-D hydraulic transport of suspended sediment, plus critical slope failure
Chase (1992); Gregory and Chase (1994)	2-D topography sculpted by precipitons which cause diffusional smoothing, then flow down steepest descent, eroding an amount proportional to QS if capacity exceeded
Anderson and Humphrey (1990)	1-D hillslope, nonlinear diffusive transport limited by soil thickness produced by bedrock weathering algorithm
Willgoose *et al.* (1991, a, b, c)	2-D nonlinear fluvial, channel initiation function, hillslope diffusion
Masek *et al.* (1994)	Chase (1992) modified to simulate orographic precipitation
Tucker and Slingerland (1994)	2-D non-uniform downstream routing of sediment and water; creep and threshold slumping; detachment- and transport-limited fluvial processes; bedrock and regolith algorithms
Howard (1994)	2-D non-uniform downstream routing of sediment and water; creep and threshold slumping; detachment- and transport-limited fluvial processes; bedrock and regolith algorithms
Kooi and Beaumont (1994); Gilchrist *et al.* (1994)	1-D and 2-D planform SPM, as Beaumont *et al.* (1992) with treatment of drainage divides
Braun and Sambridge (1997)	2-D planform on irregular self-refining grid, downstream routing of sediment and water; hillslope diffusion; fluvial carrying capacity proportional to river power; fluvial entrainment proportional to entrainment potential
TM–SPMs	
Stephenson (1984); Stephenson and Lambeck (1985)	SPM noted above coupled to elastic and viscoelastic lithospheric flexure
Flemings and Jordan (1989)	1-D diffusion of basement and sediment coupled to 2-D kinematic orogen-foreland basin model
Ellis *et al.* (1990)	2-D SPM from Beaumont *et al.* (1992) coupled to thin-sheet tectonic model
Sinclair *et al.* (1991)	1-D diffusion coupled to 2-D kinematic orogen-foreland basin model
Beaumont *et al.* (1992)	2-D (vertical section) dynamical tectonics, 2-D planform unsteady, non-uniform downstream routing of sediment and water; hillslope diffusion; fluvial carrying capacity proportional to river power; fluvial entrainment proportional to entrainment potential, orographic precipitation
Anderson (1994)	2-D self-similar diffusional hillslopes including a threshold for landslides; fluvial lowering rate proportional to QS; no capability to deposit sediment; coupled to planform kinematic tectonic model
Johnson and Beaumont (1995)	Kinematic 3-D orogen, using 2-D critical wedge concepts, flexural isostasy, planform SPM, orogen denudation and foreland basin stratigraphy
Beaumont *et al.* (1996)	2-D (vertical section) tectonics, 1-D SPM orographic precipitation, dynamical coupling
Tucker and Slingerland (1996)	Kinematic uplifts that represent elliptical hogsbacks, planform SPM modified from Tucker and Slingerland (1994)
Kooi and Beaumont (1996)	SPM of Kooi and Beaumont (1994) with kinematic tectonics of varying time scale
Avouac and Burov (1996)	1-D diffusive SPM coupled to 2-D ductile flow in lower crust and flexure

In table Q = discharge, S = slope.

the crust in which specified surface denudation or deposition causes a local isostatic response of the lithosphere. When distributed as a spatial planform and coupled together by flexural isostasy, models of this type have been used to investigate the regional isostatic response of the crust (Stephenson 1984; Stephenson and Lambeck, 1985), and more local orogenic response (Moretti and Turcotte, 1985), to diffusive denudation. Zhou and Stüwe's (1994) model is a variation on this theme that includes some aspects of the kinematics of lithospheric deformation but assumes local isostatic compensation. The advantage of these models when they are simple is that subsidence and uplift are analytical functions of initial surface elevation and densities of the crust and mantle (e.g. Turcotte and Schubert, 1982, pp. 112–132). Consequently, the models allow relatively simple relationships to be established among the first-order processes and parameters. These relationships may also be factors in the dynamics of the more complex numerical models and in natural systems.

At the next level of complexity, models incorporate geometric or kinematic assumptions concerning the tectonic process and its horizontal variation. For zones of plate divergence, these assumptions include simple or pure shear of the crust/lithosphere specified by a small number of parameters (e.g. the stretching factor of the McKenzie (1978)-type stretching models, the simple shear of the Wernicke (1985) model and other combinations noted in the introduction). For zones of convergence, the converse shortening factors (e.g. England and Thompson (1984)) are equivalent. Such models predict geometrical extension or compression of the lithosphere, the isostatic response, the vertical surface displacement and vertical motion of the rock column in response to specified surface and thermal processes. However, the lack of mechanics in the tectonic processes and lack of process in the surface processes means that these models are unsuitable for dynamical investigations of the coupled system.

Models designed to investigate the dynamics of the coupled system must include the mechanics of the tectonic process in order that the primary coupling, through gravitational and buoyancy forces owing to topography, can influence the state of stress and deformation rate. The specialized vertical (plane-strain) and planform (thin-sheet) mechanical models (Sections 3.3.1, 3.3.2) have been used.

Barr and Dahlen (1989) and Dahlen and Barr (1989) have considered the special case of a critical Coulomb wedge that maintains its uniform critical taper geometry while it is eroded; that is, the deformation within the wedge and material uplift velocity at the surface have to satisfy the Coulomb rheology and match the denudation rate of the surface processes. While these conditions can be satisfied for functionally simple spatial denudation rates, with correspondingly smooth tectonic velocity fields, this type of model cannot address the mechanics when the Coulomb wedge is not at yield everywhere (i.e. does not have uniform critical taper), or where the boundary conditions or material properties are more general. The same restriction applies to all critical wedge mechanical models that predict the geometry of critical wedges, but not their internal deformation. The fully numerical models (e.g. Willett, 1992) include both components. Recent progress, in which models of both simple and more complex surface processes have been coupled to vertical cross-section mechanical models, is discussed in Section 3.5.

Correspondingly, investigations of the response of planform TMs to SPMs that include both hillslope and fluvial processes have recently been published. Anderson (1994) has described a coupled model in which the TM is kinematic and represents flexurally compensated deformation of the crust adjacent to a bend in a strike-slip fault. The TMs employed by Johnson and Beaumont (1995), Kooi and Beaumont (1996), and Tucker and Slingerland (1996) are equally kinematic. Ellis *et al.* (1990) have described a dynamical model in which the response of a viscous thin-sheet (planform) TM subject to surface denudation by a multiprocess SPM was investigated. The limitation here is that the feedback effects of localized denudation on the deformation are not accurate when they create local sheet thickness variations that violate the conditions for thin-sheet calculations; that is, the thin-sheet approximation is only valid when horizontal gradients in model properties are small. It could be argued that only the long wavelength components of the denudation field affect the tectonics and that the low-pass filtered denudation should be coupled to the thin sheet. However, this approach is probably not valid. Complete plane-strain vertical section models (Section 3.5.2), that do not use the thin-sheet simplifications, tend to show strong denudational feedback effects on tectonics that localize at tectonic–topographic fronts where there are large gradients in the model properties.

In summary, the capabilities of the present generation of coupled numerical tectonic–surface process models are restricted by assumptions required to make the computations feasible. Two-dimensional vertical section models have addressed the full mechanical coupling for a restricted range of tectonic assumptions

(Section 3.5). There are a range of coupled, but not feedback, models in which tectonic uplift is prescribed kinematically (Section 3.3). In addition, it would certainly be possible to calculate the effects of diffuse coupling in planform models if thin-sheet approximations are made in the tectonic calculations. The heart of the problem is that surface processes fundamentally operate in planform. Only in certain circumstances is it valid to couple this planform to specialized two-dimensional tectonic models. In order to assess these circumstances, investigations of coupling between planform SPMs and three-dimensional TMs are necessary.

3.3 Denudational Response to Kinematic Tectonic Forcing

In this section results of model experiments using a surface processes model (SPM) forced by assumed simple kinematic uplift functions are presented. In these experiments denudation does not affect the uplift rate. As mentioned earlier, these are coupled models but without feedback; however, the experiments illustrate a number of fundamental concepts that can be used to understand the behaviour of the fully coupled system discussed in later sections. The section largely summarizes work published by Kooi and Beaumont (1996). We emphasize those behaviours which are thought to be robust in the sense that they do not depend critically on the particular formulation of the SPM.

3.3.1 OUTLINE OF THE SURFACE PROCESSES MODEL

The SPM calculates long-term (millions of years time scale) changes in topography, h, that result from simultaneous short- and long-range mass transport processes. Short-range transport represents the cumulative effects of hillslope processes and is modelled as linear diffusion, $\mathrm{d}h/\mathrm{d}t = K_s \nabla^2 h$. Diffusivity, K_s, implicitly includes climatic and substrate influences, and, presumably, is scale dependent.

Long-range transport represents fluvial transport and is modelled by a network of one-dimensional rivers that drain the current model topography via their respective routes of steepest descent. The equilibrium sediment carrying capacity q_f^{eqb} of the long-range transport network is considered proportional to local linear river power, $q_f^{eqb} = K_f q_r \mathrm{d}h/\mathrm{d}l$, with discharge q_r, and local downstream gradient, $\mathrm{d}h/\mathrm{d}l$. Discharge is the result of precipitation distributed over the model topography and collected by the drainage network. No distinction is made between hillslope and channel elements. The local fluvial sediment flux q_f is determined by integrating a reversible entrainment–deposition equation down the river network, $\partial h/\partial t = -\mathrm{d}q_f/\mathrm{d}l = -(1/l_f)(q_f^{eqb} - q_f)$. This entrainment equation is analogous to a first-order 'chemical' reaction in that the exchange of mass between the immobile and mobile phases is driven by the degree of disequilibrium in the sediment load, with a proportionality constant $1/l_f$. Depending on the sign of the disequilibrium, l_f is either a deposition or erosion length scale. For erosion, l_f is proportional to the detachability of the substrate. A more detailed description and justification of the model formulation can be found in Beaumont *et al.* (1992) and Kooi and Beaumont (1994).

3.3.2 RESPONSE TO TECTONIC FORCING

3.3.2.1 Steady-state landscape

Plate 3.1 shows results of an experiment in which a nearly flat surface with low-amplitude white noise topography is subjected to a step change in uplift rate at $t = 0$. Subsequently, uplift rate and all other model parameters are held constant. Two opposing boundaries (b) serve as base levels of erosion; they are held at constant elevation, and sediment delivered at these boundaries is removed from the model. The other boundaries (a) have been assigned no-flux boundary conditions.

The plate illustrates that the model landform evolves toward a steady state. In steady state, the drainage network and topographic relief are finely adjusted such that every part of the model landscape is denuded at the local rock uplift rate. In this example there is no sediment deposition in the steady-state landscape. Other experiments show that steady states exist for any uplift function when the external controls, substrate conditions and drainage network are constant.

3.3.2.2 Linear response and landscape response time

Linear system behaviour implies that the principle of superposition applies. That is, the response to a sum of inputs equals the sum of the outputs of the individual inputs. A characteristic of simple linear system behaviour is an exponential approach to a new

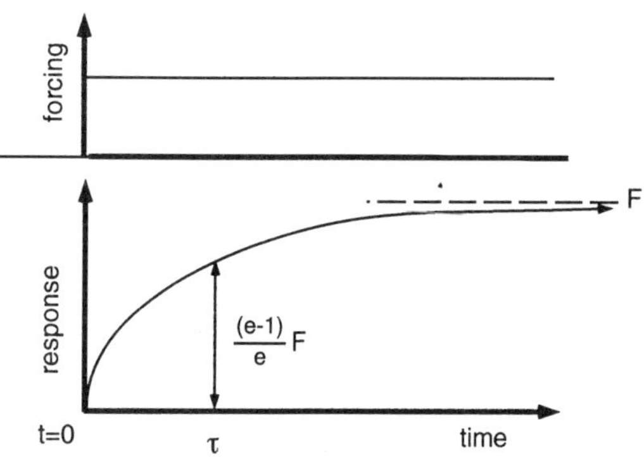

Figure 3.3 *Character of the exponential response of a linear geomorphic system to a step-function change in input (forcing). The forcing may be tectonic or climatic and the time scale of the response, the response time of the system or landscape, is measured by τ. Note that the response time will generally vary among different forcing functions*

equilibrium (steady state) following a step change in input (forcing) (Figure 3.3). The time scale of the exponential behaviour is a system constant which fully characterizes the system response to any forcing function. This time scale, referred to as the system response time, is defined by the time it takes to achieve a fraction (1 – 1/e) of the total step response, where e = 2.71828 (Figure 3.3).

The model response to kinematic tectonic forcing (specified uplift) is quasi-linear provided that the uplift geometry is approximately symmetric with respect to the base levels and tends to impose a drainage system that is fully connected with base level – that is, one that is not internally drained. For the experiment shown in Plate 3.1 a quasi-linear exponential evolution is observed for: (i) average elevation; (ii) average local relief; (iii) average sediment yield from the landscape; and (iv) average denudation rate.

The model landscape response time is a sensitive function of climate and substrate conditions, factors that influence K_s, q_r and l_f, and the spatial scale of the tectonic uplift. Conditions which lower the substrate erodibility (increase detachability) or enhance the transport capacity reduce the response time and vice versa. The response time roughly scales with the length scale of uplift, L, as L^{α}, with $1 < \alpha < 2$. For example, the response time for the experiment of Plate 3.1, with L = 50 km, is approximately 6 Ma. Correspondingly, response times of tens of millions of years are predicted for orogens with widths of 100 km (α = 2) and a Himalayan–Tibetan-scale orogen may have a response time >100 Ma.

3.3.2.3 Slow, intermediate, rapid and impulsive tectonic forcing

The concept of landscape response time is useful, because it is central to the characterization of the large-scale landscape response to arbitrary uplift histories (provided the system is linear). That is, the response depends on the ratio of the time scale of tectonic forcing and the response time, t_T/τ, where t_T is the period (or interval) for which the temporal uplift rate function operates. The model response for different forcings is schematically illustrated in Figure 3.4. For slow forcing, denoted by $t_T/\tau >> 1$, the response variables of morphology, denudation rate and sediment yield can closely track the steady state for the current uplift rate (Figure 3.4a). For rapid forcing, denoted by $t_T/\tau << 1$, the response variables cannot follow and only 'feel' the longer wavelength components in the forcing, when these exist (Figure 3.4c). For intermediate forcing, denoted by $t_T/\tau \sim 1$, the response variables do follow, but are damped in amplitude and have a phase shift or lag with respect to the forcing (Figure 3.4b). For impulsive tectonic forcing which instantaneously establishes a tectonic landscape, and which corresponds to a t_T/τ which contains all possible values (white spectrum), the response variables follow a decay curve which is approximately exponential (Figure 3.4d). The initial deviation from an exponential reflects nonlinear transient effects owing to adjustments at the scale of interfluves and river reaches, and demonstrates that the linear macroscale model behaviour is only an approximation.

The model geomorphic responses for impulsive, intermediate and slow tectonic forcing contain many characteristics of the classical conceptual geomorphic frameworks envisaged by W.M. Davis (1899), W. Penck (1953), and J.T. Hack (1960), respectively. Moreover, the model provides a tangible process-based framework in which more modern concepts such as steady-state landscapes and the scale dependence of landscape response times (e.g. Ahnert, 1984) and landscape sensitivity (Brunsden and Thornes, 1979) can be quantified and investigated. Some relationships between model behaviour and these concepts were discussed by Kooi and Beaumont (1996).

A result of significance in the present context is the predicted lag in response for intermediate tectonic forcing (Figure 3.4b). The time lag between topographic evolution and sediment yield is shown in some detail in Plate 3.2 for an experiment employing a wedge-shaped uplift geometry. The experiment shows a lag of about 6 Ma, which is similar in magnitude to

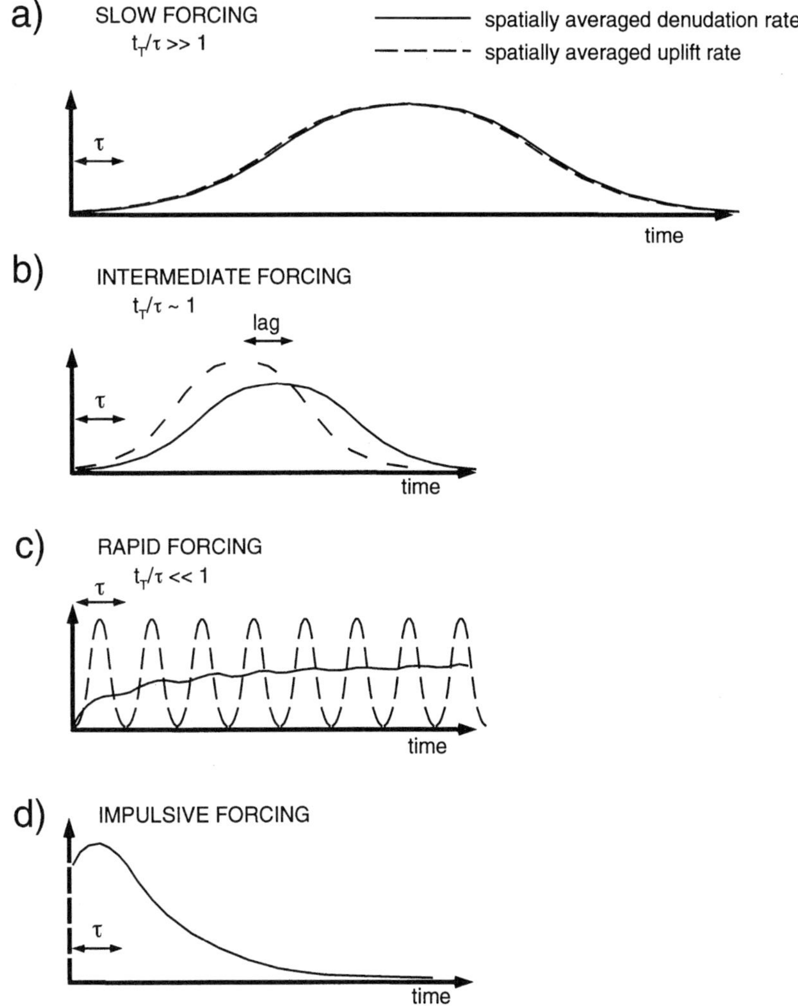

Figure 3.4 *Relationship between tectonic forcing (measured by spatially averaged uplift rate) and response (measured by spatially averaged denudation rate, or sediment yield at the boundaries) of quasi-linear SPMs. t_T is the period or interval of the tectonic forcing and τ is the response time of the landscape. For slow forcing (a) the evolution closely tracks steady-state conditions. For intermediate forcing (b) the response is attenuated and lags the forcing. For rapid forcing (c) the response is dominated by the approach to the average conditions and the rapid variations are strongly attenuated. For impulsive, that is, very rapid 'spike-like' forcing (d) the immediate response is attenuated and subsequently it decays with the characteristic response time of the landscape*

the response time of this model. Correspondingly, lags of tens of millions of years are predicted by the model for landscapes with longer response times, for example owing to larger spatial scales of uplift and/or climatic and substrate conditions corresponding to lower erosion and transport efficiency. There are at least two implications of these model outcomes if natural orogens behave in a similar way. One is that measurements of the time lag provide information on the response time, particularly if the time scale of tectonic forcing is known independently. Another is that maximum sediment yield from an orogen may lag significantly behind the tectonic maximum, and that sedimentary evidence from ancient orogens may therefore only give a lower bound for the timing of peak orogenic activity.

The phase lag for the model of Plate 3.2 is the most simple one and occurs even when substrate conditions and the drainage network are constant. More recently, Tucker and Slingerland (1996) have discussed the same basic phase lag and other reasons for more complex phase lags in the context of a different SPM formulation. These significant delays are not restricted, however, to a particular SPM formulation but are also anticipated for natural systems and should be taken into account when using the stratigraphic record to date tectonic events.

3.3.2.4 Nonlinear behaviour

The concepts of linear landscape behaviour and landscape response time are useful, but require special conditions. Some nonlinearity owing to minor reorganization of the drainage system (often associated with some temporary deposition which changes substrate properties) is a typical model behaviour, even for conditions favouring a linear response. Pronounced nonlinear behaviour occurs when: (1) tectonic forcing includes local or regional subsidence; and (2) the uplift geometry is either strongly asymmetric with respect to the base levels, or (3) tends to impose internal drainage systems (Figures 12–14, Kooi and Beaumont, 1996). Such geometrical nonlinearities occur even when the processes operating within the SPM are linear.

3.3.3 COMPARISON WITH OTHER WORK

Steady-state landscapes and landforms have been recognized in several studies using SPMs. Ahnert (1987a, b) described a steady-state form of a model interfluve when its elevation changes at the same rate as the adjacent streams incise. The same was noted by Koons (1989) in a diffusion model for the Southern Alps. In Koons' model, linear diffusion transports mass from hillslopes to imposed rivers with a fixed elevation with respect to base level, while the whole model landscape is subject to uplift. It can be shown that the response time of interfluves (to achieve a steady-state form), subject to linear diffusion, scales as width of the interfluve squared as expected for a linear diffusive system. In our model this scaling is similar for small, diffusion-dominated interfluves. However, large fluvially dominated interfluves in our model have a response time that is roughly proportional to interfluve width (Kooi and Beaumont, 1996). The difference reflects the L versus L^2 scaling of the fluvial and hillslope model processes.

As in our experiments, Koons (1989) also noted the development of steady-state macroscale landscapes for constant uplift rate in his model, although this behaviour was in part determined by the specified river profiles. Willgoose *et al.* (1991a) found the same behaviour for a single-catchment evolution model. It is likely that large-scale steady-state landscapes exist for constant external forcing (tectonics, climate) irrespective of the formulation of the transport processes. That is, denudation rate can balance uplift rate at all points within a model landscape for a wide range of SPM formulations.

Quasi-linear model behaviour, however, does depend to a certain degree on the formulation of process. Several SPMs use a fluvial transport formulation where the local fluvial sediment flux is nonlinearly dependent on stream gradient (e.g. Willgoose *et al.*, 1991b; Tucker and Slingerland, 1994, 1996). This will probably lead to a nonlinear, large-scale response in these models, reflected, for instance, in a non-exponential step-response behaviour. Nonlinearity may also be introduced when hillslope transport is divided into component processes that include thresholds, such as mass failure, and transport of the regolith boundary layer that is limited by an explicit weathering formulation. However, the concept of landscape response time would still be meaningful – even though it might be defined in a somewhat different way (e.g. Ahnert, 1987a, b). It is useful to define a response time if this time gives an indication of the time scale at which a landscape adjusts to changes in tectonic forcing, and the delays in the response that may be expected. Moreover, it remains possible that a system which consists of a great number of interacting component processes that are individually complex, self-organizes to exhibit a quasi-linear behaviour at sufficiently large spatial and temporal scales.

Nonlinearities arising from the competition between drainage basins for relatively complex uplift geometry will be discussed below. These nonlinearities are probably important for regional and continental-scale geomorphic evolution and sediment transport and, obviously, cannot be captured by single-catchment evolution models, or models with imposed and fixed drainage systems.

In addition to the geomorphic response characteristics discussed here, it is important to compare empirical relationships (commonly observed statistical properties), such as those between denudation rate and local relief, and area–slope–elevation characteristics of drainage networks, to the equivalent model properties (e.g., Willgoose, 1994a; Willgoose *et al.*, 1991a). In particular, the search for model statistics that may discriminate between different states of landscape evolution (steady-state, dynamic equilibrium, transient) is very valuable (Willgoose, 1994b). However, the problem remains that the current state of a natural landscape must be known before its properties can be correctly equated with the equivalent model statistics. Many landscapes may combine different states at different scales because the response time varies with the scale at which the landscape is analysed (Kooi and Beaumont, 1996).

3.4 Post-Rift Evolution of Rifted Margins

In this section, we discuss the model surface and denudational response to kinematic uplift functions that are designed to be in accord with the tectonics of rifted margins. A similar model approach to that of Section 3.3 is used to address the specific problem of 'great escarpments'. Current understanding of tectonics in extensional settings (Figure 3.2c) leads to the view that vertical movements in rift basins are dominated by subsidence within the rift, owing to crustal thinning and ensuing thermal contraction of the lithosphere. Uplift of a region outside the rift may be explained by a number of processes mentioned earlier. Rift flank topography may also be, in part, inherited with the rejuvenation of relief caused by base-level lowering of the rift basin within high-elevation surroundings.

Rift shoulders are often characterized by a marked large-scale asymmetry (Figure 3.2c) with a steep rift-facing escarpment and a more subdued opposite-facing flank, which may take the form of an upland plateau perhaps modified by flexural tilting during rifting. At young rift basins along the East African rift zone and at Lake Baikal the tectonic uplift geometry can still be recognized where the escarpments coincide with major graben-bounding normal faults. This coincidence implies that syn-rift tectonic uplift was fast relative to the destructive work of denudation. At mature rifted continental margins, such as southern Africa, Brazil, western India and eastern Australia, the great escarpment is found far inland from the rift basin (Ollier, 1985). Evidence exists for kilometre-scale denudation in the coastal area between the escarpment and the rifted margin, indicating considerable elevation of rift shoulders during and soon after rifting. In the case of southern Africa, substantial denudation has also been inferred inland of the great escarpment (Gilchrist *et al.*, 1994). At these margins, denudation has overprinted the original rift tectonic signature. The margins of the Gulf of Suez and Red Sea have intermediate properties: no large-scale retreat of the basement escarpments but sufficient denudation of the overlying sediments to be recorded in fission-track data (Steckler and Omar, 1994).

Backstacking techniques (Brown, 1991; van der Beek *et al.*, 1994) (see Section 4.4.1) can provide a rough indication of syn-rift morphology, by removing rift-related sediments from the basin and restoring mass on the denuded part of the margin using denudation estimates from apatite fission-track analysis and assuming isostatic adjustment to the associated loading. However, backstacking does not have enough resolution to determine unequivocally the relative role of syn-rift tectonics and syn-rift denudation. It is not clear, for example, that good criteria exist to distinguish tectonic denudation associated with normal faults from erosional denudation, solely on the basis of thermochronological data. Moreover, the importance of syn-rift erosional denudation probably depends strongly on the duration of rifting (which can comprise several phases and last as long as 100 Ma, e.g. North Atlantic margins), and on climatic conditions. In spite of these complicating factors concerning their evolution, great escarpments are one of the foremost large-scale morphological features of young and old rifted continental margins, particularly where the upland region inland of the escarpment has a high elevation.

3.4.1 ESCARPMENT EVOLUTION

The SPM of the previous section has been used to investigate some of the factors that control the evolution of model simulations of great escarpments (Gilchrist *et al.*, 1994; Kooi and Beaumont, 1994). In these models the initial morphology includes an escarpment, and represents the tectonically imposed rift-flank morphology at the end of rifting. The syn-rift evolution is not explicitly modelled. It is, therefore, implicitly assumed that the syn-rift tectonics are fast in comparison with the denudation processes. Both one- and two-dimensional (planform) models were investigated. No strike variation was included in the initial planform morphology (except for small amplitude white noise) and each model was used to investigate particular controls in isolation.

Among the principal controls inferred from the experiments was the antecedent topography inherited from rifting. In particular, the location of drainage divides relative to the tectonically imposed escarpment play a first-order role. Plate 3.3 illustrates landscape evolution in a composite model which combines the two basic styles of initial upland drainage – drainage towards and away from the initial escarpment. The motivation is to investigate the planform evolution when the tectonic escarpment coincides with the primary drainage divide in some locations, whereas in others the drainage divide is inland of the tectonic escarpment. This style of morphology must be common in rift zones. The initial escarpment is linear and located near the bottom of the first panel, which shows the morphology after 1 Ma of the model evolution. The upland surface is sinusoidal with angled sinusoidal truncations at the lower ends of the two long wavelength antiforms. The details of this artificial morphology are not important. What is important is that upland drainage is toward the escarpment in some areas and away from the escarpment in others. Reflective boundary conditions for model fluxes are used on the sides perpendicular to the escarpment. The base of the initial escarpment is kept at a constant sea-level elevation and acts as the base level for erosion through which sediment is removed from the model. The opposite, inland boundary is unerodible but allows the passage of the sediment flux. This boundary acts as an elevated local base level in the continental interior. The model includes flexural isostasy in response to the denudational unloading. The factor of five vertical exaggeration of the figure suggests more upland relief than the true low amplitude, which is typical of an old planated upland analogous to parts of the eastern Australia tableland, inland of the escarpment. The 1–10 Ma evolution shows that where the escarpment and drainage divides are widely spaced, so that a significant amount of discharge drains over the escarpment top, the upland becomes strongly incised as the river heads retreat rapidly towards the divide powered by the discharge from the large upland catchment. The initial escarpment quickly degrades, moderately long interfluves develop, and the region between the drainage divide and the original escarpment evolves mainly by downwearing of the interfluves. As the interfluves are gradually removed, a new escarpment emerges at the inland drainage divide. In contrast, escarpment retreat occurs where the top of the escarpment coincides with a drainage divide. The key to preservation of escarpment slopes and, therefore parallel retreat, is the competition between diffusive slope decline and fluvial slope steepening that occurs in the vicinity of the escarpment top when it is a drainage divide.

Both of these modes of landscape evolution, parallel retreat and incision, are analysed by Kooi and Beaumont (1994). The results suggest that an initial tectonic escarpment will evolve to an escarpment positioned at the upland drainage divide and then retreat slowly, but uniformly, into the upland (e.g., 40 Ma panel Plate 3.3).

At approximately 50 Ma (Plate 3.3), some of the upland drainage in the model is captured and flow over the escarpment causes rapid incision beyond the embayment in the escarpment (60 Ma, Plate 3.3). The same process occurs later in the other embayment. The combined capture process leads to a new phase of escarpment roughening and degrading in which the promontory is converted into a remnant erosional outlier, while the escarpment is partly re-established farther into the upland. The causes of major river capture in the model and the resulting new mode of landscape evolution by degradation and incision are the significant elevation difference between the upland and lowland catchments and the low divide between them (mostly lowered by the denudational processes in response to the elevation difference between the catchments). These causes are in agreement with the conditions of 'true river capture' listed by Bishop (1995). These capture events in the model are also associated with significant changes in the temporal and spatial distribution of sediment yield to the adjacent basin (Kooi and Beaumont, 1996).

If the two modes, (R) slow parallel retreat and (I) rapid incision of a catchment, occur in nature, the broad-scale morphology of escarpments may be a manifestation of the dynamic superposition of the two modes. Only in a few localities will either I or R create morphologies like Plate 3.3 (Gilchrist *et al.*, 1994) or Plate 3.1 (Kooi and Beaumont, 1994). The predominant pattern may be more like the 100 Ma panel in Plate 3.3.

Apart from the inherited morphology from rifting, there are a number of other controls of escarpment evolution which we have inferred from the model experiments. Details can be found in Gilchrist *et al.* (1994) and Kooi and Beaumont (1994, 1996). Figure 3.5 summarizes these controls in a conceptual planform diagram which represents one flank of a rift system that subsequently evolved into a passive margin. The master normal fault (1) is shown in the same vicinity as the coastline (2) but could equally be farther offshore. Also shown are subsidiary normal faults (3) which may control tectonic escarpments on the coastal plain. The main escarpment (4) was originally coincident with the master fault but has retreated into the upland by the R- and I-modes.

In particular, the I-mode allowed the large catchment antecedent river (5) to compete with rift-flank tectonics and continue to drain as before. That is, some existing rivers, depending on their denudational capacity, can

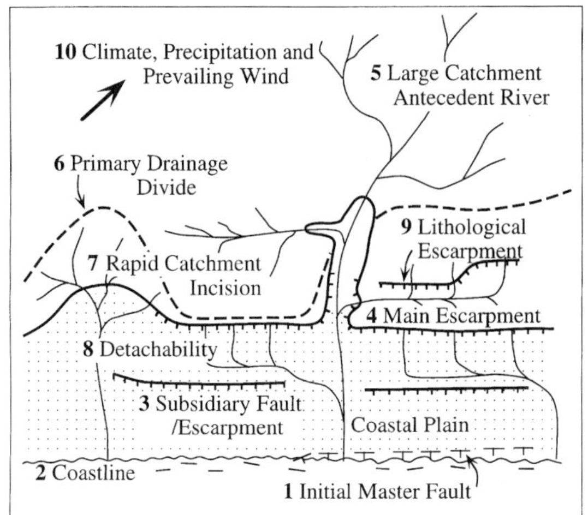

Figure 3.5 *Planform illustration of controls on evolution of rift margin tectonic escarpments inferred from model experiments with SPMs. The relationship of the evolution of the main escarpment to antecedent drainage, substrate lithology and climate is described in the text*

compete with rapid tectonic uplift of the rift flanks. Other rivers did not compete, and elsewhere a primary drainage divide was established coincident with, or somewhat inland of, the escarpment top. Where the drainage divide was coincident with the escarpment top, the escarpment retreated inland in an R-mode.

Continuous backtilting of the escarpment due to flexural isostatic uplift in response to denudational unloading helps to maintain both the escarpment top as a drainage divide and the escarpment gradient during retreat. Denudation of the coastal plain can be a factor of two to three greater than the initial elevation (e.g. Gilchrist *et al.*, 1994; Figure 11), depending on the flexural isostatic response to denudation and the counteracting effect of sediment deposition in the rift basin, which may, or may not, be flexurally coupled to the rift flanks.

The current position of the drainage divide (6) represents: I-mode retreat (area 7), like the early evolution of Plate 3.3; R-mode retreat to the left of the major river; and modified R-mode retreat to the right of the river, where escarpments above the main escarpment (one shown, 9) are controlled by the lithology of the eroding substrate.

We interpret the SPM parameters K_s and l_f (Section 3.3.1) as substrate material properties that are influenced by climate (10) (Kooi and Beaumont, 1994). A simple interpretation is that the magnitude of K_s is large for a substrate that weathers rapidly and a large l_f represents bedrock that strongly resists fluvial detachment (8). The particular combination of lithologies that are exhumed during escarpment retreat, combined with the prevailing climate, can strongly modify the I- and R-modes. For example, when lithological contrasts are present, the models can produce complex escarpment morphologies, flights of escarpments, and scarps topped with a resistant caprock form (Figure 8 in Kooi and Beaumont, 1994). However, a resistant caprock is not found to be an essential requirement for scarps to form and/or retreat. Retreat of steep model escarpments is promoted by a small (i.e. resistant) substrate detachability for fluvial entrainment (large l_f), and by a small ratio of the short-range diffusive and the long-range fluvial transport efficiencies ($K_s/K_f q_r$). A low value for this ratio is interpreted to correspond to a semi-arid climate.

In summary, a combination of inference from model results and intuition suggests that the evolution of great escarpments is controlled by a hierarchy of processes. The combination of antecedent and tectonic topography, coincident with rifting, determines the fundamental drainage divides in relation to the tectonic escarpments. The system evolves by I- and R-mode denudation with rates that are determined by climate and the ease with which material weathers or is detached during bedrock channel incision by the rivers. Spatial and temporal variations in climate and river discharge, and rock types exposed at the surface, modify the progress of the denudational modes and may, under some circumstances, cause all escarpments in the landscape to degrade. The model experiments suggest that large-scale escarpments are a dynamical feature of landscapes which require a particular combination of circumstances for retreat and/or preservation.

3.4.2 COMPARISON WITH OTHER WORK

Tucker and Slingerland (1994) inferred essentially the same requirements for escarpment evolution from their SPM: (1) bedrock channel incision on the escarpment face that causes channel steepening at the escarpment-top drainage divide; (2) a low rate of sediment production relative to sediment transport efficiency, which promotes relief-generating processes over diffusive ones; (3) a high continental elevation; and (4) processes, such as isostasy, that help to maintain a drainage divide near an escarpment crest. The similarity of the inferred controls gives some confidence that these controls are robust and do not depend critically on the particular formulation of the SPM.

Masek *et al.* (1994) proposed that orographic precipitation plays a role in the retreat of part of the eastern edge of the central Andean plateau and of the southern margin of the Tibetan Plateau (Nepal, Himalaya). The idea is that orographic precipitation impinging on a high plateau margin will generally tend to be concentrated at low elevations, eroding back the plateau while enhancing or maintaining the steep long-wavelength slope. Such orographic control may, for obvious reasons, also play a role in the evolution of major rift shoulders at rifted margins. We return to the coupling between climate and tectonics and the role of orographic precipitation in Section 3.5.

Despite these recently gained insights into the process controls of the first-order aspects of the morphological evolution of rifted continental margins, there remains ample opportunity to explore the morphological evolution of rifted margins in greater detail (see Chapters 11–15), including the role of post-rift tectonic and magmatic disturbances and the coupling between offshore sedimentation and onshore denudation through isostasy (van Balen *et al.*, 1995). The results of the simple SPMs should also be used with caution because processes involving groundwater are not included although they may be another important control (Stark, 1994), both at rifted margins and in general.

3.5 Dynamically Coupled Models of Tectonics and Denudation in Orogens

3.5.1 EVIDENCE OF COUPLING BETWEEN TECTONICS AND SURFACE PROCESSES

Orogens provide examples of plate boundary and, in some instances, plate interior convergence where the tectonics and surface processes may be strongly coupled. Partial evidence of the importance of both processes is the similar magnitudes of the tectonic uplift and denudational mass fluxes. For example, the estimates by Adams (1980) of the near equality of these mass fluxes for the Southern Alps orogen, New Zealand, led to his interpretation of most of the Southern Alps as a steady-state mountain belt (see Chapter 6). He also suggested (Adams, 1980; Table 10) that Taiwan and parts of Japan are in similar equilibrium between the rate of tectonic rock uplift and erosional denudation (see Chapters 7 and 8). Taiwan is also considered to be a steady-state orogen by Suppe (1981) and Davis *et al.* (1983). The steady-state constraint was used by Barr and Dahlen (1989) and Dahlen and Barr (1989) in their analysis of material flow and metamorphism in critical Coulomb wedges. The constraint was also assumed, but not proven, in their application of the results to Taiwan.

There are many studies in which denudational volume estimates from orogens are compared with sediment volumes accumulated in surrounding sedimentary basins (see Einsele *et al.* (1996) for the most recent application to the Himalayan system). However, a fundamental difficulty is the accurate estimation of the tectonic mass flux, in particular because the proportion of convergent lithosphere that subducts (or is consumed by any other process) is, in general, poorly known. Magmatic mass fluxes are also uncertain. For these and other reasons direct evidence for the importance of the long-term denudational feedback on tectonics is often elusive at the scale of the orogen. There is no dispute that denudation contributes to orogenic unroofing, rock uplift and exhumation (Jamieson and Beaumont, 1988; England and Molnar, 1990; Koons, 1995), but denudational control of the tectonic evolution, although demonstrated in models, remains an inference for natural orogens (e.g. Suppe, 1981; Beaumont *et al.*, 1992; Hoffman and Grozinger, 1993).

The argument that tectonics and denudation are closely linked can be demonstrated by a simple mechanical analysis. When analysed as a thin sheet, deformation of the crust or lithosphere by a combination of horizontal and basal forces is governed by two main non-dimensional parameters, the Argand (*Ar*) and Ampferer (*Am*) numbers (Figure 3.6). *Ar* relates the gravitational forces (F_2) owing to crustal (lithospheric) thickening to the corresponding horizontal compressive force (F_1). *Am* relates the integrated basal tractions (F_3) to F_1. As England and McKenzie (1982) and Ellis *et al.* (1995), among others, have shown, the distribution of crustal or lithospheric thickening in simple planform viscous thin-sheet models depends on these parameters and, in particular, *Ar* determines the tendency for deformation to spread out laterally during continued convergence and growth of the orogen (England and Houseman, 1986; Ellis, 1996). It follows that the outward propagation of deformation is also highly sensitive to denudation because F_2 (Figure 3.6) is directly related to crustal (lithospheric) thickening. While the denudation rate equals the rate of thickening, and other conditions also remain constant, an orogen will not increase in size. Instead, the horizontal tectonic mass flux into the orogen from convergence will be balanced by the denudation mass flux in such a steady-state orogen and the mechanics will favour exhumation as opposed to additional crustal thickening and outward growth (Jamieson and Beaumont, 1988).

Control Parameters: Argand and Ampferer Numbers

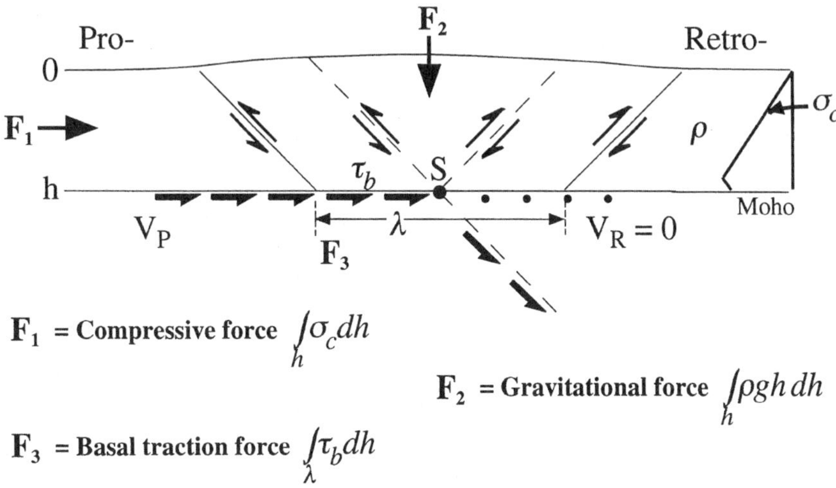

$Ar = F_2/F_1$

Emile Argand
1879-1940

$Am = F_3/F_1$

Otto Ampferer
1875-1947

Figure 3.6 *Illustration of the parameters that control deformation in simple models of a crustal layer, thickness h, that is subject to basal subduction with the mantle subducting at S. Material converges from the pro-side of the system. The integrated forces F_1, F_2 and F_3 represent measures of the strengths of the crustal layer, the gravitational forces and the integrated basal traction. The ratio F_2/F_1 is termed the Argand number (England and McKenzie, 1982) and F_3/F_1 is termed the Ampferer number (Ellis* et al.*, 1995)*

Critical Coulomb wedges (Dahlen, 1984) are a special case of this general result in which the F_1, F_2 and F_3 forces are all related through the Coulomb (frictional) yield criterion and, consequently, *Ar* and *Am* are also related. When denudation reduces the wedge taper below the critical value, an orogenic wedge becomes unstable and must deform internally, thicken and restore F_2 to the critical value before the outward propagation of deformation can return to the toe of the wedge. Erosional denudation of the surface, therefore, has an immediate and important effect on deformation when orogens are controlled by frictional mechanics. While the denudation rate equals the rock uplift rate, material will flow through a critical wedge orogen and be exhumed in its interior (Dahlen and Barr, 1989).

It is certainly true that orogen mechanics is more complicated than these simple models. However, the relationships among F_1, F_2 and F_3 (and therefore *Ar* and *Am* and the generalization to three dimensions) and their relationship to denudation are sufficiently fundamental that there is good reason to believe that at the largest scale natural orogens are controlled in the same way as the models. The more difficult problem is the coupling of tectonics and surface processes at sub-orogenic scales.

Recent work on the tectonic development of compressional orogens has not only successfully explained aspects of the large-scale structure and deformation in past and present orogenic belts (Brown *et al.*, 1993; Willett *et al.*, 1993; Beaumont and Quinlan, 1994; Beaumont *et al.*, 1994; Willett and Beaumont, 1994), but has also demonstrated the sensitivity of orogenic processes to surface redistribution of mass through denudation and sedimentation. The strength of the feedback mechanisms can also be understood physically because most denudational process rates are normally enhanced by the high elevation, and particularly the high relief, characteristic of mountain belts. Mountain belts achieve both high relief and high mean elevation when moderate to high rock uplift rates are coupled with moderate to high denudation rates at the scale of the orogen. In contrast, continental plateaus achieve a high mean elevation, but have little relief, because the long-range mass transport by rivers and glaciers from the plateau to surrounding areas is inefficient.

3.5.2 RESULTS FROM NUMERICAL MODELS

Early numerical models, that coupled simple plane-strain frictional orogen tectonics (modelled as a plastic process) with denudation predicted by an SPM, were able to demonstrate that the geometry and dynamics of a model collisional orogen are not only sensitive to the total denudation but also to the focus of the denudation on the pro- or retro-side of the orogen (Beaumont *et al.*, 1992). These particular models were motivated by the concept that crustal-scale deformation in orogens commonly involves thrusting of colliding terranes over a ramp formed by an inherited passive margin (Jamieson and Beaumont, 1988). Although we now prefer the subduction model (Willett *et al.*, 1993), the geometry of the uplift in the ramp models (Figs 9, 11 and 13 in Beaumont *et al.*, 1992) is very similar to that of the early plug evolution in subduction-controlled models where the crust is strongly coupled to the mantle (Plate 3.4A). The significance of these model results was the demonstration that an uplifting mountain belt may interact with atmospheric circulation, through orographically derived precipitation for example, to create a dynamical system. In this system the orogen tectonics is controlled by the climate and, in particular, the prevailing wind direction. A steady state is achieved when the mountain system grows dynamically to the point where it extracts sufficient precipitation to denude the orogen at a rate equal to the uplift rate of material at the surface.

A comparison of extremes – no denudation versus total denudation – demonstrates the general importance of surface processes to tectonic development, exhumation, crustal thickening, isostatic balance and foreland flexure in very simple subduction-controlled doubly vergent orogens (Plate 3.4). When denudation is absent, the mass of the orogen increases with the tectonic mass flux, there is uplift and burial, but no exhumation, and foreland basins subside as the increasing orogenic mass is flexurally compensated (Plate 3.4A). In contrast, total denudation prevents orogenic mass accumulation and no foreland basins are created (Plate 3.4B) unless other tectonic subsidence mechanisms operate. High denudation rates favour rapid exhumation and, in the limit of total denudation (Plate 3.4B) exhumation rates approach tectonic convergence rates. As is shown below, a dynamical equilibrium (equal tectonic and surface denudational mass fluxes) can be achieved for any model intermediate to those shown in Plates 3.4A and 3.4B. The primary control is the increase of the denudational mass flux with the increase in elevation, relief and precipitation within the orogen. Even such simple models may have natural analogues. For example, the Southern Alps orogen in the continental collision zone of South Island, New Zealand (Norris *et al.*, 1990), has much in common with models that are close to the total denudation limit (Beaumont *et al.*, 1996).

The effect of climate-coupled denudation on tectonics can be demonstrated by coupling the Plate 3.4A model to a simple orographic precipitation model in which precipitation occurs at a constant rate on windward-facing slopes and is zero on leeward slopes (Plate 3.5). Precipitation is collected into a one-dimensional river and the denudation rate is taken to be proportional to the discharge of water in the river times the slope of the surface (the same carrying capacity used in the fluvial component of the SPM (Section 3.3.1)). The coupling between uplift and net runoff allows the system to develop into a dynamic steady state in which the denuded mass flux is equal to the mass flux added by convergence, and there is no further change in topography or in the maximum burial depth experienced by exhumed material at the orogen surface.

The exhumation and deformation in the orogen acquire a distinct character depending on the direction of mantle subduction. If the dominant wind direction is opposite to the motion of the subducting mantle slab (Plate 3.5a), exhumation tends to be highly focused on the deformation front of the orogen whereas deformation is diffuse throughout the orogen. In contrast, if the dominant winds are in the same direction as subduction, the enhanced denudation tends to 'short-circuit' the orogen, creating a new, 'out-of-sequence' crustal shear zone in the interior of the orogen and leaving the leeward side tectonically inactive (Plate 3.5b). In this case, exhumation of the deepest (high metamorphic grade) rocks occurs in the orogen interior. Note that in both examples the exposure of higher grade metamorphic rocks does not imply higher denudation rates; the denudation rate is nearly constant across the orogen interior. The distribution of exhumation depends, rather, on the internal kinematics of deformation coupled to the surface denudation rates.

Surface slopes and morphology also reflect the changing denudational conditions; in each case the windward side of the orogen develops a long river profile that is concave upward with a nearly exponential shape, reflecting the particular form of the denudation rule. Leeward slopes are much steeper and reflect the deformational processes more than the surface processes. Although this type of model is necessarily simplistic with respect to the orogen topography, it suggests that very interesting, and likely unanticipated, results will be obtained by fully coupling the geodynamic and surface process models with all the implicit feedback mechanisms.

This predicted relationship between topography, denudation and the distribution of exhumation has important implications for the kinematics, climate and erosional history of specific orogenic belts. Although models are still too simple to provide detailed predictions or interpretations of natural orogens, some general comparisons can be made. On the largest scale, the Himalaya–Tibetan Plateau orogenic belt shows some interesting characteristics. The Himalayas represent the southern margin of the Tibetan Plateau, the formation of which is the primary response to the India–Asia collision. Although the peaks of the Himalayas are the highest in the world, the average elevation of the range is not significantly higher than that of the Tibetan Plateau (see Chapter 10) (Fielding *et al.*, (1994)). The high peaks are the reflection of the increased relief due to fluvial incision and isostatic uplift (Montgomery, 1994). The denudational incision is, in turn, a reflection of the regional climate with orographically focused precipitation on the Himalayan front and near-desert conditions in the Tibetan interior. Exhumation is also focused on the Himalayan front with the highest grade rocks exposed in the Lesser and Greater Himalaya, with only limited exhumation to the north of the high peaks (e.g. Einsele *et al.*, 1996). There is no comparable high-grade exhumation north of the Tibetan Plateau so that the orogenic belt overall shows a strong asymmetry reflecting the climatic and kinematic conditions of its formation, as do the models illustrated in Plate 3.5.

Although models are not available for large orogens like Himalaya–Tibet, Plate 3.6 illustrates some basic results obtained by coupling the small-orogen tectonic model (Plate 3.4A) with a planform surface processes model. The SPM is the one described in Section 3.3.1 and the coupling is as described by Beaumont *et al.* (1992), except that the subduction basal boundary condition has been used. The model is approximately a planform equivalent of the one shown in Plate 3.5a, but illustrated with the opposite polarity. The model result shows something of the potential for coupling full planform SPMs to plane-strain TMs with subduction basal boundary conditions. We suggest that this model orogen would be analogous to the Southern Alps, New Zealand, if convergence across the Pacific–Indian plate boundary were normal as opposed to oblique. Cross-sections through the model also have many of the characteristics of the schematic conceptual system described by Hovius in Chapter 5. They also illustrate tectonic as well as climatic–geomorphic linkages. Overall, the model serves as a prototype for the operation of the complete system.

The model orogen has achieved a large-scale steady-state mass balance (equal denudation and tectonic mass fluxes) which is close to the total denudation limit (Plate 3.4B). The up to 10 m a^{-1} orographically enhanced precipitation that falls on the windward flank of the orogen is sufficient to denude (mostly fluvial denudation) the model mountain belt at the tectonic uplift rate. The mountains have a maximum elevation of about 4000 m and are in diffusive equilibrium with the adjacent river valleys. Diffusion in this model is linear (see Section 3.3.1), but could be made nonlinear to represent the integrated effects of landsliding and give the mountains a maximum threshold slope (Anderson, 1994; Tucker and Slingerland, 1994) (see Chapter 5). Sediment produced by the model has been deposited in the flexural foreland basins formed on both sides of the model orogen. Sediment mostly bypasses the coastal plains, where it was previouly deposited by a network of avulsing rivers with braided-type behaviour, and is currently prograding into deeper water via a system of model deltas (Plate 3.6). The orogen-scale tectonics operate in a similar manner to the retro-windward model (Plate 3.5a), with maximum exhumation in the hangingwall of the retro-step-up shear. The exhumation profile across the model is also in steady state, reflecting deformation, uplift and removal of converging pro-crust on the retro-side of the orogen. Despite the steady state, the total exhumation is not directly related to either rock uplift or denudation rates. Unlike simple models that consider only vertical tectonics, the Plate 3.6 model (like the Plates 3.4 and 3.5 models) explains consequences of the conversion of horizontal convergence into a combination of uplift and convergence. The only evolving model components (Plate 3.6) are the lithospheric convergence of the pro-plate (Pacific plate) and the associated lateral motion of the topography. This motion is an important aspect because the topographic evolution is not merely a consequence of vertical movements as in the models of Sections 3.3 and 3.4. Instead, topography created at the deformation front on the pro-side (Pacific side) of the orogen is carried towards the orogen interior, crosses from the pro- to retro-side and the drainage divide (not necessarily at the same location, see Plate 3.5), and is finally destroyed at the retro-step-up shear (Alpine fault). It is not known whether steady-state landscapes (like those described in Section 3.3.2) can exist in model orogens, or in corresponding natural orogens, where the topography is carried along and modified by the convergent motions.

3.5.3 RESULTS FROM PHYSICAL MODELS

Although we have focused on numerical models, scaled laboratory experiments, particularly sandbox models, provide complementary insights into the mechanics of crustal deformation and the coupling with surface processes. The models will scale to represent the crust, although the scaling is only approximate because neither sand, nor the crust, are perfect non-cohesive Coulomb materials. In sandboxes the surface processes either occur naturally during the model experiment, or are purposely superimposed. Although combined erosion–deformation experiments are difficult to conduct and rare in the literature, sedimentation processes are easily imposed and illustrate the deformational effects of mass redistribution by sedimentation, a surface process comparable to denudation.

Sandbox experiments have successfully simulated subduction, or substrate subduction processes, similar to the numerical models of the previous section. Deformation of a uniform layer of sand under subduction boundary conditions (Malavieille, 1984; Malavieille *et al.*, 1993; Wang and Davis, 1996) or against a rigid backstop (Davis *et al.*, 1983; Liu *et al.*, 1992, Storti and McClay, 1995) creates tectonic wedges with the wedge-scale geometry predicted by critical wedge mechanics (Dahlen, 1984) and numerical models (Willett, 1992), with the caution that sand exhibits strain localization into fault-like structures that are not exhibited in analytic or numerical continuum models.

The interaction between deformation and sedimentation is illustrated well by a set of field observations, sandbox models and numerical models in the geological setting of forearc basins (Figure 3.7). Forearc basins form by ponding of sediment between a volcanic

Figure 3.7 *Effect of sedimentation on deformation in a forearc setting demonstrated by numerical models and sandbox, analogue models. (a) Numerical model with no sedimentation and velocity boundary conditions similar to Plate 3.4, but with more convergence to simulate accretionary wedge conditions. (b) Identical numerical model except for the addition of sediment to the depression between the arc and outer-arc structural high. Note that deformation is localized and does not propagate towards the arc. (c) Line drawing of principal faults formed in a sandbox experiment (modified from Wang and Davis (1996)). Boundary conditions are similar to the numerical model (Figure 3.7a). (d) Line drawing of faults and stratigraphy (inset) from sandbox experiment of Malavielle* et al. *(1993). Boundary conditions are the same as all models above, but sediment is added to depression as in Figure 3.7b. Note localization of deformation along the edge of the model basin*

a)

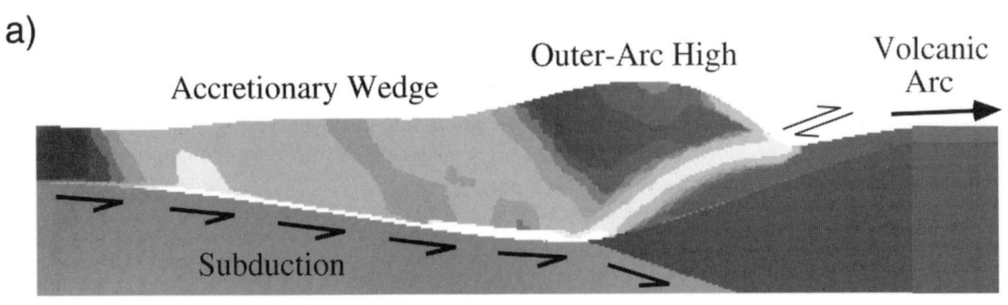
Outer-Arc High
Volcanic Arc
Accretionary Wedge
Subduction

b)

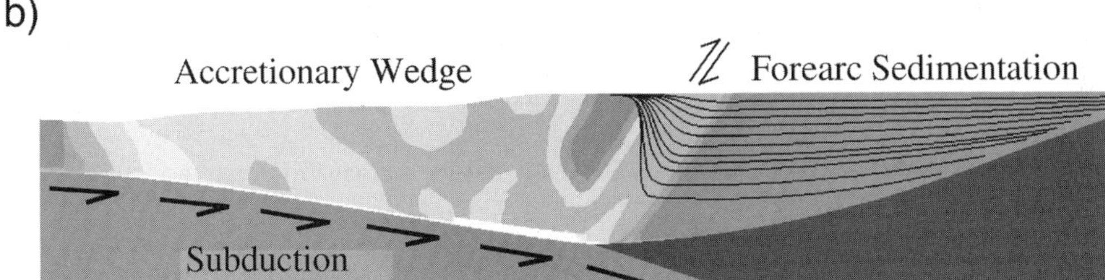
Accretionary Wedge
Forearc Sedimentation
Subduction

c)

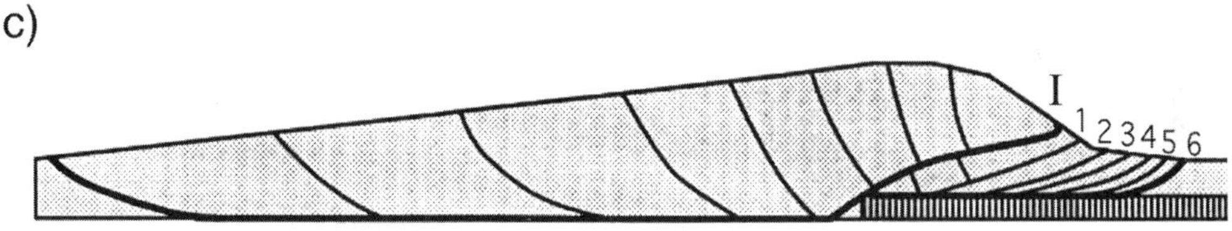
I
1 2 3 4 5 6

d)

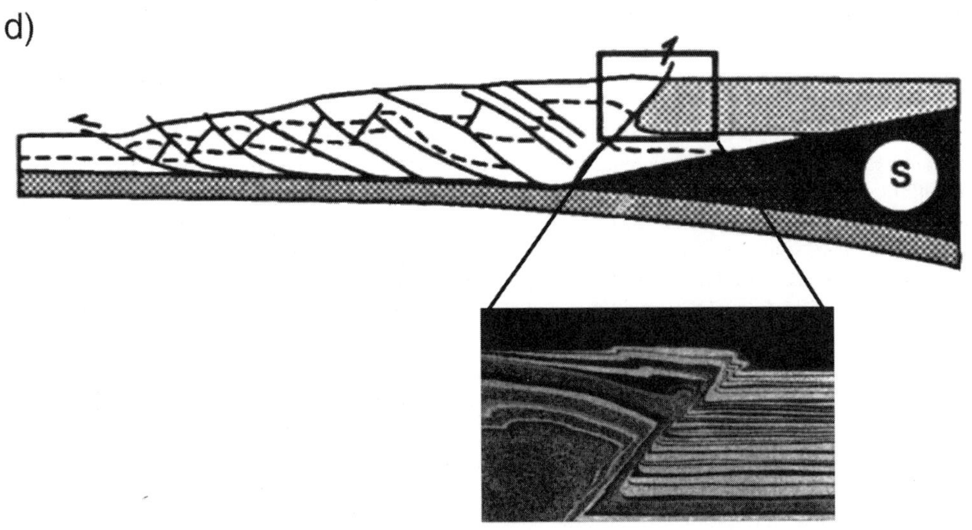
S

arc and the actively deforming accretionary wedge of material scraped off a subducting oceanic plate (e.g. Dickinson and Seely, 1979). Accretionary wedges deform at high rates and, since the basin is bounded by the accretionary wedge, stratigraphic patterns in the basin reflect this deformation. Moreover, when sedimentation advances over the accretionary wedge, the weight of these overlying sediments affects the internal deformation. The result is a boundary zone, or inner deformation belt (Byrne *et al.*, 1993), between the basin and the wedge, characterized by localized faulting, deformation and arcward tilting of young sediments (Beaudry and Moore, 1981; Silver *et al.*, 1983; Westbrook *et al.*, 1988; Byrne *et al.*, 1993).

The coupling of sedimentation and deformation in the inner deformation belt has been simulated in both sandbox and numerical models. Figures 3.7c shows a sandbox experiment from Wang and Davis (1996) that simulates subduction down to the right beneath a horizontal, or shallowly dipping overriding plate. Off-scraped sediments form a doubly vergent accretionary wedge with deformation propagating in both directions. Arcward-verging faults are numbered sequentially as they form to indicate the younging of the structures towards the arc. This sandbox deformation pattern is strongly altered when sedimentation occurs between the arc and the high point of the accretionary wedge, as demonstrated in an experiment by Malavieille *et al.* (1993) (Figure 3.7d). Sedimentation lowers surface slopes and inhibits outward propagation of faulting. The result is a strong localization of deformation on a single arcward-verging thrust structure (Figure 3.7d, inset). The same effect can be seen in numerical models of this process. Figures 3.7a and 3.7b show two experiments comparable to the sandbox models of Figures 3.7c and 3.7d; the only difference between these numerical models is the addition of sedimentation in the latter model. Sedimentation is simulated by filling the topographic low with material having the same physical properties as the accretionary wedge. As in the sandbox, sedimentation causes deformation to be localized and restricted to a single shear zone with deformation of the basin sediments adjacent to the wedge, but with no arcward propagation of this zone of deformation. However, this balance is only maintained by an adequate sedimentation rate; if sedimentation does not keep up with the topographic growth of the wedge, the deformation will propagate arcward, and progressively deform the basin sediments.

The effect of syn-deformational sedimentation on the main forward-verging pro-wedge has also been demonstrated by sandbox models (Storti and McClay, 1995) (Plate 3.7). Increased amounts of sedimentation over the toe and onto the slope of a sand wedge (as shown) suppress the internal deformation and focus thrusting closer to the backstop. Although this response, and presumably the corresponding one to broad-scale denudation, follow from the basic requirements of critical wedge mechanics, the interactions on a smaller scale are more complex and reflect the internal deformation modes of the sand wedge. If the sandbox results are extrapolated to natural systems in which fluvial and hillslope processes redistribute mass in even more complex ways, it can be seen that there are many possible ways surface processes and tectonics become coupled.

3.6 Future Research Directions

The coupling between tectonics and surface processes, within the larger-scale system comprising the atmosphere, hydrosphere, biosphere and solid Earth, is proving to be a rich field for future research, although the research reviewed here has little more than scratched the surface. Perhaps the most significant achievement to date is the recognition that the problem is a dynamical one, with a hierarchy of couplings and feedback mechanisms. This recognition should ensure integration of geomorphological, geological and geophysical studies of rift margins and orogens, for example, in directions illustrated by Figure 3.2.

Although there has been less research concerning interactions between surface processes and tectonics at rifted margins, we suggest that the general principles will apply equally to both orogens and rifts. Analogous results to those from finite element TMs coupled to SPMs (for example, Plates 3.5 and 3.6) are therefore anticipated for syn-tectonic evolution of rifts and their flanks. Such models will need to include the clastic sedimentation component as well as the denudational component of the SPM (for instance, improved versions of Johnson and Beaumont's (1995) foreland basin sedimentation models).

The surface process models are still in their infancy. Improvements are needed to include better physically based process components appropriately parameterized for model scales. In addition to improved fluvial models, chemical weathering, landsliding, and the effects of glaciation should be added. There is also the problem of spatial resolution in numerical models that use a grid or cell structure for integration. Although improved algorithms (e.g. random, self-refining tessellation techniques (Braun and Sambridge,

1997)) and increased computer speeds will increase resolution to the scale of the drainage density, it will still be necessary to develop sub-grid scale models to predict slope–channel interactions (e.g. Howard *et al.*, 1994).

Time resolution also poses problems. Mechanical SPMs designed to predict mass transport, development of topography and growth of relief, and to be coupled to mesoscale TMs need to be parameterized for time scales of 10^4 years or longer. Meteorological events, however, vary temporally 10^6–10^7 times more rapidly and also spatially. There is therefore a need to develop appropriately scaled climatic models that allow correct integration of nonlinear components in the SPMs when the model time resolution is 10^2–10^3 years.

At the scale of an orogen or rifted margin, current coupled models are, at best, as good as the TMs used to model lithospheric deformation. Significant advances have been made in the development of quantitative numerically based (and physical laboratory) thermo-mechanical models. Fully three-dimensional numerical models (e.g. Braun, 1993, 1994) will require greater computational power for truly large deformation. Most of these models are based on continuum mechanics and most applications address the deformation at the scale of the orogen. The primary role of denudational mass loss in this type of model is to modify buoyancy forces and, more generally, all forces that are linked to gravity. The effects of distributed denudation, or denudation focused on the windward flank of an orogen, are therefore relatively easy to calculate, as shown by the models of Plates 3.4 and 3.5, provided they do not couple to the model mechanics at intermediate scales.

The more significant results are those illustrated by Plate 3.5b, where the positive feedback from the spatial variation in denudation has localized deformation creating the out-of-sequence crustal-scale shear. While denudation continues with this focus, the shear remains active and can assume orogen-scale significance. This shear is the denudational equivalent of the one localized by sedimentation in numerical (Figure 3.7b) and sandbox (Figure 3.7d) experiments.

In addition, the sandbox models indicate something of the difficulties to be anticipated for the numerical models. A primary result is that a uniform layer of sand deformed under subduction boundary conditions or against a rigid backstop, exhibits internal modes of deformation that do not correspond to any simple flow of material (e.g. the Barr and Dahlen (1989) styles). Yet these models create tectonic wedges with the wedge-scale geometry predicted by critical wedge mechanics (Dahlen, 1984). Among the model results, Liu *et al.* (1992) demonstrate the style of the internal modes as a function of the internal and basal properties of the sand layer and illustrate the effect of the finite thickness of the sand layer. The modes range from the development of thrust sheets to the formation of a chain of plug uplifts, depending on the basal strength of the wedge. An explanation of the mechanics of these internal modes would provide a missing component in critical wedge mechanics – how critical wedges actually grow. The same internal modes occur in numerical models of Coulomb wedges and, consequently, they cannot be an artefact of the strain softening–hardening localization that characterizes sand deformation. The mechanics of these internal modes may be the key to an understanding of intermediate-scale deformation in orogens, in the same way that critical wedge geometries have found application at the large scale.

These results serve as a caution. We know that continuum mechanics models fail to resolve fully the self-localization of strain and deformation on discrete faults and shear zones in the manner observed in nature. It follows that any coupled process, which naturally produces and amplifies strain localization and is itself modified by localized strain, will not be accurately simulated by current continuum mechanics models. Denudation and sedimentation can force (Figure 3.7) and suppress (Plate 3.7) strain localization, and in these particular examples the interactions have been captured by the model dynamics. Under what circumstances are these effects important? Perhaps the greatest difficulty confronting the development of coupled TM–SPMs is the determination of the range of space and time scales through which such localized interactions cascade. Perfectly acceptable independent TMs and SPMs will fail when coupled unless this problem is solved. In closing, few would argue that the lot of a single sand grain on a remote drainage divide determines the destiny of the Himalayan–Tibetan system, but at what space and time scales *do* surface processes seal an orogen's fate?

Acknowledgements

HK acknowledges a Dalhousie University Killam Postdoctoral Fellowship. CB acknowledges funding through the Inco Fellowship of the CIAR and from the Natural Sciences and Engineering Research Council. SW acknowledges National Science Foundation funding through EAR. We also acknowledge and thank Alan Gilchrist for his contribution to our research on surface processes. We thank the reviewers, Kerry Gallagher, Frank Ahnert and Michael Ellis, for their constructive suggestions and Susan Ellis for her review of an earlier version of the manuscript. We also thank Mike Summerfield for his invitation to contribute this chapter. Tammy Chouinard, Juliet Hamilton and Bonny Lee are thanked for their help in manuscript preparation.

References

Adams, J. 1980. Contemporary uplift and erosion of the Southern Alps, New Zealand. *Geol. Soc. Amer. Bull.*, Part II, **91**, 1–114.

Ahnert, F. 1976. Brief description of a comprehensive process-response model for landform development. *Z. Geomorph. Suppl.*, **25**, 29–49.

Ahnert, F. 1984. Local relief and the height limits of mountain ranges. *Amer. J. Sci.*, **284**, 1035–1055.

Ahnert, F. 1987a. Approaches to dynamic equilibrium in theoretical simulations of slope development. *Earth Surf. Processes Ldfms.*, **12**, 3–15.

Ahnert, F. 1987b. Process–response models of denudation at different spatial scales. *Catena Suppl.*, **10**, 31–50.

Allen, P.A. and Allen, J.R. 1990. *Basin Analysis*, Blackwell Scientific Publications, Oxford.

Anderson, R.S. 1994. Evolution of the Santa Cruz Mountains, California through tectonic growth and geomorphic decay. *J. Geophys. Res.*, **99**, 20 161–20 180.

Anderson, R.S. and Humphrey, N.F. 1990. Interaction of weathering and transport processes in the evolution of arid landscapes. In: T.A. Cross, (ed.) *Quantitative Dynamic Stratigraphy*, Prentice-Hall, Englewood Cliffs, NJ, pp. 349–361.

Argand, E. 1924. La tectonique d'Asie. *C.R. Congr. Geol. Int. Belgique*, 1922, 171–372.

Armstrong, A.C. 1976. A three-dimensional simulation of slope forms. *Z. Geomorph. Suppl.*, **25**, 20–28.

Avouac, J.P. and Burov, E.B. 1996. Erosion as a driving mechanism of intracontinental mountain growth. *J. Geophys. Res.*, **101**, 17 747–17 769.

Barr, T.D. and Dahlen, F.A. 1989. Steady-state mountain building 2. Thermal structure and heat budget. *J. Geophys. Res.*, **94**, 3923–3947.

Barr, T.D., Dahlen, F.A. and McPhail, D.C. 1991. Brittle frictional mountain building 3. Low-grade metamorphism. *J. Geophys. Res.*, **96**, 10 319–10 338.

Bassi, G., Keen, C.E. and Potter, P. 1993. Contrasting styles of rifting: models and examples from the eastern Canadian margin. *Tectonics*, **12**, 639–655.

Beaudry, D. and Moore, G.F. 1981. Seismic–stratigraphic framework of the forearc basin off central Sumatra, Sunda Arc. *Earth Planet. Sci. Lett.*, **54**, 17–28.

Beaumont, C. and Quinlan, G. 1994. A geodynamic framework for interpreting crustal-scale seismic-reflectivity patterns in compressional orogens. *Geophys. J. Int.*, **116**, 754–783.

Beaumont, C., Fullsack, P. and Hamilton, J. 1992. Erosional control of active compressional orogens. In: K.R. McClay (ed.) *Thrust Tectonics*, Chapman & Hall, London, pp. 1–18.

Beaumont, C., Fullsack, P. and Hamilton, J. 1994. Styles of crustal deformation caused by subduction of the underlying mantle. *Tectonophysics*, **232**, 119–132.

Beaumont, C., Kamp, P.J.J., Hamilton, J. and Fullsack, P. 1996. The continental collision zone, South Island, New Zealand: comparison of geodynamical models and observations. *J. Geophys. Res.*, **101**, 3333–3360.

Bird, P. 1989. New finite element techniques for modeling deformation histories of continents with stratified temperature-dependent rheology. *J. Geophys. Res.*, **94**, 3967–3990.

Bird, P. and Piper, K. 1980. Plane-stress finite-element models of tectonic flow in California. *Phys. Earth Planet. Interiors*, **21**, 158–175.

Bishop, P. 1995. Drainage rearrangement by river capture, beheading and diversion. *Prog. Phys. Geog.*, **19**, 449–473.

Braun, J. 1993. Three dimensional numerical modeling of compressional orogenies: thrust geometry and oblique convergence. *Geology*, **21**, 153–156.

Braun, J. 1994. Three-dimensional numerical simulations of crustal-scale wrenching using a non-linear failure criterion. *J. Struct. Geol.*, **16**, 1173–1186.

Braun, J. and Beaumont, C. 1989. A physical explanation of the relation between flank uplifts and the breakup unconformity at rifted continental margins. *Geology*, **17**, 760–764.

Braun, J. and Beaumont, C. 1995. Three dimensional numerical experiments of strain partitioning at oblique plate boundaries: implications for contrasting tectonic styles in the southern Coast Ranges, California, and in central South Island, New Zealand. *J. Geophys. Res.*, **100**, 18 059–18 074.

Braun, J. and Sambridge, M. 1997. Modelling landscape evolution on geological timescales: a new method based on irregular spatial discretization. *Basin Res.*, **9**, 27–52.

Brown, R.L., Beaumont, C. and Willett, S.D. 1993. Comparison of the Selkirk Fan structure with mechanical models: implications for palinspastic reconstructions of the southern Canadian Cordillera. *Geology*, **21**, 1015–1018.

Brown, R.W. 1991. Backstacking apatite fission-track 'stratigraphy': a method for resolving the erosional and isostatic rebound components of tectonic uplift histories. *Geology*, **19**, 74–77.

Brunsden, D. and Thornes, J.B. 1979. Landscape sensitivity and change. *Trans. Inst. Brit. Geog.*, NS **4**, 463–484.

Buck, W.R. 1986. Small-scale convection induced by passive rifting: the cause for uplift of rift shoulders. *Earth Planet. Sci. Lett.*, **77**, 362–372.

Byrne, D.E., Wang, W.-H. and Davis, D.M. 1993. Mechanical role of backstops in the growth of fore-arcs. *Tectonics*, **12**, 123–144.

Carson, M.A. and Kirkby, M.J. 1972. *Hillslope Form and Process*, Cambridge University Press, Cambridge.

Chapple, W.M. 1978. Mechanics of thin-skinned fold-and-thrust belts. *Geol. Soc. Amer. Bull.*, **89**, 1189–1198.

Chase, C.G. 1992. Fluvial landsculpting and the fractal dimension of topography. *Geomorphology*, **5**, 39–57.

Chèry, J., Lacazeau, F., Daignières, M. and Vilotte, J.P. 1992. Large uplift at rift flanks, a genetic link with lithospheric rigidity? *Earth Planet. Sci. Lett.*, **122**, 195–211.

Cochran, J.R. 1983. Effects of finite rifting times on the development of sedimentary basins. *Earth Planet. Sci. Lett.*, **66**, 289–302.

Cook, F. and Varsek, J.L. 1994. Orogen-scale decollements. *Rev. Geophys.*, **32**, 37–60.

Dahlen, F.A. 1984. Noncohesive critical Coulomb wedges: an exact solution. *J. Geophys. Res.*, **89**, 10 125–10 133.

Dahlen, F.A. 1990. Critical taper model of fold-and-thrust belts and accretionary wedges. *Ann. Rev. Earth Planet. Sci.*, **18**, 55–99.

Dahlen, F.A. and Barr, T.D. 1989. Brittle frictional mountain building 1. Deformation and mechanical energy budget. *J. Geophys. Res.*, **94**, 3906–3922.

Dahlen, F.A. and Suppe, J. 1988. Mechanics, growth and erosion of mountain belts. In: S.P. Clark (ed.) *Processes in Continental Lithospheric Deformation, Geol. Soc. Amer. Spec. Pap.*, **218**, 161–171.

Davis, D., Suppe, J. and Dahlen, F.A. 1983. Mechanics of fold-and-thrust belts and accretionary wedges. *J. Geophys. Res.*, **88**, 1153–1172.

Davis, W.M. 1899. The geographical cycle. *Geogr. J.*, **14**, 481–504.

Dewey, J.F. and Bird, J.M. 1970. Mountain belts and the new global tectonics. *J. Geophys. Res.*, **75**, 2625–2647.

Dickinson, W.R. and Seely, D.R. 1979. Structure and stratigraphy of forearc regions. *Bull. Am. Assoc. Petrol. Geol.*, **63**, 2–31.

Dunbar, J.A. and Sawyer, D.S. 1989. How preexisting weaknesses control the style of rifting. *J. Geophys. Res.*, **94**, 7278–7282.

Einsele, G., Ratschbacher, L. and Wetzelm, A. 1996. The Himalaya–Bengal Fan denudation–accumulation system during the past 20 Ma. *J. Geol.*, **104**, 163–184.

Elliott, D. 1976a. The motion of thrust sheets. *J. Geophys. Res.*, **81**, 949–963.

Elliott, D. 1976b. The energy balance and deformation mechanisms of thrust sheets. *Phil. Trans. R. Soc. Lond.*, **A283**, 289–312.

Ellis, M. and Merritts, D. (eds) 1994. *Tectonics and Topography*, American Geophysical Union, Washington D.C.

Ellis, S. 1996. Forces driving continental collision: reconciling indentation and mantle subduction tectonics. *Geology*, **24**, 699–702.

Ellis, S., Fullsack, P. and Beaumont, C. 1990. Incorporation of erosion into thin sheet numerical models of continental collision. *EOS, Trans. Amer. Geophys. Un.*, **71**, 1562.

Ellis, S., Fullsack, P. and Beaumont, C. 1995. Oblique convergence of the crust driven by basal forcing: implications for length-scales of deformation and strain partitioning in orogens. *Geophys. J. Int.*, **120**, 24–44.

England, P. and Houseman, G. 1986. Finite strain calculations of continental deformation, 2, Comparison with the India–Asia collision. *J. Geophys. Res.*, **91**, 3664–3676.

England, P. and McKenzie, D.P. 1982. A thin viscous sheet model for continental deformation. *Geophys. J. R. Astr. Soc.*, **70**, 295–321.

England, P. and Molnar, P. 1990. Surface uplift, uplift of rocks and exhumation of rocks. *Geology*, **18**, 1173–1177.

England, P.C. and Richardson, S.W. 1977. The influence of erosion upon the mineral facies of rocks from different metamorphic environments. *J. Geol. Soc. Lond.*, **134**, 201–213.

England, P.C. and Thompson, A.B. 1984. Pressure–Temperature–Time paths of regional metamorphism 1. Heat transfer during the evolution of regions of thickened crust. *J. Petrol.*, **25**, 894–928.

England, P., Houseman, G. and Sonder, L. 1985. Length scales for continental deformation in convergent, divergent and strike-slip environments: analytical and approximate solutions for a thin viscous sheet model. *J. Geophys. Res.*, **90**, 4797–4810.

Fielding, E., Isacks, B., Barazangi, M. and Duncan, D. 1994. How flat is Tibet? *Geology*, **22**, 163–167.

Flemings, P.B. and Jordan, T.E. 1989. A synthetic stratigraphic model of foreland basin development. *J. Geophys. Res.*, **94**, 3851–3866.

Gilchrist, A.R. and Summerfield, M.A. 1990. Differential denudation and flexural isostasy in formation of rifted-margin upwarps. *Nature*, **346**, 739–742.

Gilchrist, A.R., Kooi, H. and Beaumont, C. 1994. Post-Gondwana geomorphic evolution of southwestern Africa: implications for the controls on landscape development from observations and numerical experiments. *J. Geophys. Res.*, **99**, 12 211–12 228.

Gregory, K.M. and Chase, G.C. 1994. Tectonic and climatic significance of a late Eocene low-relief, high level geomorphic surface, Colorado, *J. Geophys. Res.*, **99**, 20 141–20 160.

Hack, J.T. 1960. Interpretation of erosional topography in humid temperate regions. *Amer. J. Sci.*, **258A**, 80–97.

Hoffman, P.F. and Grotzinger, J.P. 1993. Orographic precipitation, erosional unloading and tectonic style. *Geology*, **21**, 195–198.

Hovius, N. 1996. Regular spacing of drainage outlets from linear mountain belts. *Basin Res.*, **8**, 29–44.

Howard, A.D. 1994. A detachment-limited model of drainage basin evolution. *Wat. Resour. Res.*, **30**, 2261–2285.

Howard, A.D., Dietrich, W.E. and Seidl, M.A. 1994. Modelling fluvial erosion on regional to continental scales. *J. Geophys. Res.*, **99**, 13 971–13 986.

Isacks, B. 1992. 'Long-term' land surface processes: erosion, tectonics and climate history of mountain belts. In: P.M. Mather (ed.) *TERRA-1: Understanding the Terrestrial*

Environment, National Environment Research Council, pp. 21–36.

Jamieson, R.A. and Beaumont, C. 1988. Orogeny and metamorphism: a model for deformation and pressure–temperature–time paths with applications to the central and southern Appalachians. *Tectonics*, **7**, 417–445.

Johnson, D.D. and Beaumont, C. 1995. Preliminary results from a planform kinematic model of orogen evolution, surface processes and the development of clastic foreland basin stratigraphy. In: S.L. Dorobek and G.M. Ross (eds) *Stratigraphic Evolution of Foreland Basins, Soc. Econ. Paleont. Mineral. Spec. Publ.*, **52**, 1–24.

Kirkby, M.J. 1986. A two-dimensional simulation model for slope and stream evolution. In: A.D. Abrahams (ed.) *Hillslope Processes*, Allen and Unwin, Boston, pp. 203–222.

Kooi, H. and Beaumont, C. 1994. Escarpment retreat on high-elevation rifted continental margins; insights derived from a surface–processes model that combines diffusion, reaction and advection. *J. Geophys. Res.*, **99**, 12191–12209.

Kooi, H. and Beaumont, C. 1996. Large-scale geomorphology: classical concepts reconciled and integrated with contemporary ideas via a surface processes model. *J. Geophys. Res.*, **101**, 3361–3386.

Kooi, H. and Cloetingh, S.A.P.L. 1992. Lithospheric necking and regional isostasy at extensional basins 2. Stress induced vertical motions and relative sea level changes. *J. Geophys. Res.*, **97**, 17573–17591.

Koons, P.O. 1989. The topographic evolution of collisional mountain belts: a numerical look at the Southern Alps, New Zealand. *Amer. J. Sci.*, **289**, 1041–1069.

Koons, P.O. 1994. Three-dimensional critical wedges: Tectonics and topography in oblique collisional orogens. *J. Geophys. Res.*, **99**, 12301–12315.

Koons, P.O. 1995. Modeling the topographic evolution of collisional orogens. *Ann. Rev. Earth Planet. Sci.*, **23**, 365–408.

Lister, G.S., Etheridge, M.A. and Symonds, P.A. 1991. Detachment models for the formation of passive continental margins. *Tectonics*, **10**, 1038–1064.

Liu, H., McClay, K.R. and Powell, D. 1992. Physical models of thrust wedges. In: K.R. McClay (ed.) *Thrust Tectonics*, Chapman and Hall, London, pp. 71–81.

Malavieille, J. 1984. Modèlisation expérimentale des chevauchements imbriqués: Application aux chaînes des montagnes. *Bull. Soc. Géol. France*, **26**, 129–138.

Malavieille, J., Larroaue, C. and Calassou, S. 1993. Modèlisation expérimentale des relations tectonique/sedimentation entre bassin avant-arc et prisme d'accretion. *C.R. Acad. Sci. Paris*, **316**, Serie II, 1131–1137.

Masek, J.G., Isacks, B.L., Gubbels, T.L. and Fielding, E.J. 1994. Erosion and tectonics at the margins of continental plateaus. *J. Geophys. Res.*, **99**, 13941–13956.

Mattauer, M. 1986. Intracontinental subduction, crust–mantle decollement and crustal-stacking wedge in the Himalayas and other collision belts. In: M.P. Coward and A.C. Ries (eds) *Collisional Tectonics, Geol. Soc. Spec. Publ.*, **19**, 37–50.

McKenzie, D.P. 1978. Some remarks on the development of sedimentary basins. *Earth Planet. Sci. Lett.*, **40**, 25–32.

Montgomery, D.R. 1994. Valley incision and the uplift of mountain peaks. *J. Geophys. Res.*, **99**, 13913–13921.

Moretti, I. and Turcotte, D.L. 1985. A model for erosion, sedimentation and flexure with applications to New Caledonia. *J. Geodynamics*, **3**, 155–168.

Musgrave, F.K., Kolb, C.E. and Mace, R.S. 1989. The synthesis and rendering of eroded fractal terrains. *Comp. Graph.*, **23**, 41–50.

Norris, R.J., Koons, P.O. and Cooper, A.F. 1990. The obliquely-convergent plate boundary in the South Island of New Zealand: implications for ancient collision zones. *J. Struct. Geol.*, **12**, 715–725.

Ollier, C.D. 1985. Morphotectonics of continental margins with great escarpments. In: M. Morisawa and J.T. Hack (eds) *Tectonic Geomorphology*, Allen and Unwin, Boston, pp. 3–25.

Penck, W. 1953. *Morphological Analysis of Land Forms*, Macmillan, London.

Price, R.A. 1981. The Cordilleran foreland thrust and fold belt in the southern Canadian Rocky Mountains. In: K.R. McClay and N.J. Price (eds) *Thrust and Nappe Tectonics, Geol. Soc. Lond. Spec. Publ.*, **9**, 427–448.

Royden, L. and Keen, C.E. 1980. Rifting process and thermal evolution of continental lithosphere, *Earth Planet. Sci. Lett.*, **51**, 343–361.

Shaw, G.H. and Mooers, H.D. 1988. Development of drainage networks: a numerical model. *EOS, Trans. Amer. Geophys. Un.*, **69**, 362.

Silver, E.A., McCaffrey, R. and Smith, R.B. 1983. Collision, rotation and the initiation of subduction in the evolution of Sulawesi, Indonesia. *J. Geophys. Res.*, **88**, 9407–9418.

Sinclair, H.D., Coakley, B.J., Allen, P.A. and Watts, A.B. 1991. Simulation of foreland basin stratigraphy using a diffusion-model of mountain belt uplift and erosion – an example from the central Alps, Switzerland. *Tectonics*, **10**, 599–620.

Sonder, L., England, P. and Houseman, G. 1986. Continuum calculations of continental deformation in transcurrent environments. *J. Geophys. Res.*, **91**, 10331–10346.

Stark, C.P. 1994. Cluster growth modeling of plateau erosion. *J. Geophys. Res.*, **99**, 13957–13969.

Steckler, M.S. and Omar, G.I. 1994. Controls on erosional retreat of the uplifted rift flanks at the Gulf of Suez and northern Red Sea. *J. Geophys. Res.*, **99**, 12159–12173.

Stephenson, R. 1984. Flexural models of continental lithosphere based on long-term erosional decay of topography. *Geophys. J.R. Astr. Soc.*, **77**, 385–413.

Stephenson, R. and Lambeck, K. 1985. Erosion-isostatic rebound models for uplift: an application to south-eastern Australia. *Geophys. J.R. Astr. Soc.*, **82**, 31–55.

Stockmal, G.S. 1983. Modelling of large-scale accretionary wedge deformation. *J. Geophys. Res.*, **88**, 8271–8281.

Storti, F. and McClay, K. 1995. Influence of syntectonic

sedimentation on thrust wedges in analogue models. *Geology*, **23**, 999–1002.

Suppe, J. 1980. A retrodeformable cross section of northern Taiwan. *Geol. Soc. China Proc.*, **23**, 46–55.

Suppe, J. 1981. Mechanics of mountain-building and metamorphism in Taiwan. *Mem. Geol. Soc. China*, **4**, 67–89.

Tucker, G.E. and Slingerland, R.L. 1994. Erosional dynamics, flexural isostasy, and long-lived escarpments: a numerical modeling study. *J. Geophys. Res.*, **99**, 12 229–12 243.

Tucker, G.E. and Slingerland, R. 1996. Predicting sediment flux from fold and thrust belts. *Basin Res.*, **8**, 329–349.

Turcotte, D. and Schubert, G. 1982. *Geodynamics: Applications of Continuum Physics to Geological Problems*, Wiley, New York.

van Balen, R.T., van der Beek, P.A. and Cloetingh, S.A.P.L. 1995. The effect of rift shoulder erosion on stratal patterns at passive margins: implications for sequence stratigraphy. *Earth Planet. Sci. Lett.*, **134**, 527–544.

van der Beek, P.A., Cloetingh, S. and Andriessen, P.A.M. 1994. Mechanisms of extensional basin formation and vertical motions at rift flanks: Constraints from tectonic modelling and fission-track thermochronology. *Earth Planet. Sci. Lett.*, **121**, 417–433.

Vilotte, J.P., Daignieres, M. and Madariaga, R. 1982. Numerical modeling of intraplate deformation: simple mechanical models of continental collision. *J. Geophys. Res.*, **87**, 10 709–10 728.

Wang., W.-H. and Davis, D.M. 1996. Sandbox model simulation of forearc evolution and noncritical wedges. *J. Geophys. Res.*, **101**, 11 329–11 340.

Wernicke, B. 1985. Uniform-sense normal simple shear of the continental lithosphere. *Can. J. Earth Sci.*, **22**, 108–125.

Westbrook, G.K., Ladd, J.W., Buhl, P., Bangs, N. and Tiley, G.J. 1988. Cross section of an accretionary wedge: Barbados Ridge complex. *Geology*, **16**, 631–635.

White, R. and McKenzie, D. 1989. Magmatism at rift zones: the generation of volcanic continental margins and flood basalts. *J. Geophys. Res.*, **94**, 7685–7729.

Willett, S.D. 1992. Dynamic and kinematic growth and change of a Coulomb wedge, In: K.R. McClay (ed.) *Thrust Tectonics*, Chapman and Hall, London, pp. 19–31.

Willett, S.D. and Beaumont, C. 1994. Subduction of Asian lithospheric mantle beneath Tibet inferred from models of continental collision. *Nature*, **369**, 642–645.

Willett, S.D., Beaumont, C. and Fullsack, P. 1993. Mechanical model for the tectonics of doubly vergent compressional orogens. *Geology*, **21**, 371–374.

Willgoose, G. 1994a. A physical explanation for an observed area–slope–elevation relationship for catchments with declining relief. *Wat. Resour. Res.*, **30**, 151–159.

Willgoose, G. 1994b. A statistic for testing the elevation characteristics of landscape simulation models. *J. Geophys. Res.*, **99**, 13 987–13 996.

Willgoose, G., Bras, R.L. and Rodriguez-Iturbe, I. 1991a. A physical explanation of an observed link area–slope relationship. *Wat. Resour. Res.*, **27**, 1697–1702.

Willgoose, G., Bras, R.L. and Rodriguez-Iturbe, I. 1991b. A physically based coupled network growth and hillslope evolution model, 1, Theory. *Wat. Resour. Res.*, **27**, 1671–1684.

Willgoose, G., Bras, R.L. and Rodriguez-Iturbe, I. 1991c. A physically based coupled network growth and hillslope evolution model, 2, Applications. *Wat. Resour. Res.*, **27**, 1685–1696.

Xiao, H.-B., Dahlen, F.A. and Suppe, J. 1991. Mechanics of extensional wedges. *J. Geophys. Res.*, **96**, 10 301–10 318.

Zhou, S. and Stüwe, K. 1994. Modeling of dynamic uplift, denudation rates, and thermomechanical consequences of erosion in isostatically compensated mountain belts. *J. Geophys. Res.*, **99**, 13 923–13 940.

4

Fission-track thermochronology and the long-term denudational response to tectonics

Andrew J. W. Gleadow and Roderick W. Brown

4.1 Introduction

Rates and patterns of denudation provide a fundamental insight into the response of landscapes to various tectonic processes and a quantitative calibration of the evolution of the Earth's surface. Estimates of present denudation rates have been derived from various sources, such as sediment yield data from large drainage basins (e.g. Pinet and Souriau, 1988; Summerfield and Hulton, 1994). Obtaining reliable estimates of long-term rates and amounts of denudation have proved more difficult, some of the most promising data coming from calculations of marine sediment volumes (e.g. Rust and Summerfield, 1990; Pazzaglia and Brandon, 1996). Such methods are mostly concerned with information averaged over large areas and thus have an inherently poor spatial resolution. Long-term denudation rates provide an important link between studies of modern surface and tectonic processes, and their continuation back into geological time.

Thermochronology provides a completely new approach to estimating amounts and rates of denudation over long periods of time. Thermochronology may be defined as the quantitative study of the thermal histories of rocks through the use of temperature-sensitive radiometric dating methods. A variety of methods are available but the most important of these for studies of surface denudation is the apatite fission-track dating system because of its sensitivity to the low temperatures found in the upper few kilometres of the Earth's crust. Apatite fission-track thermochronology is effective over time scales of millions, to hundreds of millions, of years, and has the potential to revolutionize our understanding of long-term rates of surface processes. The wide applicability of fission-track methods means that this information can be collected relatively easily from large areas of the crust to give a new insight into regional patterns of long-term denudation with a high spatial resolution compared with estimates based on offshore sediment volumes.

Methods of thermochronology are relatively recent variants of several older geological dating techniques for which retention of the products of radioactive decay in a mineral system is dependent on the temperature at which the material is held at various times. Each particular mineral used in these methods has a characteristic temperature interval over which the dating system is progressively reset. Eventually, if the temperature is great enough, the measured, or apparent, age may be reset to zero. From measurements of the degree of resetting which has occurred, and an understanding of the physical processes involved, it is possible to arrive at a quantitative reconstruction of the thermal history that a rock sample has experienced over geological time.

Thermochronology is now dominated by the $^{40}Ar/^{39}Ar$ and fission-track dating methods. Many of the minerals used in these systems have characteristic 'closure' temperatures of the order of several hundred degrees, more relevant to processes at mid- to lower crustal depths. The apatite fission-track dating system, however, is useful over a temperature range from ambient surface temperatures up to about 110°C which characterizes the upper few kilometres of the crust. At such relatively shallow depths, proximity to the surface is likely to be the dominant influence on palaeotemperature, and the denudation rate at the surface the major control on the cooling patterns obtained.

Geomorphology and Global Tectonics. Edited by Michael A. Summerfield.

Heat production within the Earth ensures that its crust is universally characterized by a thermal gradient of increasing temperature with depth. In the near-surface environment of the upper crust, denudation must inevitably have the effect of cooling the rocks that move upwards through this gradient towards the surface. Through this cooling, therefore, denudation will have a major influence on the apparent ages obtained by a low-temperature thermochronometer. In the simplest case, a measured thermal history will serve as a direct tracer of the progressive movement of the host rock towards the surface. It is important to recognize, however, that denudation is not the only process that can cause cooling in the crust and other processes also need to be considered, such as changes in heat flow.

Apatite fission-track analysis is particularly important for reconstructing the thermal histories of rocks in the low-temperature environment characteristic of the uppermost part of the crust. This has led to a variety of applications in studies of thermal and tectonic processes. Most notably this approach has been applied to young orogenic belts (e.g. Wagner *et al.*, 1977; Hurford *et al.*, 1989; Foster *et al.*, 1994a), rifted continental margins (e.g. Gleadow and Lovering, 1978; Moore *et al.*, 1986; Bohannon *et al.*, 1989; Omar *et al.*, 1989; Brown *et al.*, 1990; Dumitru *et al.*, 1991; Foster and Gleadow, 1992b; Gallagher *et al.*, 1994; Omar and Steckler, 1995), continental extension zones (e.g. Gleadow and Fitzgerald, 1987; Fitzgerald and Gleadow, 1988; Foster *et al.*, 1993; Foster and Gleadow, 1996) and sedimentary basins (e.g. Naeser, 1979; Gleadow *et al.*, 1983; Green *et al.*, 1989b).

The application of fission-track dating to tectonics arose from the important work of Wagner and Reimer (1972) in the European Alps, and the idea of using thermochronology in general to make inferences about denudation and tectonics in orogenic belts is even older (Clark and Jäger, 1969). Wagner and Reimer interpreted observed profiles of increasing apatite fission-track age with sample elevation in terms of rates of uplift. Since that time, the logical sequence involved in making such inferences has for the most part been oversimplified or confused. Indeed the very existence of such an inferential sequence has often been simply ignored, many authors preferring to talk about 'uplift-and-erosion', almost as if these constituted a single process and something that could be measured directly by the fission-track data.

The relationship between the thermal history experienced by the rocks in a particular crustal block and denudation that may have occurred at the land surface is, in fact, a complex one. Indeed the linkage may be quite tenuous in some cases and it is not surprising that some confusion has arisen in understanding the significance of fission-track results. It is the purpose of this chapter to review the principles of fission-track thermochronology, and to discuss its potential, and limitations, for studies of the long-term denudational response to tectonic processes. The following section will describe the basis of fission-track analysis and the principles upon which the interpretation of fission-track data are based.

4.2 Fission-Track Thermochronology

4.2.1 FISSION-TRACK DATING

Fission-track dating is unlike most methods of radiometric dating in that it measures the effect, rather than the product, of a radioactive decay scheme. In this case the radioactive decay is the spontaneous nuclear fission of ^{238}U which produces linear trails of radiation damage, known as fission tracks, from the passage of highly energetic fission fragments through the crystal lattice of the host mineral. The fission tracks are highly reactive compared to the surrounding undamaged lattice and can be enlarged by a simple chemical etching treatment until they can be observed under an ordinary optical microscope. When properly etched the tracks form nearly cylindrical holes which in apatite are about 1–2 μm in diameter and up to about 16 μm long. They are counted and measured on a polished surface cut through the mineral grain as illustrated in Figure 4.1.

The fission-track age is calculated from the number of etched tracks per unit area (the spontaneous track density) observed on the polished surface. For a particular uranium concentration in the apatite, the number of tracks observed in this way will steadily increase with time. The uranium concentration must also be measured and this is done via a second set of fission tracks which are induced by a thermal neutron irradiation procedure. The principles of fission-track dating have been reviewed in more detail by Brown *et al.* (1994a), and practical techniques are described by Wagner and Van den Haute (1992) and Gallagher *et al.* (1998).

Fission-track dating can be applied to a number of uranium-bearing minerals, and some natural glasses, but the most important of these is apatite, a common calcium phosphate mineral. It was discovered very early in the development of fission-track dating methods that apatites frequently give ages significantly younger than

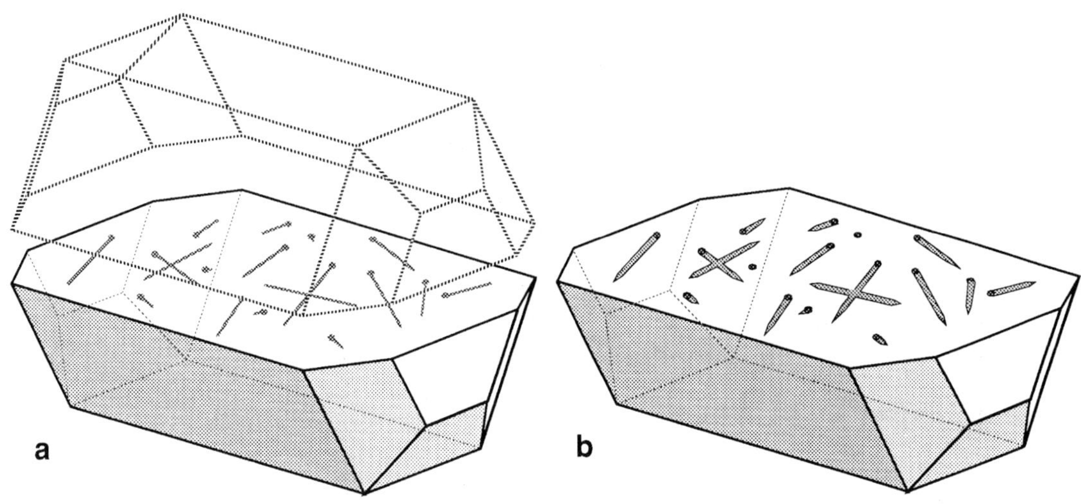

Figure 4.1 *Etching of fission tracks in an apatite crystal. In the first step (a) a polished surface is cut through the mineral grain which will intersect a number of radiation damage trails from the spontaneous nuclear fission of ^{238}U contained within the crystal lattice. The polished surface is then chemically etched (b), which selectively enlarges the fission tracks so that they can be observed under an optical microscope. Two examples of confined fission tracks, used for track length measurements, are shown, which do not intersect the surface but have been etched where they cross another track. Confined fission tracks are up to about 16 μm in length*

the independently known age of the host rock. Only very rarely will the measured fission-track age be related to the age of formation, or geological age, of the mineral as a consequence of the phenomenon of fission-track annealing.

4.2.2 FISSION-TRACK ANNEALING AND THERMAL HISTORY MODELLING

The radiation damage making up fission tracks is stable over geological time, but only at relatively low temperatures. At elevated temperatures the damage is gradually repaired, or 'annealed', as the displaced atoms along the tracks gradually diffuse back into a more ordered state. Eventually the fission tracks cease to be preferentially etchable, so that they can no longer be observed. Temperature is the only environmental factor to have any appreciable effect on the annealing of fission tracks. Other factors, such as pressure, have long been shown to have no observable effect on the annealing process (Fleischer *et al.*, 1975). Understanding the process of thermal annealing forms the basis of the interpretation of apatite fission-track data in terms of thermal histories, but it is important to be able to assess the degree of annealing which has occurred. This is achieved from the study of confined fission-track lengths (e.g. Gleadow *et al.*, 1986).

When fission tracks are first produced in a particular mineral, they are all very similar in length – approximately 16 μm in the case of apatite. During annealing they are progressively shortened from their ends and this results in a reduction in the observed area density of the tracks and therefore a reduction in the measured fission-track age. Each individual fission track shortens to a length which is characteristic of the maximum temperature to which it has been exposed. As new tracks are continually being added to the mineral by radioactive decay, a range of track lengths may be produced, with each track being shortened to a different extent depending on the temperatures experienced since it was formed. Complex distributions of spontaneous track lengths can thus result which are quite different to the simple unimodal distribution which is found if all the tracks remain thermally undisturbed (Gleadow *et al.*, 1986). The track length distribution is thus a direct record of the thermal history experienced and, when analysed together with the fission-track age, can be used to quantitatively reconstruct the thermal history.

Fission-track annealing is a kinetic process in which the effects of temperature and time are to some degree interchangeable. This means that the same annealing process that occurs in spontaneous tracks over geological time can be reproduced in the laboratory,

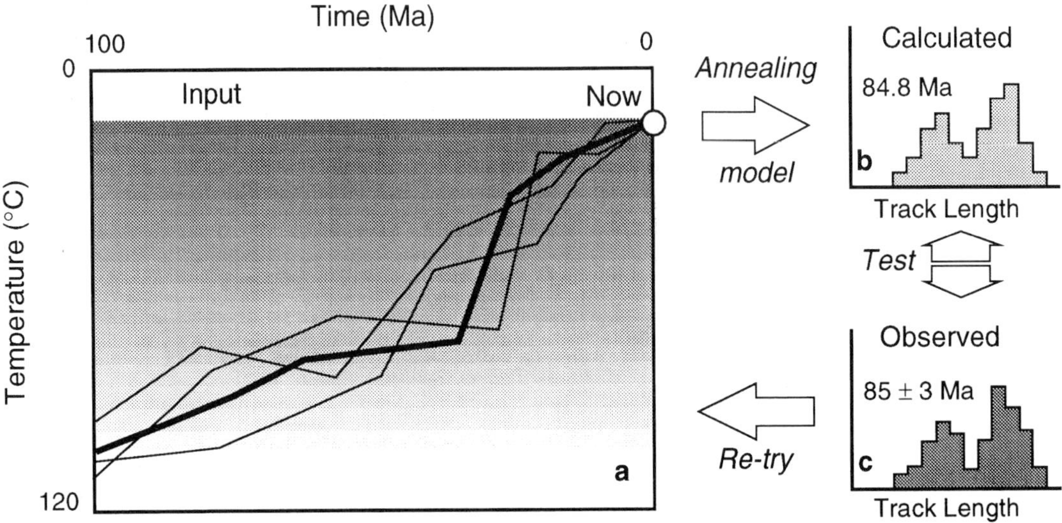

Figure 4.2 *Modelling of thermal histories from apatite fission-track data. The left-hand diagram (a) illustrates a temperature–time path from which a fission-track age and track length distribution (b) are calculated using a numerical annealing model. These are then compared with the observed fission-track data (c) until a match is obtained. Many different possible thermal histories are examined in this way until a reasonable fit to the observed data is obtained*

but at higher temperatures acting for much shorter times. Exhaustive studies of the annealing properties of fission tracks in apatite (Naeser and Faul, 1969; Wagner and Reimer, 1972; Laslett *et al.*, 1987; Green *et al.*, 1989a; Carlson, 1990; Crowley *et al.*, 1991) have produced various numerical models which can then be used to calculate the fission-track age and length distribution resulting from any given thermal history. The results of such 'forward' modelling procedures can then be tested against actual observations in well-constrained geological environments, such as deep drill holes. In this way, Green (1996) has recently shown that, of the currently available annealing models, that of Laslett *et al.* (1987) makes the most realistic predictions. This is likely to improve further once the effects of minor compositional variations in apatite are taken into account (Green *et al.*, 1985).

The reverse-modelling problem, that is, inverting a given set of fission-track observations to obtain the underlying thermal history (e.g. Corrigan, 1991), has proved more difficult. Gallagher (1995), however, has described an effective 'quasi-reversed' modelling procedure which combines a Monte-Carlo simulation of a large number of possible thermal histories with statistical testing of the outcome against observed fission-track measurements. A genetic algorithm is used to provide rapid convergence to an acceptable fit. The principles involved in this approach to thermal history modelling are illustrated in Figure 4.2. These modelling techniques reconstruct the thermal history which best matches the observed fission-track data in any individual rock sample. The approach can also now be extended to modelling the thermal histories of large numbers of samples in broad regional arrays to establish the pattern of cooling which has prevailed over an extensive area (Gleadow *et al.*, 1996). Such arrays can be used to generate images of the distribution of palaeotemperature and regional patterns of denudation at various times in the past.

4.3 Interpretation of Fission-Track Data

4.3.1 COOLING HISTORY STYLES

The preceding section showed how fission-track analysis can be used to model the thermal history of rock samples up to the temperature at which tracks are lost by thermal annealing. For typical apatites, which are close to the fluorapatite end-member in composition, this temperature range extends from ambient surface temperatures up to a maximum of around 100–110°C for geological heating times of 10^6–10^7 years. Fission tracks in more chlorine-rich apatites are stable to somewhat higher temperatures, but even for such compositions the critical temperature for complete track loss is nearly always below about 120°C. The Laslett *et al.* (1987) model is relevant to typical apatites of fluorine-rich composition. The rate of annealing increases rapidly with temperature and is extremely slow at the lower temperatures, below about 60°C. For this reason it is often assumed that most fission-track annealing in apatites occurs between 60, and 120°C, a temperature range known as the fission-track annealing zone.

Depending on the nature of the thermal history experienced, there are essentially three different kinds of 'ages' recorded by thermochronological systems – 'event ages', 'cooling ages' and 'mixed ages'. The significance of these three types can easily be confused but can be clearly resolved from the track length distributions. Figure 4.3 shows examples of three different thermal histories, which are representative of the different common styles of cooling found in the upper crust. The calculated fission-track ages and length distributions are also shown for each of these according to the Laslett *et al.* (1987) model. All three models show cooling from 120 to 20°C over a period of 100 Ma, but follow quite different paths. Most cooling patterns in the upper crustal domain fall into one or other of these three cooling styles (Wagner, 1981).

The first case (Figure 4.3a) illustrates a rock that cools rapidly through the annealing zone and resides at low, near-surface temperatures thereafter. This gives a modelled fission track age of 99.8 Ma that is essentially identical to the input age of 100 Ma, and a narrow distribution of track lengths about a mean value of 15 μm. Such a case might represent the pattern in a rapidly cooled volcanic rock, and indeed many superficial volcanic rocks display exactly this kind of track length distribution. It is only when a length distribution of this kind is observed that the apatite fission-track age can be interpreted directly as the age of a particular cooling event. We might call this kind of age an 'event' age, but it is still only indicating the time of cooling below about 110°C, and it may not therefore relate to a higher temperature phenomena such as the formation age of an igneous rock.

Such rapid cooling patterns, however, are comparatively rare and most samples of basement rocks exposed at the surface reveal fission-track age and length distributions more like (b) or (c) in Figure 4.3. Figure 4.3b shows the example of linear cooling over the 120–20°C range which gives a 'cooling' age of around 76 Ma and a much broader and shorter track length distribution. Such patterns are relatively

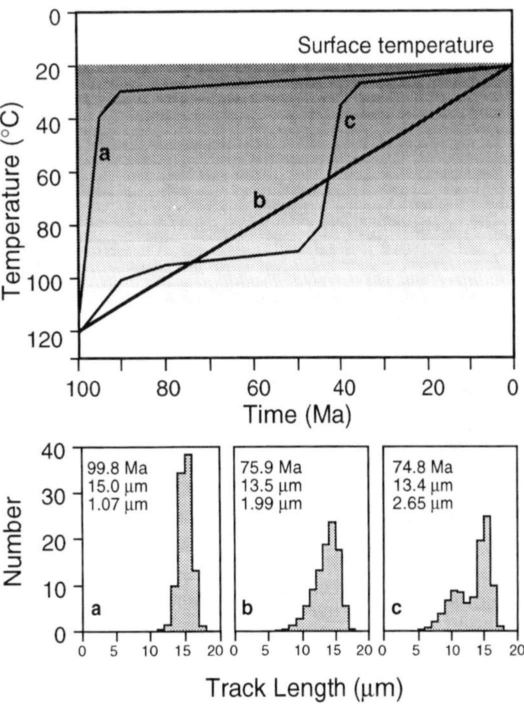

Figure 4.3 *Three different cooling history styles give rise to different fission-track ages and distinctive track length distributions. The fission-track results are calculated using the model of Laslett* et al. *(1987) and the zone of fission-track annealing in apatite is indicated by the background shading gradient. Numbers on the track length histograms give, from top to bottom, the calculated fission-track age, the mean track length and the standard deviation of the length distribution. (a) shows an example of rapid cooling which results in an event age closely approximating the time of major cooling; (b) shows a steady cooling path leading to a broader, skewed length distribution and a much younger cooling age; and (c) shows a discrete episode of rapid cooling from temperatures near the base of the annealing zone. This last case results in a mixed age of 75 Ma which is intermediate between the time of original entry into an annealing zone at 100 Ma and the rapid cooling event at about 45 Ma*

common in old basement terrains which have undergone slow cooling over very long periods of geological time. An important conclusion from such results is that the observed, or apparent, age is the cumulative effect of the slow passage of the rock through the fission-track annealing zone and does not relate to any particular event.

The third case (Figure 4.3c) shows a two-stage cooling history in which one generation of tracks resides at relative higher temperatures within the annealing zone, prior to a rapid cooling to low temperatures. The track length distribution in such cases is typically bimodal with a short peak representing the higher temperature tracks and a second, long peak being added after an episode of rapid cooling. The apparent age is a combination of the two components and is properly described as a 'mixed' age. Again, it is important to stress that the apparent age in this case has no direct significance in terms of the timing of any geological event. However, modelling of such examples reveals both the magnitude and timing of the cooling episode, the temperature before cooling, and an indication of when the sample first cooled into the annealing zone. Such fission-track patterns are therefore particularly valuable in understanding the possible thermal effects in the upper crust resulting from a tectonic disturbance and its denudational response.

Bimodal length distributions representative of mixed ages have proved to be quite common in environments of continental extension, such as along rifted margins (e.g. Moore *et al.*, 1986; Fitzgerald and Gleadow, 1988; Foster and Gleadow, 1992b). Such age and length patterns frequently form part of a regional trend which represents continuous variation from the end-member age components represented in the mixed age. These trends may be identified in a so-called 'boomerang' plot which shows the variation of mean fission-track length against apparent fission-track age (Green, 1986; Brown *et al.*, 1994a).

It is clear from this discussion that the length distribution is crucial in correctly interpreting the significance of the associated fission-track age (Gleadow *et al.*, 1986). This is particularly true when comparing cases (b) and (c) which give rise to almost identical modelled ages and mean track lengths. It is only the form and standard deviation of the track length distributions that indicate the significantly different cooling history styles involved. Most fission-track results for surface rocks represent one or other of these thermal history styles.

4.3.2 DISCRIMINATING COOLING FROM THERMAL EVENT STYLES

The previous discussion concerned the response of the fission-track dating system to patterns of cooling where it is relatively easy to discriminate different cooling styles. We now need to consider what the response will be if an apatite sample cools before being re-heated to some peak temperature and then cooling again. It is very clear that heating and cooling patterns of this kind do occur, for example in the vicinity of a high-level igneous body, but it may be difficult on the basis of fission-track evidence alone to discriminate between a 'thermal event' and a 'cooling event'. These two will actually become the same if the maximum temperature reached during the thermal event exceeds the critical temperature for total track annealing in apatite, whereupon all previous tracks will be lost. At this point the only part of the thermal history which will be preserved will be the subsequent cooling back into the fission-track annealing zone.

The response of the apatite fission-track system to a low-temperature thermal event is illustrated by three different thermal histories in Figure 4.4. One of these (a) is a purely cooling history (the same as (c) in Figure 4.3), while the other two (b) and (c) involve smaller and larger degrees of cooling, respectively, before a thermal event where temperatures increase again. The calculated fission-track parameters obtained from these thermal histories are shown, as before, in track length histograms and summary data for the three cases. It is immediately obvious that the fission-track data are almost identical for the three histories. With real data the very small differences would not be resolvable and are much less than the normal experimental uncertainties in the measurements.

The reason for this behaviour is the acceleration of annealing with increasing temperature, such that even small increases in temperature lead to significant increases in annealing. Most of the annealing experienced by any track takes place within about 15°C of the maximum temperature that it experiences. The duration of heating at that maximum is much less important. In Figure 4.4 the two thermal event histories, (b) and (c), reach a peak temperature about 10°C higher than the average temperature of (a) from 90 to 50 Ma. This brief peak temperature causes as much annealing as the entire residence at the slightly lower plateau temperature for (a) over this period.

When a sample cools, there is essentially no further effect on tracks that have already been partially annealed at higher temperatures until the temperature again rises to greater temperatures. During heating the annealing of all pre-existing tracks rapidly increases so that there is very little 'memory' of annealing at lower temperatures. Only when cooling beings again will a record of former annealing at higher temperatures be clearly preserved. Thus the fission-track system is essentially one that records cooling, and even in a history that has included some heating, most of the

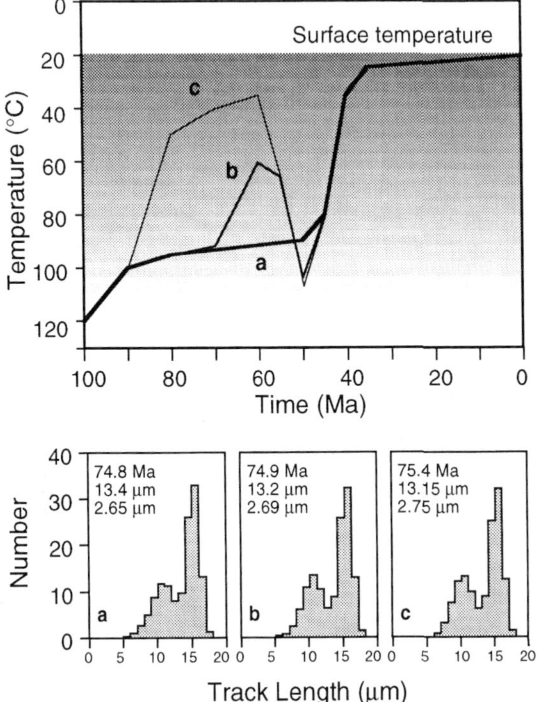

Figure 4.4 *Comparison of three different thermal histories that all lead to very similar calculated fission-track ages and track length distributions. The cooling history (a) is the same as curve (c) in Figure 4.3. (b) and (c) show two examples which cooled to different degrees between 80 and 60 Ma prior to a thermal event at 50 Ma after which cooling followed the same path as for (a). In contrast to the clear distinction between different thermal history styles in Figure 4.3, these examples are indistinguishable from each other from the fission-track data. This illustrates that fission-track thermochronology is most sensitive to the cooling component of thermal histories*

record will be from the cooling segments. Although the discussion here is concerned with the apatite fission-track system, it is worth noting that similar arguments could be made for all other systems of thermochronology.

An important conclusion is that fission-track thermochronology can most clearly establish those parts of a thermal history where cooling has occurred. The existence of a transient thermal event can be inferred, but it is difficult, on the basis of fission-track data alone, conclusively to differentiate this from a purely cooling history. However, the temperature and timing of the thermal maximum or plateau temperature prior to subsequent cooling will be similar in each case and quite well determined from modelling studies

4.3.3 VERTICAL FISSION-TRACK SAMPLING PROFILES

So far we have considered the case where fission-track modelling is applied to the thermal history experienced by a single sample. Much more information is potentially available when a suite of samples in a vertical array can be analysed, such as may be obtained by sampling from a deep drill hole or over a significant range of vertical relief. Such vertical arrays are collected across the thermal gradient through which the samples have cooled.

A consequence of fission-track annealing is that fission-track ages gradually decrease from some observed value at the Earth's surface to an apparent value of zero at the depth where no fission tracks are retained. The depth to the base of the fission-track annealing zone depends on the thermal gradient and the annealing properties of the apatites. For typical fluorapatites, as found in most granitic rocks, this usually occurs at a depth where temperatures are ~100–110°C. The shape of the profile of apatite fission-track age with sample depth below the surface reflects the thermal history of the rocks as they have cooled through the annealing zone. Gleadow (1990) and Brown *et al.* (1994a) have discussed how these profiles vary for different thermal history styles.

The vertical profile of apatite fission-track age resulting from a simple linear cooling pattern is illustrated in Figure 4.5, calculated using the Laslett *et al.* (1987) model. The left-hand part of the diagram shows the input temperature–time (T–t) path which cools from around 90°C at 100 Ma ago to the present surface temperature of 20°C. This linear cooling trend corresponds to a continuous denudation rate of 30 m Ma^{-1} over this period in a thermal gradient of 25°C km^{-1}. The calculated fission-track age for apatite resulting from this cooling path is shown on the right. Results calculated in a similar way for a series of parallel T–t paths are also shown which define an array, or profile, on the plot of fission-track age against present-day temperature.

Fission-track age–temperature profiles calculated in this way for various denudation rates in the same thermal gradient of 25°C km^{-1}, are shown in Figure 4.6, assuming that the thermal gradient remains constant. The calculated fission-track age profiles cover a range of denudation rates from 0 to 50 m Ma^{-1}. It can be seen that the form of these curves varies from nearly linear for the highest denudation rates over this time period, to concave-upwards curves at the lowest rates. The zero denudation rate indicates

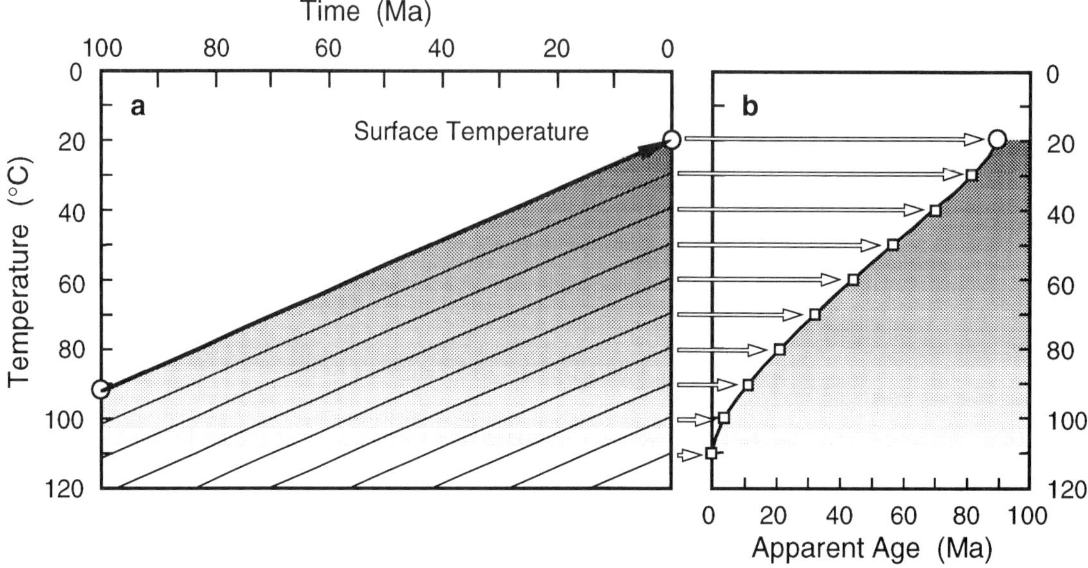

Figure 4.5 Modelling of fission-track ages in a vertical profile resulting from simple linear cooling in the upper crust, such as might be observed in a deep drill hole. The fission-track annealing zone for apatite is illustrated by the background shading gradient. Each linear temperature–time path in (a) results in a different fission-track age in (b) which together define a profile of varying fission-track age with final temperature. (b) corresponds to the profile that would be obtained from simple linear cooling in a thermal gradient of 25°C km^{-1} from a denudation rate of 30 m Ma^{-1}

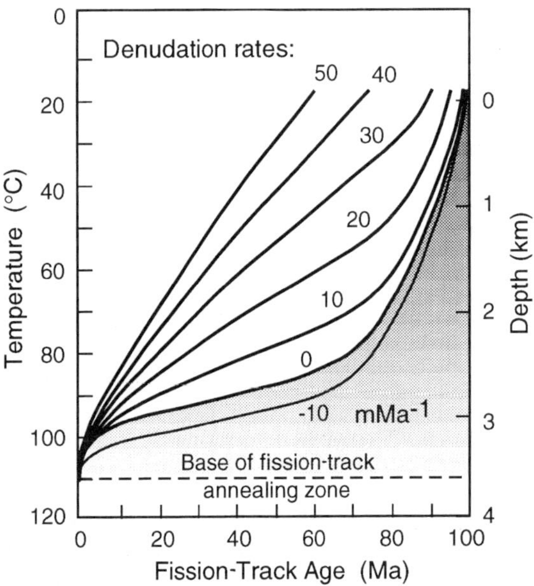

Figure 4.6 Modelled fission-track age profiles against final temperature which result from a series of linear cooling paths of the kind illustrated in Figure 4.5. The profiles represent cooling through denudation in a thermal gradient of 25°C km^{-1} at rates from −10 to 50 m Ma^{-1}

the static situation with constant temperatures being maintained throughout the modelled period of 100 Ma. Little difference from this pattern is obtained when the rate becomes negative (that is, slight heating), indicating the situation that might be found during modest burial.

For the highest denudation rates (>30 m Ma^{-1}) in Figure 4.6 the nearly linear profiles have gradients which correspond closely to the input denudation rates and reach a zero fission-track age at about 100°C. Such profiles are extremely common in active orogenic belts where samples can be collected over a significant vertical relief and they were first described in detail by Wagner and Reimer (1972) and Wagner *et al.* (1977) in the European Alps. In many cases the traditional interpretation of such linear profiles in terms of 'uplift' rates and a closure temperature of about 100°C is probably a reasonable approximation, but this will be discussed further below. Similar linear profiles have also been reported over several kilometres of depth in some deep drill holes, such as the Eielson drill hole in Alaska (Naeser, 1981). The track length distribution for each sample in such a profile is essentially the same indicating that each has experienced the same cooling history, but at slightly different times.

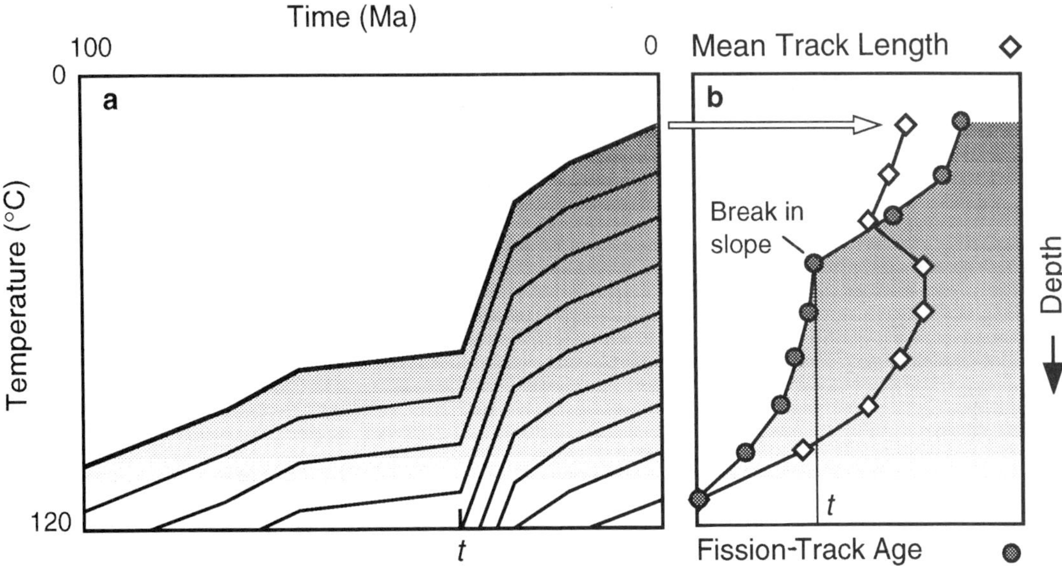

Figure 4.7 *Modelling of fission-track age and track length profiles from a two-stage thermal history incorporating a distinct cooling event at time t. The distinctive profiles observed in this case give a clear indication of the cooling history and show a break in slope separating the upper from the lower region of the curves. The break in slope represents the position that was at the base of the annealing zone prior to the rapid cooling event and has an apparent age approximately equal to the time of cooling, t*

The concave-upwards curves are characteristic of either extremely low cooling rates, such as might be found in cratonic regions, or slow heating due to progressive burial in sedimentary basins. The form of the profile in this case is primarily controlled by the increase in temperature with depth and closely approximates the annealing pattern obtained in laboratory annealing experiments at constant or increasing temperature. Such curves in nature are interpreted as signifying prolonged residence in the partial annealing zone (PAZ). In this case the variation of apparent age with elevation indicates progressive annealing due to increasing temperature and the gradient is not related in any way to cooling rate. Each individual sample in the profile has equilibrated at a slightly different temperature and will have a slightly different track length distribution indicating progressively greater degrees of track annealing.

Typical examples of these concave-upwards PAZ profiles are found in deep drill holes in sedimentary basins (e.g. Naeser, 1979, 1981; Gleadow and Duddy, 1981; Green *et al.*, 1989b). They are also observed in topography produced by broad epeirogenic uplift such as the Transantarctic Mountains (Gleadow and Fitzgerald, 1987; Fitzgerald and Gleadow, 1988) or basement relief on the flanks of the East African Rift system (Foster and Gleadow, 1992a, 1996). Track length distributions in such uplifted PAZ profiles are distinctive (Gleadow and Fitzgerald, 1987) and significantly different from those characteristic of linear profiles.

Most observed fission-track age profiles in nature are made up of either linear arrays of age with elevation or convex-upwards PAZ curves, or combinations of the two (Gleadow, 1990). The existence of such compound profiles in nature (e.g. Fitzgerald and Gleadow, 1988; Foster and Gleadow, 1992a, 1996) is highly significant as it indicates a two-stage cooling history, the upper PAZ type profile indicating a period of relative stability during which cooling rates were very low, followed by a period of accelerated cooling. The break in slope between the two is usually interpreted as signifying the onset of denudation following renewed tectonic activity.

Figure 4.7 shows the profile of apparent fission-track age and mean track length that would result from a two-stage thermal history including a distinct episode of rapid cooling. The apparent age profile shows an obvious break in slope at an age which approximates the time, *t*, of onset of the rapid cooling episode in this example. Samples below this break in slope were deeper than the annealing zone prior to the cooling event and, as a result, retain no fission tracks from before this time. Samples at higher levels record a mixture of tracks from before and after the cooling event and give intermediate apparent ages. Such compound profiles are relatively common and may be observed in deep

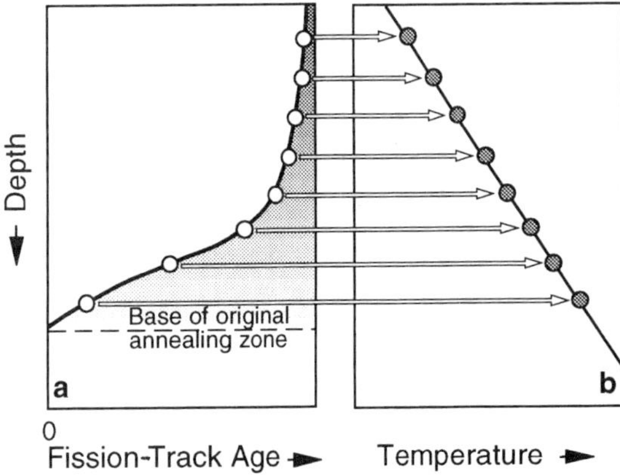

Figure 4.8 *In a vertical profile of apatite fission-track age with depth (a), reconstruction of the palaeotemperature for all of the samples, at some particular time, defines the thermal gradient at that time (b)*

drill holes (e.g. Hammerschmidt *et al.*, 1984; Gleadow and O'Brien, 1994) or, at the surface, in areas of high relief (e.g. Gleadow and Fitzgerald, 1987; Foster and Gleadow, 1996). The existence of these profiles is strong evidence that the thermal history has been characterized by an episode of sharp cooling which was rapid relative to the duration of the recorded history.

The importance of such vertical arrays of samples is that they contain more information than that which can be obtained from any individual sample alone. The samples in such an array clearly have a fixed geometric relationship to each other at the present time, and, provided they have not experienced some tectonic disruption relative to each other, this relationship will have remained constant throughout their recorded thermal history. Modelling the temperature–time paths experienced by the samples within such a fixed array means that the thermal gradient experienced at various times in the past can be directly determined, as illustrated in Figure 4.8. This is an extremely important conclusion as it means that where such profiles are available it will not be necessary to assume the thermal gradient in order to calculate denudation rates from the modelled cooling rate data.

4.3.4 TECTONIC DISRUPTION OF FISSION-TRACK PATTERNS

In areas where block uplift has disrupted an earlier cooling pattern, the observed fission-track profiles will be offset relative to each other. The apparent ages form palaeodepth markers which define an invisible fission-track 'stratigraphy' within the rocks (Gleadow, 1990; Brown, 1991). Offsets in this stratigraphy can be used to determine relative uplift between different blocks and the amount of throw on bounding faults. In this situation a common thermal history style is assumed to be responsible for the variation in fission-track parameters between all of the samples in the area. This approach gives a direct indication of tectonic activity and important insights into the underlying structural evolution of mountain ranges. Such discontinuities in apatite fission-track age patterns do not directly indicate the timing of the tectonic disruption, but they do provide a maximum age for tectonic activity in that it must have occurred at, or after, the youngest apparent age in the disturbed sequence.

Notable examples of such offset fission-track patterns have been observed in a variety of tectonic settings, but they are particularly characteristic of continental extension where a block mountain style is common. Examples include the Transantarctic Mountains (Gleadow and Fitzgerald, 1987; Fitzgerald and Gleadow, 1988), the basement mountains flanking the East African Rift system (Foster and Gleadow, 1992a, 1996), the Basin and Range Province of the southwestern United States (Foster *et al.*, 1993), the southern continental margin of Australia (Foster and Gleadow, 1992b) and the Snowy Mountains of southeastern Australia (Gleadow *et al.*, 1996).

4.4 Application to Continental Denudation

4.4.1 DERIVING ESTIMATES OF DENUDATION

The preceding sections have shown how apatite fission-track data may be used to reconstruct the variation of temperature, particularly cooling, that has been experienced by a rock sample through time. Estimation of the amount and rate of denudation from such cooling histories requires knowledge of the thermal gradient that has prevailed in the geological environment from which the sample was obtained. This involves not just the present-day thermal gradient, but also possible variations in the past. In many cases this information is not known explicitly and must be assumed on the basis of reasonable crustal values.

The logical sequence followed in deriving and using denudation data from fission-track thermochronometry is shown in Figure 4.9. First, the measured fission-track age and length data are combined with a numerical annealing model to calculate the best-fitting

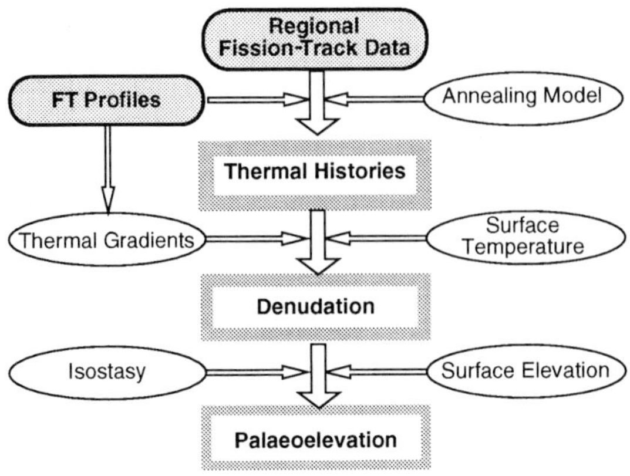

Figure 4.9 *Flow diagram illustrating the inferential sequence used in estimating patterns of denudation from fission-track data. Input from other sources of information is shown by the elliptical symbols. The bottom box shows that it is possible to correct for the amount of removed section to calculate the surface elevation prior to the denudation*

thermal history consistent with the observations. If a vertical profile is available in the area then the form of the profile contains additional constraints about the style of thermal history. The amount of cooling experienced in any given time interval within the cooling history can then be converted to an amount of denudation via the prevailing thermal gradient and the average surface temperature. The denudation rate over particular time intervals can also be calculated.

Having calculated the amount of denudation it is possible to 'backstack' the amount of removed section onto the current surface elevation at that location, as described by Brown (1991). Readjusting the backstacked column to isostatic equilibrium gives an estimate of the average surface elevation prior to the inferred denudation. Such 'palaeoelevation' estimates need to be interpreted with caution, however, as they have not allowed for any transient tectonic uplift or subsidence that may have affected the area in the past. The various sources of uncertainty involved in this inferential sequence would, of course, be cumulative so that the errors would increase with each step from the original data in Figure 4.9. One of the most important sources of uncertainty in such calculations will lie in the estimates of past thermal gradients.

4.4.2 ESTIMATING THERMAL GRADIENTS

On the basis of modern heat flow observations in a wide variety of geological environments, we can infer that changes in thermal gradient only occur within clearly defined limits. Most thermal gradients in basement regions of the continental crust lie between about 10 and 40°C km^{-1} (e.g. Cull and Conley, 1983; Pollack *et al.*, 1993), with an average of close to 30°C km^{-1}. The highest gradients that could be expected in normal continental environments are about 50°C km^{-1}, and such extremes are generally found only in areas of low thermal conductivity or those affected by the perturbing influence of a major groundwater flow system. Larger anomalies are usually extremely localized and clearly associated with a local heat source, such as magmatic activity. Thermal gradients are controlled by the heat flow from depth at a particular place and the thermal conductivity of the rocks. At the Earth's surface, heat flow reflects heat production within the underlying crust and a component which is transferred to the crust by conduction from the mantle beneath.

In crystalline terranes, such as the granitic intrusions that are most commonly used for fission-track dating, the thermal conductivities are generally higher than for sedimentary environments. This means that the thermal gradients in such rocks are usually somewhat lower than mean crustal figures, averaging 20 ± 5°C km^{-1}. In sedimentary basin environments typical gradients are higher and more variable, with typical values in the range 20 to 40°C km^{-1}. Thermal gradients also show a systematic variation with the overall age of the crust in a particular region, reaching as low as 10°C km^{-1} or less in Archaean shield areas. For the most part, thermal gradients are reasonably predictable for a particular region and only rarely do anomalous values occur which are significantly outside the normal range of variation.

Reasonable values for the thermal gradient in a particular area may thus normally be obtained from studies of surface heat flow. Applying such estimates to past thermal gradients will introduce an uncertainty into calculations of denudation. On the basis of the present-day variation in heat flow, however, the magnitude of the resulting error in denudation would be unlikely to exceed 50%, even in the worst case. It is important to recognize that where fission-track data are available from vertical sampling profiles then past variations in thermal gradient can be estimated *directly* and no such assumptions need be made.

4.4.3 TRANSIENT THERMAL EVENTS

Thermal anomalies in the crust are essentially of two kinds – those that have a deep-seated origin, due to changes in heat flow at the base of the lithosphere, and those related to local heat sources within the crust. In the first case, transfer of heat occurs by conduction, which is an extremely inefficient process due to the low thermal conductivity of rocks. Chapman and Rybach (1985) have shown that heat flow anomalies of this kind are related to plate tectonic processes, such as continental rifting, and are associated with length scales of 100–1000 km. Even in such environments thermal gradients are unlikely to exceed values of around 40°C km^{-1}. Perhaps more significantly, changes in heat flow from such deep-seated processes are necessarily slow, and their effects in the upper crust may take tens of millions of years to be felt. Thermal anomalies due to conduction of heat from the base of the lithosphere will therefore be inherently regional in scale and will be associated with only gradual temperature changes.

The consequences of such heat flow changes on thermal gradients in upper crustal environments have been discussed in detail by Brown *et al.* (1994b) and Brown and Summerfield (1997). They concluded that conductive heating from sub-crustal processes is unlikely to explain directly the kinds of relatively rapid cooling often indicated by fission-track data sets. However, the tectonic effects of such lithospheric processes are likely to be much more immediate and these may initiate episodes of cooling due to denudation at the surface.

Another factor also argues against the possibility of rapid changes in near-surface thermal gradients from deep-seated tectonic processes. This is the fact that a substantial fraction of surface heat flow is contributed from heat production within the crust itself, particularly the more radioactive upper crustal rocks. Again, this will be particularly true for the granitic rocks from which many fission-track measurements are obtained. The dominance of heat from upper crustal sources will considerably reduce the near-surface effect of deep-seated changes in heat flow.

Relatively short-lived thermal events within the upper crust must therefore be caused by a different mechanism which is much more efficient in transferring heat than conduction. Possibilities include the convective transfer of heat by magmatic fluids, or its redistribution by groundwater flow. The latter is more likely to be a characteristic of sedimentary basin environments and Chapman and Rybach (1985) have shown that these typically operate on scales of 10–100 km. Magmatic transfer of heat is usually much more local in the upper crust and generally occurs only on a kilometre scale. Geothermal systems related to near-surface magmatic activity are well known and produced extreme, but highly variable, thermal anomalies on a 1–10 km scale. A number of examples of the effects of both magmatic and fluid-flow systems on fission-track age patterns have been described (e.g. Foster *et al.*, 1994b; Gleadow and O'Brien, 1995). In areas where thermal disturbance of the upper crust has been associated with later magmatic activity, the effects are usually extremely localized to the immediate surroundings of known igneous rocks (e.g. Calk and Naeser, 1973; Gleadow and O'Brien, 1994).

The most important elements in identifying these more localized heating systems is their characteristic scale and the rocks with which they are associated. A large-scale groundwater flow system would require plausible aquifers to be identified, for example, and an appeal to magmatic heating would require an indication of the presence of suitable igneous rocks. Some fission-track studies have suggested a combination of both these effects, that is, large-scale hydrothermal circulation in sedimentary basins due to the input of heat from a magmatic source (Gleadow and Duddy, 1984; Brown *et al.*, 1994b). In most environments, however, it is difficult to see that transient 'thermal events' provide a convincing explanation of the regional-scale cooling patterns in basement crystalline rocks that are increasingly apparent from fission-track studies.

4.4.4 DENUDATION IN ACTIVE OROGENIC BELTS

One final consideration in discussing thermal gradients is the quite significant change that can be induced by the effects of denudation itself. Where denudation is extremely rapid, as occurs in active orogenic belts, large-scale advective transfer of heat occurs due to the mass transfer of rock towards the surface, and this leads to an increase in thermal gradient by a compression of isotherms towards the surface, which itself remains at the same temperature. Cooling is clearly dominated by denudation in such environments which are characterized by linear profiles of apatite fission-track age with sample elevation (Wagner and Reimer, 1972; Wagner *et al.*, 1977).

The gradient of age with elevation in young mountain belts has usually been taken as a direct indication of the 'uplift' rate, or, more correctly, the denudation rate. Parrish (1983, 1985) showed that such estimates were likely to be in error where the apparent 'uplift' rate is greater than about 300 m Ma^{-1} due to the effect of advection on the thermal structure of the crust. As denudation rates in young mountain belts reach values of at least 1000 m Ma^{-1}, this effect could be quite significant. A more detailed analysis by Brown and Summerfield (1997) broadly confirmed and extended the conclusions of Parrish but showed that, even above 300 m Ma^{-1}, the true denudation rate can be calculated via a predictable relationship with the apparent rate. Despite this effect of advection on estimates of the rate of denudation, Brown and Summerfield (1997) showed that it had little effect on estimates of the time of cooling below a particular closure temperature. It is also clear from these studies that, in areas outside of actively eroding mountain belts, the effects of advection are insignificant.

The thermal structure of active orogenic belts is also likely to be affected by the topographic relief found in such areas. The thickness of the apatite partial annealing zone is actually of the same order as the relief found in many young mountain chains. The isothermal surfaces defining the annealing zone are thus unlikely to remain horizontal, but will be perturbed to some degree towards the contours of the surface. Zeitler (1985) and Stüwe *et al.* (1994) have studied the effects of topography on critical isotherms in the upper crust in such environments. An important observation is that the deviation of the isothermal surfaces from horizontal is always much less than the relief observed on the surface, especially for the deeper isotherms (90–120°C) that are the most critical in their annealing effect in apatite. Such a departure from horizontal isotherms is unlikely to introduce significant errors into fission-track estimates of denudation at the surface.

4.4.5 THE RELATIONSHIP BETWEEN DENUDATION AND TECTONICS

We now come to the question of the relationship between denudation, which can be inferred on the basis of apatite fission-track data, and tectonic movements in the crust. Many fission-track studies in the past have interpreted the patterns of cooling observed in terms of 'uplift and erosion'. This interpretation makes an implicit assumption that 'uplift' will always be matched by an equivalent amount of 'erosion', although what is meant by these concepts is almost never defined. It has become increasingly clear that this assumption must at least be recognized and justified and, further, that in many cases it simply does not hold (Summerfield and Brown, 1998). A much more precise usage of terminology is required and recent fission-track studies have tended to provide interpretations of cooling histories in terms of denudation, by which is meant the general relative lowering of the land surface by all available processes.

Tectonic activity can have a number of effects which may include an absolute change in surface elevation, which may be defined as 'surface uplift' (or subsidence). Such a change in surface elevation, however, has no effect on the temperature of rocks in the near-surface environment, as there is no change in their position relative to the surface. Fission-track analysis will have nothing to say about such 'uplift' movements. Conversely, in young orogenic belts, where gradients of fission-track age with elevation have often been interpreted in terms of 'uplift' rates, there may actually be no net change in surface elevation. For such environments to be long-lived, there is probably a dynamic equilibrium between denudation and the upward transport of rocks towards the surface; in other words the amount of surface 'uplift' may actually be zero in this case. The terms, 'rock uplift' and 'crustal uplift' are usually used to denote the upward movement of rock with respect to the geoid. The rate of such movement will correlate with the fission-track 'uplift' rate only in the case where the change in surface elevation is zero. But again, the occurrence of such 'rock uplift' does not necessarily imply the existence of tectonic activity, and could easily be associated with a decrease in surface elevation.

The various factors involved in this discussion and their effect on the vertical distribution of fission-track ages are illustrated diagrammatically in Figure 4.10. Column (a) shows a crustal block with an average surface elevation h_0 above the geoid at a time 100 Ma ago. The distance D shows the amount of denudation required to bring a particular rock sample to the surface. Diagram 4.10b shows the profile of apatite fission-track age against sample depth in this same crustal block.

Figure 4.10c shows the same crustal block at the present day after an amount of denudation D has brought the indicated rock sample to the surface on the assumption that the tectonic uplift U_t has been zero, that is, there has been no tectonic uplift or subsidence. In this situation the elevation of the land surface, h, is actually significantly lower, by an amount Δh, than the original elevation following local isostatic adjustment, due to the thinning of the lithosphere by denudation at the surface. The amount of rock uplift, U_r, is represented by the change in elevation of the rock sample from its original position in (a) to its position at the surface in (c). Figure 4.10d shows the new apatite age profile which would result after an episode of denudation giving a form similar to that illustrated in Figure 4.7. In this case the base of the original, or fossil, annealing zone is indicated by the break in slope

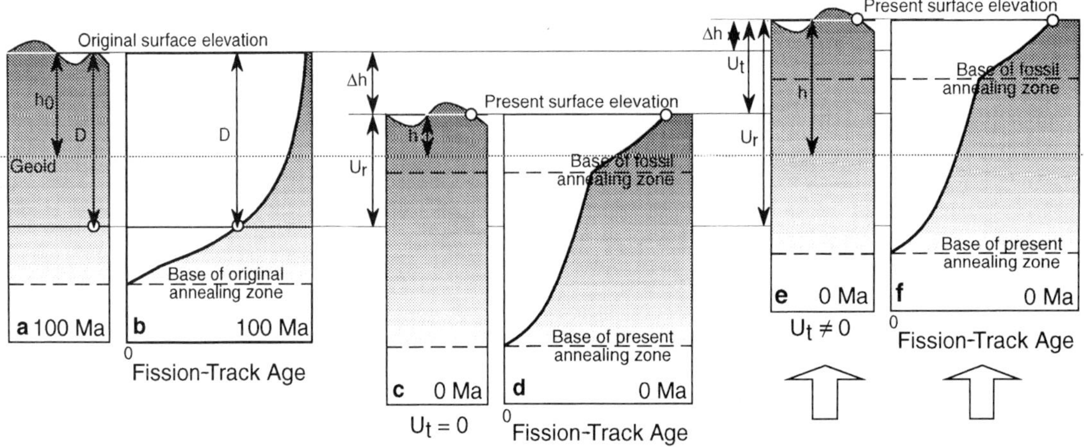

Figure 4.10 *The effects of denudation and tectonic uplift in various crustal columns (a, c, e) on profiles of apatite fission-track age (b, d, f). (a) shows the original situation at 100 Ma in a crustal column following some period of relative stability giving rise to the concave-upwards fission-track age profile illustrated in (b). The average surface elevation is defined by h_0. Column (c) shows the situation at the present following an amount of denudation, D, in the case where there is no tectonic uplift, i.e. $U_t = 0$. The cooling history has followed a path similar to that shown in Figure 4.7. In this case the land surface will have been lowered by an amount Δh to a new average elevation h. The change in elevation of a particular rock sample, the rock uplift, is indicated by U_r. A quite different fission-track age profile is now observed with a break in slope indicating the base of the fossil annealing zone prior to denudation. Column (e) shows the situation where the denudation is the same as in (c) but including the effect of some tectonic uplift, U_t. The fission-track age profile (f) is essentially identical to (d), but standing at a higher elevation*

in the profile. If the thermal gradient has remained constant through the 100 Ma period then the height of this break in slope above the base of the present-day annealing zone would also equal the amount of denudation, D.

Figure 4.10e shows the crustal block after the same amount of denudation, D, but where there has also been a finite, and positive, amount of tectonic uplift U_t. The apatite fission-track age profile (f) in this case is exactly the same as in (d) but at a higher elevation. The amount of rock uplift in (e) is greater than in (c) by the amount of tectonic uplift U_t and the height of the land surface is indicated by h in both cases. Where U_t is zero, the rock uplift U_r is equivalent to the isostatic rebound, so that in the more general case where U_t is not zero, U_r is given by:

$$U_r = \left(1 - \frac{\Delta\rho}{\rho_m}\right) D + U_t \tag{1}$$

where ρ_m is the mantle density, and $\Delta\rho = \rho_m - \rho_c$ (ρ_c being the crustal density). From Figure 4.10 the change in surface elevation, Δh, is related to the denudation and rock uplift by:

$$\Delta h = D - U_r \tag{2}$$

so that from (1):

$$\Delta h = \frac{\Delta\rho}{\rho_m} D + U_t \tag{3}$$

and, from Figure 4.10, the original height of the land surface, h_0, is given by:

$$h_0 = h - \Delta h + U_t \tag{4}$$

In each case it is obvious that the amount of tectonic uplift U_t is not a function of any of the known or measurable parameters and cannot therefore be determined directly from the denudation in any case where U_t is not zero.

We conclude that there is no necessary geometric relationship between denudation and tectonics, and yet it is clear that there is at least a broad temporal relationship between the two in many cases. This link must come via the geomorphic processes whereby denudation is initiated, principally through the creation of relief. Various authors have shown a relationship between denudation rate and local relief within a drainage basin (e.g. Ahnert, 1970; Pinet and Souriau, 1988; Summerfield and Hulton, 1994). Where tectonic activity produces an increase in relief, this will lead to denudation which in turn will cause cooling as rocks migrate towards the surface. Denudation will initiate isostatic rebound which will act to exaggerate the effects of the original denudation. Surface uplift is not necessarily involved in this process at all, as the initial relief could come from a drop in regional base level due to tectonic subsidence, or rifting (Brown *et al.*, 1994a; Gilchrist and Summerfield, 1994).

Fission-track thermochronology can thus provide powerful constraints on patterns of denudation, but the interpretation of such patterns in terms of tectonic activity cannot be inferred directly on the basis of the fission-track analysis alone. To assess the relationship to tectonic activity, fission-track data must therefore be integrated with other sources of information and can be used to test the thermal consequences of various models of the geological evolution of an area. The characteristic rates of denudation may give important clues to the underlying tectonic processes operating in different environments.

Continental margin and rift environments appear to have characteristic denudation rates of around 100 m Ma^{-1}, which mostly operate to produce total denudation of only about 1–3 km over a limited time period (e.g. Gleadow and Fitzgerald, 1987). In active orogenic belts, denudation rates are as much as ten times higher than this, typically 200–2000 m Ma^{-1}, and the total amount of denudation probably exceeds the total thickness of the apatite annealing zone several times over (e.g. Wagner *et al.*, 1977). Environments of tectonic denudation, such as observed in the metamorphic core complexes of the Basin and Range Province and the Aegean Sea, generally have even more extreme cooling rates of ~100°C Ma^{-1}, which imply denudation at 2–4 km Ma^{-1} (e.g. Lister *et al.*, 1984; Baldwin *et al.*, 1993; Foster *et al.*, 1994b). Such rates can only be fully assessed in combination with higher temperature systems such as $^{40}Ar/^{39}Ar$ thermochronology. These tectonic denudation rates clearly have quite different implications in terms of surface processes because the overlying material is removed by lateral transport on major low-angle fault systems.

4.5 Conclusions

Fission-track analysis is based on the accumulation of radiation damage in uranium-bearing minerals, of which apatite is the most important. The interpretation of apatite fission-track ages and track length distributions in terms of the thermal history experienced by the host rocks has been securely established and produces realistic and testable predictions. Understanding of the long-term stability of fission tracks has led to numerical modelling procedures which can quantify the exposure of rocks to elevated temperatures up to about 110°C over time scales of 10^6 to10^8 years. Measured fission-track ages and track length data are used to calculate the best-fitting thermal history consistent with the observations. The calculated thermal histories are essentially a response to cooling through the upper several kilometres of the crust which, in most cases, is controlled principally by denudation at the surface.

Different styles of thermal histories give rise to distinctive fission-track length distributions which show that only rarely can a fission-track age be taken as indicating the timing of some particular event. Most fission-track ages are cooling ages or mixtures of two different age components which can be misleading if interpreted at face-value. Where sampling is available over a significant vertical interval, such as from a drill hole or in areas of high surface relief, then distinctive profiles of fission-track age with sample elevation are found which are also indicators of the style of thermal history experienced. Because the samples in such a profile have a fixed geometric relationship to each other, reconstruction of the thermal histories also enables the palaeothermal gradients to be estimated directly. Knowledge of the thermal gradient is important in calculating denudation rates from the observed cooling rates. Even without such vertical sampling information, however, it is still possible to estimate denudation rates because the range of variation of thermal gradients is well known in a broad range of geological environments.

Using these fission-track techniques, thermal histories of upper crustal rocks can be reconstructed in great detail and have proved effective in the study of patterns of denudation in various tectonic settings. These patterns can also be studied on various spatial scales from regional up to continental with a high spatial resolution. The wide applicability of fission-track methods means that this information can be collected relatively easily from large areas of the crust. This type of information, together with associated studies of the natural controls on denudation, provides an extremely powerful methodology for investigating the relationship between tectonics and the evolution of topography.

4.6 Outstanding Questions and Future Directions

The development of quantitative methods for analysing the thermal histories of rocks is leading to a much more rigorous approach to interpreting the thermal evolution of the upper crust. An important shift has taken place away from explanations based on purely tectonic factors to those which incorporate a much fuller appreciation of the important geomorphic mechanisms involved. These trends open up a range of important new opportunities for testing and constraining models of landscape evolution.

Although factors such as climate, lithology and drainage evolution can affect spatial and temporal patterns of denudation, studies of modern drainage basins indicate that topographic relief primarily determines erosional energy and hence denudation rate (e.g. Ahnert, 1970; Pinet and Souriau, 1988; Summerfield and Hulton, 1994; Summerfield and Brown, 1998). A key question, however, is whether these rates, and, more critically, their dependence on the identified controlling factors, can be extrapolated to geological time scales relevant to understanding the denudational response to tectonic processes. Preliminary comparisons indicate that long-term denudation rates, derived from sediment volume calculations and thermochronologic data, combined with drainage evolution models (e.g. Rust and Summerfield, 1990), are indeed of the same order of magnitude as modern rates.

The challenge for future research is therefore to devise experiments that can further test the uniformity of these relationships and to establish that there are no additional or alternative controlling parameters operable at the longer time scales. This might be achieved by quantifying the present-day relief and discharge characteristics for well-characterized tectonic environments at a number of sites representing different stages of development. Expected sediment yields and denudation rates for each of these specific regions could then be determined using the relationships between morphometric and discharge parameters and denudation rates derived from large global data sets (e.g. Summerfield and Hulton, 1994). The predicted values could then be tested against measured values for each site. The purpose would be to test the relations based on modern data in a number of regions representing the same tectonic environment but successive stages in the geomorphic response to this style of tectonism. The validity of extrapolating the short-term relations would be enhanced if they were shown to hold for all stages and were also consistent with long-term measurements based on thermochronologic and sediment volume data.

In addition, thermochronologic data could be used to test the predictions of key geomorphic concepts. For example, an axial traverse of the Andes could be used to test the idea that the geomorphic sensitivity of mountain ranges is a function of their width (Ahnert, 1984). The morphology of the Andes changes progressively from a low, narrow range in Argentina into a high, wide and internally drained plateau in Bolivia (Summerfield and Brown, 1998). The temporal correlation between the timing of tectonic uplift and onset of denudation should itself be a function of the geomorphic sensitivity (as might the absolute denudation rates for a given location). The tectonic history of the Andes could be accurately reconstructed from plate motion data and the onset and rate of denudation measured using thermochronologic data.

Thermochronologic data, particularly apatite fission-track data, are proving ideal for testing the predictions and scaling the input parameters of the growing number of complex numerical models (e.g. Gilchrist *et al.*, 1994; Kooi and Beaumont, 1994) (see Chapter 3) which aim to simulate large-scale, long-term landscape change. These models integrate the effects of surface processes (on short temporal and length scales) with tectonic processes (on longer temporal and length scales) and make clear predictions about the thermal evolution of the rocks involved.

A major potential use of thermochronologic data that has so far been largely overlooked or neglected, primarily because of the past emphasis on interpreting results in terms of tectonically driven surface uplift, is in studies of mass and chemical recycling within the Earth. A major concern in the current global change debate is understanding the factors controlling ocean chemistry and how these have changed over various time scales. Denudation of terrestrial material clearly plays a crucial role here in that weathering rates and riverine sediment fluxes are a major control on several important chemical parameters commonly used as proxies for monitoring palaeoclimatic variables. Examples are the $^{87}Sr/^{86}Sr$ ratio (Harris, 1995; Derry and France-Lanord, 1996; Ruppel *et al.*, 1996) or CO_2 concentration (Berner, 1990, 1991; Weissert and Mohr, 1995). More generally, thermochronologic data can be used effectively to map the provenance of clastic sedimentary sequences. This would contribute to a better understanding of the spatial and temporal aspects of sediment supply which is an important element in the tectonic analysis of sedimentary basins (e.g. Cross, 1990).

Refinements of the fission-track modelling procedures are also a major challenge for future work. For example, strategies for quantitative extraction of palaeogradient estimates from vertical profiles need to be devised as well as procedures for handling variably spaced three-dimensional data sets. Techniques for integrating stratigraphic and other relevant geological data with these large data sets would also be extremely valuable. Underlying this is a need for much more and better quality data upon which to base the modelling procedures. This need is driving the development of high levels of automation in the acquisition of fission-track data and may lead in the near future to routine fully automated track counting systems. Such systems would dramatically enhance the availability of high-quality fission-track data and greatly increase the accessibility of thermochronologic constraints to researchers in landscape evolution.

Overall, a shift away from a purely tectonic emphasis in interpretations of thermochronologic data to a geomorphic emphasis will encourage a new perspective that incorporates aspects of tectonics, geomorphology and palaeoenvironmental analysis into a much wider and better integrated application of the technique. The ultimate goal might be to derive accurate, quantitative palaeogeographic reconstructions which explicitly include the combined effects of tectonics and all surface processes.

Acknowledgements

Aspects of this work have been supported by the Australian Research Council and the Australian Institute of Nuclear Science and Engineering. This paper is published by permission of the Director of the Australian Geodynamics Cooperative Research Centre. Fruitful discussions with many colleagues and students in the La Trobe University Fission Track Research Group, Mike Summerfield and Kerry Gallagher are gratefully acknowledged.

References

Ahnert, F. 1970. Functional relationships between denudation, relief and uplift in large mid-latitude drainage basins. *Amer. J. Sci.*, **268**, 243–263.

Ahnert, F. 1984. Local relief and the height limits of mountain ranges. *Amer. J. Sci.*, **284**, 1035–1055.

Baldwin, S.L., Lister, G.S., Hill, E.J., Foster, D.A. and McDougall, I. 1993. Thermochronologic constraints on the tectonic evolution of active core complexes, D'Entrecasteaux Islands, Papua-New Guinea. *Tectonics*, **12**, 611–628.

Berner, R.A. 1990. Atmospheric carbon dioxide levels over Phanerozoic time. *Science*, **249**, 1382–1386.

Berner, R.A. 1991. A model for atmospheric CO_2 over Phanerozoic time. *Amer. J. Sci.*, **291**, 339–376.

Bohannon, R.G., Naeser, C.W., Schmidt, D.L. and Zimmerman, R.A. 1989. The timing of uplift, volcanism and rifting peripheral to the Red Sea: A case for passive rifting. *J. Geophys. Res.*, **94**, 1683–1701.

Brown, R.W. 1991. Backstacking apatite fission track stratigraphy: a method for resolving the erosional and isostatic rebound components of tectonic uplift histories. *Geology*, **19**, 74–77.

Brown, R.W. and Summerfield, M.A. 1997. Some uncertainties in the derivation of rates of denudation from thermochronologic data. *Earth Surf. Processes Ldfms.*, **22**, 239–248.

Brown, R.W., Rust, D.J., Summerfield, M.A., Gleadow, A.J.W. and De Wit, M.C.J. 1990. An early Cretaceous phase of accelerated erosion on the southwestern margin of Africa: Evidence from apatite fission track analysis and the offshore sedimentary record. *Nucl. Tracks Rad. Measur.*, **17**, 339–351.

Brown, R.W., Summerfield, M.A. and Gleadow, A.J.W. 1994a. Apatite fission track analysis: Its potential for the estimation of denudation rates and implications for models of long-term landscape development. In: M.J. Kirkby (ed.) *Process Models and Theoretical Geomorphology*, Wiley, Chichester, pp. 23–53.

Brown, R.W., Gallagher, K. and Duane, M. 1994b. A quantitative assessment of the effects of magmatism on the thermal history of the Karoo sedimentary sequence. *J. Afr. Earth Sci.*, **18**, 227–243.

Calk, L.C. and Naeser, C.W. 1973. The thermal effect of a basalt intrusion on fission tracks in quartz monzonite. *J. Geol.*, **81**, 189–198.

Carlson, W.D. 1990. Mechanism and kinetics of fission track annealing. *Amer. Mineral.*, **75**, 1120–1139.

Chapman, D.S. and Rybach, L. 1985. Heat flow anomalies and their interpretation. *J. Geodynamics*, **4**, 3–37.

Clark, S.P. and Jäger E. 1969. Denudation rate in the Alps from geochronologic and heat flow data. *Amer. J. Sci.*, **267**, 1143–1160.

Corrigan, J. 1991. Inversion of apatite fission track data for thermal history information. *J. Geophys. Res.*, **96**, 10 347–10 360.

Cross, T.A. (ed.) 1990. *Quantitative Dynamic Stratigraphy*, Prentice-Hall, Englewood Cliffs, NJ.

Crowley, K.D., Cameron, M. and Schaefer, R.L. 1991. Experimental studies of annealing of etched fission tracks in apatite. *Geochim. Cosmochim. Acta*, **55**, 1449–1465.

Cull, J.P. and Conley, D. 1983. Geothermal gradients and heat flow in Australian sedimentary basins. *BMR J. Austr. Geol. Geophys.*, **8**, 329–337.

Derry, L.A. and France-Lanord, C. 1996. Neogene Himalayan weathering history and river $^{87}Sr/^{86}Sr$: impact on the marine Sr record. *Earth Planet. Sci. Lett.*, **142**, 59–74.

Duddy, I.R., Green, P.F. and Laslett, G.M. 1988. Thermal annealing of fission tracks in apatite 3. Variable temperature annealing. *Chem. Geol.* (*Isotope Geosci.*), **73**, 25–38.

Dumitru, T.A., Hill, K.C., Coyle, D.A., Duddy, I.R., Foster, D.A., Gleadow, A.J.W., Green, P.F., Laslett, G.M., Kohn, B.P. and O'Sullivan, A.B. 1991. Fission track thermochronology: application to continental rifting of southeastern Australia. *APEA J.*, **31**, 131–142.

Fitzgerald, P.F. and Gleadow, A.J.W. 1988. Fission track geochronology, tectonics and structure of the Transantarctic Mountains in northern Victoria Land, Antarctica. *Chem. Geol.* (*Isotope Geosci.*), **73**, 169–198.

Fleischer, R.L., Price, P.B. and Walker, R.M. 1975. *Nuclear Tracks in Solids*, University of California Press, Berkeley.

Foster, D.A. and Gleadow, A.J.W. 1992a. The morphotectonic evolution of rift-margin mountains in central Kenya: constraints from apatite fission track analysis. *Earth Planet. Sci. Lett.*, **113**, 157–171.

Foster, D.A. and Gleadow, A.J.W. 1992b. Reactivated tectonic boundaries and implications for the reconstruction of southeastern Australia and northern Victoria Land, Antarctica. *Geology*, **20**, 267–270.

Foster, D.A. and Gleadow, A.J.W. 1996. Structural framework and denudation history of the flanks of the Kenya and Anza Rifts, East Africa. *Tectonics*, **15**, 258–271.

Foster, D.A., Gleadow, A.J.W., Reynolds, S.J. and Fitzgerald, P.F. 1993. The denudation of metamorphic core complexes and the reconstruction of the Transition Zone, west central Arizona: constraints from apatite fission track thermochronology. *J. Geophys. Res.*, **98**, 2167–2185.

Foster, D.A., Gleadow, A.J.W. and Mortimer, G. 1994a. Rapid Pliocene exhumation in the Karakoram, revealed by fission-track thermochronology of the K2 gneiss. *Geology*, **22**, 19–22.

Foster, D.A., Murphy, J.M. and Gleadow, A.J.W. 1994b. Middle Tertiary hydrothermal activity and uplift of the northern Flinders Ranges, South Australia: insights from apatite fission track thermochronology. *Austr. J. Earth Sci.*, **41**, 11–18.

Gallagher, K. 1995. Evolving temperature histories from apatite fission-track data. *Earth Planet. Sci. Lett.*, **136**, 421–435.

Gallagher, K., Brown, R.W. and Johnson, C. 1998. Fission track analysis and its applications to geological problems. *Ann. Rev. Earth Planet. Sci.*, **26**, 519–572.

Gallagher, K., Hawkesworth, C.J. and Mantovani M.S.M. 1994. The denudational history of the onshore continental margin of SE Brazil from fission track data. *J. Geophys. Res.*, **99**, 18 117–18 145.

Gilchrist, A.R. and Summerfield, M.A. 1994. Tectonic models of passive margin evolution and their implications for theories of long-term landscape development. In: M.J. Kirkby (ed.) *Process Models and Theoretical Geomorphology*, Wiley, Chichester, pp. 55–84.

Gilchrist, A.R., Kooi, H. and Beaumont, C. 1994. Post-Gondwana geomorphic evolution of southwestern Africa: Implications for the controls on landscape development from observations and numerical experiments. *J. Geophys. Res.*, **99**, 12 211–12 228.

Gleadow, A.J.W. 1990. Fission track thermochronology – reconstructing the thermal and tectonic evolution of the crust. *Proceedings of the Pacific Rim Congress 1990*, Gold Coast, Queensland 6–12 May 1990, Vol. III, 15–21.

Gleadow, A.J.W. and Duddy, I.R. 1981. A natural long-term annealing experiment for apatite. *Nucl. Tracks Rad. Measur.*, **5**, 169–174.

Gleadow, A.J.W. and Duddy, I.R. 1984. Fission track dating and thermal history analysis of apatites from wells in the northwestern Canning Basin. In: P.G. Purcell (ed.) *The Canning Basin*, GSA and PESA, Perth, pp. 377–387.

Gleadow, A.J.W. and Fitzgerald, P.F. 1987. Uplift history and structure of the Transantarctic Mountains: new evidence from fission track dating of basement apatites in the Dry Valleys area, southern Victoria Land. *Earth Planet. Sci. Lett.*, **82**, 1–14.

Gleadow, A.J.W. and Lovering J.F. 1978. Fission track geochronology of King Island, Bass Strait, Australia: Relationship to continental rifting. *Earth Planet. Sci. Lett.*, **37**, 429–437.

Gleadow, A.J.W. and O'Brien P.E. 1994. Apatite fission track thermochronology and tectonics in the Clarence–Moreton Basin of eastern Australia. *AGSO Bull.*, **261**, 189–194.

Gleadow, A.J.W., Duddy, I.R. and Lovering, J.F. 1983. Fission track analysis: a new tool for the evaluation of thermal histories and hydrocarbon potential. *APEA J.*, **23**, 93–102.

Gleadow, A.J.W., Duddy, I.R., Green, P.F. and Lovering, J.F. 1986. Confined fission track lengths in apatite: a diagnostic tool for thermal history analysis. *Contrib. Min. Petrol.*, **94**, 405–415.

Gleadow, A.J.W., Kohn, B.P., Gallagher, K. and Cox, S.J. 1996. Imaging the thermotectonic evolution of eastern Australia during the Mesozoic from fission track dating of apatites. Mesozoic of the eastern Australia Plate, Brisbane. *Geol. Soc. Austr. Extend. Absts.*, **43**, 194–204.

Green, P.F. 1986. On the thermotectonic evolution of northern England: evidence from fission track analysis. *Geol. Mag.*, **123**, 405–415.

Green, P.F. 1996. The importance of compositional influence on fission track annealing in apatite. (Abstract) *International Fission Track Dating Workshop*, Gent, Belgium, 43.

Green, P.F., Duddy, I.R., Gleadow, A.J.W., Tingate P.R. and Laslett, G.M. 1985. Fission track annealing in apatite: track length measurements and the form of the Arrhenius plot. *Nucl. Tracks*, **10**, 323–328.

Green, P.F., Duddy, I.R., Laslett, G.M., Hegarty, K.A., Gleadow, A.J.W. and Lovering, J.F. 1989a. Thermal annealing of fission tracks in apatite: 4 – Quantitative modelling techniques and extension to geological timescales. *Chem. Geol.* (*Isotope Geosci.*), **79**, 155–182.

Green, P.F., Duddy, I.R., Gleadow, A.J.W. and Lovering J.F.

1989b. Apatite fission track analysis as a paleotemperature indicator for hydrocarbon exploration. In: N.D. Naeser and T.H. McCulloch (eds) *Thermal History of Sedimentary Basins*, Springer-Verlag, New York, pp. 181–195.

Hammerschmidt, K., Wagner, G.A. and Wagner, M. 1984. Radiometric dating on research drill core Urach III: a contribution to its geothermal history. *J. Geophys.*, **54**, 97–105.

Harris, N. 1995. Significance of weathering of Himalayan metasedimentary rocks and leucogranites for the Sr isotope evolution of seawater during the early Miocene. *Geology*, **23**, 795–798.

Hurford, A.J., Flisch, M. and Jäger E. 1989. Unravelling the thermo-tectonic evolution of the Alps: a contribution from fission track analysis and mica dating. In: M.P. Coward, D. Dietrich and R.G. Park (eds) *Alpine Tectonics, Geol. Soc. Lond. Spec. Publ.*, **45**, 369–398.

Kohn, B.P. and Eyal, M. 1981. History of uplift of the crystalline basement of Sinai and its relation to opening of the Red Sea as revealed by fission track dating of apatites. *Earth Planet. Sci. Lett.*, **52**, 129–141.

Kooi, H. and Beaumont, C. 1994. Escarpment evolution on high-elevation rifted margins: Insights derived from a surface processes model that combines diffusion, advection and reaction. *J. Geophys. Res.*, **99**, 12 191–12 209.

Laslett, G.M., Green, P.F., Duddy, I.R. and Gleadow, A.J.W. 1987. Thermal annealing of fission tracks in apatite: 2 – A quantitative analysis. *Chem. Geol. (Isotope Geosci.)*, **65**, 1–13.

Lister, G.S., Banga, G. and Feenstra, A. 1984. Metamorphic core complexes of Cordilleran type in the Cyclades, Aegean Sea, Greece. *Geology*, **12**, 221–225.

Moore, M.E., Gleadow, A.J.W. and Lovering, J.F. 1986. Thermal evolution of rifted continental margins: new evidence from fission track dating of apatites from southeastern Australia. *Earth Planet. Sci. Lett.*, **78**, 255–270.

Naeser, C.W. 1979. Thermal history of sedimentary basins by fission track dating of sub-surface rocks. In: P.A. Scholle and P.R. Schulger (eds) *Aspects of Diagenesis, Soc. Econ. Paleont. Mineral. Spec. Publ.*, **26**, 109–112.

Naeser, C.W. 1981. The fading of fission tracks in the geologic environment – data from deep drill holes. *Nucl. Tracks*, **5**, 248–250.

Naeser, C.W. and Faul, H. 1969. Fission track annealing in apatite and sphene. *J. Geophys. Res.*, **74**, 705–710.

Omar, G.I. and Steckler, M.S. 1995. Fission track evidence for the initial rifting of the Red Sea – Two pulses, no propagation. *Science*, **270**, 1341–1344.

Omar, G.I., Steckler, M.S., Buck, W.R. and Kohn, B.P. 1989. Fission-track analysis of basement apatites at the western margin of the Gulf of Suez rift, Egypt: Evidence for synchroneity of uplift and subsidence. *Earth Planet. Sci. Lett.*, **94**, 316–328.

Parrish, R.R. 1983. Cenozoic thermal evolution and tectonics of the coast mountains of British Columbia 1. Fission track dating, apparent uplift rates, and patterns of uplift. *Tectonics*, **2**, 601–631.

Parrish, R.R. 1985. Some cautions which should be exercised when interpreting fission track and other data with regard to uplift rate calculations. (Abstract) *Nucl. Tracks Rad. Measur.*, **10**, 425.

Pazzaglia, F.J. and Brandon, M.T. 1996. Macrogeomorphic evolution of the post-Triassic Appalachian mountains determined by deconvolution of the offshore basin sedimentary record. *Basin Res.*, **8**, 255–278.

Pinet, P. and Souriau, M. 1988. Continental erosion and large-scale relief. *Tectonics*, **7**, 563–582.

Pollack, H.N., Hurta, S.J. and Johnson, J.R. 1993. Heat flow from the Earth's interior: analysis of the global data set. *Rev. Geophys.*, **31**, 267–280.

Ruppel, S.C., James, E.W., Barrick, J.E., Nowlan, G. and Uyeno, T.T. 1996. High resolution $^{87}Sr/^{86}Sr$ chemostratigraphy of the Silurian: Implications for event correlation and strontium flux. *Geology*, **24**, 831–834.

Rust, D.J. and Summerfield, M.A. 1990. Isopach and borehole data as indicators of rifted margin evolution in southwestern Africa. *Mar. Petrol. Geol.*, **7**, 277–287.

Stüwe, K., White, L. and Brown, R. 1994. The influence of eroding topography on steady state isotherms. Applications to fission track analysis. *Earth Planet. Sci. Lett.*, **124**, 63–74.

Summerfield, M.A. and Brown, R.W. 1998. Geomorphic factors in the interpretation of thermochronologic data. In: P. Van den Haute and F. De Corte (eds) *Advances in Fission-Track Geochronology*, Kluwer, Dordrecht, pp. 269–284.

Summerfield, M.A. and Hulton, N.J. 1994. Natural controls of fluvial denudation rates in major world drainage basins. *J. Geophys. Res.*, **99**, 13 871–13 884.

Wagner, G.A. 1981. Correction and interpretation of fission track ages. In: E. Jäger and J.C. Hunziker (eds) *Lectures in Isotope Geology*, Springer-Verlag, Berlin, pp. 170–177.

Wagner, G.A. and Reimer, G.M. 1972. Fission track tectonics: the tectonic interpretation of fission track apatite ages. *Earth Planet. Sci. Lett.*, **14**, 263–268.

Wagner, G.A. and Van den Haute, P. 1992. *Fission Track Dating*, Kluwer, Dordrecht.

Wagner, G.A., Reimer, G.M. and Jäger, E. 1977. Cooling ages derived by apatite fission track, mica Rb–Sr and K–Ar dating: the uplift and cooling history of the central Alps. *Institute of Geology and Mineralogy, University of Padova, Memoir*, **30**, 1–27.

Weissert, H. and Mohr, H. 1995. Late Jurassic climate and its impact on carbon cycling. *Palaeogeog. climat. ecol.*, **122**, 27–43.

Zeitler, P.K. 1985. Cooling history of the NW Himalaya, Pakistan. *Tectonics*, **4**, 127–151.

5

Macroscale process systems of mountain belt erosion

Niels Hovius

5.1 Introduction

The Earth's topography reflects the interaction of tectonic processes, involving the displacement of rocks, and geomorphic processes, responsible for the redistribution of material across the resulting relief. An imbalance between the tectonic input and denudational response results in an alternation in the distribution and/or amplitude of relief. This, in turn, may affect the heat budget and atmospheric circulation to produce a regional or global climatic change. The ensuing changes in rates and modes of earth surface processes then feed back into the tectonic stress system through shifts in regional mass budgets. This complex of relations and feedbacks between tectonic, geomorphic and climatic processes culminates in the orogenic system (see Section 3.5). Mountain belts are narrow zones of tectonic convergence in which crustal material is processed rapidly due to high rates of denudation augmented by orographically enhanced precipitation: they constitute the single largest source of clastic material eroded from the present-day continents (Milliman and Syvitski, 1992).

Conceptual models developed by geomorphologists have provided a qualitative framework for the evaluation of landscape evolution in tectonically active areas. In some cases, such models have been around for almost a century (e.g. Davis, 1899; Penck, 1924) without attracting particular attention from the geophysical community. Recently, however, the interest of geophysicists in the role of surficial mass redistribution during collisional mountain building has increased markedly (e.g. Stein *et al.*, 1988; Molnar and Lyon-Caen, 1989; Merritts and Ellis, 1994; Small and Anderson, 1995). This initiative has followed an increase in the availability of information on the shape of the Earth's surface (e.g. Fielding *et al.*, 1994) (see Chapter 10) paired with the development of techniques to assess the horizontal and vertical changes of that shape (e.g. Jackson and Bilham (1994) and Chappell (1974), respectively). A further factor has been the advent of dating techniques allowing estimation of rates of geomorphic processes beyond the reach of radiocarbon and archaeologically aided studies (Cerling and Craig, 1994).

Notwithstanding these advances, current research into the topographic evolution of collisional orogens has met with considerable problems (see Chapter 3). Several orders of magnitude separate the time scales over which tectonic and geomorphic processes operate, and matters are complicated further by the superposition of the general scaling characteristics of topography, spanning a wide range of length scales, and heterogeneities in erodibility of the substratum, rates of bedrock weathering and amounts of precipitation. Thus, it is not surprising that the present generation of numerical models of the tectonic evolution of active continental margins (e.g. Koons, 1989; Beaumont *et al.*, 1992; Chase, 1992; Anderson, 1994; Masek *et al.*, 1994) are lacking in their capacity to extend realistic earth surface process model components to geophysically relevant spatial and temporal scales (cf. Koons, 1995) (see Chapter 3).

Here, I review the erosion and topographic evolution of collision zones, based on the observation that the sediment flux from mountain regions is a crucial constraint on the link between tectonic, geomorphic and atmospheric processes. Sediment flux is controlled

Geomorphology and Global Tectonics. Edited by Michael A. Summerfield. © 2000 John Wiley & Sons Ltd.

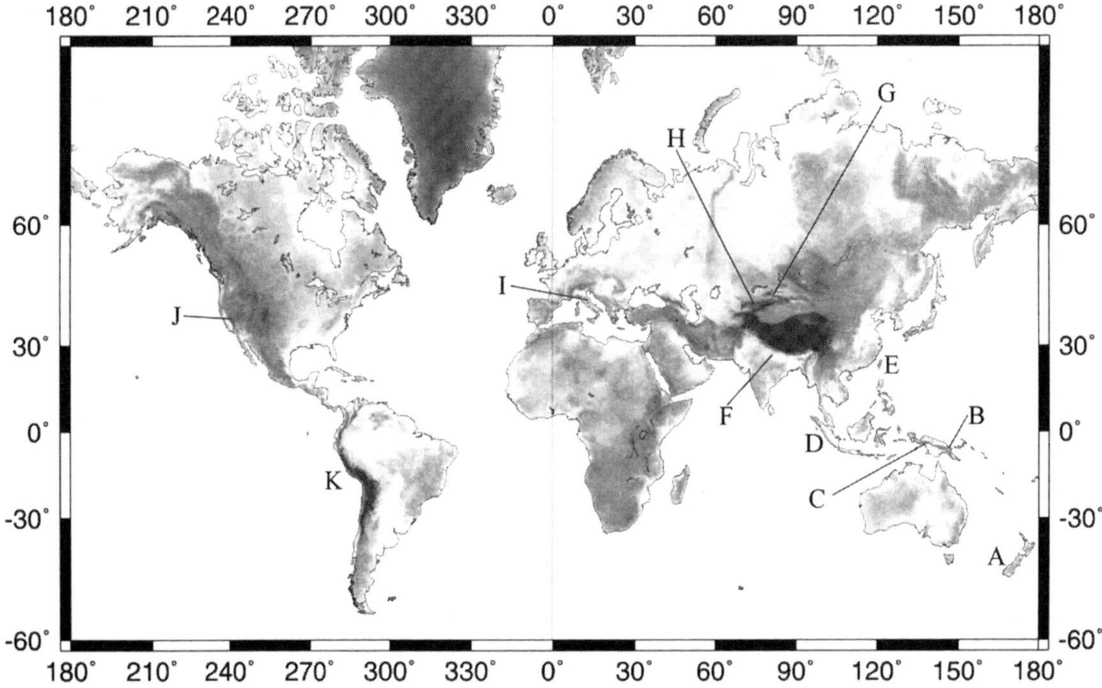

Figure 5.1 Location of mountain belts discussed herein. Names of ranges: A – Southern Alps, New Zealand; B – Finisterre Range, Papua New Guinea; C – Maoke Range, Irian Jaya; D – Barisan Range, Sumatra; E – Central Range, Taiwan; F – Himalayas, India/Nepal; G – Tien Shan, China; H – Kirgizskiy Khrebet, Kirgizstan; I – Apennines, Italy; J – Sierra Nevada, California; K – Andes*, Peru. For the two regions marked (*) only drainage spacing information is available*

by the spatial and temporal organization of the macro-scale process systems of orogenic erosion. Incision of fluvial (and, locally, glacial) systems into uplifting bedrock drives the formation of transverse valley systems, which are the principal landscape units in many mountain belts. The main source of sediment in these drainage basins derives from episodic events of slope erosion, occurring over a range of length and time scales. The debris produced by mass movement generally feeds rapidly into the fluvial system and is eventually transferred beyond the mountain front. In this chapter I will frequently use data from 11 mountain belts across a wide range of climatic and tectonic settings (Figure 5.1) that have been more comprehensively discussed in Hovius (1995, 1996, 1998) and Hovius *et al.* (1997).

5.2 Global Sediment Flux

Erosion is expressed as a relative lowering of the Earth's surface and a resulting sediment flux. While it may be difficult to make systematic observations of surface lowering, sediment fluxes are readily observable through measurements of sediment loads in rivers, which may be converted to denudation rates. The total, present-day rate of sediment delivery to the oceans has been estimated at 2×10^{10} t a^{-1} (Meybeck, 1988; Harrison, 1994). The corresponding rock volume of around 8 km^3 is dwarfed by the total volume of the Earth's crust (10^{10} km^3), thus illustrating the importance of considering mass fluxes on geological time scales in order to estimate the mass balance of the lithospheric system.

5.2.1 EMPIRICAL MODELS FOR SEDIMENT YIELD: THE ROLE OF CLIMATE AND RELIEF

Numerous workers have attempted to derive functional relationships of erosion rates from variables which are compatible with the physical driving forces of denudation. Often, this search has taken the form of statistical analyses of data sets comprising information on sediment transport by major rivers. Milliman and Meade (1983) and Milliman and Syvitski (1992) have made frequently cited compilations, while much larger data sets have been gathered by Jansson (1988) and Dedkov and Moszherin (1992). These, and other data sets (Holeman, 1968; Lisitzin, 1972; Pinet and Souriau, 1988; Enos, 1991; Leeder, 1991; Summerfield and Hulton, 1994), focus on the solid load of rivers at the interface of the continental and the marine realm, and contain little detail on the source area or the transport history of the material. The sediment load at the mouth of a river reflects the sum total of all erosional and depositional processes that occur within the drainage basin. With an increase in catchment area, there is an increase in the relative importance of depositional processes, this being reflected in the inverse relation of specific sediment yield to drainage area noted in many studies (e.g. Schumm and Hadley, 1961; Milliman and Meade, 1983).

Interpretation of these data sets is hampered by the variable reliability of the information they contain (Milliman and Meade, 1983), as well as by a possible phase lag in the variability in sediment fluxes in some rivers (Church and Slaymaker, 1989). This, however, has not stopped speculation about controls on sediment yield. Emphasis has long been on mean annual precipitation (Langbein and Schumm, 1958; Douglas, 1967; Wilson, 1973; Ohmori, 1983), while others have stressed the role of topography (Ahnert, 1970; Pinet and Souriau, 1988; Milliman and Syvitski, 1992). Models combining climatic and topographic controls have been proposed by Fournier (1960) and Summerfield and Hulton (1994).

Hovius (1998) collated data on 14 climatic and topographic variables used in previous studies for 97 major catchments around the world. None of these variables correlates well with sediment yield (Table 5.1), indicating that no single variable can be regarded as the prime control on denudation rate. Rather, a multiple regression relation combining several terms may serve as a successful predictor. Using step-wise regression, Hovius (1998) constructed a five-term model for estimating sediment yields:

$$\ln E = -0.416\ln A + 4.26\times 10^{-4}H_{max} + 0.150T + 0.095T_{range} + 0.0015R + 3.585 \quad (1)$$

where E is specific sediment yield (t km^{-2} a^{-1}), A is drainage area (km^2), H_{max} is the maximum elevation of the catchment (m), T is the mean annual temperature (°C), T_{range} is the annual temperature range (°C), and R is the specific runoff (mm a^{-1}). Equation (1) explains 49% of the observed variance in specific sediment yield. This model combines the influence of drainage area with weathering, hillslope erosion and fluvial transport components to explain the large-scale variance in specific sediment yield (Hovius, 1998). Not all model variables correlate particularly well with specific sediment yield, while some estimators with relatively high correlation coefficients do not appear in the equation. These variables are not necessarily unimportant to the variance in yield, and part of their influence may be reflected by controls included in the model (Table 5.1).

5.2.2 TECTONIC CONTROL ON SEDIMENT YIELD

Of the variance in yield, 51% is not readily explained by terms considered in this analysis, a poor but realistic result when compared with other, more limited, studies. Several reasons may combine to cause the large residual. First, poor data quality may introduce a random component. Second, variables other than those included in the analysis may be more adequate at expressing the aspects of climate and relief that control erosion rates. Third, the relationship of sediment yield with estimators may be more complex than is assumed in the regression procedure. Finally, climate and topography may not be the only controls on erosion rate. A non-horizontal trend in the residuals of the model is interpreted by Hovius (1998) to indicate that sediment yield is only in part governed by the combined potential of earth surface processes to remove material. This is reasonable as one would expect the rate of removal of material from a reference volume to be dependent not only on the erosivity of the processes acting on the surface of that volume, but also on the rate of input of new matter (Leeder, 1991). In geological terms, input is equivalent to uplift of rocks, and Hovius (1998) postulated that this process may exert an influence on sediment yield, reflected in much of the unexplained variance.

Equation (1) overestimates yields from areas with relatively low mechanical denudation rates, and underestimates yields from rapidly eroding catchments. This systematic difference between observed and predicted yields may be interpreted in terms of a rock uplift control. Over-prediction for drainage basins with small yields indicates that denudation of those catchments is limited by the input of material into the weathering–erosion–transport system. The potential for erosion, as controlled by environmental factors, and especially climate, exceeds the rate at which tectonic processes

Table 5.1 *Matrix of Pearson correlation coefficients between the estimators of specific sediment yield*

	Y	A	H	H_{max}	H_{pk}	H_r	a_{riv}	P	P_{max}	P_{pk}	T	T_{range}	R	R_{cf}	Q_{pk}
Y	1.0000														
A	–0.1493	1.0000													
H	0.4128	0.0222	1.0000												
H_{max}	0.3391	0.2917	0.6180	1.0000											
H_{pk}	–0.1054	0.4010	–0.3853	0.3342	1.0000										
H_r	0.4427	–0.3187	0.2485	0.4728	0.1026	1.0000									
a_{riv}	0.4187	–0.4812	0.1543	0.4368	0.0442	–0.1820	1.0000								
P	0.3463	0.0579	0.1958	0.1325	0.1296	–0.0143	–0.0382	1.0000							
P_{max}	0.4340	0.0909	0.2604	0.2049	0.0468	–0.0361	–0.0641	0.8649	1.0000						
P_{pk}	0.3610	–0.0440	0.1345	0.1549	–0.2251	–0.0127	0.0168	–0.0939	0.3516	1.0000					
T	0.1792	0.0027	0.0438	0.0275	0.0840	–0.0248	–0.0077	0.4847	0.5654	0.2301	1.0000				
T_{range}	–0.0575	0.0093	–0.0524	–0.0359	–0.1635	–0.0526	–0.0851	–0.5500	–0.5151	0.0174	–0.8763	1.0000			
R	0.4367	–0.0002	0.1512	0.2627	0.2560	0.2759	0.2646	0.7413	0.5507	–0.2546	0.1443	–0.2549	1.0000		
R_{cf}	–0.0707	0.1034	–0.0393	0.1762	0.2410	–0.0317	0.0470	0.0056	–0.1253	–0.2606	–0.4643	0.3124	0.5654	1.0000	
Q_{pk}	–0.0535	0.8309	–0.2189	–0.2374	–0.1609	0.1455	–0.2483	–0.0092	0.4037	–0.0251	0.1296	–0.3065	–0.0946	–0.2346	1.0000

Y is specific sediment yield (t km^{-2} a^{-1}); A is drainage area (km^2); H is mean catchment elevation (m); H_{max} is maximum catchment elevation (m); H_{pk} is the ratio of the mean elevation and the maximum elevation of the catchment; H_r is the ratio of the maximum elevation of the catchment and the basin length (m km^{-1}); A_{riv} is the slope of the river bed (m km^{-1}); P is mean annual precipitation (mm); P_{max} is mean maximum monthly precipitation (mm); P_{pk} is the ratio of the mean monthly precipitation and the maximum monthly precipitation; T is mean annual temperature (°C); T_{range} is mean annual temperature range (°C); R is specific runoff (mm a^{-1}); R_{cf} is the ratio of the mean annual precipitation and mean annual water discharge at the river mounth; Q_{pk} is the ratio of the mean flow rate and the maximum flow rate.

make material available for erosion and generate local relief. Under-prediction of yields from rapidly eroding drainage basins may be due to the fact that here the production and removal of regolith is outpaced by tectonic input of material, resulting in steepening of hillsides and slope failure involving bedrock. Eventually bedrock involved landsliding could lead to a near steady-state topography, in which sediment yield is determined by rock uplift rates. These scenarios are at variance with traditional explanations of erosional styles in flat and steep terrains, which assume transport-limited lowland erosion and weathering-limited upland erosion.

5.3 Landsliding

5.3.1 LIMITS TO LOCAL RELIEF

Landsliding is increasingly appreciated as a major control on landscape development in montane terrain (e.g. Eyles, 1983; Ahmad *et al.*, 1993; Gerrard, 1994; Greenbaum *et al.*, 1995). However, although a great body of work has documented the morphology and mechanics of individual instances of slope failure, few studies have considered the process at a larger scale.

Among the rare investigations into the long-term role of landsliding is work by Schmidt and Montgomery (1995) and Burbank *et al.* (1996a). Burbank *et al.* (1996a) analysed slope distributions in morphologically distinct regions within the Karakoram, finding that distributions are essentially indistinguishable between incised plateaus and mountain ranges, despite significant differences in denudation rates. This implies that above a certain gradient slope angles are largely unrelated to denudation and rock uplift rates, and are, instead, controlled by a common threshold. Uniformly high mean slope angles are interpreted as reflecting an average effective angle of internal friction which controls slope stability in fractured bedrock. Burbank *et al.* (1996a) postulate that, in the Nanga Parbat region of the Karakoram, rates of denudation are so high that processes which typically affect soil-mantled slopes are volumetrically unimportant. Instead, rapid mass wasting through landsliding is the primary means by which a hillslope adjusts to changes in boundary conditions resulting from river incision at its base.

Support for this model comes from a study by Schmidt and Montgomery (1995), which demonstrates that mountain-scale material strength can limit relief development in bedrock landscapes. They used a simple model for bedrock landsliding to predict the maximum gradients and heights of hillslopes and mountain fronts, and compared predicted and observed values for the western Cascade Range and the Santa Cruz Mountains, in the western United States. Pervasive landsliding in both mountain belts implies that the observed relief approaches a critical state. Agreement between bedrock material properties backcalculated from this topography, and those derived from failure tests on actual rocks, is interpreted by Schmidt and Montgomery (1995) to indicate that large-scale rock strength, rather than valley incision, may limit relief in areas of rapid rock uplift.

5.3.2 RATE AND SCALING BEHAVIOUR OF LANDSLIDES

Both Schmidt and Montgomery (1995) and Burbank *et al.* (1996a) mention pervasive landsliding, but fail to present evidence of the importance of this process to the erosional mass balance of an orogen. Such evidence is given by Hovius *et al.* (1997). From multiple sets of aerial photographs, they obtained a 60 year record of mass wasting in the western Southern Alps of New Zealand, enabling quantification of the rates and scaling of landsliding and the concomitant mass fluxes.

The Southern Alps are a linear, asymmetric mountain belt marking the oblique compressional boundary between the Australian and the Pacific plate (Walcott, 1978) (see Chapter 6). Rock uplift rates approaching 7 mm a^{-1} (Bull and Cooper, 1986; Tippett and Kamp, 1993; Simpson *et al.*, 1994) have assisted the building of 2 to 4 km of relief, which forces orographically enhanced precipitation to rates of up to 15 m a^{-1} on the steep western flank of the orogen (Griffiths and McSaveney, 1983a). Here, dissected, rectilinear slopes, with thin regolith cover, support dense, natural rainforests. These conditions are very favourable to the occurrence of rainfall-triggered rapid mass wasting, especially landsliding. Seismic activity has not been a major cause of landsliding in the central Southern Alps during historic times (Cooper and Norris, 1990; Berryman *et al.*, 1992).

In a region comprising 13 transverse catchments draining the central Southern Alps towards the west, landslides were mapped at a 1:50 000 scale, In all, 7691 landslides were recorded over a total surface area of 4970 km^2, 2670 km^2 of which are in montane terrain (Figure 5.2).

In the montane zone, east of the range-bounding Alpine fault, the sizes of observed landslides range

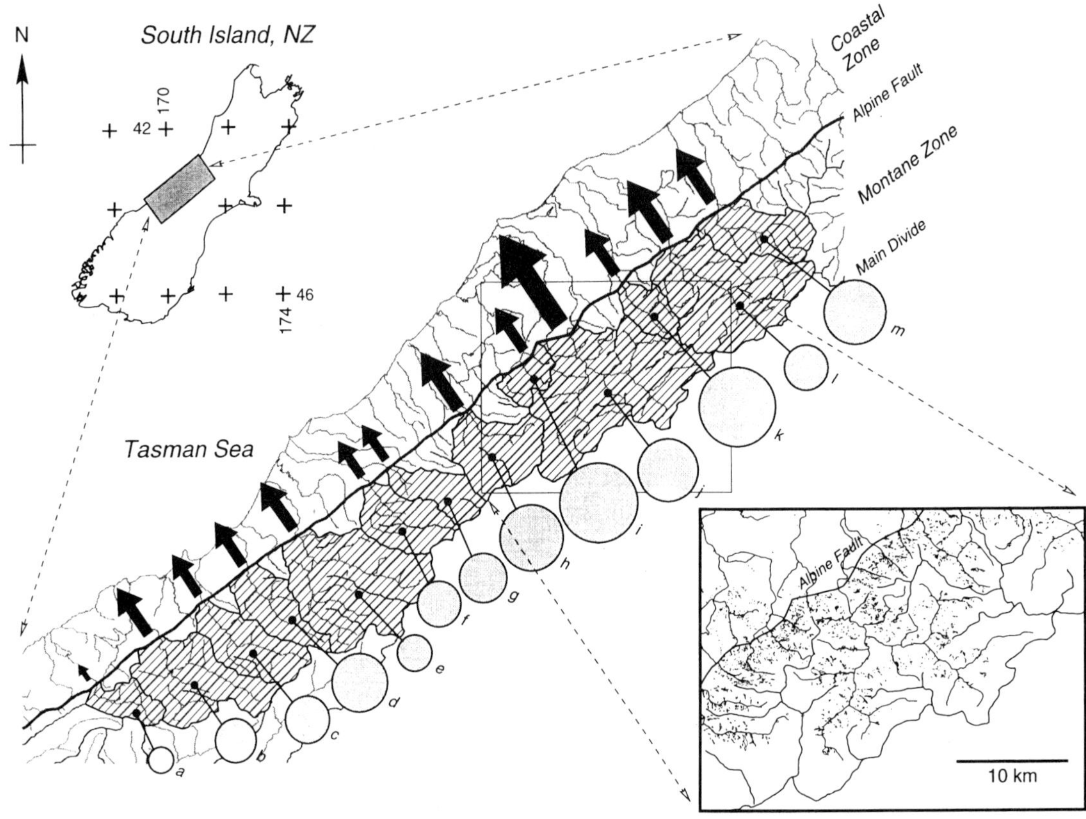

Figure 5.2 *Overview of denudation rates and sediment discharges from 13 catchments draining the western side of the Southern Alps, New Zealand, calculated from a 60 year landslide record. The inset shows a section of the landslide data base, covering the montane part of the Whataroa catchment, flanked by the Poerua and Waitangitaona basins. Denudation (E) and sediment discharges (D) are indicated respectively by circles and arrows, whose areas are proportional to each estimate, and are listed as follows: (a) Moeraki:* $E = 1.8$ mm a^{-1} ($D = 1.2\times10^5$ m^3 a^{-1})*; (b) Paringa:* $E = 5.5$ mm a^{-1} ($D = 1.3\times10^6$ m^3 a^{-1})*; (c) Mahitahi:* $E = 6.3$ mm a^{-1} ($D = 9.8\times10^5$ m^3 a^{-1})*; (d) Makawhio:* $E = 9.9$ mm a^{-1} ($D = 1.1\times10^6$ m^3 a^{-1})*; (e) Karangarua:* $E = 3.7$ mm a^{-1} ($D = 1.3\times10^6$ m^3 a^{-1})*; (f) Cook:* $E = 5.8$ mm a^{-1} ($D = 7.9\times10^5$ m^3 a^{-1})*; (g) Fox:* $E = 7.5$ mm a^{-1} ($D = 7.1\times10^5$ m^3 a^{-1})*; (h) Waiho:* $E = 12.2$ mm a^{-1} ($D = 2.0\times10^6$ m^3 a^{-1})*; (i) Waitangitaona:* $E = 18.1$ mm a^{-1} ($D = 1.1\times10^6$ m^3 a^{-1})*; (j) Whataroa:* $E = 11.4$ mm a^{-1} ($D = 1.1\times10^6$ m^3 a^{-1})*; (k) Poerua:* $E = 18.1$ mm a^{-1} ($D = 1.2\times10^6$ m^3 a^{-1})*; (l) Wanganui:* $E = 6.1$ mm a^{-1} ($D = 2.1\times10^6$ m^3 a^{-1})*; (m) Waitaha:* $E = 11.6$ mm a^{-1} ($D = 1.7\times10^6$ m^3 a^{-1})

from <100 m^2 to about 1 km^2, exhibiting a magnitude–frequency distribution as illustrated in Figure 5.3. This distribution can be described by a power law over the scale range of approximately two orders of area magnitude for which reliable measurements are available. In cumulative form it may be written as

$$n_c(A \geq A_c) = \kappa(A_c/A_r)^{-\beta} A_r \quad (2)$$

where n_c $(A \geq A_c)$ is the number of slides of magnitude greater than or equal to A_c over a reference area A_r per year, and κ is the intercept of the linear regression with $x = 0$ and is a measure of the rate of landsliding per unit area per year. β is a dimensionless scaling exponent, derived from the slope of the best-fit linear regression through the linear segment of the log–log plot of the cumulative distribution. It represents the proportional frequency of events of different magnitude. By fitting a power law model (2) over the most robust data range Hovius *et al.* (1997) obtained $\beta = 1.16$ and $\kappa = 5.4\times10^{-5}$ km^{-2} a^{-1}.

A similar scale-invariant distribution (Turcotte, 1992) of landslide magnitude–frequency has been observed by Fuyii (1969) and Sugai *et al.* (1994) in

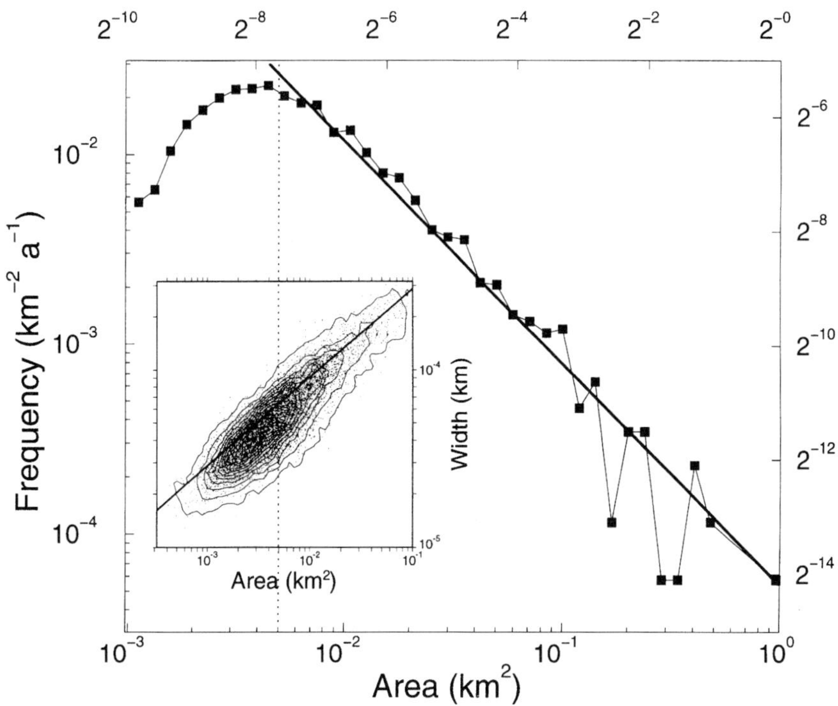

Figure 5.3 *The size distribution of mapped landslides in the central western Southern Alps, New Zealand, with best-fit linear regression. The main graph is a histogram employing logarithmic bin widths* ($\log_2 w = 1/4$). *Map resolution and digitizing accuracy put the lower area scale of reliable measurements at about 2500 m^2, which is indicated by a dashed line. Above this cutoff, a very clear power law trend is observable over two orders of area scale magnitude. The kink in the histogram may be a mapping artefact or a reflection of a size discontinuity in mass wasting mechanics. The inset graph illustrates the relationship between landslide area and width of the best-fit ellipse. This relation is employed in volumetric analysis*

Japanese upland areas. Given independent observations of power law scaling behaviour from several regions, it seems that scale invariance is a general property of the landslide mechanism.

The scaling exponent, β, may be used to evaluate the relative importance of events of different magnitudes to cumulative regional erosion. $\beta > 1$, as observed in the western Southern Alps, implies that low-magnitude events have a predominant influence on the total area affected by slope failure, whereas $\beta < 1$ would indicate a prevalence of high-magnitude events.

The existence of a power law magnitude–frequency distribution of landslides permits a simple volumetric analysis to be undertaken, providing an adequate constraint on the area–volume relation for landsliding is available. Such a relation has not been defined for the western Southern Alps, and literature on the subject is scant and diverse (e.g. Simonett, 1967; Ohmori, 1992). Ohmori (1992) proposed a linear relation of landslide length, l, to depth d, where $d(l) = \varepsilon l$. Adopting this relation, Hovius *et al.* (1997) have demonstrated that the total volume, V, eroded by landslides may be modelled as

$$V \approx \frac{2\beta\varepsilon\kappa}{3-2\beta} L_1^{3-2\beta} \quad (3)$$

where L_1 is the maximum length scale of a landslide ($L_1 \approx 1$ km in the western Southern Alps). From limited field observations, Hovius *et al.* (1997) assumed that $\varepsilon = 0.05$ to obtain an estimate for the erosion rate due to landsliding of the western Southern Alps of 9 ± 4 mm a^{-1}. Clearly, changes of ε have major implications for the outcome of this estimate.

The very important observation follows from equation (3) that when $\beta < 1.5$, as is the case in the Southern Alps, denudation due to landsliding, and the ensuing topographic evolution, are dominated by the largest events. The relation between slope instability and topography should accordingly be considered at the upper length scale of the landsliding mechanism, instead of at an arbitrary or non-specified length scale (cf. Schmidt and Montgomery, 1995; Burbank *et al.*, 1996a).

Using general values for β and ε, and local rate constants κ, drainage basins were found to have erosion rates due to landsliding of between 5.5 and 12 mm a^{-1} (Figure 5.2). Estimated mechanical denudation rates are lower in the southern half of the studied segment of the mountain belt, coinciding with a decrease in mean annual precipitation (Henderson, 1993) and possibly also in rock uplift rate towards the south. Denudation rates are highest in small catchments draining the frontal part of the mountain belt, immediately adjacent to the Alpine Fault. This may relate to an observed decrease in rock uplift rates and deformation intensity away from the fault towards the southeast.

Estimates of erosion rates in the western Southern Alps by Hovius *et al.* (1997) are comparable to rates calculated from sediment efflux measurements for some streams along the Alpine Fault, which range from 4.7 to 11.9 mm a^{-1} (Griffiths, 1979; Griffiths and McSaveney, 1983b, 1986; Hicks *et al.*, 1990). This correspondence indicates that bedrock-involved landsliding is indeed accounting for most material that is seen to leave the mountain belt, and that, at present, given little intramontane storage of eroded material, other processes are of subsidiary importance to the orogenic mass budget of the Southern Alps.

5.3.3 RESIDENCE TIME

Assumptions about the residence time of landslide material in montane catchments are somewhat problematic. Landslide material not directly deposited in river channels can remain in storage for periods ranging from months to thousands of years (Keefer, 1994). Two studies of earthquake-triggered landslides can be cited to illustrate this point. The first is a study of the magnitude 7.1 earthquake in 1970 in the Adelbert Range of Papua New Guinea by Pain and Bowler (1973), showing that landslide material was rapidly transported out of the area by rivers, with about half of the material moving into the sea within the first six months after the seismic event. The second is a study by Pearce and Watson (1986) of landslides triggered by the magnitude 7.7 earthquake in 1929 in northwestern South Island, New Zealand. They found over 50% of the landslide debris to be retained in catchments 50 years after the earthquake.

5.3.4 CAUSES OF SLOPE INSTABILITY

The causes of landslides can be divided into two categories (Selby, 1993): factors contributing to high shear stress, and factors lowering shear strength. Water is by far the most important contributor to slope failure. The general effect of rainfall on hillslope stability is simultaneously to increase porewater pressure and load. In seismically active regions, strong ground motion during earthquakes can induce slope failure over large areas (e.g. Keefer, 1984; Greenbaum *et al.*, 1995; Schmidt and Montgomery, 1995; Bull, 1996).

The proportional contribution of coseismic and aseismic landslides to regional erosion varies between areas (Keefer, 1994). In the San Francisco Bay area, the mean coseismic erosion rate is more than four times that for aseismic slope processes. This contrasts sharply with the Transverse Ranges, California, where the mean coseismic erosion rate is approximately 5% of the total erosion rate. Wieczorek and Jäger (1996) have added detail to this work. Using reports of historic slope failures in Yosemite Valley, California, they have demonstrated that here earthquake-triggered slides, although fewer than rainfall-induced slope failures (10.6% and 53.2% of events, respectively), are volumetrically more important by a factor of 1.5. This suggests that the magnitude–frequency distribution of landslides differs with triggering mechanism, either through a change in scaling coefficient β, or in upper length scale L_1. The latter is not unreasonable, as rainfall effects probably do not penetrate far into the rocky subsurface, whereas strong ground motion may activate much deeper instabilities.

The evidence presented above suggests a conceptual model for the erosion of regions of persistently high rates of bedrock uplift, not unlike that proposed by Burbank *et al.* (1996a). In their schema, processes of valley lowering interact with uplifting bedrock to form relief. Between valleys, bedrock is uplifted towards denuding hillslopes whose rate of mass transfer to valley bottoms through landsliding is controlled by the rate of valley lowering and the bulk strength of the fractured bedrock. The distribution of topography is then determined by the large-scale geometry of the drainage pattern and the mechanical properties of the rock mass in which it develops.

5.4 Drainage Patterns

Two controls compete to determine the outline of montane drainage patterns (Koons, 1995). One involves the mechanics of mountain building, the other fluvial mechanics. The balance between these two determines whether rivers are entrained by, or cut through, orogenic structures.

5.4.1 FLUVIAL ENTRAINMENT

Orogenic wedge dynamics determine the primary distribution of topographic ridges and valleys, by means of a heterogeneous uplift field over length scales of 10–50 km (Koons, 1995). Inherent to such uplift is the formation, or reactivation, of zones of strain concentration. In the complete absence of erosion, the resulting pattern of ridges and valleys has an important component parallel to the orogen (Koons, 1995). Rivers will be entrained by this structurally controlled topography if their erosive capacity does not permit them to cut through the uplifting material. Thus, river entrainment is expected where channel gradient, water discharge and/or flow confinement are low, or where the sediment load approaches carrying capacity. Such conditions are preferentially found in two different parts of an orogenic structure. In one case low channel gradients and discharge are often associated with the dry trailing (rain-shadow) side of an orogen. In the other, high sediment loads, coupled with low channel gradients are characteristic of the foothills of mountain fronts. Consequently, strike-parallel drainage is primarily found on the lee-side of the orogenic ridge pole (Koons, 1995) and in the proximity of structures related to propagation of the mountain front (Talling, 1995). Both types are well known from the Himalayas (Figure 5.4), but are also present in many other mountain belts (e.g. Abbott *et al.*, 1994; Molnar *et al.*, 1994; Hovius, 1996).

5.4.2 TRANSVERSE DRAINAGE

If rivers have the capacity to incise uplifting bedrock, their course may deviate from the structural grain of the orogen, responding instead to the regional topographic gradient. This condition is most commonly encountered in regions with steep channel gradients and large water discharges, and is associated with the steep frontal side of mountain belts (Koons, 1995), where precipitation is dominated by orographic effects.

The straight fronts of many actively uplifting mountain belts have simple patterns of drainage transverse to their main structural trend (Figure 5.4). Streams rising near, or beyond, the topographic ridgepole of such mountain belts are spaced at seemingly regular intervals. Hovius (1996) quantified this regularity, using morphometric observations from 11 mountain belts (Figure 5.1). The spacing of drainage basins can be expressed using a spacing ratio, defined as the ratio of the half-width of the mountain belt to the distance between the outlets of adjacent catchments with headwaters immediately below the main divide, as measured at the base of the mountain front (Figure 5.5). Spacing ratios between individual catchments show a gamma-type distribution around a median of 2.13. The ratio, R, of mean half-width, W, of the range to the mean outlet spacing, S, may be used to characterize drainage organization on a mountain belt scale. Ratios of means for most mountain belts are within a narrow range of values between 1.91 and 2.23 (Table 5.2), with an average of 2.07. Thus, a linear relationship exists between the spacing of catchment outlets and the average distance between the main divide and the front of the mountain belt:

$$S \approx 0.46W \tag{4}$$

Assuming that transverse catchments are approximately rectangular in shape, this may be converted to a drainage area–half-width relationship (Hovius, 1996):

$$W = 1.44\sqrt{A/p} \tag{5}$$

Where p is an expression for the percentage of the rectangular area covered by the actual catchment, and A is the surface area of the catchment.

Equation (5) is of the same general form as Hack's (1957) relation for stream length and drainage area, and broadly agrees with observations by Montgomery and Dietrich (1992). It suggests that there is a basic geometric similarity between transverse catchments in linear mountain belts in a range of different climatic and geological settings, and implies scale invariance of these drainage basins. This geometry may represent the most probable state of energy distribution in a montane river system (cf. Leopold and Langbein, 1962). Spacing ratios similar to those discussed here are also created by numerical models simulating similar conditions (Koons, 1989; Beaumont *et al.*, 1992).

5.4.3 RESPONSE OF DRAINAGE PATTERNS TO THRUST-GENERATED FOLDING

In many foreland basins growing folds result from horizontal shortening and subsequent hangingwall

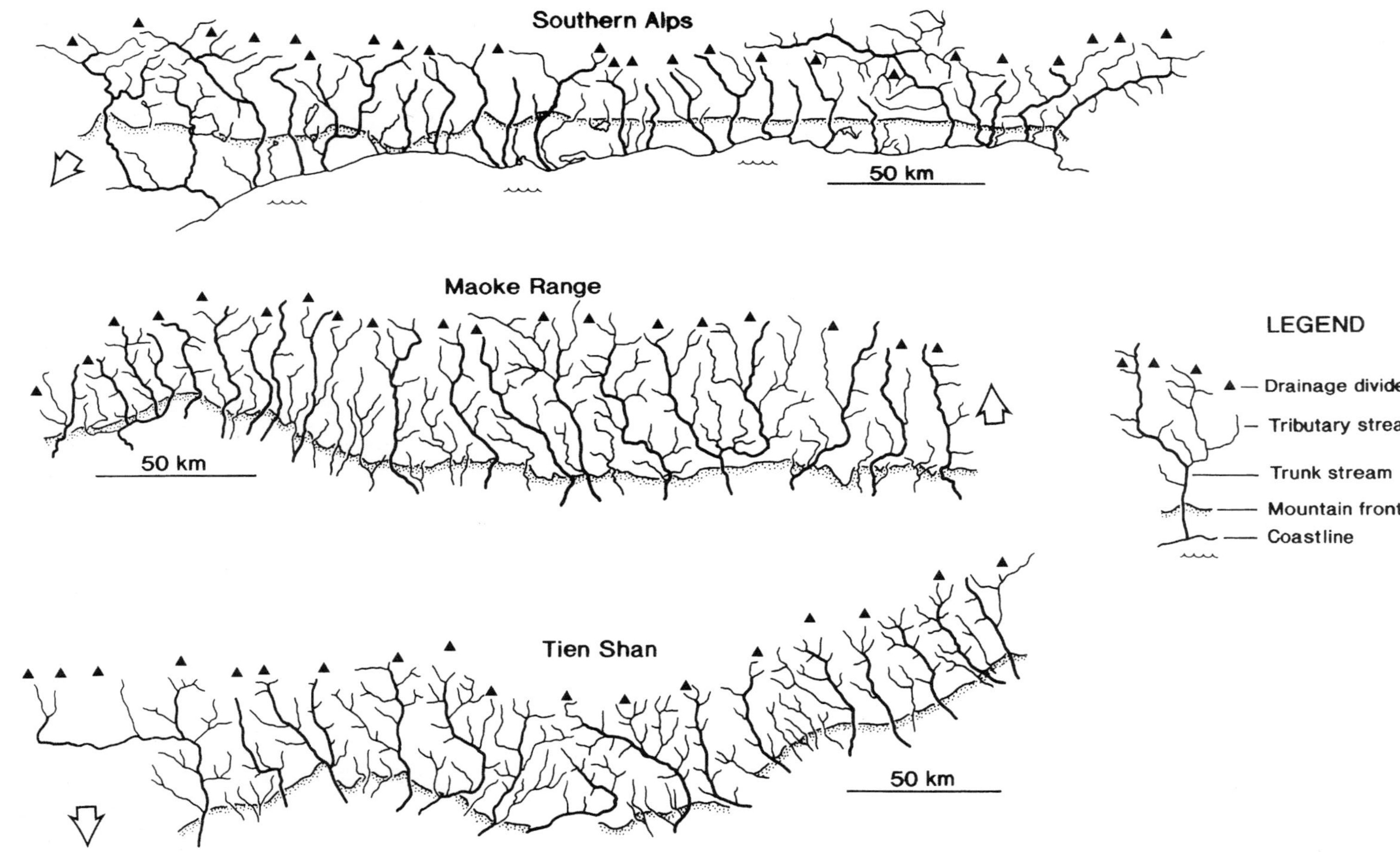

Figure 5.4 *Examples of equally spaced transverse streams in three mountain belts: the western Southern Alps between the Grey River and the Turnbull River, the southern Maoke Range between Ukemepuko River and the Lorentz River, and the northern Tien Shan between the Dzhuku River and the Agiaz River. Although these ranges have very diverse tectonic and climatic settings, the average spacing of drainage outlets relative to the width of the mountain flank is highly uniform (see Table 5.2)*

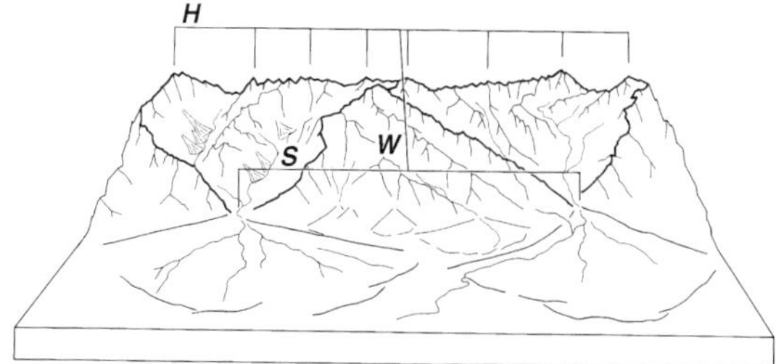

Figure 5.5 *Two montane river catchments. H is the average height of the culminations of the axial ridge of the mountain belt; W is the half-width of the mountain belt, defined as the horizontal distance between the axial ridge and the base of the mountain front; and S is the distance between the outlets of the two catchments*

uplift above thrusts that are actively propagating into the basin fill (Burbank *et al.*, 1996b). If the rate of structural uplift on such folds outpaces the rate of deposition in the surrounding area, the resulting positive topography interferes with the existing drainage pattern at the mountain front and affects the transport and deposition of sediment by transverse streams. A river must be able to erode the crest of the growing fold at a rate equal to, or greater than, the difference between uplift and aggradation rates in order to sustain its course across the edge of the developing piggyback basin (Ori and Friend, 1984). Defeated rivers (Burbank *et al.*, 1996b) may be diverted laterally, given the availability of an alternative drainage path.

During development of a fold-and-thrust belt, faults increase their displacement through repeated earthquakes. As a result, the overlying folds grow both laterally and vertically (Cowie and Scholz, 1992; Scholz *et al.*, 1993). Individual fault segments are seldom longer than 25 km, but along-strike amalgamation of segments may generate larger structures (Talling, 1995). Such composite frontal ridges have caused large-scale diversions of some Himalayan rivers (Gupta, 1997). Similar diversions at positions north of the Main Boundary Thrust (Seeber and Gornitz, 1983) suggest that this structure has had much the same effect on past drainage paths across the Himalayan front (Talling, 1995).

The response of fluvial systems to changing tectonic slopes offers a chance to monitor active fault growth where few geological observations are available (Stewart, 1995; Jackson *et al.*, 1996). Moreover, extensive preservation of syntectonic strata along fold limbs may permit a time-lapse reconstruction of the development of a fault propagation fold (Suppe *et al.*, 1992). If the resulting geomorphology and/or stratigraphy can be dated, rates of uplift, incision and structural propagation may be calculated (e.g. Medwedeff, 1992), yielding unique constraints on theoretical models of fault growth.

Table 5.2 *Mountain belt geometry and drainage spacing*

Mountain belt	H (m)	W (km)	a (m km^{-1})	L (km)	n	S (km)	R
Southern Alps	2653	21.1	123	198	19	10.99	1.92
Finisterre Range	3640	25.5	137	187	15	13.36	1.91
Maoke Range	4038	36.4	108	298	19	16.56	2.20
Barisan Range	1914	28.0	66	439	33	13.72	2.04
Central Range	3292	25.0	124	243	20	12.79	1.95
Kirgizskiy Khrebet	4101	34.7	92	202	14	15.52	2.23
Northern Tien Shan	4950	38.1	84	269	15	19.18	1.99
Apennines	1949	39.2	48	547	33	17.65	2.22
Sierra Nevada	3192	84.0	36	422	12	38.36	2.19
Peruvian Andes	5797	81.9	70	1000	25	40.00	2.05
Central Himalayas	7822	158	49	1610	12	139.2	1.17

H is the mean height of the culminations of the main drainage divide; W is the average half-width of the mountain belt, measured as the horizontal distance between the main divide and the base of the mountain front; a is the mean gradient of the enveloping slope of the range flank; L is the length of the section over which spacing of drainage outlets was measured; n is the number of streams draining the section; S is the average spacing of drainage outlets in the section; R is the characteristic aspect ratio of drainages for the section, calculated as $R = W/S$. The characteristic aspect ratio falls within a narrow range of values for all mountain belts, except in the central Himalayas. Here, drainage deflection by fault-propagation-folds in the foothills has caused amalgamation of transverse streams into larger catchments.

5.5 Longitudinal River Profiles

Within a catchment, landscape evolution is driven by the incision of the river system into the underlying bedrock. At any given point along a stream profile, the altitude of the river bed will change at a rate equal to the rate of rock uplift minus the rate of incision. Some theoretical and experimental studies (Begin *et al.*, 1981; Snow and Slingerland, 1987) suggest that a longitudinal stream profile rapidly attains a nearly 'ideal shape', even in the presence of significant tectonism. This stream profile is the backbone of the erosional landscape.

5.5.1 MATHEMATICAL MODELS

Longitudinal river profiles have the following geometric features: (1) they decrease in gradient monotonically, (2) they are concave-up with occasional local convexities, and (3) except for knickpoints they are smooth over length scales of kilometres. Such characteristics suggest that it may be possible to describe river bed elevation as a function of downstream distance. Linear (Ohmori and Shimazu, 1994), exponential (Shulits, 1941; Strahler, 1964; Ohmori and Shimazu, 1994), logarithmic (Hack, 1957; Leopold and Langbein, 1962), and power functions (Gilbert, 1877; Leopold and Maddock, 1953; Schumm, 1960; Langbein and Leopold, 1964; Snow and Slingerland, 1987; Ohmori and Shimazu, 1994) have been reported to fit measured and model-generated longitudinal profiles. These mathematical functions are theoretically explained by considering the relations between morphometric characteristics (such as channel slope and river bed form) and the downstream changes in physical parameters (such as sediment load, sediment size, and water discharge) (Morisawa, 1985; Ohmori, 1991). Often, the equations are solved according to the concept of grade (Mackin, 1948), assuming sediment continuity. However, in an active orogenic belt, most streams are actively incising into bedrock (Merritts *et al.*, 1994).

Ohmori (1991) and Ohmori and Shimazu (1994) found that Japanese streams may be divided into segments, each characterized by a different best-fitting function, and demonstrated a correlation between profile form and dominant mode of sediment transport. More often though, a single model is assumed to be an adequate approximation of an ideal river profile, and is consequently used as a reference state (e.g. Willemin and Knuepfer, 1994). Hovius (1995) matched simple exponential, logarithmic and power functions with normalized longitudinal profiles of transverse streams from nine active mountain belts (Figure 5.1). Of 171 profiles, 64 were found to have a logarithmic best fit, while 36 were most adequately described by an exponential, and 30 by a power function. The remaining 41 profiles were not particularly well fitted by any of the models, often due to tectonically induced profile convexities. Most individual mountain belts show a similar diversity in best-fitting functions. Although simple mathematical functions may provide reasonable approximations of longitudinal profile form in certain cases, they are, more often than not, incapable of producing close fits of the form in its entirety. Moreover, there is no specific reason for picking any of the functions which are often used since others may describe longitudinal profiles more accurately.

5.5.2 REGIONAL CONSISTENCIES IN PROFILE FORM

Notwithstanding the diversity in best-fitting functions, most longitudinal profiles from a mountain belt have a significant geometric overlap (Figure 5.6). It is useful to identify a shape common to all profiles within a region, in order to establish a rational starting point for developing a process-based model which describes the evolution of longitudinal river profiles. In individual profiles, this underlying geometry may be obscured by local deviations, but such effects may be suppressed by combining information from several profiles within one region.

In studies primarily concerned with the shape of river profiles, it is possible to extract each profile from its absolute coordinate system, and translate it in space in order to force a match with a reference form. Hovius (1995) defined an initial reference form by calculating the best-fit Tschebyscheff polynomial (Davis, 1975) for the combined, but untransformed, river profile data from a mountain belt (Figure 5.7a). He then demonstrated how all profiles for a region can be forced to match this curve, by applying simple horizontal and vertical shifts, while taking out short profile segments affected by faulting and/or folding (Figure 5.7c and d). These shifts result in a remarkable compression of the data set without significant loss of information, demonstrating the basic uniformity of river profiles within a region (Figure 5.7b).

One way of comparing profiles between regions is by looking at the first and second derivatives of the best-fit curves. The example in Figure 5.8 shows an upstream section over which $f(x)$, $f'(x)$ and $f''(x)$ all decrease monotonically with x. The initial rapid

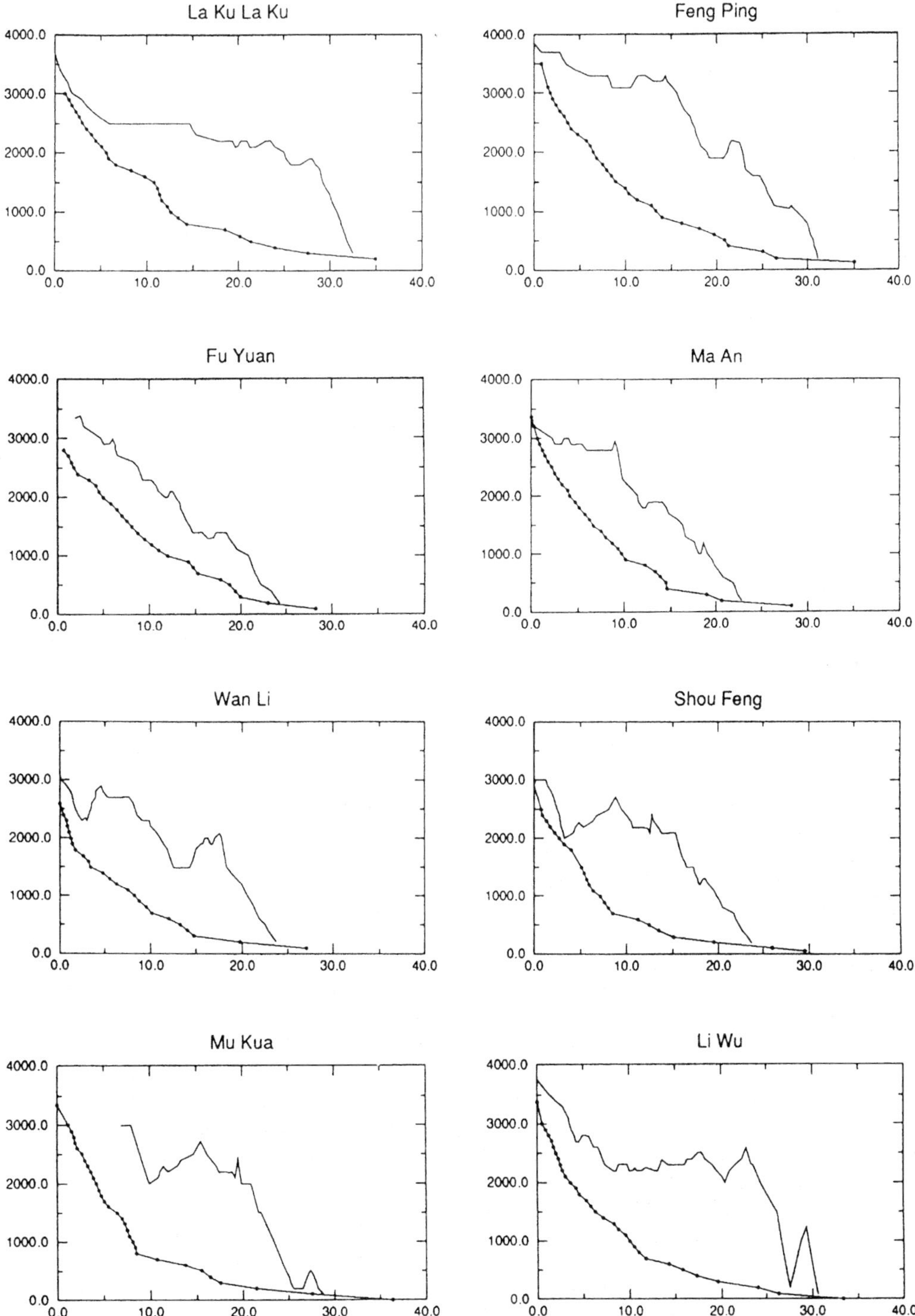

Figure 5.6 *Longitudinal profiles of trunk streams and interfluves of eight catchments in the eastern flank of the Central Range, Taiwan. Horizontal units are kilometres, vertical units metres. Underlying many small-scale irregularities is a uniform basic profile shape*

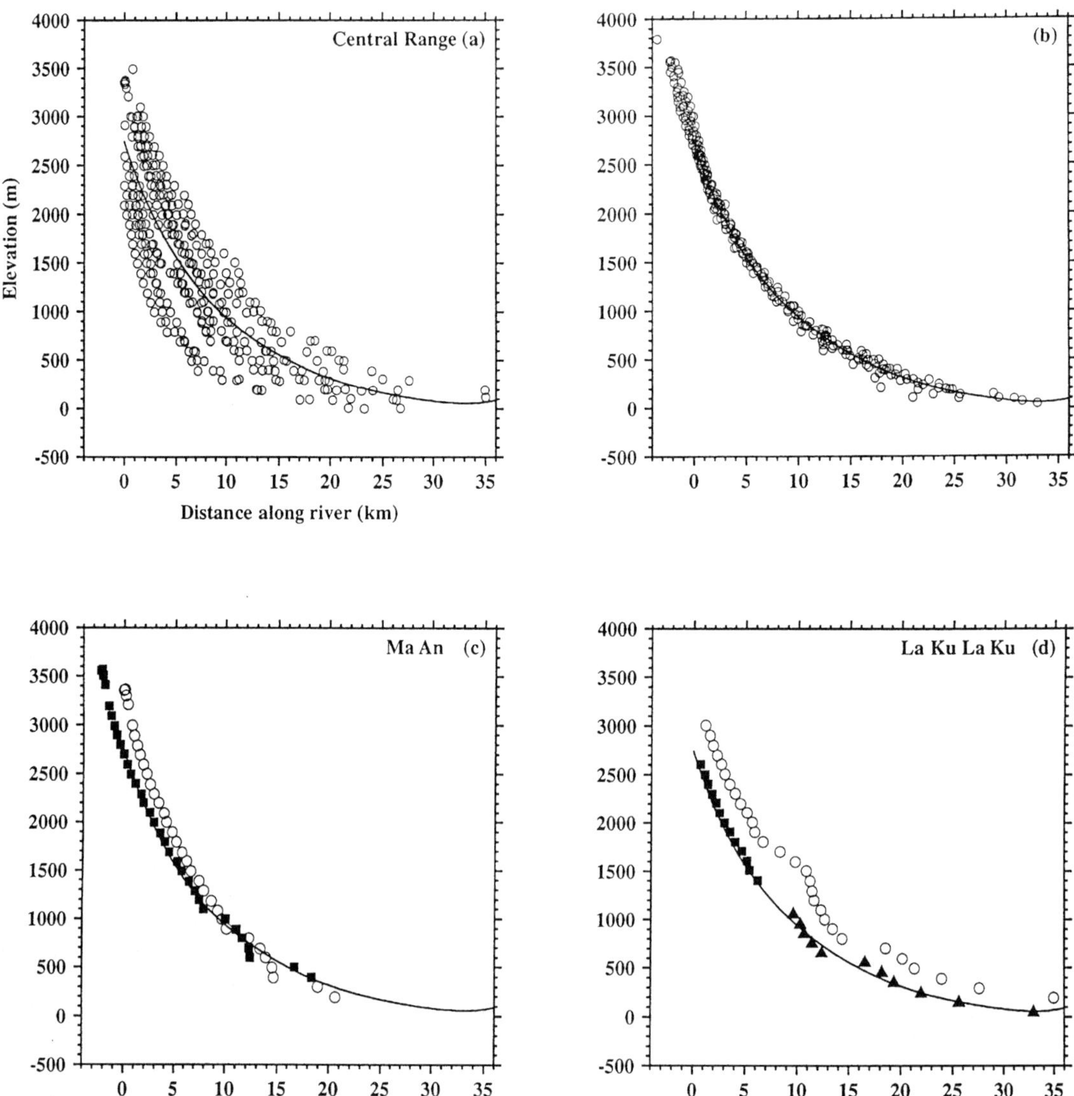

Figure 5.7 *Compression of longitudinal river profile data. (a) Original profile data for the eastern flank of the Central Range of Taiwan, with best-fit Tschebyscheff polynomial. (b) The same data set after compression. This compression has been achieved through (c) single translations, with a horizontal and a vertical component, of all data describing one profile towards the best-fitting polynomial (open circles are original data and filled squares are shifted data), or through (d) elimination of some deviating data points and separate translation of the remaining segments (open circles are original data: three data points in the convex section of the profile have been eliminated, after which the upper profile segment, indicated with filled squares, has been translated vertically to match the average profile, and the lower profile segment, indicated with filled triangles, has been shifted both downward and left laterally towards the average form)*

decline of the rate of change of slope ends in a local minimum, beyond which the curve oscillates around a gently descending or level trend. On the plot of the first derivative the corresponding point marks the transition from high, but rapidly declining gradients, to lower gradients and a more gradual decline. The persistence of these features throughout all nine regions underlines its significance as a universal characteristic of long profile shape that may well mark the transition between two profile domains.

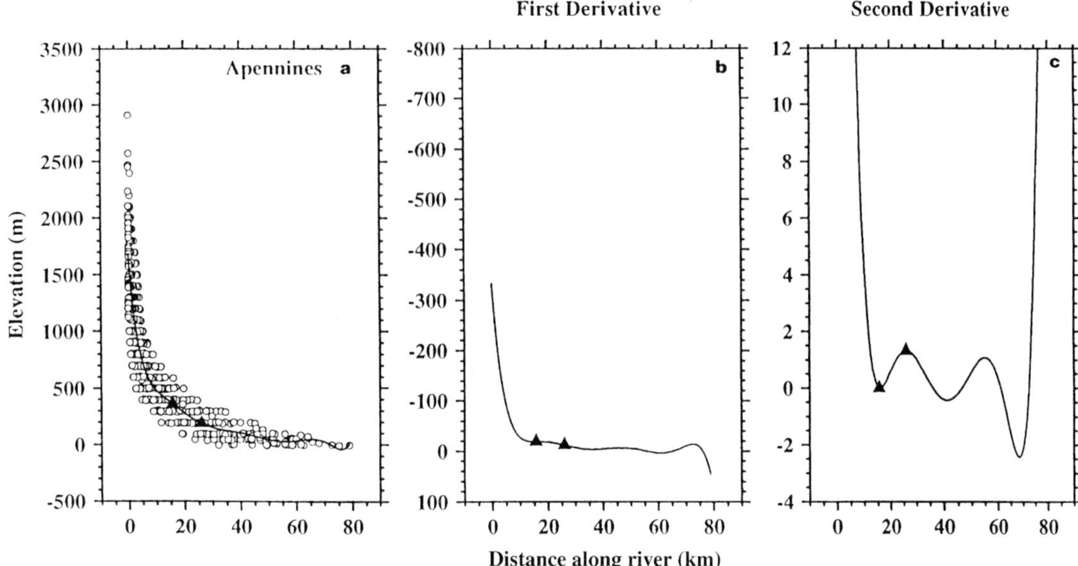

Figure 5.8 *First and second derivatives of best-fit polynomials to the combined river profile data for the eastern Apennines, Italy. The first local minimum and maximum of the second derivative are marked with triangles. The position of the corresponding points on the valley-floor elevation curve and its first derivative are also marked. The first local minimum of the second derivative marks the transition between the upstream and downstream profile segment*

In order to compare basic profile forms between regions, Hovius (1995) normalized the data sets with respect to the elevation change between the average height of the main divide of a mountain belt and the elevation of the first local minimum of the second derivative of the basic profile form. Figure 5.9 shows compressed and rescaled river profile data from nine mountain belts. In a given region, most major transverse streams fit the basic form very closely, indicating that, within each region, the spatial organization of the processes involved in fluvial downcutting is highly uniform. The basic long profile consists of an upstream component which displays very little variance between mountain belts, and a regionally uniform downstream component. A plot combining all rescaled upstream profile segments (Figure 5.10) is evidence of a globally uniform erosional river profile in the higher parts of mountain belts. Its slope declines from 0.6 m m^{-1} in the uppermost reaches to around 0.05 m m^{-1} near the transition to the downstream segment. This decrease is exponential at first, but changes character at bed angles around 0.08 m m^{-1}. Characteristic bed angles of downstream segments vary between mountain belts, while segment length varies both regionally and globally.

5.5.3 PROCESSES OF BEDROCK INCISION

From the universality of the upstream profile form it is clear that its geometry is independent of external controls such as rate of uplift of rocks and rate of precipitation, given that the duration of profile evolution exceeds the response time of the landscape to changes of such controls. Under these conditions, upstream profile form is solely a function of the mechanics of the processes acting on the river bed.

Erosion of bedrock may occur by several mechanisms, including corrosion, plucking, cavitation and abrasion. In steep channels, dominant erosion by particles due to cutting and abrasion is periodically aided by mechanisms related to macroturbulence (Matthes, 1947). Foley (1980) developed a model for bedrock incision, giving abrasion rates as complicated functions of single bed load particle mass, relative particle velocity, angle of impact, and impactor surface properties. This model is based on the assumption that erosion by bed load is proportional to bed load times single particle erosion rate divided by the characteristic saltation path length. The latter two vary with the dominant mode of sediment transport along a river profile.

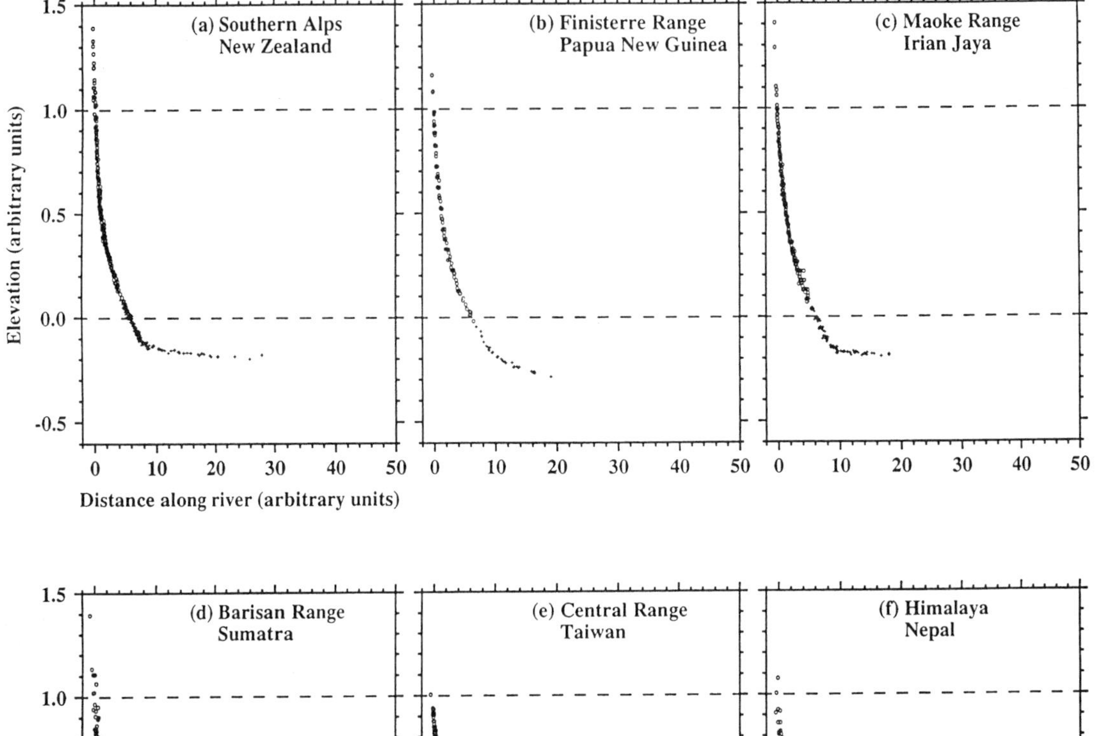

Figure 5.9 *Compressed and rescaled river profile data from nine mountain belts. After compression, each data set has been rescaled to unit length, defined as the vertical distance between the average summit elevation of the main drainage divide and the elevation of the first local minimum of the second derivative of the best-fit polynomial to the data set. Upstream and downstream data points are marked ○ and ● respectively. These data sets describe the basic river profile shape for each region. This form consists of a universal upstream component, and a regionally uniform downstream component*

The detailed mechanisms of fluvial bedrock incision remain poorly understood. Theoretical models for bedrock erosion assume that process rates vary systematically with discharge, sediment load and channel gradient. Howard and Kerby (1983) and Howard (1994) suggest that bedrock erosion may be proportional to bed shear stress. A stream-power-dependent erosion model is proposed by Seidl and Dietrich (1992), and this model is supported by some field observations (Wohl, 1992, 1993; Young and McDougall, 1993; Seidl *et al.*, 1994). However, neither model is capable of explaining profile evolution adequately, as several erosion mechanisms, each with their own mechanical properties, may operate in a single channel (Seidl *et al.*, 1994).

The beds of many montane channels contain very coarse particles resulting from *in situ* disintegration of

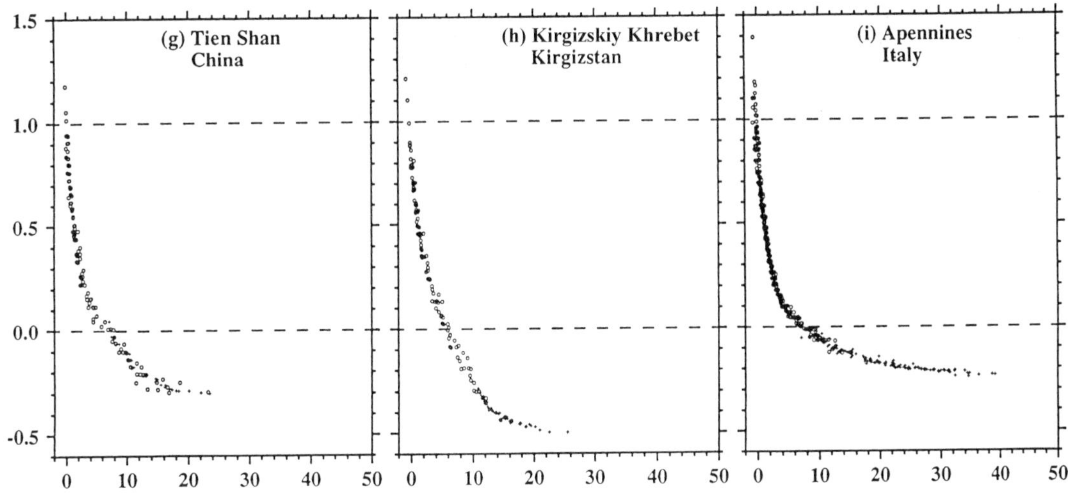

Figure 5.9 (*continued*)

bedrock or emplacement by hillslope erosion processes. These particles protect the underlying bedrock from impact abrasion under all but the most extreme high-flow conditions and, consequently, geomorphologically effective events are infrequent (e.g. Bull, 1988; Grant *et al.*, 1990). The uppermost segments of steep rivers are periodically scoured by debris flows (e.g. Benda, 1990), and in Japanese upland areas Ohmori and Shimazu (1994) and Shimazu (1994) observed a dominance of debris flow erosion in channels steeper than 0.20 m m^{-1}. In channel segments with slopes between 0.20 and 0.08 m m^{-1}, they found debris flow to be an important transport agent, but insignificant in terms of erosion. This lower limit for debris flow

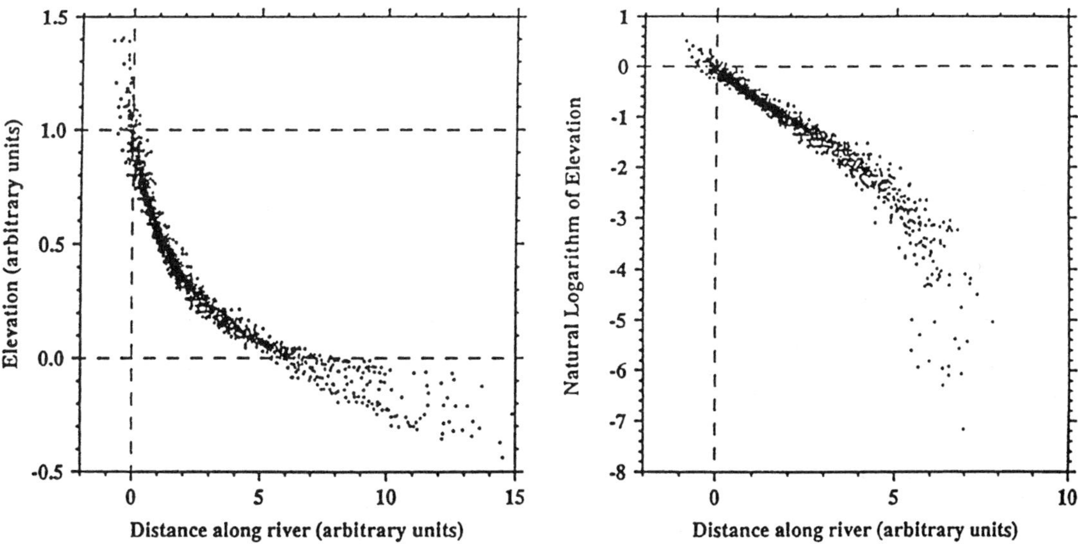

Figure 5.10 *Combined compressed and normalized upstream river profile data. The diagram on the left shows all upstream profile data collapsed, describing the universal shape of the uppermost part of erosional river profiles. The right-hand diagram shows the same data set plotted on semilogarithmic scales. This plot has an approximately linear trend in the data over the first four arbitrary length units from the main divide, indicating an exponential character of the universal profile form*

transport coincides with that of the exponential segment of the plot of combined river profiles (Figure 5.10). Seidl and Dietrich (1992) found a change from debris flow to stream flow dominated bedrock channel erosion at similar channel bed gradients of around 0.2 m m^{-1} in catchments in coastal Oregon.

With declining bed gradients, stream flow becomes incapable of moving all material provided by upstream erosion, and the substratum is covered by a sediment layer of increasing thickness. Incision into the substratum can only occur after complete removal of this mantle. It is unreasonable to assume that such removal may be achieved through fluvial erosion at peak discharge. Instead, knickpoint migration (cf. Penck, 1924; Leopold *et al.*, 1964; Pickup, 1977; Miller, 1991; Seidl and Dietrich, 1992; Merritts *et al.*, 1994) may be responsible for lowering of the downstream long profile. Vertical offset of a river profile along a major internal or range-bounding structure results in local steepening of the stream gradient. Enhanced erosivity of the fluvial system in such knickpoints may cause local removal of the sediment cover and exposure of bedrock to fluvial abrasion. The effect of a relative base-level fall propagates upstream, provided that the knickpoint form is preserved. Seidl and Dietrich (1992) suggested that steps will propagate only when the channel bed gradient is low relative to the knickpoint gradient, causing this mode of channel lowering to peter out towards the upper reaches of mountain streams.

5.5.4 TERRACE STUDIES

Incomplete erosion of an old valley floor gives rise to the formation of strath or cut terraces (Leopold and Miller, 1954). Strath (bedrock) terraces have been described for lower valley sections in many montane regions, although Merritts *et al.* (1994) suggest that their formation is limited to larger catchments, with streams powerful enough to erode laterally. In a study of the Oregon Coast Range, Personius (1995) found that valley widening primarily occurred during brief episodes of increased sediment input into streams, clogging the channel and inducing lateral incision. A subsequent decrease in sediment supply would allow the stream to resume vertical incision, causing abandonment of the former valley floor.

A fluvial terrace provides a useful datum for tectonic and geomorphic studies. If the time since abandonment, and the height of a terrace above the valley bottom are known, a mean rate of river incision can be determined. Dating of strath terraces has long been problematic, and many workers have used ages of overlying sediments as proxies for strath ages (e.g. Merritts and Vincent, 1989; Amorosi *et al.*, 1996). Recently, exposure dating techniques employing cosmogenic isotopes have become available for geomorphic applications (Cerling and Craig, 1994), allowing direct age determination of bedrock surfaces.

Burbank *et al.* (1996a) have used bedrock exposure dates to determine fluvial incision rates in the middle Indus Gorge, Pakistan. They found a sharp gradient in incision rates across the rapidly deforming Nanga Parbat/Haramosh structural axis. This gradient cannot be attributed to differences in erodibility of the bedrock, but rather seems to coincide with contrasting rates of rock uplift and a trend in channel slope (Burbank *et al.*, 1996a). A similar correlation between rate of rock uplift, rate of incision and channel gradient has been found by Merritts and Vincent (1989), in a comprehensive investigation of drainage basin morphologies across a region of differential rock uplift associated with the Mendocino triple junction in northern California. However, they found that with increasing stream order, channels are less altered by tectonism, indicating that larger rivers are capable of maintaining their longitudinal profiles against the vertical movement of bedrock, and can therefore be regarded as fixed reference frames with respect to which rock uplift can be measured. However, this assumption of a fixed longitudinal profile in order to derive rock uplift rates is fraught with difficulties. Measurements of terrace elevations convert to incision rates, but these equate with rock uplift rates only when vertical stability of the valley floor with respect to a fixed datum can be demonstrated.

Strath terrace width depends predominantly on the ratio of the rate of lateral migration to the rate of vertical incision of a stream (Merritts *et al.*, 1994). The former is a function of stream power, whereas the latter is controlled by the long-term rock uplift rate. Broad straths will result from a large ratio, but if the ratio is small, steep, narrow terraces, with little preservation potential, will be carved. Amorosi *et al.* (1996) used this concept to evaluate the relation between incision of the Pedi-Apenninic thrust front, Italy, and alluvial stratigraphy in the adjacent foreland. Using radiocarbon-derived age constraints on some of the younger terraces in the intramontane valleys and on sequence boundaries in the adjacent alluvial sediments, they demonstrated a link between rapid valley incision (and low terrace preservation) and propagation of the alluvial fan system downstream. Periods of valley planation and terrace preservation were found to correlate with retrogradational patterns of alluvial sedimentation. By relating the degree of terrace preservation directly with the sedimentary signal, Amorosi *et al.* (1996) could extend their observations beyond the temporal resolution of the dating technique.

5.5.5 LONG PROFILE DEVELOPMENT

The overall geometry of long profiles varies significantly between mountain belts. One of the most important expressions of this variance is in the relative length of low-gradient downstream sections. Differences in profile shape can be interpreted in terms of the balance between rock uplift and fluvial incision.

Rivers can only attain smooth, concave profiles when they are capable of removing bedrock at a rate equal to, or greater than, the rate of rock uplift. When rates of incision equal rates of rock uplift, the river profile will not change position, but when fluvial incision outpaces uplift of rocks, the river profile is lowered with time. Assuming that the position of the drainage outlet is pinned at the base of the mountain front, profile lowering will result in a decrease of the stream gradient until the channel bed is at the minimum slope required to transport all material transferred from upstream areas, and incision ceases. Further upstream, erosion of the channel bed continues, causing propagation of the profile form into the mountain and elongation of the minimum gradient section (cf. Howard *et al.*, 1994). Lengthening of the profile is reflected in the sedimentation downstream of the mountain front through increased sorting and rounding, a decrease in grain size, and decreased fan gradient and increased fan radius (Figure 5.11A).

The opposite happens when fluvial incision is outpaced by uplift of rocks. In this scenario not all tectonic input is removed through fluvial erosion, resulting in a net increase in elevation of the channel bed (Figure 5.11B). Persistence of this inbalance will result in steepening of the range flank, and regional slope instability removing part of the excess mass. At the range front, deposition is characterized by an alternation of coarse water-lain sediments with debris flow material and landslide breccias.

Hovius (1995) found different states of profile development in mountain belts across a range of tectonic and climatic settings. Rivers draining eastern Taiwan have longitudinal profiles consisting almost entirely of upstream segments. The smooth concavity of profiles like that of the Feng Ping (Figure 5.12A), combined with the absence of substantial downstream segments, indicates a long-term balance between rock uplift and fluvial incision. Large convexities dominate many river profiles on the north flank of the Kirgizskiy Khrebet of Kirgizstan (Figure 5.12C) and are due to the relative inefficiency of the local streams. The opposite is the case in the western Southern Alps of New Zealand, where the main transverse streams have outpaced rock uplift for a considerable amount of time, and valleys have downstream profile segments dominated by gravel deposition that reach deep into the mountain belt (Figure 5.12B).

Other regions of dominant incision are the Apennines of Italy, and the Barisan Range of Sumatra. The Maoke Range of Irian Jaya, the central Himalayas, and the Finisterre Range of Papua New Guinea have long profiles with well-defined lower sections, but relatively little intramontane sedimentation. The valleys of these regions are probably less removed from the state of balance between rock uplift and fluvial incision than in the western Southern Alps. Convexities of considerable importance are present in some of the Himalayan valleys. These profile deviations are situated over the mid-crustal ramp in the detachment fault, beneath the High Himalayas (Seeber and Gornitz, 1983; Pandey *et al.*, 1995), that coincides with a jump in uplift velocities as measured in the Kathmandu region and the Sun Kosi (Jackson and Bilham, 1994). In the northern Tien Shan the balance seems to tip slightly towards uplift dominance.

River long profile geometry may therefore be used as an indicator of the regional balance between rock uplift and incision. Quantification of rock uplift rates and rates of fluvial incision would allow the construction of a simple model of long profile development. However, at present it is virtually impossible to do this with much accuracy, due to a paucity of relevant data. Regional rock uplift rates may be derived from dated vertical displacements of laterally extensive marker surfaces (e.g. Chappell, 1974; Bull and Cooper, 1986). Alternatively, they may be inferred from fission-track data (assuming the initial elevation before uplift is constrained) (e.g. Kamp and Tippett, 1993), palaeobotanical data, or basin studies combined with geophysical modelling. Mean annual precipitation can be used as a crude proxy for erosivity of the fluvial system. In Figure 5.13 estimates of regional rock uplift rates are plotted against regionally averaged mean annual precipitation rates for nine mountain belts for which river profile information has been analysed (data in Table 5.3). The left-hand side of the diagram is occupied by profile geometries that indicate dominance of rock uplift over incision. On the right-hand side, profile geometries reflect prevalence of fluvial downcutting. The two domains are separated by a zone of

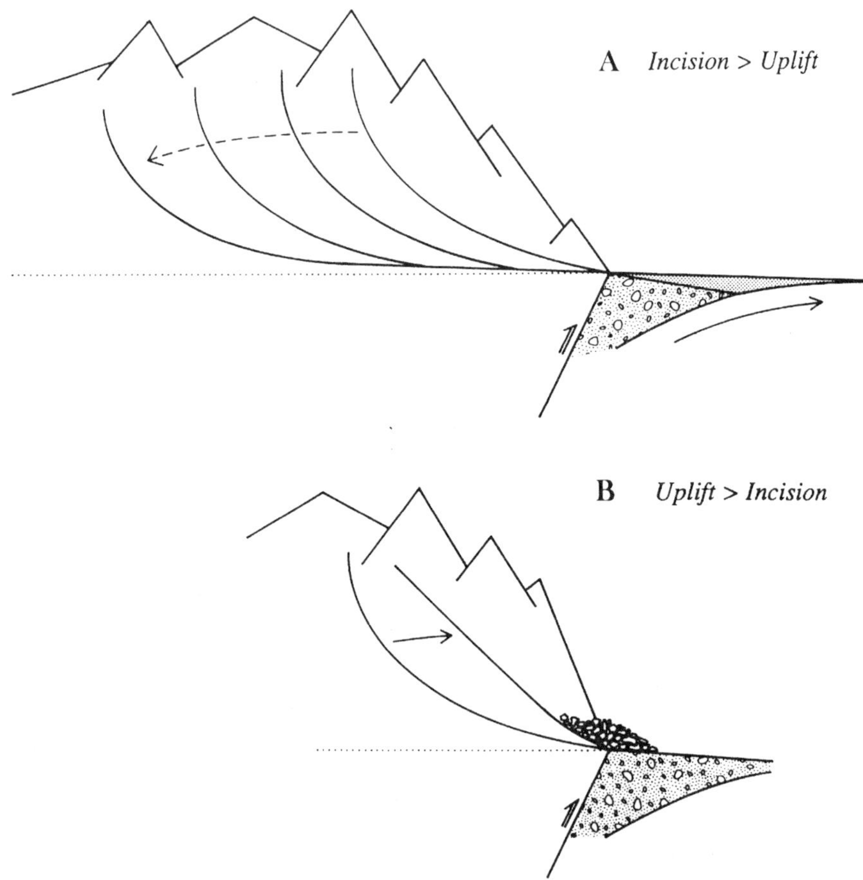

Figure 5.11 Scenarios of longitudinal river profile development. (A) When incision outpaces uplift of rocks, the river profile propagates into the montane topography. Elongation of the profile may result in fining and better sorting of the sediments leaving the catchment, and building of progressively larger, lower angle fans adjacent to the mountain front. (B) Deficiency of the fluvial system in countering uplift of rock leads to progressive steepening of the river profile and adjacent topography. Periodic slope instability may deposit important quantities of brecciated material near the mountain front, in alternation with conglomeratic sediments

balance between controls. A trend in profile form exists, perpendicular to this zone, defined by the relative importance of rock uplift and downcutting.

Another trend in profile form is parallel to the diagonal zone of balance. With increasing rates of rock uplift and precipitation, local relief grows. This can be demonstrated by comparing normalized integrals of river and divide profiles between regions (Hovius, 1995). A measure of local relief within a catchment can be obtained by subtracting the river profile integral from the divide integral. Average values for normalized integrals in the nine mountain belts are listed in Table 5.3. In regions where both the rate of rock uplift and the mean annual precipitation are low, local relief is subdued and the differential profile integral is low (Apennines and Barisan Range). High rates of rock uplift and precipitation yield extreme local relief and high differential profile integrals (Southern Alps, New Zealand).

Combined with observations on the scale invariance of montane catchments, the correlation of local relief to rock uplift and precipitation rates has some implications for models of topographic development in mountain belts. Assuming equal drainage densities, hillslope gradients are lower in catchments with smaller differential integrals. This may reflect lower rates of landsliding, and a reduced relative importance of this type of mass wasting in regional denudation. In such areas, weathering-limited hillslope erosion processes act to subdue the local relief below the rock

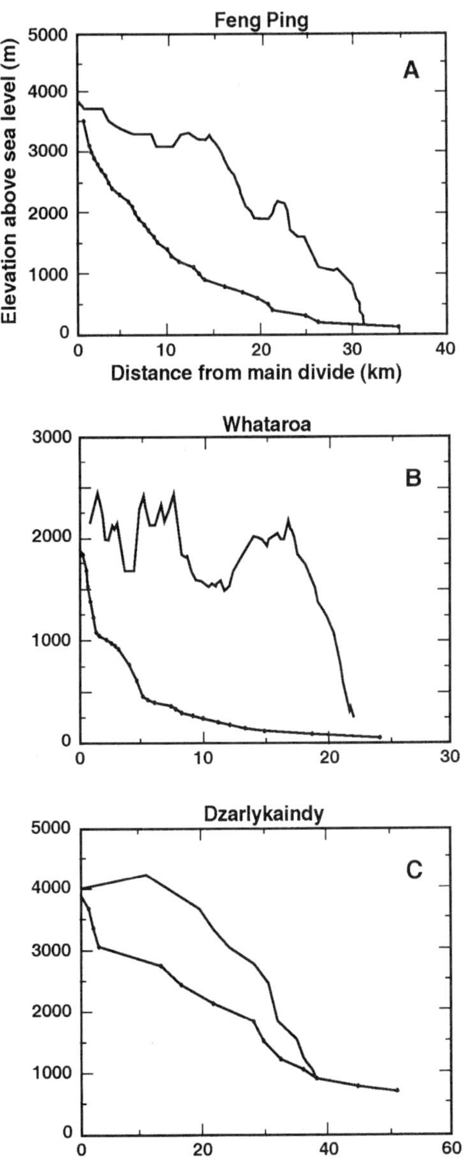

Figure 5.12 *Different states of river profile development. (A) The longitudinal profile of the Feng Ping (Central Range, Taiwan) consists almost entirely of a smooth upstream segment. This configuration indicates long-term balance between rock uplift and incision. (B) The longitudinal profile of the Whataroa River (Southern Alps, New Zealand) is characteristic of a long-term dominance of incision over rock uplift. (C) Dominant rock uplift has caused profile convexity in the Dzarlykaindy Valley (Kirgizskiy Khrebet)*

strength related limit to topography. Thus, landslide-controlled topography must be regarded as an end-member of the range of montane landscapes (cf. Anderson, 1994).

5.5.6 GLACIAL EROSION

Apart from fluvial processes, glaciers may act to lower valleys in montane regions. In most conceptual and numerical models of landscape evolution, the glacial component is systematically ignored, although valleys in many mountain belts have undergone glaciation at some stage during their Quaternary development. The formation of glacial valleys has been modelled by Harbor *et al.* (1988), who demonstrated that the conversion from a V-shaped river valley to a steady-state parabolic glacial form may be achieved in 10^5 years, given erosion rates representative of present-day glacial systems. These model predictions have been substantiated in a recent study by Kirkbride and Matthews (1997) of glacial landforms in the eastern Southern Alps of New Zealand, where 2×10^5 years of temperate glacial erosion produces a recognizable trough-and-arête topography. Kirkbride and Matthews found that mean and modal relief increase where glacial activity is confined to cirques, but decrease when trough incision by ice becomes established as the dominant valley-lowering mechanism. This implies that, in the eastern Southern Alps, glacial erosion is more vigorous than its fluvial counterpart. Raymo *et al.* (1988) and Molnar and England (1990) proposed that the initiation of late Cenozoic glaciation may have led to a global increase of denudation rates. Although the evidence of this increase is substantial (e.g. Isacks, 1992), the issue of glacial erosivity remains hotly debated. Comparison of present-day sediment yields from glaciated and non-glaciated catchments (e.g. Drewry, 1986; Hicks *et al.*, 1990; Gardner and Jones, 1993; Warburton and Beecroft, 1993) demonstrates that values for both mechanisms span a similar range of magnitudes, without one consistently outweighing the other.

5.6 Topography and Orogen Dynamics

5.6.1 TOPOGRAPHIC DEVELOPMENT AND STEADY STATE: A SYNTHESIS

The observations outlined above can be brought together in a conceptual model of the development of mountain belt topography (Figure 5.14). This model represents many aspects of the surface behaviour of a

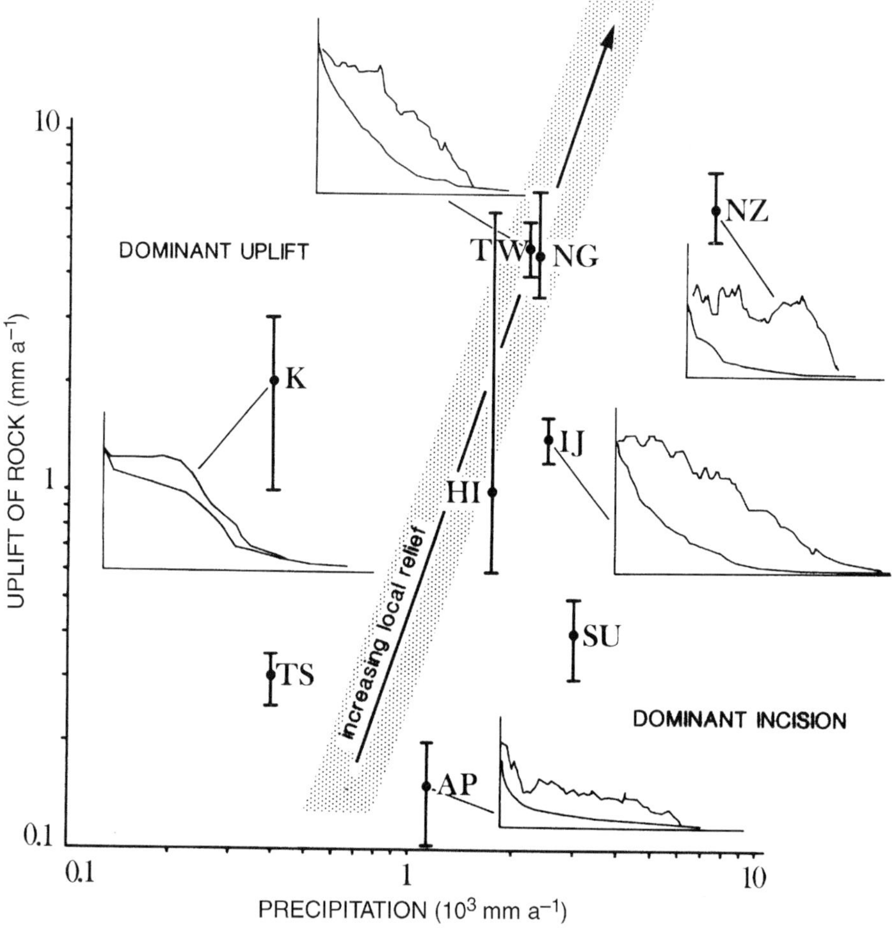

Figure 5.13 River profile geometry related to mean annual precipitation and rate of uplift of rock (data in Table 5.3). Two fields in the diagram, one of dominant rock uplift, the other of dominant incision, are separated by a diagonal zone in which the two controls are in approximate balance. Increase of local relief trends parallel to this zone of balance, in the direction of higher rates of precipitation and rock uplift. The data used to construct this diagram are in part speculative, and the diagram should be regarded accordingly

two-sided orogen that have also been investigated through numerical models such as those discussed by Beaumont *et al.* (Section 3.5).

Orogenic topography consists of several macro-morphological elements, each with their own process system and balance of forces. On the largest scale, the symmetry of a mountain range is controlled by the competition between drainage systems on either side of the main divide. Within a catchment, the balance between valley lowering and bedrock uplift determines the evolution of the longitudinal river profile, thus setting a boundary condition for the surrounding hillslopes. Hillslope evolution, in turn, is controlled by the balance between rock uplift with respect to the valley floor, and mass movement.

The regional slope of a mountain belt is at high angles with its structural grain. Rivers with a low capacity for incision are entrained in this grain, whereas actively incising streams respond to the regional topographic gradient. The steep frontal side of a mountain belt is eroded by transverse streams, set at regular intervals along the range front. Here, the elevation of the river bed is a function of the competition between fluvial incision and rock uplift. If rock uplift is not adequately countered by incision, the channel bed undergoes a net surface uplift, the persistence of which

Table 5.3 *Long profile integrals and environmental controls*

Mountain belt	River	Divide	Difference	Y (t km^{-2} a^{-1})	P (mm a^{-1})	R (mm a^{-1})	U (mm a^{-1})
Southern Alps	0.202	0.770	0.568	15000	7500	7000	5.0–7.8 (1)
Finisterre Range	0.255	0.550	0.295		2350		3.5–6.8 (2)
Maoke Range	0.226	0.493	0.267		2500		1.2–1.6 (3)
Barisan Range	0.183	0.346	0.163		3000		0.3–0.5 (4)
Central Range	0.240	0.598	0.358	14710	2300	2100	4.0–5.5 (5)
Central Himalaya	0.214	0.536	0.322	3310	1750		0.6–1.8 (6, 7)
N. Tien Shan	0.244	0.573	0.329		400		0.25–0.35 (8)
Kirgizskiy Khrebet	0.367	0.584	0.217	42	400	290	–
Apennines	0.175	0.417	0.242	1480	1150		0.1–0.2 (9)

The table lists the average values of the normalized integrals of river profiles and divide profiles, calculated for each of nine mountain belts with the exclusion of those profiles that show important tectonically induced deviations. The differential integral ('difference') is proportional to the potential energy for hillslope erosion into the river thalweg. Y is the specific sediment yield averaged over all drainage basins for which information is available; P is the mean annual precipitation averaged over the mountain belt; R is the specific runoff calculated from discharge data for a limited number of catchments; U is a regional estimate of rock uplift rate. References: (1) Cooper *et al.* (1979); Bull and Cooper (1986); Kamp and Tippett (1993); Simpson *et al.* (1994). (2) Chappell (1974); Crook (1989); Lon Abbot, pers. comm. (1994). (3) Kevin Hill, pers. comm. (1994). (4) Steven Moss, pers. comm. (1994). (5) Li (1976); Barr and Dahlen (1990); Lundberg and Dorsey (1990). (6) Iwata *et al.* (1984); Hubbard *et al.* (1991); Searle (1995). (7) Jackson and Bilham (1994). (8) Avouac and Tapponnier (1993); Hendrix *et al* (1994); An Yin, pers. comm. (1994). (9) Paul Andriessen, pers. comm. (1995).

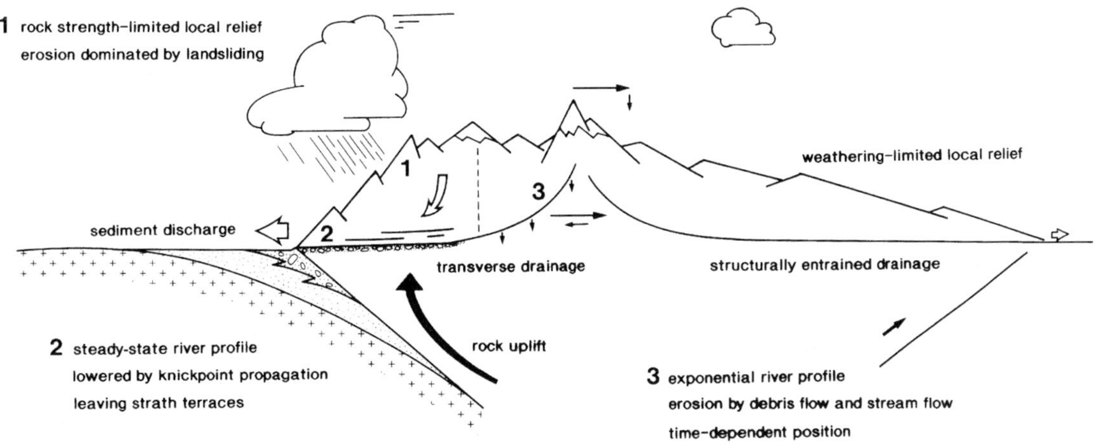

Figure 5.14 *Conceptual model of the topographic development of a two-sided orogen, wherein orographically forced precipitation is concentrated on the rapidly uplifting frontal side of the mountain belt, while the trailing side of the structure receives less rainfall. See text for explanation*

will eventually force the river back into a structural straitjacket. More often, though, rivers are capable of removing bedrock at a rate equal to, or greater than, the rate of rock uplift, resulting in a smooth concavity of their longitudinal profiles. Predominant incision results in propagation of the valley head away from the range front, and formation of a downstream profile segment at a minimum bed gradient. The vertical position of this profile segment is maintained by knickpoint propagation, following local base-level lowering. It is the only part of an intramontane river profile for which a steady-state assumption is reasonable. Further upstream, bedrock incision, due to abrasion by particles transported by debris flows and stream flow, results in an exponential decay of channel bed elevation away from the drainage divide. Here, the rate of incision is controlled by the local climate and properties of the substratum.

Between valleys, bedrock is uplifted towards denuding hillslopes, whose rate of mass transfer is primarily controlled by the rate of bedrock weathering. If this rate is slower than the local rate of valley lowering, the hillslope continues to steepen, becoming increasingly susceptible to landsliding involving bedrock. Thus, landsliding controlled by rock strength effectively places an upper limit on the amplitude of local relief, and time-independent topography may ensue in areas where this mechanism is dominant. In such areas, sediment yield is solely determined by rock uplift rates, and is, by consequence, independent of climate.

River systems on opposing sides of a drainage divide compete to determine the position of the ridge pole of a mountain range. If opposing systems are equally erosive, the divide will remain stationary, but ongoing fluvial incision will result in a lowering of the topographic boundary. Differential erosivity of opposing river systems will cause a shift of the divide away from the topographic culmination of the mountain range and into lower terrain. This shift may eventually lead to capture of drainage elements across the divide, aided, in some instances, by the advection of topography into the frontal side of a mountain belt (cf. Koons, 1990).

After Adams (1985), many workers (e.g. Summerfield, 1991) have implied a landscape evolution scheme in which time-dependent topography is replaced by time-independent, steady-state landscapes, once erosion rates have caught up with rock uplift. This assumes a perfect and rapid coupling of fluvial and hillslope systems, and ignores the possibility of denudation outpacing rock uplift. Here, I have demonstrated that steady-state topography in mountain belts is limited to low-angle, higher order stream segments, adjacent to the range front, and rock-strength-controlled valley sides. Precise steady-state topography on the scale of an orogenic flank occurs only when there is an exact and long-term balance between rock uplift and fluvial incision, combined with a perfect coupling of the hillslope and valley systems.

5.6.2 SURFACE CONTROL OF OROGEN WEDGE DYNAMICS

The form of convergent orogens is approximated by a wedge-shaped prism, the geometry of which is a function of its internal rheology and the stresses applied at its boundaries (Elliot, 1976; Chapple, 1978; Dahlen *et al.*, 1984; Platt, 1986). If the stress state exceeds the mechanical strength of the material, the wedge deforms internally, until it has attained the configuration of a critical taper, such that the gravitational forces generated by its surface slope balance the traction exerted on its base by underthrusting. Although serious problems remain with applying existing mechanical formulations of critical wedge theory to fold-thrust belts (Bombolakis, 1994), its broad conceptual basis has been validated by studies of active mountain belts (e.g. Davis *et al.*, 1983; Dahlen, 1984).

The evolution of the upper surface affects the taper of an orogenic wedge, and may drive, or modify, its outward propagation and internal deformation. Recent modelling studies (e.g. Koons, 1989, 1994) (see Chapter 3) and field studies (e.g. DeCelles and Mitra, 1995) underline this coupling of topography and tectonics. Rates and patterns of valley incision determine the distribution of topography across an orogen (Gilchrist *et al.*, 1994; Montgomery, 1994), and longitudinal river profiles may therefore be used as an indicator of surface taper development. Propagation of valley lowering in an upstream direction causes a decrease of the surface gradient of an orogenic wedge, away from the critical taper. Such a wedge is unlikely to undergo thrust-front propagation, being characterized instead by deformation styles that promote internal thickening. This type of orogen is likely to remain a narrow, linear feature. Outward propagation of deformation is more likely to occur in regions where mass removal through erosion cannot maintain the surface gradient at, or below, a critical angle. Such mountain belts are uplift dominated, which is typically expressed in significant valley profile convexity. The ensuing frontal imbrication and plateau formation are well known from rapidly uplifting or arid mountain belts like the Himalayas (Pandey *et al.*, 1995), the Andes (Masek *et al.*, 1994) (Chapter 9) and the Tien Shan (Avouac and Tapponnier, 1993; Molnar *et al.*, 1994). It may be concluded that periods of aridity have a profound influence on the geometry of compressional tectonic systems, and that in dry mountain belts topography is controlled by tectonic strain distribution. By contrast, in wet mountain belts drainage organization and rock strength properties determine the distribution of relief.

5.7 Scaling the Mountain

Progress in our understanding of the development of (montane) topography across a range of spatial and temporal scales is constrained by at least two factors. The first is our comprehension of the mechanics of tectonics and the geomorphic processes involved in landscape evolution. The second is the systematic observation of the rates at which these processes operate. Although the recent advent of comprehensive data sets on continental-scale topography has promoted the integration of disciplines, such as climatology, hydrology, structural geology and geomorphology, into a unified study of physical landsurface processes (Isacks, 1992; Merritts and Vincent, 1994), scale differences between these approaches have been only partially resolved.

Problems with the mechanics of geomorphic processes centre around fluvial and glacial valley lowering. Fluvial incision theory may be enhanced through experiments in natural and artificial bedrock channels, while new bedrock exposure dating techniques will provide temporal constraints on channel bed lowering and knickpoint propagation. Glacial erosion has not yet been integrated into models of orogen development, and difficulties in doing so will arise from the lack of relevant field observations and our fragmentary understanding of the environmental controls on the process.

The coupling of earth surface process systems and the transfer of sediment between them is ill constrained, and will have to be resolved in order to allow for a comprehensive evaluation of the geomorphic response to changes in tectonic and/or climatic controls. This evaluation also needs observations of geomorphic process characteristics which facilitate extrapolation across spatial and temporal scales. Although scale independence is assumed for many erosional processes (Isacks, 1992), few field observations are available to support this notion.

Direct estimates of rates of geomorphic processes are often problematic, and it would be helpful to link rates with other observations. It is reasonable to expect a correlation between the topography of the Earth's surface and the processes from which it results. Burbank *et al.* (1996a) have disproved a general correlation between slope gradient and rate of valley lowering, while the hope expressed by Howard (1996) that other aspects of relief, such as stream gradient or stream spacing, depend on incision rates, has been partially refuted here. However, other aspects of topography might provide the desired link: Stark and Hovius (unpublished data) for instance, have observed a coincidence of the scaling exponents of landslides and topographic curvature in the Southern Alps of New Zealand.

Further advances may be expected from the coupling of orogenic erosion with the geochemical record, and from the matching of high-resolution geodetic analyses with mechanical models of orogens. Finally, the introduction of geomorphology to the extreme rates of denudation observed during gravitational collapse of orogens (e.g. Zeck *et al.*, 1992) is a challenging prospect.

Acknowledgements

This chapter is largely based upon my doctoral work with Philip Allen at the Department of Earth Sciences of the University of Oxford. Many of the results presented here have been produced through close collaboration with Colin Stark and Mark Audet. The British Council is gratefully acknowledged for a Research Fellowship. Amerada Hess and Royal Dutch Shell Research Laboratories have provided financial support for the sediment flux work. I thank Chris Beaumont and an anonymous reviewer for many helpful comments on an earlier version of the manuscript.

References

Abbott, L.D., Silver, E.A., Thompson, P.R., Filewicz, M.V. Schneider, C. and Abdoerias 1994. Stratigraphic constraints on the timing of arc–continent collision in northern Papua New Guinea. *J. Sed. Res.*, **B64**, 169–183.

Adams, J. 1985. Large scale tectonic geomorphology of the Southern Alps, New Zealand. In: M. Morisawa and J.T. Hack (eds) *Tectonic Geomorphology*, Allen and Unwin, Boston, pp. 105–128.

Ahmad, R., Scatena, F.N. and Gupta, A. 1993. Morphology and sedimentation in Caribbean montane streams: examples from Jamaica and Puerto Rico. *Proceedings 5th International Conference on Fluvial Sedimentology, Brisbane, Australia, July 1993*.

Ahnert, F. 1970. Functional relationships between denudation, relief and uplift in large mid-latitude drainage basins. *Amer. J. Sci.*, **268**, 243–263.

Amorosi, A., Farina, M., Severi, P., Preti, D., Caporale, L. and di Dio, G. 1996. Genetically related alluvial deposits across fault zones: an example of alluvial fan–terrace correlation from the upper Quaternary of the southern Po Basin, Italy. *Sed. Geol.*, **102**, 275–295.

Anderson, R.S. 1994. Evolution of the Santa Cruz Mountains, California, through tectonic growth and geomorphic decay. *J Geophys. Res.*, **99**, 20 161–20 179.

Avouac, J.P. and Tapponnier, P. 1993. Active thrusting and folding along the northern Tien Shan and Late Cenozoic rotation of the Tarim relative to Dzungaria and Kazakhstan. *J. Geophys. Res.*, **98**, 6755–6804.

Barr, T.D. and Dahlen, F.A. 1990. Constraints on friction

and stress in the Taiwan fold-and-thrust belt from heat flow and geochronology. *Geology*, **18**, 111–115.

Beaumont, C., Fullsack, P. and Hamilton, J. 1992. Erosional controls of active compressional orogens. In: K.P. McClay (ed.) *Thrust Tectonics*, Chapman and Hall, London, pp. 1–18.

Begin, Z.B., Meyer, D.F. and Schumm, S.A. 1981. Development of longitudinal profiles of alluvial channels in response to base level lowering. *Earth Surf. Processes Ldfms.*, **6**, 49–68.

Benda, L. 1990. The influence of debris flows on channels and valley floors in the Oregon Coastal Ranges, U.S.A. *Earth Surf. Processes Ldfms.*, **15**, 457–464.

Berryman, K.R., Beanland, S., Cooper, A.F., Cutten, H.N., Norris, R.J. and Wood, P.R. 1992. The Alpine Fault, New Zealand: Variation in Quaternary structural style and geomorphic expression. *Ann. Tect. Suppl.* vol. VI, 126–163.

Bombolakis, E.G. 1994. Applicability of critical-wedge theory to foreland belts. *Geology*, **22**, 535–538.

Bull, W.B. 1988. Floods: degradation and aggradation. In: V.R. Baker, R.C. Kochel and P.C. Patton (eds) *Flood Geomorphology*, Wiley, New York, pp. 157–165.

Bull, W.B. 1996. Prehistorical earthquakes on thc Alpine Fault, New Zealand. *J. Geophys. Res.*, **101**, 6037–6050.

Bull, W.B. and Cooper, A.F. 1986. Uplifted marine terraces along the Alpine Fault, New Zealand. *Science*, **234**, 1225–1228.

Burbank, D.W., Leland, J., Fielding, E., Anderson, R.S., Brozovic, N., Reid, M.R. and Duncan, C. 1996a. Bedrock incision, rock uplift and threshold hillslopes in the northwestern Himalayas. *Nature*, **379**, 505–510.

Burbank, D.W., Meigs, A. and Brozovic, N. 1996b. Interactions of growing folds and coeval depositional systems. *Basin Res.*, **8**, 199–223.

Cerling, T.E. and Craig, H. 1994. Geomorphology and *in-situ* cosmogenic isotopes. *Ann. Rev. Earth Planet. Sci.*, **22**, 273–317.

Chappell, J. 1974. Geology of coral terraces, Huon Peninsula, New Guinea: A study of Quaternary tectonic movements and sea-level changes. *Geol. Soc. Amer. Bull.*, **85**, 553–570.

Chapple, W.M. 1978. Mechanics of thin-skinned fold-and-thrust belts. *Geol. Soc. Amer. Bull.*, **89**, 1189–1198.

Chase, C.G. 1992. Fluvial landsculpting and the fractal dimension of topography. *Geomorphology*, **5**, 39–57.

Church, M. and Slaymaker, O. 1989. Disequilibrium of Holocene sediment yield in glaciated British Columbia. *Nature*, **337**, 452–454.

Cooper, A.F., Bishop, D.G. and van der Lingen, G.J. 1979. Uplift rates and high level marine platforms associated with the Alpine Fault at Okuru River, south Westland. *Bull. R. Soc. N.Z.*, **18**, 35–43.

Cooper, A.F. and Norris, R.J. 1990. Estimates for the timing of the last coseismic displacement on the Alpine Fault, northern Fiordland, New Zealand. *N.Z. J. Geol. Geophys.*, **33**, 303–307.

Cowie, P.A. and Scholz, C.H. 1992. Growth of faults by accumulation of seismic slip. *J. Geophys. Res.*, **97**, 11 085–11 095.

Crook, K.A.W. 1989. Suturing history of an allochtonous terrane at a modern plate boundary traced by flysch-to-molasse facies transitions. *Sed. Geol.*, **61**, 49–79.

Dahlen, F.A. 1984. Noncohesive critical Coulomb wedges: An exact solution. *J. Geophys. Res.*, **89**, 10 125–10 133.

Dahlen, F.A., Suppe, J. and Davis, D. 1984. Mechanics of fold-and-thrust belts and accretionary wedges: Cohesive Coulomb theory. *J. Geophys. Res.*, **89**, 10 087–10 101.

Davis, P.J. 1975. *Interpolation and Approximation*, Dover Publications, New York.

Davis, D., Suppe, J. and Dahlen, F.A. 1983. Mechanics of fold-and-thrust belts and accretionary wedges: cohesive Coulomb theory. *J. Geophys. Res.*, **89**, 10 087–10 101.

Davis, W.M. 1899. The geographical cycle. *Geogr. J.*, **14**, 481–504.

DeCelles, P.G. and Mitra, G. 1995. History of the Sevier orogenic wedge in terms of critical taper models, northeast Utah and southwest Wyoming. *Geol. Soc. Amer. Bull.*, **107**, 454–462.

Dedkov, A.P. and Moszherin, V.I. 1992. Erosion and sediment yield in mountain regions of the world. *IAHS Publ.*, **209**, 29–36.

Douglas, I. 1967. Man, vegetation and sediment yield of rivers. *Nature*, **215**, 925–928.

Drewry, D. 1986. *Glacial Geologic Processes*. Arnold, London.

Elliott, D. 1976. The motion of thrust sheets. *J. Geophys. Res.*, **81**, 949–963.

Enos, P. 1991. Sedimentary parameters for computer modelling. In: E.K. Franseen, W.L. Watney, C.G.St.C. Kendall and W. Ross (eds) *Sedimentary Modeling: Computer Simulations and Methods for Improved Parameter Definition. Kansas Geological Survey Bulletin*, **233**, 63–99.

Eyles, G.O. 1983. The distribution and severity of present soil erosion in New Zealand. *N.Z. Geogr.*, April 1983, 12–27.

Fielding, E.J., Isacks, B.L., Barazangi, M. and Duncan, C. 1994. How flat is Tibet? *Geology*, **22**, 163–167.

Foley, M.G. 1980. Bed-rock incision by streams *Geol. Soc. Amer. Bull.*, Part II, **91**, 2189–2213.

Fournier, F. 1960. *Climat et Erosion: La Relation entre l'Erosion du sol par l'Eau et les Précipitations Atmospheriques*, Presse Universitaire de Paris, Paris.

Fuyii, Y. 1969. Frequency distribution of the magnitude of the landslides caused by heavy rainfall. *J. Seism. Soc. Japan*, **22**, 244–247.

Gardner, J.S. and Jones, N.K. 1993. Sediment transport and yield at the Raikot glacier, Nanga Parbat, Punjab Himalaya. In: J.P. Schroder (ed.) *Himalaya to the Sea*. Routledge, London, pp. 184–197.

Gerrard, J. 1994. The landslide hazard in the Himalayas: geological control and human action. *Geomorphology*, **10**, 221–230.

Gilbert, G.K. 1877. *Report on the Geology of the Henry Mountains*. US Govt. Printing Office, Washington, DC.

Gilchrist, A.R., Summerfield, M.A. and Cockburn, H.A.P.

1994. Landscape development, isostatic uplift, and the morphologic development of orogens. *Geology*, **22**, 963–966.

Grant, G.E., Swanson, F.J. and Wolman, M.G. 1990. Pattern and origin of stepped bed morphology in high-gradient streams, western Cascades, Oregon. *Geol. Soc. Amer. Bull.*, **102**, 340–352.

Greenbaum, D. *et al.* 1995. *Rapid Methods of Landslide Hazard Mapping: Papua New Guinea Case Study*. British Geological Survey Technical Report WC/95/27.

Griffiths, G.A. 1979. High sediment yields from major rivers of the western Southern Alps, New Zealand. *Nature*, **282**, 61–63.

Griffiths, G.A. and McSaveney, M.J. 1983a. Distribution of mean annual precipitation across some steepland regions of New Zealand. *N.Z. J.Sci.*, **26**, 197–209.

Griffiths, G.A. and McSaveney, M.J. 1983b. Hydrology of a basin with extreme rainfalls – Cropp River, New Zealand. *N.Z. J. Sci.*, **26**, 293–306.

Griffiths, G.A. and McSaveney, M.J. 1986. Sedimentation and river containment on Waitangitaona alluvial fan – South Westland, New Zealand. *Z. Geomorph.*, **30**, 215–230.

Gupta, S. 1997. Himalayan drainage patterns and the origin of fluvial megafans in the Ganges foreland basin. *Geology*, **25**, 11–14.

Hack, J.T. 1957. Studies of longitudinal stream profiles in Virginia and Maryland. *U.S. Geol. Surv. Prof. Pap.*, **505-B**.

Harbor, J.M., Hallet, B. and Raymond, C.F. 1988. A numerical model of landform development by glacial erosion. *Nature*, **333**, 347–349.

Harrison, C.G.A. 1994. Rates of continental erosion and mountain building. *Geol. Rund.*, **83**, 431–447.

Henderson, R.D. 1993. Extreme storm rainfall in the Southern Alps, New Zealand. *IAHS Publ.*, **213**, 113–120.

Hendrix, M.S., Dimitru, T.A. and Stephan, A.G. 1994. Late Oligocene – early Miocene unroofing in the Chinese Tien Shan: An early effect of the India – Asia collision. *Geology*, **22**, 487–490.

Hicks, D.M., McSaveney, M.J. and Chinn, T.J.H. 1990. Sedimentation in proglacial Ivory Lake, Southern Alps, New Zealand. *Arctic Alp. Res.*, **22**, 26–42.

Holeman, J.N. 1968. Sediment yield of major rivers of the world. *Wat. Resour. Res.*, **4**, 737–747.

Hovius, N. 1995. Macro scale process systems of mountain belt erosion and sediment delivery to basins. Unpublished DPhil. thesis, University of Oxford.

Hovius, N. 1996. Regular spacing of drainage outlets from linear mountain belts. *Basin Res.*, **8**, 29–44.

Hovius, N. 1998. Controls on sediment supply by large rivers. In: K.W. Shanley and P.J. McCabe (eds) *Relative Role of Eustacy, Climate and Tectonics in Continental Rocks. Soc. Econ. Paleont. Mineral. Spec. Publ.*, **59**, 3–16.

Hovius, N., Stark, C.P. and Allen, P.A. 1997. Sediment flux from a mountain belt derived by landslide mapping. *Geology*, **25**, 231–234.

Howard, A.D. 1994. A detachment-limited model of drainage basin evolution. *Wat. Resour. Res.*, **30**, 2261–2285.

Howard, A.D. 1996. The ephemeral mountains. *Nature*, **397**, 488–489.

Howard, A.D. and Kerby, G. 1983. Channel changes in badlands. *Geol. Soc. Amer. Bull.*, **94**, 739–752.

Howard, A.D., Dietrich, W.E. and Seidl, M.A. 1994. Modeling fluvial erosion on regional to continental scales. *J. Geophys. Res.*, **99**, 13 971–13 986.

Hubbard, M.S., Royden, L. and Hodges, K.V. 1991. Constraints on unroofing rates in the High Himalayas, eastern Nepal. *Tectonics*, **10**, 287–298.

Isacks, B.L. 1992. 'Long-term' land surface processes: erosion, tectonics and climate history in mountain belts. In: P.M. Mather (ed.) *TERRA-1: Understanding the Terrestrial Environment*. Taylor and Francis, London, pp. 21–36.

Iwata, S., Sharma, T. and Yamanaka, H. 1984. A preliminary report on geomorphology of central Nepal and Himalayan uplift. *J. Geol. Soc. Nepal*, **4**, 141–150.

Jackson, M. and Bilham, R. 1994. Constraints on Himalayan deformation inferred from vertical velocity fields in Nepal and Tibet. *J. Geophys. Res.*, **99**, 13 897–13 912.

Jackson, J., Norris, R.J. and Youngson, J. 1996. The structural evolution of active fault and fold systems in central Otago, New Zealand: evidence revealed by drainage patterns. *J. Struct. Geol.*, **18**, 217–234.

Jansson, M.B. 1988. A global survey of sediment yield. *Geogr. Ann.*, **70A**, 81–98.

Kamp, P.J.J. and Tippett, J.M. 1993. Dynamics of Pacific plate crust in the South Island (New Zealand) zone of oblique continent–continent convergence. *J. Geophys. Res.*, **98**, 16 105–16 118.

Keefer, D.K. 1984. Landslides caused by earthquakes. *Geol. Soc. Amer. Bull.*, **95**, 406–421.

Keefer, D.K. 1994. The importance of earthquake-induced landslides to long-term slope erosion and slope-failure hazards in seismically active regions. *Geomorphology*, **10**, 265–284.

Kirkbride, M. and Matthews, D. 1997. The role of fluvial and glacial erosion in landscape evolution: the Ben Ohau Range, New Zealand. *Earth Surf. Processes Ldfms*, **22**, 317–327.

Koons, P.O. 1989. The topographic evolution of collisional mountain belts: a numerical look at the Southern Alps, New Zealand. *Amer. J. Sci.*, **289**, 1044–1069.

Koons, P.O. 1990. The two-sided orogen: Collision and erosion from the sandbox to the Southern Alps. *Geology*, **18**, 679–682.

Koons, P.O. 1994. Three-dimensional critical wedges: Tectonics and topography in oblique collisional orogens. *J. Geophys. Res.*, **99**, 12 301–12 315.

Koons, P.O. 1995. Modeling the topographic evolution of collisional belts. *Ann. Rev. Earth Planet. Sci.*, **23**, 375–408.

Langbein, W.B. and Leopold, L.B. 1964. Quasi-equilibrium states in channel morphology. *Amer. J. Sci.*, **262**, 782–794.

Langbein, W.B. and Schumm, S.A. 1958. Yield of sediment in relation to mean annual precipitation. *Trans. Amer. Geophys. Un.*, **39**, 1076–1084.

Leeder, M.R. 1991. Denudation, vertical crustal movements and sedimentary basin infill. *Geol. Rund.*, **80**, 441–458.

Leopold, L.B. and Langbein, W.B. 1962. The concept of entropy in landscape evolution. *U.S. Geol. Surv. Prof. Pap.*, **500-A**.

Leopold, L.B. and Maddock, T.U. 1953. The hydraulic geometry of stream channels and some physiographic implications. *U.S. Geol. Surv. Prof. Pap.*, **352**.

Leopold, L.B. and Miller, J.P. 1954. A post-glacial chronology for some alluvial valleys in Wyoming. *U.S. Geol. Surv. Wat. Sup. Pap.*, **1261**, 1–90.

Leopold, L.B., Wolman, M.G. and Miller, J.P. 1964. *Fluvial Processes in Geomorphology*, Freeman, San Francisco.

Li, Y.H. 1976. Denudation of Taiwan Island since the Pliocene epoch. *Geology*, **4**, 105–107.

Lisitzin, A.P. 1972. Sedimentation in the world ocean. *Soc. Econ. Paleont. Mineral. Spec. Publ.*, **17**, 1–218.

Lundberg, N. and Dorsey, R.J. 1990. Rapid Quaternary emergence, uplift and denudation of the Coastal Range, Eastern Taiwan. *Geology*, **18**, 638–641.

Mackin, J.H. 1948. Concept of the graded river. *Geol. Soc. Amer. Bull.*, **101**, 1373–1388.

Masek, J.D., Isacks, B.L., Gubbels, T.J. and Fielding, E.J. 1994. Erosion and tectonics at the margins of continental plateaus. *J. Geophys. Res.*, **99**, 13 941–13 956.

Matthes, G.H. 1947. Macroturbulence in natural stream flow. *Trans. Amer. Geophys. Un.*, **28**, 255–262.

Medwedeff, D.A. 1992. Geometry and kinematics of an active, laterally propagating wedge thrust, Wheeler Ridge, California. In: S. Mitra and G.W. Fisher (eds) *Structural Geology of Fold and Thrust Belts*, Johns Hopkins University Press, Baltimore, pp. 3–28.

Merritts, D. and Ellis, M. 1994. Introduction to special section on tectonics and topography. *J. Geophys. Res.*, **99**, 12 135–12 141.

Merritts, D. and Vincent, K.R. 1989. Geomorphic response of coastal streams to low, intermediate and high rates of uplift. Mendocino triple junction region, northern California. *Geol. Soc. Amer. Bull.*, **101**, 1373–1388.

Merritts, D., Vincent, K.R. and Wohl, E.E. 1994. Long river profiles, tectonism, and eustasy: A guide to interpreting fluvial terraces. *J. Geophys. Res.*, **99**, 14 031–14 050.

Meybeck, M. 1988. How to establish and use world budgets of riverine materials. In: A. Lerman and M. Meybeck (eds) *Physical and Chemical Weathering in Geochemical Cycles*, Kluwer, Dordrecht, pp. 247–272.

Miller, J.R. 1991. The influence of bedrock geology on knickpoint development and channel-bed degradation along downcutting streams in south-central Indiana. *J. Geol.*, **99**, 591–605.

Milliman, J.D. and Meade, R.H. 1983. World-wide delivery of river sediment to the oceans. *J. Geol.*, **91**, 1–21.

Milliman, J.D. and Syvitski, J.P.M. 1992. Geomorphic/tectonic control of sediment discharge to the ocean: the importance of small mountainous rivers. *J. Geol.*, **100**, 525–544.

Molnar, P. and England, P. 1990. Late Cenozoic uplift of mountain ranges and global climate change: chicken or egg? *Nature*, **346**, 29–34.

Molnar, P. and Lyon-Caen, H. 1989. Fault plane solutions of earthquakes and active tectonics of the Tibetan Plateau and its margins. *Geophys. J. Int.*, **99**, 123–153.

Molnar, P., Brown, E.T., Burchfiel, B.C., Qidong, D., Xianyue, F., Jun, L., Raisbeck, G.M., Jianbang, S., Zhangming, W., Yiou, F. and Huichuan, Y. 1994. Quaternary climate change and the formation of river terraces across growing anticlines on the north flank of the Tien Shan, China. *J. Geol.*, **102**, 583–602.

Montgomery, D.R. 1994. Valley incision and the uplift of mountain peaks. *J. Geophys. Res.*, **99**, 13 931–13 921.

Montgomery, D.R. and Dietrich, W.E. 1992. Channel initiation and the problem of landscape scale. *Science*, **255**, 826–830.

Morisawa, M. 1985. *Rivers*, Longman, New York.

Ohmori, H. 1983. Erosion rates and their relation to vegetation from the viewpoint of world-wide distribution. *Bull. Dept. Geog. Univ. Tokyo*, **15**, 77–91.

Ohmori, H. 1991. Change in the mathematical function type describing the longitudinal profile of a river through an evolutionary process. *J. Geol.*, **99**, 97–110.

Ohmori, H. 1992. Morphological characteristics of the scar created by large-scale rapid mass movement. *Trans. Japan. Geomorph. Un.*, **13**, 185–202.

Ohmori, H. and Shimazu, H. 1994. Distribution of hazard types in a drainage basin and its relation to geomorphological setting. *Geomorphology*, **10**, 95–106.

Ori, G.G. and Friend, P.F. 1984. Sedimentary basins formed and carried piggyback on active thrust sheets. *Geology*, **12**, 475–479.

Pain, C.F. and Bowler, J.M. 1973. Denudation following the November 1970 earthquake at Madang, Papua New Guinea. *Z. Geomorph. Suppl.*, **18**, 92–104.

Pandey, M.R., Tandukar, J.P., Avouac, J.P., Lavé, J. and Massot, J.P. 1995. Interseismic strain accumulation on the Himalayan crustal ramp (Nepal). *Geophys. Res. Lett.*, **22**, 751–754.

Pearce, A.J. and Watson, A.J. 1986. Effects of earthquake-induced landslides on sediment budget and transport over a 50-yr period. *Geology*, **14**, 52–55.

Penck, W. 1924. *Die Morphologische Analyse: Ein Kapital der Physikalischen Geologie*, Englehorn, Stuttgart.

Penck, W. 1953. *Morphological Analysis of Land Forms*, Macmillan, London.

Personius, S.F. 1995. Late Quaternary stream incision and uplift in the forearc of the Cascadia subduction zone, western Oregon. *J. Geophys. Res.*, **100**, 20 193–20 210.

Pickup, G. 1977. Simulation modelling of river channel erosion. In: K.J. Gregory (ed.) *River Channel Changes*, Wiley, Chichester, pp. 47–60.

Pinet, P. and Souriau, M. 1988. Continental erosion and large scale relief. *Tectonics*, **7**, 563–582.

Platt, J.P. 1986. Dynamics of orogenic wedges and uplift of high-pressure metamorphic rocks. *Geol. Soc. Amer. Bull.*, **97**, 1037–1053.

Raymo, M.E., Ruddiman, W.R. and Froelich, P.N. 1988. Influence of Late Cenozoic mountain building on ocean geochemical cycles. *Geology*, **16**, 649–653.

Schmidt, K.M. and Montgomery, D.R. 1995. Limits to relief. *Science*, **270**, 617–620.

Scholz, C.H., Dawers, N.H., Yu, I.Z. and Anders, M.H. 1993. Fault growth and scaling laws: preliminary results. *J. Geophys. Res.*, **98**, 21 951–21 961.

Schumm, S.A. 1960. The shape of alluvial channels in relation to sediment type. *U.S. Geol. Surv. Prof. Pap.*, **352-B**, 17–30.

Schumm, S.A. and Hadley, R.F. 1961. Progress in the application of landform analysis in studies of semiarid erosion. *U.S. Geol. Surv. Circ.*, **437**, 1–14.

Searle, M.P. 1995. The timing of metamorphism, magmatism, and cooling in the Zanskar, Garhwal, and Nepal Himalaya. *J. Geol. Soc. Nepal*, **11**, 103–120.

Seeber, L. and Gornitz, V. 1983. River profiles along the Himalayan arc as indicators of active tectonics. *Tectonophysics*, **92**, 335–367.

Seidl, M.A. and Dietrich, W.E. 1992. The problem of channel erosion into bedrock. *Catena Suppl.*, **23**, 101–124.

Seidl, M.A., Dietrich, W.E. and Kirchner, J.W. 1994. Longitudinal profile development into bedrock: An analysis of Hawaiian channels. *J. Geol.*, **102**, 457–474.

Selby, M.J. 1993. *Hillslope Materials and Processes*, 2nd edn, Oxford University Press, Oxford.

Shimazu, H. 1994. Segmentation of Japanese mountain rivers and its causes based on gravel transport processes. *Trans. Japan. Geomorph. Un.*, **15**, 111–128.

Shulits, S. 1941. Rational equation of river bed profile *Trans. Amer. Geophys. Un.*, **22**, 622–630.

Simonett, D.S. 1967. Landslide distribution and earthquakes in the Bewani and Torricelli Mountains, New Guinea. In: J.N. Jennings and J.A. Mabbutt (eds) *Landform Studies in Australia and New Guinea*, Australian National University Press, Canberra, pp. 64–84.

Simpson, G.D.H., Cooper, A.F. and Norris, R.J. 1991. Late Quaternary evolution of the Alpine Fault zone at Paringa, South Westland. *N.Z. J. Geol. Geophys.*, **37**, 49–58.

Small, E.E. and Anderson, R.S. 1995. Geomorphologically driven Late Cenozoic rock uplift in the Sierra Nevada, California. *Science*, **270**, 277–280.

Snow, R.S. and Slingerland, R.L. 1987. Mathematical modeling of graded river profiles. *J. Geol.*, **95**, 15–33.

Stein, R.S., King, G.C.P. and Rundle, J.B. 1988. The growth of geological structures by repeated earthquakes, 2, field examples of continental dip-slip faults. *J. Geophys. Res.*, **93**, 13 219–13 331.

Stewart, M.D. 1995. The effect of fault growth and fault geometry change on drainage development and sediment flux in the active, extensional Basin and Range Province, USA. Unpublished PhD thesis, University of Leeds.

Strahler, A.N. 1964. Hypsometric (area–altitude) analysis of erosional topography. *Geol. Soc. Amer. Bull.*, **63**, 1117–1142.

Sugai, T., Ohmori, H. and Hirano, M. 1994. Rock control on magnitude–frequency distribution of landslide. *Trans. Japan. Geomorph. Un.*, **15**, 233–251.

Summerfield, M.A. 1991. *Global Geomorphology*, Longman, London.

Summerfield, M.A. and Hulton, N.J. 1994. Natural controls of fluvial denudation in major world drainage basins. *J. Geophys. Res.*, **99**, 13 871–13 884.

Suppe, J.S., Chou, G.T. and Hook, S.C. 1992. Rates of folding and faulting determined from growth strata. In K.R. McClay (ed.) *Thrust Tectonics*, Chapman and Hall, London, pp. 105–121.

Talling, P.J. 1995. Sedimentation and tectonic geomorphology in areas of active tectonic compression. Unpublished PhD thesis, University of Leeds.

Tippett, J.M. and Kamp, P.J.J. 1993. Fission track analysis of the Late Cenozoic vertical kinematics of continental Pacific crust, South Island, New Zealand. *J. Geophys. Res.*, **98**, 16 119–16 148.

Turcotte, D.L. 1992. *Fractals and Chaos in Geology and Geophysics*, Cambridge University Press, Cambridge.

Walcott, R.I. 1978. Present tectonics and Late Cenozoic evolution of New Zealand. *Geophys. J. R. Astr. Soc.*, **52**, 137–164.

Warburton, J. and Beecroft, I.1993. Use of meltwater stream material loads in the estimation of glacial erosion rates. *Z. Geomorph.*, **37**, 19–28.

Wieczorek, G.F. and Jäger, S. 1996. Triggering mechanisms and depositional rates of postglacial slope movement processes in Yosemite Valley, California. *Geomorphology*, **15**, 17–31.

Willemin, J.H. and Knuepfer, P.L.K. 1994. Kinematics of arc–continent collision in the eastern Central Range of Taiwan inferred from geomorphic analysis. *J. Geophys. Res.*, **99**, 20 267–20 280.

Wilson, L. 1973. Variations in mean annual sediment yield as a function of mean annual precipitation. *Amer. J. Sci.*, **273**, 335–349.

Wohl, E.E. 1992. Bedrock benches and boulder bars: Floods in the Burdekin Gorge of Australia. *Geol. Soc. Amer. Bull.*, **104**, 770–778.

Wohl, E.E. 1993. Bedrock channel incision along Piccaninny Creek, Australia. *J. Geol.*, **101**, 749–761.

Young, R. and McDougall, I. 1993. Long-term landscape evolution: Early Miocene and modern rivers in southern New South Wales, Australia. *J. Geol.*, **101**, 35–49.

Zeck, H.P., Monié, P., Villa, I.M. and Hansen, B.T. 1992. Very high rates of cooling and uplift in the Alpine belt of the Betic Cordilleras, southern Spain. *Geology*, **20**, 79–82.

PART III

MORPHOTECTONIC EVOLUTION IN INTERPLATE SETTINGS

6

Geodynamic processes in the Southern Alps, New Zealand

J. Mark Tippett and Niels Hovius

6.1 Introduction

The Southern Alps orogen, a result of late Cenozoic rock uplift and erosion of the Pacific plate margin in New Zealand, presents an unparalleled opportunity to understand relationships between tectonics, surface processes and topography in a collisional setting. This is because of the orogen's particular tectonic setting, its high and spatially variable rates of processes, and its small extent and accessibility. A significant feature of the Alps is the obliquely convergent nature of the late Cenozoic motion between the Pacific and Australian plates. The orogen is asymmetric with a steeper western flank, a function of tectonic convergence from the east that provides material to be eroded by runoff from precipitation driven from the west. Both crustal mass and topography are advected westwards through the orogen towards the plate boundary, with the consequence that there may exist an equilibrium erosional geometry whereby the orogen deformation and shape is maintained under conditions of constant convergence and climate. Furthermore, orientations of major topographic features within the orogen (such as large valleys) are to some extent controlled by the mechanical deformation pattern that results from oblique, rather than normal, compression.

Rates of tectonic and surface processes are high and spatially variable both along and across the Southern Alps. The high process rates mean that signals are strong and measurable, and variations offer the chance to quantify relations between tectonics, surface processes and topography over a range of values. The amount of late Cenozoic erosion varies systematically over the orogen from ~2 to 20 km, and rates of modern rock uplift and erosion vary between 2 and 12 mm a^{-1}. These spatially variable amounts and rates of rock uplift and erosion are clearly reflected in the character of the topography both across and along the orogen.

The advantage of the small size of the Southern Alps (300 × 100 km) is that landforms and processes can be measured at a relatively finer resolution than for a larger orogen, and spatial structures of tectonic and surface processes better constrained, given equal scientific resources. Due in part to its accessibility, there is a large and rapidly growing body of empirical information about the orogen derived from a wide range of geological, geophysical, thermochronological and geomorphological methods. This information has also made the Southern Alps an excellent test site for numerical models of mechanical deformation and erosion during continental collision. Nevertheless, in spite of the orogen being one of the best natural laboratories in which to examine endogenic–exogenic interactions, fundamental problems concerning its evolution remain unsolved and continue to generate intense geoscientific interest and excitement.

6.2 Framework of the South Island Collision Zone

6.2.1 TECTONIC SETTING OF THE SOUTHERN ALPS

The plate kinematics of the southwest Pacific are characterized by relative motions between the Australian,

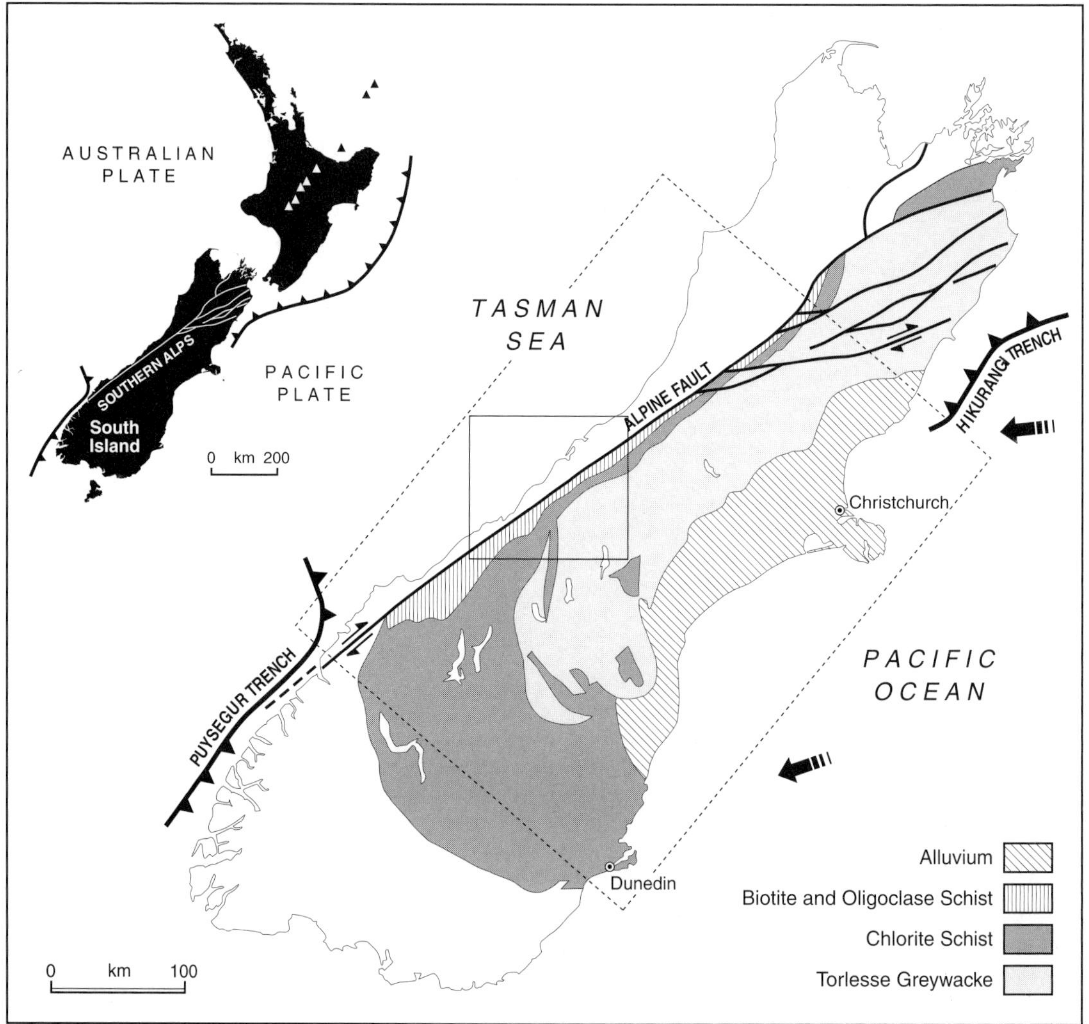

Figure 6.1 *The tectonic and geological setting of the Southern Alps. Arrows show directions of relative plate motion (Australian plate fixed). The dashed box shows the area covered by Figures 6.2 and 6.6, and the solid line box shows the area covered by Figure 6.3*

Pacific and Antarctic plates (Stock and Molnar, 1982; Kamp and Fitzgerald, 1987; Sutherland, 1995). South of New Zealand, a triple junction joins the Australian–Pacific plate boundary with the Indian Ridge and the Pacific–Antarctic Ridge. From here, the Australian–Pacific plate boundary runs northeastward and cuts through New Zealand's South Island, which contains a 100–200-km-wide zone of deformation due to late Cenozoic oblique convergence between the two plates (Walcott, 1978). The Southern Alps orogen, a product of this convergence, is located in central South Island at the leading edge of the Pacific plate (Figure 6.1).

The Alpine fault is the boundary between the Australian and Pacific plates in central South Island, and links opposite dipping subduction zones along the east coast of North Island and southern Fiordland. At its northern end, the Alpine fault branches into a number of strike-slip faults of the Marlborough system. This system converges in the Hikurangi Trench, marking active subduction of the Pacific plate under Australian plate continental crust. In the south of the

South Island, the Australian plate subducts beneath Fiordland, along the Puysegur Trench.

In the 1950s, Wellman demonstrated the dextral strike-slip character of the Alpine fault, recognizing a 480 km horizontal separation between similar Permian sequences in Nelson and West Otago. Later studies (Carter and Norris, 1976; Kamp, 1986; Cooper *et al.*, 1987) showed a late Eocene–early Oligocene inception of the plate boundary and the development of the Alpine fault as a through-going transform at around 23 Ma ago. During its initial ~10 Ma history, the Alpine fault was essentially a transcurrent structure (Kamp, 1986). Compression along the plate boundary has long been thought to have started at about anomaly 5 (9.8 Ma ago), following a change in the migration direction of the finite rotation pole of the Australian–Pacific plate pair towards the southwest (Stock and Molnar, 1982). However, Sutherland (1995, 1996) has suggested that inception of the convergent component dates back to 11–12 Ma ago, with an acceleration at anomaly 3 (5 Ma ago). This acceleration in the shortening component may be related to a change in angular velocity between the Pacific and Antarctic plates at 5.9 Ma ago, as noted by Cande *et al.* (1995). In addition, the close agreement between the anomaly 3 finite pole of Sutherland (1995) and the Australian–Pacific instantaneous pole suggests that the relative plate motion has not changed significantly since 5 Ma ago (Sutherland, 1995). The current plate displacement vector (DeMets *et al.*, 1990, 1994) for the central segment of the Alpine fault is 071 ± 3°, with an average rate of 38.5 ± 3 mm a^{-1}. This motion is obliquely convergent on the plate boundary, with a rate normal to the average fault strike (055°) of 11 ± 2 mm a^{-1}, and a fault parallel rate of 37 ± 2 mm a^{-1}.

The total amount of dextral slip on the Alpine fault since the onset of convergence has been estimated by Stock and Molnar (1982) at 330 ± 110 km, of which a minimum of 100 km has occurred since 3.6 ± 0.5 Ma ago (Sutherland *et al.*, 1995). The latter yields an average slip rate since 3.6 Ma ago of 27 mm a^{-1}. Quaternary slip rates on the Alpine fault (Berryman *et al.*, 1992; Cooper and Norris, 1995; Sutherland and Norris, 1995) converge on the same value. Thus, there is a discrepancy between the rate of relative plate motion and the slip observed on the main structure of the plate boundary. A range of studies (e.g. Walcott, 1978, 1984; Bibby *et al.*, 1986; Pearson, 1991, 1994) have demonstrated that only two thirds of the predicted (NUVEL-1A) relative plate motion is accommodated near the Alpine fault, with the remainder being distributed principally across a wide zone further east.

Since 5 Ma ago, 60 ± 30 km of shortening has occurred across the plate boundary in South Island. Rates between 10 Ma and 5 Ma ago were probably slower at 40 ± 30 km, giving a total shortening of 100 ± 40 km since convergence started (Walcott, 1994). The most recent calculated finite rotations (Sutherland, 1995) based on new Geosat data suggest that ~40 km of shortening has occurred since 5 Ma ago, and that prior to this time the motion in central South Island was almost pure strike-slip. Allis (1986) proposed that the amount of shortening varies along the plate boundary, from ~70 km in the northeast near Hurunui River, to ~100 km in the southwest near Haast. Several mechanisms have been proposed for the accommodation of this shortening. These include rock uplift, building of topography and denudation (Adams, 1980; Kamp and Tippett, 1993) (see Chapter 5), crustal thickening and formation of a lithospheric root (Woodward, 1979; Allis, 1986), underthrusting of Australian plate material, particularly towards the south of the Alps (Allis, 1986), crust–mantle or mid-crustal detachment of the Pacific plate (Wellman, 1979; Norris *et al.*, 1990; Beaumont *et al.*, 1996), flexural subsidence of flanking foreland basins (Kamp *et al.*, 1992; Beaumont *et al.*, 1996), and ductile flow (England, 1996). The relative importance of these processes in accommodating the shortening varies depending upon location in the collision zone. Many of the processes are interdependent, and they may occur simultaneously (see Chapter 3).

The Southern Alps are the surface manifestation of uplift and erosion of the leading edge of the Pacific plate during collision. They form a linear mountain belt, with a length of around 350 km, as measured from the Hurunui River in the northeast to the Arawata River in the southwest (Figure 6.2). The surface expression of the orogen is an elongated asymmetric dome, with short, steep gradients on the western flank and longer, more gentle gradients on the eastern side. In the west, the mountains rise abruptly from a coastal plain with hummocky topography of glacial origin. They form a range whose main divide runs parallel to, and 10–20 km east of, the range-bounding Alpine fault; summits reach altitudes of between 2000 and 4000 m. The present culmination of the Southern Alps is at Mt Cook (Figure 6.2). To the east of the main divide, mountain heights decrease gradually, through a series of ridges separated by broad valleys, to the ~50-km-wide eastern coastal plain. The distance between the Alpine fault and the southeastern margin of basement outcrop increases southwestwards along strike from 60 km to 180 km.

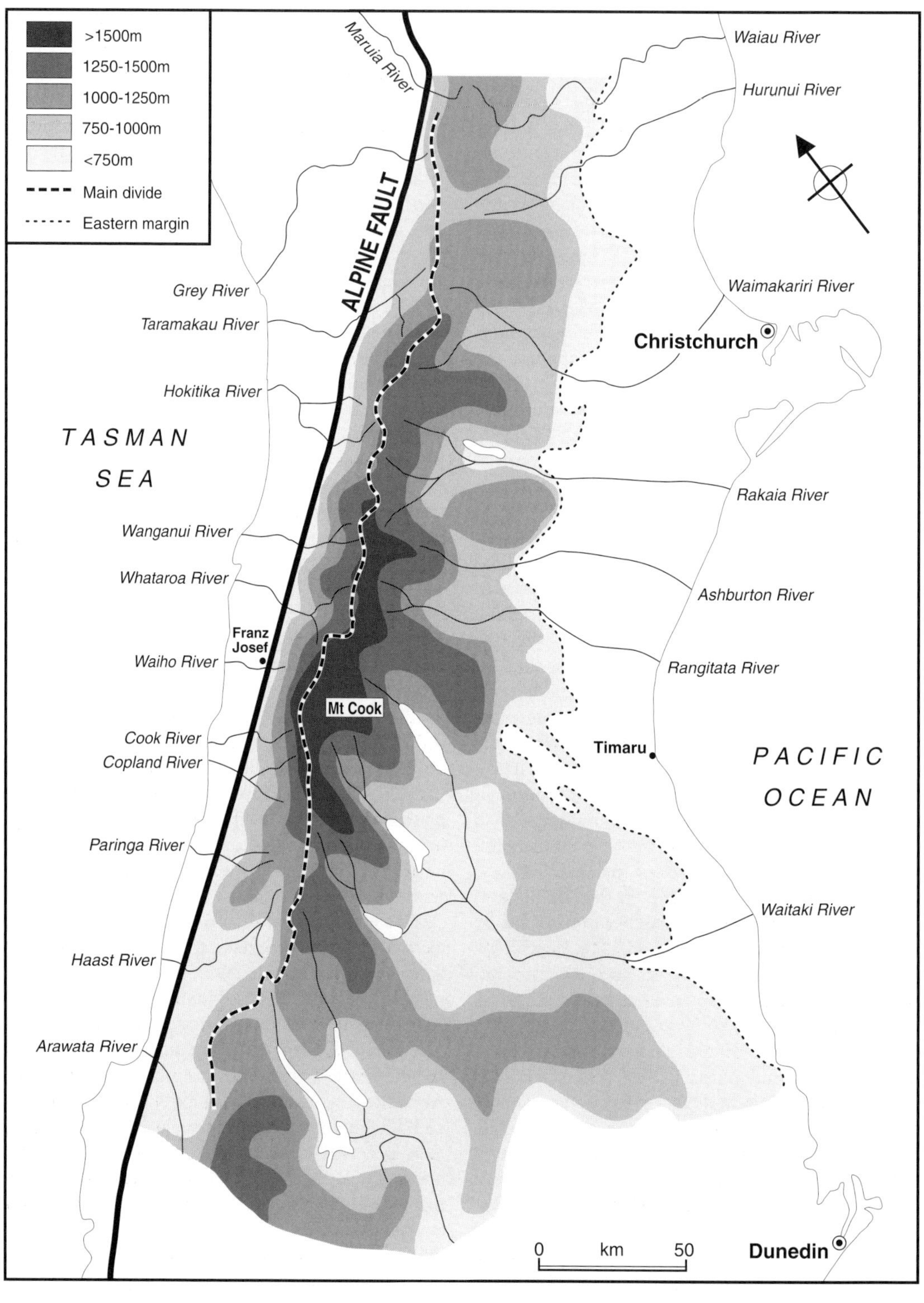
>1500m
1250-1500m
1000-1250m
750-1000m
<750m
Main divide
Eastern margin
Maruia River
ALPINE FAULT
Waiau River
Hurunui River
Grey River
Taramakau River
Hokitika River
Waimakariri River
Christchurch
TASMAN
SEA
Rakaia River
Wanganui River
Whataroa River
Ashburton River
Franz Josef
Waiho River
Rangitata River
Mt Cook
Cook River
Copland River
Timaru
PACIFIC
OCEAN
Paringa River
Waitaki River
Haast River
Arawata River
0 km 50
Dunedin

Deposits flank both sides of the orogen. A sedimentary basin lies offshore to the west of South Island, extending southwestwards from Hokitika to the Puysegur Trench. The Westland basin started to subside rapidly at around 10–15 Ma ago, with an acceleration at ~5 Ma ago (Sircombe, 1993), and it contains 2–3.5 km of shelf and slope deposits that increase in thickness close to the modern coastline. Remnants indicate that the inner margin of the basin formerly covered the area of the modern coastal strip on the edge of the Australian plate, which was uplifted and the basin inverted by reverse faulting during late Cenozoic compression (Kamp *et al.*, 1992). Adjacent to the eastern margin of the Southern Alps a smaller basin contains up to 1.5 km of fill comprising Cretaceous–early Miocene sediments that are overlain by late Miocene regressive facies and Pleistocene gravels (Field *et al.*, 1989; Hicks, 1989).

6.2.2 CLIMATE

The Southern Alps form a barrier across the prevailing northwest wind, which carries moist air masses off the Tasman Sea, forcing spectacular precipitation rates on their leading flank. Precipitation distribution over the mountains in a function of elevation and horizontal distance from the main divide (Griffiths and McSaveney, 1983a). Mean annual precipitation varies across the mountain belt, from ~5500 mm at the western edge, rising to a maximum of up to 15 000 mm on the crest of the first major range, then declining through a secondary maximum at the main divide to reach ~900 mm in the eastern foothills of the Alps (Griffiths and McSaveney, 1983a). Snow makes up part of this precipitation, the proportion increasing with altitude. Ivory Glacier, at 1400–1800 m, receives about 25% of its precipitation as snow (Anderton and Chinn, 1978). On the crest of the Southern Alps at Mt Cook, the end of summer glacier snowline lies at an altitude of about 1800–2200 m.

Along the west flank of the Southern Alps, precipitation maxima increase from 12 000 mm a^{-1} in the Hokitika catchment to around 15 000 mm a^{-1} at Franz Josef, and then decrease towards 7000 mm a^{-1} at Haast (Henderson, 1993). This side of the mountain belt is characterized by frequent, intense, rainstorms, often associated with cyclonic systems off the south of the island (Wratt *et al.*, 1996). One hour and 24 hour, five-year rainfall maxima are ~70 mm and ~600 mm, respectively. These values decrease towards the east, in a pattern similar to that of mean annual precipitation (Whitehouse, 1985). The heaviest recorded rainfall in Westland peaked at 1810 mm over three days (Thompson, 1993).

6.2.3 GEOMORPHOLOGY

Whitehouse (1987, 1988) divided the central Southern Alps into three geomorphological regions: the western, the axial and the eastern Alps, each with their own distinct assemblage of geomorphic process systems and resulting landforms (see Figure 5.14). The western Alps comprise a region of unparalleled rock uplift and precipitation. Here, steep, dissected, rectilinear slopes, with a thin regolith cover, dominate the landscape. Valley cross-sections are usually sharply V-shaped, with sides frequently steeper than 45° over much of their height. An extremely dense network of channels results in a drainage density of probably around 50 km km^{-2} (Griffiths and McSaveney, 1983a). The morphology is predominantly erosional, in a landscape not dissimilar to that of some mountain areas of the humid tropics (see Chapter 7). Ongoing fluvial incision maintains steep slopes, and the main rivers transport much of the material eroded from the valley sides out of the mountain belt without major delay. Valley sides are eroded predominantly by bedrock and debris slides of all length scales (Hovius *et al.*, 1997) (see Chapter 5). Long avalanche scars with a variety of ages of vegetation cover and soils are common, and most erosion occurs during high-intensity rainstorms (Whitehouse, 1985).

All the main valleys in the Westland region contained Pleistocene glaciers. Many of these reached below present-day sea level during the Otiran glaciation, which ended about 14 ka ago (Suggate, 1965). Remnant morphological elements are restricted mainly to glacially-shaped headwaters and low-relief crests at higher elevations, and some fluvio-glacial terraces at the outlets of mountain valleys. Present valley glaciation is minor and deposits from more extensive Holocene glaciers have been removed or covered. However, glaciers occur along the crest of the central Alps, and in the Mt Cook area large névées dominate

Figure 6.2 *The large-scale topography of the Southern Alps, showing mean surface elevation (in metres), and the locations of the main divide (from Whitehouse, 1987), the major river valleys and lakes, and the eastern margin of basement outcrop*

the upper valleys. From these, glaciers descend to an altitude of 300–400 m on the western side of the main divide. Many of these glaciers are still retreating from an advance at the end of the 19th century (Mosley *et al.*, 1984).

The terrain flanking the main divide is heavily glaciated, and consists of glacially eroded bedrock forms dominated by horns, cirques and U-shaped valleys. Holocene glacial and fluvio-glacial deposits are present in many headwaters. Rock falls, landslides and debris flows have modified many of the glacially steepened slopes, and some ridge crests bear the scars of kilometre-scale slope failure; for instance, a rock avalanche with an estimated volume of 14×10^6 m^3 lowered the summit of Mt Cook by 10 m in December 1991 (McSaveney *et al.*, 1992). Below about 2000 m, gravity transported material accumulates on rectilinear talus slopes, which are gradually being undercut by white-water streams.

The eastern Southern Alps are in a rain-shadow and have experienced lesser amounts of late Cenozoic rock uplift and erosion. Here, mountain ranges, with elevations slightly over 2000 m and lengths of several tens of kilometres, are separated by 2–10-km-wide braid plains. Pleistocene and Holocene cirques are common above 1500 m (Kirkbride and Matthews, 1997). At this elevation bedrock outcrop is interspersed with angular scree. Regolith thickness increases with decreasing height, to a maximum of several tens of metres. Streams dissect the ranges, forming open, V-shaped valleys. In between these, ridge flanks form triangular facets, commonly mantled with Pleistocene moraines and kame terraces. Inter-range valleys retain a U-shape, although this glacial imprint lessens eastward. Extensive deposits of Pleistocene moraine and alluvium, some >300 m thick, occur in all these valleys. Outlets of many tributary valleys are occupied by large, fossil, alluvial fans. Believed to have formed in the early post-glacial period, they are being entrenched at present and constitute an important source of sediment in this region.

6.2.4 LITHOLOGY AND STRUCTURE

The Alpine fault separates the Karamea terrane in the west from the Torlesse terrane in the east. The former consists of early Palaeozoic low-grade metasediments, later intruded by granites and locally metamorphosed to amphibolite facies. The latter comprises Permo-Triassic greywacke and its metamorphosed equivalent, the Haast Schist, which is subdivided geographically into Otago schist in the south and Alpine schist further north. From southeast to northwest, the basement rock exposed in the Southern Alps passes from prehnite–pumpellyite grade greywackes, through pumpellyite–actinolite facies, to greenschist facies rocks of the chlorite, biotite and garnet zones, and finally to mid-amphibolite grade rocks along the central segment of the Alpine fault (Figure 6.3) (Mason, 1962). Cenozoic cover rocks are preserved in isolated pockets around the southeast margin of the orogen, and in the adjacent Canterbury plains, while much of the region to the west of the Alpine fault is covered by glacial, fluvio-glacial and fluvial sediments.

At the large scale, the Alpine fault runs essentially as a single straight structure along the edge of the Southern Alps, striking northeastwards (055°). In detail, however, the fault is not straight, but consists of a number of 2–5-km-long, 020–050°-striking segments that are dominated by oblique overthrusting, linked by faults of a more strike-slip nature (Allis, 1981; Norris *et al.*, 1990; Norris and Cooper, 1995). The strike-slip segments strike 065–090°, close to the orientation of the plate motion vector in this area. Hangingwalls of the overthrusts expose sections through basal cataclasites grading up through fractured mylonites into fairly intact mylonites, derived at depth from overlying schists (Sibson *et al.*, 1979; Cooper and Norris, 1994). These mylonites dip fairly consistently at around 50°E, and reach a thickness of up to 1 km (Norris *et al.*, 1990).

Both the schist and the greywacke have undergone multiple phases of deformation prior to the late Cenozoic, and exhibit complex patterns of folding and faulting. Within the schists a few kilometres east of the Alpine fault, slip has occurred on numerous faults, approximately parallel to the range-bounding structure (Adams, 1979). Findlay (1979, 1987) showed that the schists display Cenozoic regional-scale folds with amplitudes that exceed 3 km. In these folds lie steeply dipping ductile shear zones that separate kilometre-scale pod-shaped domains of low-dipping limbs within which Mesozoic structures are preserved.

Cox and Findlay (1995) identified two regional-scale structural domains in the main divide area of the central Southern Alps, separated by a major fault system that runs parallel to the Alpine fault. Hangingwall rocks to the west comprise intensely sheared and faulted prehnite–pumpellyite facies greywackes and semi-schists, dipping 50–60°W. The footwall to the east is characterized by less deformed prehnite–

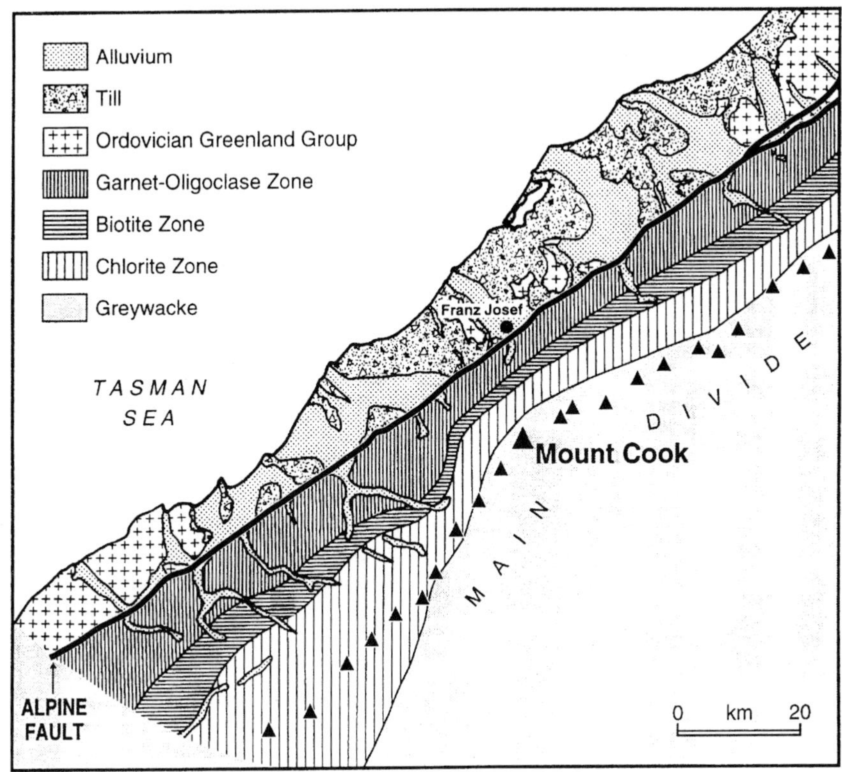

Figure 6.3 Geology of the central part of the Southern Alps in the vicinity of Franz Josef and Mt Cook

pumpellyite facies greywackes that exhibit kilometre-scale folds. On a smaller scale, the main divide is segmented by 5–15-km-long N-striking oblique-reverse faults, linked by shorter NE-striking oblique strike-slip structures.

To the east of the main divide, the greywacke strata exhibit complex patterns of folding, faulting and intense jointing (Spörli, 1979; Findlay and Spörli, 1984). In the Torlesse and Otago Schist, two sets of faults can be identified (Norris, 1979), one oriented NE–SW that dips NW and another oriented NW–SE that dips NE (Figure 6.4). The spatial distribution of the two sets of faults is characterized by areas being dominated by one or the other (Norris *et al.*, 1990). Many of these faults originated as normal faults during the Mesozoic and have been reactivated as reverse faults during late Cenozoic deformation (Norris, 1979; Spörli, 1979). Slip rates on active faults tend to be relatively low (Berryman, 1979). The eastern range front in Canterbury is marked by a zone of N-striking reverse faults which offset the basement by ~2 km. In the fold-thrust belt of Otago, the widest part of the mountain belt, basin and range topography has developed across NE-striking faults.

6.2.5 SEISMICITY

A map of shallow seismic events (Figure 6.5) for the area of the Southern Alps shows a pattern of diffuse seismicity to the east of the Alpine fault. There is subdued activity along the central segment of the Alpine fault between Lewis Pass in the north, and Jackson Bay in the south. Seismicity to the east of the fault is lowest, and mapped faults are least dense, in the central Alps where mountain elevations, uplift and erosion are highest. To the southwest and northeast, deformation appears to be more dispersed and rock uplift and erosion rates are lower.

Between 1991 and 1994, there were on average nine earthquakes per year of magnitude 2.5–3.6 along the Alpine fault (Eberhart-Phillips, 1995). However, since

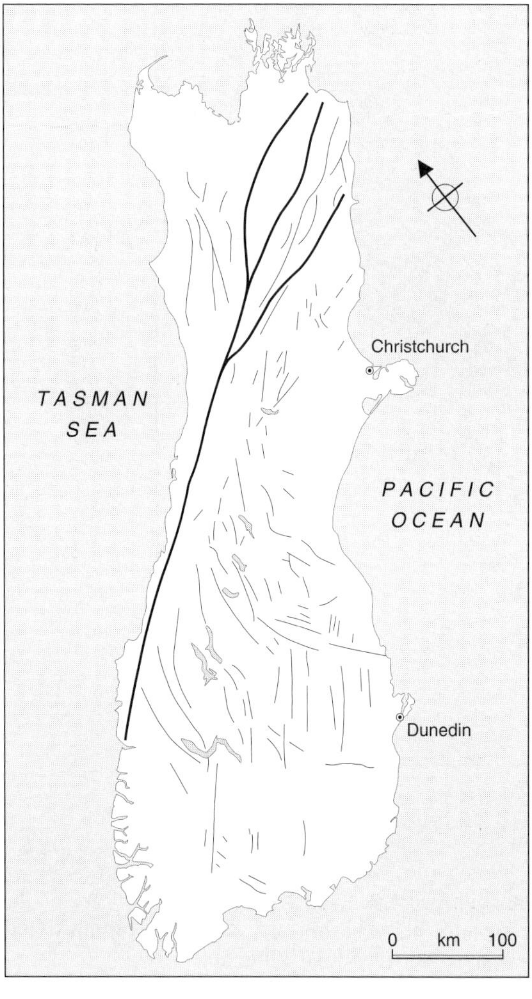

Figure 6.4 *Map of South Island showing the major mapped faults (complied using data from Norris (1979), Spörli (1979), Craw (1985), Norris* et al. *(1990), Berryman* et al. *(1992) and Cox and Findlay (1995))*

European settlement in New Zealand began in about AD 1840, there has been no rupture of the onshore trace of the fault (Berryman *et al.*, 1992). Sag pond sediments at the southern end of the fault, in northern Fiordland, indicate a major seismic event at 250 ± 50 a BP (Cooper and Norris, 1990). Offsets of geomorphic features of close to 10 m, or multiples thereof, have been found along various sections of the Alpine fault (Hull and Berryman 1986; Berryman *et al.*, 1992). Such offsets are indicative of earthquakes of magnitude >7.5.

6.2.6 DEEP LITHOSPHERIC STRUCTURE AND THERMAL REGIME

Outside the area of the Southern Alps, the crustal thickness at the northeast coast near Christchurch, as determined by microseimsic methods, is ~27 km (Reyners and Cowan, 1993). The deep structure here is interpreted as comprising an 11-km-thick seismic upper crust, a 9-km-thick seismic middle crust, and a 7-km-thick seismic lower crust. The lower crust may consist of Palaeozoic oceanic crust, upon which Torlesse greywacke sediments were deposited in the Triassic and later metamorphosed to schist. At the coast near Timaru, the crust may be ~35 km thick (Smith *et al.*, 1995). Unfortunately there are no reliable seismic determinations of crustal thickness within the zone of crustal deformation underlying the Southern Alps.

The area of the Southern Alps is characterized by a large negative Bouguer gravity anomaly (Reilly and Whiteford, 1979). The anomaly is displaced to the east of the highest elevations of the Alps, it trends away from the Alpine fault particularly in the south, and it is not present in the northern part of the Alps. The presence of the anomaly has been interpreted as thickening of continental crust into a root (Woodward, 1979; Allis, 1981, 1986). Woodward (1979) modelled a maximum crustal thickness of 50 km underneath the Southern Alps by assuming an original 30-km-thick crust, but Allis (1986) calculated a 35-km-thick crust from an initial thickness of 25 km. An ongoing major geophysical programme (Henyey *et al.*, 1993) will improve knowledge of subsurface deformation in the collision zone, and the nature of the relationships between features and processes of the mantle, crust and surface.

Rapid uplift and erosion of the Southern Alps has significantly perturbed the subsurface thermal regime. Earlier models of the thermal evolution of the crust predicted near-surface geothermal gradients of 70–200°C under uplift and erosion rates of 10 mm a^{-1} sustained for ~2 Ma (Allis *et al.*, 1979; Koons, 1987). More recent modelling (Allis and Shi, 1995; Shi *et al.*, 1996) has taken into account non-vertical uplift paths and the effects of crustal thickening. The results of this modelling indicate a more moderate thermal regime consistent with measured near-surface geothermal gradients. Measured heat flows, when corrected for topographic effects, are 190 ± 50 mW m^{-2} (60 ± 15°C) 4 km east of the Alpine fault at Franz Josef, and 90 ± 25 mW m^{-2} (30 ± 8°C) 6 km east of the fault at Haast (Shi *et al.*, 1996). These figures compare with a background

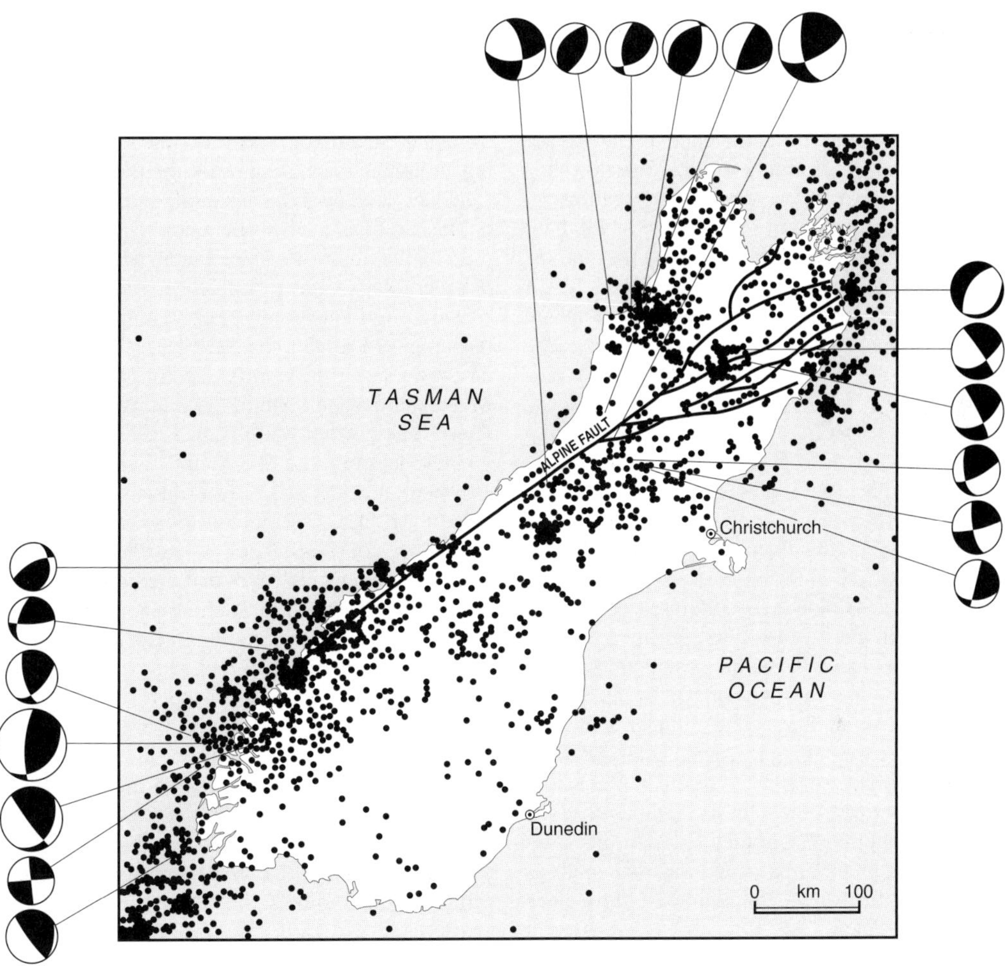

Figure 6.5 *Map of shallow (<50 km) seismic activity in South Island (from Hovius, 1995). The plot shows the epicentres of all shallow events in the ISC catalogue that occurred in the period 1964–1992. The Harvard CMT focal mechanisms for the period 1977–1994 are also shown*

heat flow of the order of 60 ± 5 mW m^{-2} measured at various localities on undisturbed crust adjacent to the Southern Alps.

The model of Shi *et al.* (1996) predicts that the 350°C isotherm is depressed to depths below ~18 km due to the effects of crustal thickening, except in the area of rapid uplift and erosion <20 km from the surface trace of the Alpine fault. This suppressed thermal regime is consistent with the pattern of subcrustal earthquakes across the central Southern Alps, which shows a cluster at depths of <20 km indicating a brittle upper/middle crust overlying an aseismic ductile lower crust (Reyners, 1987).

6.3 Uplift and Erosion of the Southern Alps

6.3.1 AMOUNT AND TIMING OF LATE CENOZOIC EROSION AND ROCK UPLIFT

A range of evidence points to a total of 18–25 km of late Cenozoic erosion adjacent to the Alpine fault in the central Southern Alps. Cooper (1980) estimated a figure of ~25 km of erosion, based on retrograde alteration of chromium kyanite in schist rocks. Within 2 km of the Alpine fault, at Franz Josef, plagioclase–garnet–biotite–muscovite geobarometry of K-feldspar zone schist implies recrystallization at, and erosion from, depths of 22–25 km (Grapes, 1995). Grapes and Watanabe (1992) had earlier estimated a slightly lower figure of 19–23 km using phengite Si geobarometry. Mineral assemblages of rocks adjacent to the Alpine fault between Franz Josef and Haast indicate that they have cooled from peak metamorphic temperatures in excess of 550°C (Grapes and Otsuki, 1983; Findlay, 1985; Grapes, 1995). However, determining the amount of erosion associated with this cooling depends upon consideration of a component of cooling before the late Cenozoic as well as the adoption of an appropriate geothermal gradient. Kamp and Tippett (1993) estimated the total amount of late Cenozoic erosion at the fault at Franz Josef to be ~18 km, by extrapolating erosion data derived from K–Ar, zircon and apatite fission-track thermochronologies (Figure 6.6).

Across the Southern Alps, sample ages derived from K–Ar, zircon and apatite fission-track thermochronometers show a pattern of decreasing age with increasing proximity to the Alpine fault (Adams, 1981; Kamp *et al.*, 1989; Tippett and Kamp, 1993). Those ages have been interpreted as representing increasing amounts of late Cenozoic cooling (by erosion) with increasing proximity to the Alpine fault. The younger ages (less than ~8 Ma) near the fault have been interpreted as reset ages, which indicate rapid and recent cooling from temperatures that exceed the 'closure' temperatures of the respective thermochronometers. Sample ages (exceeding ~8 Ma, and for apatite fission-track analysis up to ~120 Ma) further to the southeast have been interpreted as indicating cooling from various temperatures lower than the respective 'closure' temperatures (in fission-track nomenclature, from temperatures corresponding to the annealing zone) (see Chapter 4 for details on the use of thermochronology in quantifying erosional histories).

Kamp *et al.* (1989) and Tippett and Kamp (1993) used apatite and zircon thermochronology of Southern Alps basement outcrop samples to establish the spatial patterns of the amount of late Cenozoic cooling and erosion. Samples from a number of transects to the east of the main divide exhibit fission-track length distributions and older apatite ages that indicate cooling from various palaeotemperatures of between ~70°C near the eastern margin of the orogen and ~125°C just to the southeast of the main divide. Nearer the Alpine fault, zircon ages decline and the distinction between older ages and reset ages allows the identification of samples corresponding to a palaeotemperature of 265°C. These pre-erosion palaeotemperatures have been assigned to 59 samples over the Alps and, using a geothermal gradient of 27 ± 2.5°C, corresponding estimates have been made of the amount of late Cenozoic erosion at each sample location.

These estimates allow the construction of a contour map of the amount of late Cenozoic erosion over the Southern Alps (Kamp and Tippett, 1993) (Figure 6.6). The amount of erosion is greatest in the Mt Cook area, reaching ~18 km at the fault near Franz Josef, and decreases along the fault to 10 km in the northeast at Maruia River and 8 km in the southwest near Arawata River. The pattern of erosion is asymmetric, with the greatest amounts at the Alpine fault and an almost exponential decline to ~2 km at the eastern margin of the orogen. The erosion contours swing around from the 2 km contour, which is parallel to the eastern margin, to higher contours that are parallel to the plate boundary (Figure 6.6). The small amount of topography (Figure 6.2) remaining after erosion indicates that most of the uplifted rock has been removed.

Kamp *et al.* (1989) and Tippett and Kamp (1993) used fission-track data derived from the bases of exhumed apatite and zircon annealing zones to date the start of late Cenozoic cooling (and inferred uplift and erosion) of rock at various sites in the Southern Alps. The results showed that different parts of the Southern Alps started to be uplifted and eroded at different times along the orogen. Rock uplift and erosion started earlier in the Haast region to the southwest (7–8 Ma ago) than in the Arthur's region to the northeast (4–5 Ma ago).

6.3.2 RATES OF LATE CENOZOIC COOLING, EROSION AND UPLIFT NEAR THE ALPINE FAULT

Interpretations of the amount and timing of erosion from fission-track annealing zone data are unaffected by the possibility of rising geotherms as any temperature increases necessarily post-date the setting of fission-track lengths and ages. However, interpretation of cooling in terms of rates of erosion for reset fission-track and K–Ar ages is more problematic, due to rising geotherms, non-vertical uplift paths and changes in convergence rate (see Chapter 4).

Figure 6.6 *Contour map of the amount of late Cenozoic erosion (in kilometres) of the Southern Alps (from Kamp and Tippett, 1993). The thin line that cross-cuts the erosion contours represents the position of the main divide*

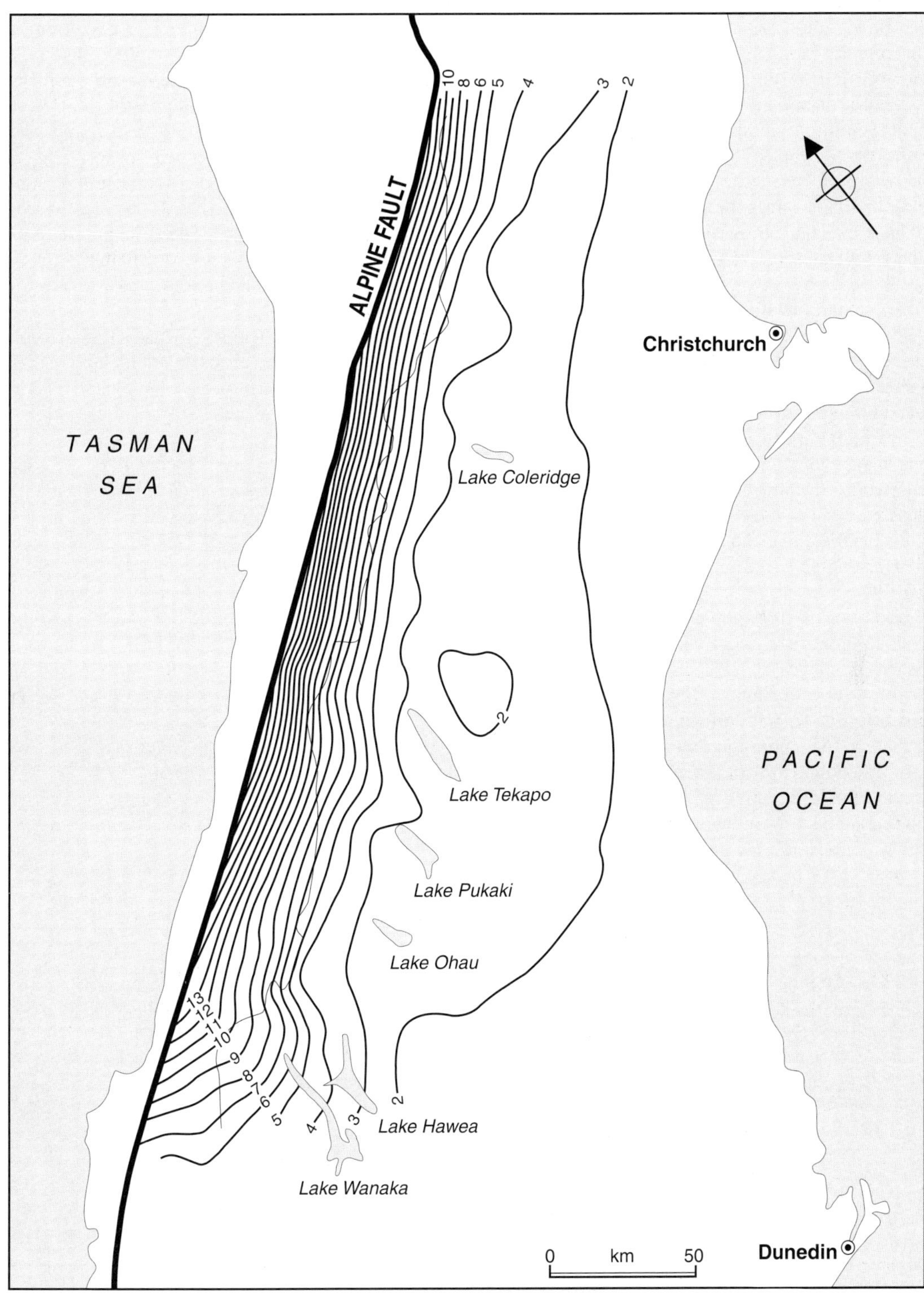
ALPINE FAULT
TASMAN SEA
PACIFIC OCEAN
Christchurch
Dunedin
Lake Coleridge
Lake Tekapo
Lake Pukaki
Lake Ohau
Lake Hawea
Lake Wanaka
0 km 50
10
8
6
5
4
3
2
13
12
11
10
9
8
7
6
5
4
3
2
2

Reset zircon fission-track ages near the Alpine fault in the central Southern Alps decrease from ~4 Ma at 17 km from the fault to <1 Ma at 1 km from the fault (Figure 6.7a). Erosion rates, calculated using these data, reach a maximum of ~10 mm a^{-1} at the Alpine fault, declining to 2–3 mm a^{-1} between 15 and 20 km from the fault (Figure 6.7b) (Tippett and Kamp, 1993). Rapid erosion has acted to enhance heat flow and to raise geotherms towards the surface for rocks <25 km from the Alpine fault trace (Allis and Shi, 1995). These erosion rates may thus be overestimates as they have probably been affected by rising geotherms, with the youngest samples (with highest erosion rates) being affected most, and the spatial trend may therefore be exaggerated (see Chapter 4). However, the thermal modelling of Shi *et al.* (1996) has suggested that the youngest zircon sample ages of ~1 Ma were at depths of ~8 km and temperatures of 240°C 1 Ma ago, giving only a slightly lower maximum erosion rate of 8 mm a^{-1} adjacent to the fault. Away from this central region of the Alps, erosion rates decrease northeastwards and southwestwards along the fault, reaching 2.5–4.5 mm a^{-1} at Hurunui and ~2 mm a^{-1} at Haast (Tippett and Kamp, 1993).

In addition to the spatial patterns of ages and inferred cooling for a single chronometer, a range of thermochronometers can be used to constrain the thermal history of a single sample or group of closely related samples. Tippett and Kamp (1993) used K–Ar muscovite, K–Ar biotite and zircon fission-track data to construct a cooling history for rocks presently exposed adjacent to the Alpine fault near Franz Josef. The curve shows rates of 22°C Ma^{-1} between 6.7 and 1.3 Ma ago, 170°C Ma^{-1} between 1.3 and 1.0 Ma ago and 240°C Ma^{-1} since 1.0 Ma ago (Figure 6.7c). Although the pattern of cooling may indicate an increase in erosion rate (driven by an increase in convergence rate) over time, it is also a function of an uplift geometry that requires faster rates of cooling nearer the fault. Rock samples have experienced uplift paths that became increasingly vertical through time as they approached the Alpine fault during westward convergence, until their exposure at the surface. The increase in cooling rate may additionally represent passage through isotherms that have become increasingly compressed at shallower depths. In this way, the pattern of cooling shown in Figure 6.7c could be consistent with erosion of

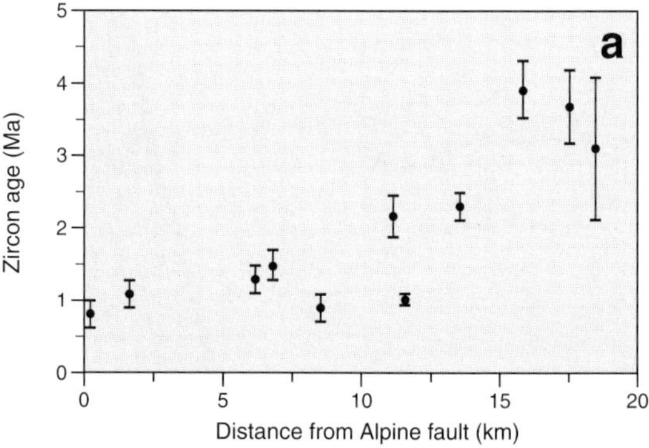

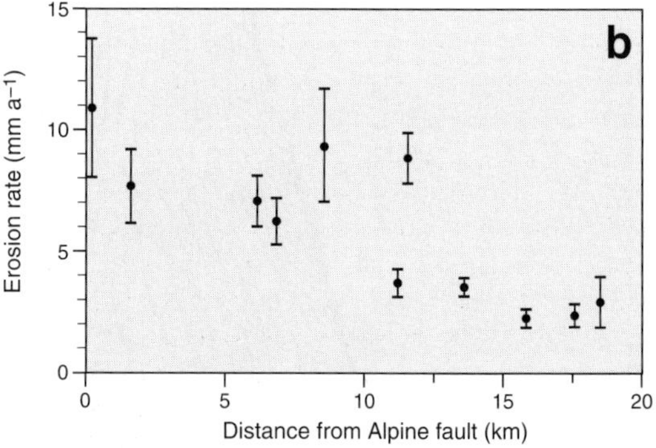

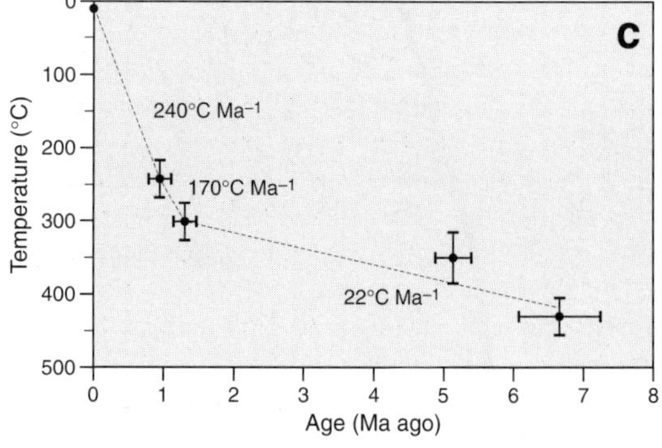

Figure 6.7 *(a) Variation in mean zircon fission-track age with distance from the Alpine fault for basement rock samples in the central Southern Alps. Data are from Kamp* et al. *(1989) and Tippett and Kamp (1993) and include data from Copland River, Franz Josef and Whataroa River. Uncertainties are 1σ. (b) Variation in erosion rate with distance from the Alpine fault in the central Southern Alps, for the same samples as in (a). The erosion rates were calculated using a geothermal gradient of 27±2.5°C and apply to the time since each sample passed upwards through the 240°C isotherm. (c) Cooling history of rocks from <5 km from the Alpine fault in the vicinity of Franz Josef and Fox Glacier (from Tippett and Kamp, 1993). The sample ages have been averaged for each thermochronometer (K–Ar muscovite, K–Ar biotite, zircon fission track). Just prior to the start of cooling at 6.5–7 Ma ago, the rocks were probably at a temperature of ~430°C (Tippett and Kamp, 1993)*

rock (at the surface location corresponding to the uplift trajectory) at a constant rapid rate rather than one that has increased over time; thermal modelling using an uplift and erosion rate of 8–10 mm a^{-1} matches the age trend of this cooling curve (Shi *et al.*, 1996).

6.3.3 MODERN RATES OF EROSION

Quantitative estimates of present-day erosion rates in the Southern Alps are limited. Basin-wide sediment yields reported by Griffiths (1979, 1981) show that rivers draining the Southern Alps to the east typically have sediment yields between ~100 t km^{-2} a^{-1} for dry intermontane areas and ~1600 t km^{-2} a^{-1} (= 0.6 mm a^{-1}) for large basins in the wetter parts of the eastern Alps (Bishop, 1984; Stirling, 1991).

Sediment yields on the western side of the main divide are greater than this by at least an order of magnitude, but few rivers in this region have been gauged systematically. The Hokitika and Haast Rivers have specific sediment yields of 17 000 and 12 700 t km^{-2} a^{-1} (= 6.5 and 5 mm a^{-1}), respectively, calculated from suspended sediment yields only. The bed load of these rivers is unknown. Sediment yield from the montane part of the Waitangitaona catchment was estimated by Griffiths and McSaveney (1986) at 12 500 ± 800 t km^{-2} a^{-1} (= 4.5–5 mm a^{-1}), from a 17-year deposition record on the alluvial fan at the range front. Measurements from Ivory basin, a small, partially glaciated cirque basin in the upper area of the Waitaha catchment, gives an average specific sediment yield of 14 900 ± 700 t km^{-2} a^{-1} (= 5–6 mm a^{-1}) for 1976–1980 (Hicks *et al.*, 1990). The nearby, non-glaciated, Cropp basin has significantly higher sediment yields, a three-year gauging campaign in this basin, which forms part of the upper Hokitika catchment, indicating suspended sediment yields of 29 600 ± 2500 t km^{-2} a^{-1} (= 10–12 mm a^{-1}) (Griffiths and McSaveney, 1983b; Basher *et al.*, 1988). Hovius *et al.* (1997) have demonstrated that the bulk of the material leaving the Southern Alps towards the west is derived from landsliding (see Chapter 5). Using a 60-year landslide record for the region between the Hokitika and Haast Rivers, they calculated catchment-wide denudation rates of 5–12 mm a^{-1}. These rates, and those calculated from river sediment loads, correspond well with estimates of current rock uplift rates near the Alpine fault. A steady-state topography has been inferred from this accordance (Adams, 1980), although care must be exercised in an unqualified acceptance of this interpretation (see Chapter 5).

6.3.4 MODERN RATES OF ROCK UPLIFT

Post-glacial rock uplift rates in the South Island have been mapped by Wellman (1979). This map reveals the highest uplift rates to be located at the Alpine fault near the Mt Cook area (>10 mm a^{-1}), decreasing along the fault to <4 mm a^{-1} at the northeastern and southwestern extremities of the Southern Alps. The uplift rates also decline to the southeast across the orogen to reach 0.1 mm a^{-1} at the southeastern limit of bedrock outcrop. Much of the evidence used to construct the map has been presented by Adams (1980). Both along and adjacent to the Alpine fault, Wellman (1979) and Adams (1980) derived uplift rates from displaced and deformed river terraces, and upthrown marine benches and shell beds. The values they found range between 2 and 15 mm a^{-1}, with the highest rates between the Hokitika and Whataroa Rivers. In the eastern flank of the Southern Alps, Wellman (1979) identified a series of post-glacial stranded lake shorelines. From the tilting of these surfaces, he calculated relative vertical displacements, increasing from 0.25 mm a^{-1} near the eastern edge of the orogen, to 2.0 mm a^{-1} closer to the main divide.

More recent work indicates post-glacial rock uplift rates approaching 7 mm a^{-1} at several locations immediately east of the Alpine fault. At Paringa River, the elevations of 13-ka-old marine silts and an 11-ka-old forest horizon, assumed to have been formed at least 5 m above sea level, imply rock uplift rates of 7 mm a^{-1} (Simpson *et al.*, 1994), and further south, at Okuru River, uplifted 10-ka-old marine silts yield a rate of 5 mm a^{-1} (Cooper and Bishop, 1979).

6.3.5 MODERN SURFACE DISPLACEMENT

Triangulation networks across the Alpine fault date from the 1870s, and reoccupations since the 1970s have been used to show the pattern of modern deformation across the plate boundary zone. These geodetic studies have shown that there are not only high rates of strain accumulation across the Alpine fault, but also significant rates adjacent to the fault both to the east and west (Walcott, 1978, 1979; Bibby *et al.*, 1986; Pearson, 1991, 1994). Pearson *et al.* (1995), using GPS determinations, reoccupied a triangulation and trilateration network established in 1978 across the plate boundary zone at the northern extremity of the Southern Alps. Results showed that about two thirds of the predicted (NUVEL-1A) relative plate motion is accommodated in the vicinity of the Alpine fault, with the remainder being taken up to the east. This supports slip data which also suggest that the Alpine fault takes up about two thirds of the calculated strike-slip component of the predicted plate boundary displacement (Berryman *et al.*, 1992). The remainder is accommodated by distributed deformation over a wide area, mainly to the east within the Pacific plate, but also in the Australian plate (Walcott, 1978; Reilly, 1986).

6.4 Evolution of Topographic Architecture

6.4.1 EASTERN FLANK

Koons (1989) numerically modelled the simultaneous rock uplift and erosion of a continental collision, using a diffusive mass transport model, and using as boundary conditions observations of rock uplift and erosion patterns from the Southern Alps. Areas of low rainfall, with stream spacings of ≥ 5 km and low erosional diffusivities of $<10^6$ m^2 Ma^{-1}, such as the eastern flank of the Southern Alps, were predicted to be unlikely to achieve a steady state between rock uplift rate and erosion rate. Elevations in these areas were predicted to be a function of the amount of rock uplift.

Tippett and Kamp (1995) explored the nature of the relations between amounts of rock uplift and measures of local elevation for 45 sites in seven transects across the eastern flank of the Southern Alps. Local mean elevation is linearly proportional to, and strongly associated with, the amount of rock uplift (Figure 6.8). Sites of equal amount of rock uplift in different transects have different mean elevations, which indicates the spatial variation in the effectiveness of late Cenozoic erosion along the orogen's eastern flank. The elevation and uplift data, when plotted, do not project through the origin, implying lower elevations than expected for amounts of rock uplift < 2.5 km. This may represent the rapid stripping of variable thicknesses of softer cover rocks that overlay the basement prior to uplift and erosion. Significant surface uplift of the eastern flank mountains may therefore not have started until after 1.5–2 km of rock uplift, that is, until ~1–2 Ma after the start of rock uplift and erosion.

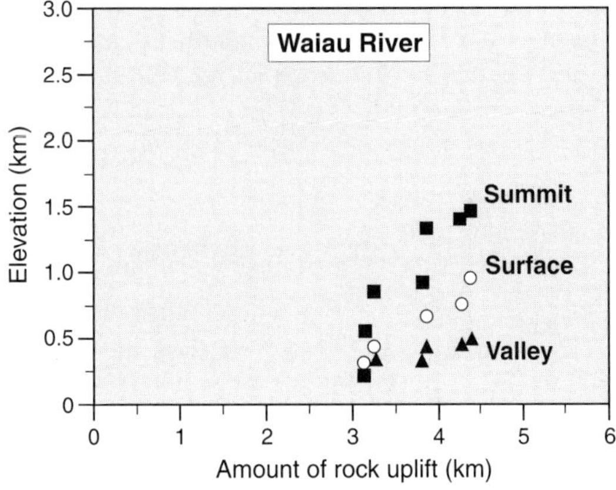

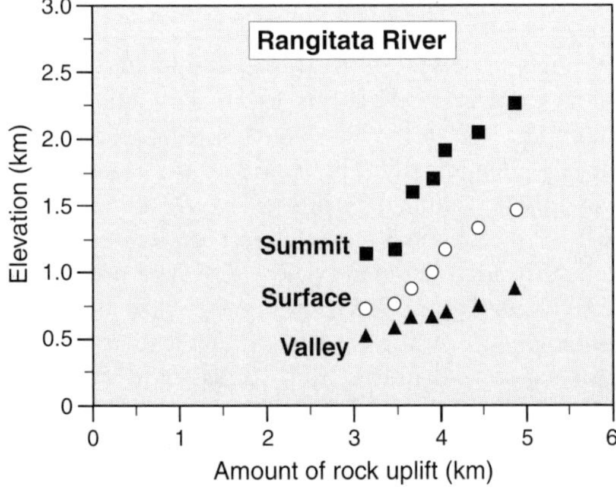

Figure 6.8 *Bivariate scatter plots of mean surface, summit and valley-floor elevations against the amount of late Cenozoic rock uplift for Waiau River and Rangitata River transects (located on Figure 6.2) to the east of the main divide in the Southern Alps (from Tippett and Kamp, 1995). Rock uplift amounts were determined using apatite fission-track thermochronology, and calculations assumed that the initial mean elevation of the surface was at sea level. The average late Cenozoic rock uplift rates for samples in these plots have been 0.5–1 mm a^{-1} since rock uplift and erosion started ~6 Ma ago (Rangitata) and ~4 Ma ago (Waiau). Elevation measures are averages for a 180 km^2 area around each fission-track sample location*

Although only tectonic uplift can contribute to an increase in mean surface elevation (England and Molnar, 1990), summit heights may additionally be raised by isostatic uplift resulting from unloading by intervening valley incision. The summit elevations of the eastern flank of the Southern Alps are identical (within error) to values for tectonic uplift (Table 1 in Tippett and Kamp, 1995). The implication is that isostatic uplift is not required in order to explain the elevations of summits – the peaks are the result of tectonic processes. These figures compare with the Himalayas and European Alps where up to 20–30% of the modern elevation of summits can be explained by isostatically compensated valley incision (Gilchrist *et al.*, 1994; Montgomery, 1994). The figures for the eastern flank of the Southern Alps are lower than these values probably on account of the low erosional intensity within valleys and their large spacing (Koons, 1989).

6.4.2 WESTERN FLANK

The results of the numerical models of Koons (1989) suggested that elevation varies in direct linear proportion to rock uplift rate, and in inverse proportion to both erosional diffusivity and the square of stream channel spacing. In one model, Koons applied an uplift pattern compatible with modern rock uplift rates (Wellman, 1979), and used west–east rainfall variations across the orogen to represent major trends in diffusivity. Steady-state elevations were predicted to be attained in the western flank where rates of rock uplift, rainfall and erosion are high.

Tippett and Kamp (1995) examined relations between the rate of erosion derived from zircon fission-track sample data and measures of local elevation for several transects across the western flank of the Southern Alps. Transects that have higher rates of erosion support higher elevations and greater relief (Figure 6.9). In transects of higher uplift and erosion rates (the Copland River, for example), elevation decreases with increasing erosion rate. However, this trend is not evident in transects of lower uplift and erosion rates, such as the Maruia River, where similar mountain elevations are associated with rates that range from 1.5 to 4.5 mm a^{-1}.

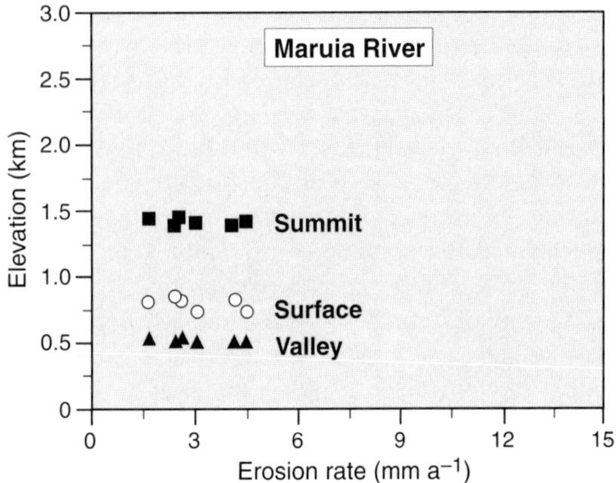

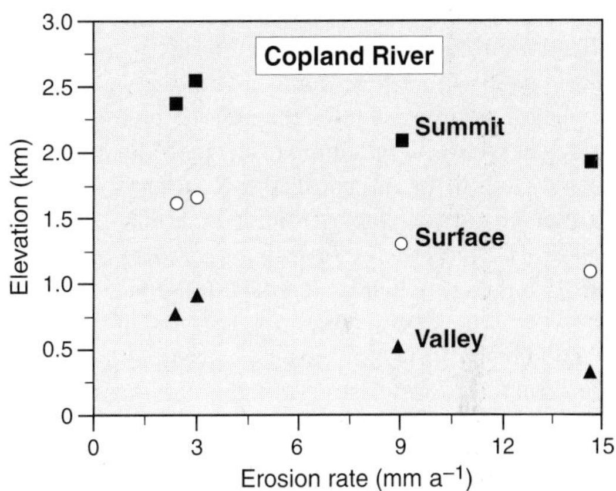

Figure 6.9 *Bivariate scatter plots of mean surface, summit and valley-floor elevations against erosion rate for Maruia River and Copland River transects (located on Figure 6.2) to the west of the main divide (from Tippett and Kamp, 1995). Erosion rates were determined using zircon fission-track thermochronology and apply to the time since each sample passed upwards through the 240°C isotherm. Elevation measures are averages for a 180 km² area around each fission-track sample location*

6.4.3 CROSS-SECTIONAL ARCHITECTURE

Adams (1980, 1985) envisaged the Southern Alps as a dynamic and self-perpetuating 'cuesta' that is constantly being eroded and simultaneously renewed by convergence (shortening) and uplift. In this model, low mountains in the east are moved closer to the fault and become higher, spikier mountains, and are themselves replaced by new lower mountains. In this scenario, the lithology and elevation of the mountains at any one location are constant, and therefore the topographic profile across the orogen remains constant over time given constant tectonic and climatic conditions. The distance from the plate boundary to the main divide, and by implication the cross-sectional asymmetry of the range, was seen as the result of a dynamic balance between the rate of eastward headward erosion of valleys on the western flank and the rate of westward crustal shortening.

Whitehouse (1988) examined the variation in geomorphology of the Southern Alps in terms of the asymmetric pattern of rates of erosion and rock uplift across the orogen. The implications of this asymmetric erosion pattern for the mechanical behaviour of a deforming collisional orogen have subsequently been explored with reference to the Southern Alps by Koons (1990), Norris *et al.* (1990) and Beaumont *et al.* (1992, 1996) (see Chapter 3 and Plate 3.6). In a departure from previous approaches to the modelling of crustal deformation during convergence, Koons (1990) assumed the indentor to be of similar elevation to the deforming plate, thereby allowing transport of material both towards (over) the indentor and away from it. In this case, a critical wedge treatment of crustal material under compression predicts that two wedges are formed with opposite-dipping topographic slopes, one facing the indentor (the 'inboard' wedge corresponding to the western flank of the Southern Alps) and one facing away (the 'outboard' wedge corresponding to the eastern flank) and separated by the main divide (Figure 6.10). The eastern wedge forms by thrusting away from the indentor, whereas the western wedge forms from movement up a basal thrust (Alpine fault) that joins the indentor surface with a lower crustal décollement. Uplifting mountains intercept moisture-laden winds and enhance the asymmetry of rainfall and erosion, thereby producing an asymmetric eroding range that has a steep western flank where rainfall and erosion rates are high, and a shallower and drier eastern flank. Continued convergence moves material through the orogen to replace the mass that is eroded. In this two-sided orogen model, the high erosion rates of the western wedge control the mechanical deformation pattern of the orogen and its shape.

Beaumont *et al.* (1992) have numerically modelled a deforming crustal wedge that is simultaneously eroded by surface processes under the rules of cellular automata, and have compared the results with the tectonics, erosion and topography of the Southern Alps (see Chapter 3 for an account of coupled tectonic–erosional models). Differences in orogen morphology are produced by the model under identical tectonic conditions but with contrasting erosion scenarios of westerly-prevailing wind/rain, and easterly-prevailing wind/rain. The gross characteristics of the Southern Alps, including the metamorphic rock outcrop pattern, the position of the main divide and the height of mountains, bear more resemblance to the first scenario, with the implication that the shape and deformation style of the orogen is controlled by the rapid denudation of its western flank by westerly-driven precipitation.

The topographic asymmetry across the Southern Alps therefore reflects the asymmetric distribution of rock uplift, precipitation and erosion rates. Hovius (1995) has used some simple geometric observations to explore the development of this asymmetry, and its variation along the orogen, focusing on the erosion of the western flank. The western flank of the central Southern Alps, between the Hokitika and Turnbull Rivers, has a more or less constant width and peak elevation (Figure 6.11). This highly linear segment is drained by a number of regularly spaced, transverse streams (Hovius, 1996), running orthogonal to the structural grain. Their valleys are characterized by steep headwaters, with narrow bedrock stream beds, and gradients decaying in a quasi-exponential fashion away from the main divide. Below this, low-gradient valley sections 5–15 km in length have slightly widened valley floors and are experiencing periodic intramontane sedimentation alternately with episodes of fluvial incision. This profile configuration may be indicative of a prolonged outpacing of rock uplift by valley lowering processes, resulting in valley head propagation (see Chapter 5). In this scenario, the position of the main divide has been shifted to the east and into the drier trailing flank of the mountain belt. From the near constant width of the western mountain front it follows that the length scale of this shift is homogeneous along the central Southern Alps. There are, however, two exceptions to this pattern.

In one location, near the southern end of the central range segment, the Haast River has breached the main divide to capture the east coast drainage of the Landsborough valley. The other deviation from the regional pattern occurs where the Southern Alps culminate in the high summits of Mt Cook and Mt Tasman, the area of which coincides with that of the minimum distance between the Alpine fault and main divide. The Mt Cook region is unique in terms of the extensive glaciation of western transverse valleys. The two long profiles in this region (Figure 6.11), for the Waiho and Fox valleys, differ substantially from the other profiles in the west flank. Both valleys have short, low-gradient sections, below overall convex upper reaches, even when corrected for glacier thickness. Such convexities indicate that the glaciers in these valleys are incapable of removing all uplifted material. At least two reasons may be identified for the difference between glaciated and non-glaciated valley long profiles in the region. One is that rock uplift rates are higher near Mt Cook; another is that the glaciers protect the valley floors from the impact of peak discharge events that cause most channel bed lowering in the fluvial valleys, implying that the local rate of glacial erosion is lower than that of fluvial incision.

The varying degree of asymmetry of the Southern Alps is illustrated in Figure 6.12, which shows the mean topography of three 10-km-wide swaths across the mountain belt. Section A, traversing the Haast-Landsborough area, shows a western flank of ~45 km wide and a mean main divide elevation of ~1650 m, giving an average gradient of 0.037 m m^{-1}. Section B, across the Mt Cook region, contains a much narrower western flank of ~15 km, with a main divide elevation of 2600 m, and yields a gradient of 0.173 m m^{-1}. The Wanganui region, crossed by section C, is an intermediate case, representative of much of the Southern Alps. The western flank here has a width of ~25 km, and a main divide elevation of ~1900 m, giving a flank gradient of 0.076 m m^{-1}. The average gradient of the eastern flank is similar in sections B and C, with values

Inboard Wedge
Outboard Wedge
MAIN DIVIDE
Erosion slope
Outcrops of deep crustal material
Zone of synthetic and antithetic thrusting
Surface geology is dominantly upper crustal material
Undeformed crust
σ_1
σ_3
$\phi = \alpha$
ϕ_b
TRANSITION ZONE
Principal stresses change orientation from outboard to inboard
Nappe emplacement and out of sequence thrusting
ϕ
ϕ_b
α
σ_1
σ_3
Indentor Plane
Décollement
Frontal Fault

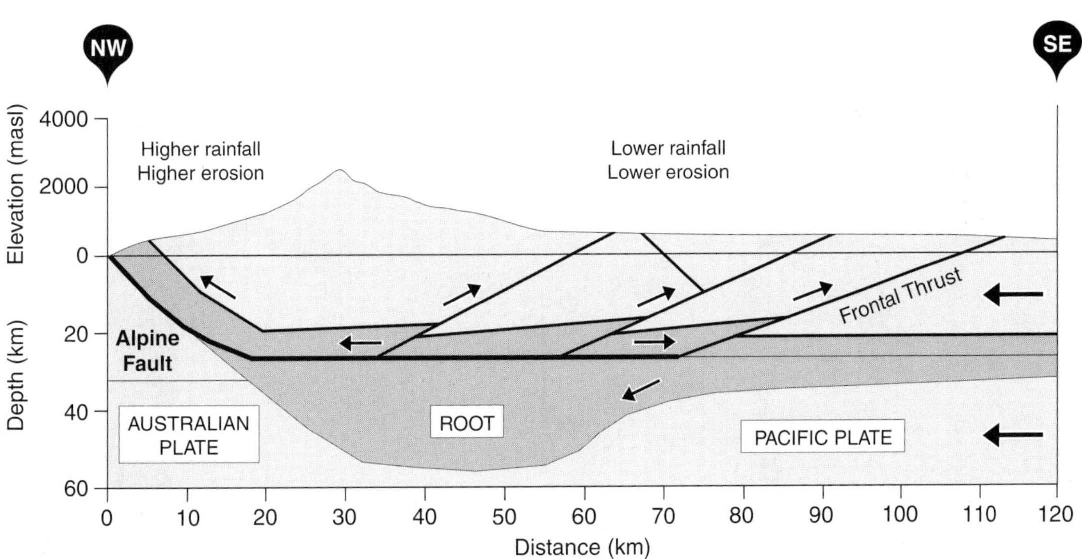

Figure 6.10 *Geometry of two-dimensional deforming zone of a two-sided orogen (top), assuming the stress and failure conditions of Coulomb wedges (from Koons, 1990). Single arrows show slip directions, and coupled arrows show the relative sense of shear at the base of the orogen. The Mohr circles show the orientations within the two wedges of the principal stress directions, which are defined by the local relations between* Φ *(the friction angle of the deforming crust),* Φ_b *(the friction angle of the décollement) and* α *(the topographic slope). Under critical wedge theory, a crustal wedge under compression deforms internally and builds a topographic slope until a critical slope is reached. At this point the entire (critical) wedge is at incipient failure and sliding occurs along the décollement. The bottom diagram is a schematic section across the Southern Alps showing its two-sided character with a lower crustal décollement (adapted from Norris* et al.*, 1990). Material above the décollement travels through the orogen from east to west and is uplifted and eroded, while material below accumulates as a root. Darker shading represents amphibolite facies rocks*

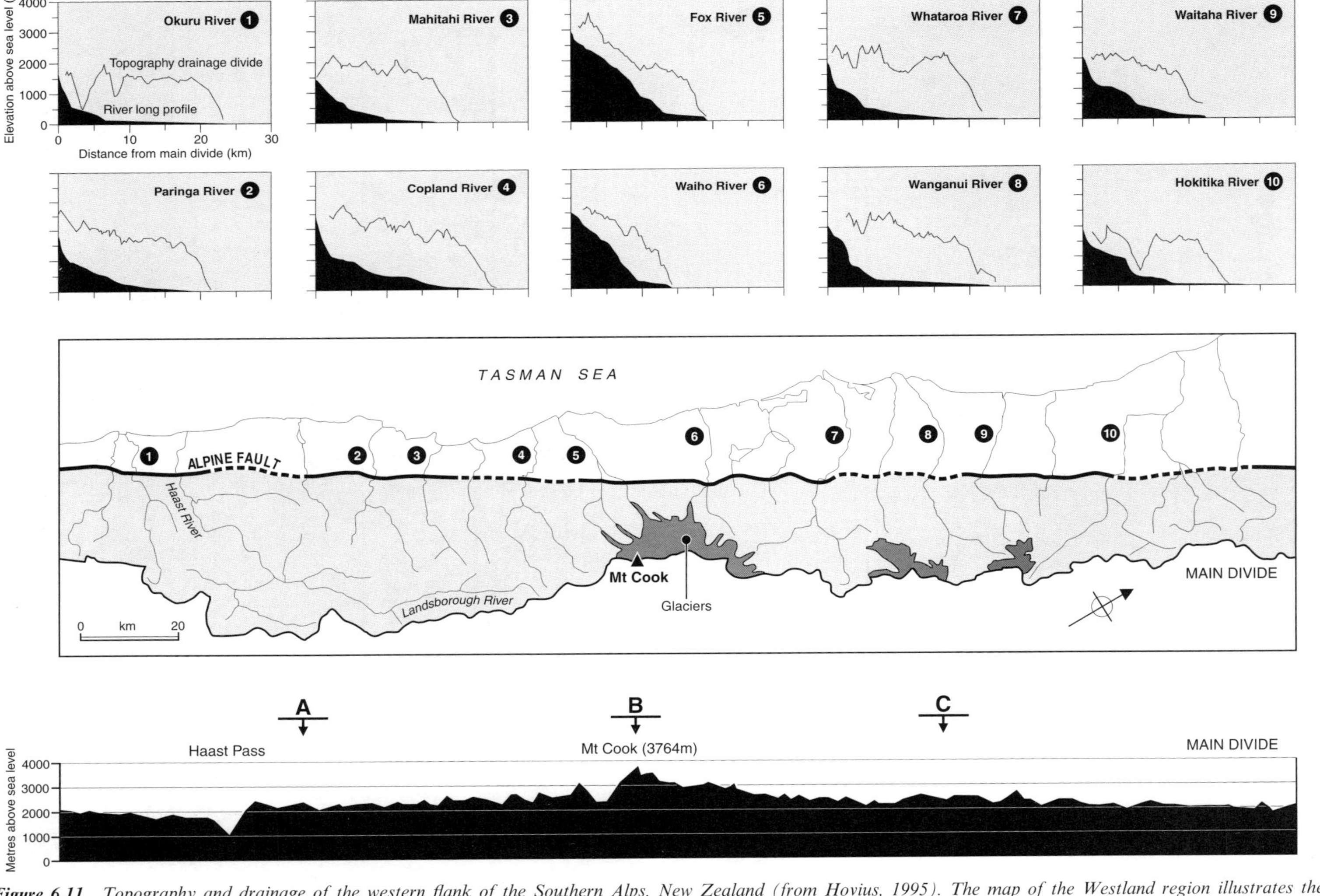

Figure 6.11 *Topography and drainage of the western flank of the Southern Alps, New Zealand (from Hovius, 1995). The map of the Westland region illustrates the uniformity of spacing between the main divide and the Alpine fault, with the two exceptions of the glaciated Mt Cook area and the Haast–Landsborough catchment. Most valleys in this region have similar longitudinal profiles, characterized by elongated minimum gradient sections. In contrast, the glaciated Waiho and Fox valleys have short and steep convex long profiles*

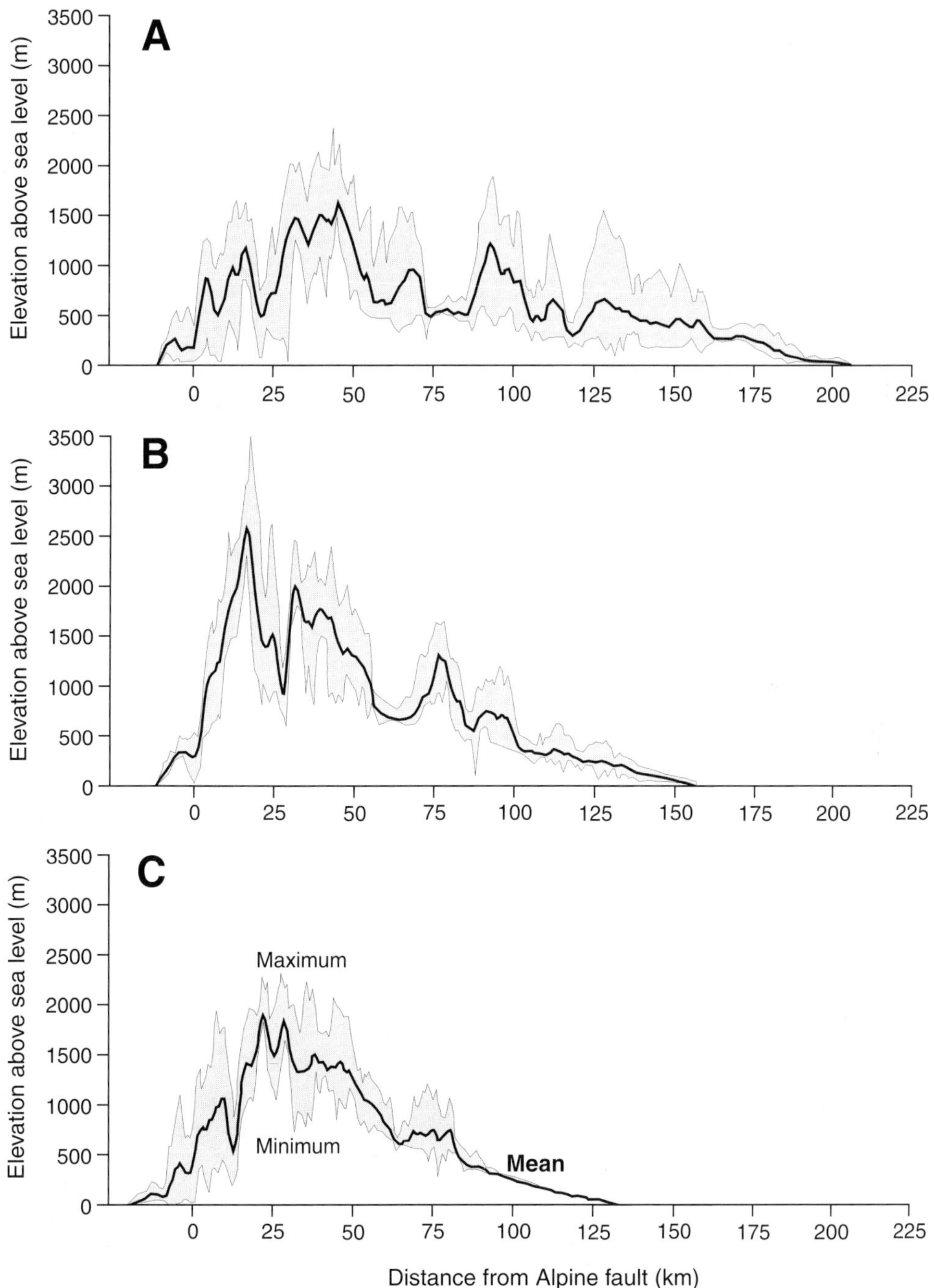

Figure 6.12 *Mean, maximum and minimum elevations of three 10-km-wide swaths across the Southern Alps, calculated from a 500 m spacing grid digital elevation model (from Hovius, 1995). The locations of the swaths are shown in Figure 6.11. Values for each elevation line represent the average of 20 data points orthogonal to the swath trend, calculated at a 500 m interval along the swath, and give reasonable approximations of the distribution of topographic mass across the orogen. The three cross-sections illustrate the varying degree of topographic asymmetry of the orogen*

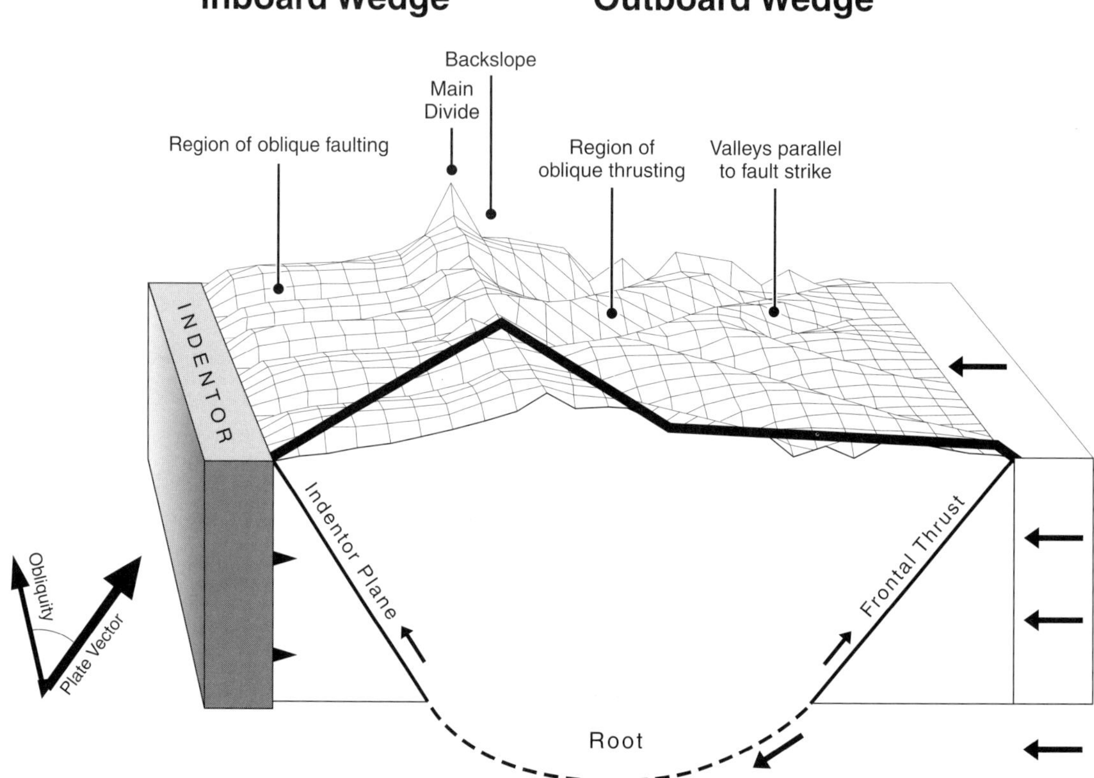

Figure 6.13 *Schematic geometry of the oblique two-sided orogen showing characteristic structural and topographic features (adapted from Koons, 1994). Mean topography (heavy line) and first harmonic of topography (wire-framed surface) are shown. Note the oblique angles between eastern flank valleys and the plate boundary*

of 0.019 m m^{-1} and 0.017 m m^{-1}, respectively. The total width of the mountain belt in the sections is around 150 km. This width increases south of section B, where the main structural trend swings from NW–SE to N–S, resulting in imbrication and folding in northwest and central Otago (Norris *et al.*, 1990). It is impossible to distinguish a proper eastern flank of the orogen in this region.

The asymmetry of the mountain belt is primarily determined by the interaction of erosion and uplift on its western side (Koons, 1990). In areas of enhanced erosion, the crest of the mountain belt is pushed eastward into progressively lower parts of the eastern flank, resulting in a lowering of the topographic asymmetry, and of the main divide elevation. Where erosion cannot keep up with rock uplift, a buildup of mass and topography occurs near the mountain front. Consequently, the main divide is heightened and shifted in the direction of the Alpine fault, steepening the western flank of the range and enhancing the topographic asymmetry (Figure 6.12).

6.4.4 IMPORTANCE OF OBLIQUE CONVERGENCE

Two-dimensional (i.e. normally convergent) mechanical models (Figure 6.10) do not predict some of the observed characteristics of the Southern Alps, particularly the oblique orientations of rivers on both flanks. Koons (1994) extended the principles developed in his earlier two-dimensional model to produce a three-dimensional critical wedge model (Figure 6.13), in which the component of horizontal shear stress reduces the topographic slope and width of an obliquely convergent orogen compared with a normally convergent orogen. In the oblique two-sided orogen, orientations of valleys are predicted to be controlled by crustal structures whose orientations are a function of convergence obliquity. In the eastern flank, intersecting sets of ridges and valleys are structurally controlled by uplifting fault blocks that entrain long subcritical rivers. In the western flank, supercritical streams incise weakened rock along steeply dipping fault planes that trend at angles to the indentor. These topographic patterns bear greater resemblance to the Southern Alps than those of two-dimensional models (Koons, 1990, 1995; Beaumont *et al.*, 1992).

Observations of the drainage of the eastern flank indicate that the valley pattern varies systematically along the orogen. Major valleys in the north and central Southern Alps, such as that of the Rakaia River, trend normal to the plate boundary. Southwestwards of the Rangitata River, the orientations of river valleys progressively swing round to become close to parallel to the plate boundary (Figure 6.2). It is not clear whether the variation in obliquity as predicted by DeMets *et al.* (1990) is sufficient to produce this orientational pattern of major river valleys. In the southwest of the Southern Alps (44.5°S, 169°E) the predicted obliquity (according to DeMets *et al.*, 1990) is 13.5°, whereas in the northeast (43.5°S, 172°E) it is 19°. Furthermore, much of the geological structure of the Southern Alps is inherited from the Mesozoic and cannot be explained purely in terms of late Cenozoic deformation. An alternative interpretation is that the drainage may be controlled by large-scale topographic gradients (Figure 6.2), which are functions of variations in late Cenozoic erosion (Figure 6.6) along the Southern Alps that are unrelated to changing obliquity.

Continued convergence, with mass moved horizontally towards the plate boundary, allows the westward advection of topography obliquely across the collision zone (Adams, 1985; Beaumont *et al.*, 1992; Koons, 1995). West-draining catchments retain their length by headward erosion and intermittent stream capture of east-draining rivers during advection. The Landsborough River, for instance, flows subparallel to the main divide in the south part of the orogen and is in the process of being transferred from the eastern to the western flank (Figure 6.11). It is one of several rivers whose trends are thought to be structurally controlled (Koons, 1994). Other longitudinal valleys, such as the Dobson and Tasman valleys, are still part of larger, east-draining catchments. They are probably locked in the structural grain, due to the low erosivity of the eastern rivers. It appears that these valleys, and the topographic and mechanical grain established in the eastern flank, are being carried obliquely through the orogen on the back of the westward-moving Pacific plate towards the erosional domain of the western flank.

6.5 Discussion and Outstanding Research Questions

6.5.1 TOPOGRAPHY

The topographic asymmetry of the orogen represents a shape determined by the interaction of tectonics and climate. The main divide is of critical significance, in that its position determines the degree of asymmetry of the orogen and therefore the proportion of the topography that is subject to the erosional processes of the western flank. The location of the main divide is a function primarily of erosion rates of the western flank, and is closest to the Alpine fault near Franz Josef where magnitudes and rates of uplift and erosion are at a maximum (Figure 6.6). The main divide may be a structurally controlled feature in the central Southern Alps (Cox and Findlay, 1995), and further investigation of its structural and geometric significance is needed. Recent work has provided much-needed constraints on erosion rates to the west of the main divide (Hovius, 1997), but although catchment-averaged techniques may indicate differences in erosion rates along the orogen, they do not reveal the spatial structure of erosion rates across the orogen between the Alpine fault and the main divide.

Knowledge and understanding of the evolution of the topography of the Southern Alps will be improved by analysis of digital elevation data, now available at a contour interval of 20 m. Measures of elevation and of its spatial structure (e.g. fractal dimension and amplitude measured at a range of scales and orientations) should yield important information about interactions between tectonics and surface processes (Chase, 1992; Koons, 1995). Topography is the result of the action of different processes at different scales (e.g. Lifton and Chase, 1992), and such an analysis may help resolve the relative importance of tectonic, structural and climatic controls on morphology. Quantification of the topography will enable the testing of the more quantitative predictions of numerical mechanical–erosional models of the Southern Alps, assuming that appropriate goodness-of-fit statistical tests can be applied.

6.5.2 UPLIFT AND EROSION

Rock uplift and erosion of the Pacific plate has accommodated only a proportion of the late Cenozoic shortening (Allis, 1986; Kamp *et al.*, 1992; Beaumont *et al.*, 1996). However, the absence of any deeper crustal rocks exposed at the Alpine fault, despite 40–100 km of shortening, has led Norris *et al.* (1990) to infer the existence of a lower crustal décollement that allows the upper ~25 km of crust to be delaminated and upthrusted. If this geometric interpretation is correct, then the actual total amount of erosion at the fault in the central Southern Alps may exceed the 18–25 km recorded by geobarometric and thermochronologic methods due to the advection of mass through the convergence zone. In a similar way, the pattern of erosion across the Southern Alps (Figure 6.6) faithfully reflects the upward passage of rocks from inferred subsurface depths, but may be an underestimate of the actual total amount of late Cenozoic erosion at any one location. According to this view, rock eroded is replaced from the east by new rock derived from the same depth, the amount of erosion at any one location as recorded by thermochronologic (or geobarometric) methods remains constant, and the pattern of erosion as shown in Figure 6.6 remains constant regardless of how much convergence occurs. The question is whether or not sufficient convergence has occurred to bring about this steady-state erosion pattern. If so, then the erosion pattern could be seen as representing an equilibrium geometry of erosional deformation under the particular conditions of tectonic and surface processes experienced by the Southern Alps during the late Cenozoic. Further, if this is the case, then the geological and topographic patterns of the orogen are a direct function of this equilibrium erosional geometry.

Shi *et al.* (1996) have identified a need for further thermochronological data (with reset ages) from the area <20 km from the Alpine fault, in order to quantify more precisely cooling and inferred erosion rates in this critical region. Further zircon and apatite samples could be obtained from this area, but the low ages and tight precisions of the dates that are required for the determination of thermal histories as suggested by Shi *et al.* (1996), of 0.2 ± 0.05 Ma for apatite samples, would be very difficult to attain in practice. In addition to generating more data, a major challenge is to disentangle the three competing effects that complicate interpretations of erosion rates using cooling curves; that is, non-vertical uplift paths, perturbations of the thermal structure of the crust, and temporal changes in convergence rates. Thermo-mechanical modelling will help to constrain the range of erosional histories consistent with cooling curves, such as the one presented in Figure 6.7c. The difficulties of inferring changes in erosion rates over time using cooling curves may hamper attempts to test proposed convergence histories derived from tectonic data (e.g. Sutherland, 1995), erosional histories postulated from sedimentological evidence (Sutherland, 1996), and the hypothesis of high erosion and uplift rates forced by global climatic change (Molnar, 1996).

6.5.3 EVOLUTION OF THE SOUTHERN ALPS

Molnar and England (1990) have suggested that the erosional evidence commonly used to infer tectonic changes in mountain belts could alternatively be caused by climatic change (cooling) of a global nature unrelated to the regional-scale changes in mountain climate induced by rising topography. In a recent discussion paper, Molnar (1996) has proposed that the rapid erosion of the Southern Alps during late Cenozoic times could be a result of a change in global climate. He suggested that early in the convergence low rates of shortening produced hills, which later were rapidly eroded due to climatic change to form high peaks and deep valleys, thereby dramatically increasing the vertical component of motion. This hypothesis demands a critical analysis of the timing of late Cenozoic tectonic, climatic, erosional and depositional changes.

In the Pacific plate, the earliest signs of erosion of the Southern Alps, as assessed using fission-track thermochronology, are 7–8 Ma ago (Kamp *et al.*, 1989; Tippett and Kamp, 1993), some 3–4 Ma after convergence started. Conglomerates containing schist and greywacke clasts derived from rocks of the western flank of the Southern Alps appear in the Westland terrestrial sedimentary record 2–3 Ma ago (e.g. Nathan *et al.*, 1986). In Waiho-1 borehole, a deep exploration well located at the mouth of the Waiho River, clasts of schist and greywacke, presumably derived from the western flank of the Southern Alps, first appear 4–4.4 Ma ago, a timing that post-dates the start of erosion. This reflects the borehole's dextral transport northeastwards past Fiordland, which, only since 4–4.5 Ma ago, enabled it to capture sediment sourced from rivers and submarine canyons that drained the southwestern Southern Alps. Clearly, however, more information is likely to be gained from other boreholes, which need to be located up to 150 km further northeast along the coast from Waiho-1 if they are to record the start of erosion of the Southern Alps. Changes in sediment provenance, as assessed from several boreholes along the Australian plate margin, could be combined with reconstructions of relative plate motions to constrain both the initiation and the development of the erosion of the western flank of the Southern Alps.

Climatic change is unlikely to have initiated erosion and rock uplift, because the start of erosion along the Southern Alps is diachronous (Kamp *et al.*, 1989; Tippett and Kamp, 1993). However, this does not preclude a later acceleration of erosion and rock uplift due to climatic change. The first marine faunal evidence of cooling in the Plio-Pleistocene sequences of central New Zealand occurs ~2.4 Ma ago (Pillans, 1991). Pliocene glaciation in northwest South Island started 2.4–2.6 Ma ago, as recorded in the Taramakau River area by tills of the Ross Glaciation that contain rock fragments derived from both east and west of the Alpine fault (Suggate, 1990). Currently, however, there is no unequivocal evidence showing that rates of erosion on the western flank of the Southern Alps since 2.5 Ma ago were any different to rates prior to that time. Moreover, it is not clear whether rates of erosion under glacial conditions are any greater than in non-glacial conditions (Whitehouse, 1987; Summerfield and Kirkbride, 1992). Further knowledge of the relative erosional efficiencies of glacial versus fluvial processes is needed, otherwise understanding of the nature of mountain evolution under conditions of changing climate is likely to remain limited (Koons, 1995). Clearly, there remain major challenges in quantifying the evolution of the Southern Alps in terms of the interactions between tectonics and surface processes.

References

Adams, C. 1981. Uplift rates and thermal structure in the Alpine fault zone and alpine schists, Southern Alps, New Zealand. *Geol. Soc. Lond. Spec. Publ.*, **9**, 211–222.

Adams, J. 1979. Vertical drag on the Alpine fault, New Zealand. *Bull. R. Soc. N.Z.*, **18**, 47–54.

Adams, J. 1980. Contemporary uplift and erosion of the Southern Alps, New Zealand. *Geol. Soc. Amer. Bull.*, **91**, 1–114.

Adams, J. 1985. Large-scale tectonic geomorphology of the Southern Alps, New Zealand. In: M. Morisawa and J.T. Hack (eds) *Tectonic Geomorphology*, Allen and Unwin, Boston, pp. 105–128.

Allis, R.G. 1981. Continental underthrusting beneath the Southern Alps of New Zealand. *Geology*, **9**, 303–307.

Allis, R.G. 1986. Mode of crustal shortening adjacent to the Alpine fault, New Zealand. *Tectonics*, **5**, 15–32.

Allis, R.G. and Shi, Y. 1995. New insights to temperature and pressure beneath the central Southern Alps, New Zealand. *N.Z. J.Geol. Geophys.*, **38**, 585–592.

Allis, R.G., Henley, R.W. and Carman, A.F. 1979. The thermal regime beneath the Southern Alps. *Bull. R. Soc. N.Z.*, **18**, 79–85.

Anderson, H. and Webb, T. 1994. New Zealand seismicity: patterns revealed by an upgraded network. *N.Z. J. Geol. Geophys.*, **37**, 477–493.

Anderton, P.W. and Chinn, T.J. 1978. Ivory Glacier, New Zealand, an I.H.D. representative basin study. *J. Glaciol.*, **20**, 67–84.

Basher, L.R., Tonkin, P.J. and McSaveney, M.J. 1988. Geomorphic history of a rapidly uplifting area on a compressional plate boundary: Cropp River, New Zealand. *Z. Geomorph. Suppl.*, **69**, 117–131.

Beaumont, C., Fullsack, P. and Hamilton, J. 1992. Erosional control of active compressional orogens. In: K.R. McClay (ed.) *Thrust Tectonics*, Chapman and Hall, London, pp. 1–18.

Beaumont, C., Kamp, P.J.J., Hamilton, J. and Fullsack, P. 1996. The continental collision zone, South Island, New Zealand: Comparison of geodynamical models and observations. *J. Geophys. Res.*, **101**, 3333–3359.

Berryman, K. 1979. Active faulting and derived PHS directions in the South Island, New Zealand. *Bull. R. Soc. N.Z.*, **18**, 29–34.

Berryman, K.R., Beanland, S., Cooper, A.F., Cutten, H.N., Norris, R.J. and Wood, P.R. 1992. The Alpine fault, New Zealand: Variation in Quaternary structural style and geomorphic expression. *Ann. Tect.*, **6** (Suppl.), 126–163.

Bibby, H.M., Haines, A.J. and Walcott, R.I. 1986. Geodetic strain and the present day plate boundary zone through New Zealand. *Bull. R. Soc. N.Z.*, **24**, 427–438.

Bishop, D.G. 1984. Modern sedimentation at Falls Dam, upper Manuherikia River, Central Otago, New Zealand. *N.Z. J. Geol. Geophys.*, **27**, 305–312.

Bishop, D.G. and Laird, M.G. 1976. Stratigraphy and depositional environment of the Kyeburn formation (Cretaceous), a wedge of coarse terrestrial sediments in central Otago. *J. R. Soc. N.Z.*, **6**, 55–71.

Cande, S.C., Raymond, C.A., Stock, J. and Haxby, W.F. 1995. Geophysics of the Pitman fracture zone and Pacific–Antarctic plate motions during the Cenozoic. *Science*, **270**, 947–953.

Carter, R.M. and Norris, R.J. 1976. Cainozoic history of southern New Zealand: An accord between geological observations and plate tectonic predictions. *Earth Planet. Sci. Lett.*, **31**, 85–94.

Chase, C.G. 1992. Fluvial landsculpting and the fractal dimension of topography. *Geomorphology*, **5**, 39–57.

Cooper, A.F. 1980. Retrograde alteration of chromian kyanite in metachert and amphibolite whiteschist from the Southern Alps, New Zealand, with implications for uplift on the Alpine fault. *Contrib. Min. Petrol.*, **75**, 153–164.

Cooper, A.F. and Bishop, D.G. 1979. Uplift rates and high level marine platforms associated with the Alpine fault at Okuru River, South Westland. *Bull. R. Soc. N.Z.*, **18**, 35–43.

Cooper, A.F. and Norris, R.J. 1990. Estimates for the timing of the last coseismic displacement on the Alpine fault, northern Fiordland, New Zealand. *N.Z. J. Geol. Geophys.*, **33**, 303–307.

Cooper, A.F. and Norris, R.J. 1994. Anatomy, structural evolution, and slip rate of a plate-boundary thrust: The Alpine fault at Gaunt Creek, Westland, New Zealand. *Geol. Soc. Amer. Bull.*, **106**, 627–633.

Cooper, A.F. and Norris, R.J. 1995. Displacement on the Alpine fault at Haast River, South Westland. *N.Z. J. Geol. Geophys.*, **38**, 509–514.

Cooper, A.F., Barreiro, B.A., Kimbrough, D.L. and Mattinson, J.M. 1987. Lamprophyre dyke intrusion and the age of the Alpine fault, New Zealand. *Geology*, **15**, 941–944.

Cox, S.C. and Findlay, R.H. 1995. The Main Divide Fault Zone and its role in formation of the Southern Alps, New Zealand. *N.Z. J. Geol. Geophys.*, **38**, 489–499.

Craw, D. 1985. Structure of schist in the Mt Aspiring region, northwest Otago, New Zealand. *N.Z. J. Geol. Geophys.*, **28**, 55–75.

DeMets, C., Gordon, R.G., Argus, D.F. and Stein, S. 1990. Current plate motions. *Geophys. J. Int.*, **101**, 425–478.

DeMets, C., Gordon, R.G., Argus, D.F. and Stein, S. 1994. Effect of recent revisions to the geomagnetic reversal time scale on estimates of current plate motions. *Geophys. Res. Lett.*, **21**, 2191–2194.

Eberhart-Phillips, D. 1995. Examination of seismicity in the central Alpine fault region, South Island, New Zealand. *N.Z. J. Geol. Geophys.*, **38**, 571–578.

England, P. 1996. The mountains will flow. *Nature*, **381**, 23–24.

England, P. and Molnar, P. 1990. Surface uplift, uplift of rocks, and exhumation of rocks. *Geology*, **18**, 1173–1177.

Field, B.D., Browne, G.H. *et al.* 1989. Cretaceous and Cenozoic sedimentary basins and geological evolution of the Canterbury region, South Island, New Zealand. *New Zealand Geological Survey Basin Studies*, **2**.

Findlay, R.H. 1979. Summary of structural geology of Haast Schist Terrain, central Southern Alps, N.Z.: implications of structures for uplift of and deformation within Southern Alps. *Bull. R. Soc. N.Z.*, **18**, 113–120.

Findlay, R.H. 1985. Metamorphic low to high 2Vx K-feldspar in kyanite zone rocks within the Alpine schists of Copland Valley, South Westland, New Zealand. *N.Z. J. Geol. Geophys.*, **28**, 77–83.

Findlay, R.H. 1987. Structure and interpretation of the Alpine schists in Copland and Cook River valleys, South Island, New Zealand. *N.Z. J. Geol. Geophys.*, **30**, 117–138.

Findlay, R.H. and Spörli, K.B. 1984. Structural geology of the Mount Cook Range and Main Divide, Hooker Valley region, New Zealand. *N.Z. J. Geol. Geophys.*, **27**, 257–276.

Gilchrist, A.R., Summerfield, M.A. and Cockburn, H.A.P. 1994. Landscape dissection, isosatic uplift, and the morphologic development of orogens. *Geology*, **22**, 963–966.

Grapes, R. 1995. Uplift and exhumation of the Alpine schist, Southern Alps, New Zealand: thermobarometric constraints. *N.Z. J. Geol. Geophys.*, **38**, 525–533.

Grapes, R. and Otsuki, H. 1983. Peristerite compositions in quartzofeldspathic schists, Franz Josef – Fox Glacier area, New Zealand. *J. Metamorph. Geol.*, **1**, 47–61.

Grapes, R. and Watanabe, T. 1992. Metamorphism and uplift of the Alpine schist, Southern Alps, New Zealand. *J. Metamorph. Geol.*, **10**, 171–180.

Griffiths, G.A. 1979. High sediment yields from major rivers of the Western Southern Alps, New Zealand. *Nature*, **282**, 61–63.

Griffiths, G.A. 1981. Some suspended sediment yields from South Island catchments, New Zealand. *Wat. Resour. Bull.*, **17**, 662–671.

Griffiths, G.A. and McSaveney, M.J. 1983a. Distribution of mean annual precipitation across some steepland regions of New Zealand. *N.Z. J. Sci.*, **26**, 197–209.

Griffiths, G.A. and McSaveney, M.J. 1983b. Hydrology of a basin with extreme rainfalls – Cropp River, New Zealand. *N.Z. J. Sci.*, **26**, 293–306.

Griffiths, G.A. & McSaveney, M.J. 1986. Sedimentation and river containment on Waitangitoaona alluvial fan – South Westland, New Zealand. *Z. Geomorph.*, **30**, 215–230.

Henderson, R.D. 1993. Extreme storm rainfalls in the Southern Alps, New Zealand. *IAHS Publ.*, **213**, 113–120.

Henyey, T., Stern, T. and Molnar, P. 1993. Continent–continent collision in Southern Alps studied. *EOS Trans. Amer. Geophys. Un.*, **74**, 316.

Hicks, D.M., McSaveney, M.J. and Chinn, T.J.H. 1990. Sedimentation in proglacial Ivory Lake, Southern Alps, New Zealand. *Arctic Alp. Res.*, **22**, 26–42.

Hicks, S.R. 1989. Structure of the Canterbury Plains, New Zealand, from gravity modelling. *New Zealand SSIR Geophysics Division Research Report*, **222**.

Hovius, N. 1995. Macro scale process systems of mountain belt erosion and sediment delivery to basins. Unpublished PhD thesis, University of Oxford.

Hovius, N. 1996. Regular spacing of drainage outlets from linear mountain belts. *Basin Res.*, **8**, 29–44.

Hovius, N., Stark, C.P. and Allen, P.A. 1997. Sediment flux from a mountain belt derived by landslide mapping. *Geology*, **25**, 231–234.

Hull, A.G. and Berryman, K.R. 1986. Holocene tectonism in the region of the Alpine fault at Lake McKerrow, Fiordland, New Zealand. *Bull. R. Soc. N.Z.*, **24**, 317–331.

Kamp, P.J.J. 1986. Mid-Cenozoic Challenger Rift System of western New Zealand and its implications for the age of Alpine fault inception. *Geol. Soc. Amer. Bull.*, **97**, 255–281.

Kamp, P.J.J. 1987. Age and origin of the New Zealand orocline in relation to Alpine fault movement. *J. Geol. Soc. Lond.*, **144**, 641–652.

Kamp, P.J.J. and Fitzgerald, P.G. 1987. Geologic constraints on the Cenozoic Antarctica–Australia–Pacific relative plate motion circuit. *Geology*, **15**, 694–697.

Kamp, P.J.J. and Tippett, J.M. 1993. Dynamics of Pacific plate crust in the South Island (New Zealand) zone of oblique continent–continent convergence. *J. Geophys. Res.*, **98**, 16 105–16 118.

Kamp, P.J.J., Green, P.F. and White, S.H. 1989. Fission track analysis reveals character of collisional tectonics in New Zealand. *Tectonics*, **8**, 169–195.

Kamp, P.J.J., Green, P.F. and Tippett, J.M. 1992. Tectonic architecture of the mountain front–foreland basin transition, South Island, New Zealand, based on fission track analysis. *Tectonics*, **11**, 98–113.

Kirkbride, M. and Matthews, D. 1997. The role of fluvial and glacial erosion in landscape evolution: the Ben Ohau Range, New Zealand. *Earth Surf. Processes Ldfms.*, **22**, 317–327.

Kneupfer, P.L.K. and Coleman, P.C. 1994. Interactions between topography and tectonics in the Southern Alps. Abstract for *Geophysical Symposium on Origin of the Southern Alps*, Wellington, New Zealand.

Koons, P.O. 1987. Some thermal and mechanical consequences of rapid uplift: an example from the Southern Alps, New Zealand. *Earth Planet. Sci. Lett.*, **86**, 307–319.

Koons, P.O. 1989. The topographic evolution of collisional mountain belts: a numerical look at the Southern Alps, New Zealand. *Amer. J. Sci.*, **289**, 1041–1069.

Koons, P.O. 1990. Two-sided orogen: collision and erosion from the sandbox to the Southern Alps, New Zealand. *Geology*, **18**, 679–682.

Koons, P.O. 1994. Three-dimensional critical wedges: Tectonics and topography in oblique collisional orogens. *J. Geophys. Res.*, **99**, 12 301–12 315.

Koons, P.O. 1995. Modeling the topographic evolution of collisional belts. *Ann. Rev. Earth Planet. Sci.*, **23**, 375–408.

Lifton, N.A. and Chase, C.G. 1992. Tectonic, climatic and lithologic influences on landscape fractal dimension and hypsometry: implications for landscape evolution in the San Gabriel Mountains, California. *Geomorphology*, **5**, 77–113.

Mason, B. 1962. Metamorphism in the Southern Alps of New Zealand. *Bull. Amer. Mus. Nat. Hist.*, **123**, 217–248.

McSaveney, M.J., Chinn, T.J. and Hancox, G.T. 1992. Mount Cook rock avalanche of 14 December 1991, New Zealand. *Landslide News*, August, 32–32.

Molnar, P. 1996. Summary of studies of active deformation, regional structure, and other issues related to the geodynamic processes in the South Island of New Zealand. Unpublished discussion paper.

Molnar, P. and England, P. 1990. Late Cenozoic uplift of mountain ranges and global climate change: chicken or egg? *Nature*, **346**, 29–34.

Montgomery, D.R. 1994. Valley incision and the uplift of mountain peaks. *J. Geophys. Res.*, **99**, 13 913–13 921.

Mosley, M.P., Blakely, R.J. and Chinn, T.J.H. 1984. Predicting the behaviour of a glacier fed river. *Streamland*, **20**.

Nathan, S., Anderson, H.J., Cook, R.A., Herzer, R.H., Hoskins, R.H., Raine, J.I. and Smale, D. 1986. Cretaceous and Cenozoic sedimentary basins of the West coast region, South Island, New Zealand. *N.Z. Geol. Surv. Basin Studs.*, **1**.

Norris, R.J. 1979. A geometrical study of finite strain and bending in South Island. *Bull. R. Soc. N.Z.*, **18**, 21–28.

Norris, R.J. and Cooper, A.F. 1995. Origin of small-scale segmentation and transpressional thrusting along the Alpine fault, New Zealand. *Geol. Soc. Amer. Bull.*, **107**, 231–240.

Norris, R.J., Koons, P.O. and Cooper, A.F. 1990. The obliquely-convergent plate boundary in the South Island of New Zealand: implications for ancient collision zones. *J. Struct. Geol.*, **12**, 715–725.

Pearson, C.F. 1991. Distribution of geodetic shear strains across the Alpine fault zone: Haast, New Zealand. *J. Geophys. Res.*, **96**, 8465–8473.

Pearson, C.F. 1994. Geodetic strain determinations from the Okarito and Godley-Tekapo regions, central South Island, New Zealand. *N.Z. J. Geol. Geophys.*, **37**, 307–318.

Pearson, C.F., Beavan, J., Darby, D.J., Blick, G.H. and Walcott, R.I. 1995. Strain distribution across the Australian–Pacific plate boundary in the central South Island, New Zealand, from 1992 GPS and earlier terrestrial observations. *J. Geophys. Res.*, **100**, 22 071–22 081.

Pillans, B.J. 1991. New Zealand Quaternary stratigraphy: an overview. *Quat. Sci Rev.*, **10**, 405–418.

Reilly, W.I. 1986. Crustal bending in Otago, New Zealand, from the evidence of geodetic measurements. *Bull. R. Soc. N.Z.*, **24**, 65–73.

Reilly, W.I. and Whiteford, C.M. 1979. *Gravity Map of New Zealand, 1:1000000; Bouguer and Isostatic Anomalies, South Island.* Department of Scientific and Industrial Research, New Zealand.

Reyners, M. 1987. Subcrustal earthquakes in the central South Island, New Zealand, and the root of the Southern Alps. *Geology*, **15**, 1168–1171.

Reyners, M. and Cowan, H. 1993. The transition from subduction to continental collision: crustal structure in the North Canterbury region, New Zealand. *Geophys. J. Int.*, **115**, 1124–1136.

Shi, Y., Allis, R.G. and Davey, F.J. 1996. Thermal modelling of the Southern Alps. *Pure Appl. Geophys.*, **146**, 469–501.

Simpson, G.D.H., Cooper, A.F. and Norris, R.J. 1994. Late Quaternary evolution of the Alpine fault zone at Paringa, South Westland, New Zealand. *N.Z. J. Geol. Geophys.*, **37**, 49–58.

Sibson, R.H., White, S.H. and Atkinson, B.K. 1979. Fault rock distribution and structure within the Alpine fault zone: Preliminary account. *Bull. R. Soc. N.Z.*, **18**, 55–65.

Simpson, G.D.H., Cooper, A.F. and Norris, R.J. 1994. Late Quaternary evolution of the Alpine fault zone at Paringa, South Westland, New Zealand. *N.Z. J. Geol. Geophys.*, **37**, 49–58.

Sircombe, K.N. 1993. Analysis of the South Westland Basin, New Zealand. Unpublished MSc thesis, University of Waikato.

Smith, E.G.C., Stern, T.A. and O'Brien, B. 1995. A seismic velocity profile across the central South Island, New Zealand. *N.Z. J. Geol. Geophys.*, **38**, 565–570.

Spörli, K.B. 1979. Structure of the South Island Torlesse in relation to the origin of the Southern Alps. *Bull. R. Soc. N.Z.*, **18**, 99–104.

Stern, T.A. 1995. Gravity anomalies and crustal loading at and adjacent to the Alpine fault, New Zealand. *N.Z. J. Geol. Geophys.*, **38**, 593–600.

Stirling, M.W. 1991. Peneplain modification in an alpine environment of Central Otago, New Zealand. *N.Z. J. Geol. Geophys.*, **34**, 195–201.

Stock, J. and Molnar, P. 1982. Uncertainties in the relative positions of the Australia, Antarctica, Lord Howe, and Pacific plates since the late Cretaceous. *J. Geophys. Res.*, **87**, 4697–4714.

Suggate, R.P. 1965. Late Pleistocene geology of the northern part of the South Island, New Zealand. *N.Z. Geol. Surv. Bull.*, **77**.

Suggate, R.P. 1990. Late Pliocene and Quaternary glaciations of New Zealand. *Quat. Sci. Rev.*, **9**, 175–197.

Summerfield, M.A. and Kirkbride, M.P. 1992. Climate and landscape response. *Nature*, **355**, 306.

Sutherland, R. 1994. Displacement since the Pliocene along the southern section of the Alpine fault, New Zealand. *Geology*, **22**, 327–330.

Sutherland, R. 1995. The Australia–Pacific boundary and Cenozoic plate motions in the SW Pacific: some constraints from Geosat data. *Tectonics*, **14**, 819–831.

Sutherland, R. 1996. Transpressional development of the Australia–Pacific boundary through southern South Island, New Zealand: constraints from Miocene–Pliocene sediments, Waiho-1 borehole, South Westland. *N.Z. J. Geol. Geophys.*, **39**, 251–264.

Sutherland, R. and Norris, R.J. 1995. Late Quaternary displacement rate, paleoseismicity, and geomorphic evolution of the Alpine fault: evidence from Hokuri Creek, South Westland, New Zealand. *N.Z. J. Geol. Geophys.*, **38**, 419–430.

Sutherland, R., Nathan, S. and Turnbull, I.M. 1995. Pliocene–Quaternary sedimentation and Alpine fault related tectonics in the lower Cascade valley, South Westland, New Zealand. *N.Z. J. Geol. Geophys.*, **38**, 431–450.

Thompson, S.M. 1993. Estimation of probable maximum floods from the Southern Alps, New Zealand. *IAHS Publ.*, **213**, 299–305.

Tippett, J.M. and Kamp, P.J.J. 1993. Fission track analysis of the late Cenozoic vertical kinematics of continental Pacific crust, South Island, New Zealand. *J. Geophys. Res.*, **98**, 16 119–16 148.

Tippett, J.M. and Kamp, P.J.J. 1995. Quantitative relationships between uplift and relief parameters in the Southern Alps, New Zealand, as determined by fission track analysis. *Earth Surf. Processes Ldfms.*, **20**, 153–176.

Walcott, R.I. 1978. Present tectonics and late Cenozoic evolution of New Zealand. *Geophys. J. R. Astr. Soc.*, **52**, 137–164.

Walcott, R.I. 1979. Plate motion and shear strain rates in the vicinity of the Southern Alps. *Bull. R. Soc. N.Z.*, **18**, 5–12.

Walcott, R.I. 1984. The kinematics of the plate boundary zone through New Zealand: a comparison of short- and long-term deformations. *Geophys. J. R. Astr. Soc.*, **52**, 137–164,

Walcott, R.I. 1994. The origin – 15 years on. Abstract for *Geophysical Symposium on Origin of the Southern Alps*, Wellington, New Zealand.

Wellman, H.W. 1955. New Zealand Quaternary tectonics. *Geol. Rund.*, **43**, 248–257.

Wellman, H.W. 1979. An uplift map for the South Island of New Zealand, and a model for the uplift of the Southern Alps. *Bull. R. Soc. N.Z.*, **18**, 13–20.

Whitehouse, I.E. 1985. The frequency of high-intensity rainfalls in the central Southern Alps, New Zealand. *J. R. Soc. N.Z.*, **15**, 213–226.

Whitehouse, I.E. 1987. Geomorphology of a compressional plate boundary, Southern Alps, New Zealand. In: V. Gardiner (ed.) *International Geomorphology*, Wiley, Chichester, pp. 897–924.

Whitehouse, I.E. 1988. Geomorphology of the central Southern Alps, New Zealand: the interaction of plate collision and atmospheric circulation. *Z. Geomorph. Suppl.*, **69**, 105–116.

Woodward, D.J. 1979. The crustal structure of the Southern Alps, New Zealand, as determined by gravity. *Bull. R. Soc. N.Z.*, **18**, 95–98.

Wratt, D.S. *et al.* 1996. The New Zealand Southern Alps experiment. *Bull. Amer. Met. Soc.*, **77**, 683–692.

7

Morphotectonic evolution of Taiwan

Jiun-Chuan Lin

7.1 Introduction

The aim of this chapter is to assess the relationships between tectonics and landscape development in Taiwan. Structural and seismic data are used to place Taiwan in its plate tectonic context, and an uplift–denudation interaction model for the Central Range and Coastal Range is presented. Uplift rates of 2–5 mm a^{-1} have been established for several areas in Taiwan using a variety of methods, and there appears to be a relationship between uplift and denudation rates. Landscape development in Taiwan clearly reflects tectonics and lithology, and has occurred under the influence of a monsoonal, sub-tropical climate.

7.2 Regional Tectonic Background

Taiwan is an area of active arc–continent collision lying between the Philippine Sea and Eurasian plates, with the Philippine Sea plate moving northwest at approximately 70 km Ma^{-1} (Seno, 1977). An instantaneous pole of rotation of 1.2° Ma^{-1} at 45.5°N, 150.2°E estimated by Seno (1977) predicts a zone of plate convergence near Taiwan involving the Philippine Sea plate (Figure 7.1). The Luzon arc is oriented N16°E and the continental margin N60°E; the angle between the arc and the compression is ~71°, and between the continental margin and compression ~65°. Solving the triangle of forces yields a rate of propagation of the collision zone of 90 km Ma^{-1} with the arc, and 95 km Ma^{-1} with the continental margin (Lin, 1994).

The ancient sedimentary and tectonic environment in the region of Taiwan was marine and geosynclinal with the resulting stratigraphy comprising a varied sequence of shale, siltstone, sandstone and limestone units ranging in age from late Palaeozoic to late Mesozoic. During orogenesis, thick Tertiary sediments in a western sedimentary basin were raised, deformed, and partly metamorphosed (Ho, 1982).

The main tectonic elements of the Philippine Sea plate margin form significant bathymetric morphological features and tectonic structures and control tectonic movements in Taiwan. They can be summarized as follows (Figure 7.2):

1. Ryukyu arc: The Ryukyu arc system lies to the southeast of the East China Sea continental slope and includes the Okinawa trough, the Ryukyu ridge and the Ryukyu trench. This island arc joins the northeastern part of Taiwan. Tsai (1978) has pointed out that the boundary of the Philippine Sea plate bends along an east–west line at latitude 24°N. The plate subducts northward beneath the Ryukyu arc which has been constructed on oceanic crust seaward of the Eurasian continental margin. A thin planar seismic zone, about 50 km thick, dips 45° northward beneath the Ryukyu arc to a depth of 150 km. The profile of the Benioff zone dips beneath the Ryukyu arc north of 24°N. Some authors (e.g. Lin and Tsai, 1981) have drawn attention to the complicated nature of the junction of the arc with the Eurasian plate close to Taiwan. It is possible that a small right-lateral fault was developed at the junction by a northward-moving fragment of crust prior to subduction. Wu and Lu (1976) consider that this fault may be a transform fault. This has important implications when considering the direction of plate convergence and landform development.

Geomorphology and Global Tectonics. Edited by Michael A. Summerfield.

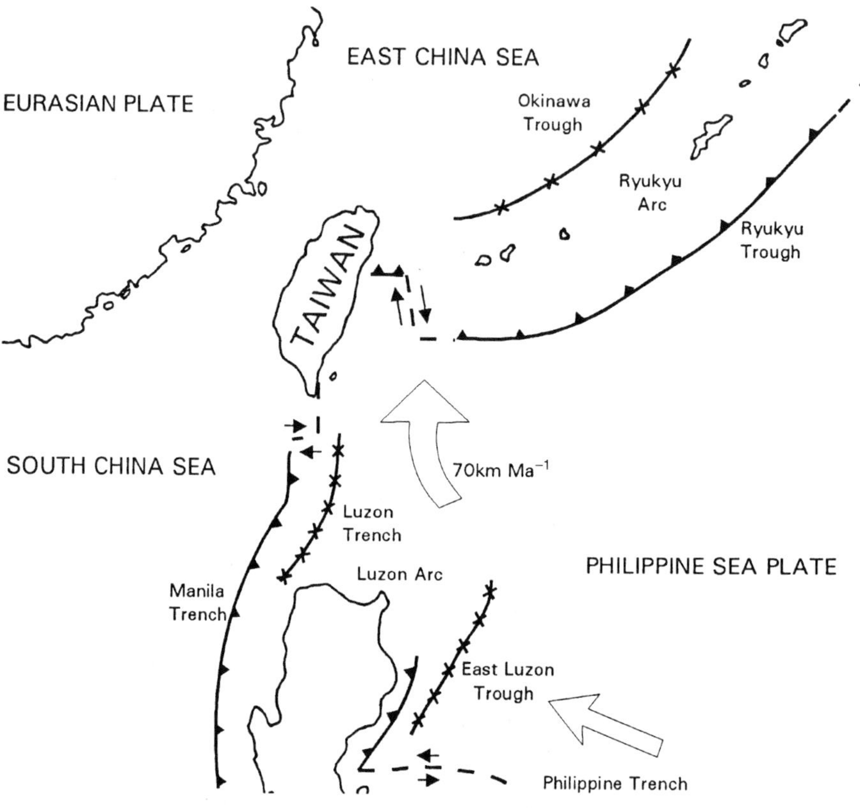

Figure 7.1 *Main tectonic elements of the western Philippine Sea plate margin showing direction and rate of convergence (modified from Seno, 1977; Lin and Tsai, 1981; Ho, 1982)*

2. Luzon arc: The Luzon arc lies to the south of Taiwan between the South China Sea and the West Philippine basin. Here there is a close correspondence between the submarine and volcanic features of eastern Taiwan and those of northern Luzon Island. In this region, seismicity is generally concentrated along two narrow zones parallel to the western and eastern boundaries of Luzon Island. These seismic zones are associated with the Manila trench which represents the subduction zone along which the Eurasian plate, underlying the South China Sea, underthrusts the Luzon arc. The Philippine trench forms the subduction zone for the Philippine plate and is marked by a very clear and active Benioff zone. It should also be noted that immediately to the south of Taiwan another right-lateral fault occurs between the Manila trench subduction zone and the seaward extension of the Taiwan Longitudinal Valley fault.

7.3 The Nature of the Plate Collision Zone in Taiwan

7.3.1 REGIONAL TECTONICS

The collision between the Luzon arc of the Philippine Sea plate and the Eurasian continental margin has been the most significant tectonic event in the morphotectonic evolution of Taiwan. The collision, known as the Penglai Orogeny, probably began in the late Pliocene and reached its climax in the middle Pleistocene (Ho, 1975). Several hundred kilometres of crustal shortening were accomplished in only a few million years, the resulting compression leading to the accretion of sedimentary units to create the island of Taiwan. According to Suppe (1984), arc–continent collision has already stopped in the northernmost quarter of Taiwan because of the southwestward propagation of the Ryukyu trench subduction zone.

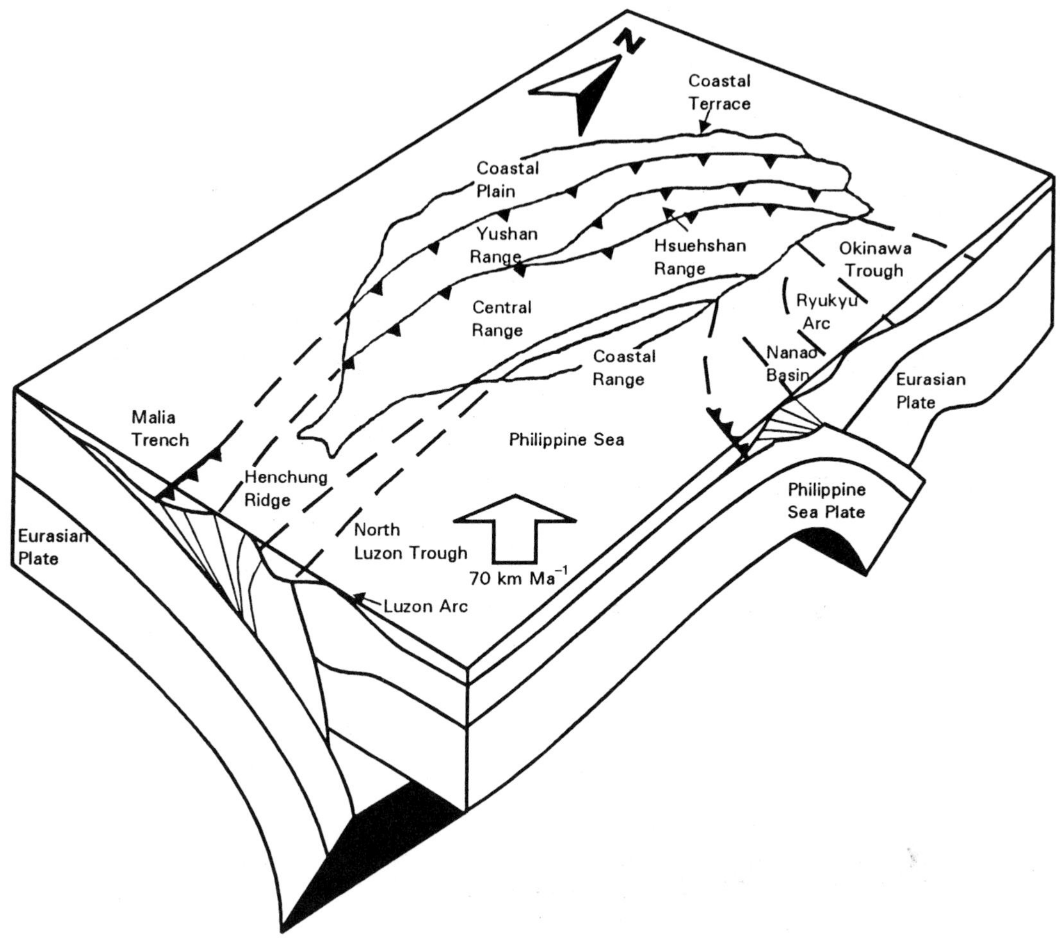

Figure 7.2 *Three-dimensional schematic plate tectonic setting of Taiwan (modified from Ho, 1982)*

The compression due to plate convergence has been reconstructed from field analyses of the fold and fault patterns of Taiwan (Angelier *et al.*, 1986; Wang, 1987). These have revealed that palaeostress trajectories can be distinguished and that two compressional events appear to have contributed to the present structure and topography (Figure 7.3). Both of Plio-Pleistocene age, these two events yield similar fan-shaped distributions of maximum compressive stress trajectories, but with an anticlockwise shift of 35–50° between them (Tan, 1977). Current interpretations of plate motion indicate that the Philippine Sea plate close to Taiwan is moving north; however, the collision of the Luzon arc is oblique and the main stress trajectory is more northwestward. Palaeomagnetic data for the Coastal Range indicate that a clockwise rotation of about 30° occurred during the Pleistocene. No significant rotation is apparent for a period of at least 3 Ma prior to this time. The rotation was very rapid, first occurring in the northern part of the Coastal Range and subsequently in the middle sector 2.3–1.9 Ma ago. The time of collision is therefore about 0.9 Ma younger in the south than in the middle of the Coastal Range (Lee, 1989).

During the collision, compressional forces first became evident in the northern area of Taiwan and then propagated southward. The indentation zone has been gradually enlarged through time (Lee and Wang, 1987). In central and southern Taiwan the direction of greatest principal stress has remained constant from the early stage of folding. The regional stress pattern derived from the analysis of the focal mechanisms of

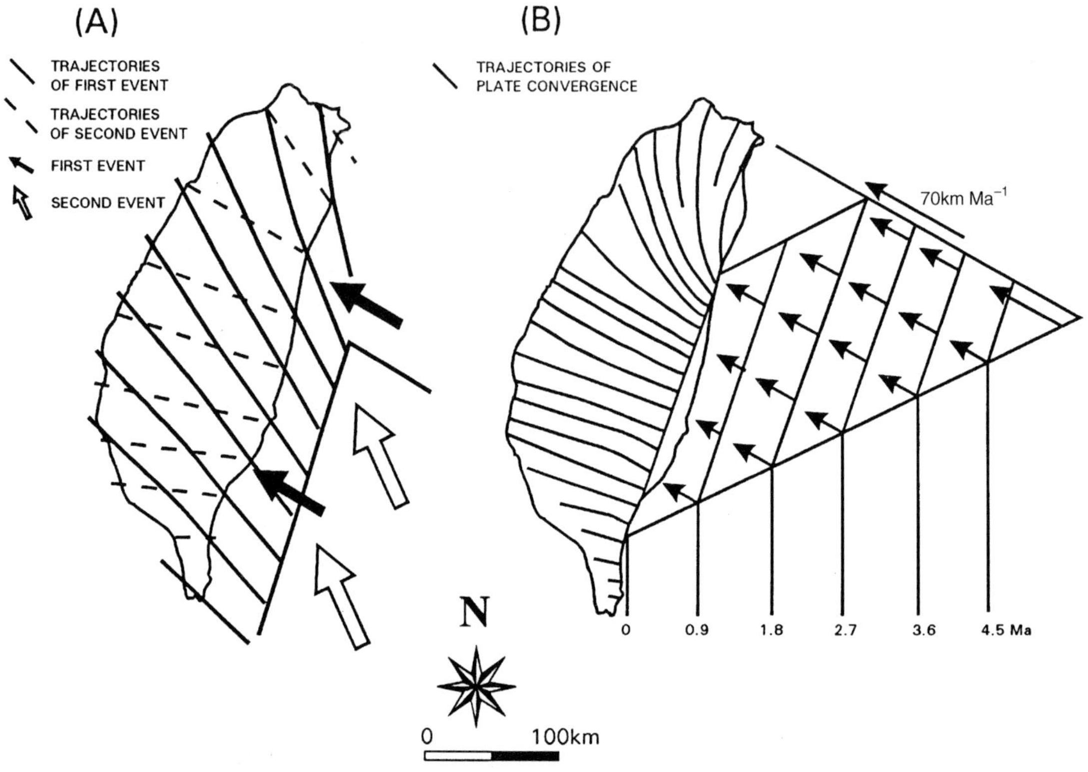

Figure 7.3 Stress trajectories and plate convergence. (A) First and second compressional events. Trajectories of the two events show a rotation of 40° (modified from Angelier et al., *1986). (B) Pliocene–Quaternary arc–continent collision (after Lee and Wang, 1987). This diagram shows trajectories of plate convergence at 70 km Ma^{-1} over the past 4.5 Ma*

principal stress directions of major earthquakes and composite focal mechanisms of micro-earthquakes demonstrates mainly shear stresses at the western end of the Ryukyu trench, with predominantly tensile stresses in the Central Range (Figure 7.4) (Lin *et al.*, 1985). The direction of maximum principal stress in the Taiwan area is SE-NW and lies in, or close to, the horizontal plane (Lin *et al.*, 1985). The fan-shaped distribution of the maximum principal horizontal stress in western Taiwan is quite consistent with the direction of the Philippine Sea plate–Eurasian plate convergence. Lin (1984) used SLAR imagery to analyse the lineaments of Taiwan and found that the geometrical solution could be explained by a collision commencing 5.4 Ma ago. Following the opening of the Okinawa trough at 1.9 Ma, the collision gradually retreated southward and this movement has continued to the present.

A large portion of the strain is accommodated by the Yushan–Hsuehshan megashear zone, the thrust faults of the Central Range and the Coastal Range (Figure 7.4), large-scale strike-slip motion, especially along the Longitudinal Valley, and by the movement of fault-wedge basins. The greatest lateral and vertical upthrust movements cumulate in the Yushan Range at nearly 4000 m above sea level. This is the highest of the western North Pacific coastal ranges. The Taiwan arc is unusually high, thick and short when compared to the other island arcs of the circum-Pacific zone (Figure 7.4). It also appears to have entered a stage of block tectonic units modified in part by the orogenic framework. This imposes recent transcurrent buckling on the landforms of the Coastal Ranges and the coastline (Figure 7.4), and the details of this process are now more clearly understood compared with earlier interpretations (Hsu, 1956) (Figure 7.5). There is thus an inconsistency in slip vectors along the Ryukyu and Philippine trenches (Figure 7.2). It should also be noted that a major fault system, the Philippine fault, extends through this area. This is equivalent in scale to the San

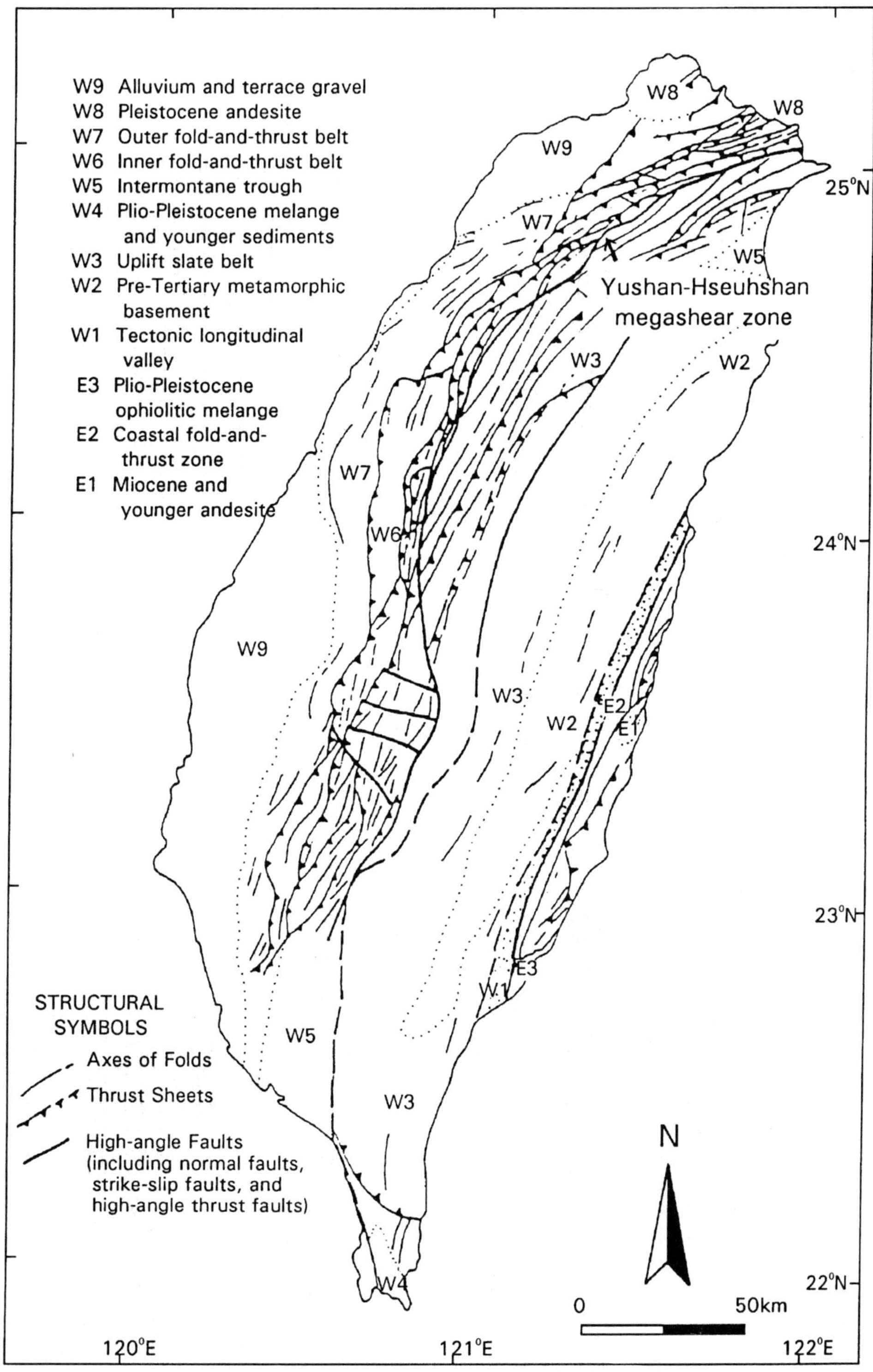

Figure 7.4 Generalized tectonic provinces of Taiwan (modified from Ho, 1982)

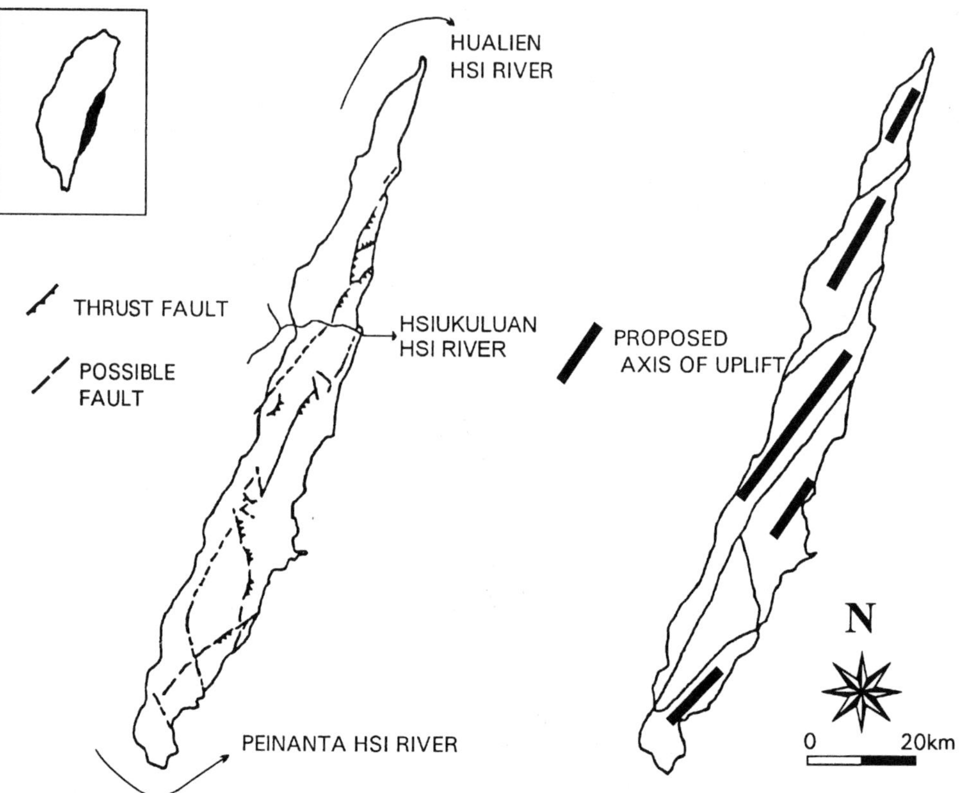

Figure 7.5 *(A) Main fault of the Coastal Range (modified from Hsu, 1956); (B) proposed axis of uplift within each tectonic block in the Coastal Range*

Andreas (USA), Alpine (New Zealand) or Atacama faults and reflects the dynamic tectonics of the region.

The two right-lateral (transform?) faults at the termination of the Ryukyu and Manila trenches are very important because they mean that subduction is not at present taking place beneath Taiwan, as some authors have suggested (Wu and Lu, 1976). Tsai *et al.* (1987) believe that the Philippine Sea plate subducts to form the Coastal Range with a boundary dipping at 50–55° towards the east. Between the two subduction zones motion also appears to be occurring along a left-lateral fault. This is the reason why deep earthquakes tend to occur south of 21°N and north of 24.5°N, whereas mainly shallow earthquakes occur in the region of the Coastal Range.

7.3.2 SEISMICITY

Seismically, Taiwan can be divided into three zones. The northeastern seismic zone occupies the area east of longitude 121.5°E and north of latitude 24.0°N. Seismicity in this zone is primarily associated with the northward subduction of the Philippine Sea plate under the Eurasian plate. The eastern seismic zone includes the Longitudinal Valley, the Coastal Range of eastern Taiwan, and the offshore area to the east. The western limit of this seismic zone lies west of, but parallel to, the Longitudinal Valley and its boundary surface dips at an angle of 50–55° eastward down to a depth of about 50 km. The western seismic zone includes most of the Hsuehshan Range, the foothill region, and the coastal plain area of western Taiwan.

Seismic epicentre and geodetic survey data suggest that present movement in Taiwan is concentrated mainly in the Chengkung area. However, the gradient of vertical movement is concentrated in the Tafu area which may be in accordance with the compression movements. Therefore, the uplift models of Hsu (1954) and Biq (1972) can be modified, with the angle of the uplift axis being rotated by 45° from a W–E direction to NNE–SSW (Figure 7.5b). The Changping, Chengkung and Tuluan areas can be thought of as having a replicated sequence of landforms from north to south. This suggests that the landforms of the Coastal Range may have developed by similar mechanisms but at different times.

Global Positioning System measurements of crustal deformation in the Taiwan arc–continent collision zone reveal a fan-shaped pattern consistent with the directions of maximum compressional tectonic stress inferred from borehole breakout data, earthquake focal mechanisms and Quaternary geological data (Yu and Chen, 1994). Palaeomagnetic studies of the collision-related bending of the fold-thrust belt show a 20° counterclockwise rotation on average across whole island (Lue *et al.*, 1995).

Taiwan has been very tectonically active since its formation during the Palaeogene. The spatial distribution of the combined seismicity data indicates that the plate boundary in the Taiwan area should be located along the Longitudinal Valley fault from Hualien to Tawu, turning southward and overlapping the 121°E meridian just off the east coast of the Hengchun Peninsula. According to Hsu (1962, 1976), Biq (1974), Tsai *et al.* (1977), Tsai (1978, 1985) and Hsu and Chang (1979), eastern Taiwan is still moving with obviously active faults and earthquakes. For instance, during the Hualien earthquake of 22 October 1951 vertical and horizontal displacements on the Meilun fault were 1.2 m and 2 m, respectively. Geodetic data show that eastern Taiwan is still moving in a direction of N16°E and N40°E at a rate of about 60–130 mm a^{-1}.

7.3.3 UPLIFT AND DENUDATION RATES

Uplift rates over the past 9000 years for the Hengchun Peninsula, the Taiwan area and the Coastal Range have been of the order of 5.0 ± 0.4 mm a^{-1}. The rate in the north was 2 mm a^{-1} from 5500 to 1500 a BP (Peng *et al.*, 1977), while Konishi *et al.* (1968) have estimated rates of 1.8–4.8 mm a^{-1} from north to south in the Central Range and 6–9.7 mm a^{-1} in the Coastal Range. Bonilla (1975) published figures of 0.3–8.7 mm a^{-1} but these originate from other, less reliable, data sources. Overall these records suggest that uplift rates increased over the past 3 Ma to a maximum mean rate of ~7.0 mm a^{-1} in the late Pleistocene, with the majority of uplift occurring in the last ~2 Ma.

These very high uplift rates are apparently matched by comparably high denudation rates. Denudation rates have been calculated by Li (1976) from sediment and solute loads from stream gauging stations (Table 7.1). There is an extensive coverage of stations for suspended load throughout Taiwan (Committee for Water Resource Control, 1973), and solute load data are available from bi-monthly sampling and water flow records for more than 30 stations (Hsu, 1964; Committee for Water Resource Control, 1970). The data show that in the Central Range, where relief is high and pre-Miocene sediments crop out, the mean mechanical denudation rate is ~4.8 mm a^{-1}, with a maximum of 5.06 mm a^{-1}. These rates are comparable to those for some catchments in the Southern Alps of New Zealand (see Chapter 6). Fission-track data from apatite, sphene and zircon from the Central Range have yielded similarly high long-term denudation rates of 5–9 mm a^{-1} (Liu, 1982).

Table 7.1 *Denudation rates for Taiwan based on data from Li (1976)*

Area	Mechanical denudation rate (mm a^{-1})	Chemical denudation rate (mm a^{-1})
West Taiwan	1204	141
Central Range	4815	241
Central Range (Pliocene)	356	–
Central Range (Pleistocene)	1426–2852	–
Coastal Range	2815	237
Taiwan maximum	5056	337

Bedrock density of 2700 kg m^{-3} assumed.

7.3.4 CLIMATIC CONTROLS: TROPICAL STORMS AND EPISODIC FORMATIVE EVENTS

The landscape of Taiwan has evolved under the influence of a monsoonal sub-tropical climate, except in the extreme south which lies within the tropics (the Hengchun Peninsula). Mean annual precipitation varies across the island from 1500 mm in the west to 4000–5000 mm in the Central Range. It rises to approximately 4000 mm in the north due to the winter monsoon. During the cold phases of the Pleistocene the climate was modified and became much cooler, but there is little information indicating the precise nature of these climatic changes. An argument has been suggested that small glaciers occurred in the Central Range, but this has never been substantiated (Tsan, 1960).

Perhaps the most important morphogenetic factor affecting the climate is that the region comes under the

Table 7.2 Maximum rainfall record in Taiwan

Duration (hr)	World record (mm)	Taiwan rainfall (mm), year, location
1	305 (42 mins)	300, 1972, Wushi (W Taiwan)
2	483 (130 mins)	560, 1972, Wushi (W Taiwan)
3	559 (165 mins)	614, 1972, Wushi (W Taiwan)
4	782 (4.5 hr)	670, 1972, Wushi (W Taiwan)
6	–	760, 1972, Wushi (W Taiwan)
12	1340	1157.5, 1996, Alishan (SW Taiwan)
18	1689 (18.5 hr)	1537.5, 1996, Alishan (SW Taiwan)
24	1870	1748.5, 1996, Alishan (SW Taiwan)
48	2500	2260.2, 1967, Ilan (NE Taiwan)

influence of severe tropical storms and typhoons (Table 7.2). In their study of the effectiveness of climate in landscape evolution, Wolman and Gerson (1978) showed that high-magnitude events are apparently of high frequency in humid tropical regions. This idea suggests that big events dominate the geomorphic work carried out and that landscape change is episodic. Amounts of denudation associated with a major storm in such an area may even approach the mean annual erosion rate. Tropical storms should therefore be regarded as the major landscape-forming events (Brunsden and Thornes, 1979).

7.4 Tectonic Landforms of Taiwan

Active arc–continent collision in Taiwan represents a typical orogen, with a deformed continental margin (Foothills Belt, Hsueshan Range and Central Range) separated from an accreted island arc (Coastal Range) by a collapsed forearc basin (Longitudinal Valley) (Willemin and Kneupfer, 1994). The maximum elevation of the Central Range is 3805 m, decreasing north and south. The island is asymmetrical, rising steeply in the east from a narrow longitudinal tectonic valley. Relief decreases more gradually to the west through subsidiary ranges, basins and tablelands to the coastal plain. Hillslopes in Taiwan are generally steep (>20°), forested and unstable. The Central Range forms the main divide, and rivers flowing to the east are therefore very steep and form spectacular gorges and terraces. Throughout the island, and particularly to the east, tectonic landforms, both active and relict, are common. In addition 'fabric relief' dominates every slope and valley. These fabrics are either planar, linear, or mixed planar–linear, and represent a pure space-constant mechanical strain form over large regions (Sander, 1970) (Figure 7.5b).

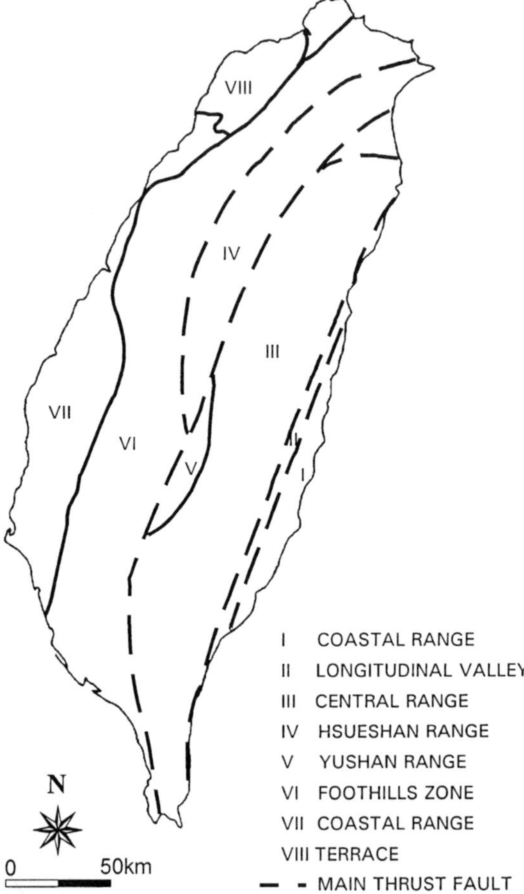

Figure 7.6 Generalized landform provinces of Taiwan

The distribution of Quarternary and active faults has been summarized by Hsu (1962, 1976), York (1976), Hsu and Chang (1979) and Shih *et al.* (1986). Some faults have moved dramatically; for example, a maximum horizontal displacement of 2.4 m occurred on the Meishan fault in southwest Taiwan in 1906 (Omori, 1907), while a maximum vertical displacement of 3 m occurred along the Szetan fault in western Taiwan in 1935.

The landforms of Taiwan closely reflect its tectonic history and stratigraphy and suggest the following morphotectonic classification developed by Hsu (1955), Lin (1957) and Ho (1975, 1987) (Figure 7.6).

1. The Coastal Range situated between Hualien and Taitung is approximately 150 km in length and 20 km wide. It is composed of a sequence of units from north to south each with a distinctive rock

type, structure, age and relief. It is crossed by only one river (Figure 7.5A).

2. The Longitudinal Valley runs parallel to the Coastal and Central Ranges and is drained by the Hualien Chi, Shiukuluan Chi and Pinanta Chi Rivers. Only the Shiukuluan Chi crosses the Coastal Range, the other two draining north and south. The valley is dominated by active tectonic landforms, faults, terraces and alluvial fans.
3. The Eastern Central Mountain Ranges are mostly made up of schist and crystalline limestone with subordinate gneiss. Deep valleys are associated with extremely high denudation rates.
4. The Western Central Mountain Range, the Hsuehshan Range and Yushan Range are mainly composed of Paleogene argillites with subordinate sandstone, pyroclastics, and limestone lenses. The rocks are indurated and deformed by very low-grade regional metamorphism (Chai, 1972).
5. The Western Foothills correspond to a major part of the passive Eurasian margin and comprise a fold-and-thrust belt. Most of the folds are asymmetric or even overturned, and the axial planes dip southeast. The folds are generally bounded by thrust faults forming a series of imbricated thrust blocks (Chai, 1972). The height of the foothills is between 100 and 1000 m and they are densely dissected by rivers.
6 The Coastal Plain and terrace region has a subdued topography with only a few rolling hills and terraces, and is mainly covered by late Pleistocene and Holocene sediments (Chai, 1972).

7.5 Morphotectonic Evolution of Taiwan

7.5.1 KINEMATICS OF ARC–CONTINENT COLLISION

Collision began in the north of Taiwan approximately 4 Ma ago and is just beginning in the south. The landforms are thus a maximum of 4 Ma old in the north, becoming progressively younger to the south, and of present-day origin on the south coast. Suppe (1981, 1984, 1987) has suggested that if we assume that the collision rate has been reasonably constant, then moving north in Taiwan by 90 km is equivalent to moving back 1 Ma in time. This space–time equivalence is a powerful tool in tectonic analysis and has been used by Page and Suppe (1981) to analyse the palaeogeography of the upper Pliocene Lichi Melange and to develop schematic models of progressive propagation, and of the flipping mechanism from subduction to collision termination (Suppe, 1984). Most importantly in the present context, Suppe drew attention to the remarkable steady-state topography of Taiwan. He argued that the major bathymetric and topographic effect of collision is the expansion to the south of the accretionary wedge of the inter-non-volcanic arc in both width and height. The mountain belt is similarly growing in height and width with time towards the north. The space–time equivalence indicates that mountain growth is steady until 120 km to the north (1.33 Ma old) where a constant width of 87 ± 4 km, a cross-sectional area of 118 ± 24 km^2 and a mean elevation of 1350 m are attained, with this elevation being maintained for another 170 km. This region of constant topography in Taiwan from 120 to 290 km north is therefore "a region of steady state topography in which the rate of growth of the mountain belt as a result of plate-boundary compression is equal to the rate of erosion. This implies that erosion places an upper limit to the size of this mountain belt and that this maximum size is reached in about 1.33 Ma 120 km north of the southern tip of Taiwan" (Suppe, 1981). Suppe (1981) further suggested that if the mountain belt is in a steady state then the rate of uplift should balance the rate of erosion and he used the comparison of a mean denudation rate of 5.5 km Ma^{-1} (Li, 1976) and uplift rate of 5.0 km Ma^{-1} for the early Holocene (Peng *et al.*, 1977) in support of this idea.

7.5.2 AN UPLIFT–DENUDATION INTERACTION MODEL FOR THE CENTRAL RANGE

The uplift and erosion data for Taiwan can be used to produce a quantitative uplift–denudation interaction model. Previous studies by Yoshikawa (1970, 1974, 1985), Ohmori (1978), Whitehouse (1987, 1988), and particularly Brunsden and Thornes (1979), Ahnert (1987, 1988) and Brunsden and Lin (1991) provide support for this general approach. It appears that the relief of the Central Range of Taiwan developed very slowly for the first 1–2 Ma of its history in the northern part of the island, but grew very rapidly during the mid–late Pleistocene. Uplift rates tend to exceed denudation rates in the south, but these rates become approximately equal in the central half of the island. However, in the northern Central Range, denudation exceeds uplift. In the north, the landforms represent a relaxation from the slow relief growth at the beginning of Taiwan's geomorphological history to the current slower uplift of the Ryukyu tectonic domain (Brunsden and Lin, 1991).

This model illustrates how a landscape achieves a characteristic form and then how it begins to adjust to a new equilibrium when one of the boundary conditions (tectonic setting and uplift rate) is changed (Brunsden and Thornes, 1979). In northern Taiwan the Ryukyu domain landscape seems to be settling to a new equilibrium for rates of denudation and uplift around 2 mm a^{-1} (Brunsden and Lin, 1991). In terms of width, height and cross-sectional flux, the central mountains of Taiwan appear to be in an approximate topographic equilibrium.

7.5.3 APPLICATION TO THE COASTAL RANGE

Because of the lack of denudation data, which are only available for the largest rivers, it is not possible to confirm a topographic steady state for the Coastal Range. Nevertheless, there are a number of indications that a steady state may exist. For instance, the rate of arc–continent collision and the propagation of collision appears to be continuing at a constant rate, and each tectonic slice of the Coastal Range between the main thrust faults appears to maintain a constant width. Moreover, the Coastal Range has a similar topographic form to that of the Central Range, and there is a similarly remarkable accordance of summit heights, range width and topographic forms.

The evolution of the Coastal Range can be divided into at least three stages from north to south, beginning approximately 1.1 Ma ago (Lin, 1994), this interpretation being consistent with recent sedimentological and palaeomagnetic studies (Lee, 1989). The rate of arc–continent collision means that the northern tip of the Coastal Range is older than the southern part. The length of the Coastal Range collision boundary is ~100 km. Dividing this length by the collision rate of 90 km Ma^{-1} means that collision took place ~1.1 Ma ago and moving north 10 km is equivalent to moving back in time by 0.1 Ma at the inner collision edge. The main tectonic units therefore appear to have been produced by progressive collisions from south to north at 0, 0.22, 0.77 and 1.16 Ma BP. This interpretation is similar to the suggestion from Lundberg and Dorsey (1990) that parts of the Coastal Range have been uplifted for up to 1 Ma.

7.6 Discussion and Conclusions

According to Ollier and Pain (1988) any plate tectonic model should be internally consistent and should explain all data, whether structural, lithological, geological, sedimentological or geomorphological. The morphotectonic interpretations presented here are proposed as examples of the value of this strategy. If they are substantiated by further research it is also hoped that they will form a basis for a future seismic hazard programme for Taiwan.

An uplift rate of about 2–6 mm a^{-1} has been established for several areas of Taiwan using a range of methods and approaches. In the coastal area along the Coastal Range minimum Quaternary uplift rates range from 4.0 to 9.7 mm a^{-1}, and are higher than those for the other coastal areas of Taiwan. Contrasts in relief from north to south, and changes in the number and height of marine terraces along the coast, suggest that different structural units have experienced differential movement.

As suggested by Lundberg and Dorsey (1990), long-term denudation rates are in approximate accord with uplift rates in the Coastal Range, and such an approximate equivalence in rates has also been suggested for the Central Range by Peng *et al.* (1977) and Suppe (1981). Current calculations of denudation and uplift rates should be thought of as long-term process rates, and the steady-state uplift–denudation model should be corrected whenever new denudation and uplift data are obtained.

Uplift appears to be irregular within the Coastal Range and indicates a block structure for the orogen. The boundaries of each block are generally demarcated by major structural lines including thrust faults, strike-slip faults and synclinal/anticlinal axes. These boundaries are often seen as lineaments on remote sensing imagery, but more confirmation from field surveys coupled with detailed geological mapping is required. Nevertheless, the main lineaments may provide a useful guide to the main structural lines and potentially seismically hazardous areas.

An uplift–denudation model, based on published uplift and denudation rate data, has been proposed for the Central Range. The model shows how the present landforms reflect the evolution of the Central Range, and can be used as a framework both for future research on landscape evolution in Taiwan and for the more precise identification of areas with a potentially high risk of seismic events.

Acknowledgements

The author wishes to express thanks to all those who have assisted and advised during this research programme, in particular Professor D. Brunsden of the Department of Geography, King's College, London.

References

Ahnert, F. 1987. Process–response models of denudation at different spatial scales. *Catena Suppl.*, **10**, 31–50.

Ahnert, F. 1988. Modelling landform change. In M.G. Anderson (ed.) *Modelling Geomorphological Systems*, Wiley, Chichester, pp. 375–400.

Angelier, J., Barrier, E. and Chu, H.T. 1986. Plate collision and paleostress trajectories in a fold-thrust belt: the foothills of Taiwan. *Tectonophysics*, **125**, 161–178.

Biq, C.C. 1972. Transcurrent buckling, transform faulting and transgression: their relevance in eastern Taiwan kinematics. *Petrol. Geol. Taiwan*, **10**, 1–10.

Biq, C.C. 1974. Metallogeny in Taiwan: a plate-tectonic approach. *Bull. Geol. Surv. Taiwan*, **24**, 139–156.

Bonilla, M.G. 1975. A review of recently active faults in Taiwan. *U.S. Geol. Surv. Open File Rep.*, 41–75.

Brunsden, D. and Lin, J.C. 1991. The concept of topographic equilibrium in neotectonic terrains. In: J. Cosgrove and M. Jones (eds) *Neotectonics and Resources*, Belhaven Press, London, pp. 120–143.

Brunsden, D. and Thornes, J.B. 1979. Landscape sensitivity and change. *Trans. Inst. Brit. Geog.*, N.S. **4**, 463–484.

Chai, B.H.T. 1972. Structure and tectonic evolution of Taiwan. *Amer. J. Sci.*, **272**, 389–422.

Committee for Water Resource Control. 1970. *Taiwan Hydrological Year Book*, Ministry of Economy.

Committee for Water Resource Control. 1973. *The Preliminary Estimation of Economy* (in Chinese).

Ho, C.S. 1975. *An Introduction to the Geology of Taiwan, Explanatory Text of the Geologic Map of Taiwan*, The Ministry of Economic Affairs, Taipei.

Ho, C.S. 1982. *Tectonic Evolution of Taiwan, Explanatory Text of the Tectonic Map of Taiwan*. The Ministry of Economic Affairs, Taipei.

Ho, C.S. 1987. A synthesis of the geologic evaluation of Taiwan. *Mem. Geol. Soc. Taiwan*, **9**, 1–18.

Hsu, T.L. 1954. On the geomorphic features and the recent uplifting movement of the Coastal Range, eastern Taiwan. *Bull. Geol. Surv. Taiwan*, **7**, 51–57.

Hsu, T.L. 1955. Landform of Taiwan. *Quart. J. Taiwan Bank*, **7**(2).

Hsu, T.L. 1956. Geology of the Coastal Range, eastern Taiwan. *Bull. Geol. Surv. Taiwan*, **8**, 15–41.

Hsu, T.L. 1962. Recent faulting in the Longitudinal Valley of eastern Taiwan. *Mem. Geol. Soc. China*, **1**, 95–102.

Hsu, T.L. 1976. Neotectonics of the Longitudinal Valley, eastern Taiwan. *Bull. Geol. Surv. Taiwan*, **25**, 53–62.

Hsu, T.L. and Chang, H.C. 1979. Quaternary faulting in Taiwan. *Mem. Geol. Soc. China*, **3**, 155–156.

Hsu, Y.P. 1964. *The Water Quality of Irrigation Water in Taiwan*, National Taiwan University, Taipei (in Chinese with English summary).

Konishi, K., Omura, A. and Kimura, T. 1968. ^{234}U–^{230}Th dating of some late Quaternary coralline limestones from southern Taiwan (Formosa). *Geol. Palaeont. SE Asia*, **5**, 211–224.

Lee, C.T. and Wang, Y. 1987. Palaeostress change due to the Pliocene–Quaternary arc–continent collision in Taiwan. *Mem. Geol. Soc. China*, **9**, 63–86.

Lee, T.Q. 1989. Evolution tectonique et Géodynamique Néogene et Quaternaire de la chaîne côtiere de Taiwan: apport du Paleomagnetisme. Thèse de doctorat de Univ. Paris, **6**.

Li, Y.H. 1976. Denudation of Taiwan Island since the Pliocene Epoch. *Geology*, **4**, 105–107.

Lin, C.C. 1957. *The Landform of Taiwan*, The Historical Research Commission of Taiwan Province (in Chinese).

Lin, C.H., Yeh, Y.H. and Tsai, Y.B. 1985. Determination of regional principal stress directions in Taiwan from plate plane solutions. *Bull. Inst. Earth Sci. Acad. Sinica*, **5**, 67–85.

Lin, J.C. 1991. The structural landforms of the Coastal Range of eastern Taiwan. In: J. Cosgrove and M. Jones (eds) *Neotectonics and Resources*, Belhaven Press, London, pp. 65–74.

Lin, J.C. 1994. An evolutionary model for the Coastal Range, Eastern Taiwan. *Chinese Environment and Development*, **5**, 7–27.

Lin, M.T. and Tsai, Y.B. 1981. Seismotectonics in Taiwan–Luzon area. *Bull. Inst. Earth Sci. Acad. Sinica*, **1**, 51–82.

Liu, J.K. 1984. Two newly-discovered tectonic patterns in Taiwan region – the circular pattern and the NW–SE shear zone – as interpreted from small-scale remote sensing images. *Int. Soc. Photogrammetry and Remote Sensing*, **25**, 293–302.

Liu, T.K. 1982. Tectonic implication of fission track ages from the Central Range, Taiwan. *Proc. Geol. Soc. China*, **25**, 22–37.

Lue, Y.T., Lee, T.Q. and Wang, Y. 1995. Paleomagnetic study on the collision-related bending of the fold-thrust belt, northern Taiwan. *J. Geol. Soc. China*, **38**, 215–227.

Lundberg, N. and Dorsey, R.J. 1990. Rapid Quaternary emergence, uplift, and denudation of the Coastal Range, eastern Taiwan. *Geology*, **18**, 638–641.

Ohmori, H. 1978. Relief structures of the Japanese mountains and their stages in geomorphic development. *Bull. Dept. Geog. Univ. Tokyo*, **10**, 31–85.

Ollier, C.D. and Pain, C.F. 1988. Morphotectonics of passive continental margins. *Z. Geomorph. Suppl.*, **69**, 1–16.

Omori, F.K. 1907. Earthquake of the Chiayi area, Taiwan, 1906. *Introduction of Earthquake*, 103–147 (in Japanese).

Page, B.M. and Suppe, J. 1981. The Pliocene Lichi Melange of Taiwan: its plate tectonic and olistostromal origin. *Amer. J. Sci.*, **281**, 193–227.

Peng, T.H., Li, Y.H. and Wu, F.T. 1977. Tectonic uplift rates of the Taiwan Island since the early Holocene. *Mem. Geol. Soc. China*, **2**, 57–69.

Pirazzoli, P.A., Arnold, M., Giresse, P., Hsieh, M.L. and Liew, P.M. 1993. Marine deposits of late glacial times exposed by tectonic uplift on the east coast of Taiwan. *Mar. Geol.*, **110**, 1–6.

Sander, B. 1970. *An Introduction to the Study of the Fabric of Geological Bodies*, Pergamon, Oxford.

Seno, T. 1977. The instantaneous rotation vector of the Philippine Sea Plate relative to the Eurasian Plate. *Tectonophysics*, **42**, 209–226.

Shih, T.T., Teng, K.H., Chang, J.C., Shih, C.D. and Yang, G.S. 1986. Geomorphological study of active fault in Taiwan. *Geographical Studies*, **12**, The Department of Geography, National Taiwan Normal University (in Chinese).

Suppe, J. 1981. Mechanics of mountain building and metamorphism in Taiwan. *Mem. Geol. Soc. China*, **4**, 67–90.

Suppe, J. 1984. Kinematics of arc–continent collision, flipping of subduction, and back-arc spreading near Taiwan. *Mem. Geol. Soc. China*, **6**, 21–33.

Suppe, J. 1987. The active mountain belt. In: S.P. Schaer and J. Rodgers (eds) *The Anatomy of Mountain Ranges*, Princeton University Press, New Jersey, pp. 277–293.

Tan, L.P. 1977. Pleistocene eastward bending of the Taiwan Arc. *Mem. Geol. Soc. China*, **2**, 77–83.

Tsai, C.C., Loh, C.H. and Yeh, Y.T. 1987. Analysis of earthquake risk in Taiwan based on seismotectonic zones. *Mem. Geol. Soc. China*, **9**, 413–446.

Tsai, Y.B. 1978. Plate subduction and the Plio-Pleistocene Orogeny in Taiwan. *Petrol. Geol. Taiwan*. **15**, 1–10.

Tsai, Y.B. 1985. A study of disastrous earthquakes in Taiwan, 1683–1895. *Bull. Inst. Earth Sci. Acad. Sinica*, **5**, 1–44.

Tsai, Y.B., Teng, T.L., Chiu, J.M. and Liu, H.L. 1977. Tectonic implications of the seismicity in the Taiwan region. *Mem. Geol. Soc. China*, **2**, 13–41.

Tsan, S.F. 1960. On the problem of the glaciated topography of the Nanhutashan. *Mem. Geol. Soc. China*, **3**, 109–111.

Wang, Y. 1987. Continental margin rifting and Cenozoic tectonics around Taiwan. *Mem. Geol. Soc. China*, **9**, 227–240.

Whitehouse, I.E. 1987. Geomorphology of a compressional plate boundary, Southern Alps, New Zealand. In: V. Gardiner, (ed.) *International Geomorphology 1986, Part I*, Wiley, Chichester, pp. 897–924.

Whitehouse, I.E. 1988. Geomorphology of the central Southern Alps, New Zealand: the interaction of plate collision and atmospheric circulation. *Z. Geomorph. Suppl.*, **69**, 105–116.

Willemin, J.H. and Knuepfer, L.K. 1994. Kinematics of arc–continent collision in the eastern Central Range of Taiwan inferred from geomorphic analysis. *J. Geophys. Res.*, **99**, 20 267–20 280.

Wolman, M.G. and Gerson, R. 1978. Relative scales of time and effectiveness of climate in watershed geomorphology. *Earth Surf. Processes*, **3**, 189–208.

Wu, F.T. and Lu, C.D. 1976. Recent tectonics of Taiwan. *Bull. Geol. Surv. Taiwan*, **25**, 97–111.

York, J.E. 1976. Quaternary faulting in eastern Taiwan. *Bull. Geol. Surv. Taiwan*, **25**, 63–72.

Yoshikawa, T. 1970. On the relations between Quaternary tectonic movement and seismic crustal deformation in Japan. *Bull. Dept. Geog. Univ. Tokyo*, **2**, 1–24.

Yoshikawa, T. 1974. Denudation and tectonic movement in contemporary Japan. *Bull. Dept. Geol. Univ. Tokyo*, **6**, 1–14.

Yoshikawa, T. 1985. Landform development by tectonics and denudation. In: A. Pitty (ed.) *Themes in Geomorphology*, Croom Helm, London, pp. 194–210.

Yu, S.B. and Chen, H.Y. 1994. Global positioning system measurements of crustal deformation in the Taiwan arc–continent collision zone. *Terrestrial, Atmospheric and Oceanic Sciences*, **5**, 477–498.

8

Morphotectonic evolution of Japan

Hiroo Ohmori

8.1 Introduction

Japan forms part of the Circum-Pacific series of island arcs. It is characterized by violent crustal movement, and is located in a warm-humid climatic region with high precipitation that promotes active mass movement. Landforms in Japan have experienced rapid and continuous change due to active tectonics and intense erosion. From an anthropogenic perspective it can also be said that the frequent occurrence of earthquakes, volcanic eruptions, landslides, debris flows and many other kinds of natural hazard has had a significant impact on human activities. In order to predict and prevent these events, contemporary crustal movements and other natural hazards have been quantitatively surveyed, especially during the past three decades when residential, industrial and commercial areas have markedly expanded over the lowlands as a result of economic development. For instance, digital altitude data have been published and analysed in order to define the regional characteristics of landforms in Japan (Sakaguchi, 1964; Hagiwara, 1967; Ohmori, 1978; Geographical Survey Institute of Japan, 1979, 1980). Vertical displacement during the Quaternary has also been investigated (Research Group for Quaternary Tectonic Map, 1968, 1969, 1973), with most of the active faults and folds being precisely surveyed in order to clarify their level of activity as part of an earthquake prediction programme (Research Group for Active Faults, 1980a, b). Levelling surveys have also been carried out at short intervals, especially along the Pacific coast, under a national project operated by the Geographical Survey Institute of Japan. Damage caused by mass movements such as landslides, slope avalanches and debris flows has been intensively surveyed in Japan every year since 1975 by the Ministry of Construction.

From these observations of contemporary and recent process rates, the frequency and magnitude of both tectonic events and mass movement have been evaluated. Through analyses of their spatial and temporal characteristics, the investigation of Cenozoic geological structures, Quaternary tectonics and geomorphic processes has markedly advanced, revealing the complexity and regional characteristics of morphotectonics and landscape-forming processes in Japan (Matsuda *et al.*, 1967; Yoshikawa, 1974, 1984, 1985a, b; Ohmori, 1978; Yoshikawa *et al.*, 1981; Kaizuka, 1987; Yonekura *et al.*, 1990; Kaizuka and Suzuki, 1993).

This chapter first introduces the tectonic framework of Japan, then examines the relationship between contemporary crustal displacements and neotectonic movements, and provides estimates of the amount of denudation in the Quaternary. The relationship between denudation rate and altitude is then assessed, and the change in mountain altitude resulting from concurrent tectonics and denudation is modelled. The chapter concludes with an evaluation of the stages of geomorphic development.

8.2 Tectonic Framework of Japan

The landscape of Japan was reduced to subdued relief in the late Tertiary, but was then uplifted in the Quaternary (Matsuda *et al.*, 1967; Research Group for Quaternary Tectonic Map, 1968, 1969, 1973; Yoshikawa *et al.*, 1981). Rates of Quaternary deformation of geological structures are an order of magnitude larger than those of the late Tertiary (Matsuda, 1981), and the present landforms of Japan have been formed dominantly in relation to Quaternary crustal movements controlled by recent plate tectonics.

Geomorphology and Global Tectonics. Edited by Michael A. Summerfield.

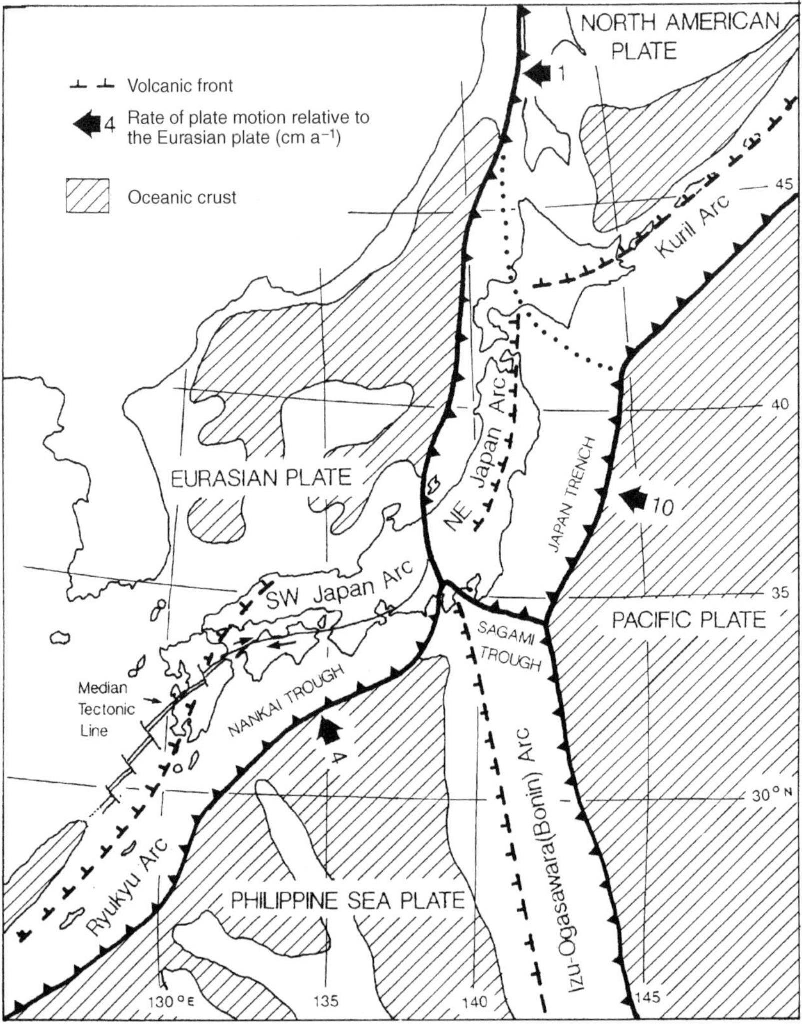

Figure 8.1 *Tectonic framework in and around Japan. A dotted line crossing the northern part of Japan shows an alternative boundary between the North American and Eurasian plates. Compiled from Research Group for Active Faults (1980a), Geological Survey of Japan (1982), Kaizuka (1987) and Kanaori and Kawakami (1996)*

Japan is located at the junction of four plates; namely the North American, Pacific, Philippine Sea and Eurasian plates (Figure 8.1) (Research Group for Active Faults, 1980a; Geological Survey of Japan, 1982; Nakamura, 1983; Nakamura *et al.*, 1984; Seno, 1985; Tamaki and Honza, 1985; Kaizuka, 1987; Kanaori *et al.*, 1992; Kanaori and Kawakami, 1996). To the east, the Japan Trench extends along the subduction zone of the Pacific plate, and to the southwest the Nankai Trough stretches along the subduction zone of the Philippine Sea plate. At the northern edge of the Philippine Sea plate collisional contact occurs against the land mass of central Japan, and from there the Izu-Ogasawara (Bonin) arc extends along the eastern margin of the Philippine Sea plate. The boundary between the North American and Eurasian plates is a convergent, collisional contact which crosses the central part of Japan, dividing the Japanese islands into the Northeast and Southwest Japan arcs across which the distribution of pre-Neogene formations shows a marked contrast (Figure 8.2A) (Geological Survey of Japan, 1982).

Due to the relative direction and rate of movement of the Pacific plate with respect to the North American

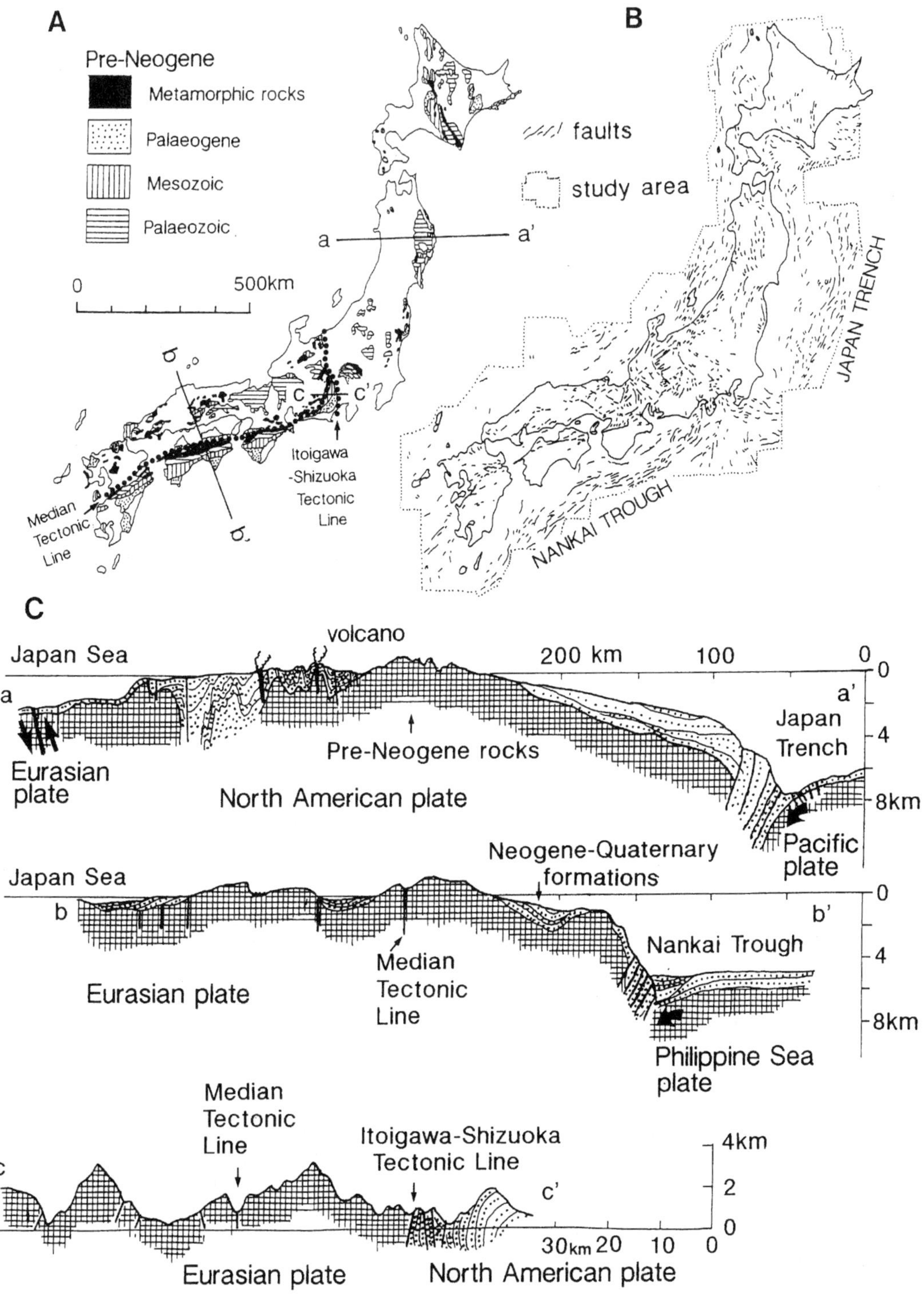

Figure 8.2 *(A) Distribution of pre-Neogene strata (Geological Survey of Japan, 1982) and (B) faults (simplified from Research Group for Active Faults, 1980a) over the Japanese islands; and (C) schematic cross-sections showing geological structures (a–a': Ishiwada* et al.*, 1977; b–b': Huzita, 1980; c–c': compiled from Matsuda, 1961, and Geological Survey of Japan, 1982)*

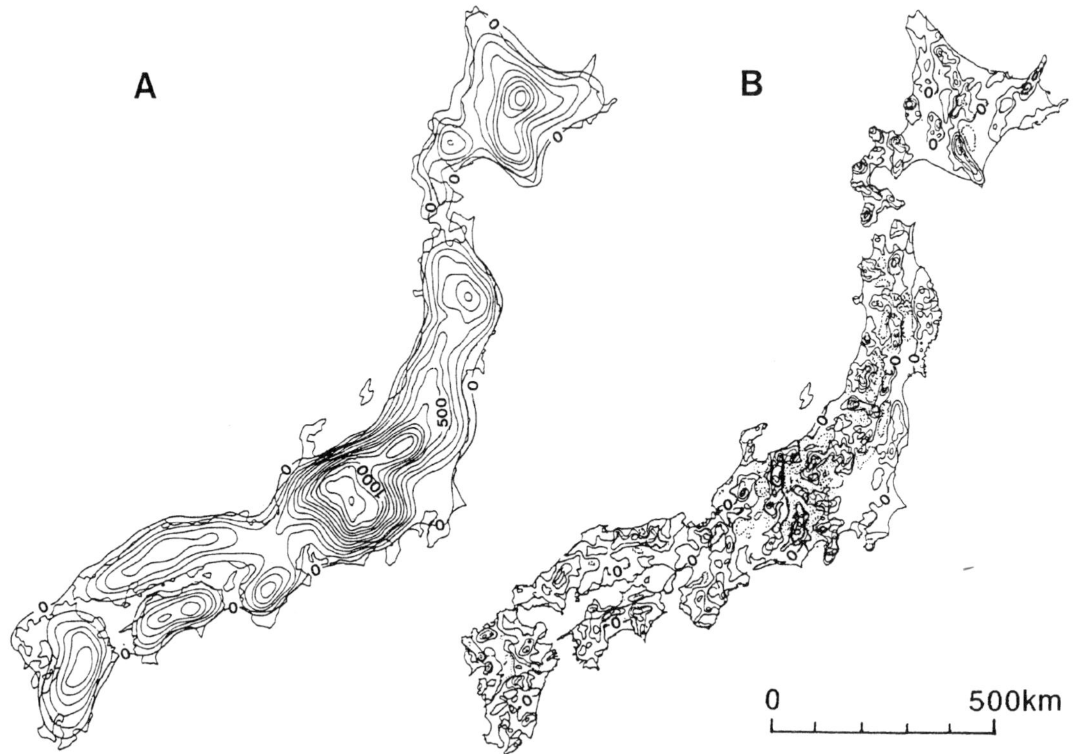

Figure 8.3 *(A) Distribution of altitude of long wavelength topography (>150 km) (contour interval: 100 m), and (B) of short wavelength topography (contour interval: 200 m), calculated by a running mean method (Hagiwara, 1967)*

plate, and of the Philippine Sea plate with respect to the Eurasian plate, the stress field of Japan is dominantly compressional in an east–west or southeast–northwest direction with complex associated fault and fold systems (Figure 8.2B) (Research Group for Active Faults, 1980a). The Northeast Japan arc, especially on its Japan Sea side, is characterized by extensive Neogene–Quaternary formations and volcanic rocks. They are markedly deformed by N–S-trending thrust faults and folds (Figure 8.2C, a–a') (Ishiwada *et al.*, 1977). The Southwest Japan arc consists of pre-Neogene sedimentary and metamorphic rocks with relatively well-preserved structures (Figure 8.2C, b–b') (Huzita, 1980). The boundary between the Northeast and Southwest Japan arcs is geologically indicated by the Itoigawa–Shizuoka Tectonic Line which is composed of active and non-active faults running between the western pre-Neogene and eastern Neogene formations (Figure 8.2C, c–c') (Matsuda, 1961; Geological Survey of Japan, 1982).

Because the direction of the stress related to plate motion varies within a plate and the mode of crustal movement observed on the surface often extends over plate boundaries, and because the mountain block which forms the basic unit of crustal movement inherits pre-Neogene geological structures, surface altitude across Japan is distributed relatively independently of the distribution of plates (Figure 8.3A) (Hagiwara, 1967). Japan can be divided into five terrains on the basis of distribution of altitude, the composition of highlands and lowlands and other landform properties such as degree of dissection and form of slopes, and structural and tectonic characteristics. These are Southwest Japan, Central Japan, Northeast Japan, the mainland of Hokkaido and the Izu Peninsula (Figure 8.4) (Okayama, 1953; Research Group for Quaternary Tectonic Map, 1968; Research Group for Active Faults, 1980a; Yoshikawa *et al.*, 1981; Ohmori, 1995). Each terrain is divided into mountain blocks which consist of tilt blocks bounded by faults and up-warped blocks (Figure 8.3B) (Hagiwara, 1967).

Southwest Japan has inherited a pre-Neogene structure to a significant extent (Figure 8.2A). The terrain is divided into an inner belt (Japan Sea side) and an outer

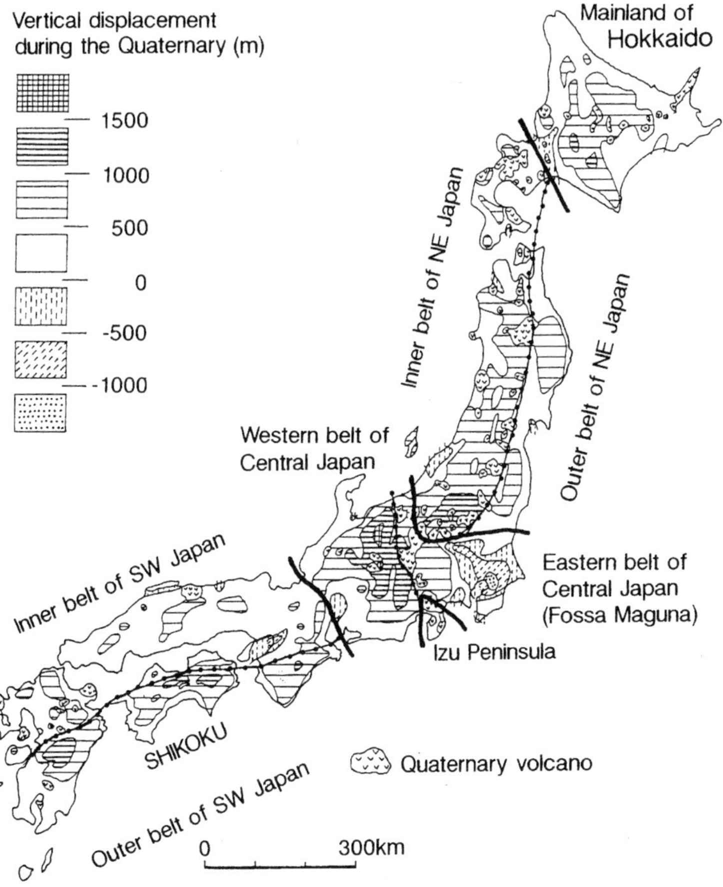

Figure 8.4 *Morphotectonic divisions of Japan and vertical displacement during the Quaternary. Compiled from Okayama (1953), Research Group for Quaternary Tectonic Map (1968), Research Group for Active Faults (1980a), Yoshikawa* et al. *(1981) and Ohmori (1995)*

zone (Pacific Ocean side) by the Median Tectonic Line which is a right-lateral strike-slip fault. The mountains in the outer zone have been built by upwarping with high uplift rates, resulting in a high, rugged landscape in sharp contrast to the low mountains in the relatively stable inner belt. Central Japan is divided into two sub-terrains, the western and eastern belts, by the Itoigawa–Shizuoka Tectonic Line. The western belt occupies the eastern part of the Southeast Japan arc where pre-Neogene sedimentary and metamorphic rocks predominate (Figure 8.2A). Due to high compression, N–S-trending thrust faults and folds are conspicuous along the eastern margin, creating the highest mountains in Japan with peaks of around 3000 m above sea level. In sharp contrast to this, plateau-like mountains occur in the western part where NW–SE and NE–SW-trending strike-slip faulting is dominant. The pairing of high mountains fronting a plateau is topographically somewhat similar to the Himalayan–Tibetan Plateau system. The eastern belt of Central Japan is an area called the 'Fossa Maguna' and comprises basins with Neogene formations. Its western margin is bounded by the Itoigawa–Shizuoka Tectonic Line and its eastern margin is demarcated by a line connecting the eastern margins of the present tectonic basins, such as the Kanto plain, which have experienced the most subsidence during the Quaternary (Figure 8.4). The Neogene and Quaternary formations in the Fossa Maguna have been strongly deformed by faulting and folding with a N–S trend (Figure 8.2C, c–c').

The boundary between Northeast Japan and the mainland of Hokkaido runs through a depression zone in the western part of Hokkaido where a possible alternative boundary between the North American and Eurasian plates is located (Figure 8.1). Both terrains have volcanic zones in their continental portions which divide them into a volcanic inner belt and a non-volcanic outer zone (Figures 8.1, 8.2C, a–a', and 8.4). The inner belts are characterized by active volcanoes, and the Neogene–Quaternary sedimentary rocks are deformed by thrust faulting and folding. The outer belts, however, are covered by pre-Neogene sedimentary, metamorphic and granitic rocks whose geological structures were formed mainly in the Neogene or before.

Izu Peninsula is an independent morphotectonic terrain at the northern edge of the Izu–Ogasawara arc. It is characterized by many active volcanoes in a N–S-trending chain. Right-lateral NW–SE-trending and left-lateral NE–SW-trending strike-slip faults are densely distributed. The Ryukyu arc, extending to the southwest from Kyushu, is divided into inner and outer belts by a volcanic front. The inner belt consists of volcanic islands and the outer belt is characterized by Quaternary limestones originating from coral reefs covering pre-Tertiary sedimentary rocks.

8.3 Seismicity and Crustal Displacement

Japan has frequently experienced great earthquakes of a magnitude larger than M 7 associated with abrupt crustal deformation. Great earthquakes affecting an area larger than 1000 km^2 have recurrence intervals of 100 years or more, often over several thousand years in the same region. Repeated precise levelling surveys have revealed slow and long-lasting vertical displacement even when no earthquakes have occurred. Moreover, coseismic movement often shows a reverse direction to interseismic movement. Thus, when long-term crustal deformation is evaluated, the resultant movement of both abrupt coseismic and continuous, slow interseismic movements should be considered.

8.3.1 RECENT AND QUATERNARY CRUSTAL MOVEMENTS

The duration of precise levelling surveys carried out in Japan is about 100 years or less as the first set of bench marks was established in the 1890s. The complete sequence of changes in crustal movements for one cycle of interseismic and coseismic events has not yet been observed in Japan. Consequently it is necessary to estimate the resultant movement by analysing the data from precise levelling for short periods including the occurrence of earthquakes. In order to estimate earthquake recurrence intervals, it is necessary to have historical records of several earthquakes that have affected a particular area. Areas with data from both repeated precise levelling surveys including seismic displacements, and historical records of earthquakes, are limited even in Japan.

The eastern part of Shikoku, an island in the outer belt of southwest Japan (Figure 8.4), is located near the old capital cities of Nara, Kyoto and Osaka. Nine earthquakes have affected this region since about 700 AD, indicating a recurrence interval ~140 a. Precise levelling surveys, which do not include measurements of horizontal deformation, have been carried out seven or eight times since the first establishment of bench marks in 1895 to 1897 along the highways running along the coast, although many of the initial set of bench marks are now missing. This period of observation includes a preseismic phase, the M 8.1 Nankai earthquake of 1946, and postseismic movements. In addition, along the southern coast of Shikoku Pleistocene marine terraces are well developed and these provide comparative evidence of relatively long-term crustal deformation. Shikoku is therefore the optimal location in Japan for the examination of displacements associated with coseismic and interseismic movements, and the resultant movements have previously been discussed on the basis of levelling data and the altitude of Pleistocene marine terraces (Yoshikawa *et al.*, 1964; Yoshikawa, 1968, 1970; Ohmori, 1990a, b).

Examples of vertical displacements of selected bench marks are shown in Figure 8.5 (Ohmori, 1990a, b). Modes of the movement differ with levelling routes, and a conspicuous characteristic is that the coseismic movement has a reverse direction to the preseismic movement. Analysis of pre- and postseismic movements from levelling surveys shows that interseismic movements consist of two components with different directions of movement, termed 'accessory' and 'recovery' (Figure 8.6) (Ohmori, 1990b). Accessory movement occurs just after an earthquake and has the same direction as the coseismic movement. Recovery movement reverts to the same direction as the preseismic movement. It is characterized by three periods of different rates with time elapsed after an earthquake, from slow to rapid and then to slow again. Interseismic movement has two components – an after-effect phase characterized by changes in both rate and direction of movement, and a secular phase with a constant rate. On the basis of the trends of movements peculiar to each levelling route, the resultant rates of

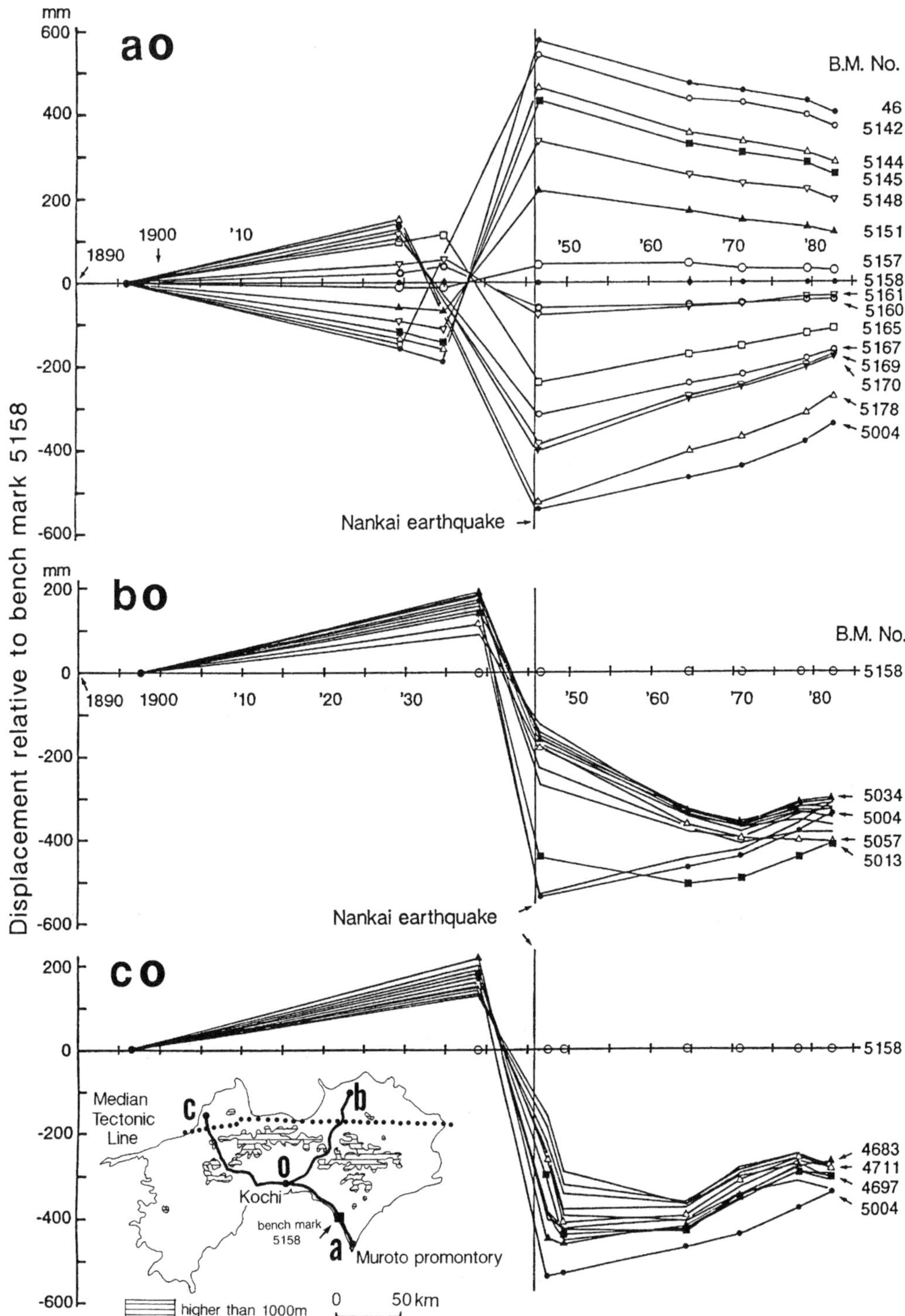

Figure 8.5 Examples of displacement of bench marks relative to reference bench mark no. 5158, showing differences in mode with levelling route (Ohmori, 1990a, b). The Nankai earthquake M 8.1 occurred on 21 December 1946

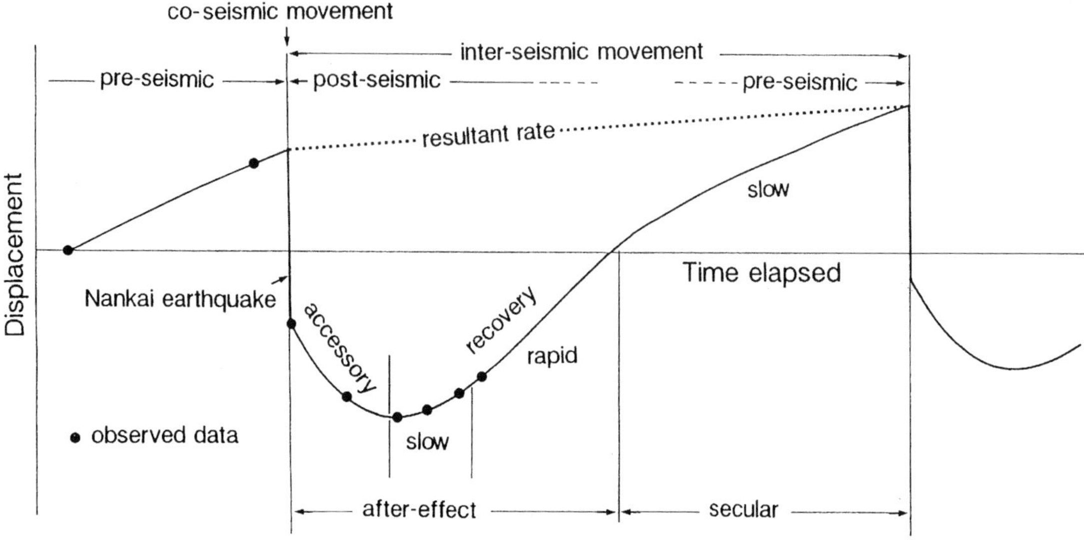

Figure 8.6 *Generalized sequential change in displacement of pre-seismic, co-seismic and post-seismic movements, showing differences in rate, mode and period of movement (Ohmori, 1990b)*

one cycle of interseismic and coseismic activity have been estimated for individual bench marks (Ohmori, 1990a, b).

Pleistocene marine terraces of various ages are well developed along the coast from Kochi to the promontory of Muroto (Figure 8.7) (Yoshikawa *et al.*, 1964). The longest continuous terrace is the 'Muroto-saki terrace I', which was formed during the last interglacial high stand about 120 ka BP, when sea level is estimated to have been about 5 m higher than at present (Miyoshi, 1983; Bloom and Yonekura, 1985; Kikuchi, 1988). The altitude of the Muroto-saki terrace I just behind bench mark no. 5158, the standard bench mark for the area, is 82 m above present sea level. Thus, the mean uplift rate of the bench mark relative to sea level is estimated to be 0.68 mm a^{-1} over the past 120 ka, a rate which is almost the same as the recent rate of uplift calculated from the change in altitude of datum lines for the tidal stations in Shikoku (Ohmori, 1990a, b). The resultant rate of crustal movement relative to sea level at each bench mark has been estimated by using the resultant rate relative to bench mark no. 5158 + 0.68 mm a^{-1}. The resultant rates at the bench marks are in close agreement with rates estimated from the adjacent raised beaches of Muroto-saki terrace I (Figure 8.7), indicating that the mode and rate of recent crustal movement have persisted from at least the late Pleistocene (Yoshikawa *et al.*, 1964; Yoshikawa, 1968, 1970; Ohmori, 1990a, b). Because resultant movements accumulate over the sequence of interseismic and coseismic movement cycles and cause cumulative landform displacement, they have been termed 'morphogenetic crustal movements'; in order to distinguish them from individual coseismic and interseismic movements (Yoshikawa, 1970, 1974, 1985a, b).

8.3.2 INLAND QUATERNARY VERTICAL DISPLACEMENTS

Tectonic movements along the Japanese coast have been discussed on the basis of the altitude of raised former shorelines of marine terraces formed since the late Pleistocene. The tectonic movements inferred from the deformation of these marine terraces indicate the cumulative character of tectonic movements including displacements by historically documented earthquakes (Ota, 1975, 1985, 1986; Matsuda *et al.*, 1978; Ota and Yoshikawa, 1978; Nakata *et al.*, 1979).

Planation surfaces developed extensively over Japan in the late Tertiary are now distributed on mountain ridges as low-relief erosion surface remnants. Quaternary vertical displacements of these uplifted surfaces, some of which are associated with Neogene marine sedimentary rocks, can be estimated from their present altitudes (Figure 8.4) (Research Group for Quaternary Tectonic Map, 1968, 1969, 1973; Yoshikawa, 1968). Sugai (1990) has estimated uplift of up to ~2000 m in central Japan since the early Quaternary on the basis of

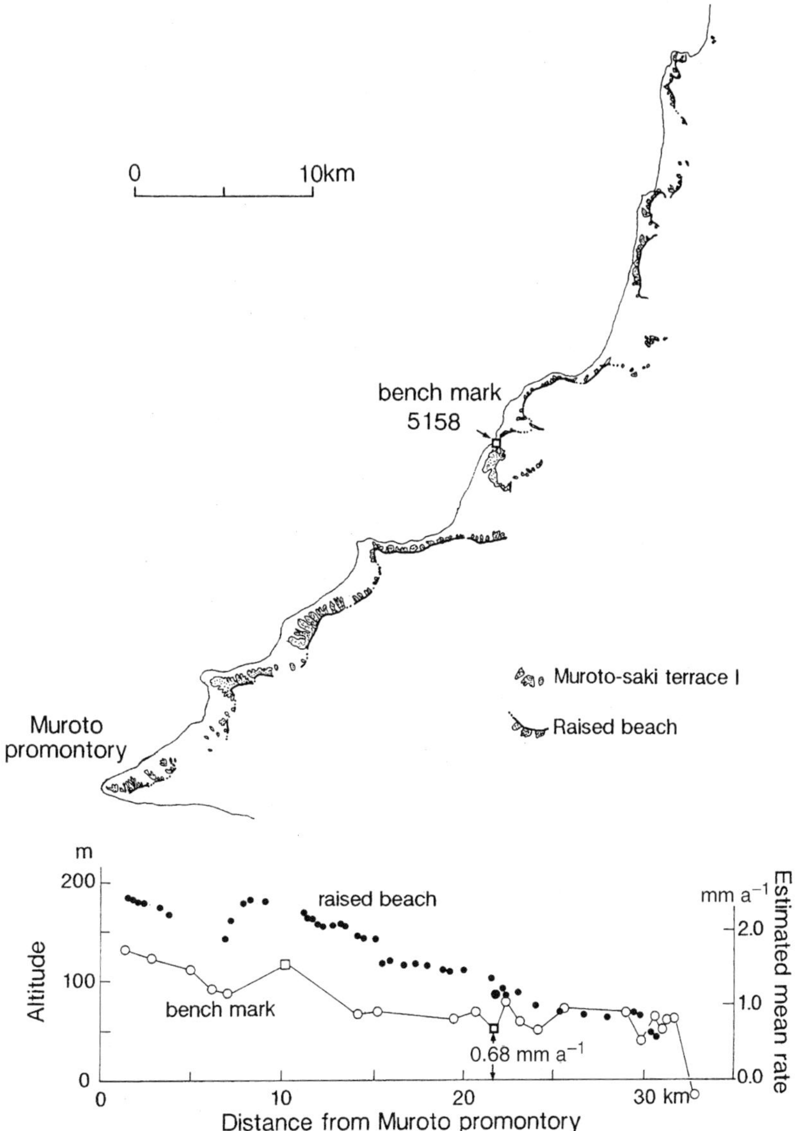

Figure 8.7 *Distribution of the Muroto-saki terrace I formed about 120 ka BP (simplified from Yoshikawa* et al., *1964), showing the relationship between the altitude of the raised beach of the Muroto-saki terrace I and the resultant rates of bench mark displacements (Ohmori, 1990b). The resultant rates are shown by the rates of movement relative to sea level*

evidence from Neogene and Quaternary sediments, periglacial deposits, weathering materials and river channel morphology. These uplifted surfaces are progressively more dissected at higher elevations, and they consist only of fragmentary remnants at altitudes approaching 2000 m. Periglacial activity is particularly evident above 2000 m and estimates of uplift based on erosion surface elevations are minimum estimates because of the possibility of erosional lowering (Yoshikawa, 1984, 1985b; Ohmori, 1987, 1990a, b; Sugai, 1990). At lower elevations, however, the presence of Neogene–Quaternary sediments provides better constraints on uplift and there is a closer correspondence with rates of vertical displacement derived from raised shorelines (Yoshikawa, 1968, 1970). More generally, since modes and rates of crustal movement at time scales ranging from the present to the Quaternary are similar, it can be inferred that landforms created during the Quaternary have been developed under the same tectonic regime as that of the present.

8.4 Denudation in the Quaternary

Various landscape development models have been proposed (Summerfield, 1991), although in many cases tectonic and denudational processes have been treated relatively independently of each other. Recently, however, the coupling of tectonic and denudational processes has received greater attention (Beaumont *et al.*, 1992; Bishop and Brown, 1992; Isacks, 1992; Hoffman and Grotzinger, 1993; Gilchrist *et al.*, 1994) (see Chapter 3). From the perspective of contemporary and recent tectonic movements and denudational processes in Japan, orogen morphology has clearly evolved as the product of a coupling of tectonics and denudation. This coupling involves complex feedbacks, with tectonic uplift and orogen development inducing both high crustal stress and deformation, and orographic precipitation. High denudation rates result from this combination of high rates of runoff and crustal deformation and are associated with the formation of deeply incised landscapes which promote active mass movement. Denudational unloading affects the tectonic stress field and causes new tectonic movements, including isostatic rebound, which contributes further to mountain uplift.

For Japan, the relationship between mean mountain altitude $\bar{H}$ (m) and mean Quaternary uplift rate over the past 1 Ma, $\bar{U}$ (ma^{-1}), is expressed by

$$\bar{H} = \mu \bar{U}^{\nu} \quad (1)$$

where μ and ν are constants ($\mu = 0.217 \times 10^6$ and $\nu = 0.84$ for the best-fit function with a correlation coefficient of 0.95) (Figure 8.8). The mean altitude is apparently controlled by the mean uplift rate. Because $\bar{U}$ is the mean uplift rate over 1 Ma and $\bar{H}$ is the present mean altitude, the total amount of denudation over the period since onset of uplift, M (m), is calculated by

$$M = \bar{U} \times 10^6 - \bar{H} \quad (2)$$

Where $\bar{U} \times 10^6$ is the altitude which the mountain can attain without denudation. Then the mean denudation rate $\bar{R}$ (ma^{-1}), is given by

$$\bar{R} = M/10^6 = (\bar{U} \times 10^6 - \bar{H})/10^6 \quad (3)$$

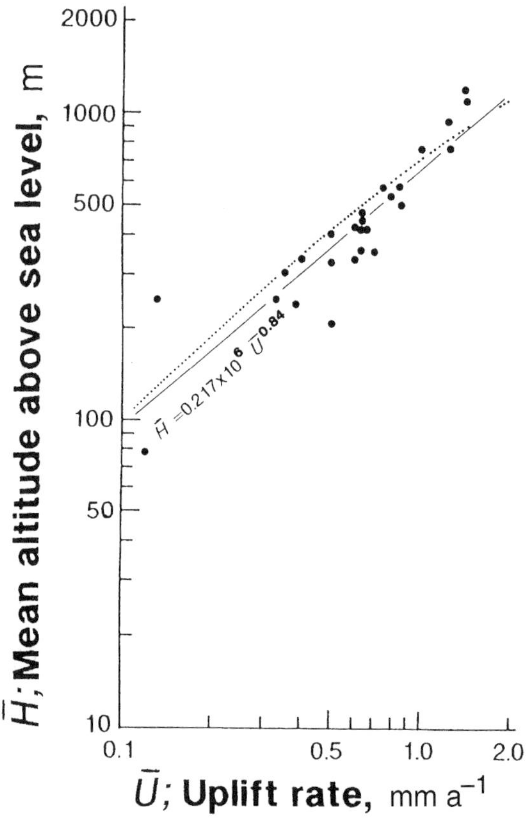

Figure 8.8 Relationship between present mean altitude and mean Quaternary uplift rate for the Japanese mountain ranges. The mean Quaternary uplift rates were estimated by dividing mean vertical displacement during the Quaternary of individual mountain ranges by 1 Ma. The dotted line shows the mean altitude which would have been attained if the ranges had been uplifted at the rate shown by the abscissa from sea level in 1 Ma derived from Figure 8.15 (based on equation (12))

By substituting equation (1) into equation (3), we obtain the equation

$$\bar{R} = (\bar{U} \times 10^6 - \mu \bar{U}^{\nu})/10^6 = \bar{U}(1 - \mu \times 10^{-6} \bar{U}^{\nu - 1}) \quad (4)$$

Equation (4) indicates the relationship between mean uplift rate and mean denudation rate over the period since the onset of uplift. From equation (4), the ratio of denudation to uplift, r_d, can be evaluated. It is given by

$$r_d = \bar{R}/\bar{U} = 1 - \mu \times 10^{-6} \bar{U}^{\nu - 1} \quad (5)$$

From equation (5), for example, r_d is 0.41 (41%) for an uplift rate of 2 mm a^{-1}, r_d is 0.34 (34%) for an uplift rate of 1 mm a^{-1} and r_d is 0.27 (27%) for an uplift rate of 0.5 mm a^{-1}.

It should be pointed out here that both rates are estimated by dividing the total amount of displacement and/or denudation by the elapsed time. They do not demonstrate a dynamic relationship, which will be discussed in the following sections, but rather show a static one which simply indicates the mean condition resulting from mountain building by concurrent tectonics and denudation. In particular, it should be noted that denudation rates are low at low altitudes but significantly increase with altitude in response to higher uplift rates.

8.5 Denudation Rates in Relation to Morphotectonics

The Japanese mountains are covered with dense vegetation mainly due to the warm-humid climate. Nevertheless, their denudation rates are high because the denudational processes in Japan are dominated by landslides and debris avalanches in response to heavy rainfall, with the erosional processes overcoming the protective effect of vegetation and sometimes being triggered by seismic and volcanic activity (Yoshikawa, 1974; Tanaka, 1976; Ohmori, 1983b, c; Machida, 1984; Japan Landslide Society, 1988; Fujita *et al.*, 1989). Under these conditions, the Japanese mountains have been deeply dissected, resulting in a landscape of V-shaped valleys and steep slopes.

Present-day denudation rates in Japan have been estimated mainly on the basis of sediment delivery rates to reservoirs (Tanaka and Ishigai, 1951; Tanaka, 1955; Namba and Kawaguchi, 1965; Ishigai, 1966; Yoshikawa, 1974; Ohmori, 1978, 1983b, c; Mizutani, 1981; Tanaka, 1982). In contrast to short-term variations as a result of individual flood events, rates calculated from total sediment supply for reasonably long periods show stable values for individual reservoirs, ranging from 10^1 to 10^3 m^3 km^{-2} a^{-1} (10–1000 mm ka^{-1}) (Figure 8.9) (Yoshikawa, 1974; Ohmori, 1978). These rates show good agreement with rates inferred from the sediment volume of alluvial cones formed since the Last Glacial Maximum (Iso *et al.*, 1980; Akojima, 1983).

Local variations in denudation rate depend predominantly not on local differences in precipitation and rock type but on local variations in relief (Ohmori, 1978; Tokunaga and Ohmori, 1989). The dispersion of altitude, defined as the standard deviation of the frequency distribution of surface altitude in an area of specific size (Ohmori, 1978), is a measure of local relief

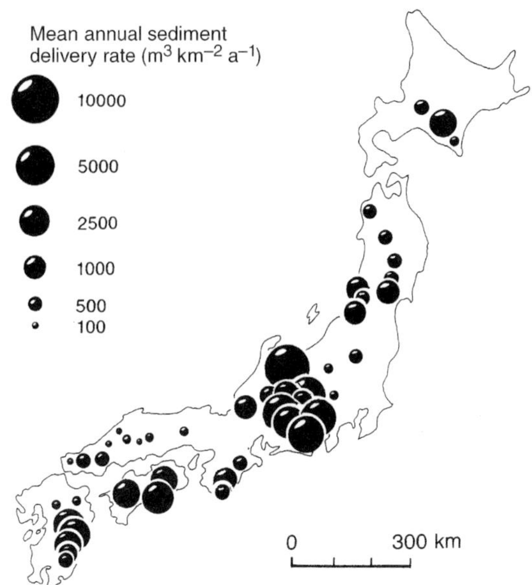

Figure 8.9 *Mean annual sediment delivery rates to reservoirs in Japan. The rate indicates the value calculated by dividing the total sediment volume accumulated in a reservoir by the area of drainage basin of the reservoir and by the observation duration. Compiled from Yoshikawa (1974) and Ohmori (1978)*

(Evans, 1972; Ohmori, 1978). It is directly proportional to mean hillslope gradient (Ohmori and Sohma, 1983; Ohmori and Hirano, 1984) and indicates the dispersion of potential energy of the surface materials. Using the altitude dispersion of unit squares of 1 km^2, the relationship between denudation rate E (m a^{-1}) and mean altitude dispersion D (m) of drainage basins has been expressed empirically for the Japanese mountains using a function for an area larger than several tens of km^2 (Ohmori, 1978, 1982),

$$E = \alpha D^{\beta} \qquad (6)$$

where α and β are constants. The correlation coefficient is 0.77. For a mean density of reservoir sediment of 1750 kg m^{-3} and mean crustal density of 2500 kg m^{-3}, $\alpha = 0.35 \times 10^{-9}$ and $\beta = 3.2$ for the best-fit function (Figure 8.10) (Ohmori, 1978, 1982, 1983b, 1993). Denudation rates estimated by the function correspond well with those calculated from the morphometric analysis of mountain slopes and volume of alluvial fans formed since the last glacial (Oguchi, 1991).

Using the mean altitude dispersion over a whole mountain range, denudation rates of individual mountains in Japan have been calculated from equation (6)

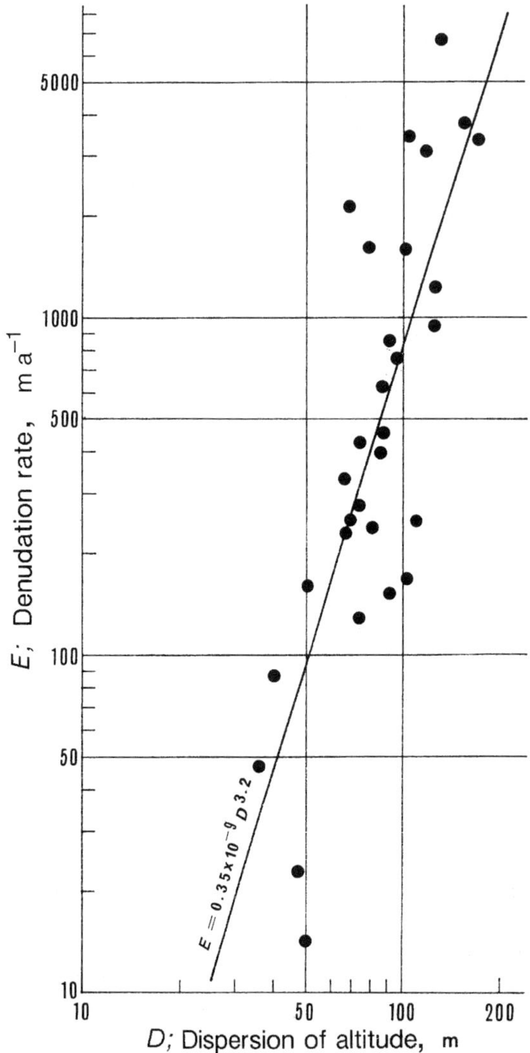

Figure 8.10 *Relationship between altitude dispersion and denudation rate for reservoirs in Japan (Ohmori, 1978). The dispersion of altitude indicates the arithmetical mean of altitude dispersion of a specific area of 1 km² over the drainage area of individual reservoirs. The denudation rate is derived from the sediment accumulation rate assuming a mean sediment density of 1750 kg m⁻³ and a crustal density of 2500 kg m⁻³*

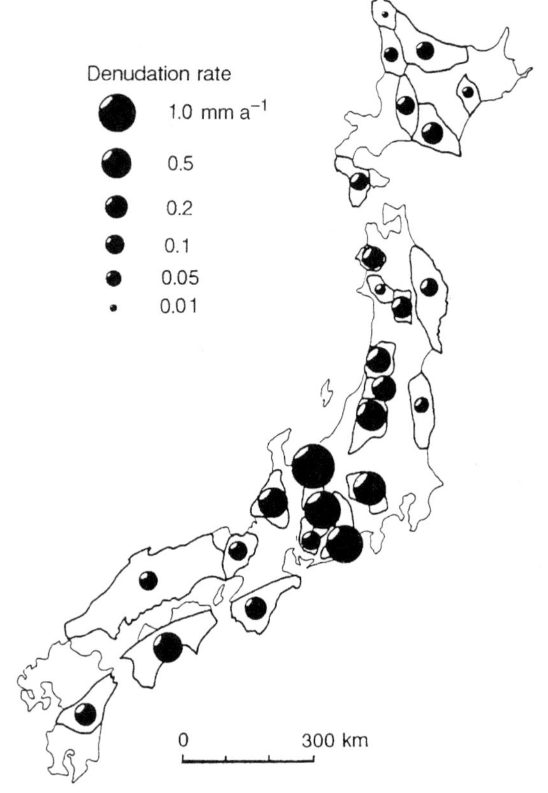

Figure 8.11 *Denudation rates in individual mountain ranges in Japan estimated from equation (6) where the dispersion of altitude is the arithmetical mean of altitude dispersion of a specific area of 1 km² for individual mountains (Ohmori, 1983b)*

(Figure 8.11) (Ohmori, 1983b). The resulting variations in estimated denudation rate show contrasts between specific morphotectonic terrains, with high rates in Central Japan, the outer belt of southwest Japan and the inner belt of northeast Japan, and low rates in the inner belt of southwest Japan, the outer belt of northeast Japan and the mainland of Hokkaido. In particular, values greater than 1 mm a^{-1} are indicated for the central Japanese mountains where mean altitudes are about 1000 m, and peaks reach 3000 m above sea level. These rates are almost equal to mean uplift rates during the Quaternary inferred from the vertical displacements illustrated in Figure 8.4.

The relationship between altitude dispersion and mean altitude H (m) (Ohmori, 1978, 1982; Oguchi, 1991; Sugai *et al.*, 1994; Ohmori and Sugai, 1995), is expressed by:

$$D = aH^b \tag{7}$$

where a and b are constants, and $a = 0.9219$ and $b = 0.6967$ for the Japanese mountains. The correlation coefficient is 0.95 (Figure 8.12) (Ohmori, 1978, 1982, 1983b; Ohmori and Hirano, 1984). By substituting (7) into (6), the following equation is obtained:

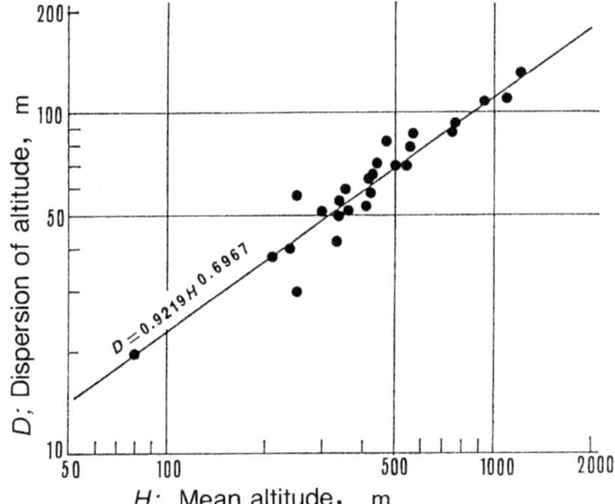

Figure 8.12 *Relationship between mean altitude and dispersion of altitude for Japanese mountain ranges (Ohmori, 1978). The mean altitude indicates the mean of surface altitude calculated from digital altitude data for each mountain range in Japan, and the dispersion of altitude indicates the arithmetical mean of altitude dispersion of a specific area of 1 km² for individual ranges*

$$E = \alpha(aH^b)^\beta = \gamma H^\delta \tag{8}$$

where $\gamma(=\alpha a^\beta)$ is 0.27×10^{-9} and $\delta(=b\beta)$ is 2.23.

Equation (8) indicates that denudation rates increase with an increase in altitude. From equation (8), the denudation rate is only 0.008 mm a^{-1} when the mean altitude is 100 m but 1.3 mm a^{-1} when the mean altitude is 1000 m, indicating that the intensity of denudation processes is related to landsurface elevation even in the areas of high precipitation such as Japan. Indeed, landslides are very scarce at lower altitudes and increase in volume with an increase in altitude dispersion which, in turn, is functionally related to mean elevation (Ohmori and Sugai, 1995). Altitude is related to tectonic movements, as discussed in the previous section. We can thus draw the kinematic inference that earth materials in areas with a large altitude dispersion (which is related to high altitude due to high uplift rate) are highly stressed by active tectonic movements and that denudational processes act to release this stored stress.

8.6 Change in Altitude Resulting from Concurrent Tectonics and Denudation

The relationships between local relief (altitude dispersion), tectonics and denudation represented by equations (6), (7) and (8) are illustrated in Figure 8.13. The relationship between precipitation and altitude is not included here because it has not yet been quantified for the Japanese mountains. The component of denudation affecting tectonic uplift is also difficult to include in the functional scheme of Figure 8.13, since tectonic movements associated with plate motion may be relatively independent of the reduction in load through denudation. The sequence of the change in altitude resulting from concurrent uplift and denudation has previously been discussed on the basis of the relationships expressed in equations (6), (7) and (8) (Ohmori, 1978, 1983a, 1984, 1985, 1993; Yoshikawa, 1984, 1985a, b). Temporal change in altitude is given by the difference between uplift rate and denudation rate. It is expressed by

$$dH/dt = U - E \tag{9}$$

From equation (9), the time necessary for mean altitude to increase from x_1 to x_2 is given by:

$$t = \int_{x_1}^{x_2} [1/(U - E)]dx \tag{10}$$

Substituting equation (8) into equation (10):

$$t = \int_{x_1}^{x_2} [1/(U - \gamma H^\delta)]dx \tag{11}$$

where the measurement units of U, H and t are m a^{-1}, m, and a, respectively. Assuming that the uplift rate is constant, and that the presently observed relationship between denudation rate and altitude dispersion is valid over time, the change in mean altitude can be modelled (Figure 8.14) (Ohmori, 1978, 1985). The resulting landscape development can be divided into developing, culminating and declining stages (Ohmori, 1978, 1985, 1993; Yoshikawa, 1984, 1985a, b).

In the developing stage, mean altitude increases due to the uplift rate exceeding the denudation rate. Denudation rate increases with an increase in mean elevation as altitude dispersion becomes greater. As it approaches the uplift rate there is a progressively slower rate of increase in mean elevation. In the culminating stage when the denudation rate equals the uplift rate, mean altitude attains a maximum and then remains constant, in spite of continuous uplift. This stage represents a steady state between uplift and denudation, with altitude dispersion reaching a maximum and maintaining a constant sediment yield. The maximum elevation becomes greater with a higher

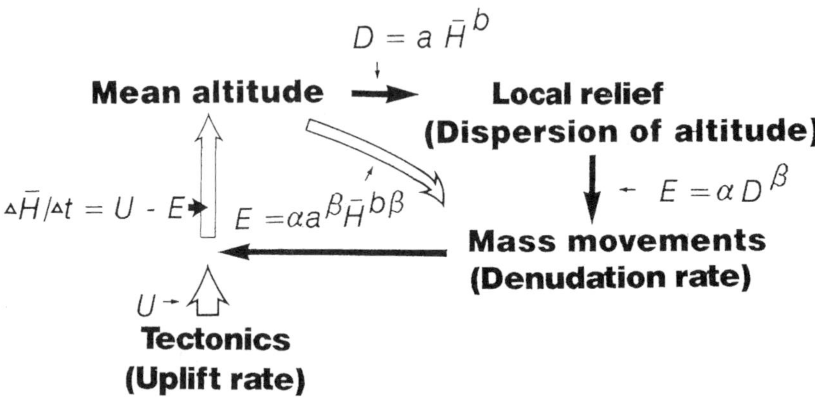

Figure 8.13 *Schematic relations among uplift rate, morphometry (altitude and altitude dispersion) and denudation. The relationships are empirically expressed by mathematical functions for Japanese mountain ranges except for the effect of denudation on tectonics*

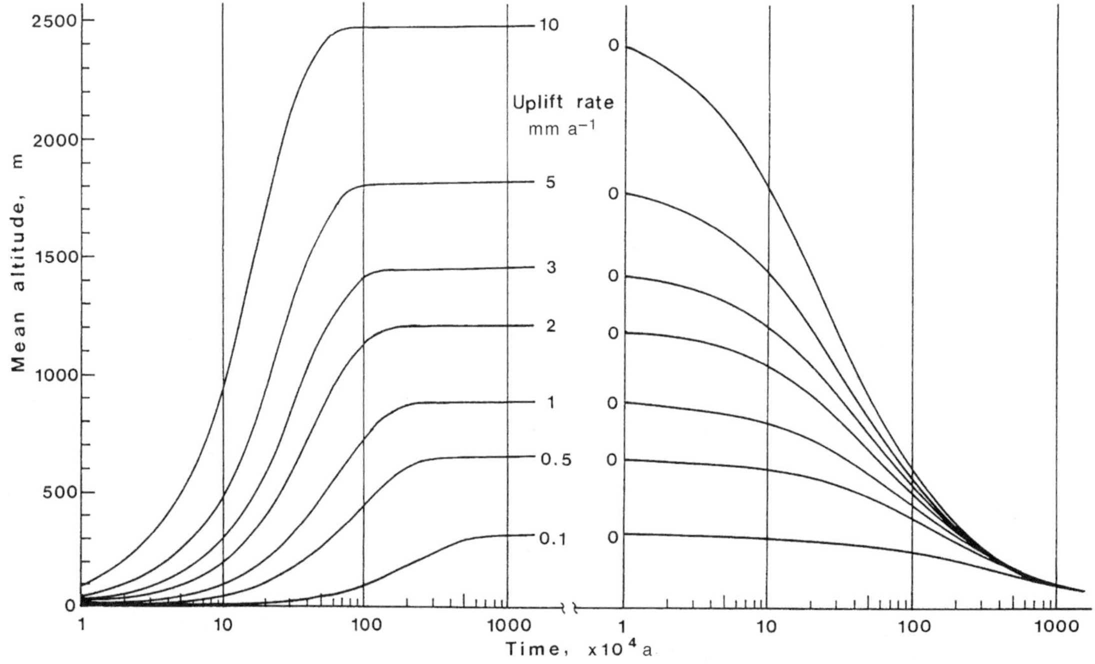

Figure 8.14 *Changes in mean altitude resulting from concurrent tectonics and denudation from an initial altitude at sea level (Ohmori, 1978, 1987)*

uplift rate, and the higher the uplift rate, the earlier the mountains attain their maximum altitude. During the declining stage, when the uplift rate declines, the mean elevation falls and there is a decrease in altitude dispersion and sediment yield. Higher areas with a large altitude dispersion are lowered more rapidly than lower areas characterized by a lower altitude dispersion. This suggests that a decline in slope gradient plays a significant part in the decrease in altitude in comparison with the parallel retreat of slopes (Ohmori, 1983a, 1984). It takes about 10 Ma for the mean altitude to decrease to 100 m above sea level, regardless of the initial altitude.

In the developing stage, the uplift rate depends on plate motion activity, with uplift being promoted not only by compression but also by isostasy. In the culminating stage, uplift is also maintained essentially by plate tectonics. A mountain range cannot maintain a constant altitude by isostatic rebound alone, since the mass isostatically uplifted is less than that of the denuded load. When a mountain range enters the declining stage, the force supporting it at a high level disappears, and subsidence may occur until an isostatic balance is achieved. Subsequently, isostatic rebound due to denudational unloading may be the main cause of uplift. In this stage the uplift rate decreases with a decrease in altitude because of a fall in denudation rate.

8.7 Morphological Evolution of the Japanese Mountain Ranges

On the basis of the model represented by equation (11), it is possible to assess the impact of a change in uplift rate. This produces a sequential change in mean altitude composed of several phases with different uplift rates. In such a case, there is no significant difference between a mean altitude calculated by the integrated uplift of individual uplift rates over ~1 Ma or longer, and a mean altitude calculated on the basis of the mean uplift rate, u (the mean weighted uplift rate):

$$u = \left(\sum_{i=1}^{n} u_i t_i\right) / \sum t_i \qquad (12)$$

where t_i is the duration of each uplift rate u_i, n is the number of the changes in uplift rate, and Σt_i is the total duration of the simulation (Ohmori, 1987). The model shows that when mountain ranges are uplifted at the same rate, their mean elevations converge to a 'critical altitude' restricted by the uplift rate, in spite of differences between their initial altitudes (Figure 8.15) (Ohmori, 1978, 1985, 1987, 1990a, b). The mean altitude attained by a mountain range uplifted in 1 Ma has a relatively narrow range, especially for high uplift rates, regardless of its initial altitude. The relationship between uplift rate and mean altitude attained by a mountain range uplifted from sea level in 1 Ma has been estimated from Figure 8.15 and is shown as a dotted line in Figure 8.8. It shows a close correspondence with the regression line expressed by a solid line for the present altitude of mountain ranges in Japan. Using Figure 8.15, the highest and lowest uplift rates have been estimated for individual ranges in Japan and are shown as Rate 1 in Figure 8.16. These show good agreement with rates estimated from Quaternary vertical displacements (Rate 2) (Figure 8.4). The highest values of Rate 2 were estimated assuming that the duration of displacement was 1 Ma, and the lowest assume a duration of 2 Ma.

The developing stage can be further divided into substages which are expressed by two parameters (Ohmori, 1978). One parameter is the ratio r_{H_n} which is given by:

$$r_{H_n} = \bar{h}/H_n \qquad (13)$$

and the other is ratio r_{H_c} which is given by:

$$r_{H_c} = \bar{h}/H_c \qquad (14)$$

where $\bar{h}$ is the present mean altitude of a mountain range, H_n is the mean altitude without denudation which is given by $H_n = U_t$, and H_c is the critical altitude which is given by $H_c = (U/\gamma)^{1/\delta}$, where γ and δ are the coefficients from equation (8), indicating the steady-state mean altitude where the denudation rate is balanced by uplift. The ratio r_{H_n} indicates the relative degree of dissection of the landscape and the ratio r_{H_c} shows the relative position with respect to the maximum mean altitude that the mountain range can attain. At the start of uplift, r_{H_n} is 1.0 but moves close to zero with advancing stage, whereas r_{H_c} starts from zero and moves towards 1.0 with an advance in stage towards culmination. The loci of the pairs of the ratios are expressed by a curve-like arc, despite the difference in uplift rate (Figure 8.17).

From the mean uplift rate in the Quaternary and present mean altitudes, the stages of individual Japanese ranges can be identified and mapped (Figure 8.17) (Ohmori, 1978, 1993). The earliest and early substages are found in the inner belts of northern Japan and Hokkaido, whereas those of the late and latest substages occur in central Japan. Landscapes of the middle substages are mainly distributed in southwest and northeast Japan, although terrain in the outer belt of southwest Japan and the inner belt of northeast Japan is at a more advanced stage than in the inner belt of southwest, and the outer belt of northeast Japan. The former is representative of the later part of the middle substage, whereas the latter represents the earlier part of the middle substage.

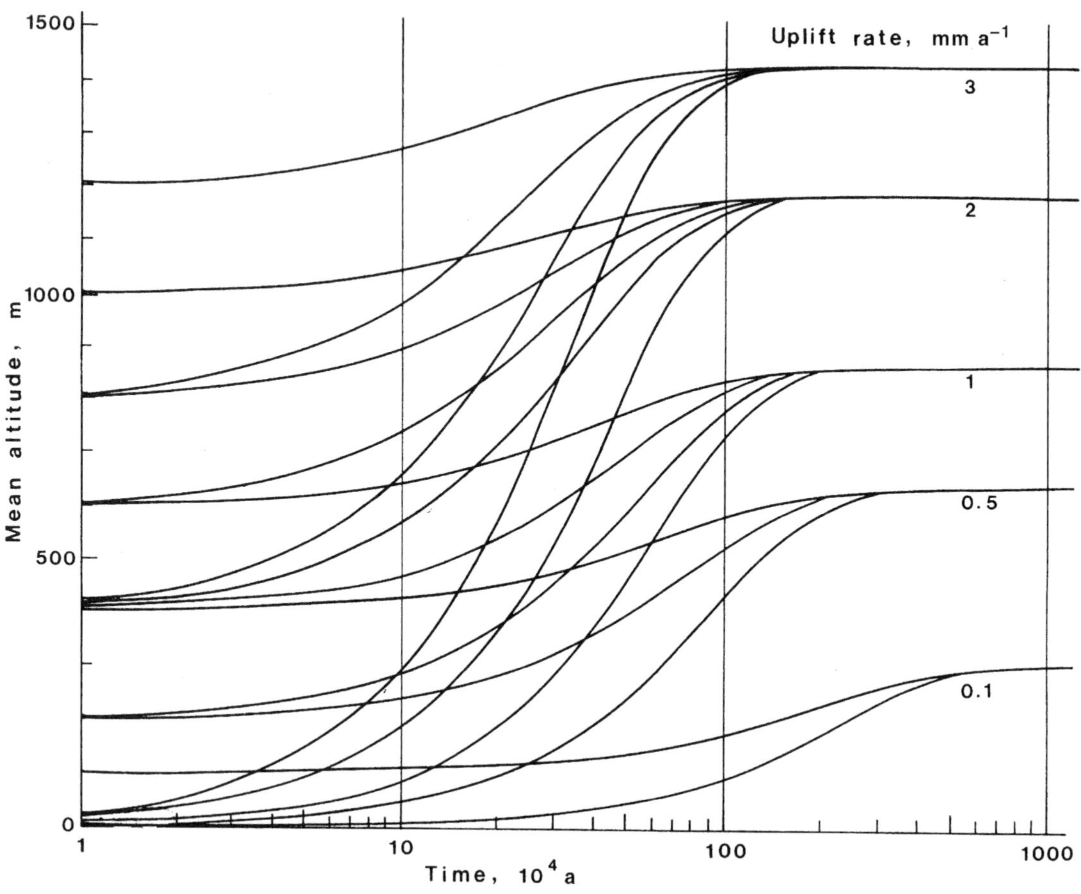

Figure 8.15 *Changes in mean altitude resulting from concurrent tectonics and denudation for different initial altitudes, showing that the mean altitude of mountain ranges with the same uplift rate converges to a similar altitude in about 1 Ma (Ohmori, 1978, 1985)*

8.8 Discussion and Conclusions

The landforms of Japan have been created in the context of Quaternary crustal movements associated with recent plate motions. Regional variations in the stress caused by plate motion and pre-Neogene geological structures have given rise to five morphotectonic terrains in Japan, each of which is divided into two sub-terrains by a major tectonic boundary or volcanic front. Resultant rates of contemporary interseismic and coseismic movement, rates of uplift calculated from the elevation of late Quaternary raised shorelines, and rates estimated from Quaternary vertical displacements, show a good correspondence indicating that present rates and modes of movement typify those throughout the Quaternary.

Mean altitudes and mean Quaternary uplift rates are related. Ratios of denudation to uplift show that Japanese mountain ranges have experienced considerable denudation. Relationships between present denudation rates and mean altitude dispersion, and between mean altitude dispersion and mean altitude, indicate that denudation rates increase with mean altitude; landslide frequency similarly increases with altitude dispersion. Areas with a large altitude dispersion are inferred to be highly stressed by active tectonic movements, and denudational processes may be a means of releasing the stress stored in the crust.

The modelled sequence of changes in mean altitude resulting from concurrent uplift and denudation based on the above relationships indicates that landform development is divided into three stages – developing,

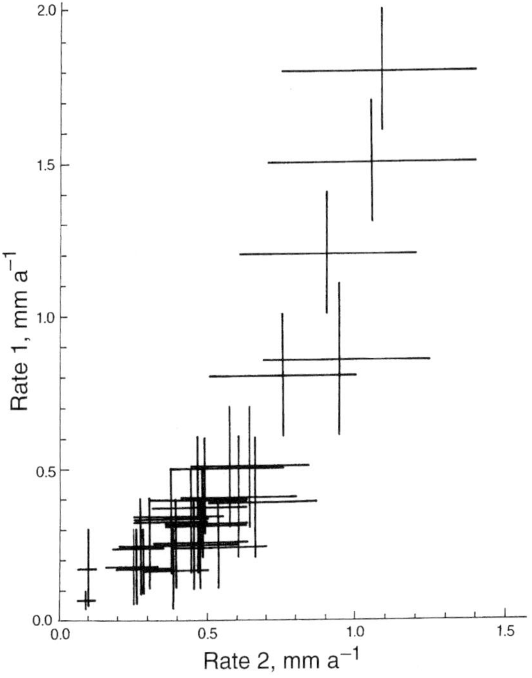

Figure 8.16 *Relationship between Rate 1 and Rate 2 (Ohmori, 1978)*

culminating and declining. Japanese mountain ranges represent various stages from the earliest to the latest substages of the developing stage. The stages depend on the tectonic uplift rates which reflect plate motion and geological structure; in general, the higher the uplift rate, the more advanced the stage.

In terms of the coupling of tectonic and denudational processes, denudation rates are intensified not only by an increase in altitude dispersion which increases with elevation, but also by orographic precipitation which may also be enhanced by the increase in elevation induced by tectonic uplift. Such intensification of orographically enhanced precipitation raises more complex problems for the interpretation of landform development than effects due to climatic change, such as glacial–interglacial cycles.

Acknowledgements

The author would like to thank M.A. Summerfield for his kind and helpful comments on this chapter.

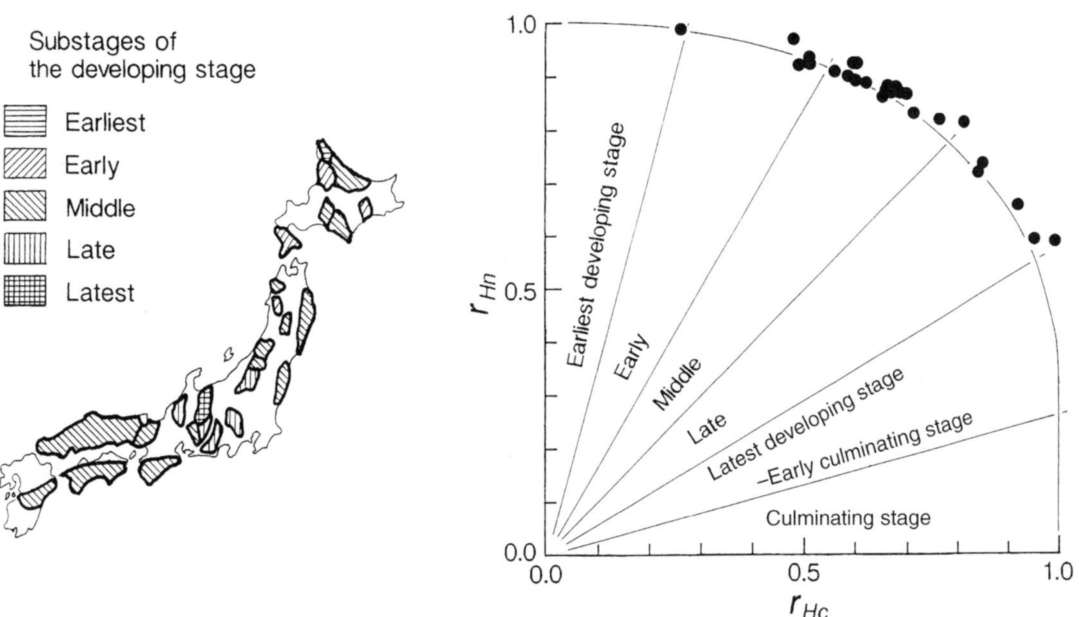

Figure 8.17 *Stages of Japanese mountain range development derived from equations (13) and (14) (Ohmori, 1978, 1993)*

References

Akojima, I. 1983. Comparison between the past and the present denudation rate of mountains around the Yamagata Basin, Northeast Japan. *Trans. Japan. Geomorph. Un.*, **4**, 97–106 (in Japanese with English abstract).

Beaumont, C., Fullsack, P. and Hamilton, J. 1992. Erosional control of active compressional orogens. In: K. McClay (ed.) *Thrust Tectonics*, Chapman and Hall, London, pp. 1–18.

Bishop, P. and Brown, R. 1992. Denudational isostatic rebound of intraplate highlands: The Lachlan River valley, Australia. *Earth Surf. Processes Ldfms.*, **17**, 345–360.

Bloom, A.L. and Yonekura, N. 1985. Coastal terraces generated by sea-level change and tectonic uplift. In: M.J. Woldenberg (ed.) *Models in Geomorphology*, Allen and Unwin, Boston, pp. 139–154.

Evans, I.S. 1972. General geomorphometry, derivatives of altitude, and descriptive statistics. In: R.J. Chorley (ed.) *Spatial Analysis in Geomorphology*, Methuen, London, pp. 19–90.

Fujita, T., Suwa, H. and Okuda, S. 1989. Mass movement. *Trans. Japan. Geomorph. Un.*, **10-A**, 23–34.

Geographical Survey Institute of Japan. 1979. *Explanatory Text of the Numerical Terrestrial Data of Japan*, Tsukuba, 279 pp. (in Japanese).

Geographical Survey Institute of Japan. 1980. *Abstract of the Numerical Terrestrial Data of Japan*, Tsukuba (in Japanese).

Geological Survey of Japan. 1982. *Geological Atlas of Japan*, Tsukuba, 119 pp. (in Japanese with English abstract).

Gilchrist, A.R., Summerfield, M.A. and Cockburn, H.A.P. 1994. Landscape dissection, isostatic uplift, and the morphologic development of orogens. *Geology*, **22**, 963–966.

Hagiwara, Y. 1967. Analyses of gravity values in Japan. *Bull. Earthquake Res. Inst.*, **45**, 1091–1228.

Hoffman, P.F. and Grotzinger, J.P. 1993. Orographic precipitation, erosional unloading, and tectonic style. *Geology*, **21**, 195–198.

Huzita, K. 1980. Role of the Median Tectonic Line in the Quaternary tectonics of the Japanese Islands. *Mem. Geol. Soc. Japan*, **18**, 129–153.

Isacks, B.L. 1992. Long-term land surface processes: erosion, tectonics and climate history in mountain belts. In: P.M. Mather (ed.) *TERRA-1: Understanding the Terrestrial Environment: The Role of Earth Observation from Space*. Taylor and Francis, London, pp. 21–36.

Ishigai, H. 1966. Study on sedimentation of clastic sediments in reservoirs. *Report of Technical Research Laboratory, Central Institute of Electric Power Industry*, 66010, 1–95 (in Japanese with English abstract).

Ishiwada, Y., Ikebe, Y., Ogawa, K. and Onitsuka, T. 1977. A consideration on scheme of sedimentary basins of Northeast Japan. *Professor Kazuo Huzioka Memorial Volume*, Akita, 1–7 (in Japanese with English abstract).

Iso, N., Yamakawa, K., Yonezawa, H. and Matsubara, T. 1980. Accumulation rates of alluvial cones, constructed by debris-flow deposits, in the drainage basin of the Takahashi river, Gifu Prefecture, central Japan. *Geogr. Rev. Japan*, **53**, 699–720 (in Japanese with English abstract).

Japan Landslide Society. 1988. *Landslides in Japan*, Tokyo, 54 pp.

Kaizuka, S. 1987. Quaternary morphogenesis and tectogenesis of Japan. *Z. Geomorph. Suppl.*, **63**, 61–73.

Kaizuka, S. and Suzuki, T. 1993. Geomorphology in Japan. In: H.J. Walker and W.E. Grabau (eds) *The Evolution of Geomorphology – A Nation-by-Nation Summary of Development*, Wiley, Chichester. pp. 255–271.

Kanaori, Y. and Kawakami, S. 1996. Microplate model and large inland earthquakes of Southwest Japan: Simplification for generation of the 1995 M7.2 Hyogo-ken-nanbu earthquake. *J. Seism. Soc. Japan*, **49**, 125–139 (in Japanese with English abstract).

Kanaori, Y., Kawakami, S. and Yairi, K. 1992. The block structure and Quaternary strike-slip block rotation of central Japan. *Tectonics*, **11**, 47–56.

Kikuchi, T. 1988. Paleo sea-levels since the Last Interglacial – A mathematical method estimating for uplift rate of marine terraces and paleo sea-levels. *Trans. Japan. Geomorph. Un.*, **9**, 81–104 (in Japanese with English abstract).

Machida, H. 1984. Large-scale rockslides, avalanches and related phenomena: a short review. *Trans. Japan. Geomorph. Un.*, **5**, 155–178 (in Japanese with English abstract).

Matsuda, T. 1961. The Miocene stratigraphy of the Fuji River valley, central Japan. *J. Geol. Soc. Japan*, **69**, 79–96 (in Japanese with English abstract).

Matsuda, T. 1981. Quaternary tectonics. In: Science Council of Japan (ed.) *Recent Progress of Natural Sciences in Japan*, **6**, 233–243.

Matsuda, T., Nakamura, K. and Sugimura, A. 1967. Late Cenozoic orogeny in Japan. *Tectonophysics*, **4**, 349–366.

Matsuda, T., Ota, Y., Ando, M. and Yonekura, N. 1978. Fault mechanism and recurrence time of major earthquakes in southern Kanto district, Japan, as deduced from coastal terrace data. *Geol. Soc. Amer. Bull.*, **89**, 1610–1618.

Miyoshi, M. 1983. Estimated ages of late Pleistocene marine terraces in Japan, deduced from uplift rate. *Geogr. Rev. Japan*, **56**, 819–834 (in Japanese with English abstract).

Mizutani, T. 1981. Drainage basin characteristics affecting sediment yield from steep mountain drainage basin. *Shin-sabo*, No. 119, 1–9 (in Japanese with English abstract).

Nakamura, K. 1983. Possible nascent trench along the eastern Japan Sea as the convergent boundary between Eurasian and North American plates. *Bull. Earthquake Res. Inst.*, **58**, 711–722 (in Japanese with English abstract).

Nakamura, K., Shimazaki, K. and Yonekura, Y. 1984. Subduction, bending and eduction. Present and Quaternary tectonics of the northern border of the Philippine Sea plate. *Bull. Soc. Géol. France*, **26**, 221–243.

Nakata, T., Koba, M., Jo, W., Imaizumi, T., Matsumoto, H. and Suganuma, T. 1979. Holocene marine terraces and seismic crustal movement. *Scientific Report of Tohoku University, 7th series (Geography)*, **29**, 195–204.

Namba, S. and Kawaguchi, T. 1965. Influences of some factors upon soil losses from large mountain watersheds. *Bulletin of Government Forest Experiment Station*, **173**, 93–116 (in Japanese with English abstract).

Oguchi, T. 1991. Quantitative study of sediment transport in mountain drainage basins since the Late Glacial. *Trans. Japan. Geomorph. Un.*, **12**, 25–39 (in Japanese with English abstract).

Ohmori, H. 1978. Relief structure of the Japanese mountains and their stages in geomorphic development. *Bull. Dept. Geog. Univ. Tokyo*, **10**, 31–85.

Ohmori, H. 1982. Functional relationship between the erosion rate and the relief structure in the Japanese mountains. *Bull. Dept. Geog. Univ. Tokyo*, **14**, 65–74.

Ohmori, H. 1983a. A three-dimensional model for the erosional development of mountain on the basis of relief structure. *Trans. Japan. Geomorph. Un.*, **4**, 107–120.

Ohmori, H. 1983b. Characteristics of the erosion rate in the Japanese mountains from the viewpoint of climatic geomorphology. *Z. Geomorph. Suppl.*, **46**, 1–14.

Ohmori, H. 1983c. Erosion rates and their relation to vegetation from the viewpoint of world-wide distribution. *Bull. Dept. Geog. Univ. Tokyo*, **15**, 77–91.

Ohmori, H. 1984. Change in the earth's surface altitude with absolute time simulated from the relation between mean altitude and dispersion, and between dispersion and denudation rate. *Bull. Dept. Geog. Univ. Tokyo*, **16**, 5–22.

Ohmori, H. 1985. A comparison between the Davisian scheme and landform development by concurrent tectonics and denudation. *Bull. Dept. Geog. Univ. Tokyo*, **17**, 18–28.

Ohmori, H. 1987. Mean Quaternary uplift rates in the central Japanese mountains estimated by means of geomorphological analysis. *Bull. Dept. Geog. Univ. Tokyo*, **19**, 29–36.

Ohmori, H. 1990a. Quaternary uplift rate and its relation to landforms of Mts. Shikoku, Japan. In: N. Yonekura, A. Okada and A. Moriyama (eds) *Tectonic Landforms*, Kokon-Shoin, Tokyo, pp. 60–86 (in Japanese).

Ohmori, H. 1990b. Geomorphogenetic crustal movement and the altitudinal limitation of peneplain remnants of the Shikoku Mountains. Japan. *Bull. Dept. Geog. Univ. Tokyo*, **22**, 17–34.

Ohmori, H. 1993. Changes in the hypsometric curve through mountain building resulting from concurrent tectonics and denudation. *Geomorphology*, **8**, 263–277.

Ohmori, H. 1995. Geological and geomorphological characteristics of Japan. In: Editorial Committee of the History of Civil Engineering in Japan (ed.) *The History of Japanese Civil Engineering from 1966 to 1990: Memorial Publication of the Eightieth Anniversary of the Japan Society of Civil Engineers*, Japan Society Civil Engineers, Tokyo, pp. 16–20 (in Japanese).

Ohmori, H. and Hirano, M. 1984. Mathematical explanation of some characteristics of altitude distributions of landforms in an equilibrium state. *Trans. Japan. Geomorph. Un.*, **5**, 293–310.

Ohmori, H. and Sohma, H. 1983. Landform classification in mountain region and geomorphic characteristics values. *J. Japan. Cart. Ass.*, **21**(3), 1–12 (in Japanese with English abstract).

Ohmori, H. and Sugai, T. 1995. Toward geomorphometric models for estimating landslide dynamics and forecasting landslide occurrence in Japanese mountains. *Z. Geomorph. Suppl.*, **101**, 149–164.

Okayama, T. 1953. Geomorphic structure of Japan – as a starting point of regional geomorphology. *Journal of Historical Association of Meiji University*, **3**, 28–38 (in Japanese).

Ota, Y. 1975. Late Quaternary vertical movement in Japan estimated from deformed shorelines. *Bull. R. Soc. N.Z.*, **13**, 231–239.

Ota, Y. 1985. Marine terraces and active faults in Japan with special reference to co-seismic events. In: M. Morisawa and J.T. Hack (eds) *Tectonic Geomorphology*, Allen and Unwin, Boston, pp. 345–366.

Ota, Y. 1986. Marine terraces as reference surfaces in late Quaternary tectonic studies: examples from the Pacific rim. *Bull. R. Soc. N.Z.*, **24**, 357–375.

Ota, Y. and Yoshikawa, T. 1978. Regional characteristics and their geodynamic implications of late Quaternary tectonic movement deduced from deformed former shorelines in Japan. *J. Phys. Earth*, **26**, Suppl., s379–389.

Research Group for Active Faults. 1980a. *Active Faults in Japan – Sheet Maps and Inventories*, University of Tokyo Press (in Japanese with English summary).

Research Group for Active Faults. 1980b. Active faults in and around Japan: the distribution and the degree of activity. *J. Nat. Disaster Sci.*, **2**, 61–99.

Research Group for Quaternary Tectonic Map. 1968. Quaternary tectonic map of Japan. *Quat. Res.*, **7**, 182–187 (in Japanese with English abstract).

Research Group for Quaternary Tectonic Map. 1969. *Quaternary Tectonic Map of Japan*, National Research Center of Disaster Prevention, Tokyo, 6 sheets.

Research Group for Quaternary Tectonic Map. 1973. *Explanatory Text of the Quaternary Tectonic Map of Japan*, National Research Center of Disaster Prevention, Tokyo.

Saito, K. 1988. *Alluvial Fans in Japan*, Kokon-Shoin, Tokyo (in Japanese).

Saito, K. 1989a. Dominant factors influencing the distribution of alluvial fans in the Taiwan Island. *Journal of Hokkai-Gakuen University*, No. 53, 19–36 (in Japanese with English abstract).

Saito, K. 1989b. Geomorphic development of mountains in the Taiwan Island. *Annals of Hokkaido Geographical Society*, No. 63, 9–16 (in Japanese with English abstract).

Sakaguchi, Y. 1964. On the geomorphic history of Japan. *Geogr. Rev. Japan*, **37**, 387–390 (in Japanese with English abstract).

Seno, T. 1985. Is northern Honshu a micro plate? *Tectonophysics*, **115**, 177–196.

Sugai, T. 1990. The origin and geomorphic characteristics of the erosional low-relief surfaces in the Akaishi Mountains and southern part of the Mikawa Plateau, central Japan. *Geogr. Rev. Japan*, **63A**, 793–813 (in Japanese with English abstract).

Sugai, T., Ohmori, H. and Hirano, M. 1994. Rock control on magnitude–frequency distribution of landslide. *Trans. Japan. Geomorph. Un.*, **15**, 233–251.

Summerfield, M.A. 1991. *Global Geomorphology*, Longman, London.

Tamaki, K. and Honza, E. 1985. Incipient subduction and obduction along the eastern margin of the Japan Sea. *Tectonophysics*, **119**, 381–406.

Tanaka, H. 1955. Geological and topographical studies on the sedimentation of reservoirs in Japan. *J. Tech. Res. Lab.*, **5**(2), 163–198.

Tanaka, H. and Ishigai, H. 1951. On the relation of sedimentation of reservoirs to configuration and nature of rocks of catchment area (1st report). *J. Japan. Soc. Civil Engineers*, **36**, 173–177 (in Japanese with English abstract).

Tanaka, M. 1976. Rate of erosion in the Tanzawa Mountains, Central Japan. *Geogr. Ann.*, **58A**, 155–163.

Tanaka, M. 1982. A map of regional denudation rate in the Japanese mountains. *Trans. Japan. Geomorph. Un.*, **3**, 159–167.

Tokunaga, E. and Ohmori, H. 1989. Drainage basin geomorphology. *Trans. Japan. Geomorph. Un.*, **10-A**, 35–46.

Yonekura, N., Okada, A. and Moriyama, A. (eds) 1990. *Tectonic Landforms*, Kokon-Shoin, Tokyo (in Japanese).

Yoshikawa, T. 1968. Seismic crustal deformation and its relation to Quaternary tectonic movement on the Pacific coast of Southwest Japan. *Quat. Res.*, **7**, 157–170 (in Japanese with English abstract).

Yoshikawa, T. 1970. On the relations between Quaternary tectonic movement and seismic crustal deformation in Japan. *Bull. Dept. Geog. Univ. Tokyo*, **2**, 1–24.

Yoshikawa, T. 1974. Denudation and tectonic movement in contemporary Japan. *Bull. Dept. Geog. Univ. Tokyo*, **6**, 1–14.

Yoshikawa, T. 1984. Geomorphology of tectonically active and intensely denuded regions. *Geogr. Rev. Japan*, **57A**, 691–702 (in Japanese with English abstract).

Yoshikawa, T. 1985a. Landform development by tectonics and denudation. In: A. Pitty (ed.) *Themes in Geomorphology*, Croom Helm, London, pp. 194–210.

Yoshikawa, T. 1985b. *Geomorphology of Tectonically Active and Intensely Denuded Regions*, University of Tokyo Press, Tokyo (in Japanese).

Yoshikawa, T., Kaizuka, S. and Ota, Y. 1964. Mode of crustal movement in the late Quaternary on the southeast coast of Shikoku, Southwest Japan. *Geogr. Rev. Japan*, **37**, 627–648 (in Japanese with English abstract).

Yoshikawa, T., Kaizuka, S. and Ota, Y. 1981. *The Landforms of Japan*, University of Tokyo Press, Tokyo.

9

Large-scale geomorphology of the Andes: interrelationships of tectonics, magmatism and climate

Lorcan Kennan

9.1 Introduction

The Andes stretch 9000 km along the western edge of South America, rise to nearly 7000 m at their highest and reach 700 km across at their widest (Plate 9.1). Pacific oceanic crust is being actively subducted beneath the western margin of the continent and is driving active volcanism and crustal deformation. Along the orogen there are striking changes in structure, geomorphology, kinematics of active deformation, distribution of active volcanism and climate, making the region an ideal natural laboratory for the study of the numerous processes which form an ocean–continent convergent margin orogen. There is a considerable literature on Andean stratigraphy, structure and magmatism, but relatively little on geomorphology, and there are few published papers using modern digital elevation models to analyse topography systematically (Isacks, 1988; Masek *et al.*, 1994). Zeil (1979), among others, has provided a broad regional review of the bedrock geology, while Clapperton (1993) has reviewed South American Quaternary geomorphology. In this chapter I will attempt to describe the morphology of the Andes at a very large scale.

One of the most striking features of the range is the near mirror symmetry about an axis trending ENE through central Bolivia (Gephart, pers. comm.). The width of the central Andean plateau region clearly drops to the north and south of the axis. Foreland shortening is greatest in central Bolivia and is less in southern Peru and northernmost Argentina. In north-central Peru and Argentina the width of the main Andean range is markedly lower and there are important N–S-trending thick-skinned basement uplifts (e.g. the Shira, Moa and Pampeanas ranges). These broad divisions correlate with dip changes in the subducting oceanic plate (e.g. Jordan *et al.*, 1983a, b), with elastic thickness of the continental plate (Watts *et al.*, 1995), and also with the presence of thick sequences of Palaeozoic sediments (e.g. Allmendinger *et al.*, 1993; Baby *et al.*, 1993). The chain of cause and effect between these large-scale features is still poorly known. Superimposed on this mega-symmetry is a marked north–south asymmetry in present-day climate of the eastern Andes (e.g. Masek *et al.*, 1994), which has resulted in dramatically different long-term denudation rates and preservation of transient features such as erosion surfaces. Contrasts in sedimentation history from northern Peru to Argentina suggest that this asymmetry is at least 50 Ma old.

I will review the evidence for the morphological development of the Andes, concentrating in particular on palaeogeographic developments since the late Cretaceous, on the influence of pre-existing structures, and on the relationship of uplift to plate tectonic processes. I have attempted to include some material from throughout the Andes, but I have concentrated on the central Andes because more relevant evidence seems concentrated in this region. This bias also in part reflects my own field work in the region since 1990.

Geomorphology and Global Tectonics. Edited by Michael A. Summerfield. © 2000 John Wiley & Sons Ltd.

9.2 Reconstructing Andean Uplift: Methods

The main methods I have used to reconstruct the development of Andean relief are basin analysis combined with accurate chronological control and regional structural studies. These data have been integrated with quantitative denudation rate data (fission-track dating), and with very rare data on actual surface altitude change through time (palaeoflora and palaeosurface studies). The Andes appear to be actively growing and present-day topographic relief is not in a 'steady state', and is not applicable further back in time than possibly the late Pliocene.

The uplift and erosion history of the Andes is recorded in forearc, intramontane and foreland basin sediments. Provenance and palaeoflow data can be integrated with radiometric and palaeontological ages to map out uplifting and actively eroding regions. Because much Andean sediment is continental, only a low-resolution chronostratigraphy is available, from dated volcanic intercalations (e.g. Kennan *et al.*, 1995) and a few mammal faunas (e.g. Gayet *et al.*, 1991; Marshall and Sémpere, 1991).

Palaeogeographic reconstructions in this chapter are based only on reliably dated sedimentary sequences (e.g. Jordan *et al.*, 1983a, b; Beer *et al.*, 1990; Kennan *et al.*, 1995; Lamb *et al.*, 1997). In many areas no direct dating is available and there is a danger that superficially similar red beds have often been correlated using the invalid assumption that bounding unconformities are time surfaces traceable over vast distances (e.g. Sébrier *et al.*, 1988a). In reality, both erosion and sedimentation may have been quasi-continuous in many areas. Seismic sections (unpublished oil company data) show widespread synsedimentary thrusting, with associated localized intraformational unconformities.

Fission-track ages in apatites and zircons (see Andriessen (1995) and Chapter 4 for reviews) are commonly used to infer the timing and rate of erosional denudation. Closure temperatures for apatite and zircon are generally taken as ~100–110°C and 200–250°C, respectively. Denudation rates can be calculated using either samples collected over a wide altitude range, or assuming a reasonable geothermal gradient (typically about 25°C km^{-1} in the central Andes (e.g. Henry and Pollack, 1988). Although the term 'uplift' is usually used when discussing the results, the data only give bulk erosion rates, or movement of a rock body towards the eroding topographic surface. They say nothing about surface uplift, or rate of uplift of a body of rock with respect to a fixed datum such as sea level (see England and Molnar (1990) for a discussion of the misuse of 'uplift' and MacFadden *et al.* (1994) for a misuse of denudation as surface uplift in the Andes). While there is a general correlation in the Andes between greatest relief and highest inferred denudation rates, the data cannot be used to infer palaeoaltitude or the rate at which average orogen altitude was increasing. Independent, palaeobotanical data from other mountain belts (e.g. the Colorado plateau (Gregory and Chase, 1992)) clearly indicate that surface uplift and erosion may be separated by more than 30 Ma. However, because fission-track data record the time and rate of erosion, and because erosion will occur during or following surface uplift, we can infer a minimum surface uplift age and minimum age of clast influx into surrounding basins where there may be no datable tuffs or fossils. Thus the data can be integrated into palaeogeographic reconstructions.

Palaeosurfaces due to marine or fluvial erosion are widespread, especially in the central Andes, and may be used to infer changes in surface altitude. While the presence of marine rocks gives a direct indication of uplift, palaeoaltitude can only be reconstructed from fluvial terraces or pediments if drainage gradient and base-level altitude are known. For instance, in the central Andes of Peru and Bolivia, recent rapid uplift was inferred from the dissection of erosion surfaces now at ~4000 m (e.g. Walker, 1949). However, base level for these surfaces was probably the perched, internally draining Altiplano (author's own observations) which could have been well above sea level. I have only quantified uplift of palaeosurfaces where base level can be clearly inferred (examples discussed below). Molnar and England (1990) have suggested that uplift of such palaeosurfaces could in part be due to isostatic rebound during, or following, deep dissections. However, Gilchrist *et al.* (1994), giving an example from the Alps, suggest that in real orogens this effect can only account for a small fraction of the apparent uplift, and incision clearly cannot account at all for the considerable surface uplift of undissected or internally draining regions such as the Andean Altiplano or Tibetan Plateau.

Surface uplift of mountain belts is commonly inferred from changes in flora. Plants are sensitive to temperature, and because temperature varies with altitude palaeofloras in mountain belts can, in principle, be used to deduce the altitude at which they grew. Palaeoflora species or communities can be compared with nearest living relatives, although there is the possibility of adaptation to a new environment.

Alternatively, potentially temperature-dependent leaf characteristics such as smoothness of margin may be used to infer palaeotemperatures (e.g. Gregory and Chase, 1992; Forest *et al.*, 1995). Unfortunately, palaeoaltitude can only be inferred from palaeotemperatures if the change of temperature with altitude at the time the plants were deposited is known.

Climatic change can also potentially explain the evidence for a late Cenozoic increase in denudation and incision rates (e.g. Ruddiman *et al.*, 1989). Deep Pliocene dissection is seen in the Himalayas, in the Colorado Plateau and in certain parts of the Andes (e.g. Kennan *et al.*, 1995) and many authors have used it to infer Plio-Pleistocene uplift of 1500–4000 m (e.g. Harrison, 1943; Walker, 1949; Servant *et al.*, 1989). However, although both dissection and floral changes could result from surface uplift, it is possible that climatic change involving cooling and an increase in precipitation could also be responsible (Molnar and England, 1990). The middle Pliocene (~3 Ma BP) was apparently a time of major climatic change; oxygen isotope trends around the world show a dramatic decrease in $\delta^{18}O$ and the first northern hemisphere sea ice is noted (e.g. Raymo and Ruddiman, 1992). Closer to the Andes, Pliocene changes in $\delta^{18}O$ in eastern Pacific sediments (Shackleton and Hall, 1995) indicate more local temperature changes as ocean currents changed, possibly in response to the completion of the Panama land bridge (Marshall *et al.*, 1979). It has been proposed that widespread orogenic uplift could have changed world climate (e.g. Raymo *et al.*, 1988; Raymo and Ruddiman, 1992), but much of the floral and geomorphological evidence used to infer uplift could equally be a result of climatic change. Andean examples are discussed below (see Molnar and England (1990) for examples from other orogens).

9.3 Overview of Plate Tectonic Setting and Kinematics

The Andes define the boundary between the South American continent and three distinct oceanic plates (Figure 9.1). Relative motions between the Nazca, Antarctic and South America plates between ~60 Ma and the present are relatively well known (Pilger, 1984; Pardo-Casas and Molnar, 1987; DeMets *et al.*, 1990). Since ~50 Ma BP convergence between the plates has been towards ~080°, but there have been considerable changes in convergence rate (Figure 9.2) which may broadly correlate with periods of more intense volcanism and shortening in the continental margin. Present-day Nazca–South America convergence is towards 080° at ~78 mm a^{-1} at 10°S and ~84 mm a^{-1} at 25–40°S, while south of the Chile Ridge spreading centre the Antarctic plate is subducting towards 100° at ~20 mm a^{-1}. Major changes in continental margin orientation with respect to plate convergence have produced very different strain patterns in different parts of the Andes (e.g. Dewey and Lamb, 1992).

The E–W-trending Andes of northernmost Colombia and Venezuela are dominated by subduction of oceanic Caribbean crust under the South American plate, and by large-magnitude (displacement >~100 km) dextral faults which link subduction on the Caribbean and central American (Panama) trenches (Burke *et al.*, 1984; Mattson, 1984; Kellogg and Vega, 1995). In Colombia and Ecuador, the very oblique plate convergence is highly partitioned between subduction zone slip, active dextral slip on NE–SW-trending faults within the Andes (Pennington, 1981; Winter and Lavenu, 1989; Dewey and Lamb, 1992), and foreland thrusting (Figure 9.1). Recent plate reconstructions indicate a prolonged complex history for the Panama triple junction (Kennerley, 1980; Burke *et al.*, 1984; Aspden and McCourt, 1986; Restrepo and Toussaint, 1988; Kellogg and Vega, 1995; Pindell and Tabutt, 1995). Cretaceous and later collision and accretion of island arc and oceanic crust fragments was followed by ~1500 km northeastward migration of Panama towards its present position.

In contrast, central Andean kinematics is dominated by ENE-directed shortening. Present-day seismicity suggests approximately 90% of inter-plate motion is taken up by slip on the subduction zone and 10% by shortening and thickening in the overlying continental margin (Stauder, 1975; Chinn and Isacks, 1983; Suarez *et al.*, 1983; Dorbath *et al.*, 1986, 1990, 1991; Assumpção, 1988, 1992; Dewey and Lamb, 1992; Harvard and ISC earthquake catalogues). Shortening directions of most subduction and Andean earthquakes are close to 080°, indicating that little strain partitioning is taking place. There is no evidence of terrain pile-up in the Arica Bend region at 19°S to suggest any long-term partitioning of trench slip. Studies of young faulting and active continental margin seismicity between 6°S and 36°S show that only a small fraction of total continental margin strain is partitioned (Lavenu, 1979; Lavenu and Ballivian, 1979; Sébrier *et al.*, 1985, 1988b; Cabrera *et al.*, 1987, 1991; Dewey and Lamb, 1992; Kennan, 1993, 1994). Displacements on orogen-parallel strike-slip faults are modest (all <<100 km) and relate to accommodation of local changes in fold-thrust directions. The N–S-

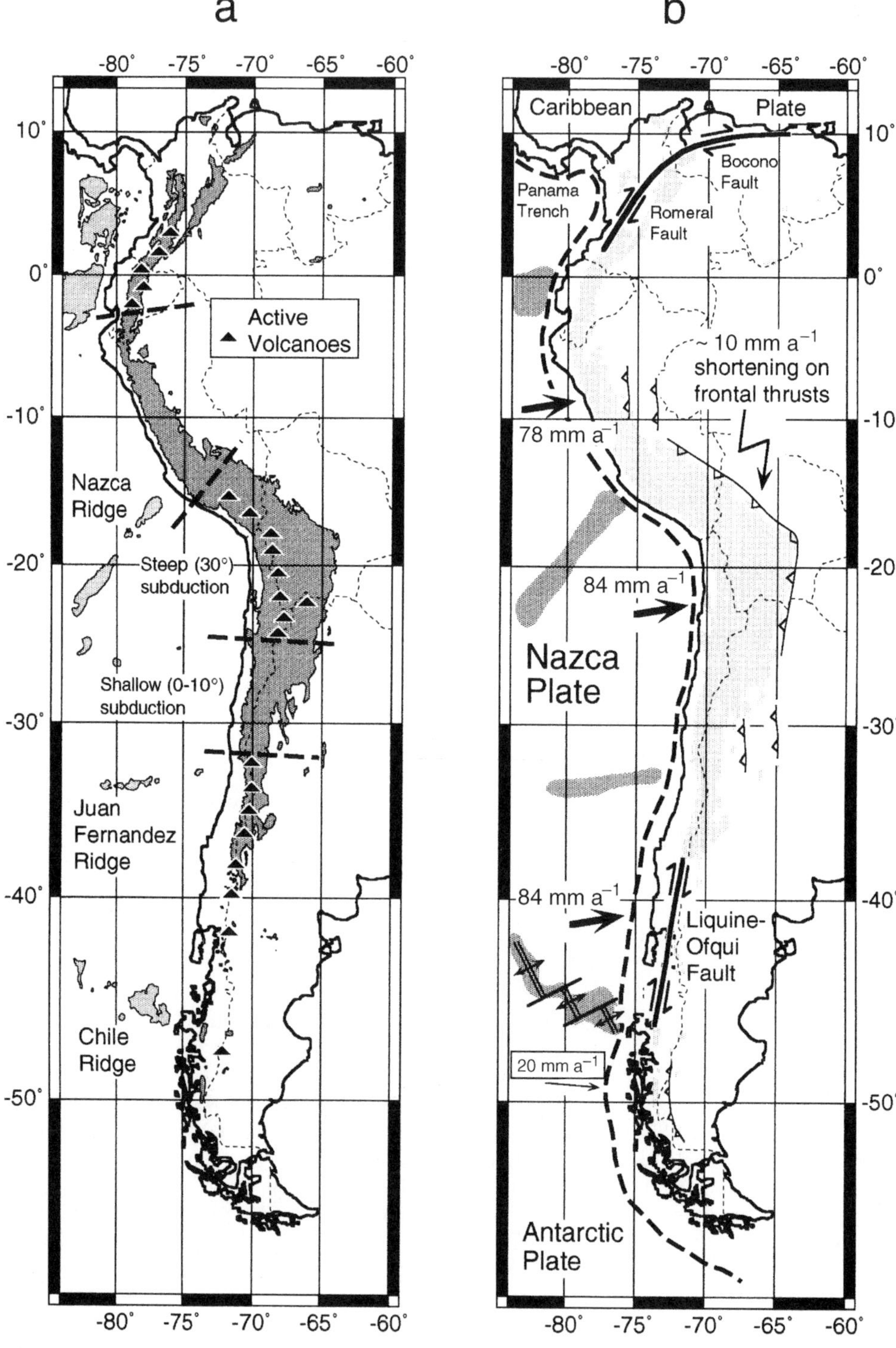
a
b
Active Volcanoes
Nazca Ridge
Steep (30°) subduction
Shallow (0-10°) subduction
Juan Fernandez Ridge
Chile Ridge
Caribbean Plate
Panama Trench
Romeral Fault
Bocono Fault
~ 10 mm a^{-1} shortening on frontal thrusts
78 mm a^{-1}
84 mm a^{-1}
Nazca Plate
84 mm a^{-1}
Liquine-Ofqui Fault
20 mm a^{-1}
Antarctic Plate

trending Atacama fault of northern Chile accommodated significant sinistral and dextral slip during the Cretaceous (Brown *et al.*, 1993) but at present shows only small normal displacements (Dewey and Lamb, 1992).

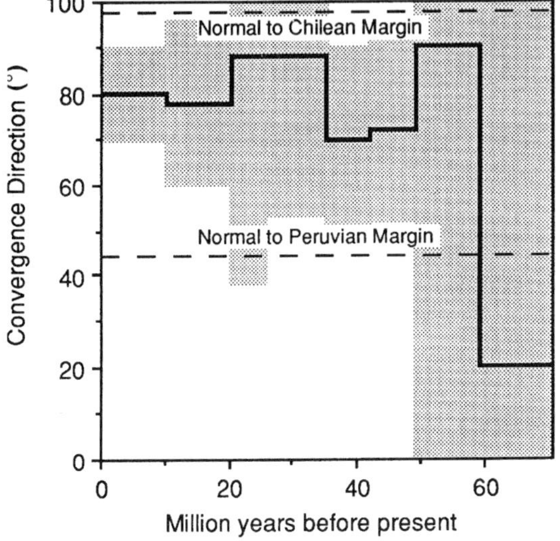

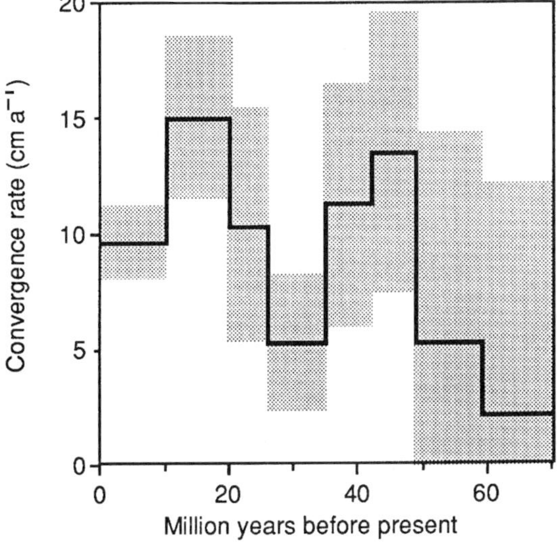

From about 36°S to 46°S the active Liquine-Ofqui fault appears to be partitioning subduction zone slip at present, resulting in maximum dextral slip at rates of < 3 cm a^{-1} (Dewey and Lamb, 1992) and E–W-directed shortening. The long-term (~50 Ma) behaviour of the Liquine-Ofqui fault is not well known, but Eocene and Miocene basins, possibly pull-aparts, are present along the fault zone (Forsythe and Prior, 1992; Pankhurst *et al.*, 1992). To the south of 46°S, intraplate convergence drops to only ~20 mm a^{-1} normal to the subduction zone, with no continental margin strike-slip (Cande and Leslie, 1986; Murdie *et al.*, 1993; Flint *et al.*, 1994). The area is not very seismically active, but Miocene to recent deformation (e.g. Ramos, 1989) suggests a similar ratio of trench-slip to continental margin shortening to further north.

Along the Andes, present-day subduction is segregated into zones of shallow (0–10°) and moderate (~30°) dip (e.g. Grange *et al.*, 1984; Isacks, 1988; Cahill and Isacks, 1992; Comte *et al.*, 1992; Lindo *et al.*, 1992). The changes in slab-dip are gradual and do not reflect sharp tears in the subducting slab. Zones of shallow subduction may be due to the buoyancy of younger oceanic crust, or of the Carnegie, Nazca and Juan Fernandez aseismic ridges. Zones of steeper dip correlate with active volcanism (Figure 9.1) and the presence of an asthenospheric wedge. Dehydration of sediments on the subducting slab promotes melting within this wedge and during ascent the resulting basaltic magma is modified by fractionation and assimilation of crustal material to a more andesitic

Figure 9.2 *Graphs of rate of Nazca–South America plate convergence and convergence direction plotted against time for 20°S (data from Pardo-Casas and Molnar, 1987). Note that since the early Cenozoic convergence direction has been nearly constant at ~080° but there were important increases in convergence rate in the Eocene and early Miocene which appear to correspond to periods of increased deformation of the continental margin. However, convergence rate appears to have been dropping as major late Cenozoic shortening occurred in the Subandean fold-thrust belt in the eastern Andes. Convergence has been oblique in Peru and Chile, but there is no clear major strike-slip in these regions*

Figure 9.1 *Outline maps of western South America. (a) Sea floor –3000 m and surface +1500 m contours highlighting the main bathymetric and topographic features of western South America. Offshore ridges are named. Note that the locations of recent volcanoes correlate with zones of moderate subduction angle (~30°). The shallow subduction zone between 3 and 15°S may reflect difficulty subducting the more buoyant Nazca (south) and Carnegie (north) ridges. (b) The major plate kinematic features of the plate margin. Note that partitioning of intraplate slip is strong in the northern Andes and nearly absent in the central Andes. Approximately 10% of Nazca–South America slip is taken up by shortening in the continental margin, and 90% by subduction zone slip. Earthquakes suggest active continental deformation is concentrated in the eastern Andes*

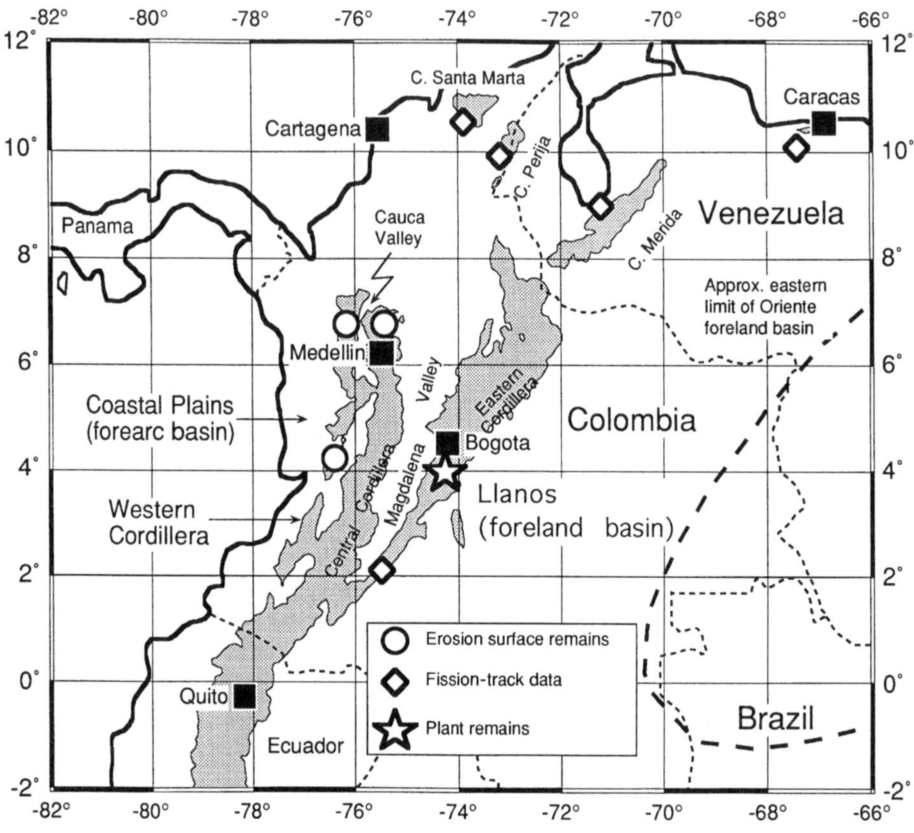

Figure 9.3 *Outline topographic map of the northern Andes of Ecuador, Colombia and Venezuela showing areas above 1500 m (grey fill), major cities and sites discussed in the text. Here the Andes consist of several parallel cordilleras separated by deep intramontane valleys with thick Cenozoic sediment fills. These valleys are bounded by active thrusts and by large-displacement dextral faults. The Eastern Cordillera/Perijá block was uplifted most recently (~10 Ma BP), separating the Magdalena valley from the foreland and diverting palaeo-Amazon/Orinoco flow from the north towards the east*

average composition (Thorpe, 1982; Davies and Bickle, 1991; Peacock, 1993).

9.4 Uplift of the Northern Andes (11°N to 4°S)

The Andes of central Ecuador and south-central Colombia comprise distinct parallel cordilleras with typical elevations of 2500–3500 m, and with the highest peaks reaching ~5000–6000 m. These ranges are separated by the deep intermontane Cauca and Magdalena valleys and flanked by coastal and foreland plains (Figure 9.3). To the northeast, the Eastern Cordillera splits into the Perijá and Merida ranges which ring Lake Maracaibo, while the isolated triangular Santa Marta range lies further west. The narrow, E–W-trending Caribbean Andes of Venezuela locally reach 2000 m.

The Western Cordillera and coastal plains of Ecuador and Colombia are an oceanic crust and island arc terrane accreted to the continental margin in the middle Cretaceous. The other cordilleras are cored with continental crust, and share a similar thick, deformed Mesozoic cover indicating that they were accreted no later than the Jurassic, and possibly during the Precambrian or Palaeozoic (e.g. Irving, 1975; Kennerley, 1980; Aspden and McCourt, 1986; Restrepo and Toussaint, 1988). Cenozoic to recent volcanics are restricted to the Western and Central Cordilleras between 2°S and 6°N, where stratovolcanoes form all the higher peaks (e.g. Arango *et al.*, 1976; CGMW, 1978). The higher peaks in the Santa Marta,

Perijá and Merida ranges consist of unroofed metamorphic basement with no volcanic cover.

9.4.1 PALAEOGEOGRAPHICAL DEVELOPMENT

Sedimentary sequences up to 6000 m thick in forearc, intramontane and foreland basins record the uplift and erosion of the distinct cordilleras. The earliest uplift of the Western and Central Cordilleras was caused by middle Cretaceous arc accretion. Major rejuvenation of the Central Cordillera in the Eocene is reflected in the first appearance of conglomerates in the Cauca and Magdalena basins. Further north, there was some Eocene erosion from the Santa Marta massif (Irving, 1975). Only in central Ecuador is Eocene uplift of the Eastern Cordillera apparent. Palaeogene K–Ar ages record metamorphism in Eastern Cordillera schists (Feininger, 1982) and conglomerates appear in the 600-m-thick Tiyuyacu Formation in the foreland basin (of presumed Palaeocene to Eocene age (Kennerley, 1980; Dashwood and Abbotts, 1990)). At about the same time oil was expelled from organic-rich sediments in the Eastern Cordillera (now graphitic schists (Feininger, 1975)) into actively growing structures.

By the Oligocene, relief and erosion rates in the Colombian Western and Central Cordilleras were low. Fine-grained facies with marine incursions are noted in surrounding basins and transgress across the Eastern Cordillera region. Major rejuvenation by ~18 Ma BP of the Central Cordillera resulted in conglomerates being deposited across the Magdalena basin (Honda Formation). The first influence of a rising Eastern Cordillera is noted at ~12 Ma BP (van der Wiel and van den Bergh, 1992a, b). Major middle-to-late Miocene thrusting of the Eastern Cordillera to the west over the Magdalena and east over the foreland (Roeder and Chamberlain, 1995) was accompanied by deposition of thick sedimentary wedges (e.g. Cooper *et al.*, 1995). This uplift caused redirection of palaeo-Amazon or Orinoco drainage away from the western Caribbean towards their present easterly courses (Hoorn *et al.*, 1995). Folding and thrusting in the easternmost ranges of Ecuador and Colombia is still active. Major thin-skinned shortening, deforming thick middle to latest Miocene sequences, is also noted in the Cordillera de Perijá further north (Kellogg, 1984).

9.4.2 FISSION-TRACK DATA

Limited data from the Eastern Cordillera (sites shown on Figure 9.3) are consistent with the palaeogeographic outline, indicating that erosion accelerated markedly in the mid–late Miocene. Zircon (Andriessen, 1995) and apatite ages (van der Wiel, 1990) from southern Colombia indicate average 100–12 Ma BP denudation rates of ~60 m Ma^{-1}, and post-12 Ma BP rates of ~300 m Ma^{-1}. Samples from the Santa Marta, Perijá, Santander and Merida ranges ringing Lake Maracaibo (Kohn *et al.*, 1984a; Shagam *et al.*, 1984; Kroonenberg *et al.*, 1990) show similar low rates of denudation until the middle Miocene. Subsequent rates were ~300–500 m Ma^{-1} over wide areas of these ranges, with very localized rates of >800 m Ma^{-1} on narrow fault slivers. Further east, the coastal ranges near Caracas have much younger zircon ages of 17–24 Ma suggesting the erosion rates of >300 m Ma^{-1} were sustained for longer, since the earliest Miocene (Kohn *et al.*, 1984b).

9.4.3 PALAEOBOTANICAL EVIDENCE

Pollen from Pliocene sediments in the Sabana de Bogota (Figure 9.3), an intramontane basin now at ~2500 m above sea level in the Colombia Eastern Cordillera show tropical floras at ~ 4 Ma BP changing rapidly to high-altitude floras by 3 Ma BP (Kroonenberg *et al.*, 1990; Andriessen *et al.*, 1993). Tropical floras are now found below 500 m but ~2000 m of rapid mid-Pliocene surface uplift can only be inferred if there was no regional change in climate at the time. However, there is no apparent sudden overdeepening of the foreland basin at this time which might result from greater topographic load on the Guyana shield (e.g. Cooper *et al.*, 1995), which would be expected if rapid surface uplift had occurred. There are no reports of palaeofloras from the region that pre-date possible Pliocene climatic change.

9.4.4 HIGH-ALTITUDE PALAEOSURFACES IN COLOMBIA

Low-relief erosion surfaces between ~2200 m and 3000 m have been reported from the Western and Central Cordilleras of Colombia (Padilla, 1981; Page and James, 1981) (see Figure 9.3). The higher ~3000 m surfaces near Medellin appear to underlie the early to middle Miocene Honda Formation (e.g. van der Wiel and van den Bergh,1992a) which lies at ~6000 m below sea level in the Magdalena valley, indicating vertical displacement of 8000–9000 m. Uplift of the surfaces from near sea level (present Magdalena valley-floor altitude is only ~100 m) must have started at ~18–15 Ma BP because the rejuvenated Central Cordillera sourced the Honda Formation (van der Wiel and van

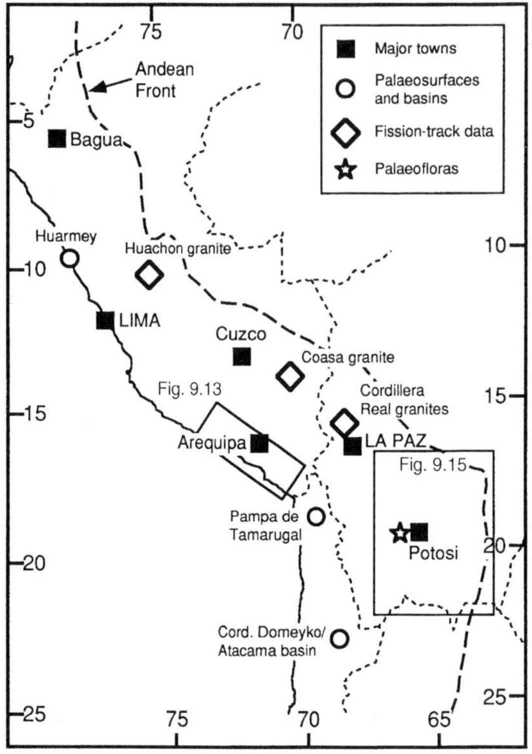

Figure 9.4 Location map for the central Andes. Sites for fission-track data, palaeoflora, palaeosurfaces and major towns are shown, in addition to box outlines of the locations of Figures 9.13 and 9.15

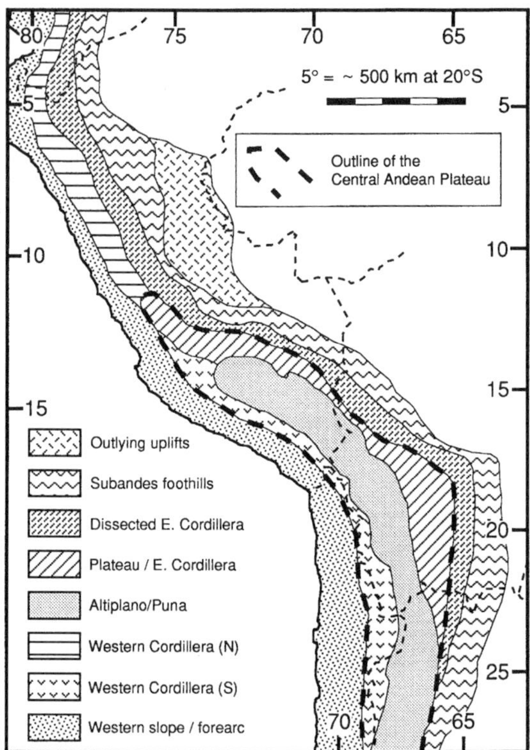

Figure 9.5 Geomorphological zonation of the central Andes. The most prominent feature of the region is the central Andean plateau, the centre of which is a compound internal drainage basin. The eastern and northern margins are low relief, little dissected, undulating surfaces cutting across Palaeozoic to Cenozoic bedrock, while the western margin is defined by the active volcanic arc. The steep eastern slope consists of deeply dissected lower Palaeozoic rocks, whereas the lower gradient Subandes is a deforming fold-thrust belt. Rainfall is high in these regions north of ~18°S. In northern Peru the Western Cordillera consists of deeply dissected Mesozoic rocks with a reduced volume of Cenozoic volcanics. The western slopes are extremely arid throughout the region, and comprise salt pans, local ranges of hills, and deep gorges cut by ephemeral streams draining from the Western Cordillera

den Bergh, 1992a), suggesting an average surface uplift rate of ~200 m Ma^{-1} and basin subsidence rate of ~400 m Ma^{-1}. The presence of a lower plain at ~2200 m may indicate that this uplift was pulsed. Deep gorges dissected the surfaces in Pliocene to recent time. The isostatic rebound due to this unloading is probably no more than ~100 m.

9.5 Uplift of the Central Andes (4°S to 46°S)

The large-scale morphology of the central Andes is strikingly different from further north. The most prominent feature is the 'Bolivian Orocline', an abrupt change in topographic trend from NE–SW in Peru to N–S in northern Argentina and Chile. In the core of this orocline lies the central Andean Plateau (11°N to 28°S), up to 500 km wide and with a remarkably constant average altitude of ~4000 m (Plate 9.1 and Figure 9.4). Plateau width, but not elevation, drops symmetrically to the northwest and south. Average elevations along the western and eastern margins of the plateau are slightly higher, with peaks locally reaching near 7000 m (glaciated above ~4800 m). Much of the plateau is occupied by the Altiplano (Peru/Bolivia) and the Puna (Argentina), a compound internal drainage basin interrupted by low-relief bedrock ridges (Figure 9.5), and much of the distinctive very flat relief of the region consists of the top surface of the youngest basin fills. The surrounding regions are low-relief undulating

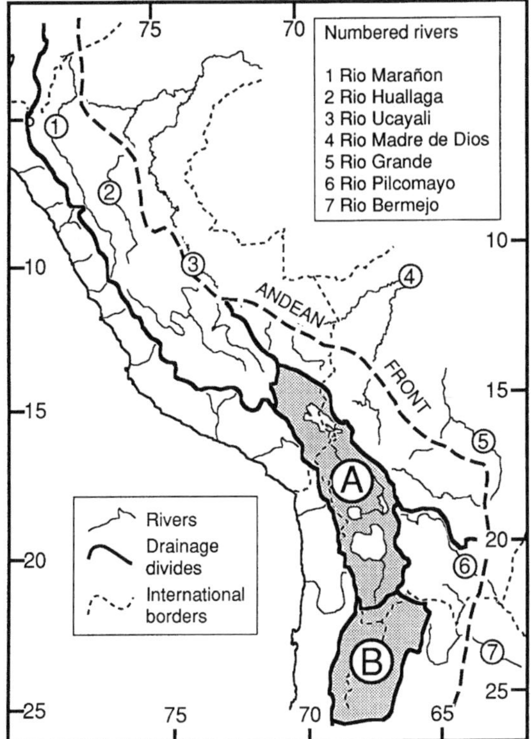

Figure 9.6 *Major rivers of the central Andes. Note that much of the central Andean plateau region is an area of internal drainage, with ephemeral rivers draining into lakes and salt pans in the (A) Altiplano and (B) Puna–Atacama basins. North of this region the main watershed coincides approximately with the volcanic Western Cordillera. Northern parts of the plateau are externally drained and deeply dissected. In Subandean regions three main watersheds divide waters draining directly into the upper Amazon (rivers 1, 2, 3), joining the Amazon in central Brazil (rivers 4, 5) and draining into the Rio de la Plata in Argentina (rivers 6, 7). The main east–west watershed and the boundaries of the internal drainage basins have been fairly constant for at least the past 25 Ma*

highlands cutting across Palaeozoic–Mesozoic bedrock (Figure 9.5). The plateau margins have very steep topographic gradients (up to 1 in 15) and are dissected by rivers draining into the Pacific and Atlantic, but there is almost no dissection deep into the plateau. In northern Chile, the western Andean slope is disrupted by a narrow coastal range reaching ~2500 m, a 50-km-wide sediment-filled 'central depression', and the discontinuous 'Precordillera' (comprising the Cordilleras de Morena and Domeyko). Dissection of the eastern slopes is strikingly more intense north of 18°S than to the south. This change is coincident with a drop in present-day mean annual precipitation from ~3000 mm in northern Bolivia to about 700 mm in southern Bolivia (Masek *et al.*, 1994). The deeply dissected plateau lip north of 18°S, the Cordillera, is slightly higher, possibly as a result of isostatic rebound.

Average elevation and width of the Andes drops dramatically north and south of the plateau region. By ~5°N and 45°S elevations are typically 500–1000 m with only isolated peaks reaching 2000–3000 m. These regions are entirely externally drained (Figure 9.6) and river valleys are typically very deep and narrow. For example, the bed of the Marañon valley in Peru is up to 3000 m below surrounding peaks for much of its ~500 km length. In contrast to the plateau region, there are distinct outlying ranges up to 1000–2000 m high up to 250 km east of the main Andean range. In both northern Peru and northern Argentina these outer ranges trend N–S.

Bedrock geology of the central Andes can be conveniently described in terms of orogen-parallel zones which broadly parallel the topography (Figure 9.7). Coastal regions of Peru and Chile are dominated by the granites of the coastal batholith (e.g. Coira *et al.*, 1982; Pitcher *et al.*, 1985; Mukasa, 1986), which intrude the Precambrian basement of the Arequipa Massif (Shackleton *et al.*, 1979), and deep extensional marginal basins formed during the Mesozoic prior to earliest Andean uplift (e.g. Cobbing, 1985; Flint and Turner, 1988).

Offshore, between the coastal batholith and the Peru–Chile Trench, there are elongate overlapping forearc basins filled with Eocene to recent sediments (e.g. von Huene *et al.*, 1987). Where the Carnegie and Nazca ridges collide with the continent, these forearc basins have been uplifted above sea level (e.g. Dunbar *et al.*, 1990). In northern Chile, the Cordilleras de Domeyko and Morena comprise deformed basement underlying Mesozoic rocks flanked by the deep Tamarugal (central depression) and Atacama basins (e.g. Jolley *et al.*, 1990; Buddin *et al.*, 1993; Flint *et al.*, 1993). The western edge of the central Andean Plateau is dominated by Eocene to recent volcanic rocks, mainly andesite flows and rhyolitic ignimbrites related to subduction zone dehydration and melting (e.g. Thorpe *et al.*, 1982). The prominent peaks of the Western Cordillera the Pleistocene stratovolcanoes spaced fairly regularly at ~50 km intervals. East of the Western Cordillera the Altiplano–Puna is a complex of Cenozoic sedimentary basins filled with up to ~10 km of fine- and coarse-grained red beds, with intermittent volcanic intercalations most common in the west (e.g. Kennan *et al.*, 1995; Lamb *et al.*, 1997).

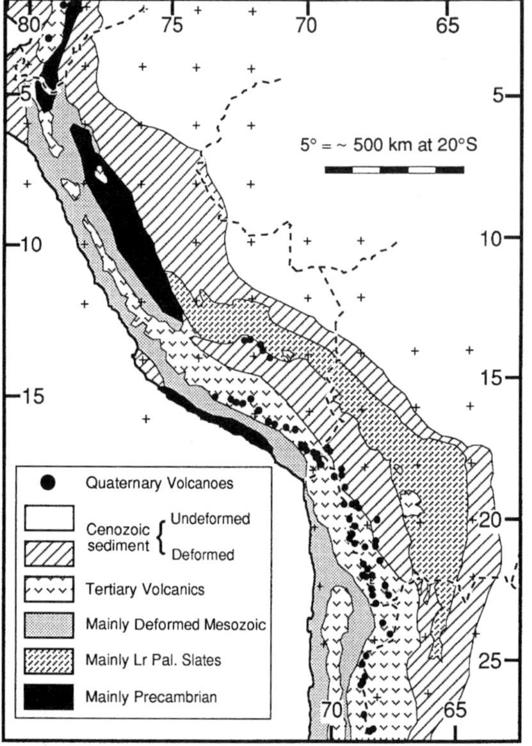

Figure 9.7 *Simplified geological map of the central Andes. The deep Mesozoic volcanic and sediment filled basins of the western Andes are intruded by coastal batholith granites (not shown). The eastern Andes are mainly composed of Precambrian schists and granites (north of 12°S) and low-grade Palaeozoic sediments (south of 12°S). Mesozoic–Cenozoic sediments are preserved locally in syncline cores. The thick Cenozoic sediments of the Altiplano and Subandean basins were deformed mainly in the last 10 Ma. Note that Tertiary volcanics are more widespread than Quaternary volcanoes. This may indicate the shallowing of the subduction angle is relatively young (<5 Ma)*

In contrast, the dominant rocks of the Eastern Cordilleras are Precambrian and Palaeozoic sediments and greenschist to amphibolite facies metamorphic rocks with Mesozoic–Cenozoic strata widely preserved in thrust footwalls and syncline cores (e.g. INGEOMIN, 1978; Pareja *et al.*, 1978). North of ~12°S, lower Palaeozoic rocks are largely absent and Carboniferous and younger strata lie directly on late Precambrian schists and gneisses (Dalmayrac, 1978). In the Macusani (14°S), Morococala (18°S) and Los Frailes (19–20°S), distinctive ignimbrite shields with highly peraluminous composition were erupted along the western edge of the Eastern Cordillera at ~12 Ma to 4 Ma BP (Grant *et al.*, 1979; Bonhomme *et al.*, 1988; Ericksen *et al.*, 1990; Cheillitz *et al.*, 1992). In contrast to the main volcanic arc these rocks are not related directly to the subduction zone, but possibly reflect melting of crustal material. Seismic sections (unpublished oil company data) indicate that, throughout the Cenozoic, the Eastern Cordillera has been overthrusting the eastern basins of the Altiplano, although average elevations in the Cordillera are now only slightly higher.

The eastern margin of the Cordillera, approximately coincident with the edge of the central Andean plateau, is also defined geologically by major overthrusting over the Subandean ranges, a thin-skinned fold-and-thrust belt involving Palaeozoic, Mesozoic and late Cenozoic strata (e.g. Jordan *et al.*, 1988; Roeder, 1988; Baby *et al.*, 1992). These Cenozoic strata are a deformed foreland basin fill, continuous with sequences in the present-day foreland. Active deformation in the central Andes is concentrated in this foreland zone. The importance of thrusting and folding in the eastern foothills declines to the north and south of the plateau, where basement-cored anticlines, bounded by steep reverse faults (Allmendinger *et al.*, 1983, 1990) define the outlying uplifts of the Sierras Pampeanas (Argentina) and the Sierras de Shira, Contaya and Moa (northern Peru).

9.5.1 PALAEOGEOGRAPHICAL DEVELOPMENT

Middle to late Cretaceous marine sediments cover much of the central Andes (Riccardi, 1988; Jaillard, 1994) except coastal regions (Figure 9.8). In Peru, a narrow fold-thrust belt emerged and shed coarse sediments east into basins in the western Altiplano (e.g. Vicente *et al.*, 1979; Noble *et al.*, 1990), while transpression on N–S faults in northern Chile uplifted the Cordillera de Domeyko, and localized extension or transtension formed the Atacama Basin (Flint *et al.*, 1993). In the late Cretaceous to Paleocene the sediments being deposited in these early clastic basins became coarser and clastic sediments became more common than carbonates further east (e.g. Gayet *et al.*, 1991; Flint *et al.*, 1993; Jaillard, 1994). During the Paleocene and Eocene deformation in this western belt intensified and resulting unconformities are sealed by a major volcanic flare-up (after a 50 Ma period with little eruption) dated at ~42–36 Ma BP in west-central Peru (e.g. Noble *et al.*, 1979a, 1990). The unconformities die out rapidly to the east into the Altiplano where a sharp influx of sands derived from the west is recorded (e.g. Bagua basin, Naeser *et al.*, 1991; Cuzco and Sicuani

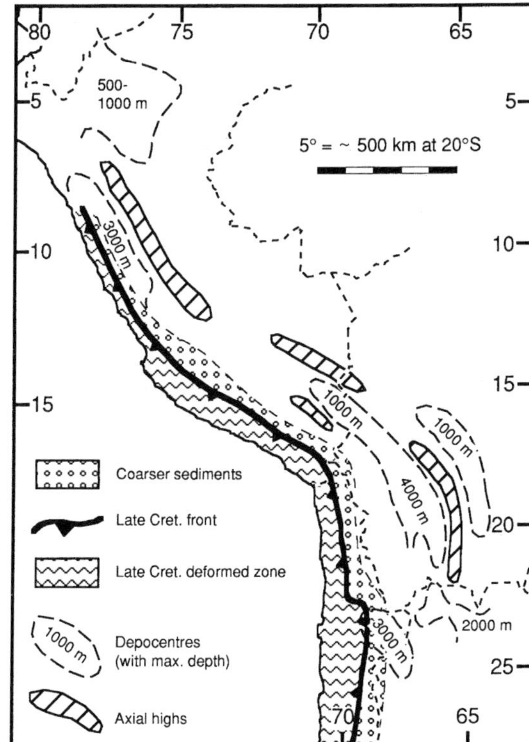

Figure 9.8 Late Cretaceous–early Cenozoic palaeogeography generalized from sources quoted in the text. No attempt has been made to palinspastically restore the effects of Andean E–W shortening. Note that deformation was restricted to immediate coastal regions with a narrow apron of coarse sediments to the east. Over almost all the Andes and foreland there are widely correlatable sequences of fine-grained clastic and carbonate rocks. Also shown are the main Cretaceous depocentres and axial highs. These highs were zones of reduced or no subsidence from the Albian onwards where locally the Cretaceous sequence is condensed to a few metres of nodular limestones (e.g. near Aiquile, Bolivia). The depocentres were bounded by normal faults overlapped in most cases by the latest Cretaceous

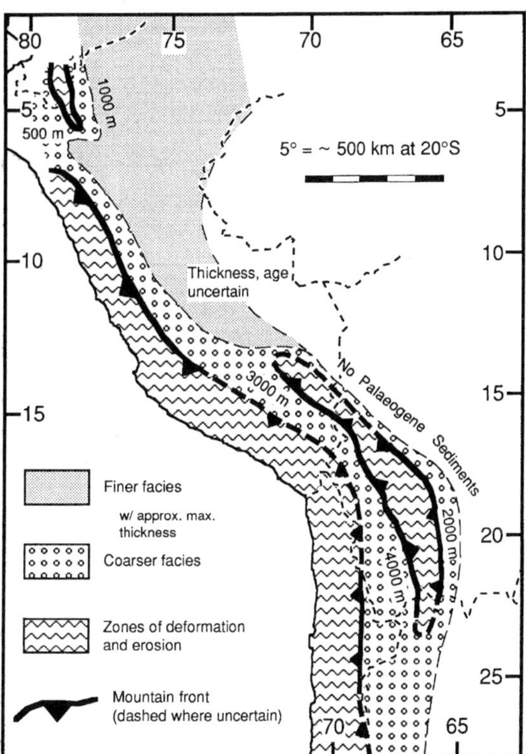

Figure 9.9 Paleocene–Eocene (~50–35 Ma BP) palaeogeography. Note that the limits of observed eroding regions coincide very closely with the Cretaceous depocentres, suggesting a basin inversion origin. The entire western Cordillera region was deformed and shed sediment to the west into forearc basins and east into the Altiplano. Eruption of thick andesite and dacite flows at ~40 Ma BP sealed many associated unconformities. The Eastern Cordillera between 14°S and ?23°S started to be eroded at about 50 Ma BP, and was bounded to east and west by active reverse faults. The eastern foreland basin was very narrow, and no Palaeogene sediments are seen in the Bolivian Subandes. Although much of the Peruvian Eastern Cordillera was not being eroded at this time, sediments up to 1000 m thick are widespread in the foreland. The northernmost uplift (5°S) resulted in coarse Precambrian clasts being shed to the east in the earliest Cenozoic

basins, Noblet *et al.*, 1987; Lopez and Cordova, 1988; Jaillard *et al.*, 1993; Chavez *et al.*, 1994). Pre-late Eocene sands were also shed west into forearc basins (e.g. von Huene *et al.*, 1987). In northern Chile, the coarser upper members of the Purilactis group in the Atacama basin record continued erosion in the Cordillera Domeyko (Hartley, 1993).

The Eastern Cordillera of Bolivia also first rose during the Palaeogene. From southern Peru to northernmost Argentina a proto-Cordillera (Figure 9.9) rose bounded to east and west by active thrust belts, and the Mesozoic cover of the region was eroded and transported into the Altiplano (e.g. Rodrigo and Castaños, 1975; Ellison *et al.*, 1989; Kennan *et al.*, 1995). Remnants of a narrow foreland basin, which pinched out west of the present-day Subandes, are preserved in synclines in the eastern part of the present Cordillera (e.g. Kennan, 1994; Kennan *et al.*, 1995). Sediments in Altiplano and foreland basins show no sign of significant Eastern Cordillera denudation between ~6°S and 10°S (e.g.

Koch and Blissenbach, 1962; Noblet *et al.*, 1987). Comparison of foreland basin sequences from southern Ecuador to northern Argentina reveal a striking asymmetry in sediment yield and foreland basin subsidence as old as >55 Ma. No Palaeogene sediments at all are known from the thrust belts and foreland basins of north-central Argentina (Precordillera and Bermejo basin, for example (Allmendinger, 1986; Jordan *et al.*, 1988)) or Bolivia (Sempere *et al.*, 1990; Marshall and Sempere, 1991), and any sediment eroded during early deformation was trapped within what is now the high plateau region. In northern Peru and southern Ecuador, rapid subsidence started in the latest Cretaceous due to thrusting in the Eastern Cordillera and at least 2000 m of fine to coarse sediments were deposited (e.g. Rosenzweig, 1953; Dashwood and Abbotts, 1990). The pinch-out of this sediment wedge to the southeast is very poorly defined. Clasts in these sediments are Eastern Cordillera schists and also volcanics and metavolcanic clasts (Tafur, 1991) probably from accreted Jurassic terranes of central Ecuador.

Between ~36 Ma and 26 Ma BP, coincident with a drop in intraplate convergence rate (Figure 9.2), there was little or no deposition in onshore or forearc basins, suggesting a reduction of palaeo-Andean relief, and there was almost no volcanic activity. In the latest Oligocene to middle Miocene (Figure 9.10), ignimbrites and andesites were erupted throughout the entire Western Cordillera (e.g. Noble *et al.*, 1979b, 1985, 1990; Lahsen, 1982; McKee and Noble, 1982). Multiple episodes of coarse-grained sedimentation, folding or tilting, and erosion are recorded in the Western Cordillera of Peru and Chile (e.g. Flint, 1985; Mortimer and Saric, 1975), the Altiplano basins (Mégard *et al.*, 1984; Naeser *et al.*, 1991; Kennan *et al.*, 1995; Lamb *et al.*, 1997) and in the forearc (e.g. von Huene *et al.*, 1987). By the early Miocene the entire Eastern Cordillera from ~5°S to >23°S was emergent and being eroded. Early middle Miocene subsidence of >>1000 m in central Peru (Koch and Blissenbach, 1962) was probably driven by overthrusting and flexure. In the Bolivian Subandes, thin clastic sequences (400 m deposited from ~27 Ma to 10 Ma BP) ended a 40 Ma hiatus as sedimentation spread east of the Eastern Cordillera (Sanjines and Jimenez, 1976; Marshall and Sémpere, 1991). As deformation progressed, sedimentation in the Altiplano and parts of the Eastern Cordillera tended to be concentrated in relatively narrow fault-bounded basins, some of them very deep (e.g. Laubacher *et al.*, 1988; Hérail *et al.*, 1993; Chavez *et al.*, 1994; Sémpere *et al.*, 1994; Lamb *et al.*, 1997). Lower Palaeozoic debris is seen for the first time in many basins (author's own observations). Subsidence and sedimentation in the Aubandes–Precordillera and foreland of northern Argentina (~25–30°S) started at 15–10 Ma BP, coincident with the onset of thrusting (Jordan *et al.*, 1988) and the first ponding of sediment within piggyback basins (Beer *et al.*, 1990).

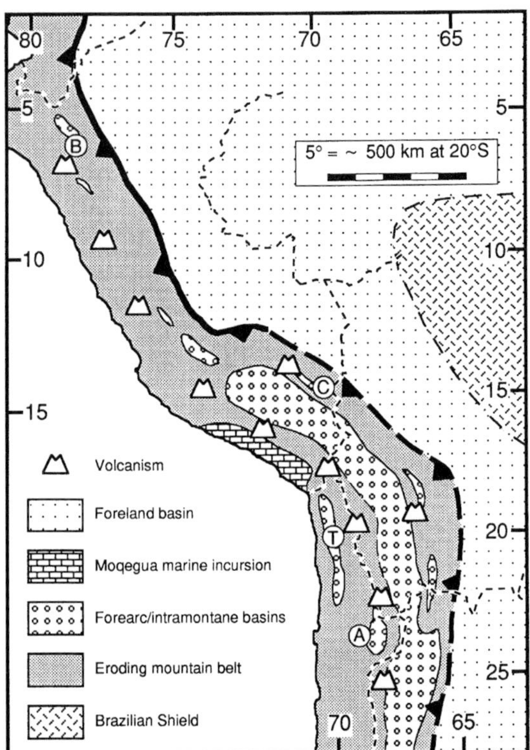

Figure 9.10 *Early to middle Miocene (~25–10 Ma BP) palaeogeography. By about 20 Ma BP the entire Eastern Cordillera was being uplifted and eroded and shedding metamorphic and granite clasts into the foreland basin. The Altiplano had become a closed internal drainage basin with volcaniclastic sediment sourced from the active Western Cordillera volcanoes and Palaeozoic and Mesozoic clasts coming from the Eastern Cordillera. The Eastern Cordillera spread to the east and sedimentation started in the Bolivian Subandes. Within the Eastern Cordillera there were numerous narrow fault-bounded basins (e.g. C. Crucero basin and B, Bagua basin). The Atacama basin (A) and Tamarugal (T) forearc basins in northern Chile filled with volcanics and clasts derived from the Western Cordillera and older uplifted blocks, such as the Cordillera Domeyko. In southern Peru, a marine incursion at 25 Ma BP deposited limestones up to 40 km inland. Until ~12 Ma BP volcanism was restricted to the west, but from 12 Ma BP several ignimbrite shields formed along the Eastern Cordillera–Altiplano boundary*

Shortening in the Eastern Cordillera continued until ~10 Ma BP when several, flat-lying ignimbrite shields were erupted. Sedimentation and deformation continued in the Altiplano basins and rapid (>600 m Ma^{-1}) deposition of sandstones and conglomerates started in the Subandes, as the Eastern Cordillera thrust east onto the Brazilian shield (Figure 9.11). By ~5 Ma BP internal shortening had more or less ceased on the Altiplano (Kennan *et al.*, 1995), and thin-skinned deformation of the foreland basin sequences in the Subandes started (e.g. Allmendinger, 1986; Allmendinger *et al.*, 1990; Baby *et al.*, 1993; Marrett *et al.*, 1994). At present the eastern range front of the Andes is a major watershed. Erosion is concentrated in this region and almost absent from the plateau region to the west.

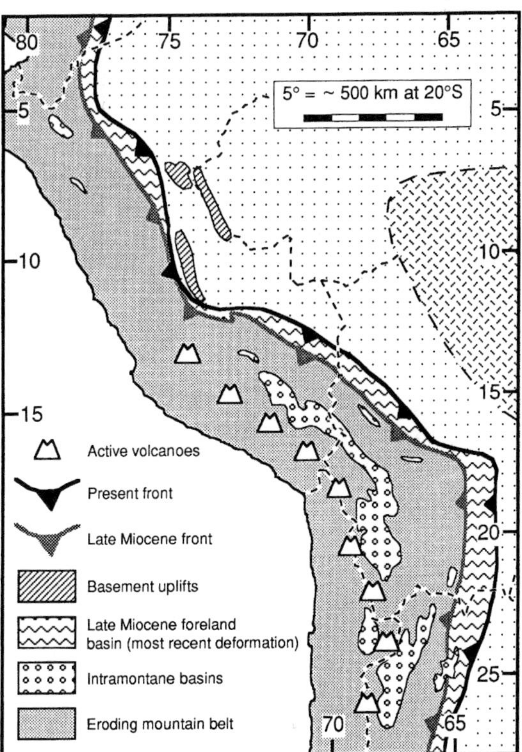

Figure 9.11 Late Miocene (~10 Ma BP) to recent palaeogeography. By the late Miocene active volcanism became restricted to the western margin of the Andean plateau. The Eastern Cordillera was thrust eastward onto the Brazilian Shield, producing a deep flexural foreland basin. Between 10 Ma and 6 Ma BP shortening in parts of the Altiplano reduced considerably the area still accumulating sediment, and during the late Miocene to Pliocene deformation advanced into the foreland basin, producing an imbricate fold-thrust belt. Thin-skinned tectonics became less important north of the plateau, where thick-skinned uplifts occur. During the middle Pliocene much of the northern Andes and eastern plateau region became externally drained and the area of intramontane basins decreased

9.5.2 FISSION-TRACK AND OTHER COOLING AGES IN PERU AND BOLIVIA

A few fission-track and K–Ar/Ar–Ar mica ages are available from three areas of the Eastern Cordillera (see Figure 9.4). In the Huachon region (~10°S), Triassic and older granites intrude probable Precambrian schists. Apatite fission-track ages (Laubacher and Naeser, 1994) suggest that denudation rates were ~70 m Ma^{-1} between 12 Ma and 22 Ma BP, and about 300 m Ma^{-1} between 12 Ma ago and the present. Much older zircon ages suggest average cooling rates of less than 1.25°C Ma^{-1} between 100 and 20 Ma BP, equivalent to denudation of only 5 m Ma^{-1} assuming a 25°C km^{-1} geothermal gradient. From the data given, there is no clear evidence of a significant early Cenozoic event, consistent with the thin, fine-grained sequences seen in the foreland.

Data from southern Peru (~14°S) are harder to interpret. Here, the Triassic Coasa batholith intrudes low-grade Palaeozoic schists (Kontak *et al.*, 1990a). The easternmost granites show clear 40 Ma K–Ar and Ar–Ar biotite ages, and younger, 20–30 Ma apatite fission-track ages. Kontak *et al.* (1990b) have interpreted these ages as due to a cryptic thermal perturbation at ~40 Ma which reset the biotites. However, such a long-lived upper crustal thermal anomaly seems unlikely. For instance, a major high-level vein-forming and stock intrusion episode only 25 km to the southwest shows concordant K–Ar and apatite fission-track ages of ~20 Ma BP indicating rapid cooling (Kontak *et al.*, 1987). Alternatively, the granite age data could indicate cooling from ~250 to 100°C between 40 and ~25 Ma due to denudation. At typical regional geothermal gradients this would require ~5–6 km of erosion. This hypothesis could be tested by looking for distinctive clasts in adjacent basin sediments of this age (Kennan, work in progress). Miocene and younger cooling rates may have averaged no more than ~3–6°C Ma^{-1}, equivalent to denudation rates of 100–200 m Ma^{-1} if the geothermal gradient was ~25°C km^{-1}. No zircon data are available for this region.

The most detailed, and most easily interpreted, data are available from the Cordillera Real of northern Bolivia. Apatite and zircon ages (Crough, 1983;

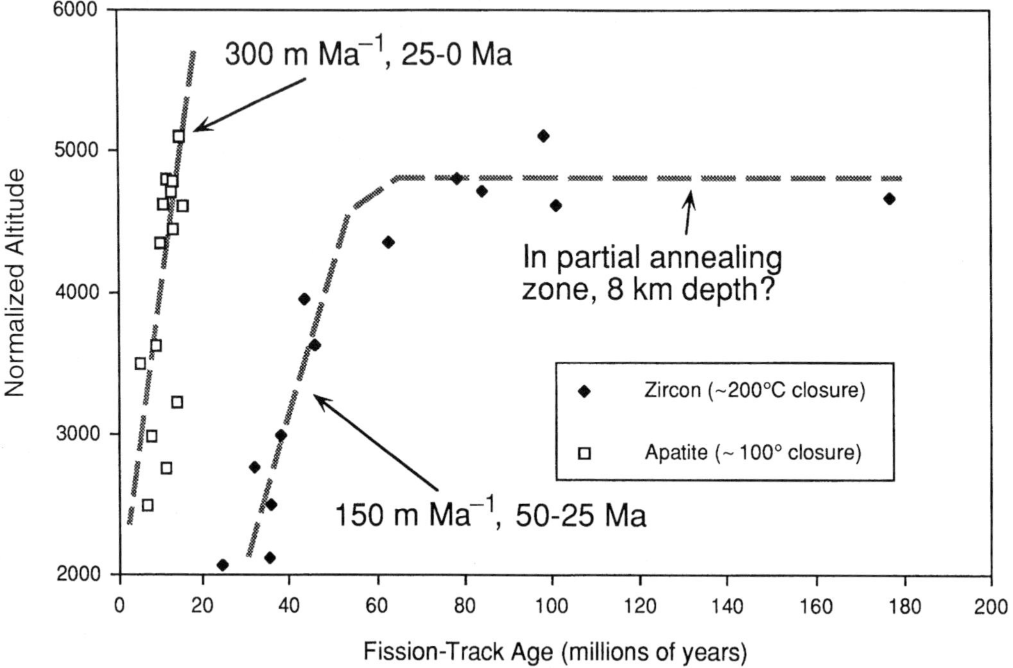

Figure 9.12 *Fission-track ages versus normalized altitude for the Cordillera Real, Bolivia (data points from Crough (1983) and Benjamin* et al. *(1987)). Because mean altitude, and thus depth to a given geotherm, drops to the northeast along the sampling profile, I have normalized altitudes to 4025 m (sample depth below mean surface altitude, calculated using digital elevation data). This result provides a more conservative estimate of denudation rate and prevents an apparent exponential increase in recent denudation rates being inferred in the younger samples, which lie deepest below the mean surface (e.g. Benjamin* et al., *1987, Figure 9.3). Unroofing in the Cordillera Real started at about 50 Ma BP during the inversion of Cretaceous depocentres. The dramatic increase in denudation at ~25 ± 5 Ma coincides with the main period of surface uplift in the central Andean plateau region, but the deduced rates give no direct indication of rate of increase of mean surface elevation. The high Miocene to present denudation rates in this region, which have led to the exposure of granites and andalusite–cordierite schists, contrast with the very low rates south of 18°S where Cretaceous and Cenozoic sediments are widely preserved. Note that the youngest apatites being eroded and transported into the foreland at present are as young as 5–4 Ma as a result of deep dissection*

Benjamin *et al.*, 1987) indicate slow denudation rates before ~50 Ma BP. Between 50 Ma and 30 Ma BP denudation rates were ~100–150 m Ma^{-1}, increasing to 300–400 m Ma^{-1} between 30 Ma BP and the present (Figure 9.12). This denudational record is entirely consistent with the palaeogeography discussed above. No fission-track data are available from further south.

9.5.3 PALAEOBOTANICAL EVIDENCE

Berry (1939) (and references therein) has described some palaeofloras from several localities in Ecuador, Peru and Bolivia and has compared them to their nearest living relatives. Unfortunately, many of the localities are poorly described and cannot be dated accurately using more modern isotopic data. Plants found at Potosi, Bolivia (~4500 m above sea level in the Eastern Cordillera), may be as old as 22 Ma and they compare with living floras now found at altitudes of 2000–3000 m. Some of these plants are intolerant of freezing (Lena Stranks, pers. comm.) and are not now seen above ~2500 m in Bolivia. In the absence of climatic change, ~2000 m of Miocene to recent uplift could be inferred, but the true figure may be somewhat lower, since oxygen isotope data suggest continental-scale climate cooling since ~3.5 Ma BP. There are no published leaf morphology studies on Andean palaeofloras from which to infer palaeoelevations.

9.5.4 HIGH-ALTITUDE PALAEOSURFACES IN PERU, CHILE AND BOLIVIA

Palaeosurfaces seem to be more widely preserved, and better described, in the central Andean plateau region

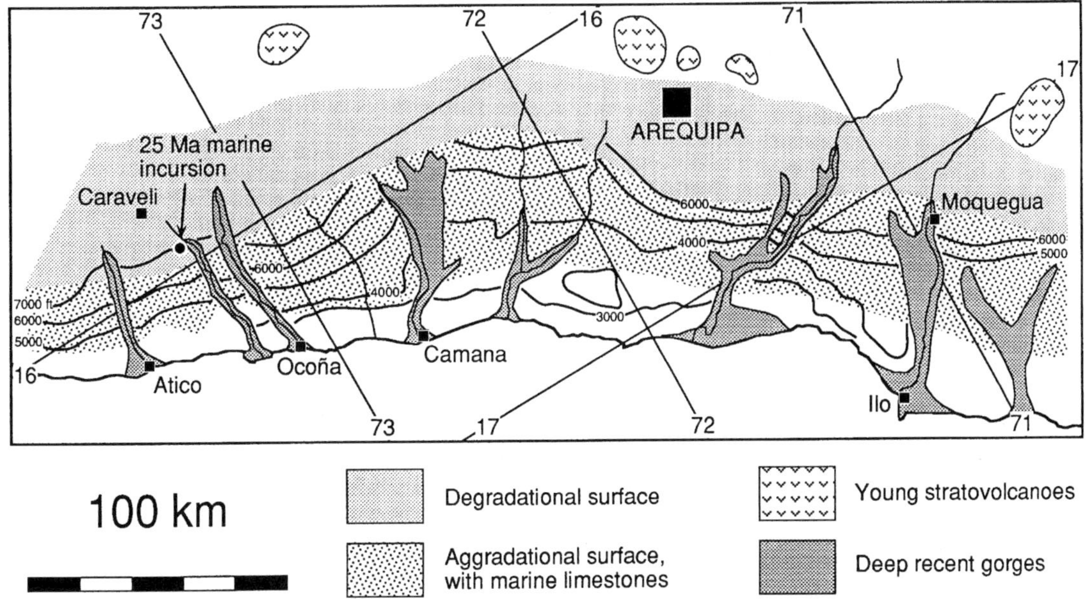

Figure 9.13 *Map of the coastal region of southern Peru showing the known limits of a late Oligocene erosion surface transgressed by marine limestones. The contours (data source Operational Navigation Chart sheet P-26 scale 1 : 1 000 000, contours in feet) show the interpolated present elevation of the marine transgression, which is now tilted to the southeast at 2–3°. To the northeast a degradational surface rose to an altitude of perhaps 1500 m above sea level and is the basement on which the Miocene arc erupted. Lower palaeosurfaces (not shown) and deep canyon incision have not removed enough material to account for more than a few per cent of the uplift of the surface through isostatic rebound*

than elsewhere. Many can be traced laterally to marine deposits so that absolute surface uplift can be reliably estimated.

In southern Peru (15–18°S, Figures 9.4 and 9.13) a prominent seaward dipping erosion surface is covered by a few metres of marine limestone dated at 25 Ma (Tosdal *et al.*, 1981, 1984; Noble *et al.*, 1985). These overlie gently folded Eocene conglomerates and are overlain by conglomerates and ignimbrites. This marine erosion surface is now at about 400–1000 m from the coast and can be traced at least 40 km inland to an altitude of ~2000 m. This essentially unbroken tilt slope has a seaward dip of 2–2.5°, but was probably at or near horizontal at 25 Ma. Northwest of the marine incursion a slightly steeper degradational part of that surface rises to the Western Cordillera watershed, mantled by Miocene to recent volcanic rocks. There seems to be little offset of the aggradational and degradational parts of this surface. Thus at ~25 Ma BP it seems likely that a low-relief forearc rose northeast to a watershed at perhaps 1000–1500 m. Subsequent simple seaward tilting suggests about 2000–3000 m of uplift of the Western Cordillera since then. As uplift progressed a series of nested lower surfaces and valleys, with successively lower present-day slopes, were cut (Figure 9.14a). Although it is not possible to quantify tilting rates from such dynamic features as river profiles, it is interesting that the most rapid change in slope seems to have occurred between about 14 Ma and 9 Ma (using valley fills and incised tuffs). Myers (1976) has described a superficially similar series of surfaces from near Huarmey (10°S, central Peru) which were also nested during the Miocene with the most recent phase of valley incision being post 5 Ma BP.

Similar seaward tilting also appears to have occurred in northern Chile (19–22°S). The Pampa de Tamarugal is a half-graben basin bounded by the Atacama fault and Coastal Cordillera (Figure 9.14b). The floor of the basin is an erosion surface of early–middle Cenozoic age mantled by earliest Miocene ignimbrites which define a simple monocline rising east to the volcanic arc (Galli-Olivier, 1967). Faulting and tilting started in the early Miocene (Mortimer *et al.*, 1974; Mortimer and Saric, 1975; Paskoff and Naranjo, 1983) and post-Pliocene offsets are known (e.g. Fig. 6 of Dewey and

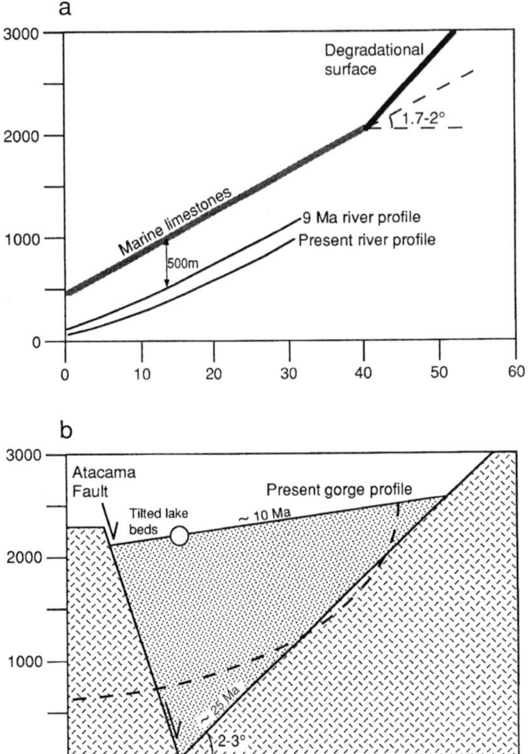

Figure 9.14 *Schematic cross-sections illustrating inferred uplift magnitude and timing in western Peru and Chile. (a) Moquegua region (see Figure 9.13) showing the attitude of the 25 Ma marine incursion surface and subsequent valley profiles (data from Tosdal* et al., *1984). By ~9 Ma BP limestone outcrops, now at 1000 m, were elevated at least 500 m above dated river profiles near Locumba, at, or above, sea level, suggesting that at least half the tilting pre-dates ~9–10 Ma BP. (b) Pampa de Tamarugal region, northern Chile. The Pampa overlies a half-graben basin, bounded by the Atacama fault. Tilting and sedimentation started at ~20–25 Ma BP as the Andes started to rise in this region. The marine incursion surface was probably close to horizontal, while river profiles were always inclined to the west. Hence, although apparently parallel, substantial tilting occurred between 25 Ma and 9 Ma BP. By ~9 Ma BP the basin was filled and drainage gained a direct outlet to the sea, initiating deep gorge dissection. Dissection was thus a result of base-level change and not directly of uplift. In both regions Plio-Pleistocene lake terraces show evidence of continued tilting*

Lamb (1992)). Sediments ponded against the Coastal Cordillera as the basin deepened, but by about 10 Ma BP the basin had filled. Rivers breached the Cordillera, gaining a direct connection with the sea. The subsequent incision of gorges (Mortimer, 1980) is thus related to a change from internal to external drainage, and not to rapid uplift of the coastal block. South of ~22°S, drainage had a direct marine connection and low-altitude late Miocene pediments (Mortimer, 1973) suggest little coastal uplift of this region which is separated from the Cordillera Domeyko by young thrusts (e.g. Jolley *et al.*, 1990). In both southern Peru (Tosdal *et al.*, 1984) and northern Chile (Hollingworth, 1964; Mortimer and Saric, 1975), substantial seaward tilts are seen in Quaternary lacustrine deposits which suggest that tilting is still active. Pleistocene coastal terraces are uplifted in northern Peru (De Vries, 1988) and in southern Peru (Macharé and Ortlieb, 1992), but are probably local transient effects related to the subduction of more buoyant aseismic ridges, and inferred uplift rates cannot be applied regionally.

By contrast, erosion surfaces related to drainage into the foreland basin in the eastern Andes are well preserved in south-central Bolivia (Figure 9.15). Servant *et al.* (1989), Gubbels *et al.* (1993) and Kennan *et al.* (1997) have described extremely flat surface remnants up to 10 km × 30 km in size over an area of ~600 × 100 km, in all cases located east of the Altiplano–foreland drainage divide. These surfaces lie at 2200–3800 m and are cut directly into folded Palaeozoic to Cenozoic bedrock. Clastic cover is usually thin or absent but reaches ~100 m where sediments have banked against protruding ridges. The surfaces define flat-bottomed valleys, commonly with steep side scarps. Valley floors typically lie up to 1000 m beneath surrounding highlands. These N–S-trending valleys are controlled by regional structures, and valley-bottom altitude systematically drops as they meander towards the east. Valley gradients drop downstream from ~1 in 125 to as low as 1 in 200–250. The lowest surfaces, at ~2200 m, are found immediately west of the Subandes fold-and-thrust belt. The surfaces cut rocks as young as 15 Ma old and are mantled by 9 to 3-Ma-old tuffs (Kennan *et al.*, 1997 and references therein), suggesting that they were cut in the mid–late Miocene. Headward erosion by gorges up to 1000 m deep post-dates 3 Ma BP. The volume of sediment eroded during surface cutting must have been ~$1–2 \times 10^4$ km^3 (Kennan *et al.*, 1997), but no sufficiently large sediment sinks exist in the Eastern Cordillera. This suggests that the ultimate sink and base level was the late Miocene foreland basin (now deformed as the Subandes), which was at, or near, sea level (the ~10-Ma-old Yecua Formation contains marine fossils). Even allowing for significant structural shortening, the lowest remnants are unlikely to have been more than 500 m above sea

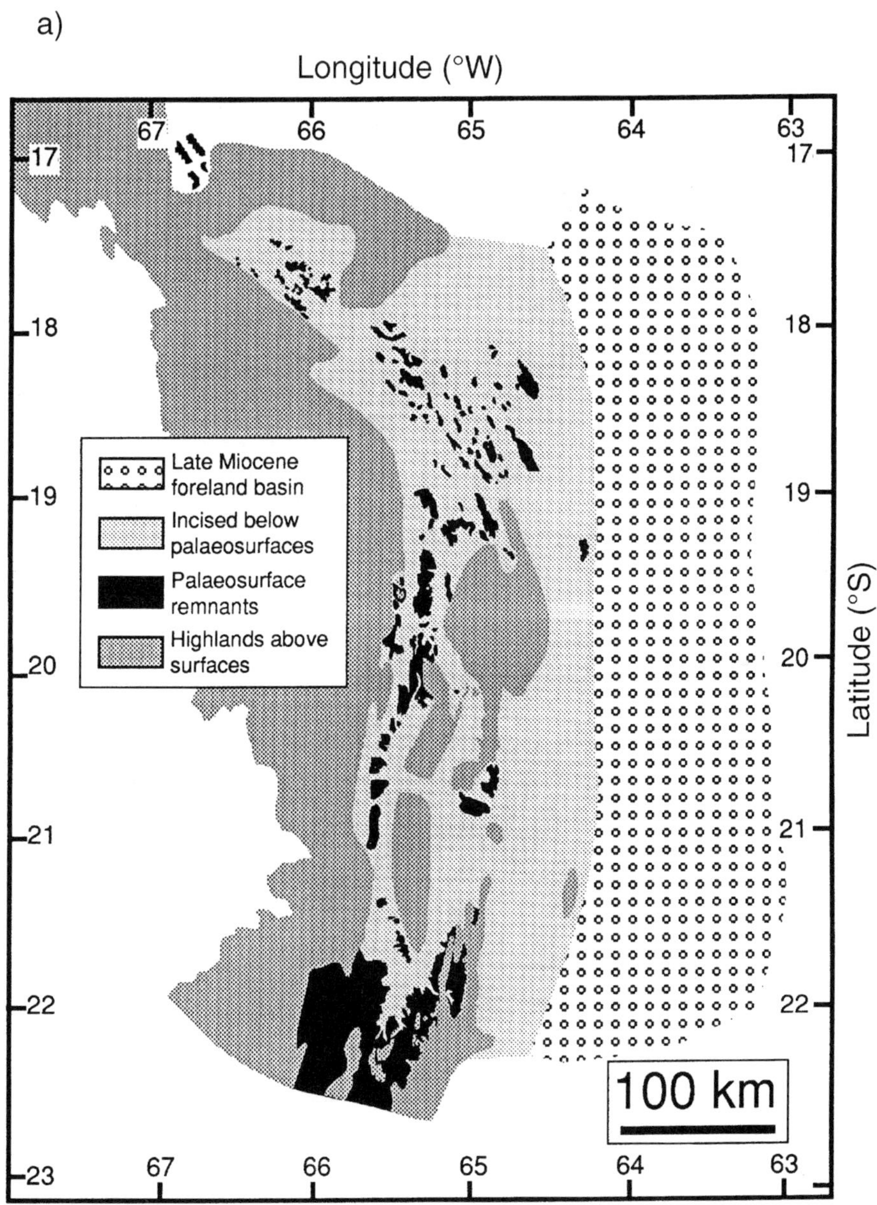

Figure 9.15 *High-altitude late Miocene palaeosurfaces in the Eastern Cordillera of south-central Bolivia. (a) The location of all mapped surface remnants, areas of higher altitude and the position of the foreland basin at ~10 Ma, now the Subandes fold-thrust belt*

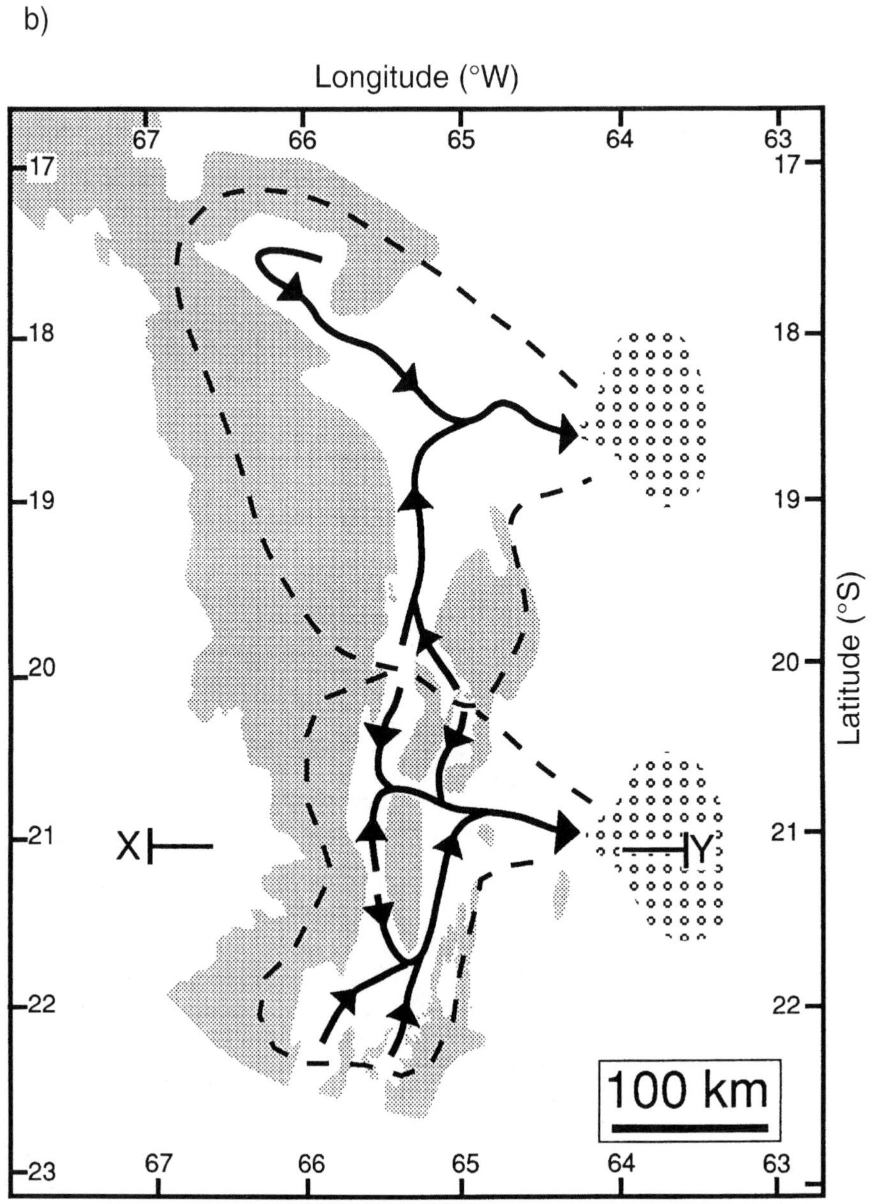

Figure 9.15 *(continued) (b) Interpretation of palaeodrainage basins with lines of drainage inferred from palaeosurface attitude and elevation. The palaeosurfaces define a mature low-gradient system of broad, flat-bottomed valleys which drained into the late Miocene foreland basin, which was at, or near, sea level. Thus the present surface altitude is the result of ~2000 m of vertical uplift since ~10 Ma BP. The lack of significant regional tilts on any surface remnants suggests there is no major crustal ramp beneath this area which could have been active since the late Miocene during Subandean thrusting. Any such ramp must be located west of ~67° W*

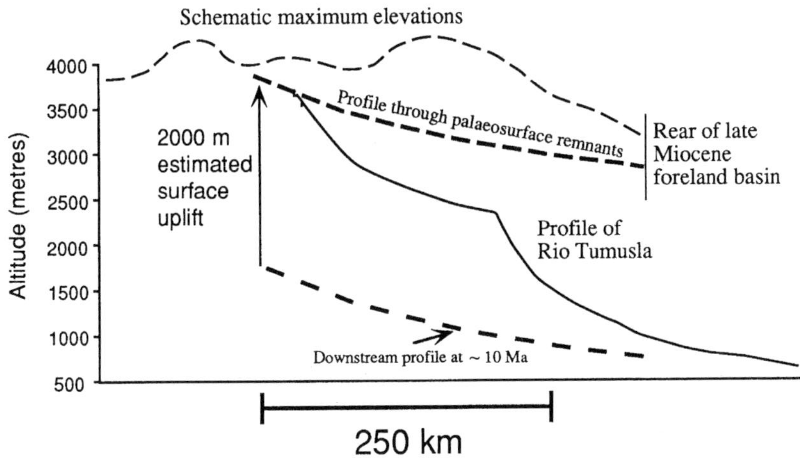

Figure 9.16 *Schematic cross-section through the high-altitude palaeosurfaces (approximately X–Y on Figure 9.15b) showing the inferred uplift of the surfaces since 10 Ma BP. Also shown are approximate maximum elevations above the palaeosurfaces, suggesting that prior to ~10 Ma BP the Eastern Cordillera was probably not much higher than 1000–2000 m. The present drainage profile of the Rio Tumusla is also shown. Although dissection has occurred all along the length of the river, it is most prominent downstream of a major knickpoint, which is not coincident with a marked lithological change and may be migrating upstream*

level, suggesting that ~2000 m of surface uplift has occurred in the Eastern Cordillera of Bolivia since ~10 Ma BP (Figure 9.16).

9.5.5 SOUTHERN PUNA AND EASTERN THRUST BELT OF NORTHERN ARGENTINA

The structure and stratigraphy of the southern Puna and eastern thrust belt of northern Argentina (25–33°S) have been extensively studied. There are no quantitative palaeobotanical or fission-track data from which to reconstruct surface altitude or denudation rates, but the geological history of the area supports a similar uplift history to that of southern Bolivia (Figure 9.16). Following an early period of shortening in the west during the Eocene, deformation advanced to the east, where the most important structure is the Precordillera fold-thrust belt which accounts for >65% of the total shortening observed at 30°S (Allmendinger *et al.*, 1990). The onset of deposition in piggyback basins and in the foreland indicates the start of thrusting at about 16 Ma BP (Allmendinger, 1986; Jordan *et al.*, 1988; Beer *et al.*, 1990). The first evaporites in the closed basins of the Puna date from this time (Vandervoort *et al.*, 1995). Enhanced erosion and sedimentation rates are observed since about 8.5 Ma BP and rapid shortening (up to 0.5–1.5 cm a^{-1}) has been recorded on individual fold-thrust structures (Sarewitz, 1988). This is essentially similar to the shortening history of southern Bolivia. Quaternary extension (Allmendinger, 1986; Marrett *et al.*, 1994) is observed in the high Puna (~3800 m) and may reflect a recent increase in altitude as a consequence of recent lithospheric delamination (e.g. Kay *et al.*, 1994).

9.6 Uplift of the Southern Andes (46°S–57°S)

South of the intersection of the Chile Ridge with the coast at 46°S there is a dramatic change in plate kinematics, geomorphology and geology (see Plate 9.1 and Figure 9.1). The mountain belt is somewhat wider and higher than to the north of 46°S, although only local peaks exceed 1500–2000 m. Core bedrock (the Mesozoic Patagonian Batholith extends from ~40°S to 54°S) is the same to the north and south of the change. Uplift and deformation can be dated from the Cenozoic marine to continental transition, and from the youngest deformed rocks, and clearly shows that deformation moved from south to north. In southernmost Chile, the Cretaceous Rocas Verdes back-arc basin, floored with oceanic crust, was inverted between 100 and 80 Ma BP (De Wit and Stern, 1978; Dalziel, 1986). Initial denudation rates were high, but dropped by ~60 Ma BP (Nelson, 1982). The youngest deformation in the extreme south is Eocene (Winslow, 1981). In contrast, further north significant deformation post-dates 15-Ma-old marine rocks, and very dramatic crustal shortening occurred in the late

Miocene (Ramos, 1989; Flint *et al.*, 1994). The highest marine rocks are found at ~500–1000 m above sea level. Sediment age and subsidence rates in the Patagonian foreland basin show a similar N–S diachroneity (e.g. Biddle *et al.*, 1986). There is little Cenozoic to recent andesitic arc-volcanism south of ~44°S (e.g. Thorpe *et al.*, 1982). In foreland regions, however, alkaline plateau basalts are common (e.g. Ramos and Kay, 1992), and these young from south to north. The intersection of the Chile Ridge with South America has also moved north through the Cenozoic (e.g. Cande and Leslie, 1986). The subcrustal presence of an active spreading ridge may have resulted in a higher heat flow and a softer upper crust, so allowing easier shortening, although the plate convergence rate is lower than to the north (e.g. Ramos, 1989). Overall crustal shortening estimates (~50 km) (Ramos, 1989) suggest a similar continental/subduction-zone slip ratio to that further north.

9.7 Discussion

Andean topography is clearly correlated with a thick crustal root. Geophysical evidence (e.g. James, 1971; Lyon-Caen *et al.*, 1985; Wilson, 1985; Bahlburg *et al.*, 1988; Wigger, 1988; Kono *et al.*, 1989; Schmitz *et al.*, 1990; De Matos and Brown, 1992; Introcaso *et al.*, 1992; James and Snoke, 1994; Whitman, 1994; Zandt *et al.*, 1994; Watts *et al.*, 1995) suggests that the crust under the highest parts of the Andes reaches a thickness of 60–70 km, or nearly twice the thickness in forearc and foreland regions (35–40 km), and that large-scale Andean geomorphology is likely to be intimately linked with crustal thickening processes such as magmatic accretion or horizontal shortening. Lavas younger than ~25 Ma BP show more crustal contamination than older lavas (Kay *et al.*, 1987, 1994; Boily *et al.*, 1990), suggesting that the crust through which they rose became markedly thicker from the early Miocene.

Widespread evidence of simple tilting or vertical uplift of erosion surfaces (Figures 9.13–9.16), and the preservation of flat-lying thick Cenozoic volcanic and sedimentary sequences over much of the Altiplano, with only localized shortening, has prompted many authors to suggest that uplift in the Andes has been caused by vertical block-movements (e.g. Myers, 1975; Cobbing *et al.*, 1981) with the deep crustal root mainly a result of magma addition from below (e.g. James, 1971; Gough, 1973; Thorpe *et al.*, 1981).

We can test this idea with a mass balance calculation for the central Andes at ~18°S where the total cross-sectional area is about 45 000– 50 000 km^2, and the trench to foreland width about 900 km (Kennan, 1994). This is an ideal area to choose because the cover of middle to late Cenozoic rocks over much of the plateau indicates that overall late Cenozoic erosion rates have been very low, and that the topography is a primary signal of tectonic processes. The significant denudation of the Cordillera Real of Bolivia accounts for less than 5% of the crustal cross-section, and much of this may have been recycled into Altiplano or Subandes basins included in the calculation. A cursory examination of the digital elevation data (Plate 9.1) for much of the central Andes shows that deeply dissected regions are relatively narrow, so the assumption of little overall erosion is reasonably valid.

The total volume of central Andean extrusive rocks is apparently relatively small. Although locally thick, total volume added is probably not more than 100 km^3 km^{-1} arc length (Baker and Francis, 1978), with production mainly over the last 25 Ma. The volume of related cumulates left within the crust due to high-level fractionation of primary basaltic melt may be two to four times the extruded volume (e.g. Thorpe *et al.*, 1982; Boily *et al.*, 1990), so extrusive volcanism can account for perhaps 500 km^2 of the cross-section. It is not so clear how to estimate the volume of magma intruded or accreted to the base of the crust. Magma addition (including cumulates) related to widespread granite intrusion (Figure 9.17) may account for 4000–8000 km^3 km^{-1} arc length (estimate from Thorpe *et al.*, (1981)). Most were intruded between ~180 and 80 Ma ago during a major period of extension, high heat flow and space creation, and therefore pre-date Andean surface uplift (e.g. Capdevila *et al.*, 1977; Lancelot *et al.*, 1978; Noble *et al.*, 1984; Beckinsale *et al.*, 1985; Mukasa, 1986; Soler and Bonhomme, 1988; Bonhomme and Carlier, 1990; Farrar *et al.*, 1990; Kontak *et al.*, 1990a; Petford and Atherton, 1992; Atherton, 1993). Assuming that magma production rates have been similar since then, perhaps 4000 km^3 km^{-1} arc length basaltic magma has ponded at the base of the crust since about 50 Ma ago. Thus, total magma addition since 50 Ma BP may be ~5000 km^3 km^{-1} arc length, accounting for only ~10% of the crust at 18°S.

Seismic velocities and densities in the lower crust (references given above) support the accretion of a basaltic root mainly beneath the Western Cordillera, while the lower crust in the Altiplano and further east has a more felsic composition. Helium isotope studies, however, indicate that modest mantle melting is occurring right across the Altiplano, and that lithospheric mantle is thin in these regions (Hoke *et al.*,

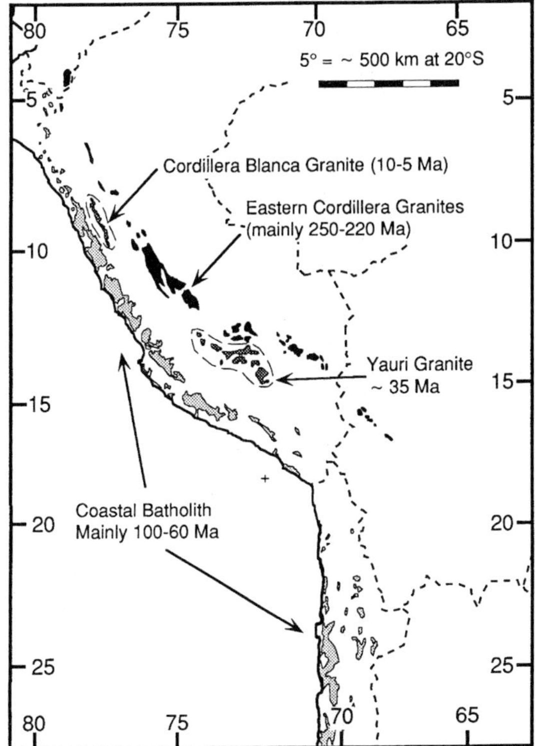

Figure 9.17 *Granitoid bodies in the central Andes (generalized from CGMW, 1978). Although large volumes of granite occur in the central Andes, most pre-date significant uplift. The Coastal batholith was mostly intruded between 100 and 60 Ma BP and the Eastern Cordillera granites are older at ~250–220 Ma BP. In both cases intrusion at a very high-level (8–4 km depth) accompanied major basin extension, injection of basaltic magma at a high level in the crust and high heat flow. Almost no large granites have been intruded since the onset of compression, crustal thickening and uplift; later felsic intrusions are mainly small, high-level stocks. The exceptions are the late Oligocene Yauri batholith and the late Miocene Cordillera Blanca batholith. The former lies in a poorly known zone of complex strike-slip faulting which may have allowed space creation, and the latter lies in the footwall of a major extensional/sinistral fault system. Because erosion is not sufficiently deep we do not know if granites are being intruded beneath the Miocene to Recent arc or whether basaltic magma is ponding at the base of the crust due to a lack of space creation mechanisms*

1994). Ponding of mantle melt near the base of the crust may have contributed to crustal melting and extrusion of peraluminous ignimbrites in this region.

At 18°S the shortening required to explain the observed crustal thickening is about 250–350 km, depending on the initial crustal thickness (~35–40 km). In addition to surface observations, seismic information (unpublished oil company data) shows significant shortening within apparently undeformed basins in the Altiplano. Altogether ~150 ± 30 km of shortening can be accounted for across the relatively gentle structures of northern Chile, the Bolivian Altiplano and the Eastern Cordillera at 18°S (Kennan, 1994; Lamb *et al.*, 1997), so leaving perhaps 100–200 km which must be concentrated in the narrow Subandean fold-thrust belt in the east. Seismic reflection studies (unpublished oil company data) clearly show a well-developed thin-skinned style of deformation between 12°S and 25°S. Palaeozoic to Cenozoic sequences are imbricated above lower Palaeozoic shale décollements. Total shortening at ~18°S is estimated to be at least 150 km (e.g. Roeder, 1988; Baby *et al.*, 1992). The low altitude of the Subandes clearly indicates that large-magnitude shortening of the middle and lower crust beneath the décollement is not occurring in that region. Instead, essentially undeformed Brazilian shield is being thrust beneath the Eastern Cordillera and Altiplano (Figure 9.18). Isacks (1988) has proposed that this produces distributed ductile shortening and thickening beneath the central Andean plateau, with passive uplift of a less deformed upper crustal 'lid'. How much uplift occurs, and when, depends closely on the thickness and width of the deforming layer. As a preliminary estimate, the ~200 × 30 km of Brazilian shield thrust beneath the central Andes would thicken the crust beneath the ~400-km-wide plateau by an average 15 km, which would produce approximately 2 km of late Miocene to recent surface uplift, consistent with geomorphological observations. This model is supported by gravity studies, which show clear flexure of the Brazilian shield beneath the load of the Eastern Cordillera (Lyon-Caen *et al.*, 1985; Watts *et al.*, 1995). Strong lithosphere extends 150–200 km east beneath the Eastern Cordillera. This underthrusting is constrained to have occurred since 10 Ma ago because only then did significant flexural subsidence start in central Bolivia. Allmendinger *et al.* (1990) report that at ~30°S, where late Miocene magmatic addition to the crust is negligible, measured crustal shortening can account for the entire observed crustal cross-section.

In addition to modest surface erosion, the subducting oceanic plate may scrape off lower crustal slices from the forearc region (e.g. Karig, 1974), with this crustal thinning causing dramatic subsidence of forearc basins below sea level (e.g. von Huene and Lallemand, 1990). Estimated maximum erosion rates are ~100 km^3 km^{-1} arc length (Sosson *et al.*, 1994), but it remains unclear whether this material is accreted to the base of the crust deeper down (resulting in high-pressure

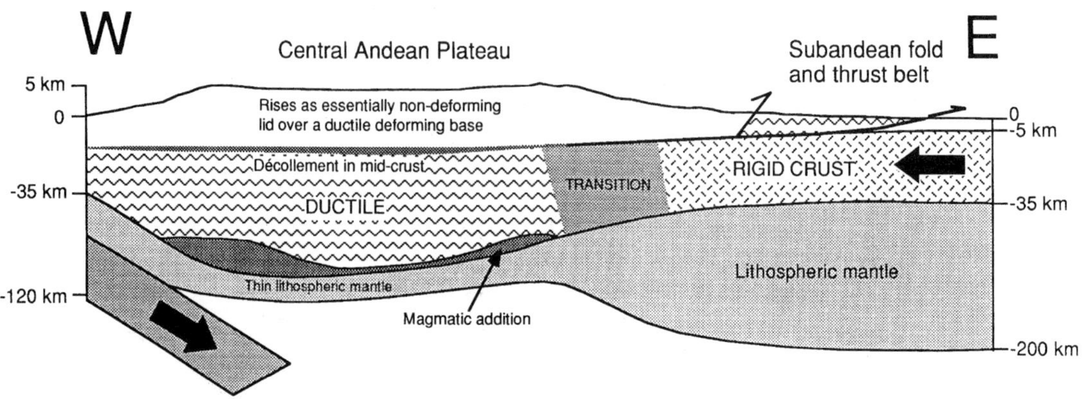

Figure 9.18 *Proposed Andean plateau schematic crustal thickening model (see Isacks, 1988). The apparent vertical uplift over much of the plateau is offset by about 300–500 km from the major shortening in the Subandes. The two are reconciled by using a mid-crustal décollement to transfer lower crustal shortening beneath the Andean plateau. This is consistent with numerical models of crustal shortening. Also limited deep seismic data show no sign of the Brazilian shield subducting steeply to the west in the mantle. In this model almost all the bulk of the crust can be accounted for using measured surface shortening, with perhaps 10% due to magma addition, mainly below the Western Cordillera volcanic arc. Note that the vertical scale is not constant*

metamorphism) or carried deep into the mantle. The error introduced in the shortening calculations above is unlikely to be greater than ~10%.

Although crustal shortening explains the crustal volume, it does not explain why shortening is distributed as it is, or why the resulting topographic form is a plateau. Numerical modelling of Andean-type scenarios provides some insights into possible preconditions for the formation of a plateau region with steep margins (e.g. Wdowinski and Bock, 1994a, b; Harry *et al.*, 1995). In order to focus deformation relatively close to the subduction zone, a thermal anomaly and/or a thick upper crustal sedimentary column are required, both consistent with the observation that early Andean shortening was focused into regions of Mesozoic extension. Assuming initially thickened or uniform crust produces a much broader, wedge-shaped orogen, the presence of strength discontinuities in the models also results in décollement formation as shortening increases, so that maximum shortening in the upper crust shifts towards the foreland while lower crust and mantle shortening remains closer to the trench. This results in the offset of regions of maximum upper crustal shortening and surface uplift as seen in the Andes.

Surface uplift in the central Andes started at different times and proceeded at different rates, and uplift apparently has not been 'plateau-like' throughout the Cenozoic (Figure 9.19a, b). However, through a combination of shortening and magmatic addition (Western and Eastern Cordilleras) and locally thick sedimentation (Altiplano), the region seems to have converged towards the plateau height of ~4000 m relatively recently (?<5 Ma BP). Numerical models also show this convergence, with the plateau height depending on the strength of the thickened crust. Essentially, vertical buoyancy forces reach a balance with horizontal compressional forces (e.g. Froidveaux and Isacks, 1984; England, 1996) and additional shortening is accommodated by widening of the plateau region. If bounding horizontal stresses drop, plateau collapse towards a slightly lower equilibrium height may occur, and this may account for observed high-altitude normal faulting in north central Peru (e.g. Cordillera Blanca fault, Dalmayrac and Molnar, 1981; Deverchère *et al.*, 1989: Quiches fault, Doser, 1987; Bellier *et al.*, 1991) where average altitudes are slightly higher than in the plateau region in Bolivia. Extension since ~2 Ma BP has also occurred in the Puna of northern Argentina (Allmendinger, 1986; Marrett *et al.*, 1994), between 22 and 39°S where average elevation is also higher than in the Bolivian Altiplano. Instead of reduced horizontal compressive stress it is possible that delamination of the lithosphere, with possible replacement by hotter asthenosphere, could provide a modest thermal component of uplift sufficient to drive the Puna over its equilibrium height (Kay *et al.*, 1994). Boundary stresses are likely to depend on plate convergence rate, subduction zone dip, age of subducting ocean crust and other factors. The

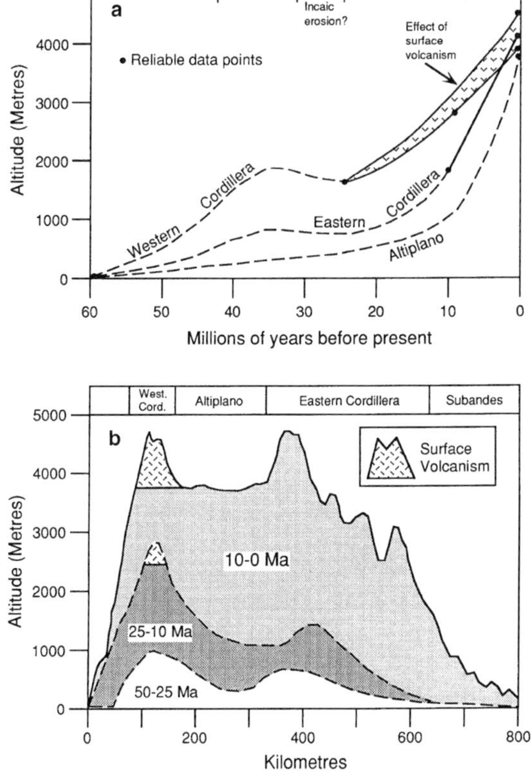

Figure 9.19 *Central Andean uplift at ~18°S in space and time. (a) Estimated elevation through time for the Western Cordillera, Altiplano and Eastern Cordillera, showing reliable data points. (b) Schematic west–east cross-section across the Andes at ~18°S. The few available data suggest that Andean uplift was not 'plateau-like' and that different regions uplifted at different times, only recently converging towards ~4000 m. This altitude may represent a limit imposed by overall lithospheric strength*

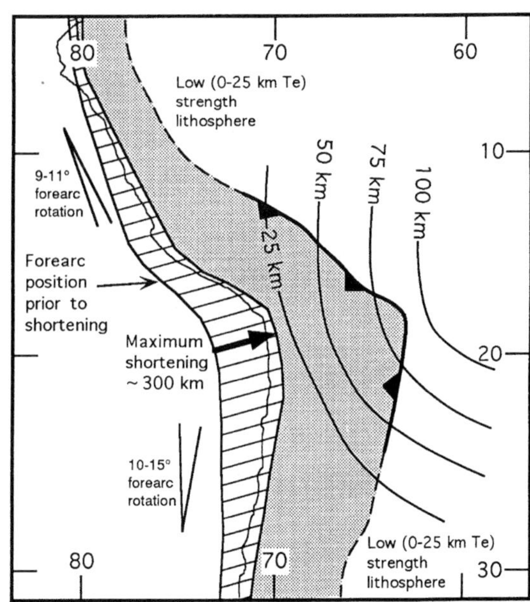

Figure 9.20 *Outline map of the central Andes showing interpreted forearc position prior to significant shortening (e.g. Isacks, 1988). Because observed shortening appears to increase dramatically into the core of the Bolivian Orocline, forearc rotations have been anticlockwise in Peru and clockwise in north Chile. The area 'removed' by shortening is shown hatched. The increased shortening is thus responsible for both the curvature of the Orocline and for the central Andean plateau. The contours east of the Andean thrust front show effective elastic thickness of the lithosphere (T_e) (after Watts* et al.*, 1995). There is a close coincidence of strong lithosphere with well-developed foreland fold-thrust belts (barbed line) and wide foreland basins. To north and south, where the effective elastic thickness of the lithosphere is 0–25 km, foreland basins show steep narrow depocentres close to the mountain front, and basement cored uplifts, which accommodate little shortening, dominate foreland structures*

mechanisms for maintaining a plateau region with constant height, despite remarkable differences in observed surface shortening and thickness of young sediments, remain unclear. Surface smoothing mechanisms such as landsliding, erosion and redeposition, extensional faulting (e.g. Green and Wernicke, 1986), and also some adjustments in the thickness of the crustal root must be occurring.

Plateau evolution appears also to be intimately linked with the development of the 'Bolivian Orocline' (Figure 9.20). Isacks (1988) has pointed out that crustal cross-sectional area drops to north and south, possibly as a result of reduced shortening. Greater shortening in the core of the orocline could result in forearc rotations of 10–15° anticlockwise in Peru, and clockwise in northern Chile, motions which are consistent with palaeomagnetic data (e.g. Macedo-Sanchez *et al.*, 1992). This is consistent with the relatively few accurate shortening estimates across the central Andes. Subandean shortening of >150 km at ~18°S drops to ~70 km at 22°S (Baby *et al.*, 1992), and perhaps as little as 15–30 km at ~12°S (author's own field observations). At ~30°S total shortening has dropped to about 150 km, of which about 100 km is concentrated in the Neogene Precordillera thrust belt (Allmendinger *et al.*, 1990). It is still not clear why shortening should be more intense in the core of the orocline. There is a close spatial correlation with the dip of the subduction zone. The restriction of the volcanic arc to areas of steeper subduction dates from the late Miocene, and may date the shallowing of the subduction zone (e.g. Jordan *et al.*, 1983a, b). However, this is exactly the time of intensified shortening, and it is not clear which is cause and which is effect. There is also a spatial correlation with better developed shale décollements in the core of the orocline (e.g. Allmendinger *et al.*, 1983, 1993) which may allow greater thin-skinned shortening. Recently, Watts *et al.* (1995) have pointed out a strong correlation with a much larger scale, and longer lived feature – lithospheric strength. Studies of flexure clearly show that the lithosphere is strongest in the core of the orocline at 18°S. It is possible that strong lithosphere is a prerequisite for developing a thin-skinned belt which can take up large amounts of shortening. By contrast, weaker lithosphere leads to steeper décollement angles, and thick-skinned structures, which cannot take up large-magnitude shortening.

I have already pointed out that an understanding of climatic change through time is vital in order to interpret correctly evidence of palaeofloral change and geomorphological features such as dissection. Cause–effect links between climate and Andean tectonics may be important but are as yet poorly known (see Chapters 3 and 5 for a general discussion of climate–tectonics interrelationships). Simulations of climate (e.g. Lenters and Cook, 1996) suggest that Andean topography plays a critical role in concentrating precipitation on the eastern slopes of the Andes, thus keeping regions to the west arid north of about 30°S. This concentration of precipitation may be why, in contrast to other parts of the Andes, these regions are very deeply eroded. For instance, from ~0°S to 18°S Precambrian and Palaeozoic metamorphic rocks of the Eastern Cordillera have had up to ~8 km of cover removed during the Cenozoic, and significant shortening appears to have been focused into these regions. Masek *et al.* (1994) have pointed out the close coincidence between this denudation and present-day higher precipitation rates and note that, in addition to focusing deformation and erosion into narrow regions, deeply cut valleys at the margin of the central Andean plateau will lead to some isostatic rebound and upwarping of the lip of the plateau, so raising the peaks of the Cordillera Real. To the south of 18°S the climate is more arid, dissection is much lower, and without this isostatic rebound the plateau edge drops more gently to the Subandes. Peaks raised in this manner may become glaciated and erosion and discharge patterns may change. This would not, however, indicate the raising of mean elevation above the snowline and illustrates that major mean elevation change, and time of change, cannot be definitively inferred from the first appearance of local glaciation.

Studies of foreland basin sequences and how they relate to erosion in the hinterland may play a vital role in understanding large-scale morphology of the Andes, in the absence of detailed palaeoaltitude data. For instance, Kooi and Beaumont (1996) have pointed out that mountain belt morphology is strongly sensitive to time lags between uplift and erosion processes (see Chapter 3), and it seems unlikely that Andean uplift and erosion have reached any kind of steady state. Thus flexural subsidence, due to the load of the orogen on the Brazilian shield, and sediment supply, have probably not been in equilibrium. Burbank (1992) has suggested that if loading outstrips sediment supply, deposition will migrate in towards the mountain front, and if erosion becomes dominant, mean elevation and flexural loading may decrease, resulting in basin-wide rebound and unconformities. Although a lot of seismic data exist for the Andes, no-one has yet tested whether this has happened in the past. It may be happening at the present. A brief examination of Landsat MSS images held at Oxford shows that almost the entire Peruvian foreland region is slightly dissected, and that sediment is being eroded to further downstream in the Amazon Basin (see also Rasanen *et al.*, 1992). Only close to the thrust front in northern Bolivia (14–18°S) is there evidence of foreland subsidence. Here, meanders in tributaries of the Beni River migrate into the mountain front and there are enormous semi-permanently flooded areas (P. Burgess, pers. comm.). This is the more intensely shortened part of the Cordillera Real and Subandes, where the topographic load is higher and narrower, and where there is a deeper foreland basin and steeper basal décollement. Although erosion rates are high in the Cordillera Real region, the continued foreland subsidence close to the mountain front suggests that mean elevation of the load is increasing. Sediment supply, although high, is not high enough to maintain a continuous positive slope downstream (as, for example, in the Pastaza region of Ecuador and northern Peru).

In addition to the provenance studies of foreland basin sediments outlined above, fission-track dating of detrital grains in the sediments may also give important clues to palaeorelief in the orogen (see Cerveny *et al.*, 1988, for an example from the Himalayas). Consider the Cordillera Real of Bolivia (see Figure 9.12); here, late Cenozoic denudation rates of ~300 m Ma^{-1} suggest it would take apatites about 10–13 Ma to reach the surface after track annealing stopped if there was no dissection. However, dissection extends well below the mean surface, and as a result the youngest apatites being released into the foreland are 5–4 Ma old. This shorter time lag depends directly on the depth of relief. In older foreland basin sediments we may also expect to see a time lag between depositional (if known) and detrital apatite age. If denudation rates can be well constrained in source regions it should be possible to estimate directly palaeorelief, which in turn yields a minimum estimate for maximum palaeoaltitudes. This requires a careful integration of fission-track dating, sediment-provenance analysis and regional structural studies.

9.8 Conclusions and Research Questions

1. The large-scale geomorphology of the Andes, especially the central Andes, reflects crustal thickening, mainly as a result of crustal shortening. Although surface volcanism is widespread and defines prominent smaller-scale morphological features, the contribution of magmatism to crustal thickening is relatively small (probably not more than 10% of the total Andean volume).
2. The earliest Andean uplifts were defined by major terrane boundaries or by inverted Mesozoic extensional basins, indicating that pre-existing weak crust and higher heat flow were probably important. Continued deformation merged these early uplifts into a single mass in the central Andes, while northern Andean uplifts have remained separated by externally drained intermontane basins.
3. Over much of the interior of the Central Andes, net Cenozoic erosion rates have been relatively low. Large-scale topography represents the primary signal of crustal thickening, and can be modelled successfully using numerical methods in which erosion is not considered. Initially thinned crust with a weak sedimentary cover and a modest thermal anomaly appear to be prerequisites for plateau development. The models also predict that strain in the crust is vertically partitioned by major décollements which allow the sites of major surface shortening and of surface uplift to be separated, accounting for the apparent vertical uplift seen in the interior of all Andean ranges.
4. The mechanical properties of the Precambrian shield regions to the east of the Andes also seem to be important. In areas of high lithospheric strength, such as Bolivia, a foreland fold-and-thrust-belt with major shortening above low-angle décollement, can develop. Where the lithosphere is weak, thick-skinned basement uplifts, which can accommodate little shortening, form. Lithospheric strength may also have had a long-term control on deposition of thick sedimentary sequences, which also correlate spatially with regions of thin-skinned shortening. For instance, subandean thrusting is less intense to the north and south where Devonian shales are much thinner.
5. In the central Andean plateau, poorly understood processes of lithospheric thinning have maintained a thermal anomaly to the present. Helium isotope studies suggest that mantle melting is occurring beneath the whole plateau region. Higher heat flow may be aiding flow in the lower crust or lithosphere, focusing intense shortening into the region.
6. As shortening increased, separate areas which were uplifted at different rates converged towards a plateau height of ~4000 m, which appears to be an upper limit defined by crustal strength. Localized drops in bounding horizontal stresses in northern Peru and northern Argentina may have allowed some collapse of topography above the threshold elevation. The reason for the stress drop is not clear.
7. Erosion plays an important role in cutting smaller scale (10–100 km wavelength) topographic features. Over large areas of the Andes erosion is concentrated in narrow valleys, and volume removed is unlikely to cause much isostatic rebound at typical lithospheric elastic thicknesses of 25–75 km. The timing of this erosion seems to coincide with a major global? climatic change. Uplift may have occurred well before dissection, which may be a result of increased precipitation, and, possibly, the onset of late Cenozoic glaciation.
8. However, climatic models suggest that growing Andean topography may have focused rainfall in the Andes eastern flank. By ~10 Ma BP, the Andes may have been high enough to cause the onset of aridity in the plateau and in the coastal deserts of Peru and Chile. Thus there seems to be a positive feedback between uplift and concentration of precipitation which may lead to more efficient dissection concentrated on the edge of the central Andean plateau.
9. The focusing of precipitation and dissection on the Andes eastern flank north of ~18°S may have kept topographic gradients steep in this region,

enhanced peak uplift of the Cordillera Real in Bolivia, and even focused increased long-term (~10–20 Ma) deformation into a narrow region.

10. Most of the links proposed between observed deformation, crustal structure and history, geomorphology and climate remain somewhat speculative. Any approach to understanding Andean topography needs to be multidisciplinary, and to be aware of potential shortcomings of any one line of evidence. For instance, climatic change, palaeofloral change, and mountain uplift or erosion are not independent of each other. Key issues for future research also include better constraints on surface and deep crustal structure, and surface field studies which can define how important are surface smoothing mechanisms such as sediment redistribution, or local faulting. Fission-track studies of dissected mountain regions and basin sediments may give direct independent estimates of palaeorelief, and thereby constrain altitude, but inferred denudation rates do not relate directly to surface uplift rates. Analysis of Andean digital terrain models may provide important insights into lithosphere-scale processes and smaller-scale geomorphic processes. Analysis of foreland basin seismic data may reveal that basin-wide rebound unconformities and valley incision are related to periods during Andean uplift when denudation was greater than rock uplift, and the load on the flexed shield was decreasing.

References

Allmendinger, R.W. 1986. Tectonic development, southeastern border of the Puna Plateau, northwestern Argentine Andes. *Geol. Soc. Amer. Bull.*, **97**, 1070–1082.

Allmendinger, R.W., Ramos, R.A., Jordan, T.E., Palma, M. and Isacks, B.L. 1983. Palaeogeography and Andean structural geometry, northwest Argentina. *Tectonics*, **2**, 1–16.

Allmendinger, R.W., Figueroa, D., Snyder, D., Beer, J., Mpodozis, C. and Isacks, B.L. 1990. Foreland shortening and crustal balancing in the Andes at 30°S latitude. *Tectonics*, **9**, 789–809.

Allmendinger, R.W., Gubbels, T., Isacks, B.L. and Cladouhos, T. 1993. Lateral variations in Late Cenozoic deformation, central Andes, 20–28°S. *International Symposium on Andean Geodynamics*, Oxford, pp. 155–158.

Andriessen, P.A.M. 1995. Fission-track analysis: principles, methodology and implications for tectono-thermal histories of sedimentary basins, orogenic belts and continental margins. *Geol. Mijnb.*, **74**, 1–12.

Andriessen, P.A.M., Helmens, K.F., Hooghiemstra, H., Riezebos, P.A. and van der Hammen, T. 1993. Absolute chronology of the Pliocene–Quaternary sediment sequence of the Bogota area, Colombia. *Quat. Sci. Rev.*, **12**, 483–501.

Arango, J.L., Kassem, T. and Duque, H. (1976) *Mapa Geologico de Colombia*, Instituto Nacional de Invertigaciones Geologico-Mineras, Bogota, Colombia.

Aspden, J.A. and McCourt, W.J. 1986. Mesozoic oceanic terrain in the central Andes of Colombia. *Geology*, **14**, 415–418.

Assumpção, M. 1992. The regional intraplate stress field in South America. *J. Geophys. Res.*, **97**, 11 889–11 903.

Assumpção, M. and Suarez, G. 1988. Source mechanisms of moderate-size earthquakes and stress orientation in midplate South America. *Geophys. J.*, **92**, 253–267.

Atherton, M.P. 1993. Granite magmatism. *J. Geol. Soc. Lond.*, **150**, 1009–1023.

Baby, P., Herail, G., Salinas, R. and Sempere, T. 1992. Geometry and kinematic evolution of passive roof duplexes deduced from cross-section balancing: example from the foreland thrust system of the southern Bolivian Subandean zone. *Tectonics* **11**, 523–536.

Baby, P., Guiller, B., Oller, J., Herail, G., Montemurro, G. and Zubieta, D. 1993. Structural synthesis of the Bolivian Subandean Zone. *International Symposium on Andean Geodynamics*, Oxford, pp. 159–162.

Bahlburg, H., Breitkreuz, C.H. and Geise, P. (eds) 1988. Southern Central Andes. *Lecture Notes in Earth Science*, **17**, Springer-Verlag, Berlin.

Baker, M.C.W. and Francis, P.W. 1978. Upper Cenozoic volcanism in the central Andes – Ages and volumes. *Earth Planet. Sci. Lett.*, **41**, 175–187.

Beckinsale, R.D., Sanchez-Fernandez, A.W., Brook, M., Cobbing, E.J., Taylor, W.P. and Moore, N.D. 1985. Rb–Sr whole rock isochron and K–Ar age determinations for the coastal batholith of Peru. In: W.S. Pitcher, M.P. Atherton, E.J. Cobbing and R.D. Beckinsale (eds) *Magmatism at a Plate Edge*, Blackie, London, pp. 239–240.

Beer, J.A., Allmendinger, R.W., Figueroa, D.E. and Jordan, T.E. 1990. Seismic stratigraphy of a Neogene piggyback basin, Argentina. *Amer. Ass. Petrol. Geol. Bull.*, **74**, 1183–1202.

Bellier, O., Dumont, J.F., Sébrier, M. and Mercier, J.L. 1991. Geological constraints on the kinematics and fault-plane solution of the Quiches fault zone reactivated during the 10 November 1946 Ancash Earthquake, Northern Peru. *Sesimol. Soc. Amer. Bull.*, **81**, 468–490.

Benjamin, M.T., Johnson, M.N. and Naeser, C.W. 1987. Recent rapid uplift in the Bolivian Andes: Evidence from fission track dating. *Geology*, **15**, 680–683.

Berry, E.W. 1939. The fossil flora of Potosi, Bolivia. *Johns Hopkins University Studies in Geology*, **13**, 9–67.

Biddle, K.T., Uliana, M.A., Mitchum, R.M., Fitzgerald, M.G. and Wright, R.C. 1986. The stratigraphic and structural evolution of the central and eastern Magallanes Basin, southern South America. In: P.A. Allen and P.

Homewood (eds) *Foreland Basins, Int. Ass. Sediment. Spec. Publ.*, **8**, Blackwell, Oxford, pp. 41–62.

Boily, M., Ludden, J.N. and Brooks, C. 1990. Geochemical constraints on the magmatic evolution of pre- and post-Oligocene volcanic suites of southern Peru: Implications for the tectonic evolution of the Central Volcanic Zone. *Geol. Soc. Amer. Bull.*, **102**, 1565–1579.

Bonhomme, M.G. and Carlier, G. 1990. Relation entre magmatisme et mineralisations dans le batholithe d'Andahualyas-Yauri (sud-Pérou): données géochronologique. *Andean Geodynamics*, ORSTOM, Grenoble, pp. 329–332.

Bonhomme, M.G., Fornari, M., Laubacher, G., Sébrier, M. and Vivier, G. 1988. New Cenozoic K–Ar ages on volcanic rocks from the eastern High Andes, southern Peru. *J. S. Amer. Earth Sci.*, **1**, 179–183.

Brown, M., Diaz, F. and Grocott, J. 1993. Displacement history of the Atacama fault system 25°00'S–27°00'S, northern Chile. *Geol. Soc. Amer. Bull.*, **105**, 1165–1174.

Buddin, T.S., Stimpson, I.G. and Williams, G.D. 1993. North Chilean forearc tectonics and Cenozoic plate kinematics. *Tectonophysics*, **220**, 193–203.

Burbank, D.W. 1992. Causes of recent Himalayan uplift deduced from deposited patterns in the Ganges basin. *Nature*, **357**, 680–683.

Burke, K., Cooper, C., Dewey, J.F., Mann, P. and Pindell, J. 1984. Caribbean tectonics and relative plate motions. In: W.E. Bonini, R.B. Hargreaves and R. Shagam (eds), *The Caribbean–South America Plate Boundary and Regional Tectonics, Geol. Soc. Amer. Mem.*, **162**, 31–64.

Cabrera, J., Sébrier, M. and Mercier, J.L. 1987. Active normal faulting in High Plateaus of Central Andes: the Cuzco Region (Peru), *Ann. Tect.*, **1**, 116–138.

Cabrera, J., Sébrier, M. and Mercier, J.L. 1991. Plio-Quaternary geodynamic evolution of a segment of the Peruvian Andean Cordillera located above the change in the subduction geometry: the Cuzco region. *Tectonophysics*, **190**, 331–362.

Cahill, T. and Isacks, B.L. 1992. Seismicity and shape of the subducted Nazca plate. *J. Geophys. Res.*, **97**, 17 503–17 529.

Cande, S.C. and Leslie, R.B. 1986. Late Cenozoic tectonics of the southern Chile trench. *J. Geophys. Res.*, **91**, 471–496.

Capdevila, R., Mégard, F., Paredes, J. and Vidal, P. 1977. Le batholithe de San Ramon (Cordillère Orientale du Pérou Central). Un granite tardihercynian mis en place à la limite Permien–Trias. Données géologiques et radiométriques. *Geol. Rund.*, **66**, 434–446.

Cerveny, P.F., Naeser, N.D., Zeitler, P.K., Naeser, C.W. and Johnson, N.M. 1988. History of uplift and relief of the Himalaya during the past 18 million years: evidence from fission-track ages of detrital zircons from sandstones of the Siwalik Group. In: K.L. Kleinspehn and C. Paola (eds) *New Perspectives in Basin Analysis*, Springer-Verlag, New York, pp. 43–61.

CGMW. 1978. *Tectonic Map of South America*, Commission for the Geological Map of the World. Geological Society of America Map and Chart Series MC-32.

Chavez, R., Gil, W., Mamani, S., Sotomayor, M., Cardenas, J. and Carlotto, V. 1994. Sedimentologia y estratigrafia de la formacion Punacancha (Eoceno?) en la region de Cusco. *VIII Congreso Peruano de Geologia*, pp. 171–173.

Cheillitz, A., Clark, A.H., Farrar, E., Arroyo, G., Pichivant, M. and Sandeman, H.A. 1992. Volcano-stratigraphy and $^{40}Ar/^{39}Ar$ geochronology of the Macusani ignimbrite field: monitor of the geodynamic evolution of the Andes of southeast Peru. *Tectonophysics*, **205**, 307–327.

Chinn, D.S. and Isacks, B.L. 1983. Accurate source depths and focal mechanisms of shallow earthquakes in western South America and in the New Hebrides island arc. *Tectonics*, **2**, 529–563.

Clapperton, C. 1993. *Quaternary Geology and Geomorphology of South America*, Elsevier, Amsterdam.

Cobbing, E.J. 1985. The tectonic setting of the Peruvian Andes. In: W.S. Pitcher, M.P. Atherton, E.J. Cobbing and R.D. Beckinsale (eds) *Magmatism at a Plate Edge*, Blackie, London, pp. 3–12.

Cobbing, E.J., Pitcher, W.S., Wilson, J., Baldock, J., Taylor, W., McCourt, W. and Snelling, N.J. 1981. *The Geology of the Western Cordillera of Northern Peru, Over. Mem. Inst. Geol. Sci.*, **5**, 143 pp.

Coira, B., Davidson, J., Mpodosis, C. and Ramos, V. 1982. Tectonic and magmatic evolution of northern Argentina and Chile. *Earth-Sci. Rev.*, **18**, 303–332.

Comte, D., Pardo, M., Dorbath, C., Haessler, H., Rivera, L., Cisternas, A. and Ponce, L. 1992. Crustal seismicity and subduction morphology around Antofagasta, Chile: preliminary results from a microearthquake survey. *Tectonophysics*, **205**, 13–22.

Cooper, M.A., Addison, F.T., Alvarez, R., Hayward, A.B., Howe, S., Pulham, A.J. and Taborda, A. 1995. Basin development and tectonic history of the Llanos Basin, Colombia. In: A.J. Tankard, R. Suarez and H.J. Welsink (eds) *Petroleum Basins of South America, Amer. Ass. Petrol. Geol. Mem.*, **62**, 659–665.

Crough, S.T. 1983. Apatite fission-track dating of erosion in the eastern Andes, Bolivia. *Earth Planet. Sci. Lett.*, **64**, 396–397.

Dalmayrac, B. 1978. Géologie de la Cordillère orientale de la région de Huanuco: sa place dans une transversale des Andes du Pérou central (9°S à 10°30'S). *Travaux et Documents de l'ORSTOM*, **93**, 161 pp.

Dalmayrac, B. and Molnar, P. 1981. Parallel thrust and normal faulting in Peru and constraints on the state of stress. *Earth Planet. Sci. Lett.*, **55**, 473–481.

Dalziel, I.W.D. 1986. Collision and cordilleran orogenesis: an Andean perspective. In: M.P. Coward and A.C. Ries (eds) *Collision Tectonics, Geol. Soc. Lond. Spec. Publ.*, **19**, 389–404.

Dashwood, M.F. and Abbotts, I.L. 1990. Aspects of the petroleum geology of the Oriente Basin, Ecuador. In: J. Brooks (ed.) *Classic Petroleum Provinces, Geol. Soc. Spec. Publ.*, **50**, 89–118.

Davies, J.H. and Bickle, M.J. 1991. A physical model for the volume and composition of melt produced by hydrous fluxing above subduction zones. *Phil. Trans. R. Soc. Lond.*, **A335**, 355–364.

de Matos, R.M.D. and Brown, L.D. 1992. Deep seismic profile of the Amazonian craton (northern Brazil). *Tectonics*, **11**, 621–633.

DeMets, C., Gordon, R.G., Argus, D.F. and Stein, S. 1990. Current plate motions. *Geophys. J. Int.*, **101**, 425–478.

DeVries, T.J. 1988. The geology of late Cenozoic marine terraces (tablazos) in northwestern Peru. *J. S. Amer. Earth Sci.*, **1**, 121–136.

De Wit, M. and Stern, C. 1978. Pillow talk. *J. Volcan. Geotherm. Res.*, **4**, 55–80.

Deverchère, J., Dorbath, C. and Dorbath, L. 1989. Extension related to a high topography: results from a microseismic survey in the Andes of Peru and tectonic implications. *Geophys. J. Int.*, **98**, 281–292.

Dewey, J.F. and Lamb, S.H. 1992. Active tectonics of the Andes. *Tectonophysics*, **205**, 79–95.

Dorbath, C., Dorbath, L., Cisternas, A., Deverchere, J., Diament, M., Ocola, L. and Morales, M. 1986. On crustal seismicity of the Amazonian foothill of the central Peruvian Andes. *Geophys. Res. Lett.*, **13**, 1023–1026.

Dorbath, C., Dorbath, L., Cisternas, A., Deverchere, J. and Sebrier, M. 1990. Seismicity of the Huancayo basin (central Peru) and the Huaytapallana fault. *J. S. Amer. Earth Sci.*, **3**, 21–29.

Dorbath, L., Dorbath, C., Jimenez, E. and Rivera, L. 1991. Seismicity and tectonic deformation in the Eastern Cordillera and the sub-Andean zone of central Peru. *J. S. Amer. Earth Sci.*, **4**, 13–24.

Doser, D.I. 1987. The Ancash, Peru, earthquake of 1946 November 10: evidence for low-angle normal faulting in the high Andes of northern Peru. *Geophys. J. R. Astr. Soc.*, **91**, 57–71.

Dunbar, R.B., Marty, R.C. and Baker, P.A. 1990. Cenozoic marine sedimentation in the Sechura and Pisco basins, Peru. *Palaeogeog. climat. ecol.*, **77**, 235–261.

Ellison, R.A., Klinck, B.A. and Hawkins, M.P. 1989. Deformation events in the Andean orogenic cycle in the Altiplano and Western Cordillera, southern Peru. *J. S. Amer. Earth Sci.*, **2**, 263–276.

England, P. 1996. The mountains will flow. *Nature*, **381**, 23–24.

England, P. and Molnar, P. 1990. Surface uplift, uplift of rocks and exhumation of rocks. *Geology*, **18**, 1173–1177.

Ericksen, G.E., Luedke, R.G., Smith, R.L., Koeppen, R.P. and Urquidi, F. 1990. Peraluminous igneous rocks of the Bolivian tin belt. *Episodes*, **13**, 3–8.

Farrar, E., Yamamura, B.K., Clark, A.H. and Taipe, A.J. 1990. (40)Ar/(39)Ar ages of magmatism and tungsten-polymetallic mineralization, Palca 11, Choquene District, southeastern Peru, *Econ. Geol.*, **85**, 1669–1676.

Feininger, T. 1975. Origin of petroleum in the Oriente of Ecuador. *Amer. Ass. Petrol. Geol. Bull.*, **59**, 1166–1175.

Feininger, T. 1982. The metamorphic "basement" of Ecuador. *Geol. Soc. Amer. Bull.*, **93**, 87–92.

Flint, S. 1985. Alluvial fans and playa sedimentation in an Andean arid closed basin: the Paciencia Group, Antofagasta Province, Chile. *J. Geol. Soc. Lond.*, **142**, 553–546.

Flint, S. and Turner, P. 1988. Alluvial fan and fan-delta sedimentation in a fore-arc extensional setting: the Cretaceous Coloso Basin of northern Chile. In: W. Nemec and R.J. Steel (eds) *Fan Deltas: Sedimentology and Tectonic Setting*, Blackie, London, pp. 387–399.

Flint, S., Turner, P., Jolley, E.J. and Hartley, A.J. 1993. Extensional tectonics in convergent margin basins: an example from the Salar de Atacama, Chilean Andes. *Geol. Soc. Amer. Bull.*, **105**, 603–617.

Flint, S.S., Prior, D.J., Agar, S.M. and Turner, P. 1994. Stratigraphic and structural evolution of the Tertiary Cosmelli Basin and its relationship to the Chile Triple Junction. *J. Geol. Soc. Lond.*, **151**, 251–268.

Forest, C.E., Molnar, P. and Emanuel, K.E. 1995. Palaeoaltimetry from energy conservation principles. *Nature*, **374**, 347–350.

Forsythe, R.D. and Prior, D.J. 1992. Cenozoic continental geology of South America and its relations to the evolution of the Chile triple junction. *Proc. Ocean Drilling Prog. Initial Rep.*, **141**, 23–31.

Froidveaux, C. and Isacks, B.L. 1984. The mechanical state of the lithosphere in the Altiplano-Puna segment of the Andes. *Earth Planet. Sci. Lett.*, **71**, 305–314.

Galli-Olivier, C. 1967. Pediplain in northern Chile and Andean uplift. *Science*, **158**, 653–655.

Gayet, M., Marshall, L.G. and Sempere, T. 1991. The Mesozoic and Palaeocene vertebrates of Bolivia and their stratigraphic context: a review. In R. Suarez (ed.) *Fossiles y Facies de Bolivia, Volume 1, Rev. Tec. Yacim. Petrol, Fisc. Bolivianos*, **12**, 393–434.

Gilchrist, A.R., Summerfield, M.A. and Cockburn, H.A.P. 1994. Landscape dissection, isostatic uplift, and the morphologic development of orogens. *Geology*, **22**, 963–966.

Gough, D.I. 1973. Dynamic uplift of Andean mountains and island arcs. *Nature Phys. Sci.*, **242**, 39–41.

Grange, F., Cunningham, P., Gagnepain, J., Hatzfeld, D., Molnar, P., Ocala, L., Rodrigues, A., Roecker, S.W., Stock, J.M. and Suarez, G. 1984. The configuration of the seismic zone and the downgoing slab in southern Peru. *Geophys. Res. Lett.*, **11**, 38–41.

Grant, J.N., Halls, C., Avila Salinas, W. and Snelling, N.J. 1979. K–Ar ages of igneous rocks and mineralisation in part of the Bolivian tin belt. *Econ. Geol.*, **74**, 838–851.

Green, A. and Wernicke, B. 1986. Possible large-magnitude Neogene extension on the southern Peruvian Altiplano: implications for the dynamics of mountain building. *AGU Fall Meeting Abstracts – EOS*, **67**, 1241.

Gregory, K.M. and Chase, C.G. 1992. Tectonic significance of palaeobotanically estimated climate and altitude of late Eocene erosion surface, Colorado. *Geology*, **20**, 581–585.

Gubbels, T.L., Isacks, B.L. and Farrar, E. 1993. High-level surfaces, plateau uplift, and foreland basin development, Bolivian central Andes. *Geology*, **21**, 695–698.

Harrison, J.V. 1943. The geology of the central Andes in part of the province of Junin, Peru. *Quart. J. Geol. Soc. Lond.*, **99**, 1–36.

Harry, D.L., Oldow, J.S. and Sawyer, D.S. 1995. The growth of orogenic belts and the role of crustal heterogeneities in decollement tectonics. *Geol. Soc. Amer. Bull.*, **107**, 1411–1426.

Hartley, A.J. 1993. Sedimentological response of an alluvial system to source area tectonism: the Seilao Member of the Late Cretaceous to Eocene Purilactis Formation of northern Chile. *Int. Ass. Sediment. Spec. Publ.*, **17**, 489–500.

Henry, S.G. and Pollack, H.N. 1988. Terrestrial heat flow above the Andean Subduction Zone in Bolivia and Peru. *J. Geophys. Res.*, **93**, 15 153–15 162.

Hérail, G., Oller, J., Baby, P., Blanco, J., Bonhomme, M.G. and Soler, P. 1993. The Tupiza, Nazareño and Estarca Basins (Bolivia): strike-slip faulting and thrusting during the Cenozoic evolution of the southern branch of the Bolivian Orocline. *Third International Symposium on Andean Geodynamics*, Oxford, pp. 191–194.

Hoke, L., Hilton, D.R., Lamb, S.H. Hammerschmidt, K. and Friedrichsen, H. 1994. ^{3}He evidence for a wide zone of active mantle melting beneath the Central Andes. *Earth Planet. Sci. Lett.*, **128**, 341–355.

Hollingworth, S.E. 1964. Dating the uplift of the Andes of northern Chile. *Nature*, **201**, 17–20.

Hoorn, C., Guerrero, J., Sarmiento, G.A. and Lorente, M.A. 1995. Andean tectonics as a cause for changing drainage patterns in Miocene South America. *Geology*, **23**, 237–240.

Hsu, J.T. 1989. Aminostratigraphy of Peruvian and Chilean Quaternary marine terraces. *Quat. Sci. Rev.*, **8**, 255–262.

INGEOMIN. 1978. *Sinopsis explicitiva del mapa geologico del Peru escala 1:1 000 000*. Instituto de Geologia y Mineria, Boletin No. 28, 41 pp.

Introcaso, A., Pacino, M.C. and Fraga, H. 1992. Gravity, isostasy and Andean crustal shortening between latitudes 30 and 35°S. *Tectonophysics*, **205**, 31–48.

Irving, E.M. 1975. Structural evolution of the northernmost Andes, Colombia. *U.S. Geol. Surv. Prof. Pap.*, **846**, 47 pp.

Isacks, B.L. 1988. Uplift of the Andean Plateau and bending of the Bolivian Orocline. *J. Geophys. Res.*, **93**, 3211–3231.

Jaillard, E. 1994. Kimmeridgian to Palaeocene tectonic and geodynamic evolution of the Peruvian (and Ecuadorian) margin. In: J. Salfity (ed.) *Cretaceous Tectonics of the Andes*. Vieweg, Braunschweig, pp. 101–167.

Jaillard, E., Capetta, H., Ellenberger, P., Feist, M., Grambast-Fessard, N., Lefranc, J.P. and Sige, B. 1993. Sedimentology, palaeontology, biostratigraphy and correlation of the late Cretaceous Vilquechico Group of southern Peru. *Cretaceous Res.*, **14**, 623–661.

James, D.E. 1971. Andean crustal and upper mantle structure. *J. Geophys. Res.*, **76**, 3246–3271.

James, D.E. and Snoke, J.A. 1994. Structure and tectonics in the region of flat-slab subduction beneath central Peru: crust and uppermost mantle structure. *J. Geophys. Res.*, **99**, 6899–6912.

Jolley, E.J., Turner, P., Williams, G.D., Hartley, A.J. and Flint, S. 1990. Sedimentological response of an alluvial system to Neogene thrust tectonics, Atacama Desert, northern Chile. *J. Geol. Soc. Lond.*, **147**, 769–784.

Jordan, T.E., Isacks, B.L., Allmendinger, R.W., Brewer, J.A., Ramos, V.A. and Ando, C.J. 1983a. Andean tectonics related to geometry of subducted Nazca plate. *Geol. Soc. Amer. Bull.*, **94**, 341–361.

Jordan, T.E., Isacks, B.L., Ramos, V.A. and Allmendinger, R.W. 1983b. Mountain building in the central Andes. *Episodes*, **3**, 20–26.

Jordan, T.E., Flemmings, P.B. and Beer, J.A. 1988. Dating of thrust-fault activity by use of foreland basin strata. In: K. Kleinspehn and C. Paola (eds) *New Perspectives in Basin Analysis*, Springer-Verlag, New York, pp. 307–330.

Karig, D. 1974. Tectonic erosion at trenches. *Earth Planet. Sci. Lett.*, **21**, 209–212.

Kay, S.M., Maksaev, V., Moscoso, R. and Mpodozis, C. 1987. Probing the evolving Andean lithosphere; mid–late Tertiary magmatism in Chile (29°–30° 30') over the modern zone of subhorizontal subduction. *J. Geophys. Res.*, **92**, 6173–6189.

Kay, S.M., Coira, B. and Viramonte, J. 1994. Young mafic back arc volcanic rocks as indicators of continental lithospheric delamination beneath the Argentine Puna Plateau, Central Andes. *J. Geophys. Res.*, **99**, 24 323–24 339.

Kellogg, J.N. 1984. Cenozoic tectonic history of the Sierra de Perijá, Venezuela–Colombia, and adjacent basins. In: W.E. Bonini, R.B. Hargreaves and R. Shagam (eds) *The Caribbean–South America Plate Boundary and Regional Tectonics, Geol. Soc. Amer. Mem.*, **162**, 239–261.

Kellogg, J.N. and Vega, V. 1995. Tectonic development of Panama, Costa Rica, and the Colombian Andes: constraints from Global Positioning System geodetic studies and gravity. In: P. Mann (ed.) *Geologic and Tectonic Development of the Caribbean Plate Boundary in Southern Central America, Geol. Soc. Amer. Mem.*, **295**, 75–90.

Kennan, L. 1993. Cenozoic evolution of the Cochabamba area, Bolivia. *Third International Symposium on Andean Geodynamics*, Oxford, pp. 199–202.

Kennan, L.J.G. 1994. Cenozoic tectonics of the central Bolivian Andes. DPhil thesis, Oxford University.

Kennan, L., Lamb, S.H. and Rundle, C.C. 1995. K–Ar dates from the Altiplano and Cordillera Oriental of Bolivia: Implications for the Cenozoic stratigraphy and tectonics. *J. S. Amer. Earth Sci.*, **8**, 163–186.

Kennan, L., Lamb, S.H. and Hoke, L. 1997. High altitude palaeosurfaces in the Bolivian Andes: evidence for late Cenozoic surface uplift. In: M. Widdowson (ed.) *Palaeosurfaces: Recognition, Reconstruction and Interpretation, Geol. Soc. Lond. Spec. Publ.*, **120**, 307–324.

Kennerley, J.B. 1980. *Outline of the Geology of Ecuador*, Overseas Geology and Mineral Resources, **55**, 17 pp.

Koch, E. and Blissenbach, E. 1962. Las Capas Rojas del Cretacico superior – Terciario en la region del curso medio del Rio Ucayali, Oriente del Peru. *Bol. Soc. Geol. Peru*, **39**, 7–141.

Kohn, B.P., Shagam, R. and Subieta, T. 1984a. Results and preliminary implications of sixteen fission-track ages from rocks of the western Caribbean mountains, Venezuela. In: W.E. Bonini, R.B. Hargreaves and R. Shagam (eds) *The Caribbean–South America Plate Boundary and Regional Tectonics, Geol. Soc. Amer. Mem.*, **162**, 415–421.

Kohn, B.P., Shagam, R., Banks, P.O. and Burkley, L.A. 1984b. Mesozoic–Pleistocene fission-track ages on rocks of the Venezuelan Andes and their tectonic implications. In: W.E. Bonini, R.B. Hargreaves and R. Shagam (eds) *The Caribbean–South America Plate Boundary and Regional Tectonics, Geol. Soc. Amer. Mem.*, **162**, 365–384.

Kono, M., Fukao, Y. and Yamamoto, A. 1989. Mountain building in the Central Andes. *J. Geophys. Res.*, **94**, 3891–3905.

Kontak, D.J., Clark, A.H., Farrar, E., Archibald, D.A. and Baadsgaard, H. 1987. Geochronologial data for Tertiary granites of the Southeast Peru segment of the Central Andean tin belt. *Econ. Geol.*, **82**, 1611–1618.

Kontak, D.J., Clark, A.H., Farrar, E., Archibald, D.A. and Baadsgaard, H. 1990a. Late Palaeozoic–early Mesozoic magmatism in the Cordillera de Carabaya, Puno, southeastern Peru: Geochronology and petrochemistry. *J.S. Amer. Earth Sci.*, **3**, 213–230.

Kontak, D.J., Farrar, E., Clark, A.H. and Archibald, D.A. 1990b. Eocene tectono-thermal rejuvenation of an upper Palaeozoic–lower Mesozoic terrane in the Cordillera de Carabaya, Puno, southeastern Peru, revealed by K–Ar and $^{40}Ar/^{39}Ar$ dating. *J. S. Amer. Earth Sci.*, **3**, 231–246.

Kooi, H. and Beaumont, C. 1996. Large-scale geomorphology: classical concepts reconciled and integrated with contemporary ideas via a surface processes model. *J. Geophys. Res.*, **101**, 3361–3386.

Kroonenberg, S.B., Bakker, J.G.M. & van der Wiel, M. 1990. Late Cenozoic uplift and palaeogeography of the Colombian Andes: constraints on the development of high-Andean biota. *Geol. Mijnb.*, **69**, 279–290.

Lahsen, A. 1982. Upper Cenozoic volcanism and tectonism in the Andes of northern Chile. *Earth-Sci. Rev.*, **18**, 285–302.

Lamb, S.H., Hoke, L. and Kennan, L. 1997. Cenozoic evolution of the central Andes in Bolivia and northern Chile. In: J.-P. Burg and M. Ford (eds) *Orogens Through Time, Geol. Soc. Lond. Spec. Publ.*, **121**, 237–264.

Lancelot, J.R., Laubacher, G., Marocco, R. and Renaud, U. 1978. U/Pb radiochronology of two granitic plutons from the Eastern Cordillera (Peru) – Extent of Permian magmatic activity and consequences. *Geol. Rund.*, **67**, 236–243.

Laubacher, G. and Naeser, C.W. 1994. Fission-track dating of granitic rocks from the Eastern Cordillera of Peru: evidence for late Jurassic and Cenozoic cooling. *J. Geol. Soc. Lond.*, **151**, 473–483.

Laubacher, G., Sérbrier, M., Fornari, M. and Carlier, G. 1988. Oligocene and Miocene continental sedimentation, tectonics, and S-type magmatism in the southeastern Andes of Peru (Crucero Basin): Geodynamic implications. *J. S. Amer. Earth Sci.*, **1**, 225–238.

Lavenu, A. 1979. Neotectonica de los sedimentos plio-cuaternarios del norte del Altiplano. *VI Congreso Nacional de Geologia*, Oruro, pp. 449–463.

Lavenu, A. and Ballivian, O. 1979. Estudios neotectonicos de las regiones de Cochabamba, Sucre, Tarija – Cordillera Oriental Boliviana. *Rev. Acad. Nac. Cienc. Bolivia*, **3**, 107–129.

Lenters, J.D. and Cook, K.H. 1996. Simulation and diagnosis of the regional summertime precipitation climatology of South America. *J. Climate*. **8**, 2988–3005.

Lindo, R., Dorbath, C., Cisternas, A., Dorbath, L., Ocala, L. and Morales, M. 1992. Subduction geometry in central Peru from a microseismicity survey: first results. *Tectonophysics*, **205**, 23–29.

Lopez, R.L. and Cordova, E.H. 1988. Estratigrafia y sedimentacion de la serie continental 'capas rojas' (Maestrichtiano Paleoceno) entre Cuzco y Ccorao. *Bol. Soc. Geol. Peru*, **78**, 149–164.

Lyon-Caen, H., Molnar, P. and Suarez, G. 1985. Gravity anomalies and flexure of the Brazilian shield beneath the Bolivian Andes. *Earth Planet. Sci. Lett.*, **75**, 81–92.

Macedo-Sanchez, O., Surmont, J., Kissel, C., Mitouard, P. and Laj, C. 1992. Late Cainozoic rotation of the Peruvian Western Cordillera and the uplift of the Central Andes. *Tectonophysics*, **205**, 65–77.

MacFadden, B.J., Wang, Y., Cerling, T.E. and Anaya, F. 1994. South American fossil mammals and carbon isotopes: a 25 million year sequence from the Bolivian Andes. *Palaeogeog. climat. ecol.*, **107**, 257–268.

Macharé, J. and Ortlieb, L. 1992. Plio-Quaternary vertical motions and the subduction of the Nazca Ridge, central coast of Peru. *Tectonophysics*, **205**, 97–108.

Marrett, R.A., Allmendinger, R.W., Alonso, R.N. and Drake, R.E. 1994. Late Cenozoic tectonic evolution of the Puna Plateau and adjacent foreland, northwestern Argentine Andes. *J. S. Amer. Earth Sci.*, **7**, 179–207.

Marshall, L.G. and Sémpere, T. 1991. The Eocene to Pleistocene vertebrates of Bolivia and their stratigraphic context: a review. In: R. Suarez (ed.) *Fossiles y Facies de Bolivia, Volume 1, Rev. Tec. Yacim. Petrol. Fisc. Bolivianos*, **12**, 631–652.

Marshall, L.G., Butler, R.F., Drake, R.E., Curtis, G.H. and Tedford, R.H. 1979. Calibration of the Great American Interchange. A radioisotope chronology for the late Tertiary interchange of terrestrial faunas between the Americas. *Science*, **204**, 272–279.

Masek, J.G., Isacks, B.L., Gubbels, T.L. and Fielding, E.J. 1994. Erosion and tectonics at the margins of continental plateaus. *J. Geophys. Res.*, **99**, 13 941–13 956.

Mattson, P.H. 1984. Caribbean structural breaks and plate movements. In: W.E. Bonini, R.B. Hargreaves and R. Shagam (eds), *The Caribbean–South America Plate Boundary and Regional Tectonics, Geol. Soc. Amer. Mem.*, **162**, 131–152.

McKee, E.H. and Noble, D.C. 1982. Miocene volcanism and

deformation in the western Cordillera and high plateaus of south-central Peru. *Geol. Soc. Amer. Bull.*, **93**, 657–662.

Mégard, F., Noble, D.C., McKee, E.H. and Bellon, H. 1984. Multiple pulses of Neogene compressive deformation in the Ayachucho intramontane basin, Andes of central Peru. *Geol. Soc. Amer. Bull.*, **95**, 1108–1117.

Molnar, P. and England, P. 1990. Late Cenozoic uplift of mountain ranges and global climate change: chicken or egg? *Nature*, **346**, 29–34.

Mortimer, C. 1973. The Cenozoic history of the southern Atacama desert, Chile. *J. Geol. Soc. Lond.*, **129**, 505–526.

Mortimer, C. 1980. Drainage evolution in the Atacama desert of northernmost Chile. *Rev. Geol. Chile*, **11**, 3–28.

Mortimer, C. and Saric, N. 1975. Cenozoic studies in northernmost Chile. *Geol. Rund.*, **64**, 395–420.

Mortimer, C., Farrar, E. and Saric, N. 1974. K–Ar ages from Tertiary lavas of the northernmost Chilean Andes. *Geol. Rund.*, **63**, 484–490.

Mukasa, S.B. 1986. Zircon U–Pb ages of super-units in the Coastal batholith, Peru: Implications for magmatic and tectonic processes. *Geol. Soc. Amer. Bull.*, **97**, 241–254.

Murdie, R.E., Prior, D.J., Styles, P., Flint, S.S., Pearce, R.G. and Agar, S.M. 1993. Seismic responses to ridge-transform subduction: Chile triple junction. *Geology*, **21**, 1095–1098.

Myers, J.S. 1975. Vertical crustal movements of the Andes in Peru. *Nature*, **254**, 672–674.

Myers, J.S. 1976. Erosion surfaces and ignimbrite eruption, measures of Andean uplift in northern Peru. *Geol. J.*, **11**, 29–44.

Naeser, C.W., Crochet, J.-Y., Jaillard, E., Laubacher, G., Mourier, T. and Sigé, B. 1991. Tertiary fission-track ages from the Bagua syncline (northern Peru): stratigraphic and tectonic implications. *J. S. Amer. Earth Sci.*, **4**, 61–71.

Nelson, E.P. 1982. Post-tectonic uplift of the Cordillera Darwin orogenic core complex: evidence from fission track geochronology and closing temperature–time relationships. *J. Geol. Soc. Lond.*, **139**, 755–761.

Noble, D.C., McKee, E.H. and Megard, F. 1979a. Early Tertiary "Incaic" tectonism, uplift and volcanic activity, Andes of central Peru. *Geol. Soc. Amer. Bull.*, **90**, 903–907.

Noble, C.R., Farrar, E. and Cobbing, E.J. 1979b. The Nazca group of south-central Peru: age, source and regional volcanic and tectonic significance. *Earth Planet. Sci. Lett.*, **45**, 80–86.

Noble, D.C., McKee, E.H., Eyzaguirre, V.R. and Marocco, R. 1984. Age and regional tectonic and metallogenetic implications of igneous activity and mineralisation in the Andahuayllas–Yauri belt of southern Peru. *Econ. Geol.*, **79**, 172–176.

Noble, D.C., Sebrier, M., Megard, F. and McKee, E.H. 1985. Demonstration of two pulses of Paleogene deformation in the Andes of Peru. *Earth Planet. Sci. Lett.*, **73**, 345–349.

Noble, D.C., McKee, E.H., Mourier, T. and Megard, F. 1990. Cenozoic stratigraphy, magmatic activity, compressive deformation, and uplift in northern Peru. *Geol. Soc. Amer. Bull.*, **102**, 1105–1113.

Noblet, C., Marocco, R. and Delfaud, J. 1987. Analyse sedimentologique des "couches rouges" du bassin intramonagneux de Sicuani (sud du Perou). *Bull. Inst. Fran. Etudes Andines*, **16**, 55–78.

Padilla, L.E. 1981. Geomorfologia de posibles areas peneplanizadas en la Cordillera occidental de Colombia. *Revista CIAF*, **6**, 391–402.

Page, W.D. and James, M.E. 1981. The antiquity of the erosion surfaces and late Cenozoic deposits near Medellin, Colombia: implications to tectonics and erosion rates. *Revista CIAF*, **6**, 421–454.

Pankhurst, R.J., Hervé, F., Rojas, L. and Cembrano, J. 1992. Magmatism and tectonics in central Chiloé, Chile (42°–42°30'S). *Tectonophysics*, **205**, 283–294.

Pardo-Cases, F. and Molnar, P. 1987. Relative motion of the Nazca (Farallon) and South American plates since late Cretaceous time. *Tectonics*, **6**, 233–248.

Pareja, J., Vargas, C., Suarez, R., Ballon R., Carrasco, R. and Villaroel, C. 1978. *Mapa Geologico de Bolivia – Memoria Explicitiva*. Yacimientos Petroliferos Fiscales Bolivianos – Servicio Geologico de Bolivia, 27 pp.

Paskoff, R. and Naranjo, J.A. 1983. Formation et évolution du piemont andin dans le desert nord du Chili (18–21° latitud Sud) pendant le Cénozoique superior. *C.R. Acad. Sci. Paris*, **297 II**, 743–748.

Peacock, S.M. 1993. Large-scale hydration of the lithosphere above subducting slabs. *Chem. Geol.*, **108**, 49–59.

Pennington, W.D. 1981. Subduction of the Eastern Panama Basin and seismotectonics of northwestern South America. *J. Geophys. Res.*, **86**, 10 753–10 770.

Petford, N. and Atherton, M.P. 1992. Granitoid emplacement and deformation along a major crustal lineament: the Cordillera Blanca, Peru. *Tectonophysics*, **205**, 171–186.

Pilger, R.H. 1984. Cenozoic plate kinematics, subduction and magmatism: South American Andes. *J. Geol. Soc. Lond.*, **141**, 793–802.

Pindell, J.L. and Tabutt, K.D. 1995. Mesozoic–Cenozoic Andean palaeogeography and regional controls on hydrocarbon systems. In: A.J. Tankard, R. Suarez and H.J. Welsink (eds), *Petroleum Basins of South America*, *Amer. Ass. Petrol. Geol. Mem.*, **62**, 101–128.

Pitcher, W.S., Atherton, M.P., Cobbing, E.J. and Beckinsale, R.D. (eds) 1985. *Magmatism at a Plate Edge*, Blackie, London.

Ramos, V.A. 1989. Andean foothills structures in northern Magallanes Basin, Argentina. *Amer. Ass. Petrol. Geol. Bull.*, **73**, 887–903.

Ramos, V.A. and Kay, S.M. 1992. Southern Patagonia plateau basalts and deformation: backarc testimony of ridge collision. *Tectonophysics*, **205**, 261–282.

Rasanen, M., Neller, R., Salo, J. and Jungner, H. 1992. Recent and ancient fluvial deposition systems in the Amazonian foreland basin, Peru. *Geol. Mag.*, **129**, 293–306.

Raymo, M.E. and Ruddiman, W.F. 1992. Tectonic forcing of late Cenozoic climate. *Nature*, **359**, 117–122.

Raymo, M.E., Ruddiman, W.F. and Froelich, P.N. 1988.

Influence of late Cenozoic mountain building on ocean geochemical cycles. *Geology*, **16**, 649–653.

Restrepo, J.J. and Toussaint, J.F. 1988. Terranes and continental accretion in the Colombian Andes. *Episodes*, **11**, 189–193.

Riccardi, A.C. 1988. *The Cretaceous System of Southern South America. Geol. Soc. Amer. Mem.*, **168**.

Rodrigo, L.A. and Castaños, A. 1975. Estudios sedimentologicos de las formaciones Tiwanaku, Coniri y Kollu-Kollu del Altiplano Septentrional Boliviano. *Bol. Soc. Geol. Boliviano*, **22**, 85–126.

Roeder, D. 1988. Andean age structure of Eastern Cordillera (Province of La Paz, Bolivia) *Tectonics*, **7**, 23–39.

Roeder, D. and Chamberlain, R.L. 1995. Eastern Cordillera of Colombia: Jurrasic–Neogene crustal evolution. In: A.J. Tankard, R. Suarez and H.J. Welsink (eds) *Petroleum Basins of South America, Amer. Ass. Petrol. Geol. Mem.*, **62**, 633–645.

Rosenzweig, A. 1953. Reconocimiento geologico en el curso medio del Rio Huallaga. *Bol. Soc. Geol. Peru*, **26**, 155–189.

Ruddiman, W.F., Prell, W.L. and Raymo, M.E. 1989. Late Cenozoic uplift in southern Asia and American west: rationale of general circulation modelling experiments. *J. Geophys. Res.*, **94**, 18 379–18 391.

Sanjines, G. and Jimenez, F. 1976. Communicacion preliminar acerca de la presencia de fossiles vertebrados en la formacion Petaca del area de Santa cruz. *Rev. Tec. Yacim. Petrol. Fisc. Bolivianos*, **4**, 147–156.

Sarewitz, D. 1988. High rates of late Cenozoic crustal shortening in the Andean foreland, Mendoza Province, Argentina. *Geology*, **16**, 1138–1142.

Schmitz, M., Baldzuhn, S., Giese, P., Wigger, P., Araneda, M., Martinez, E. and Ricaldi, E. 1990. Variation of crustal structure between the Coastal Cordillera (N. Chile) and Chaco plain. *Final Andes Workshop*, Berlin, 98.

Sébrier, M., Mercier, J.L., Megard, F., Laubacher, G. and Carey-Gailhardis, E. 1985. Quaternary normal and reverse faulting and the state of stress in the central Andes of south Peru. *Tectonics*, **4**, 739–780.

Sébrier, M., Lavenu, A., Fornari, M. and Soulas, J.P. 1988a. Tectonics and uplift in Central Andes (Peru, Bolivia and northern Chile) from Eocene to present. *Géodynamique*, **3**, 85–106.

Sébrier, M., Mercier, J.L., Machare, J., Bonnot, D., Cabrera, J. and Blanc, J.L. 1988b. The state of stress in an overriding plate situated above a flat-slab: the Andes of central Peru. *Tectonics*, **7**, 895–928.

Sémpere, T., Herail, G., Oller, J. and Bonhomme, M.G. 1990. Late Oligocene–early Miocene major tectonic crisis and related basins in Bolivia. *Geology*, **18**, 946–949.

Sémpere, T., Marshall, L.G., Rivano, S. and Godoy, E. 1994. Late Oligocene–Early Miocene compressional tectosedimentary episode and associated land-mammal faunas in the Andes of central Chile and adjacent Argentina (32–37°S), *Tectonophysics*, **229**, 251–264.

Servant, M., Sempere, T., Argollo, J., Bernat, M., Feraud, G. and Lo Bello, P. 1989. Morphogenese et soulevement des Andes de Bolivie au Cénozoique. *C.R. Acad. Sci. Paris*, **309**, 417–422.

Shackleton, N.J. and Hall, M.A. 1995. Stable isotope records in bulk sediments (leg 138). *Proc. Ocean Drilling Prog. Sci. Results*, **138**, 797–805.

Shackleton, R.M., Ries, A.C., Coward, M.P. and Cobbold, P.R. 1979. Structure, metamorphism and geochronology of the Arequipa Massif of coastal Peru. *J. Geol. Soc. Lond.*, **136**, 195–214.

Shagam, R., Kohn, B.P., Banks, P.O., Dasch, L.E., Vargas, R., Rodriguez, G.I. and Pimental, N. 1984. Tectonic implications of Cretaceous–Pliocene fission-track ages from rocks of the circum-Maracaibo basin region of western Venezuela and eastern Colombia. In: W.E. Bonini, R.B. Hargreaves and R. Shagam (eds) *The Caribbean–South America Plate Boundary and Regional Tectonics, Geol. Soc. Amer. Mem.*, **162**, 385–412.

Soler, P. and Bonhomme, M.G. 1988. New K–Ar age determinations of intrusive rocks from the Cordillera Occidental and Altiplano of central Peru; identification of magmatic pulses and episodes of mineralization. *J. S. Amer. Earth Sci.*, **1**, 169–177.

Sosson, M., Bourgois, J. and Mercier de Lepinay, B. 1994. SeaBeam and deep-sea submersible *Nautile* surveys in the Chiclayo canyon off Peru (7°S): subsidence and subduction–erosion of an Andean-type convergent margin since Pliocene times. *Mar. Geol.*, **118**, 237–256.

Stauder, W. 1975. Subduction of the Nazca Plate under Peru as evidenced by focal mechanisms and by seismicity. *J. Geophys. Res.*, **80**, 1053–1064.

Suarez, G., Molnar, P. and Burchfiel, B.C. 1983. Seismicity, fault plane solutions and depth of faulting, and active tectonics of the Andes of Peru, Ecuador and Southern Colombia. *J. Geophys. Res.*, **88**, 10 403–10 428.

Tafur, H.I.A. 1991. Estratigrafia geologica de la cuenca del alto Maranon. *Bol. Soc. Geol. Peru*, **82**, 73–94.

Thorpe, R.S. 1982. Introduction. In: R.S. Thorpe (ed.) *Andesites*, Wiley, Chichester, pp. 1–7.

Thorpe, R.S., Francis, P.W., Hammill, M. and Baker, M.C.W. 1982. The Andes. In: R.S. Thorpe (ed.) *Andesites*, Wiley, Chichester, pp. 187–205.

Thorpe, R.S., Francis, P.W. and Harmon, R.S. 1981. Andean andesites and crustal growth. *Phil. Trans. R. Soc. Lond.*, **A 301**, 305–320.

Tosdal, R.M., Farrar, E. and Clark, A.H. 1981. K–Ar geochronology of the late Cenozoic volcanic rocks of the Cordillera Occidental, southernmost Peru. *J. Volcan. Geotherm. Res.*, **10**, 157–173.

Tosdal, R.M., Farrar, A.E. and Clark, A.H. 1984. Cenozoic polyphase landscape and tectonic evolution of the Cordillera Occidental, southernmost Peru. *Geol. Soc. Amer. Bull.*, **95**, 1318–1332.

Vandervoort, D.S., Jordan, T.E., Zeitler, P.K. and Alonso, R.N. 1995. Chronology of internal drainage development and uplift, southern Puna plateau, Argentine central Andes. *Geology*, **23**, 145–148.

van der Wiel, A.M. 1990. Uplift age of the Garzon massif

(Eastern Cordillera, S. Colombia) in relation to the infill of the adjacent S. Neiva basin. *First International Symposium on Andean Geodynamics*, Grenoble, pp. 217–218.

van der Wiel, A.M. and van den Bergh, G.D. 1992a. Uplift, subsidence and volcanism in the southern Neiva Basin, Colombia, Part 1: Influence on fluvial deposition in the Miocene Honda Formation. *J. S. Amer. Earth Sci.*, **5**, 157–173.

van der Wiel, A.M. and van den Bergh, G.D. 1992b. Uplift, subsidence and volcanism in the southern Neiva Basin, Colombia, Part 2: Influence on fluvial deposition in the Miocene Gigante Formation. *J. S. Amer. Earth Sci.*, **5**, 175–196.

Vicente, J.C., Sequieros, F., Valdivia, M.A. and Zavala, J. 1979. El sobre-escurrimiento de Cincha-Lluta: elemento del accidente mayor Andino al NW de Arequipa. *Bol. Soc. Geol. Peru*, **61**, 67–99.

von Huene, R. and Lallemand, S. 1990. Tectonic erosion along the Japan and Peru convergent margins. *Geol. Soc. Amer. Bull.*, **102**, 704–720.

von Huene, R., Suess, E. and Emeis, K.-J. 1987. Convergent tectonics and coastal upwelling: a history of the Peru continental margin. *Episodes*, **10**, 87–93.

Walker, E.H. 1949. Andean uplift and erosion surfaces near Uncia, Bolivia. *Amer. J. Sci.*, **247**, 646–663.

Watts, A.B., Lamb, S.H., Fairhead, J.D. and Dewey, J.F. 1995. Lithospheric flexure and bending of the central Andes. *Earth Planet. Sci. Lett.*, **134**, 9–21.

Wdowinski, S. and Bock, Y. 1994a. The evolution of deformation and topography of high elevated plateaus 1. Model, numerical analysis and general results. *J. Geophys. Res.*, **99**, 7103–7119.

Wdowinski, S. and Bock, Y. 1994b. The evolution of deformation and topography of high elevated plateaus 2. Application to the central Andes. *J. Geophys. Res.*, **99**, 7121–7130.

Whitman, D. 1994. Moho geometry beneath the eastern margin of the Andes, northwest Argentina, and its implications for the effective elastic thickness of the Andean foreland. *J. Geophys. Res.*, **99**, 15 277–15 289.

Wigger, P.J. 1988. Seismicity and crustal structure of the central Andes. In: H. Bahlburg, C. Breitkreuz and P. Giese (eds) *The Southern Central Andes, Lecture Notes in Earth Sciences*, **17**, Springer-Verlag, Berlin, 209–229.

Wilson, D.V. 1985. The deeper structure of the Central Andes and some geophysical constraints. In: W.S. Pitcher, M.P. Atherton, E.J. Cobbing and R.D. Beckinsale (eds) *Magmatism at a Plate Edge*, Blackie, London, pp. 13–18.

Winslow, M.A. 1981. Mechanisms for basement shortening in the Andean foreland of southern South America. In: K, McClay and N.J. Price (eds) *Thrust and Nappe Tectonics, Geol. Soc. Lond. Spec. Publ.*, **9**, 513–528.

Winter, T. and Lavenu, A. 1989. Morphological and microtectonic evidence for a major active right-lateral strike-slip fault across central Ecuador (South America). *Ann. Tect.*, **3**, 123–139.

Zandt, G., Velasco, A.A. and Beck, S.L. 1994. Composition and thickness of the southern Altiplano crust, Bolivia. *Geology*, **22**, 1003–1006.

Zeil, W. 1979. *The Andes. A Geological Review*, Gebrüder Borntraeger, Berlin.

10

Morphotectonic evolution of the Himalayas and Tibetan Plateau

Eric J. Fielding

10.1 Introduction

Tibet is the largest and highest plateau on Earth, a continuously elevated area nearly 3500 km by 1500 km in extent. It is perhaps the biggest land uplift formed in the last billion years (Harrison *et al.*, 1992a), and has resulted from the convergence of the Indian and Asian continental plates (Figure 10.1) (Argand, 1924; Gansser, 1964; Powell and Conaghan, 1973; Molnar and Tapponnier, 1975). The plateau includes 82% of the world's land surface area that is >4 km above sea level and has an average elevation (for the internally drained portion) of 5023 m above sea level (Fielding *et al.*, 1994). The vast volume of the Tibetan Plateau is much larger than all the other presently active orogenic belts on the Earth (such as the Altiplano–Puna plateau of South America (see Chapter 9), the Southern Alps of New Zealand (see Chapter 6), and Taiwan (see Chapter 7)) as can be seen in the topographic cross-sections in Figure 10.2 (Fielding, 1996). The high topography of Tibet strongly influences climate by affecting atmospheric circulation, not only locally but also regionally and perhaps over the entire northern hemisphere (Ruddiman and Kutzbach, 1989; Ruddiman *et al.*, 1989; Kutzbach *et al.*, 1993). The plateau is arid to semi-arid over most of its area and central Tibet is internally drained (Figure 10.1). Conversely, the edges of Tibet are wet, externally drained, and the site of intense erosion that may have caused a decrease of atmospheric carbon dioxide and global cooling (e.g. Raymo and Ruddiman, 1992).

The Himalayan and Karakoram ranges contain the highest mountain peaks on the Earth today, including all 14 peaks over 8 km above sea level. The Himalayas form the southern edge of the Tibetan Plateau (Figure 10.1), where the intense precipitation of the Indian monsoon impinges on the elevated mass and cuts deep canyons between the mountains. The Karakoram is a narrow spine connecting the main part of Tibet with the Pamirs (Figure 10.1) and has extreme relief due to spectacular glacial erosion. The height of the peaks in these ranges is isostatically balanced by the depth of the canyons so the average elevation is only slightly higher than the elevation of the uneroded plateau (e.g. Wager, 1933, 1937). The Himalayas and Karakoram have traditionally been studied as separate tectonic and geomorphic provinces from the Tibetan Plateau, in part because of political boundaries that limited access by Western scientists to Tibet from the 1920s to the 1980s, and also because their geomorphic and tectonic characters are distinct.

Several excellent reviews of the geological, geophysical and thermochronometric constraints on the various proposed mechanisms and consequences of the uplift of the Tibetan Plateau have appeared in the last decade (e.g. Dewey *et al.*, 1988; Molnar, 1988; Burchfiel and Royden, 1991; Harrison *et al.*, 1992a; Molnar *et al.*, 1993). Fielding (1996) has reviewed some of the newer work on uplift and erosion, distinguishing between the evidence for bedrock uplift and denudation with histories covering the development of the plateau since the Mesozoic. This chapter will focus on the evolution of the present geomorphology of the Tibetan Plateau and its margins, including the Himalayan mountain range.

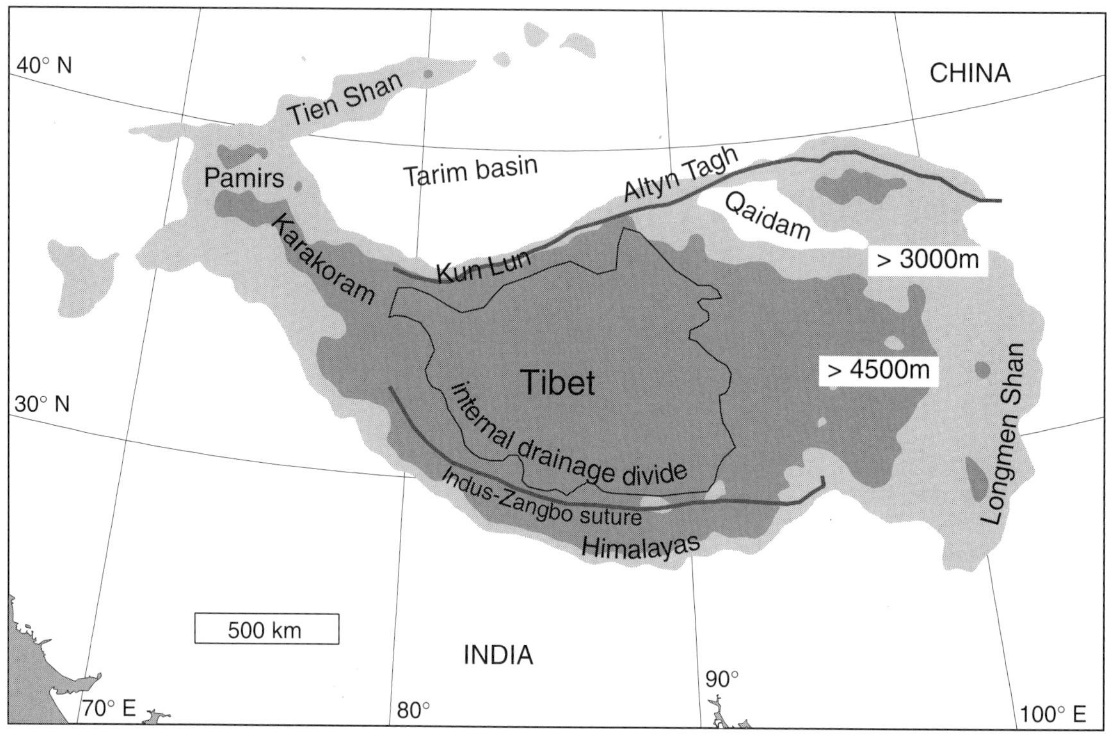

Figure 10.1 Overview map of Tibet showing major features over smoothed topography (darker grey shading >4500 m and lighter grey 3000–4500 m). Thin line encloses zone of internal drainage used for hypsometry illustrated in Figure 10.8. Lambert conformal conic projection is used for Figures 10.1, 10.3, 10.6, 10.9, 10.10 and Plates 10.1–10.3. Modified from Fielding et al. (1994)

10.2 Tectonic Models

Geologists, extending back to Argand (1924), have proposed a wide variety of tectonic mechanisms for the uplift of the Tibetan Plateau (e.g. Powell and Conaghan, 1973; Dewey *et al.*, 1988; Molnar, 1988; Burchfiel and Royden, 1991; Harrison *et al.*, 1992a; Molnar *et al.*, 1993). Most of these models have attempted to answer three basic tectonic questions about the plateau: (1) how did the crust of Tibet become 50–75 km thick, twice the normal thickness of continental crust (e.g. Hirn, 1988; Molnar, 1988); (2) what mechanisms absorbed the ~2500 km of convergence between the Indian and Asian plates since 40–50 Ma BP (Patriat and Achache, 1984; Besse and Courtillot, 1988; Dewey *et al.*, 1989; Le Pichon *et al.*, 1992); and (3) when did the plateau surface reach its present 5 km elevation above sea level? A complete Tibet uplift theory should answer all three of these related questions. The tectonic mechanisms proposed to answer the first two questions will largely determine the answer to the third question because constructional topography is linked to lithospheric structure (thickness and density) by isostasy.

Some hypotheses propose that the double thickness of crust beneath Tibet has formed by continual north–south shortening of the entire Asian crustal column (e.g. Dewey and Burke, 1973; England and Houseman, 1986), while other hypotheses posit formation of the double crust by underthrusting of the Indian crust (e.g. Argand, 1924; Powell and Conaghan, 1973; Barazangi and Ni, 1982). A variation on the underthrusting model proposed by Zhao and Morgan (1985) argues for the Indian crust being injected into a viscous, ductile lower crust of Tibet. All of these mechanisms would absorb a major (perhaps 1000 km or more) portion of the convergence between India and Asia within, or beneath, Tibet. Another set of hypotheses suggests that a significant part of the convergence between India and Asia was accommodated by motion on strike-slip faults that extruded blocks laterally (Molnar and Tapponnier, 1975; Tapponnier *et al.*, 1982; Peltzer and Tapponnier, 1988). This mechanism does not contribute to the thickening of Tibetan crust. Early versions of these

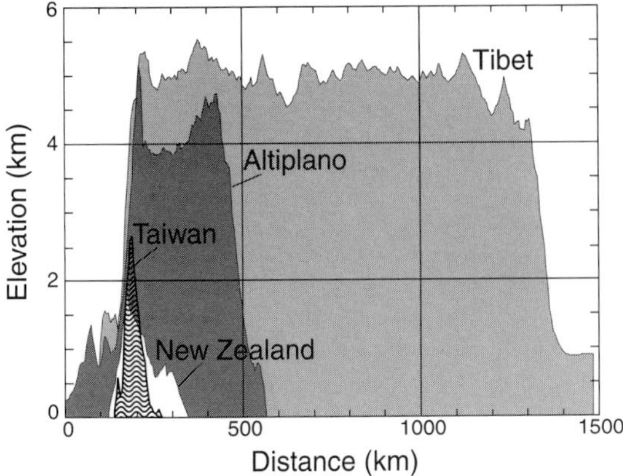

Figure 10.2 *Profiles of four active orogenic belts, taken perpendicular to strike. Each profile represents the mean elevation from 100-km-wide swaths. Profiles are approximately aligned at their steepest mountain front and are filled with shades and patterns. Tibet profile (light grey) runs roughly S–N (left to right), Altiplano profile (medium grey) runs NE–SW, New Zealand profile (white) is NW–SE, and Taiwan profile (wavy lines) runs E–W. Reprinted from* Tectonophysics, *v. 260, E.J. Fielding, Tibet Uplift and Erosion, pp. 55–84, Copyright (1996), with permission from Elsevier Science*

hypotheses assumed that the given mechanism operated more or less continuously since the beginning of the collision, but many variations in timing and duration have been proposed. Another mechanism, first modelled by Houseman *et al.* (1981), can cause uplift of the plateau without changing the crustal thickness. Houseman *et al.* (1981) and England and Houseman (1988, 1989) propose that part of the lithospheric mantle of Tibet has been removed and replaced with hotter, less dense asthenosphere. This would have the effect of decreasing the mass of the lithospheric column and could result in isostatic uplift of the surface by 1–3 km.

Many recent theories propose that different mechanisms have operated at various times since the first contact of the Indian and Asian continental plates. The exact timing of the initial 'collision' has little effect on the tectonic and erosional processes that have shaped the modern plateau, so I will arbitrarily use ~50 Ma BP. The current increase in data on both the structure and history of Tibet and adjoining areas requires more complex theories to match all the newly available facts. In particular, there is now strong evidence that certain structures were active for subsets of the time since ~50 Ma BP, which suggests that the deformation mechanisms expressed by those structures were similarly restricted in time (e.g. Harrison *et al.*, 1992a, b).

Several studies using satellite and seismic data sets have contributed to our understanding of the neotectonics and present structure of the crust and upper mantle for the vast interior of Tibet and surrounding areas. Earthquake source mechanisms provide information on the modern deformation of the crust (and mantle), while propagation of the seismic waves provides data on the present velocity and attenuation structure of the crust and mantle through which they pass (Figure 10.3). The present tectonic regime of the Tibetan Plateau clearly involves some component of east–west extension as shown by earthquake focal mechanisms (e.g. Molnar, 1988), field mapping and the interpretation of airphotos and Landsat imagery of normal and strike-slip faults (Rothery and Drury, 1984; Armijo *et al.*, 1986, 1989; Kidd and Molnar, 1988).

The southern part of Tibet, south of the Indus–Zangbo Suture (Figures 10.1 and 10.3), is made up of the deformed edge of the Indian plate. Palaeozoic to Mesozoic sedimentary rocks (the Tethyan or Tibetan sedimentary series) cover much of the region between the suture and the High Himalaya peaks (Figure 10.4). These rocks are strongly deformed but unmetamorphosed. The Tethyan sedimentary strata on the northern passive margin of the Indian plate show that it was undeformed and below sea level until collision with the Asian plate closed the Tethyan Ocean (e.g. Searle *et al.*, 1987). The contact between the Tethyan sedimentary series and the High Himalaya Crystalline rocks to the south is a faulted contact, first described by Burg (1983; Burg *et al.*, 1983) who called it the North Himalayan Normal Fault. In most places, this is the low-angle, north-dipping South Tibetan Detachment System (STDS), which had down-to-the-north or normal-fault displacement in the Miocene (Burchfiel *et al.*, 1992).

The High Himalaya Crystalline series are high-grade metamorphic rocks with Miocene leucogranite intrusions and form the main range of the Greater Himalaya. The Main Central Thrust (MCT) places these high-grade rocks over the low-grade metamorphosed sedimentary and volcanic units of the Lesser Himalaya sequence which form the Lesser Himalayan ranges (Figures 10.4 and 10.5). The Lesser Himalaya sequence is then thrust over the late Tertiary sedimentary rocks, largely the Siwalik molasse, by the Main Boundary Thrust (MBT). The Miocene–Holocene foreland basin sediments of the Siwaliks are deformed by mostly buried thrust faults propagating out into the Indo-Gangetic foredeep. Structural cross-sections indicate that the Himalayan thrusts from the Siwaliks to the MCT root into a major décollement at depth (e.g. Molnar, 1988; Schelling and Arita, 1991;

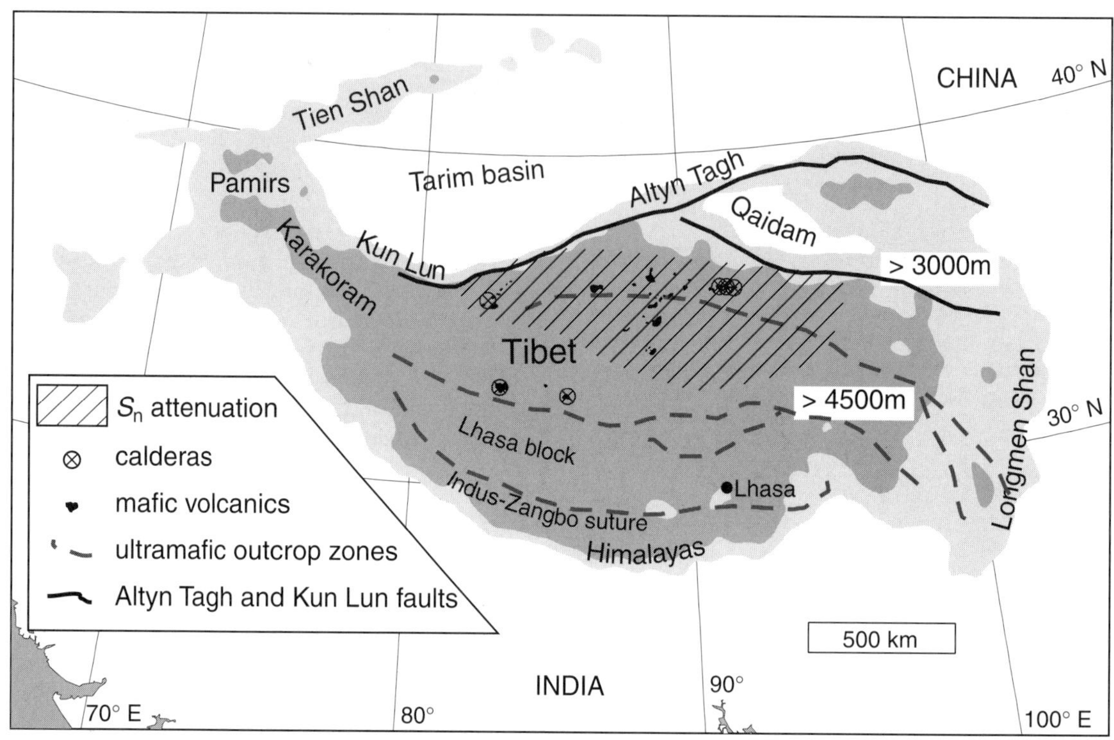

Figure 10.3 *Highly simplified tectonic map of Tibet showing late Cenozoic volcanic rocks, anomalous mantle, and ultramafic outcrop zones. Late Cenozoic calderas and mafic volcanics adapted from Landsat map by Rothery and Drury (1984). Anomalous mantle S_n attenuation zone reinterpreted from Barazangi and Ni (1982) and Ni and Barazangi (1983). Zones of ultramafic outcrops that are likely to be associated with suture zones were generalized from geologic map of Tibet (Zhou* et al.*, 1986). Other symbols as in Figure 10.1. Reprinted from* Tectonophysics, *v. 260, E.J. Fielding, Tibet Uplift and Erosion, pp. 55–84, Copyright (1996), with permission from Elsevier Science*

Schelling, 1992), called the Main Himalayan Thrust (MHT). The MHT is the active structure along which the Indian plate underthrusts the Himalaya and southern Tibet, absorbing a large part of the convergence between India and Asia. Schelling (1992) interprets a major footwall ramp on the MHT below the present outcrop of the MCT based on his structural data from eastern Nepal. Seismic reflection data in southern Tibet (southwest of Lhasa) indicate that the MHT continues down to a depth of at least 40 km and the Indian plate continues northward until at least the position of the Indus–Zangbo Suture zone (Zhao *et al.*, 1993; Nelson *et al.*, 1996).

The tectonic chronology of the major late Cenozoic structures of the Himalayas (STDS, MCT, MBT, and MHT) is clearly important for understanding the development of the modern Himalayan landforms (Figure 10.5). Each of these features has local complexities and is probably diachronous, but they can be traced along most of the length of the Himalayas. Most of the geochronological work carried out to date on these structures has been in the central part of the belt in Nepal and southern Tibet. The Main Central Thrust is responsible for uplifting the High Himalaya Crystalline rocks from mid- and lower crustal levels. Early movement on the MCT is reflected by metamorphic ages before ~22 Ma BP (Hubbard and Harrison, 1989; Guillot *et al.*, 1994), but it was reactivated in the late Miocene and Pliocene (Hodges *et al.*, 1996; Harrison *et al.*, 1997). Even later deformation on brittle faults in the MCT zone is supported by Late Pliocene ages (Macfarlane, 1993).

Geodetic measurements indicate ongoing uplift of the High Himalaya on a structure close to the Main Central Thrust in Nepal (Jackson and Bilham, 1994). The wavelength of this deformation is consistent with displacement on a deep structure that Jackson and Bilham (1994) interpret as a footwall ramp in the MHT similar to that described by Schelling (1992). Seismicity is also concentrated in this zone (Pandey *et al.*, 1995).

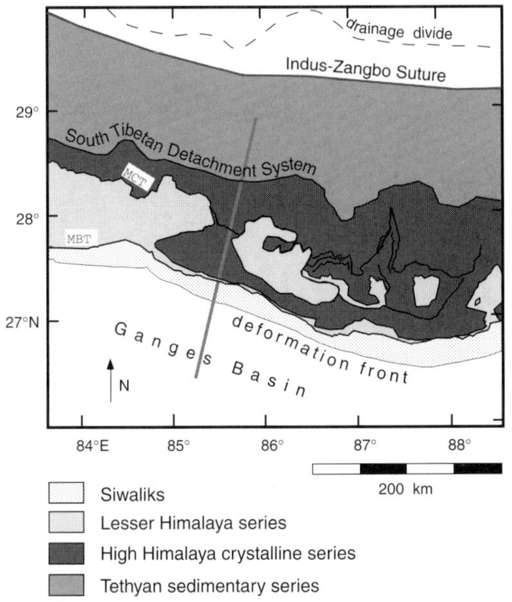

Figure 10.4 *Generalized geological map for central Nepal Himalaya. The location of the section in Figure 10.5 is shown as a thick grey line. Major fault systems are shown as medium black lines including MBT (Main Boundary Thrust) and MCT (Main Central Thrust). Modified from Masek* et al. *(1994a)*

These observations suggest that modern uplift of the Himalayas in Nepal may be occurring as a type of 'ramp anticline' over a ramp in the MHT where the Indian plate bends to a steeper dip as it underthrusts the Himalayas.

The South Tibetan Detachment System has recently become recognized as a major system of low-angle normal faults that affects most, if not all, of the North Himalaya. It is closely associated with the Miocene leucogranites that intrude the uppermost part of the High Himalaya Crystalline series, both spatially and temporally (Burchfiel *et al.*, 1992). In some locations the STDS cuts the leucogranites and in other places the leucogranites cut the STDS, so ages on the granites constrain STDS motion. It is clear that the STDS is not the same age everywhere and also that its normal-fault displacement was simultaneous with thrust movement on the MCT in some areas, including the Everest area (Hodges *et al.*, 1992). In some areas, the STDS moved prior to ~22 Ma BP (Guillot *et al.*, 1994; Hodges *et al.*, 1996), while at other locations the last movement was during or after the middle Miocene. Further east, in the Himalayan range north of Bhutan, the STDS was apparently active in the early Pliocene (Edwards *et al.*, 1996). Burchfiel and Royden (1985) and Burchfiel *et al.* (1992) have suggested that the movements on the MCT and STDS represent a gravitational collapse of high topographic relief between southern Tibet and the Indian plains. Another possibility is that erosion of the southern margin of the plateau during the Miocene removed enough mass to cause the Indian continent to begin underthrusting Tibet, as proposed for the Altiplano–Puna by Isacks (1988) and Gubbels *et al.* (1993). Both theories would predict that the Himalayas and Tibet were already at a high elevation in the Miocene.

Thrusting on the Main Boundary Thrust must have occurred sometime after formation of the Main Central Thrust, because it folds the MCT in several places. The MBT also cuts Quaternary units of the Siwalik group, so its motion must have continued into the Quaternary. Unfortunately, no intrusive or metamorphic ages can be used to date the MBT, nor are there overlapping or cross-cutting relations that closely bracket its deformation. Meigs *et al.* (1995) used stratigraphic and geochronologic data to interpret formation of the MBT before 10 Ma BP at two sites in the western Himalayas and Burbank and Raynolds (1988) have reported stratigraphic evidence of thrust displacement on the MBT at 5–4 Ma BP in Pakistan and northwestern India. Pliocene ages have been suggested elsewhere. Because the MBT is not a single thrust, but a system of thrusts that root in the Main Himalayan Thrust, it may not be possible to define an end to MBT motion. Both geological and geodetic evidence show Holocene deformation on structures south of the MBT where the MHT décollement terminates in the fold-thrust belt of the Siwaliks. It is likely that the ongoing motion of the MHT has been transferred to the surface via the MCT, MBT and other thrusts at different times, diachronously along the 2500-km-long Himalayan front.

10.3 Previous Geomorphological Studies

Early geomorphological studies of the Himalaya and Tibet were made by geodetic and geological mapping expeditions. One long-running controversy concerns the development of the Himalayan river drainages between those who supported consequent drainage and those who proposed antecedent drainage. Nearly all of the major rivers in the Himalaya start north of the High Himalayan peaks in southern Tibet and cut deep gorges through the range. The consequent drainage theory

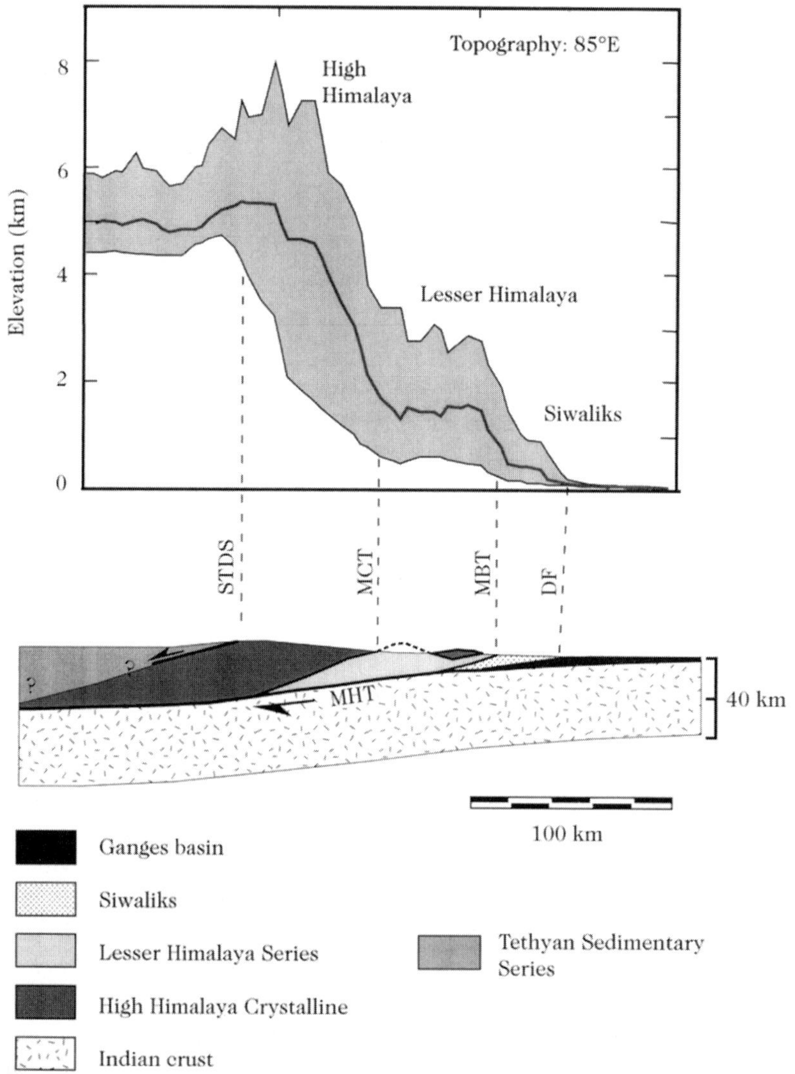

Figure 10.5 *Topographic and geological sections across central Nepal Himalaya at about 85°E. Location shown in Figure 10.4. Top figure shows vertically exaggerated topographic profile with 100-km-wide swath similar to Figure 10.7. Thick grey line is mean elevation at each point in the swath, with light grey fills between maximum and minimum elevations. Bottom figure shows a 1:1 scale crustal geological section. Major tectonic boundaries are STDS (South Tibetan Detachment System), MCT (Main Central Thrust), MBT (Main Boundary Thrust), DF (deformation front) and MHT (Main Himalayan Thrust). Modified from Masek* et al. *(1994a)*

asserts that these major rivers have cut back through the Himalayas by headward erosion and captured Tibetan rivers (Hayden, 1907; Heron, 1922; Burrard and Heron, 1933). Medlicott (1868) first suggested antecedent drainage, arguing that the rivers were present before the uplift of the present Himalayan range and maintained their courses by vigorous down-cutting. Fox (1922) and Odell (1925) also advanced this view of the evolution of the Himalayan rivers. Antecedent origins for the Himalayan rivers would provide important constraints on what topographic relief existed before formation of the modern High Himalaya, presumably by Miocene uplift on the Main Central Thrust.

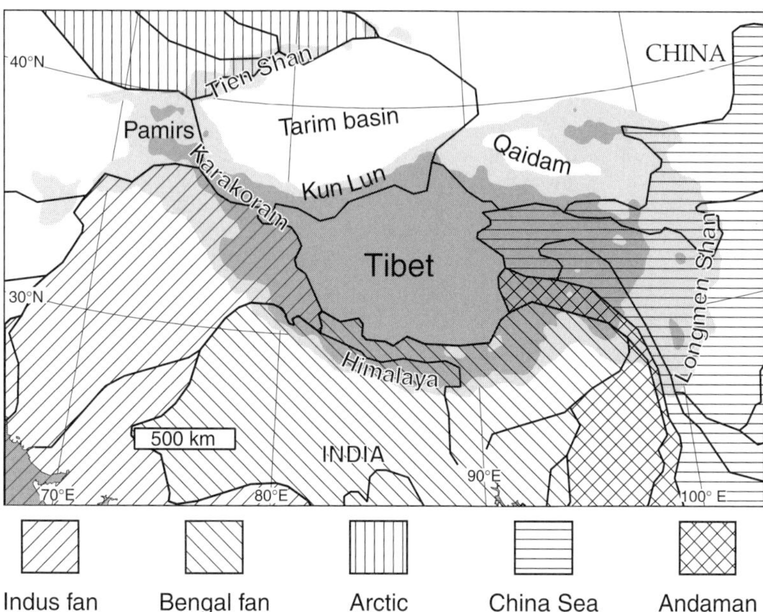

Figure 10.6 *Central Asia drainage divides and basins. Hatching patterns (see key) show destination of rivers in each drainage basin. Unhatched basins represent internal drainage. Major divides are marked with solid lines. Base map is similar to Figure 10.1*

Wager (1937) supported the antecedent drainage theory, based on his investigation of the Arun River which cuts through the Himalayas east of Mount Everest in eastern Nepal. The Arun cuts a narrow gorge in high-grade crystalline rocks of a range just north of the High Himalayan peaks that he called Nyonno Ri. A low pass in soft schist crosses the Nyonno Ri slightly east of the Arun gorge. It seems likely that a river cutting through this range by headward erosion would have avoided the more resistant gneisses, so it was inferred that the gorge in the gneisses must be following a river course antecedent to the uplift of the Nyonno Ri range. The main gorge of the Arun carves through the High Himalayan range starting about 15 km south of the Nyonno Ri. Wager (1937) concluded that the Arun, and probably other major rivers that originate north of the Himalayas, must be antecedent to the uplift of the main Himalayas, However, the Nyonno Ri range uplift may be part of the Quaternary graben structures in Tibet and therefore younger than the uplift of the main Himalayas, so antecedent drainage across the Nyonno Ri may not prove that the main gorge of the Arun and other Himalayan rivers are all antecedent.

Seeber and Gornitz (1983) also supported an antecedent origin for the major Himalayan rivers that have headwaters in southern Tibet. One of their main arguments was that the drainage divide between the Himalayan rivers of southern Tibet and the two main longitudinal rivers (Indus and Zangbo) parallels the High Himalayan range. Figure 10.6 shows the location of most of this drainage divide, interpreted from small-scale maps. When mapped in more detail, however, the divide is not so regular in shape. The locations of several trans-Himalayan rivers appear to be controlled by the north–south graben systems of southern Tibet (Seeber and Gornitz, 1983), but these structures are now known to be younger than those responsible for uplift of the Himalayas (e.g. Harrison *et al.*, 1995). I believe the question of consequent or antecedent drainage in the Himalayas is still open (Oberlander, 1985).

Wager (1937) was also the first to propose that the uplift of the highest peaks of the Himalayas above the elevation of the Tibetan Plateau is caused in part by isostatic compensation in response to the reduced load on the crust due to the eroded volume of the deep valleys. Wager (1933) noted that if the elevation of the Himalayas in Sikkim were levelled by transferring the tops of the mountains into the adjacent valleys, the average elevation would be approximately the same as that of the Tibetan Plateau to the north. This same

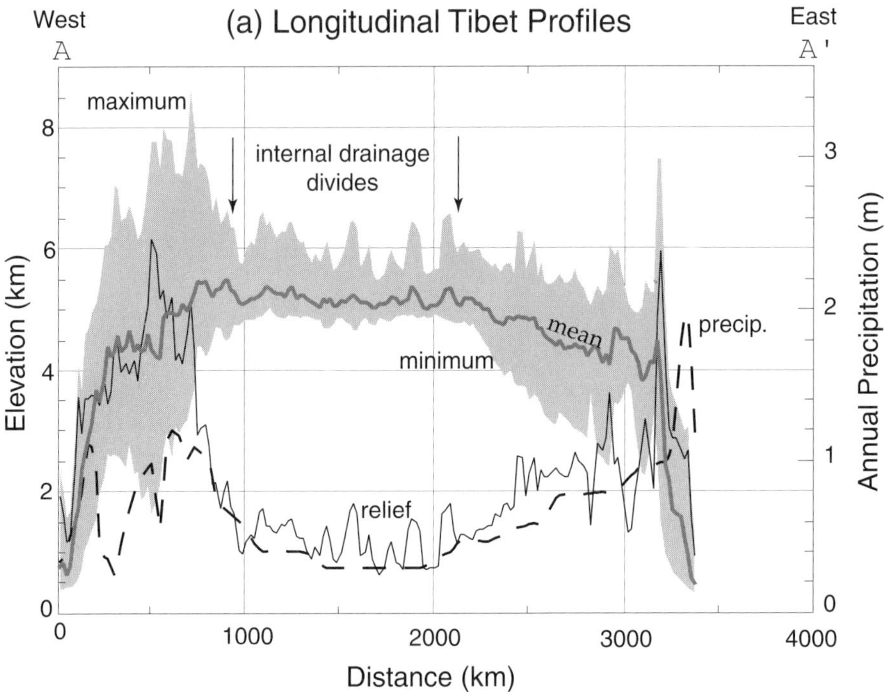

Figure 10.7 *Profiles of Tibetan Plateau, taken along 100-km-wide swaths. (a) West–east longitudinal profile A–A' across Tibet; (b) northeast–southwest profile B–B' across the western Himalayas (Nanga Parbat) and Karakoram; (c) north–south C–C' profile across the west-central Himalayas; (d) north-trending profile D–D' across widest part of central Tibet near longitude 85°E; (e) north-trending profile E–E' across east-central Tibet; and (f) north-trending profile F–F' across eastern end of the Himalayas and southern Tibet. See Plate 10.3 for profile locations. Each profile shows maximum, minimum (shading), and mean elevations (thick solid lines) of the topography within 5 x 100 km segments, along with topographic relief (thin lines). Thick dashed lines indicate annual precipitation adapted from* Climatic Atlas of Asia *(1981); scale is on right (see also Figure 10.10). Parts (b) and (d) modified from Fielding* et al. *(1994)*

feature can be seen in swath profiles through the Himalayas where the elevations are averaged along strike for some distance (Bird, 1978; Fielding *et al.*, 1994; Masek *et al.*, 1994a; Duncan, 1997) (Figure 10.7). Gilchrist *et al.* (1994) have calculated the amount of uplift to be expected from isostatic compensation of eroded valleys. Using an example from the Alps, they estimated that the isostatic uplift component can only account for a quarter of the elevation of the peaks with local compensation and less with flexural compensation. If the 8-km-high peaks of the Himalayas have a similar topographic configuration, this would correspond to ~2 km of isostatic uplift.

Seeber and Gornitz (1983) studied the profiles of the major rivers that cross the Himalayas. Their profiles were primarily based on the 1:1 000 000 Operational Navigation Charts (ONC) which generally have 1000 foot (305 m) contours. Almost every Himalayan river has an unusually steep reach that ends downstream where it crosses what they call the 'basement thrust front' marked by a narrow zone of thrust earthquakes. This feature is probably the same as the ramp in the Main Himalayan Thrust described above. Many rivers also have sharp knickpoints where they cross the MCT, this being consistent with young displacement. The MHT ramp is also beneath the surface trace of the MCT in much of the Himalaya, so it is difficult to separate their effects. Recent analysis of medium-resolution digital topography by Duncan (1997) also indicates a strong control of the Himalayan front topography by active faults.

Shackleton and Chang (1988) studied the geomorphic features along the 1985 Royal Society–Academica Sinica Geotraverse across east-central

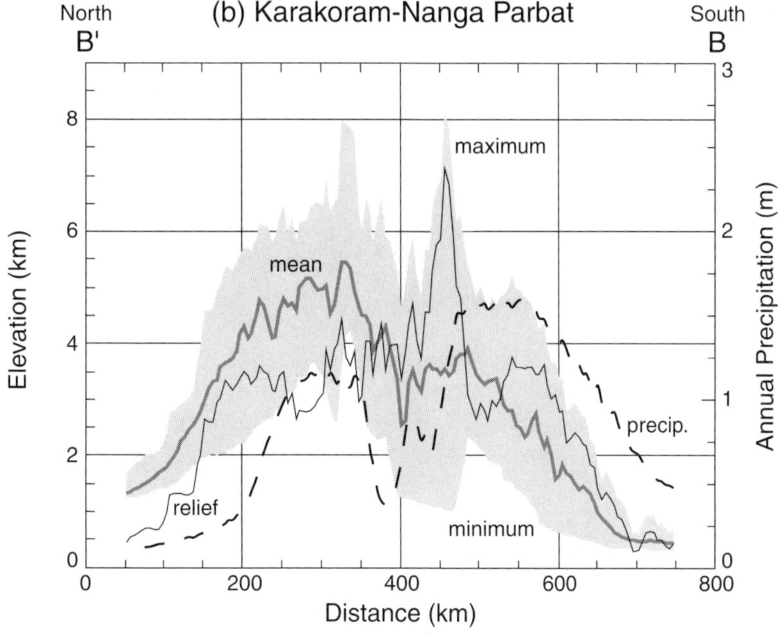

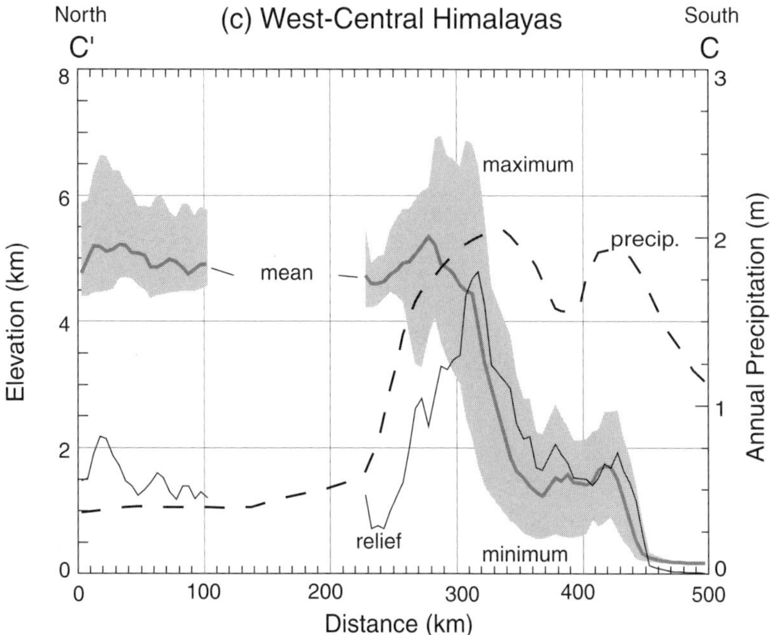

Figure 10.7 *(continued)*

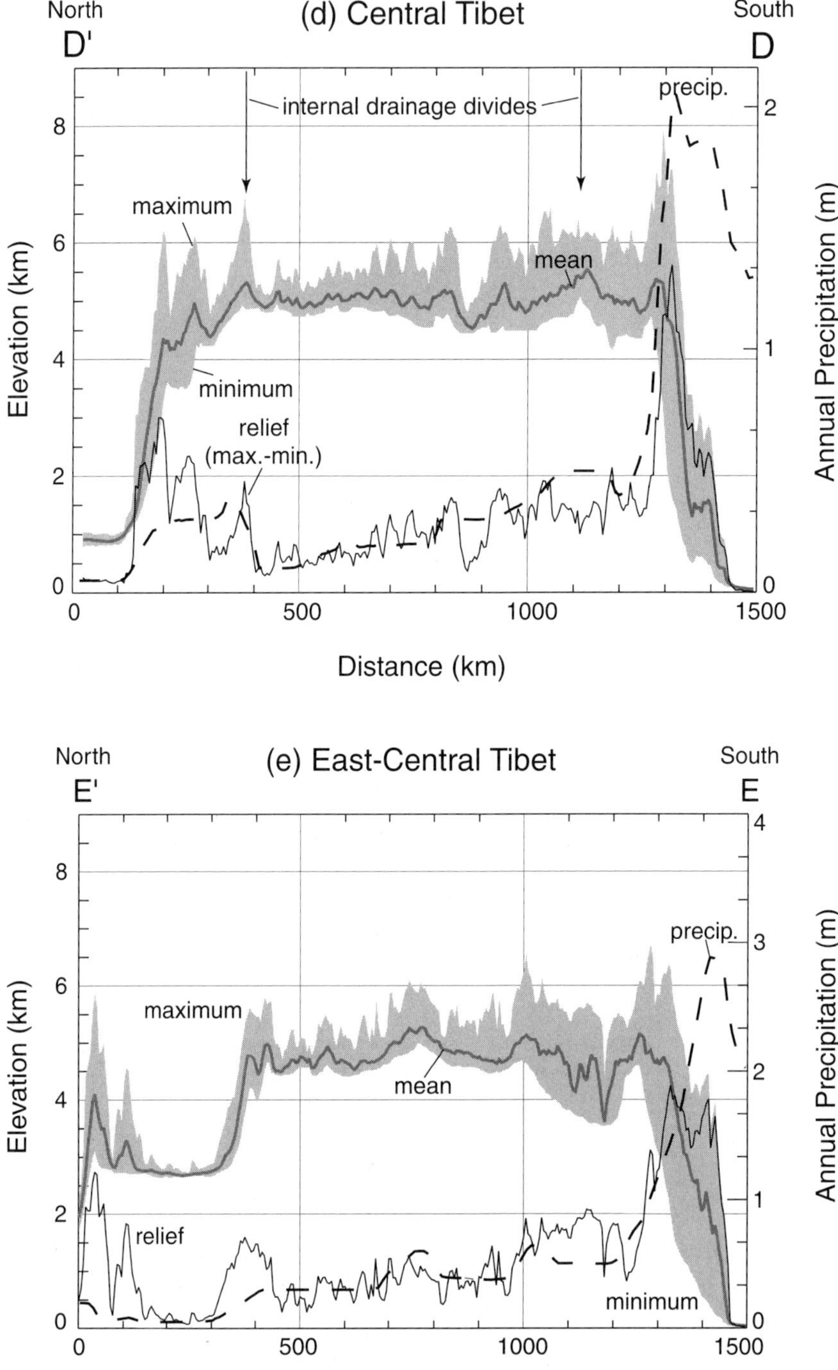

Figure 10.7 *(continued)*

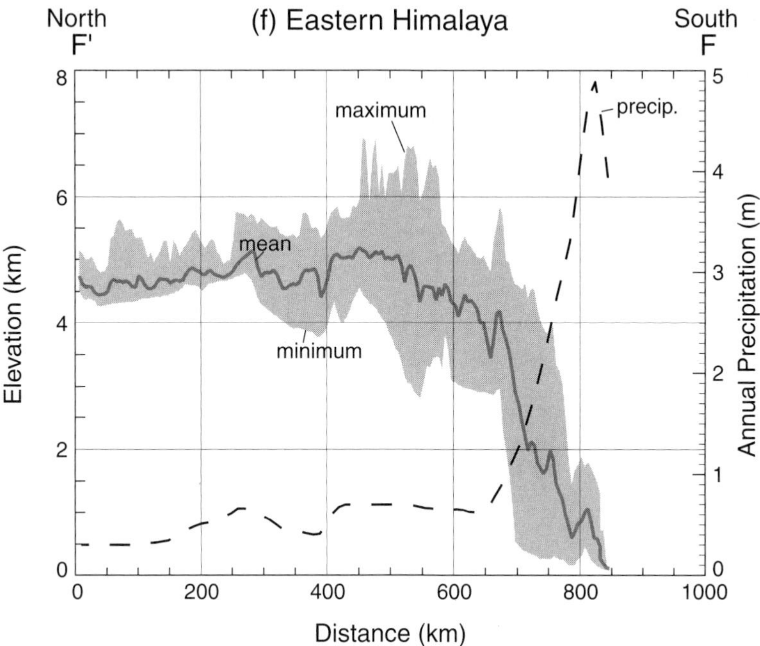

Figure 10.7 *(continued)*

Tibet from Lhasa to Golmud (in the Qaidam basin), close to the boundary between internal and external drainage (Figures 10.1 and 10.3). They described an extensive erosion surface cutting across the tops of the ranges that they interpreted as a middle to late Miocene pediplanation surface. The pediplanation cuts across deformed Eocene strata and mid-Miocene granites, but the surface has been deformed by the late Cenozoic graben structures. The surface is highly eroded in some areas, and it is generally expressed as summit accordances. It is most visible in the southern half of the plateau, but a similar surface cuts across the Kun Lun between Tibet and the Qaidam basin.

10.4 Morphological Characteristics

The recent production of digital elevation models (DEMs) for various areas of the Earth, including mountain belts, has enabled the quantification of morphology at scales that are impractical or impossible to measure in the field (Plate 10.1) (see Chapter 2). Fielding *et al.* (1994) used a 3-arcsecond (~90 m) grid DEM (generated by the US Defense Mapping Agency – recently renamed the National Imaging and Mapping Agency – with unfortunately restricted access) to measure slopes by least-squares fitting of a plane to a 4×4 point window approximately 250 m wide (Plate 10.2). The measurement of hillslope angles from DEMs depends on the length scale or size of the analysis window used. Thus, it is very important to specify the analysis window size or length scale used to measure slopes when comparing numerical values, but the spatial distribution of slopes measured with different window sizes is highly correlated, so the spatial pattern of steep and shallow slopes does not change.

Masek *et al.* (1994a, b), Burbank *et al.* (1996) and Duncan (1997) and Duncan *et al.* (1998) also analysed portions of the 3-arcsecond topographic data set. The Eros Data Center has recently released a reduced-resolution version of the 3-arcsecond data set with a 30-arcsecond (~0.9 km) grid that has complete coverage for all of the land area of Eurasia (Plate 10.3; Eros Data Center, 1996). To produce this data set, they filled regions where the 3-arcsecond data were not available with an interpolation of the contour lines from the Digital Chart of the World or other data sources.

Fielding *et al.* (1994) presented analyses that quantify many of the modern topographic characteristics of Tibet, the Himalayas, and surrounding areas, especially the extraordinary flatness of the plateau at a wide range of scales. The abrupt topographic boundaries of the Tibetan Plateau to the north and south (Plate 10.1, Figure 10.7) are, respectively, the Himalayan ranges and Kun Lun mountains. The western part of Tibet is continuous with the Karakoram and Pamir mountains and the eastern part of Tibet slopes downward until it reaches the Longmen Shan at the edge of the Szechwan basin (Figure 10.7a).

The central and northern parts of the plateau have the least relief, on both short and long wavelengths. The dark blue colour on Plate 10.2 shows the vast areas with shallow slopes of less than 10°, measured with ~250 m windows (Fielding *et al.*, 1994). The area of lowest relief also corresponds to the area of anomalous (hot?) mantle lithosphere and contains most of the young volcanic extrusions (Figure 10.3). The slopes in Tibet steepen only near a few glaciated peaks and near the Pliocene–Quaternary grabens in the south and south-central part of the plateau (e.g. Armijo *et al.*, 1986). Average slopes over ~250 m wavelengths are about 5° in central Tibet. At moderate wavelengths of the 100-km-wide swaths of Figure 10.7, relief is about 1 km or less for most of Tibet, as opposed to the much higher relief of up to 7 km on the plateau edges (Fielding *et al.*, 1994). The shape of the Himalayan front changes along strike (Figures 10.7b–f, Plate 10.3), due to both structural and erosional complexities (Duncan, 1997).

At the longest wavelengths, Tibet hypsometry demonstrates that more than half of the internally drained area lies within ±200 m of the mean elevation of 5023 m above sea level (plotted on Figure 10.8) (Fielding *et al.*, 1994). The total internally drained area of Tibet is 5.5×10^5 km^2, although the entire Tibetan Plateau is more than twice as large (Figures 10.1 and 10.6). The modal elevation (between 4920 and 4960 m), where the area within a single 10 m elevation bin reaches 1.01×10^4 km^2 or 1.8% of the total area, is slightly less than the mean indicating a small degree of skew, although it is surprisingly close to a symmetric distribution. This long wavelength flatness cuts across the different accreted terranes and changes in mantle lithospheric properties (Figure 10.3).

Synthetic aperture radar (SAR) interferometry is an innovative technique for mapping surface elevations using a pair of SAR images with a suitable geometry (e.g. Zebker *et al.*, 1994). I have produced experimental high-resolution DEMs for three sites in the Himalaya and Tibet by interferometric analysis of data collected by the NASA/JPL Shuttle Imaging Radar–C (SIR-C) flight in October 1994 (Figure 10.9). These DEMs have 10- to 20-m grid spacings but cover only limited 25 km by 100 km areas imaged by SIR-C. The SIR-C tracks allow a high-resolution sampling of some of the major topographic features.

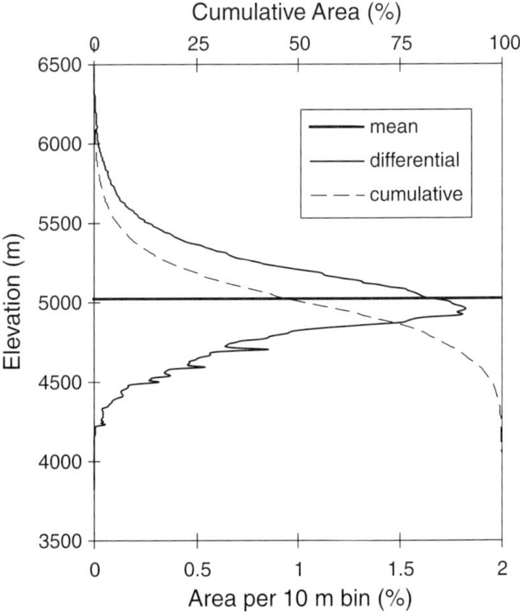

Figure 10.8 *Hypsometry of internally drained area of central Tibet (see Figure 10.1 for location). Differential (solid) and cumulative (dashed) hypsometry have the same vertical (elevation) axis and different horizontal axes (top and bottom). Thick solid line indicates mean elevation. Modified from Fielding* et al. *(1994)*

The Karakax River in northwestern Tibet is one of the largest rivers flowing out into the Tarim basin. One stretch of the Karakax flows along a valley that is controlled by the active Altyn Tagh fault (Figures 10.1 and 10.9). I have generated a SIR-C interferometric DEM covering a portion of this valley just beyond the entrance of the Karakax River (Plate 10.4). At the ~20 m resolution of this DEM, the active strand of the Altyn Tagh fault stands out as a sharp break in slope part way up the valley side (Plate 10.5). While the Altyn Tagh is primarily a left-lateral strike-slip fault, there is apparently some dip-slip component to its long-term displacement which has formed the Kun Lun range to the north. Several uplifted terrace remnants are also visible in the three-dimensional perspective view.

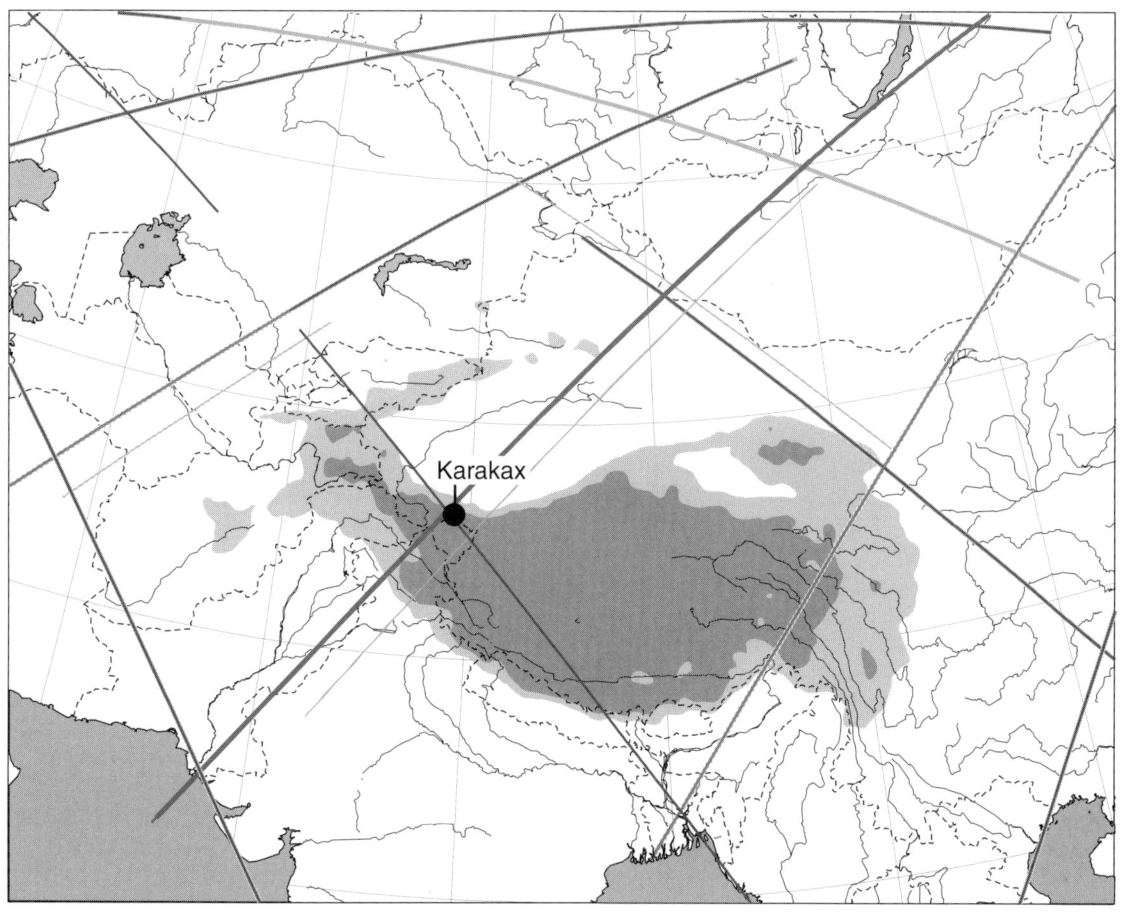

Figure 10.9 *Location of high-resolution experimental digital elevation model (DEM) derived from interferometric analysis of Shuttle Imaging Radar-C (SIR-C) data. Base map is similar to Figure 10.1, with major rivers as thin black lines. SIR-C interferometric data takes are shown as grey lines. SIR-C DEM location is marked with a filled circle*

10.5 Landform Development

The present morphology of the plateau has been shaped by erosional and depositional processes acting on the topography constructed by tectonic mechanisms over a long period of time. As more recent erosion and deposition has tended to erase the evidence of earlier events, I will cover here only the landform development since the middle Cenozoic. Landform development is covered in chronological order, with the modern erosional regime, for which we have the most information, being covered last.

Deposition of sedimentary rocks in the Tibetan orogen during the Cenozoic has been restricted to relatively small local basins within the plateau and to the large foreland basins adjacent to the plateau, so erosion has been the dominant landform-shaping process for most of the plateau. (See Fielding (1996) for a discussion of the entire Mesozoic through Quaternary uplift and erosion history.) The total amount of erosion in an area is the integral of the erosion rate over a given period. The rate of erosion and dissection of an area is a function of (1) the relief that provides potential energy, (2) the amount and type of precipitation that, as runoff, does the geomorphic work, and (3) the erodibility of the materials exposed at the surface. The configuration of the drainage network strongly controls the distribution of erosion.

Areas of internal drainage, such as the vast interior of Tibet, cannot lose mass through fluvial or glacial erosion so the net denudation will be approximately zero (except for aeolian erosion), and medium wavelength relief between highs and lows generally will be reduced by the redistribution of material.

The geological evidence for the denudational history of a given area generally falls into three categories: unconformities and the truncation of deformed structures in the area, sedimentary deposits containing clasts eroded from the area, and thermochronological measurements of the cooling of rocks. The former two techniques are traditional geological methods of documenting erosion, but they do not provide direct measurements of amounts of erosion, and precise time control is sometimes difficult to obtain. Thermochronology represents a relatively recent set of techniques that, in some cases, can provide a detailed thermal history of a rock sample, or samples, by measuring the time since various minerals (or mineral domains) cooled enough to retain isotopes or fission tracks (e.g. Harrison *et al.*, 1992a; Copeland *et al.*, 1995) (see Chapter 4). Mineral cooling ages can then be combined with assumptions (or measurements) of the geothermal gradient at those dates to estimate at what depth below the surface the samples were located at those times. Erosional or tectonic denudation of the overlying rocks decreases the depth below the surface and causes the sample to cool and retain markers. Unfortunately, high rates of denudation, high local relief, or underthrusting of colder rocks can strongly affect or even temporarily reverse the normal geothermal gradient, and these effects can greatly complicate data interpretation (e.g. Stüwe *et al.*, 1994; Copeland *et al.*, 1995).

10.5.1 MID-TERTIARY

An episode of rapid cooling, and presumably rapid erosion, affected many of the Gangdese plutons of southern Tibet in the Miocene (Harrison *et al.*, 1992a; Harrison *et al.*, 1993; Pan *et al.*, 1993; Copeland *et al.*, 1987, 1995). Across east-central Tibet along the Geotraverse route, Shackleton and Chang (1988) describe evidence for extensive mid–late Miocene planation, although their timing control is less precise. Unfortunately, detailed thermal and stratigraphic chronologies have not yet been obtained for the enormous interior of Tibet, so little can be determined about erosion there.

In addition, the Miocene extensional movement on the STDS is likely to have modified, and possibly greatly increased, the gradient of the rivers draining southward from the Gangdese across the Himalayas and STDS by lowering the regional base level. This would require that the Gangdese was already at a high elevation. If the STDS also acted to lower the average elevation of the Himalayas, it would have allowed more precipitation to reach the Gangdese by reducing the rain-shadow effect. Increased river gradients and more precipitation would produce more rapid erosion. The Gangdese is now at an average elevation of 5–5.5 km above sea level (Figure 10.7), but it is not being rapidly eroded due to low amounts of precipitation and high regional base level that are both caused by the Himalayas to the south (Copeland and Harrison, 1990). Thus, the rapid erosion of the Gangdese during and after the Miocene may have been controlled more by the complex history of uplift in the Himalayas than by regional uplift of southern Tibet in the Miocene.

Rapid denudation also affected the Himalayas during the mid-Tertiary. Some of the Himalayan denudation resulted from tectonic denudation associated with movement of the STDS, but erosional denudation also must have contributed. Thermobarometric data in several places require the removal of at least 20–25 km, and perhaps as much as 35 km, of rock from the high-grade High Himalaya Crystalline series (e.g. Hubbard *et al.*, 1991; Vannay and Hodges, 1996). Combining the thermobarometry with geochronology partially constrains timing and rates. Typical average denudation rates are ~1 mm a^{-1} since the early Miocene (~20 Ma BP), although there is evidence for much more rapid denudation in some locations and time periods. Details of the cooling history depend on the assumed thermal structure (Hubbard *et al.*, 1991). The sediments deposited in the Ganges and Indus foreland basins and subsea fans record a large amount of erosion starting in the mid-Miocene (Harrison *et al.*,1993), and it seems likely that most of this material came from the Himalayas (Curray, 1994; Johnson, 1994). The fact that the Bengal fan had prograded far out into the Indian Ocean by the early Miocene suggests that a significant amount of material had already been eroded from the Himalayas by that time (Curray, 1994). Sorting out the relative contributions from Tibet and the Himalayas to the sedimentary deposits would help to constrain the evolution of the drainage divide with time.

10.5.2 LATE TERTIARY–QUATERNARY

The present morphology reflects the late Cenozoic erosional history of Tibet to a large extent. For most of the plateau, especially the northern and central parts,

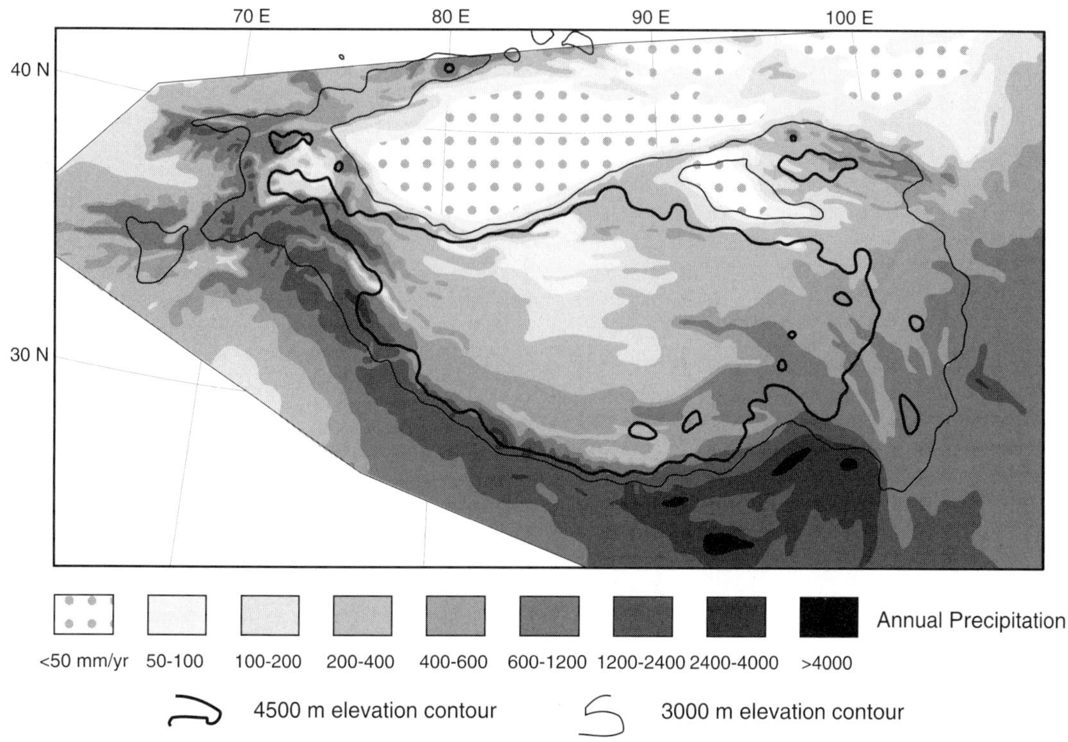

Figure 10.10 *Mean annual precipitation for Tibet and surrounding areas adapted from* Climatic Atlas of Asia *(1981). Key shows shading patterns for precipitation. Outline of Tibet topography (3000 m and 4500 m contours, similar to Figure 10.1) shown as lines for geographic reference. Reprinted from* Tectonophysics, *v. 260, E.J. Fielding, Tibet Uplift and Erosion, pp. 55–84, Copyright (1996), with permission from Elsevier Science*

erosion has apparently been very slow for much of the late Cenozoic. The degree of dissection is minimal, as shown by the predominance of slopes of less than 5° (Plate 10.2). Many of the scarps of the graben systems in Tibet are only slightly eroded (Masek *et al.*, 1994b). The drainage network is very poorly developed in the dry climate of interior Tibet (Figure 10.10). In contrast, the deep canyons and steep slopes in the externally drained eastern Tibet, the Himalayas and the Karakoram probably reflect rapid erosion during all of the Quaternary and perhaps much of the Tertiary.

A comprehensive review of thermochronologies in Tibet, the Himalayas and the Karakoram is beyond the scope of this chapter, but a few examples will serve to illustrate the evidence for rapid denudation in some areas. (See Fielding (1996) and Sorkhabi and Stump (1993) for recent reviews of thermochronologic histories for Tibet and the Himalayas, respectively.) One site where there is clear evidence of extremely swift late Tertiary–Quaternary denudation is the Nanga Parbat–Haramosh Massif in the western Himalayas where rates of 1–8 mm a^{-1} have been calculated for various intervals since 4 Ma BP from fission track and other isotopic ages (e.g. Zeitler, 1985; George *et al.*, 1995). Denudation rates as high as this must certainly be interpreted with caution (Stüwe *et al.*, 1994), but rates >2 mm a^{-1} are required by the data, which include an apatite fission-track age as young as 0.4 Ma where the Indus River gorge cuts the massif (Zeitler, 1985).

Holocene incision by the Indus River has been measured by Burbank *et al.* (1996) using cosmogenic nuclide dates on strath terraces. Straths cut by the former Indus into bedrock are now far above the present river, requiring incision and almost certainly uplift at rates of 8–12 mm a^{-1}. This area has very steep slopes (averaging >30°) (Plate 10.2) (Burbank *et al.*, 1996) that are both a consequence and a cause of the rapid denudation, and it has a moderately high mean annual precipitation (>1200 mm) (Figure 10.10). The apparent similarity between the hillslope angles adjacent to sections of the Indus that differ in incision rates by a factor of two suggests that the slopes are limited by rock strength in this area of rapid erosion. The extreme rates in this area are probably related to the structural syntaxis at Nanga Parbat, combined with the abundant erosional power of the Indus River.

Detailed thermochronologies are not yet available from the interior of Tibet, and because the expected erosion rates there are low, measurements would be unlikely to provide much information on Pliocene–Quaternary denudation. One location at the northern edge of Tibet in the Kun Lun close to Golmud was studied by Lewis (1990), who measured apatite fission-track ages of ~20 Ma and track length evidence that the rocks were in the partial annealing zone (50–100°C) before 20 Ma BP. It is possible that as little as 2 km has been eroded there in the past 20 Ma at an average rate of 0.1 mm a^{-1}. There is a good chance that the rest of the internally drained, dry interior of Tibet has experienced even slower late Cenozoic erosion rates than the Kun Lun. An upper-bound average erosion rate of 0.1 mm a^{-1} would remove only 1.5 km from the mean surface of Tibet in the past 15 Ma, which represents roughly the same amount of crustal thinning as 2% extension of the 70 km thick crust.

10.5.3 PRESENT LANDSCAPE DEVELOPMENT

The present erosion of Tibet is concentrated at the edges of the plateau, where there is high relief (2000–6000 m) (Figure 10.7), steep slopes (Plate 10.2), and external drainage. Intense fluvial and glacial dissection maintains high local relief. The steepest slopes are in the Karakoram and Himalayas where they average about 35° and reach up to 78° (at a 250 m wavelength; Fielding *et al.*, 1994). The along-strike averaged elevation of the High Himalaya is only slightly greater than that of the plateau as a result of the combination of very high peaks and deep river valleys (Figures 10.5 and 10.7); (e.g. Wager, 1937; Bird, 1978). Dissection is more limited on most of the northern margin of the plateau in the Kun Lun Shan where the topography drops 3–4 km over some 50–100 km into the Tarim and Qaidam basins, but there are few deep valleys. In those regions, there is only a narrow band of steep slopes (Plate 10.2) and relief within a 100-km-wide swath profile (Figure 10.7a, c–e) reaches 3 km compared to the 5.5 km relief in the Himalayas. The areas of high relief (and steep slopes) clearly correlate with the areas of high precipitation (Figure 10.10).

The Karakoram is deeply dissected by intense Quaternary and continuing modern glaciation, producing deep valleys between high peaks, with an average range elevation slightly higher than that of Tibet (Figure 10.7a, b). The elevation of the peaks in the Pamirs is about the same as in the central part of Tibet, but the average elevation (3.9 km) is lower by about 900 m due to a much higher degree of dissection, removal of mass by erosion, or thinner crust (Figure 10.7a). The western Pamirs are much more dissected than the eastern portion that includes a relatively flat area at about 4.2 km elevation (Plates 10.1 and 10.2). The increase in relief towards the west in the Pamirs matches an increase in annual precipitation (Figures 10.7a and 10.10).

As described in Section 10.4, the eastern part (nearly a third) of the Tibetan Plateau slopes gradually downward from the centre of the plateau beginning roughly at the external drainage divide (Figure 10.7a). Both the mean and maximum elevations decrease, but the minimum elevations drop more rapidly due to the deep canyons that eventually drain into many of the major rivers of south and east Asia, including the Brahmaputra, Mekong, Yangtze, and Huang He (Yellow) Rivers (Figure 10.6). This produces the gradual eastward increase in relief shown in Figure 10.7a. The more gradual long wavelength rise of eastern Tibet also allows precipitation to penetrate further onto the plateau (Figure 10.10). The greater precipitation and external drainage are then responsible for the greater dissection and relief. Since eastern Tibet is likely to have been formed by extrusion, the external drainage of this area is probably an old feature, perhaps dating back to the Miocene or earlier.

Southern Tibet includes a substantial area of external drainage, where the Indus, Ganges, and Zangbo/Brahmaputra and tributary rivers leave the plateau (Figures 10.1 and 10.6). This zone is also somewhat wetter (mostly 400–600 mm a^{-1}) than the central and northern parts of Tibet, but much drier than the Himalayas just to the south (Figure 10.10). The Zangbo River that drains southern Tibet is now at elevations of 3.5–4.5 km above sea level, so the regional base level is very high. Sand grains from a dune next to the Zangbo (Tsangpo) River were analysed by Copeland and Harrison (1990). They found that the K-feldspar grains in the sand had age minima with a maximum at ~17 Ma and no grains younger than 8 Ma, and suggested that all of these grains could have come from the erosion of the Gangdese batholith, which is north of the river. While this sample of 26 grains may not be representative of the large region of southern Tibet and the north slope of the Himalayas that drains through this point, the lack of very young ages suggests there is little or no area of rapid, deep erosion in that region, unlike the southern slope of the Himalayas (Copeland and Harrison, 1990).

The climate of the central part of Tibet is now arid (less than 400 mm precipitation per annum) (Figure 10.10) and cold (average annual temperature below freezing) (*Climatic Atlas of Asia*, 1981), so the rate of fluvial erosion is presumably low. The highest peaks (above about 6500 m) are presently glaciated. However, due to the aridity of Tibet, Pleistocene glaciers did not extend much below this altitude (Wissmann, 1959; Derbyshire *et al.*, 1991). In this cold, dry desert regime, erosional processes presumably are less efficient at transporting material over long distances and thus slowly remove topography uplifted by tectonic deformation. The disconnected internal drainage basins of Tibet (Figure 10.6) also prevent long-distance transport of material (except by aeolian erosion). In contrast, the Karakoram, Himalaya and eastern Tibet receive a much greater amount of precipitation than central Tibet, 600–4000 mm mm a^{-1} over much of the area (Figures 10.7 and 10.10); the spectacular and extensive glaciers of the Karakoram are maintained by this moisture. External drainage (Figure 10.6) is carrying away the material eroded from these regions. The clear spatial correlation between the areas of steep slopes, deep dissection, and present high precipitation (Plate 10.2, Figures 10.7 and 10.10) indicate that this climatic distribution has continued for a substantial period, at least throughout the Quaternary, and perhaps the late Tertiary.

10.6 Summary of Morphotectonic History

The maximum elevation and consequent change to an extensional regime of the Tibetan Plateau are likely to have resulted from some type of lithospheric thinning process, such as convective removal of the thermal boundary later (e.g. England and Houseman, 1989). Extension and volcanism, which were presumably related to this lithospheric thinning, indicate that it probably happened by 8 Ma BP and may have happened before 14 Ma BP. The apparently sudden development of the Gangdese Thrust System, Main Central Thrust, and South Tibetan Detachment in southernmost Tibet and the Himalayas at ~25 Ma BP record a change in the deformational regime of Tibet that suggests that lithospheric thinning may have begun then. Fielding (1996) has speculated that the crust of the interior of Tibet between 25 and 8 Ma BP remained about the same thickness, since it appears that deformation was concentrated in southern Tibet and the Himalayas between ~25 and ~8 Ma BP, while the mantle lithosphere was thinned dramatically sometime during the Miocene. If this scenario is correct, then this amount of thinning would result in about 3.2 km of uplift of the surface of Tibet between 25 and 8 Ma BP without a change in crustal thickness (Fielding, 1996), similar to the calculations of England and Houseman (1989).

How does the present low local relief of Tibet match with this proposed history? At moderate-to-long wavelengths from tens to hundreds of kilometres, central Tibet is extremely flat (Plate 10.2, Figures 10.7 and 10.8) (Fielding *et al.*, 1994). This lack of long wavelength relief may be caused by flow in the lower crust due to the extreme thickness of, and heat production within, the Tibetan crust (Gaudemer *et al.*, 1988; Bird, 1991), or other processes that result in lithosphere with a low effective viscosity (England and Houseman, 1986). Flexural modelling of the rift-flank uplift on some of the graben in Tibet indicates lithosphere with a very thin effective elastic thickness (<10 km) that would be consistent with a low viscosity zone in the Tibetan crust (Masek *et al.*, 1994b). The low rates of fluvial cut-and-fill processes for central Tibet cannot erase tectonic relief at moderate wavelengths, so either there has been little significant deformation of the upper crust in the late Cenozoic or some other process, such as viscous flow in the lower crust, has removed relief at these wavelengths.

The Himalayas and Karakoram, unlike central Tibet, have not been protected from precipitation and have been the site of much denudation during the Neogene (20–25 km in many places), and perhaps as much as 35 km of rock have been removed from the metamorphic core of the Himalayas (Vannay and Hodges, 1996). The concentration of precipitation on the edges of the plateau concentrates erosion there, and the external drainage provides the means for removing mass from the orogen. The tectonic input to the orogen due to continued convergence can be partially balanced by the erosional outflow (Figure 10.11). The ongoing shortening of the Himalayas provides new material to replace the kilometres of rock that have been removed and maintains the high elevation of the range, so the total uplift of rock with respect to sea level is much greater in the Himalayas than in Tibet.

Another major factor is time of erosion since the last construction of tectonic relief. The short wavelength

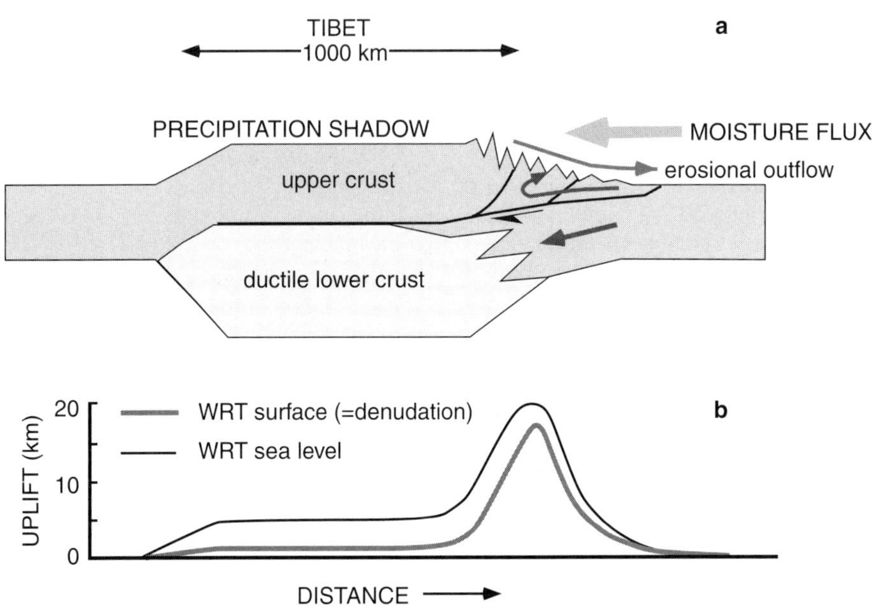

Figure 10.11 *(a) Schematic diagram (not to scale) illustrating interaction between tectonic inflow and erosional outflow of mass for continental plateaus. High rates of erosion are concentrated at the edge of the plateau by orographic precipitation, and the interior of the plateau is shielded from erosion. (b) Uplift curves for the hypothetical plateau shown in (a). Denudation or uplift of rock with respect to (WRT) the surface is concentrated at the plateau margin (thick grey line). Net uplift of the surface with respect to sea level is approximately the elevation of the modern surface in (a). Total uplift of rock with respect to sea level is the sum of net surface uplift and denudation (black line). Vertical scale is an approximate lower bound for Cenozoic denudation in the Himalayas and Tibet. Adapted from Masek* et al. *(1994a)*

topographic slopes of central Tibet are generally shallow, but locally steep close to the young normal faults of southern Tibet (Plate 10.2) (Armijo *et al.*, 1986). The relatively small scale of these graben and their scattered distribution suggests that they involve only minor upper crustal deformation, and do not indicate significant lithospheric deformation. Fault scarps formed by youthful deformation, if they were present in the large smooth areas of northern Tibet, would probably be well preserved and visible at the surface, as they are in other regions of similar climate such as the thrust belt on the southern margin of the Tien Shan, southern Tibet or the Basin and Range of the western USA. The late Cenozoic deformation regime of Tibet cannot involve significant deformation, especially shortening, of the surface of northern Tibet. Early Cenozoic or Cretaceous shortening of the upper crust of Tibet is consistent with the present relief since there has been enough time for even slow erosion to have reduced the constructional topography.

Some models of denudation have made a simplifying assumption that erosion rates are directly proportional to elevation. This may be reasonable in a wet, narrow, externally drained mountain range such as the Southern Alps or Taiwan, but it clearly does not apply to wide plateaus such as Tibet or the Altiplano–Puna (Figure 10.2) (Isacks, 1992). The great interior of Tibet is now (and probably has been for millions of years) at a very high elevation (~5 km), but has a low erosion rate. Since the distribution of precipitation strongly affects erosion rates, the intense focusing of precipitation at moderate elevations (~2 km) on a mountain front such as the Himalayas may cause the most rapid erosion at elevations below the peak of the range (Figure 10.11) (Masek *et al.*, 1994a). In addition to the spatial variations of climate and drainage network that produce variations in erosion rates at a given time, there have been major temporal changes in global climate, especially during the Cenozoic, that surely have affected erosion rates (Molnar and England, 1990). These considerations (and others) prevent a direct relationship between erosion and uplift of the surface with respect to sea level (England and Molnar, 1990).

10.7 Conclusions

The surface of Tibet records the integrated effects of tectonic and erosional processes over at least the past 20 Ma but it is more indicative of more recent effects that have tended to erase evidence of earlier processes. Some early studies suggested that the Tibetan Plateau is continuing to shorten in a N–S direction by 'accordion-style' thrust faulting and folding (e.g. Dewey and Burke, 1973). The present smooth topographic surface cuts across upper Tertiary folded strata without noticeable relief, despite apparently low erosion rates. Folding and thrust faulting of the interior of Tibet, therefore, has not been active during latest Tertiary and Quaternary time. Uplift of the plateau during the late Cenozoic must have been through a mechanism that does not involve significant shortening of the uppermost crust, such as the underthrusting of India beneath Tibet (e.g. Barazangi and Ni, 1982; Beghoul *et al.*, 1993), injection of Indian crust into the Tibetan lower crust (Zhao and Morgan, 1985), other mechanisms of flow of the lower crust (Royden, 1996), or lithospheric thinning (e.g. England and Houseman, 1989). Distributed shortening may describe the early evolution of the Tibetan Plateau, before the change to an extensional regime at 8 Ma BP (or perhaps earlier). The symmetrical and extremely narrow peak of the hypsometry over an area of more than 0.5×10^6 km^2 in north- and south-central Tibet (Figure 10.8) suggests some kind of fluid crustal process has been operating recently to level the elevation (England and Houseman, 1986), perhaps involving lower crustal flow as proposed by Bird (1991) and Royden (1996).

Fielding (1996) reviewed the presently available constraints on the uplift and erosion history of the Tibetan Plateau. Given the relatively small amount of geological and geophysical research that has so far been carried out over the immense area (more than 10^6 km^2, equal to a large part of Europe) of the plateau, we can expect some surprises out of future tectonic studies. Further work should be undertaken on the erosional history of the Tibetan Plateau and the Himalayas. The evolution of the present drainage system (Figure 10.6) over time is an important topic for constraining the destinations of the sediments eroded from the Tibetan Plateau, Himalaya, and related mountain ranges of central Asia. The modern river systems are carrying sediment to widely separated ocean basins. Mass balance calculations of eroded and deposited material must estimate where the major drainage divides have been located in the past. New high-resolution DEMs are becoming available for portions of Tibet and will allow much more detailed geomorphological measurements.

Acknowledgements

Part of this research was performed at the Jet Propulsion Laboratory, California Institute of Technology, under contract with the US National Aeronautics and Space Administration.

References

Argand, E. 1924. La tectonique de L'Asie. *Comptes Rendus, Congrès Géologique International*, **1**, 171–372.

Armijo, R., Tapponnier, P., Mercier, J.L. and Tong-Lin, H. 1986. Quaternary extension in southern Tibet: Field observations and tectonic implications. *J. Geophys. Res.*, **91**, 13 803–13 872.

Armijo, R.,Tapponnier, P. and Tonglin, H. 1989. Late Cenozoic right-lateral strike-slip faulting in Southern Tibet. *J. Geophys. Res.*, **94**, 2787–2838.

Barazangi, M. and Ni, J. 1982. Velocities and propagation characteristics of Pn and Sn beneath the Himalayan arc and Tibetan plateau: Possible evidence for underthrusting of Indian continental lithosphere beneath Tibet. *Geology*, **10**, 179–185.

Beghoul, N., Barazangi, M. and Isacks, B.L. 1993. Lithospheric structure of Tibet and western North America: Mechanisms of uplift and a comparative study. *J. Geophys. Res.*, **98**, 1997–2016.

Besse, J. and Courtillot, V. 1988. Paleogeographic maps of the continents bordering the Indian Ocean since the early Jurassic. *J. Geophys. Res.*, **93**, 11 791–11 808.

Bird, P. 1978. Initiation of intracontinental subduction in the Himalaya. *J. Geophys. Res.*, **83**, 4975–4987.

Bird, P. 1991. Lateral extrusion of lower crust from under high topography, in the isostatic limit. *J. Geophys. Res.*, **96**, 10 275–10 286.

Burbank, D.W. and Raynolds, R.G.H. 1988. Stratigraphic keys to the timing of thrusting in terrestrial foreland basins: Applications to the Northwestern Himalaya. In: K.L. Kleinspehn and C. Paola (eds) *New Perspectives in Basin Analysis*, Springer-Verlag, New York, pp. 331–351.

Burbank, D.W., Leland, J. Fielding, E.J., Anderson, R.S., Brozovic, N., Reid, M. and Duncan, C.C. 1996. Fluvial incision, rock uplift, and threshold hillslopes in the northwestern Himalaya. *Nature*, **379**, 505–510.

Burchfiel, B.C. and Royden, L.H. 1985. North–south extension within the convergent Himalayan region. *Geology*, **13**, 679–682.

Burchfiel, B.C. and Royden, L.H. 1991. Tectonics of Asia 50

years after the death of Emile Argand. *Eclogae Geol. Helv.*, **84**, 599–629.

Burchfiel, B.C., Chen, Z., Hodges, K.V., Liu, Y., Royden, L.H., Deng, C. and Xu, J. 1992. The South Tibetan Detachment System, Himalayan orogen: Extension contemporaneous with and parallel to shortening in a collisional mountain belt. *Geol. Soc. Amer. Spec. Pap.*, **269**, 41 pp.

Burg, J.-P. 1983. Tectogénèse comparée de deux segments de chaîne de collision: le sud du Tibet (suture du Tsanpo), et la chaîne hercyienne en Europe (suture de Massif-Central). Thesis, Montpellier.

Burg, J.-P., Proust, F., Tapponnier, P. and Chen, G.M. 1983. Deformation phases and tectonic evolution of the Lhasa Block (southern Tibet, China). *Eclogae Geol. Helv.*, **76**, 643–665.

Burrard, S.G. and Heron, A.M. 1933. *A Sketch of the Geography and Geology of the Himalaya Mountains and Tibet*, revised: Dehli, Manager of Publications, pp. 261–266 and 347–349.

Climatic Atlas of Asia. 1981. *Maps of Mean Temperature and Precipitation*. World Meteorology Organization, Geneva, 28 pp.

Coleman, M. and Hodges, K. 1995. Evidence for Tibetan plateau uplift before 14 Myr ago from a new minimum age for east–west extension. *Nature*, **374**, 49–52.

Copeland, P. and Harrison, T.M. 1990. Episodic rapid uplift in the Himalaya revealed by Ar-40/Ar-39 analysis of detrital K-feldspar and muscovite, Bengal Fan. *Geology*, **18**, 354–357.

Copeland, P., Harrison, T.M., Kidd, W.S.F., Ronghua, X. and Yuquan, Z. 1987. Rapid early Miocene acceleration of uplift in the Gangdese Belt, Xizang-southern Tibet, and its bearing on accommodation mechanisms of the India–Asia collision. *Earth Planet. Sci. Lett.*, **86**, 240–252.

Copeland, P., Harrison, T.M., Pan, Y., Kidd, W.S.F., Roden, M. and Zhang, Y. 1995. Thermal evolution of the Gangdese batholith, southern Tibet: A history of episodic unroofing. *Tectonics*, **14**, 223–236.

Curray, J.R. 1994. Sediment volume and mass beneath the Bay of Bengal. *Earth Planet. Sci. Lett.*, **125**, 371–383.

Derbyshire, E., Shi, Y.F., Li, J.J., Zheng, B.X., Li, S.J. and Wang, J.T. 1991. Quaternary glaciation of Tibet – the geological evidence. *Quat. Sci. Rev.*, **10**, 485–510.

Dewey, J.F. and Burke, K.C.A. 1973. Tibetan, Variscan and Precambrian reactivation: Products of continental collision. *J. Geol.*, **81**, 683–692.

Dewey, J.F., Shackleton, R.M., Chengfa, C. and Yiyin, S. 1988. The tectonic evolution of the Tibetan Plateau. *Phil. Trans. R. Soc. Lond.*, *A*, **327**, 379–413.

Dewey, J.F., Cande, S. and Pitman, W.C. 1989. Tectonic evolution of the India–Eurasia collision zone. *Eclogae Geol. Helv.*, **82**, 717–734.

Duncan, C.C. 1997. Topographic signatures of tectonics and erosion in the Himalayas from digital elevation data and numerical modeling. PhD dissertation, Cornell University.

Duncan, C.C., Klein, A., Blodgett, T., Masek, J. and Isacks, B. 1998. Modern and last glacial maximum equilibrium lines of central Nepal from digital elevation model analysis. *Quat. Res.*, **49**, 241–254.

Edwards, M.A., Kidd, W.S.F., Li, J., Yue, Y. and Clark, M. 1996. Multi-stage development of the southern Tibet detachment system near Khula Kangri: New data from Gonto La. *Tectonophysics*, **260**, 1–19.

England, P. and Houseman, G. 1986. Finite strain calculations of continental deformation, 2, Comparison with the India–Asia collision zone. *J. Geophys. Res.*, **91**, 3664–3676.

England, P. and Houseman, G. 1988. The mechanics of the Tibetan Plateau. *Phil. Trans. R. Soc. Lond.*, **A**, **326**, 301–320.

England, P. and Houseman, G. 1989. Extension during continental convergence, with application to the Tibetan plateau. *J. Geophys. Res.*, **94**, 17 561–17 579.

England, P. and Molnar, P. 1990. Surface uplift, uplift of rocks, and exhumation of rocks. *Geology*, **18**, 1173–1177.

Eros Data Center. 1996. Global 30 arc-second Elevation Data Set: http://edcwww.cr.usgs.gov/landdaac/30asdewdem/30asdcwdem.html.

Fielding, E.J. 1996. Tibet uplift and erosion. *Tectonophysics*, **260**, 55–84.

Fielding, E.J., Isacks, B.L., Barazangi, M. and Duncan, C.C. 1994. How flat is Tibet? *Geology*, **22**, 163–167.

Fox, C.S. 1922. Discussion of geological results of the Mount Everest Reconnaissance Expedition. *Geogr. J.*, **59**, 431–436.

Gansser, A. 1964. *The Geology of the Himalayas*. Interscience, New York.

Gaudemer, Y., Jaupart, C. and Tapponnier, P. 1988. Thermal control on post-orogenic extension in collision belts. *Earth Planet. Sci. Lett.*, **89**, 48–62.

George, M., Reddy, S. and Harris, N. 1995. Isotopic constraints on the cooling history of the Nanga Parbat–Haramosh Massif and Kohistan arc, western Himalaya. *Tectonics*, **14**, 237–252.

Gilchrist, A.R., Summerfield, M.A. and Cockburn, H.A.P. 1994. Landscape dissection, isostatic uplift, and the morphologic development of orogens. *Geology*, **22**, 963–966.

Gubbels, T.L., Isacks, B.L. and Farrar, E. 1993. High-level surfaces, plateau uplift, and foreland development, Bolivian Central Andes. *Geology*, **21**, 695–698.

Guillot, S., Hodges, K.V., LeFort, P. and Pécher, A. 1994. New constraints on the age of the Manaslu leucogranite: Evidence for episodic tectonic denudation in the central Himalayas. *Geology*, **22**, 559–562.

Harrison, T.M., Copeland, P., Kidd, W.S.F. and Yin, A. 1992a. Raising Tibet. *Science*, **255**, 1663–1670.

Harrison, T.M., Chen, W., Leloup, P.H., Ryerson, F.J. and Tapponnier, P. 1992b. An early Miocene transition in deformation regime within the Red River fault zone, Yunnan, and its significance for Indo-Asian tectonics. *J. Geophys. Res.*, **97**, 7159–7182.

Harrison, T.M., Copeland, P., Hall, S.A., Quade, J., Burner,

S., Ojha, T.P. and Kidd, W.S.F. 1993. Isotopic preservation of Himalayan/Tibetan uplift, denudation, and climatic histories of two molasse deposits. *J. Geol.*, **101**, 157–175.

Harrison, T.M., Copeland, P., Kidd, W.S.F. and Lovera, O.M. 1995. Activation of the Nyainqentanghla Shear Zone: Implications for uplift of the southern Tibetan Plateau. *Tectonics*, **14**, 658–676.

Harrison, T.M., Ryerson, F.J., Le Fort, P., Yin, A., Lovera, O.M. and Catlos, E.J. 1997. A Late Miocene/Pliocene age for the Central Himalayan inverted metamorphism. *Earth Planet. Sci. Lett.*, **146**, E1–E7.

Hayden, H.H. 1907. The geology of the provinces of Tsang and U in Central Tibet. *Mem. Geol. Surv. India*, **36**(2), 5.

Heron, A.M. 1922. Geological results of the Mount Everest Reconnaissance Expedition. *Geogr. J.*, **59**, 418–431.

Hirn, A. 1988. Features of the crust–mantle structure of the Himalayas–Tibet: A comparison with seismic traverses of the Alpine, Pyrenean and Variscan orogenic belts. *Phil. Trans. R. Soc. Lond.*, **A**, **326**, 17–32.

Hodges, K.V., Parrish, R.P., Housh, T.B., Lux, D.R., Burchfiel, B.C., Royden, L.H. and Chen, Z. 1992. Simultaneous Miocene extension and shortening in the Himalayan Orogen. *Science*, **258**, 1466–1470.

Hodges, K.V., Parrish, R. and Searle, M. 1996. Tectonic evolution of the central Annapurna Range, Nepalese Himalayas. *Tectonics*, **15**, 1264–1291.

Houseman, G.A., McKenzie, D.P. and Molnar, P. 1981. Convective instability of a thickened boundary layer and its relevance for the thermal evolution of continental convergent belts. *J. Geophys. Res.*, **86**, 6155–6132.

Hubbard, M.S. and Harrison, T.M. 1989. ^{40}Ar/^{39}Ar age constraints on deformation and metamorphism in the MCT Zone and Tibetan Slab, eastern Nepal Himalaya. *Tectonics*, **8**, 865–880.

Hubbard, M., Royden, L. and Hodges, K. 1991. Constraints on unroofing rates in the high Himalaya, eastern Nepal. *Tectonics*, **10**, 287–298.

Isacks, B.L. 1988. Uplift of the Central Andean Plateau and bending of the Bolivian Orocline. *J. Geophys. Res.*, **93**, 3211–3231.

Isacks, B.L. 1992. 'Long-term' land surface processes: Erosion, tectonics and climate history in mountain belts. In: P.M. Mather (ed.) *Terra-1: Understanding the Terrestrial Environment*. Taylor and Francis, London, pp. 21–36.

Jackson, M. and Bilham, R. 1994. Constraints on Himalayan deformation inferred from vertical velocity-fields in Nepal and Tibet. *J. Geophys. Res.*, **99**, 13 897–13 912.

Johnson, M.R.W. 1994. Volume balance of erosional loss and sediment deposition related to Himalayan uplifts. *J. Geol. Soc. Lond.*, **151**, 217–220.

Kidd, W.S.F. and Molnar, P. 1988. Quaternary and active faulting observed on the 1985 Academica Sinica–Royal Society Geotraverse of Tibet. *Phil. Trans. R. Soc. Lond.*, **A**, **327**, 337–363.

Kutzbach, J.E., Prell, W.L. and Ruddiman, W.M. 1993. Sensitivity of Eurasian climate to surface uplift of the Tibetan Plateau. *J. Geol.*, **101**, 177–190.

Le Pichon, X., Fournier, M. and Jolivet, L. 1992. Kinematics, topography, shortening, and extrusion in the India–Eurasia collision. *Tectonics*, **11**, 1085–1098.

Lewis, C.L.E. 1990. Thermal history of the Kunlun Batholith, N. Tibet, and implications for uplift of the Tibetan Plateau. *Nucl. Tracks Rad. Measur.*, **17**, 301–307.

Macfarlane, A.M. 1993. Chronology of tectonic events in the crystalline core of the Himalaya, Langtang National Park, central Nepal. *Tectonics*, **12**, 1004–1025.

Masek, J., Isacks, B.L., Gubbels, T.L. and Fielding, E.J. 1994a. Erosion and tectonics at the margins of continental plateaus. *J. Geophys. Res.*, **99**, 13 941–13 956.

Masek, J.G., Isacks, B.L., Fielding, E.J. and Browaeys, J. 1994b. Rift-flank uplift in Tibet: Evidence for a viscous lower crust. *Tectonics*, **13**, 659–667.

Medlicott, H.B. 1868. The Alps and the Himalaya, a geological comparison. *Quart. J. Geol. Soc. Lond.*, **24**, 34–52.

Meigs, A.J., Burbank, D.W. and Beck, R.A. 1995. Middle–late Miocene (>10 Ma) formation of the Main Boundary thrust in the western Himalaya. *Geology*, **23**, 423–426.

Molnar, P. 1988. A review of geophysical constraints on the deep structure of the Tibetan Plateau, the Himalaya and the Karakoram, and their tectonic implications. *Phil. Trans. R. Soc. Lond.*, **A**, **326**, 33–88.

Molnar, P. and England, P. 1990. Late Cenozoic uplift of mountain ranges and global climate change – chicken or egg? *Nature*, **346**, 29–34.

Molnar, P. and Tapponnier, P. 1975. Cenozoic tectonics of Asia: Effects of a continental collision. *Science*, **189**, 419–426.

Molnar. P., England, P. and Martinod, J. 1993. Mantle dynamics, uplift of the Tibetan Plateau, and the Indian monsoon. *Rev. Geophys.*, **31**, 357–396.

Nelson, K.D., Zhao, W.J., Brown, L.D., Kuo, J., Che, J.K., Liu, X.W., Klemperer, S.L., Makovsky, Y. and Meissner, R. *et al.* 1996. Partially molten middle crust beneath southern Tibet – synthesis of project INDEPTH results. *Science*, **274**, 1684–1688.

Oberlander, T.M. 1985. Origin of drainage transverse to structures in orogens. In: M. Morizawa and J.T. Hack (eds) *Tectonic Geomorphology*. Allen and Unwin, Boston, pp. 155–182.

Odell, N.E. 1925. Observations of the rocks and glaciers of Mount Everest. *Geogr. J.*, **66**, 300.

Pan, Y. and Kidd, W.S.F. 1992. Nyainqentanglha shear zone: A late Miocene extensional detachment in the southern Tibetan Plateau. *Geology*, **20**, 775–778.

Pan, Y., Copeland, P., Roden, M.R., Kidd, W.S.F. and Harrison, T.M. 1993. Thermal and unroofing history of the Lhasa area, southern Tibet – Evidence from apatite fission track thermochronology. *Nucl. Tracks Rad. Measur.*, **21**, 543–554.

Pandey, M.R., Tandukar, R.P., Avouac, J.P., Lave, J. and Massot, J.P. 1995. Interseismic strain accumulation on the

Himalayan crustal ramp (Nepal). *Geophys. Res. Lett.*, **22**, 751–754.

Patriat, P. and Achache, J. 1984. India–Eurasia collision chronology has implications for crustal shortening and driving mechanism of plates. *Nature*, **311**, 615–621.

Peltzer, G. and Tapponnier, P. 1988. Formation and the evolution of strike-slip faults, rifts, and basins during the India–Asia collision: An experimental approach. *J. Geophys. Res.*, **93**, 15 085–15 117.

Powell, C.M. and Conaghan, P.J. 1973. Plate tectonics and the Himalayas. *Earth Planet. Sci. Lett.*, **20**, 1–12.

Ratschbacher, L., Frisch, W., Liu, G. and Chen, C. 1994. Distributed deformation in southern and western Tibet during and after the India–Asia collision. *J. Geophys. Res.*, **99**, 19 917–19 945.

Raymo, M.E. and Ruddiman, W.F. 1992. Tectonic forcing of late Cenozoic climate. *Nature*, **359**, 117–122.

Rothery, D.A. and Drury, S.A. 1984. The neotectonics of the Tibetan Plateau. *Tectonics*, **3**, 19–26.

Royden, L. 1996. Coupling and decoupling of crust and mantle in convergent orogens – implications for strain partitioning in the crust. *J. Geophys. Res.*, **101**, 17 679–17 705.

Ruddiman, W.F. and Kutzbach, J.E. 1989. Forcing of late Cenozoic northern hemisphere climate by plateau uplift in southern Asia and the American West. *J. Geophys. Res.*, **94**, 18 409–18 427.

Ruddiman, W.F., Prell, W.L. and Raymo, M.E. 1989. Late Cenozoic uplift in southern Asia and the American West: rationale for general circulation modeling experiments. *J. Geophys. Res.*, **94**, 18 379–18 391.

Schelling, D. 1992. The tectonostratigraphy and structure of the eastern Nepal Himalaya. *Tectonics*, **11**, 925–943.

Schelling, D. and Arita, K. 1991. Thrust tectonics, crustal shortening, and the structure of the far-eastern Nepal Himalaya. *Tectonics*, **10**, 851–862.

Searle, M. 1986. Structural evolution and sequence of thrusting in the High Himalaya, Tibetan–Tethys and Indus suture zones of Zanskar and Ladakh, Western Himalaya. *J. Struct. Geol.*, **8**, 923–936.

Searle, M. 1995. The rise and fall of Tibet. *Nature*, **374**, 17–18.

Searle, M.P., Windley, B.F., Coward, M.P., Cooper, D.J.W., Rex, A.J., Rex, D., Li, T.D., Xiao, X.C., Jan, M.Q., Thakur, V.C. and Kumar, S. 1987. The closing of the Tethys and the tectonics of the Himalaya. *Geol. Soc. Amer. Bull.*, **98**, 678–701.

Seeber, L. and Gornitz, V. 1983. River profiles along the Himalayan arc as indicators of active tectonics. *Tectonophysics*, **92**, 335–367.

Shackleton, R.M. and Chang, C. 1988. Cainozoic uplift and deformation of the Tibetan Plateau: the geomorphological evidence. *Phil. Trans. R. Soc. Lond.*, **A**, **327**, 365–378.

Sorkhabi, R.B. and Stump, E. 1993. Rise of the Himalaya: a geochronologic approach. *GSA Today*, **3**, 85–92.

Stüwe, K., White, L. and Brown, R. 1994. The influence of eroding topography on steady-state isotherms. Application to fission track analysis. *Earth Planet. Sci. Lett.*, **124**, 63–74.

Tapponnier, P., Peltzer, G., Ledain, A.Y., Armijo, R. and Cobbold, P. 1982. Propagating extrusion tectonics in Asia – new insights from simple experiments with plasticine. *Geology*, **10**, 611–616.

Vannay, J.C. and Hodges, K.V. 1996. Tectonometamorphic evolution of the Himalayan metamorphic core between Annapurna and Dhaulagiri, central Nepal. *J. Metamorph. Geol.*, **14**, 635–656.

Wager, L.R. 1933. Elevation of the Himalayas in Sikkim. *Nature*, **132**, 28.

Wager, L.R. 1937. The Arun River drainage pattern and the rise of the Himalaya. *Geogr. J.*, **89**, 239–250.

Wissmann, H. von, 1959. Die heutige Vergletscherung und Schneergrenze in Hochasien. *Mainz, Akademie Wissenschaften und der Literatur, Abhandlungen der Mathematisch-Naturwissenschaftlichen Klasse*, **14**, 5–307.

Zebker, H.A., Werner, C.L., Rosen, P.A. and Hensley, S. 1994. Accuracy of topographic maps derived from ERS-1 radar interferometry. *IEEE Trans. Geosci. Remote Sensing*, **32**, 823–836.

Zeitler, P.K. 1985. Cooling history of the NW Himalaya, Pakistan. *Tectonics*, **4**, 127–151.

Zhao, W. and Morgan, W.J. 1985. Uplift of Tibetan Plateau. *Tectonics*, **4**, 359–369.

Zhao, W.J., Nelson, K.D. and Project INDEPTH Team. 1993. Deep seismic-reflection evidence for continental underthrusting beneath southern Tibet. *Nature*, **366**, 557–559.

PART IV

MORPHOTECTONIC EVOLUTION IN INTRAPLATE SETTINGS

11

Geomorphological evolution of the East Australian continental margin

Paul Bishop and Geoff Goldrick

11.1 Introduction

The geomorphological evolution of the East Australian continental margin has been the subject of research since the early 20th century, and a considerable literature has accumulated in recent decades, much of it with an implicit or explicit global tectonic context (Young, 1977; Wellman, 1979a, 1987; Ollier, 1982; Veevers, 1984; Stephenson and Lambeck, 1985; Bishop, 1986, 1988; Lambeck and Stephenson, 1986; Lister and Etheridge, 1989). The major focus of this research has been the evolution of the highlands that parallel the East Australian continental margin. This margin is of particular interest because it is an example of perhaps the less common form of high-elevation passive continental margin, namely, one on which the major highland drainage divide lies inland of the crest of the great escarpment (Seidl *et al.*, 1996). It is also of considerable interest because the general tectonic stability of the margin means that its generally low rates of denudation enable the clarification of long-term margin evolution in the absence of the 'complications' of ongoing active tectonism. Indeed, this tectonic stability has enabled exploration of rates and styles, and assessment of the major models of, long-term landscape evolution.

Much of the insight into the evolution of the East Australian margin has been derived from Cenozoic valley-filling and plateau basaltic lavas that are widespread throughout the East Australian highland belt on, and about, the continental divide and the highland flanks. These lavas provide excellent control on long-term landscape evolution and probably yield the best such data for any high-elevation passive continental margin. Moreover, there is a wealth of apatite fission-track data for the margin. Despite (or because of?) this wealth of data, however, a wide range of highlands histories has been suggested and a full consensus on this matter is yet to emerge (Bishop, 1988; Ollier and Pain, 1996, and their discussants).

This chapter examines the evolution of the East Australian continental margin in the light of field evidence and the findings of numerical modelling of high-elevation passive continental margins. We focus on the tectonic and denudational histories of the principal elements of the macrogeomorphology of the margin, namely, the escarpment, the continental drainage divide, and drainage systems. We first provide a brief introduction to its geology and geomorphology before proceeding with a discussion of the overall denudation of the margin, the evolution of the escarpment, and the history of the drainage systems in plan view and long profile. We conclude with a synthesis of these data and a comparison between them and the findings of numerical modelling of the evolution of high-elevation passive continental margins (see Chapter 3).

11.2 Character of the East Australian Continental Margin

The East Australian passive continental margin formed by continental extension, rifting and break-up in the late Mesozoic and early Cenozoic during the fragmentation of Gondwana (Falvey and Mutter, 1981; Veevers, 1984; Colwell *et al.*, 1993). Break-up in the southeast, along the Tasman margin, occurred about 80 Ma BP and Tasman Sea opening lasted until

Geomorphology and Global Tectonics. Edited by Michael A. Summerfield. © 2000 John Wiley & Sons Ltd.

about 60 Ma BP when the Tasman spreading ridge became defunct; the opening of the Coral Sea, in the northeast, lasted for no more than about the first 10 Ma of the Tertiary (Weissel and Hayes, 1977; Falvey and Mutter, 1981; Veevers, 1984; Colwell *et al.*, 1993).

The margin is composed of a diversity of geological domains, consisting essentially of multiply deformed, folded and faulted Palaeozoic basement (granites and metasedimentary sequences), and intervening mildly deformed Mesozoic sedimentary basins. Generally thin Cenozoic sequences are found in inland sedimentary basins (e.g. the Murray Basin and the Great Australian Basin) and the thicker Otway and Gippsland Basin sequences on the Victorian coast, in the far southeast. Minor Cenozoic deposits are widespread throughout the East Australian highlands.

The margin is over 3000 km long and is characterized by the high elevations of the East Australian highlands close to the present coastline. The highlands take one of two general forms (Wellman, 1987) (Figure 11.1). More commonly (north of 18°S and south of 26°S (Wellman, 1987)), the highlands are of the form of a broad asymmetrical arch that rises gently from the interior lowlands to the continental drainage divide which is an area of generally low relief and low denudation rates. Low relief continues east of the divide until the crest of the great escarpment, across which the easterly (exterior or Tasman Sea) drainage falls to an erosional coastal region with relief that generally declines dramatically towards the coastline. We emphasize that the continental drainage divide lies well inland of the crest of the escarpment which can generally be recognized for the length of the continental margin (Ollier, 1982) (Figure 11.1). The continental drainage divide coincides with the top of the escarpment for only very short distances representing only a minor proportion of the total length of the highlands. The second, and less common, form of the highlands (18°S to 26°S) is more symmetrical and lies further inland (Figure 11.1).

These two morphologically distinct sectors are broadly mirrored by the continental shelf morphology (Figure 11.1). In the northern sector, a broad continental shelf–marginal plateau complex, discontinuously underlain by rift basins, lies between the shoreline and the continent–ocean boundary (COB), whereas in the southern sector the shelf is narrow and steep. The southern sector has a generally thin post-break-up sedimentary cover of presumably late Mesozoic and Cenozoic sediments (maximum of 900 m off southern New South Wales (NSW); generally less than ~600 m in the central part of the southern sector, off NSW) (Davies, 1975; Falvey and Mutter, 1981; Veevers, 1984; Colwell *et al.*, 1993). The rift valley sequence is absent from the shelf in the southern sector, and is interpreted to be attached to the opposing continental margin, the Lord Howe Rise (Jongsma and Mutter, 1978). This configuration was initially interpreted as the result of 'anomalous' asymmetrical break-up of the rift (Jongsma and Mutter, 1978), but later interpretations have incorporated this feature as a 'normal' component of the 'upper plate–lower plate' model for the development of this margin (Lister *et al.*, 1986; Lister and Etheridge, 1989).

The highlands in the southern sector are high, narrow and close to the coast and the COB; the highest elevations on the Australian continent (just over 2 km) occur in the far southeast. The highlands in the northern sector are broader, more diffuse and of lower elevation than in the south; highlands width and the distance inland from the coast of the continental drainage divide therefore broadly form a mirror image of the trend of the COB (Figure 11.1). Interestingly, the highest parts of the Queensland highlands (~1 km) are in the north where the shelf is narrowest and the COB closest to the shoreline. The continental drainage divide in Queensland is generally much lower than this maximum elevation of 1 km.

Late Mesozoic and Cenozoic volcanism was a widespread (if volumetrically insignificant) feature of the margin (Wellman, 1974; Wellman and McDougall, 1974; Johnson, 1989) (Figure 11.2). The youngest volcanism (Quaternary) occurs in the far north and far south, with the intervening region of southern Queensland, NSW and eastern Victoria exhibiting extensive evidence of Tertiary basaltic and minor intermediate

Figure 11.1 *Eastern Australia showing generalized elevations. Continental altitudes from an 11 × 11 km grid derived from gravity station altitudes with smoothed contours at 0.2, 0.6, 1.0 and 1.2 km intervals (based on Wellman, 1987, Fig. 1), and bathymetric contours at –1 –4 km depths. The location of the Great Escarpment and the continental drainage divide in East Australia are based on Ollier's (1982) Fig. 5, and the sea-floor spreading ridges and transforms are based on data given by Falvey and Mutter (1981), Veevers (1984) and Colwell* et al. *(1993). The narrow strip of oceanic crust shown between the Tasman and Coral Seas is based on Falvey and Mutter's (1981) interpretation. Veevers (1984) does not indicate this strip of oceanic crust; hence, the question marks adjoining this strip of ocean. The four guyots in the Tasman Sea are indicative of the many such southward-younging Tasmantid Guyots that are found in the Tasman Sea basin (Duncan and McDougall, 1989)*

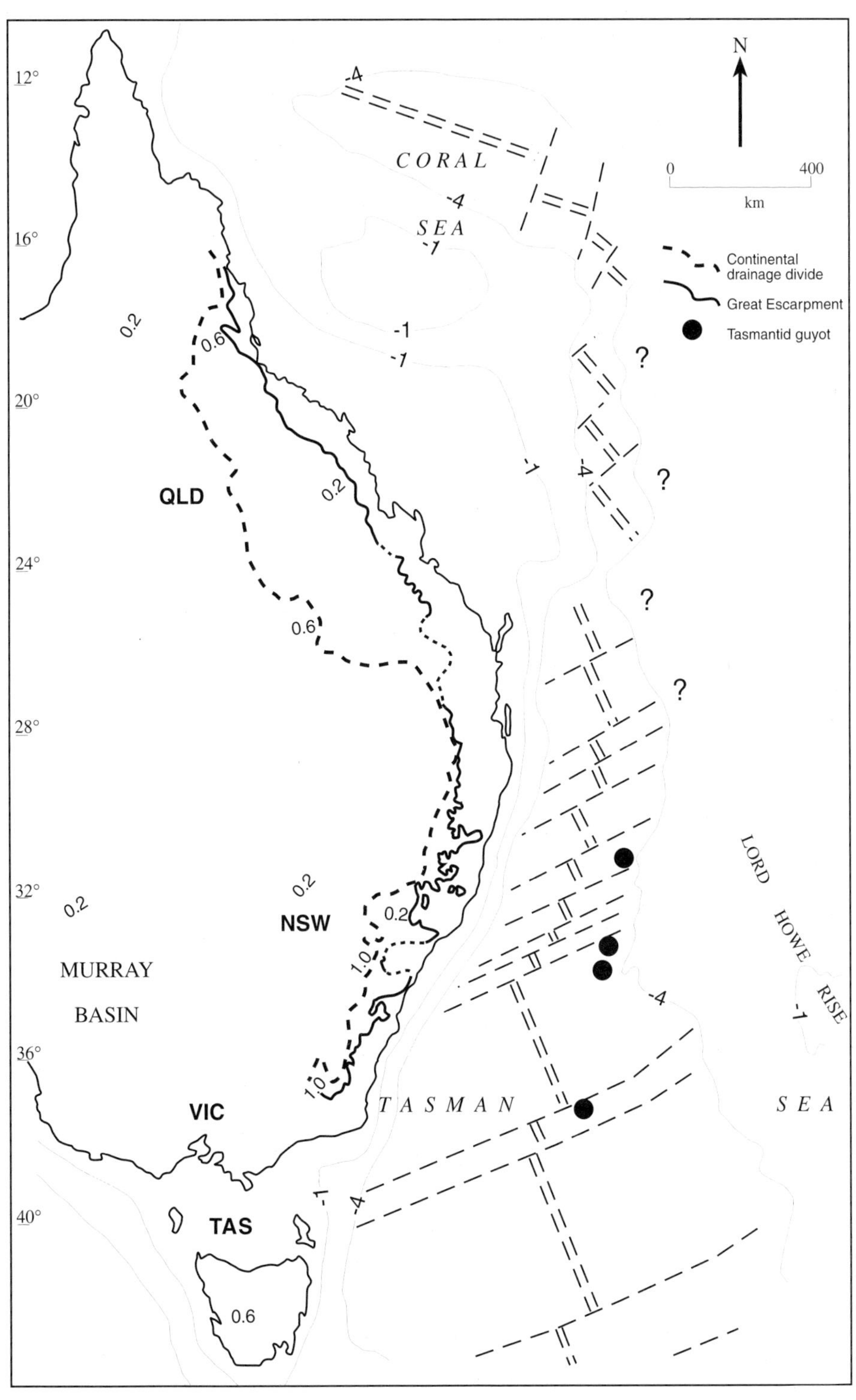
N
0
400
km
Continental drainage divide
Great Escarpment
Tasmantid guyot
12°
16°
20°
24°
28°
32°
36°
40°
CORAL
SEA
QLD
NSW
MURRAY
BASIN
VIC
TAS
TASMAN
SEA
LORD HOWE RISE
0.2
0.6
1.0
?

volcanic activity. Two broad types of eruptive style have been identified. These are the basaltic lava field provinces, the result of diffuse fissure eruptions now preserved as plateau lava fields and long, narrow, valley-filling lava flows, and the central volcanoes which consist of basaltic and more minor intermediate eruptive centres. The latter are now preserved as the dissected volcanic edifices that 'young' towards the south and reflect, as do seamounts in the Tasman Sea (Figure 11.1), the northward movement of the Indo-Australian plate over one or more hot spots (Wellman and McDougall, 1974; McDougall and Duncan, 1988; Duncan and McDougall, 1989; Sun *et al.*, 1989). Upper mantle and lower crustal xenoliths from the volcanic rocks, and seismic and gravity data from along the margin, indicate the presence beneath the highland belt of thickened crust which is generally interpreted to have resulted from underplating, either in association with rifting along the East Australian margin, or with the later (Cenozoic) volcanism (Wellman, 1987; Ewart, 1989; Lister and Etheridge, 1989; O'Reilly, 1989). The isostatic equilibrium of the highlands indicates that they are isostatically supported by, and genetically related to, this crustal 'root' (Wellman, 1979b; Young, 1989); magmatic underplating is therefore a prime candidate for the mechanism of highlands uplift, but whether this underplating is related to rifting and Tasman Sea formation or to Cenozoic volcanism (or both) remains unresolved (Wellman, 1987). Estimates of the effective elastic thickness of the southeast Australian highlands continental lithosphere are of the order of 15 km (Zuber *et al.*, 1989).

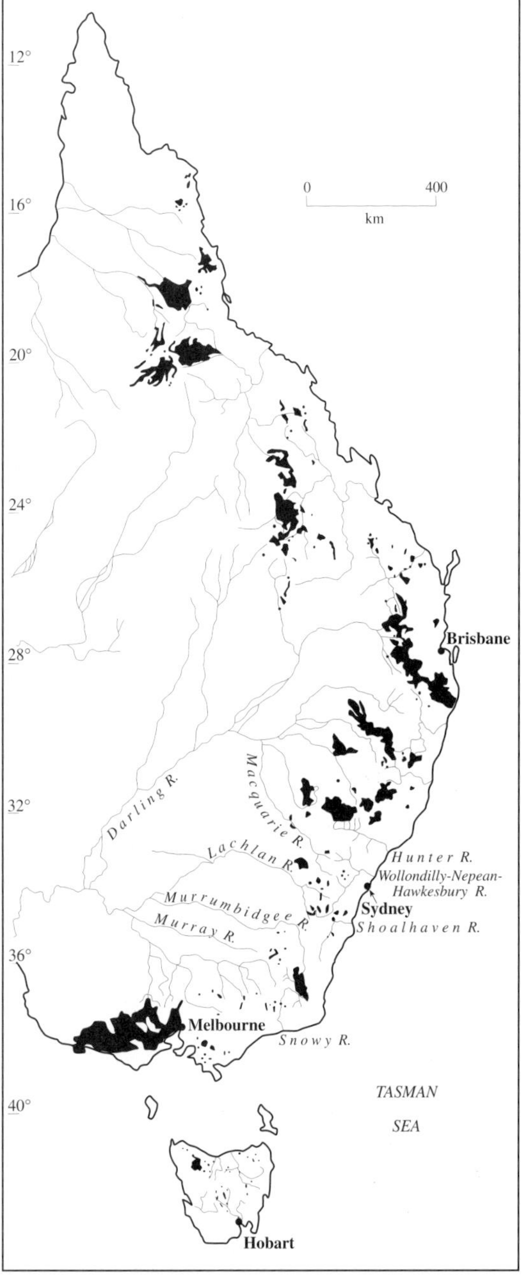

Figure 11.2 Eastern Australia showing main rivers and principal Cenozoic basaltic provinces (black) (after Johnson, 1989, Fig. 1.1.5)

The age of highlands uplift has yet to be satisfactorily resolved; it even remains unclear whether the whole of the highlands experienced the same history. It has been assumed by some (explicitly or implicitly) that the highlands have acted as a single morphotectonic entity with a single history (e.g. Veevers, 1984; Wellman, 1987), whereas others, including Jennings (1972), Bishop (1988) and Ollier and Pain (1996), have argued that such unity cannot be assumed. In any event, post mid-Triassic uplift of the highlands in NSW is demonstrated by mid-Triassic paralic sediments at 1000 m above sea level in central-eastern NSW (Wellman, 1987). Post-early Cretaceous uplift of the Queensland highlands is demonstrated by early Cretaceous sediments that were deposited close to sea level and that now cross, or are close to, the highlands crest for a significant length of the Queensland highlands (Wellman, 1987). This interpretation is consistent with the transport of large volumes of volcanogenic sediments from present offshore Queensland across the present highlands area to interior basins in the middle Cretaceous (Veevers, 1984; Wellman, 1987). Veevers (1984) has suggested that these sediments were derived from a volcanic arc that stretched the length of now-offshore East Australia prior to break-up along the Tasman and Coral Seas margins. The only evidence of this volcanic arc in the southern part of the margin is in the far southeast (eastern Victoria) in the Gippsland and Otway Basins where large volumes of late Jurassic to middle Cretaceous volcanics and volcanogenic sediments with an easterly provenance are found. Veevers (1984) was unable to decide whether the source of these sediments was a volcanic centre of restricted extent at the eastern end of the Gippsland Basin, or a centre of more general extent along present offshore NSW. Nonetheless, he argued by analogy with the Queensland margin that the major (continental) early Cretaceous drainage divide lay in the present offshore, essentially coincident with this arc system: "the Eastern Highlands were initiated around 90 Ma ago, and the crestline subsequently migrated west from an initial location at the present coastline" (Veevers, 1984, p. 126). Wellman (1987) and Ollier and Pain (1994) likewise pointed to 95–90 Ma BP as the period of establishment of the East Australian highlands in their present gross configuration. The issue remains unresolved and we return to it below (see Sections 11.4.2 and 11.6.1).

It is currently generally accepted, therefore, that highland uplift pre-dated the Cenozoic. This is consistent with the topographic positions of the highlands basaltic lavas which show that highlands relief in the southeast of the continent was at least several hundred metres, and probably higher, by the early Cenozoic (Bishop, 1988 and references therein; Taylor *et al.*, 1990; Young and McDougall, 1993; Nott *et al.*, 1996). The far southeast (the Monaro), where Paleocene lavas preserve a high-relief, cool-climate landscape, provides the best evidence for pre-Cenozoic uplift (Taylor *et al.*, 1990). Modelling by van der Beek *et al.* (1995) has indicated the need for Miocene uplift to explain the present configuration of the highlands, and Wellman (1987, 1996) argued for the possibility of some Cenozoic active uplift caused by underplating in association with the Cenozoic volcanic activity. Many authors, however, find no evidence for active Cenozoic uplift, and the most recent modelling of van der Beek and Braun (1999) does not require Cenozoic uplift.

The evolution of the highlands in the northeast of the continent has received considerably less attention. Uplift is certainly post-Cretaceous (Wellman, 1987), and Grimes (1980) has argued that the Queensland highlands experienced a series of alternating pulses of uplift and quiescence (planation) throughout the Cenozoic. Veevers (1984) finds the same cycles of minor uplift and quiescence in the southeast highlands (superimposed on overall highlands settling). It would be fair to say, however, that the interpretation of Veevers (1984) has received little support, not least because the pulses of flanking basin sedimentation that Veevers correlated with pulses of highlands uplift reflect eustatic sea level, notwithstanding his attempt to make a link between tectonism in the southeast Australian highlands and eustatic sea level (C.M. Brown, 1983).

The considerable paucity of data for the northeast part of the highlands relative to southeast Australia (related in part to the relative youthfulness of the Cenozoic volcanism in Queensland), the marked morphological differences between these two sectors of the whole East Australian highlands, and the different plate tectonic histories of their respective margins, make it inappropriate to treat the highlands necessarily as a morphological unity; it is therefore appropriate to focus this chapter on the southeastern part of the highlands fronting the Tasman margin. We therefore concentrate on southeast Australia, but make some more general closing comments about the Queensland part of the highlands.

11.3 Southeast Australian Denudation

The long-term denudation of southeast Australia is well constrained by fission-track data, the volumes of sediment in Cenozoic and Mesozoic sedimentary basins, the ages and topographic positions of weathering profiles, and the valley-filling lavas that are widespread throughout the highlands.

Rates of stream incision throughout the Cenozoic across wide areas of the highlands have been uniformly low (of the order of 1–10 m Ma^{-1}). There is order-of-magnitude correspondence between these rates of stream incision and the rates of sediment supply from the highlands to the interior sedimentary basins (Figure 11.3), meaning that Cenozoic rates of river incision may be thought of as broadly equivalent to rates of landscape denudation (Bishop, 1985). In detail, however, Cenozoic denudation has not been distributed uniformly across the landscape. Plateau stream incision and plateau denudation are notably very low in areas distant from base levels, particularly in headwater regions, even in the major river systems such as the Murray, Murrumbidgee, Lachlan and Maquarie (all interior drainage) and the Snowy, Shoalhaven and Wollondilly–Nepean–Hawkesbury (exterior drainage to the Tasman Sea) (Young, 1981; Bishop *et al.*, 1985; Taylor *et al.*, 1985; Young and McDougall, 1993).

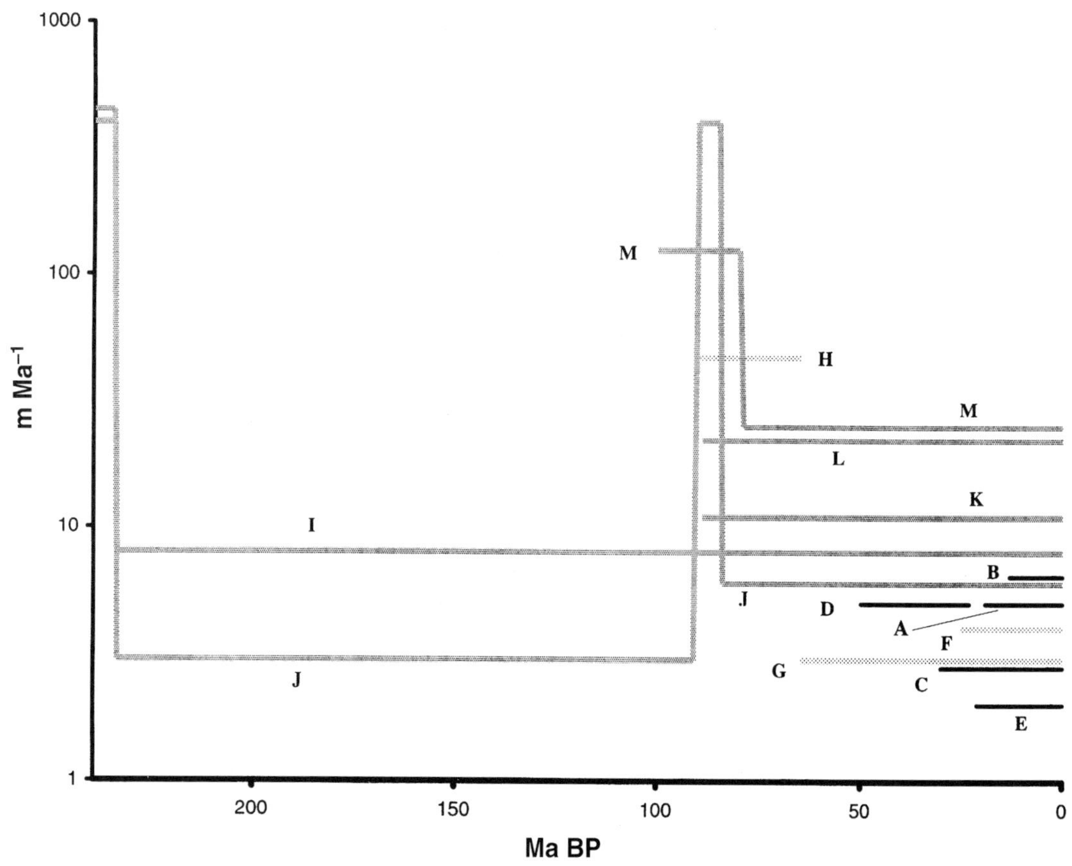

Figure 11.3 *Rates of denudation and river incision in southeast Australia. Only representative Cenozoic rates of post-basaltic incision (heavy lines A, B, C, D and E) and basin sedimentation (lightly stippled lines F, G and H) are presented from the extensive data and compilations of Young (1983), Bishop (1985), Wellman (1987), Fabel and Finlayson (1992), Fried and Smith (1992), Gale (1992), Nott (1992), Young and McDougall (1993) and Nott* et al. *(1996). See Table 11.1 for rates of gorge retreat and cliff and escarpment retreat (not included here). Denudation rates based on AFTT data (heavily stippled lines I, J, K, L and M) are derived from the discussions of Brown* et al. *(1994) and O'Sullivan* et al. *(1995, 1996)*

Conversely, rates of gorge extension into the highland mass, particularly along its seaward edge, are considerably higher, and are the most visible form of highlands denudation (of the order of 2.5 km Ma^{-1} – Young, 1983; Nott *et al.*, 1996). These rates are up to orders of magnitude greater than the rates of gorge and valley sidewall retreat (<<1 to ~30 m Ma^{-1}) (Young, 1983; Young and McDougall, 1985, 1993; Nott *et al.*, 1996). They are also substantially greater than the 1 to 10 m Ma^{-1} of stream incision and plateau lowering.

Rates of escarpment retreat in southeast Australia have been constrained by various data related to the volume of the 'wedge' eroded to produce the escarpment, and/or the distance of escarpment retreat from dated earlier positions, such as those marked by dated lavas, sediments and weathering profiles on and about the escarpment, and the escarpment's presumed starting point (at the edge of the continental shelf) (Table 11.1). It is noteworthy that the longer-term average rates, which are comparable to average rates from other passive continental margins (Ollier, 1985), hide the variability revealed by the more detailed rates from the south coast of NSW (Table 11.1).

In summary: "The major process of highlands denudation [in the Shoalhaven catchment] is gorge extension, not scarp retreat or upland surface lowering" (Nott *et al.*, 1996, p. 230). This conclusion may be applied to much of southeast Australia: all three of the major exterior drainage systems in southern NSW exhibit gorges cutting back into the highlands, as do

Table 11.1 *Rates of escarpment retreat and land lowering below the escarpment for various lengths of the escarpment in New South Wales over various time intervals since 100 Ma BP (approximating the age of Tasman rift margin formation)*

Erosion type	Region	Reference	Method	Age range	Rate
(i) Escarpment retreat	Southern NSW	Young and McDougall (1982); Nott *et al.* (1991, 1996)	Distance of escarpment retreat from dated rocks on coastal plain	Neogene	<200 m Ma^{-1}
	Northern NSW	Seidl *et al.* (1996); Weissel and Seidl (1997)	Distance from the escarpment to the continental edge and age of break-up	100 Ma	2 km Ma^{-1}
	NSW	Based on data in Ollier (1982, Fig. 5)	Distance from the escarpment to the continental edge and age of break-up	100 Ma	1–2 km Ma^{-1}
(ii) Land lowering seaward of escarpment	Central NSW	Based on data in Li *et al.* (1996); Ollier and Pain (1996)	Volume of eroded prism between escarpment and coastal plain	100 Ma	1–5 m Ma^{-1}
	Southern NSW	Based on data in Seidl *et al.* (1996)	Volume of eroded prism between escarpment, former plateau surface and coastal plain (to continental edge)	100 Ma	5–10 m Ma^{-1}

Rates were calculated on the basis of (i) the distance that the escarpment has retreated beyond a datable feature, and (ii) the volume of the prism eroded to form the escarpment. In effect, these are two different ways of considering the same quantity. Note that volumes of offshore sediment (either on the Tasman Sea continental shelf (Ollier and Pain, 1994) or on the continental shelf and in the Tasman Sea abyssal plain (Seidl *et al.*, 1996)) were not used for these calculations because of uncertainties as to the source area for these sediments, both in terms of the proportion of the sediments that was derived from catchments landward of the escarpment, and the proportion of sediment transferred by marine processes to the Tasman Sea abyssal plain from adjacent ocean basins. Likewise, the amounts of continental margin denudation that are inferred from apatite fission-track thermochronology (AFTT) are not considered in this table (see Figure 11.3, and discussion of AFTT data in text)

the major systems dissecting the escarpment in northern NSW (Seidl *et al.*, 1996; Weissel and Seidl, 1997). Thus, gorge extension accounts for the major dissection of the highland mass in southeast Australia, but it should be noted that widespread stream incision and plateau lowering, albeit at relatively low rates, may produce larger volumes of sediment than does spatially restricted gorge extension, even though it occurs at much higher rates. That is, gorge extension is associated with the highest local erosion rates but not necessarily the highest rates of sediment production and delivery to sedimentary basins, which is a central concern in understanding sedimentary basin development on passive continental margins (e.g. Rust and Summerfield, 1990; Seidl *et al.*, 1996).

The consistency of the southeast Australian data on rates of denudation and escarpment retreat should not be taken to indicate unanimity on the details of the denudational development of southeast Australian margin coastal strip. This debate is currently concerned with two major and related issues: (i) did escarpment retreat involve dissection of an upland plateau surface (an elevated rift shoulder) (Young and McDougall, 1982, 1985; Nott and Purvis, 1995, 1996) or a downwarped rift shoulder, with the coastal plain incorporating portions of this downwarped rift shoulder (M.C. Brown, 1983, 1996; Ollier and Pain, 1994; Orr, 1996; Seidl *et al.*, 1996) (Figure 11.4) and (ii) what is the total depth of Mesozoic and post-Mesozoic denudation along the continental margin (Moore *et al.*, 1986; Dumitru *et al.*, 1991; Nott and Purvis, 1996)? We turn to these two points now.

11.3.1 ESCARPMENT RETREAT ACROSS A DOWNWARPED COASTAL STRIP?

The implications of Nott and Purvis's (1996) reporting of early Cretaceous (~100 Ma) lavas close to sea level below the escarpment on the NSW south coast are clear: if the lavas were erupted close to their present elevations, they must post-date escarpment retreat across this part of the margin, meaning that escarpment

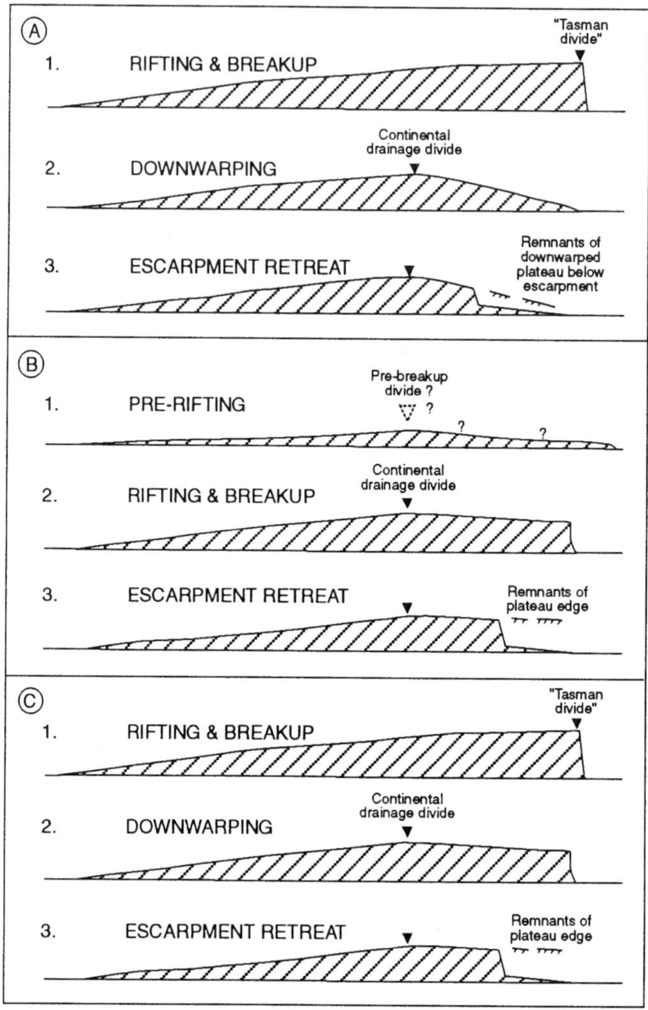

Figure 11.4 *Three possible evolutionary histories for the southeast Australian continental margin. (A) is the history suggested by Ollier and Pain (1994, their Fig. 5); (B) is an alternative history that does not include downwarping of the continental margin and associated westward migration of the continental drainage divide; and (C) is a combination of the two interpretations. See text for discussion*

formation and retreat significantly pre-dated break-up along the Tasman margin (~80 Ma BP). If, on the other hand, the lavas were erupted at higher altitudes and subsequently downwarped to their present elevations, the chronology of break-up and escarpment retreat seems more reasonable in terms of the accepted understandings of rifting, break-up and escarpment initiation.

The dips required to downwarp the plateau surface close to sea level are low and are not unreasonable (M.C. Brown, 1983, 1996; Orr, 1996). Support for the interpretation that the escarpment has been eroded into a downwarped rift shoulder (Figure 11.4A) comes from the fall in the highest elevations (interfluves) from the escarpment front seawards across the dissected coastal strip to the present coast (Pain and Ollier, 1986; Orr, 1996; Seidl *et al.*, 1996), and from modelling of the post-break-up subsidence of the Tasman margin due to flexural coupling across the continental margin between the continental lithosphere and the subsiding oceanic lithosphere (Seidl *et al.*, 1996). These data pertain particularly to the NSW mid-south and mid-north coast parts of the margin formed largely in folded Palaeozoic basement rocks.

Important data that indicate that there has been very little or no downwarping of the continental margin (Figure 11.4B) include the essentially horizontal or gently westerly dipping attitude of the very prominent middle Triassic Hawkesbury Sandstone in the more northerly part of the NSW south coast (Young, 1977; Young and McDougall, 1985; Nott and Purvis, 1996), and weathering profiles that have been dated to the late Mesozoic on the south coast (Young et al., 1996). The dating of these weathering profiles is not universally accepted (e.g. Paton *et al.*, 1989) and, in any case, their topographic position below the escarpment means that the profiles can be interpreted as remnants of the downwarped plateau surface (Ollier and Pain, 1994).

This issue remains unresolved and it is possible that a combination of the two evolutionary sequences occurred (Figure 11.4C) and/or that the evolutionary sequence has varied along the margin. Elevations are nowhere particularly great and quite small dips (1–2°) can account for substantial changes in elevation across the coastal strip, as would be argued by the proponents of the coastal downwarping hypothesis (M.C. Brown, 1983, 1996; Ollier and Pain, 1994; Seidl *et al.*, 1996). However, the essentially horizontal, or gently westerly dipping, attitude of the upper Sydney Basin strata (middle Triassic) as they cross the plateau surface and approach the escarpment top poses considerable problems for interpretations favouring downwarping. Young and McDougall (1983, p. 249) showed that the dip of the plateau-forming Hawkesbury Sandstone near the mouth of the Shoalhaven River (Figure 11.2) is essentially horizontal, with the plateau terminating close to the present coast (see also Young and McDougall, 1985). Indeed, for the 175 km of central NSW coastline between the Shoalhaven and Hunter Rivers (Figure 11.2), the upper Sydney Basin strata (middle Triassic) are either essentially horizontal or dip very gently to the west (Department of Mineral Resources, 1983, 1985, 1997). These dips are almost certainly not syn-depositional (none of these strata exhibits palaeocurrent directions from the eastern hemisphere – Conaghan *et al.*, 1982) and they are therefore firm evidence against the coastal downwarping claimed by Ollier and Pain (1994) and Brown (1996) in central coastal NSW. Moreover, these dips are almost certainly more reliable indicators of crustal structure than are the elevations of interfluves across the coastal plain (Orr, 1996; Seidl *et al.*, 1996), but it should be noted that the arguments used by Orr (1996) and Seidl *et al.* (1996) are for areas outside the Sydney Basin, leaving final resolution of the issue impossible at this stage and allowing for varying post-break-up subsidence histories along the coast.

In summary, rates of escarpment retreat in southeast Australia have been deduced from theoretical considerations and some limited field data, but the morphology of the rift flank into which the escarpment retreated is still being debated. Data from apatite fission-track thermochronology (AFTT) (see Chapter 4) in southeast Australia are another important element in understanding the evolution of the margin, especially in light of the Mesozoic weathering profiles and 100-Ma-old sub-aerially erupted lavas reported on the coastal plain of southern NSW (Nott and Purvis, 1995, 1996). We turn to these AFTT data now.

11.3.2 APATITE FISSION-TRACK THERMOCHRONOLOGY DATA AND MARGIN DENUDATION

The large southeast Australian AFTT data set that is available indicates kilometre-scale denudation of the southeast Australian highlands and adjacent coastal region at varying rates from the late Mesozoic to the present (Figure 11.3) (Moore *et al.*, 1986; Dumitru *et al.*, 1991; Brown *et al.*, 1994; O'Sullivan *et al.*, 1995, 1996). The late Mesozoic pulse in denudation that is indicated by both basin sedimentation data (Figure 11.3, line H) and the AFTT cooling age data (Figure 11.3, lines M and J) is taken to indicate major uplift and denudation of southeast Australia coincident with pre-break-up extension and rifting along the Tasman margin. Coal vitrinite reflectance data from the Sydney Basin and palaeomagnetic overprints in southeast NSW are also consistent with this interpretation (Schmidt and Embleton, 1981; Middleton and Schmidt, 1982).

The regional pattern of AFTT ages across the highland belt to the present coastline is taken to indicate that late Mesozoic denudation was deepest (up to 4 km) along the coastal strip (Moore *et al.*, 1986; Dumitru *et al.*, 1991; Brown *et al.*, 1994) (Figure 11.3, lines K, L and M). However, deeper regional denudation along the coastal strip than along the highlands proper is difficult to reconcile with the 100-Ma-old lavas and Mesozoic weathering profiles on the NSW south coast and the essentially horizontal attitude of the middle Triassic Sydney Basin strata as they pass from the plateau surface across the escarpment and coastal strip to the coastline. And this generally horizontal attitude of the Sydney Basin strata is certainly inconsistent with the denudational rebound that would be expected of such kilometre-scale differential denudation along the coastal strip. The latest modelling of the evolution of the Tasman margin by van der Beek and Braun (1999) predicts only about 400 m of isostatic rebound at the coast, increasing to a maximum of ~800 m at ~80 km inland, which in turn predicts sub-horizontal dips in the middle Triassic Sydney Basin strata. This interpretation is also consistent with the arguments of Branagan (1983).

It should also be noted that the regional pattern of AFTT ages, and the denudation that they imply, cannot be taken to provide a detailed history of any particular locality among the margin. The 100-Ma-old lavas and Mesozoic weathering profiles on the NSW south coast, for example, are inconsistent with deep denudation having occurred along and across the whole of the coastal strip below the escarpment, and there are many areas throughout the southeast Australian highlands where landscape elements that date from the Mesozoic have been identified (Young 1981; Bird and Chivas, 1993; Twidale, 1994; Twidale and Campbell, 1995; Hill, 1999). Hill (1999) has argued that variations in relief in the highlands and across the coastal strip below the escarpment, and the associated differential preservation of Mesozoic landscape elements, may explain the apparent conundrum of kilometre-scale denudation in the late Mesozoic and the preservation of Mesozoic landscape elements. Elevated geothermal gradients would also assist in minimizing the amounts of denudation required to expose late Mesozoic fission-track ages at the present ground surface, but there is currently a strong consensus among AFTT researchers that geothermal gradients along the southeast Australian continental margin were probably not significantly greater than 25°C km^{-1} to 30°C km^{-1} in the late Mesozoic (Dumitru *et al.*, 1991; Brown *et al.*, 1994).

11.3.3 DENUDATION AND MARGIN REBOUND

Results of investigations of flexural isostatic rebound of the East Australian margin in response to escarpment retreat (e.g. Gilchrist and Summerfield, 1990) are yet to be published. The very low angle westerly dip of the middle Triassic Hawkesbury Sandstone (see Section 11.3.1) is perhaps consistent with such flexural isostasy. For the Cenozoic, rates of escarpment retreat have been low (see Section 11.3) (Table 11.1) suggesting that amounts of Cenozoic flexural isostasy would, in any case, be expected to be low. Any flexural isostasy must be consistent with middle Miocene estuarine deposits about 30 m above present sea level on the coast near Sydney (Pickett *et al.*, 1997) and the ages and elevations of the Cenozoic sub-aerial lavas and other deposits on the coastal plain of southern NSW (Young and McDougall, 1982; Nott *et al.*, 1991, 1996). Unfortunately, the ages of these latter (non-estuarine) materials, and/or their relationship to sea level at the times of formation or deposition, may be quite imprecise, diminishing somewhat their value for testing of models of flexural isostasy due to escarpment retreat.

Loading of the continental shelf by sediment can, in principle, also lead to flexural tectonics on the sub-aerial part of a margin (Summerfield, 1991; Pazzaglia and Gardner, 1994). There seems to be little evidence of this effect on the southeast Australian margin, probably reflecting the southeast Australian continental shelf's low flexural moment (the result of the narrowness of the continental shelf and the generally low Cenozoic rates of sediment loading of the shelf).

Cenozoic (Neogene) isostatic rebound via faulting and/or flexure in response to denudation of the highlands as a whole (and not just that due to escarpment retreat) has been inferred from both modelling and field evidence from the western edge of the highlands in central NSW (Bishop and Brown, 1992). It was concluded that passive denudational rebound (rather than active uplift) has been operating because modelling of flexural behaviour using the known rates of highland denudation and reasonable values of lithospheric properties can reproduce the amounts of tectonic offset observed in the field (Bishop and Brown, 1992). It is noteworthy, in terms of the discussion in the preceding paragraphs, that the Neogene denudation has a grossly regional character, as indicated by rates of river incision and sediment supply to adjacent basins. These rates of denudation are low, relative to denudation rates in glaciated regions or more tectonically active areas (Young, 1983), but they result in rebound where lithospheric properties are appropriate.

11.4 Southeast Australian Rivers: Planform

The evolution of river systems on passive continental margins is important for several reasons. Continental margin drainage networks (river systems in plan view) may provide data on the broad palaeogeographic history and gross tectonic framework of the margin, including the history of the continental drainage divide (e.g. Gardner and Sevon, 1989), and detail on the history of sediment supply to the margin and to interior basins (e.g. Dingle and Hendey, 1984; Bishop, 1985;

Table 11.2 *Selected East Australia localities showing parallelism between modern and ancient drainage directions as deduced from Cenozoic valley-filling lavas*

Locality and drainage (noting Tasman Sea or interior)	Age of ancient drainage (from K–Ar dates on valley-filling lavas)	Reference
Central eastern Victoria (Tasman)	Oligocene	Wellman and McDougall (1974)
Far eastern Victoria (Tasman)	Eocene	Li (1995)
Southern NSW (interior)	Late Oligocene to early Miocene	Young and McDougall (1993)
Southern NSW (Tasman)	Paleocene	Taylor *et al.* (1985, 1990)
South central NSW (interior)	Late Oligocene to early Miocene	Bishop *et al.* (1985); Bishop (1986); Bishop and Brown (1992)
South central NSW (Tasman)	Middle and late Oligocene	Young (1977); Ruxton and Taylor (1982); Bishop (1986); Nott (1992); Nott *et al.* (1996)
Central eastern NSW (interior)	Early to middle Miocene	Dulhunty (1973)
Central eastern NSW (Blue Mountains) (Tasman)	Early to middle Miocene	Wellman and McDougall (1974)
Northeastern NSW (interior)	Oligocene to Miocene	Fried and Smith (1992)
Northeastern NSW (Tasman)	Oligocene to Miocene	Fried and Smith (1992)

Rust and Summerfield, 1990; Gilchrist *et al.*, 1994). The continental margin drainage systems might also be expected to record post-break-up flexural tectonics if these deformations are sufficient to reorganize the drainage net (e.g. Summerfield, 1991; Driscoll and Karner, 1994). Many numerical models of the long-term evolution of high-elevation passive continental margins incorporate flexural rebound tectonics that modify drainage systems in plan view and generate migration of major drainage divides, including the continental divide (e.g. Gilchrist *et al.*, 1994; Kooi and Beaumont, 1994). The various models for Tasman margin rifting, break-up and sea-floor spreading have different implications for the long-term history of the continental divide and the major river systems (Ollier, 1978, 1982; Bishop, 1986, 1988; Young, 1989; Ollier and Pain, 1994, 1996). The bedrock long profiles of continental margin drainage systems and their deposits may also be used to interpret tectonic history via the ways in which relative changes in base level, reflecting sea-level changes or tectonics, are reflected in the long profiles (e.g. Pazzaglia and Gardner, 1994).

11.4.1 CENOZOIC VALLEY-FILLING LAVAS AND SOUTHEAST AUSTRALIAN DRAINAGE HISTORY

Parallelism between Cenozoic and modern flow directions is virtually universally reported from throughout the East Australian highlands in southeast Australia (Table 11.2); the few exceptions occur in situations where complete filling of the pre-basaltic valleys has resulted in broad lava fields that have significantly modified the drainage patterns, as in, for example, parts of northern and southern NSW (Taylor *et al.*, 1990; Fried and Smith, 1992).

It is also clear from the topographic positions of the valley-filling basalts, deep within their valleys, that many of the valleys were established long before the lavas flowed into them. The late Oligocene/early Miocene lavas in the Lachlan and Wollondilly valleys, for example, lie up to ~100 m above the present valley bottom and about 400 m below the interfluves of the catchment boundary (Bishop *et al.*, 1985; Bishop, 1986). This means that the valleys must date from the early Cenozoic at least, and probably the late Mesozoic (Bishop, 1986; Ollier and Pain, 1994, 1996). The topographic positions of Cenozoic valley-filling lavas throughout southeast Australia show that essentially the same conclusion can be drawn for the whole of southeast Australia (see references in Table 11.2, and the sub-basaltic data used by Wellman (1979a)). In short, the valley-filling lavas throughout southeast Australia indicate persistence of drainage directions from the latest Mesozoic or early Cenozoic.

11.4.2 MESOZOIC DRAINAGE

The flow directions of Mesozoic drainage are more controversial than those from the Cenozoic. As we have seen, Veevers (1984) argued that the Mesozoic drainage divide lay to the east of the present continental margin, coincident with a volcanic arc. Early in the 20th century, Taylor (1911) hypothesized from planform river channel patterns that the continental drainage divide in southeast Australia formerly lay to the east (offshore) of the present coastline. This divide migrated inland to the present position of the continental divide as a great 'crustal wave' that disrupted and rearranged the river systems that now lie east of the continental divide (the Tasman drainage). The main support for this interpretation came from the classic evidence of drainage rearrangement, including elbows of capture, wind gaps, high-level fluvial gravels and the so-called 'barbed' drainage. Taylor's (1911) interpretations were incorporated by Smith (1979) into early Atlantic-type passive margin models of continental rifting and break-up along the East Australian margin (e.g. Dewey and Bird, 1970; Falvey, 1974). These early interpretations used the Atlantic-style models' hypothesized active uplift along the rift axis, and the subsequent widening of the rift valley and post-break-up migration of drainage divides away from the rift axis, to account for the supposed inland migration of the continental drainage divide in East Australia.

Ollier and co-workers have argued similarly that the major drainage divide (equivalent to a 'continental' drainage divide) lay to the east of the present continental margin prior to rifting and break-up (Ollier and Pain, 1994) (Figure 11.4A). In this scenario (Ollier and Pain, 1994, 1996), all southeast Australian rivers, including the present easterly flowing (Tasman) rivers, flowed to the west. Downwarping of the new continental margin in association with rifting and break-up led to the formation of a new continental drainage divide inland of the margin and the onset of escarpment retreat into the downwarped plateau. Ollier and Pain 1994, 1996) argued that the easterly drainage to the new continental margin (the new Tasman drainage) was formed by the diversion, capture and reversal of the systems that formerly flowed to the west. This produced the various planform characteristics, especially elbows of capture and barbed drainage, that are taken to be evidence of drainage re-arrangement.

Ollier and Pain (1994, 1996) argued that these former drainage directions, hypothesized on the basis of the planform characteristics of the modern drainage networks, must be of Mesozoic age because they must pre-date the Cenozoic drainage directions recorded by the valley-filling lavas. They claimed that Mesozoic westerly flow from a continental divide to the east is demanded by the presence of dry land (the Lord Howe Rise) to the east of the present southeast Australian continental margin, prior to break-up, and the notion that rivers could not therefore have drained to the east (Ollier and Pain, 1996) (Figure 11.1). Palaeocurrent data from the pre-break-up middle Triassic Sydney Basin in central eastern NSW (Conaghan *et al.*, 1982) demonstrate unequivocally, however, that easterly drainage operated prior to continental break-up and Tasman Sea formation, and that the presence of land to the east of the present continental margin is not an *a priori* argument against easterly drainage.

The sole fluvial deposit of possible Mesozoic age that lies east of the divide and has the potential to elucidate this debate is the poorly dated Rickabys Creek Gravel which drapes the Lapstone Monocline, a major tectonic structure at the eastern edge of the highlands inland from Sydney (Bishop, 1982, 1986). Palaeocurrent and compositional data demonstrate that this gravel was deposited by easterly flow from the area of the present continental divide in an ancient precursor of the major east-flowing Wollondilly–Nepean–Hawkesbury river system (see Figure 11.2 for location of Wollondilly–Nepean–Hawkesbury River). Unfortunately, the gravel's age(s) and its structural relationships with the underlying Lapstone Monocline remain equivocal (Bishop, 1982, 1986; Bishop and Hunter, 1990; Pickett and Bishop, 1992). Palaeomagnetic data demonstrate that there was significant movement on the monocline between ~90 Ma and ~15 Ma BP (Bishop *et al.*, 1982; Schmidt *et al.*, 1995). Schmidt *et al.* (1995) have argued that this movement was related to precursor events leading to rifting in the Tasman Sea and that it must therefore have occurred soon after 90 Ma BP. If the gravel is deformed by this monoclinal folding, the gravel must pre-date the Mesozoic break-up of the Tasman Sea. And even if the gravel makes an angular unconformity with (i.e. pre-dates) the monocline, a Mesozoic age of the gravel's oldest units is entirely possible (Pickett and Bishop, 1992).

The debate about the Mesozoic drainage systems of southeast Australia remains polarized and unresolved, partly because of the meagreness of good chronological control. Many are unwilling to accept the supposed evidence of barbed drainage and elbows of capture, and the very multiplicity of former connections between easterly and westerly drainage that have been hypothesized to have been disrupted by the supposed drainage re-arrangement (e.g. Taylor, 1911; Ollier, 1978; Ollier and Pain, 1994) points to the difficulties of assessing whether the reconstructions can be trusted (Young, 1977; Bishop, 1982, 1995). Moreover, the supposed downwarping of the southeast Australian margin said to be responsible for the disruption and reorganization of the drainage east of the continental divide seems unlikely on the basis of the structural dips in the middle Triassic Sydney Basin strata (see Section 11.3.1). It is not even unequivocal that Pain and Ollier's (1986) supposedly downwarped surfaces capped by Cenozoic basalts east of the escarpment on the NSW north coast actually represent a remnant of the former plateau surface. Seidl *et al.*'s (1996) 'confirmation' of plateau downwarping by best-fitting of a surface to interfluves (supposed plateau remnants) east of the escarpment reflects to a considerable extent the very high statistical weighting given to these basalt-capped surfaces.

The uncertainties concerning the now-disrupted Mesozoic drainage directions are compounded when the considerable thickness of crustal section that AFTT indicates has been removed by post-break-up denudation is considered. The details of these uncertainties have been discussed more fully elsewhere (Bishop, 1995); in summary, it seems unlikely to us that the details of drainage directions would necessarily have been preserved during the differential denudation of up to several kilometres of crustal section. The requirement that subtle geomorphological features, such as wind gaps and divide-crossing gravels, be preserved from the Mesozoic drainage systems (e.g. Haworth and Ollier, 1992), seems difficult to meet given the depths of post-Mesozoic denudation indicated for the margin (see Section 11.3.2), unless, of course, these features represent fragments of Mesozoic landscape that have escaped denudation. More detailed (and opposing viewpoints) on this unresolved issue are given by Bishop (1995) and Ollier and Pain (1994, 1996).

11.5 Southeast Australian Rivers: Long Profiles

11.5.1 CENOZOIC PROFILES

The middle reaches of Cenozoic (generally sub-basaltic) river profiles in southeast Australia generally lie above the modern river systems as a result of post-basaltic incision (Figure 11.5). A marked decrease in the amount of Cenozoic (post-basaltic) incision is found in the headwaters of the major systems, with the headwater reaches rising on broad lava fields that are locally associated with the continental divide (Figure 11.5).

Where sub-basaltic surfaces have been examined in detail, the Cenozoic and modern long profiles are commonly grossly similar in form. Bishop *et al.* (1985) and Young and McDougall (1993) interpreted these similarities to be indicative of slow post-basaltic incision with only minor, if any, tectonic disturbance; ongoing active highlands uplift, such as was invoked by Wellman (1979a), was argued to be unnecessary to account for the observed post-basaltic incision. Indeed, whether the considerable similarities between the modern and sub-basaltic long profiles are interpreted in terms of general adjustment of long profiles to lithology (Bishop *et al.*, 1985), or in terms of a less specific persistence of steeper and gentler reaches (Young and McDougall, 1993), the similarities have been interpreted to indicate a lack of tectonic disturbance in post-basaltic times. It therefore remains to clarify the role in long profile development of denudational rebound, such as was inferred by Bishop and Brown (1992) for the Lachlan catchment. We address this issue now.

11.5.2 MODERN RIVER LONG PROFILES

11.5.2.1 Introduction

The identification of deformation and anomalous gradients in fluvial long profiles is a well-established technique for the analysis of tectonism, both local and regional, long-term and neotectonic (Seeber and Gornitz, 1983; Keller and Rockwell, 1984; Merritts *et al.*, 1994; Pazzaglia and Gardner, 1994; Goldrick and Bishop, 1995). For our present purposes, long profile characteristics provide a most useful way of assessing the impact of Cenozoic passive tectonism on the drainage networks of southeast Australia and the rates at which such tectonism is accommodated by the drainage system.

The key to this type of analysis is the identification of the equilibrium profile, with departures from this equilibrium (generally steepening) being interpreted as the result of relative base-level change (generally in the form of tectonic activity). The general concave-up form of the equilibrium profile has been modelled based on a 'theory-free' least-squares best-fit to the elevation–distance long profile (Jones, 1924) or, more commonly, based on an understanding of the theoretical form of the equilibrium long profile (e.g. Hack, 1975). Hack (1975) modelled the equilibrium long profile as a semi-logarithmic relationship between elevation (normal) and distance (logarithmic), and this is the most common form of the equilibrium long profile in current use (e.g. Reed, 1981; Seeber and Gornitz, 1983; Keller and Rockwell, 1984; Bishop *et al.*, 1985; McKeown *et al.*, 1988; Goldrick and Bishop, 1995). Goldrick and

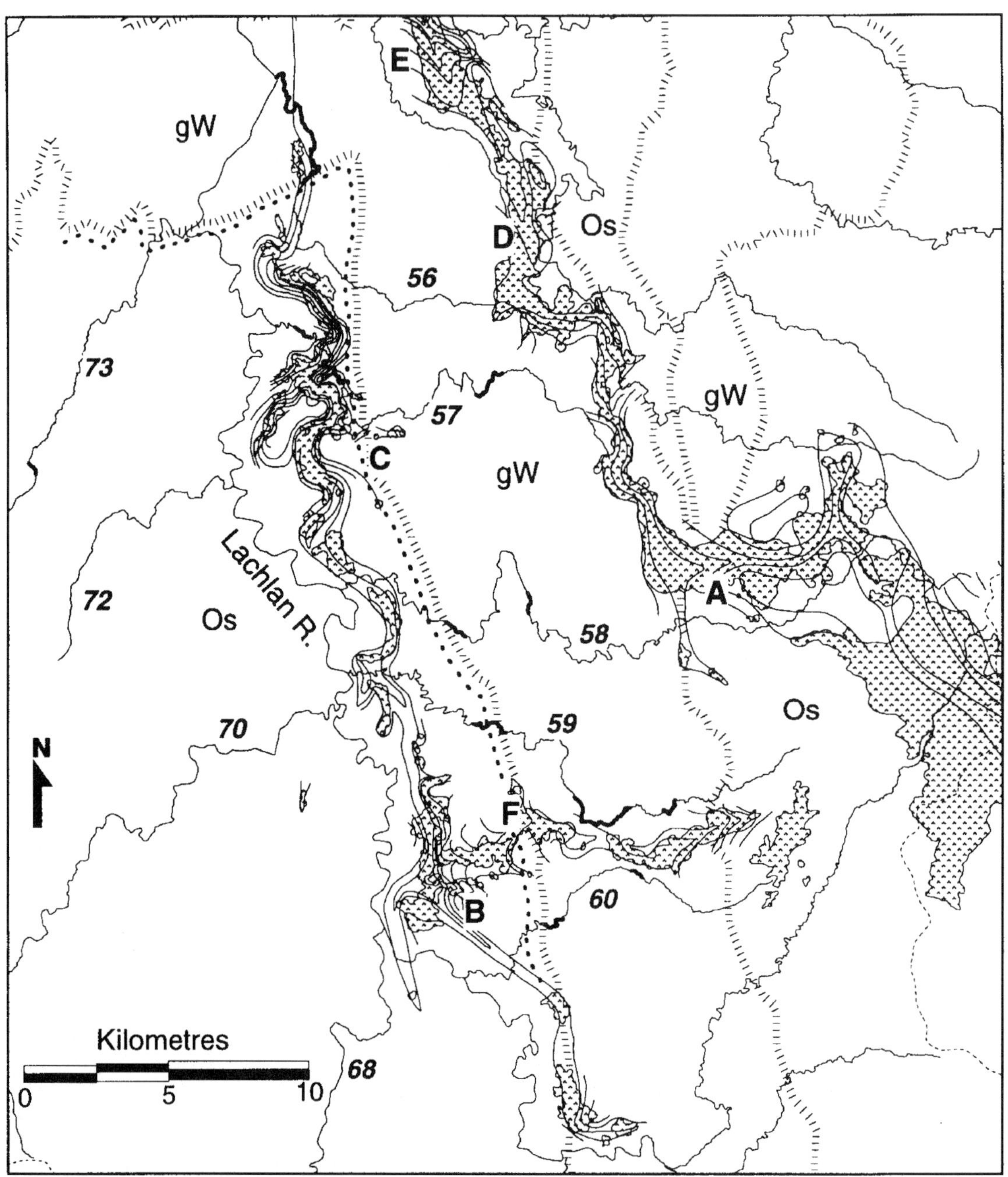

Figure 11.5 Upper Lachlan early Miocene basalts, reconstructed sub-basaltic surfaces with 20 m contour interval, and country rock lithologies (Bishop et al., 1985). Country rock lithologies: gW: granitic rocks of the Wyangala Batholith; Os: Ordovician metasediments. Dots along the geological boundary indicate the prominent hornfels ridge (see text). Numbers are stream numbers for modern drainage net. The reaches highlighted by a thickened stream line are disequilibrium reaches corresponding to peaks in the DS plots (Figure 11.8) and the upper case letters (A to F) are the localities of steepening of the sub-basaltic surface. Location of this figure is given in Figure 11.7

Bishop (1995) used this Hack form to identify disequilibria (oversteepening) in profiles that have been deformed by denudational rebound.

A more general equilibrium long profile than that used by Hack (1975) is possible and is used here to assess long profile disequilibria in the Lachlan valley, central NSW, the only system in the southeast Australian highlands in which Cenozoic tectonism has so far been investigated in detail (Bishop and Brown, 1992; Goldrick and Bishop, 1995). The passive rebound uplift reported by Bishop and Brown (1992) and Goldrick and Bishop (1995), on the basis of modelling and field data from the transition zone between the highlands and the interior alluviated lowlands, is also evident in the detail of long profiles of the Lachlan's Cenozoic terraces in this transition zone (Figure 11.6). These data suggest ongoing uplift into the Holocene.

11.5.2.2 Summary of a more general equilibrium river long profile form

We present in the Appendix to this chapter the theoretical derivation of the equilibrium form of the long profile that is more general than the semi-logarithmic form proposed by Hack (1975). In summary, it is a condition of equilibrium or grade in this more general form that a plot of the logarithm of slope against the logarithm of distance downstream should describe a straight line which we name a DS plot. For stream reaches which plot as a straight line on a DS plot, the following formulation of the long profile can be deduced (see Appendix):

$$H = H_0 - k\frac{L^{1-\lambda}}{1-\lambda} \tag{1}$$

where H and L are channel elevation and distance downstream, respectively, and H_0, λ and k are constants. The equilibrium long profile can be projected downstream to deduce the amount of offset of the long profile that can be attributed to tectonic activity (see Appendix for a more detailed explanation).

11.5.2.3 The long profile characteristics of the Lachlan Basin drainage net

The long profiles of 110 streams in the Lachlan's highland (bedrock) tract (Figure 11.7) were digitized from the 'blue lines' on 1:100 000 and 1:50 000 topographic sheets with a 20 m contour interval. Of the 110 streams digitized, only 23 exhibit linear DS plots with essentially no peaks or no changes of slope or intercept. All 23 uniformly graded streams cross geological boundaries, indicating that lithology is, at most, only a second-order control on long profile form, a conclusion that is consistent with findings on the lithological control of the cross-sectional dimensions of valleys (Bishop and Cowell, 1997).

Sixteen of these 23 streams, which we term here uniformly graded to convey the notion that their profiles exhibit a linear DS plot irrespective of changes in substrate lithology, are concentrated in the west of the catchment, largely downstream of the zone of passive tectonic activity marking the transition from the highlands to the interior (Bishop and Brown, 1992; Goldrick and Bishop, 1995). The remaining seven uniformly graded streams are in the upper catchment, with the majority (five) of these lying to the west of a prominent hornfels ridge at the granite–metasediment contact east of the Lachlan River (Figures 11.5 and 11.8). The widespread occurrence of disequilibrium, and the preferential location of uniformly graded streams downstream of the passive tectonic offset, are consistent with the hypothesis of ongoing rejuvenation at the western edge of the highlands and upstream propagation of knickpoints.

In the upper Lachlan, where the Miocene valley-filling basalts permit further elucidation of the response of the drainage net to rejuvenation, there is a marked left-bank–right-bank asymmetry in the long profiles of the Lachlan's modern tributaries (Figure 11.8). The left-bank tributaries are either uniformly graded (linear DS plot) or characterized by one disequilibrium reach. In the left-bank streams that exhibit a disequilibrium reach (streams 68 and 73), projection of the upstream graded reach to the modern Lachlan, using the method of Goldrick and Bishop (1995) and the more general DS model, indicates a maximum long profile offset of about 15 m (Figure 11.9A).

All right-bank tributaries exhibit marked discontinuities in their DS plots with at least one major disequilibrium reach, upstream of which are one or more equilibrium reaches (Figure 11.8). The sub-basaltic contouring of the Miocene tributaries is sufficiently detailed in the case of the basalt-filled tributary that lies between streams 59 and 60 (Figure 11.5) to allow reconstruction of a DS plot; this DS plot also shows a disequilibrium reach in the Miocene tributary (Figure 11.8). Complications which will be explored in detail elsewhere prevent the construction of DS plots for other basalt flows, but the anomalously close spacing of the sub-basaltic contours in these Miocene streams indicates that discontinuities in the equilibrium long

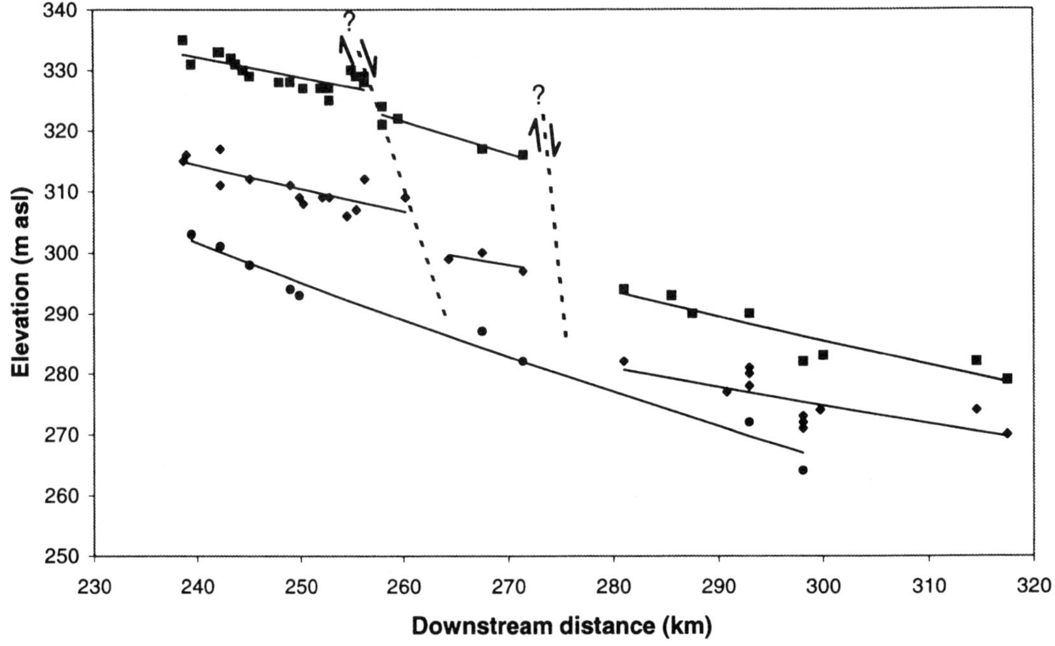

Figure 11.6 *Long profiles of the present water surface (●) and Cenozoic terraces (■,◆) of the Lachlan River in the highlands–interior transition zone between Wyangala Dam and Cowra (W and C in Figure 11.7). Elevation data were derived using the precision barometric survey method of Goldrick (1994), and best-fit lines fitted using the equilibrium profile form introduced in this chapter (Appendix). No lithological changes are associated with the gradient discontinuities between 260 and 280 km; the relative amounts of deformation of the terrace and modern profiles therefore indicate ongoing tectonism into the Holocene. It has not yet been possible to determine the age of the upper terrace (a sample of this terrace on the right bank at Darbys Falls (D in Figure 11.7) has returned an infinite thermoluminescence age (Goldrick, unpublished data)). The upper surface of the lower terrace immediately downstream of Wyangala Dam has returned a radiocarbon age of 2 ka BP (Bishop, unpublished data)*

profiles (steepening) are also present in these Miocene tributaries (points A to F in Figure 11.5).

This steepening of the modern and Miocene profiles was interpreted by Bishop *et al.* (1985), using a qualitative Hack-type semi-logarithmic long profile analysis, as an equilibrium response to the high lithological resistance of the hornfels ridge at the granite–Ordovician metasediments contact (Figure 11.5). The quantitative analysis that is now possible using the DS plots shows, however, that the steepening represents disequilibrium in the long profile. Moreover, the disequilibria and/or the greatest steepening are not always precisely coincident with the hornfels ridge (compare streams 58, 59 and 60 on Figures 11.5 and 11.8), and streams 56 and 57 flow across this contact with no evident discontinuities in the DS plots (Figures 11.5 and 11.8).

We suggest, therefore, that, rather than being an equilibrium response to high lithological resistance, the discontinuities in the DS plots indicate long profile disequilibrium resulting from rejuvenation of the Lachlan drainage net. For all the right-bank tributaries in the upper Lachlan, including the sub-basaltic Miocene tributary for which a DS plot can be constructed, downstream projections of the most upstream graded reach to the modern or Miocene Lachlan indicate long profile offsets due to rejuvenation of between 60 and 120 m (e.g. Figure 11.9B and C). The disequilibria in the Miocene profiles indicate that this tectonic activity is time-transgressive. The approximate, but by no means universal, coincidence between the hornfels and the steepened reaches indicates that the greater resistance of the hornfels generally slows the passage of knickpoints, retarding the rate at which the streams can reattain equilibrium. Thus, the knickpoints are preferentially, but not always, located in the vicinity of the hornfels ridge. This retarding effect of the hornfels ridge would also explain the left-bank–right-bank asymmetry in tributary long profiles (Figure 11.8). Knickpoint passage and the reattain-

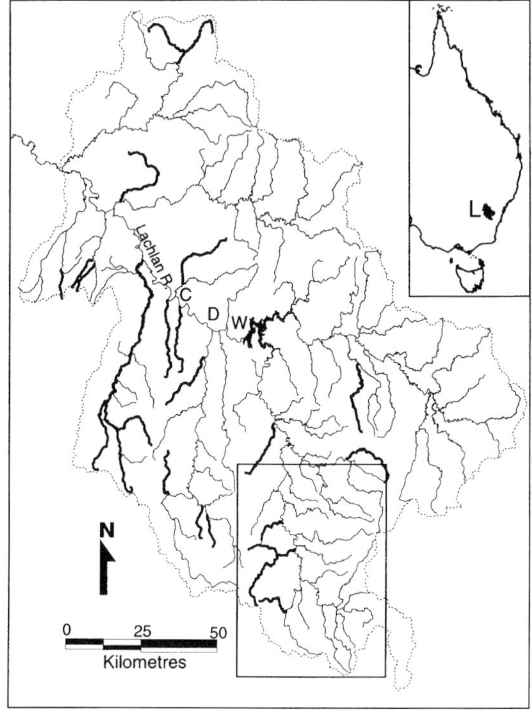

Figure 11.7 *The Lachlan drainage net highlighting in bold the streams that exhibit linear DS plots ('uniformly graded streams'). C: Cowra; D: Darbys Falls; W: Wyangala Dam. The lower box shows the location of Figure 11.5 and the inset the location of the Lachlan catchment (L)*

ment of equilibrium are evidently more rapid in the left-bank tributaries, which flow over relatively low resistance Ordovician metasediments that have not experienced the contact metamorphism associated with intrusion of the granites.

The broad pattern is consistent with the headward migration of one or more rejuvenation knickpoints along the Lachlan and its left-bank tributaries, and retardation of the passage of the knickpoints along the right-bank tributaries where they cross the hornfels ridge. The 60 to 120 m of long profile offset indicated by projection of the right-bank long profiles is comparable to the amount of Neogene tectonic offset in the transition zone between the highlands and the interior (Bishop and Brown, 1992; Goldrick and Bishop, 1995). In summary, therefore, comparisons between the modern and Miocene profiles in the upper Lachlan suggest time-transgressive tectonism consistent with denudational isostatic rebound of the highlands. Two final points warrant emphasis. First, the modelling of Bishop and Brown (1992) showed that the uplift we interpret from the long profiles and other data is passive denudational rebound and that active uplift does not have to be invoked to account for the long profile disequilibria and other field data indicative of uplift. Secondly, the identification of long profile disequilibria via the use of the DS plot (which derives from the more general form of the equilibrium long profile which itself encompasses the Hack-type semi-logarithmic long profile as a special case (see Appendix)) means that the qualitative identification of a Hack-type dynamic equilibrium in the upper Lachlan (Bishop *et al.*, 1985) cannot now be sustained. The minimal incision in the headwaters of the drainage net (see modern headwaters rising on basalt in Figure 11.5) was also noted by Bishop *et al.* (1985) to be inconsistent with a Hack-type dynamic equilibrium with catchment-wide downwasting.

11.6 Synthesis and Discussion

11.6.1 SYNTHESIS

Our current interpretation of rifting and break-up along the southeast Australian margin is that it followed the upper-plate/lower-plate model proposed by Lister *et al.* (1986) and Lister and Etheridge (1989). This model synthesizes the asymmetric break-up in the Tasman Sea, and underplating and uplift inboard of the margin. This uplift occurred between 100 and 90 Ma BP and was evidently associated in turn with a major denudational episode reflected in the crustal cooling that is expressed in apatite fission-track ages throughout the region. The preservation of Mesozoic landscape elements throughout the highlands and below the escarpment means, however, that this denudation cannot be thought of as the regional stripping of a uniform 'slice' of the upper crust.

The pre-Cenozoic location and form of the continental drainage divide remain contentious. The ?Mesozoic Rickabys Creek Gravel points to a Mesozoic origin for the divide, related to Tasman rifting. This viewpoint was espoused more than 65 years ago by Craft (1931), who argued that the present rivers flowing eastwards from the continental drainage divide had their antecedence in the Mesozoic palaeogeography in which streams delivered sediment to the Mesozoic Sydney Basin from the areas of the present uplands, north, southwest and south of the basin (Bishop, 1982; Conaghan *et al.*, 1982). Even today on these flanks of the Sydney Basin, Palaeozoic rocks rise up above the basin, much as would have been the case

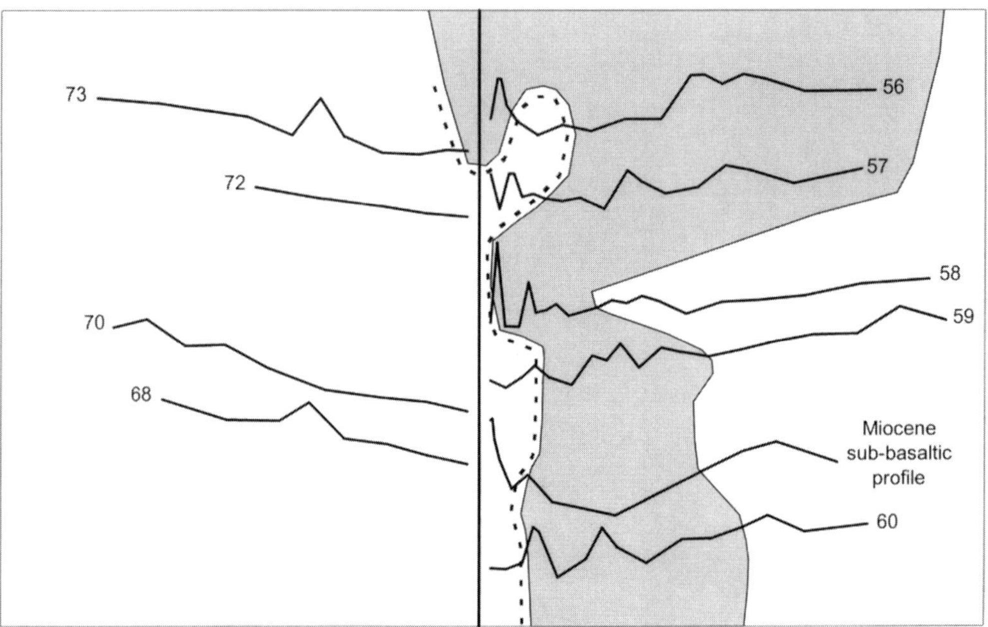

Figure 11.8 DS plots of the left- and right-bank tributaries of the upper Lachlan and country rock geology (see Figure 11.5 for stream locations). Shading indicates granitic rocks of the Wyangala batholith and unshaded areas denote Ordovician meta-sediments. Dashes along the geological boundary indicate the prominent hornfels ridge. The diagram combines map representation of the geology and DS plots on XY axes. The logarithmic scale on the DS plot's X axis distorts the locations of the geological boundaries which are therefore diagrammatic

when it was a depositional centre. It is clear, however, that parts of the continental divide demonstrably must have been uplifted since Sydney Basin times because Sydney Basin strata form the continental divide in central eastern NSW. As we have seen, however, Ollier and Pain (1994) have a different viewpoint on the Mesozoic drainage history. Veevers (1984) also "speculate[d] that the lineaments [the geomorphic framework or structure] of the current regime took shape over eastern Australia in the interval 95–90 Ma" (p. 138 and his Fig. 72). In any event, the valley-filling lavas throughout the highlands demonstrate that the continental divide dates from at least the latest Mesozoic or earliest Cenozoic, and the coincidence of a deep crustal root beneath the continental divide (Young, 1989) and the accepted link between Tasman Sea rifting and crustal underplating beneath the highlands (e.g. Johnson, 1989; but see Wellman, 1987, 1996) indicate that some divide uplift must have been associated with Tasman Sea rifting and break-up (90 Ma BP). Craft's (1931) conjectural dating of the continental divide from Sydney Basin times, at least in the southern part of the highlands, would mean that the rifting-related uplift of the highlands 90–100 Ma ago reactivated a pre-existing highland belt in southern NSW and northern NSW (see Fried and Smith (1992) on the latter) (Figure 11.4B).

Where the evidence is apparently strongest (in the Sydney Basin, where the sub-horizontal to horizontal middle Triassic strata are found), the highlands seem to have taken the form of an uplifted plateau into which the escarpment retreated (and not a downwarped plateau surface). This point also remains highly contentious, however, and the literature also records strong support for the downwarped plateau model.

There appears to have been minimal active uplift since the highlands received their final form, but ongoing denudational rebound has been an important feature of the highlands' post-uplift history (although Wellman (1987, 1996) has a different viewpoint). The lack of deformation of the Sydney Basin strata and

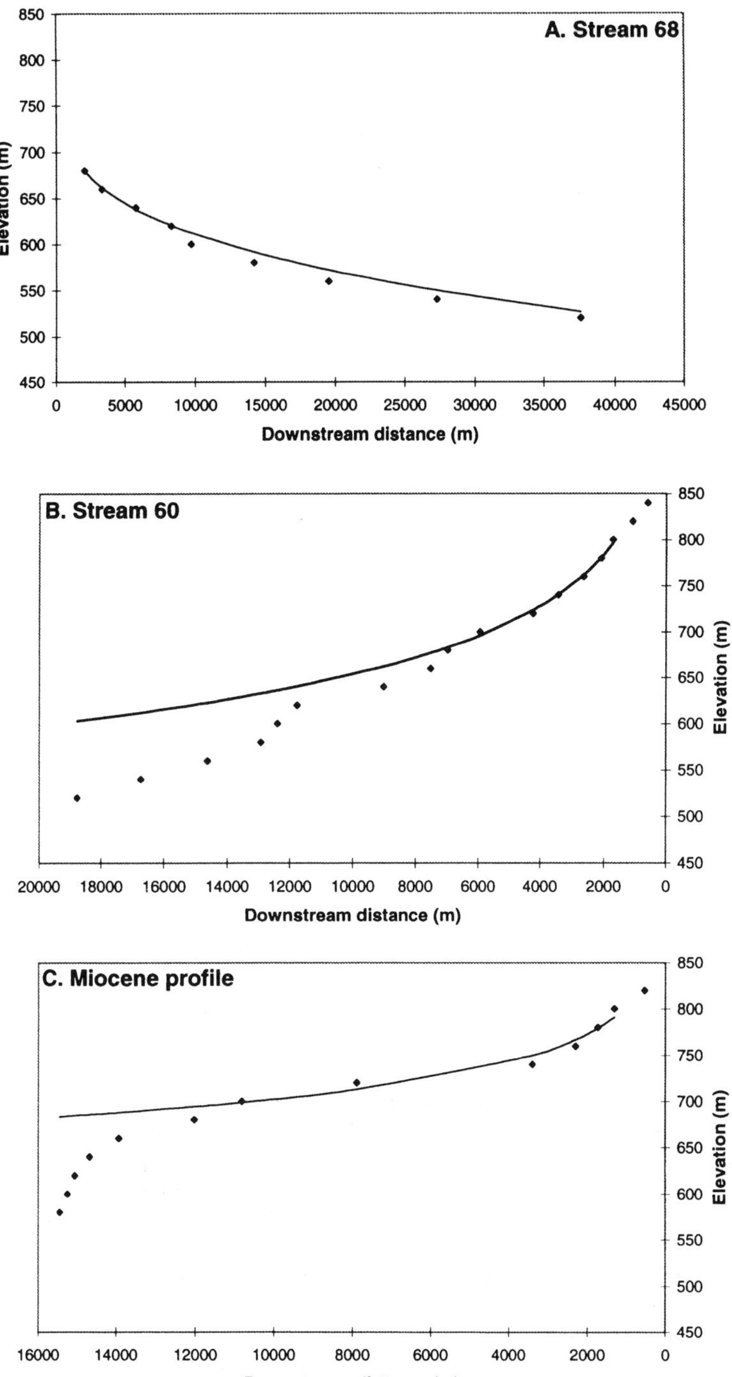

Figure 11.9 *Projections of upstream graded reaches of selected upper Lachlan tributaries to their confluences with the modern and Miocene Lachlan. A and B are modern left- and right-bank tributaries, and C is the sub-basaltic Miocene tributary between modern streams 59 and 60. See Figure 11.5 for locations*

the evident lack of an upwarped seaward edge of the highlands (Gilchrist and Summerfield, 1990) mean that denudational rebound of the highlands as a whole has apparently been of a greater magnitude than any flexural rebound due to escarpment retreat. The apparent insignificance of flexural denudational rebound due to escarpment retreat probably reflects the narrowness of the coastal strip across which the escarpment has retreated relative to the broad regionality of denudation, albeit at generally low rates. As we have already noted in relation to the amount of sediment produced by landscape denudation relative to much higher, but spatially more restricted, gorge retreat (see Section 11.3), the regionality of denudation is often a much more important issue than other more spectacular, but spatially restricted, forms of erosion (including escarpment retreat).

Ongoing denudational rebound of the highlands is recorded in the highland rivers' long profiles as a series of disequilibria that move through the drainage net at different rates probably dependent on the discharge, slope and lithological characteristics of individual streams. The upper Lachlan, even though it is remote from the rebound-associated base-level changes, demonstrates the time-transgressive movement of knickpoints through the drainage net, with both the Miocene and modern drainage nets exhibiting steepened reaches that appear to be genuine disequilibria and not equilibrium adjustments to more resistant lithologies. Young and McDougall (1993) also identified persistent and headward migrating knickpoints in the interior Murray system, and we see no reason to think that the same knickpoint persistence will not be identified in other systems as they are investigated in detail. These relatively subtle rejuvenation features cannot generally be identified in the exterior drainage systems because the major knickpoints associated with the exterior rivers' fall from the plateau through the gorges eroding back into the highland mass dominate the long profiles of these systems.

The Lachlan, like the other major catchments throughout East Australia, has exhibited increasing relief through time (Bishop *et al.*, 1985; Young and McDougall, 1993; Nott *et al.*, 1996). This feature of the landscape means that neither the Davisian nor the Hack model of long-term landscape evolution is appropriate to the southeast Australian highlands (Young and McDougall, 1993; Nott *et al.*, 1996). Twidale (1991) has stressed the inapplicability of the Davis and Hack models in a wide range of settings, highlighting the roles of structure and lithology in generating and maintaining, and even increasing, relief differences in the landscape. These influences may operate in a somewhat different way where the continental divide is mantled by basalts. The Tertiary basalts in the southeast Australian highlands generally exhibit smooth, well-rounded forms and the very low (to zero) drainage densities indicative of a dominance of infiltration over runoff. The columnar structure of the flows and their well-weathered, strongly pedal regolith ensure this dominance of infiltration and perhaps, in turn, the relatively low rates of incision where divides are characterized by low relief and basaltic mantles.

11.6.2 EAST AUSTRALIA AND THE MODELLING OF THE EVOLUTION OF HIGH-ELEVATION PASSIVE CONTINENTAL MARGINS

Modelling of the interplay between global tectonics and geomorphology in the evolution of passive continental margins has generally been investigated at two temporal and spatial scales (see Chapter 3). The first scale considers the genesis of the gross topographic character of the margin, broadly encompassing the rifting and break-up that leads to the formation of the new margin, whereas the focus at the second scale is the post-break-up denudational and tectonic (isostatic) development of the margin. The split between the two spatial and temporal scales of investigation is, of course, only a device to assist understanding, and should not be taken so far as to obscure interactions that are essential in understanding continental margin evolution (e.g. van der Beek *et al.*, 1995). Nonetheless, such a conceptual split is useful for our present purpose of understanding of the evolution of the East Australian continental margin, and for comparing the reasonably detailed understanding that now exists for southeast Australia and the models of passive margin evolution.

The geomorphological and tectonic development at the first conceptual scale has been elucidated for passive margins in general by conceptual and numerical modelling of the development of young rifted margins (e.g. the Red Sea – Steckler and Omar, 1994), and by data from pre-break-up, syn-rift and post-break-up sedimentary basins (e.g. Veevers, 1984; Wellman, 1987; Brown *et al.*, 1990; Rust and Summerfield, 1990) and apatite fission-track thermochronology of passive continental margins (e.g. Dumitru *et al.*, 1991; Steckler *et al.*, 1993). Direct geomorphological input to this first-order understanding has generally been limited to general comparisons of modelled and actual topographic profiles of young continental margins because of the antiquity of most passive continental margins and the difficulty of identifying landscape elements that can be dated confidently to the time of rifting and break-up. Indeed, the depths of syn-rift and post-rift passive margin denudation indicated by AFTT data should preclude identification of rift stage elements in the landscape, at least close to the continental margin (Brown *et al.*, 1990, 1994; Dumitru *et al.*, 1991; Gallagher *et al.*, 1994). The data from the southeast Australian margin presented in this chapter indicate, however, that fragments of this pre-break-up landscape elements may persist in the landscape and it is now appropriate to move beyond the blanket application of long-term, supposedly regional rates of denudation derived from AFTT and sedimentary basin analysis (see Section 11.3.2).

The second scale of investigation, that of post-break-up margin evolution, has been elucidated by conceptual and numerical models that are variably constrained by geomorphological data. The geomorphological control has generally consisted of an assessment of the amount of post-break-up continental margin denudation (as indicated again by the volumes of sediment in offshore and/or interior basins, and by AFTT) and/or a comparison of the topographic profiles of actual continental margins to those generated by the numerical modelling of tens of millions of years of post-break-up margin evolution (e.g. Gilchrist and Summerfield, 1990; Arvidson *et al.*, 1994; Gilchrist *et al.*, 1994; Kooi and Beaumont, 1994; Pazzaglia and Gardner, 1994; Steckler and Omar, 1994; Tucker and Slingerland, 1994) (see Chapter 3). In some cases (e.g. Pazzaglia and Gardner, 1994), more detailed geomorphological data, such as dated river and terrace long profiles, have been incorporated (see Chapter 13).

A major aim of this modelling has been an understanding of the maintenance and ongoing evolution of the highland belt and the associated history of the escarpment (e.g. Gilchrist and Summerfield, 1990; Gilchrist *et al.*, 1994; Kooi and Beaumont, 1994; Tucker and Slingerland, 1994; van der Beek *et al.*, 1995). Tucker and Slingerland (1994) and van der Beek *et al.* (1995) concluded that the escarpment is more readily generated and maintained when the models incorporate regional plateau uplift inland of the rift and new continental margin, as well as the usual 'tectonic' uplift associated with the rifting. Such a conclusion is consistent with the tentative suggestions that substantial elevation existed along the trace of at least parts of the southeast Australian highlands prior to their further uplift in association with Tasman Sea rifting.

Modelling also indicates that isostatic uplift of the escarpment lip as a result of erosional unloading accompanying escarpment retreat is generally significant in maintaining the escarpment (see, for example, Gilchrist and Summerfield (1990)). Sediment loading on the continental shelf may either damp such denudational uplift (when the wavelength and timing of the flexural response to the shelf loading results in downwarping of the exterior region (e.g. Tucker and Slingerland, 1994)) or enhance the denudational uplift (as when the wavelengths of loading and flexural response have a positive interaction (e.g., Summerfield, 1991, Fig. 4.20; Pazzaglia and Gardner, 1994)). Maintenance of escarpments is also enhanced in these modelling investigations by resistant lithologies in the escarpment zone (Kooi and Beaumont, 1984; Pazzaglia and Gardner, 1994), and by evolution under supply-limited conditions, such as greater aridity (e.g. Gilchrist *et al.*, 1994; Kooi and Beaumont, 1994); Tucker and Slingerland's (1994) modelling results can also have the same implication although they explicitly noted that aridity is not the only way in which to interpret their sediment supply-limited case.

None of these effects appears to be of particular significance in the maintenance of the escarpment in southeast Australia. The escarpment crosses many different lithologies and humid climatic types ranging from tropical monsoonal to humid temperate. Palaeobotanical evidence indicates that aridity, or even semi-aridity, has been unimportant throughout the history of the Tasman margin which has generally been characterized by humid climates (Taylor *et al.*, 1990; Nott and Owen, 1992). Nor can denudational rebound due to escarpment retreat or flexure associated with sediment loading of the continental shelf be implicated in escarpment maintenance. In short (and this remains one of the striking outcomes of much of the research on the evolution of the southeast Australian margin), low rates of denudation and persistence of geomorphic forms are very characteristic of this margin (Young, 1983; Nott *et al.*, 1996).

The behaviour of drainage divides is also an important element of various models of high-elevation passive continental margin evolution (see Chapter 3). The most commonly (and most easily?) modelled case, in which the escarpment lip and the continental drainage divide coincide, is generally associated with enhanced escarpment preservation, the divide being maintained by backtilting of the escarpment as a result of flexural denudational isostasy associated with escarpment retreat (Kooi and Beaumont, 1994). Even in the case of modelling in which the continental drainage divide lies inland of the escarpment lip (e.g. Gilchrist *et al.*, 1994; Kooi and Beaumont, 1994), the original escarpment is destroyed and a new one created inland so that escarpment lip and continental divide coincide after about 30 Ma (see, for instance, Gilchrist *et al.*'s (1994) Model 3 and their Plate 3). Alternatively, backtilting of the plateau edge as a result of flexural uplift in response to escarpment retreat produces reversal of the exterior drainage between the continental divide and the escarpment lip so that the escarpment lip becomes the continental drainage divide (Kooi and Beaumont, 1994). The only way in which Kooi and Beaumont (1994), for example, were able to maintain the exterior drainage was to run the model with lithosphere with high flexural rigidity (effective elastic thickness = 90 km). The rates of evolution of the southeast Australian highlands are clearly much slower than in such modelling. To reiterate: the highland mass is being consumed only very slowly by gorge retreat; flexural denudational isostasy as a result of escarpment retreat is evidently negligible; the continental divide has been horizontally stable for 65 Ma at least; and the effective elastic thickness of the southeast Australian highlands lithosphere is of the order of only 15 km (Zuber *et al.*, 1989). Denudational uplift of the highlands as a whole is consistent with these observations, but this form of highlands evolution appears not to have been modelled yet. Moreover, the southeast Australian data show that if fluvial long profiles are included in numerical models of highlands evolution, they would be expected to exhibit persistent disequilibria indicative of slow adjustment to ongoing denudational rebound.

The stability of the continental drainage divide in southeast Australia also means that the model of drainage development in southern Africa, in which the highlands are breached and a large interior drainage system is diverted to the exterior (Rust and Summerfield, 1990; Gilchrist *et al.*, 1994), cannot be applied to southeast Australia. The southeast Australian continental divide is the locus of extremely low denudation rates and tectonic activity, and river long profiles record ongoing disequilibria of (presumably) considerable antiquity.

11.7 Future Directions and Research Issues

The large amount of data from a wide range of disciplines enables a reasonably clear general understanding of the development of the southeast Australian margin, but it also introduces significant complexity into interpretations of some of the detail of the margin's history. A case in point is the convergence of data from AFTT, palaeomagnetism and coal vitrinite reflectance studies, indicating a significant cooling event in the late Mesozoic (see Section 11.3.2). The various thermochronological approaches and their data suggest that the present land surface is broadly the result of kilometre-scale denudation in the late Mesozoic, whereas geomorphological interpretations indicate the preservation of Mesozoic landscape elements at the present land surface. The apparent conflict between these data will be resolved by combined geomorphological and thermochronological studies of critical sites where supposed Mesozoic landscape elements are preserved. If the southern NSW coastal strip indeed dates from 100 Ma ago, it is striking that there is no evidence in this area for the Cretaceous marine transgression (M. Orr, 1996, pers. comm.; S. Hill, 1996, pers. comm.). This issue needs to be resolved as part of the foregoing investigation but could simply reflect the fact that the 100 Ma lavas on the NSW south coast (Nott and Purvis, 1995) are not truly sub-aerial but were emplaced at very shallow depths (Peter van der Beek, 1996, pers. comm.). Such shallow emplacement would not substantially affect Nott and Purvis's (1995) interpretation.

Such investigations should also continue the recent welcome shift in emphasis of AFTT studies from the coastal strip to the highland crest (e.g. O'Sullivan *et al.*, 1995, 1996). This renewed interest should clarify the Mesozoic tectonic history of the highlands' crest and perhaps the associated long-term history of Mesozoic drainage directions and the continental drainage divide. Key issues will be whether major tectonic reorganization of the highland crest and downwarping of the continental margin are indicated (Figure 11.4). It is also to be hoped that modelling studies, which should include structural data from the Sydney Basin sediments, will clarify whether downwarping is likely. Finally, modelling studies are yet to reproduce satisfactorily the known rates of landscape evolution in southeast Australia, and this issue also requires attention because of the importance of these rates in determining rates of rebound and the likelihood of major drainage re-arrangement. Rates of denudation have both direct and indirect influence on drainage rearrangement: direct via the possibility of headward retreat and drainage disruption, and indirect via possible passive tectonic (denudational) impacts on drainage systems.

Finally, the late Mesozoic–Cenozoic evolution of the East Australian margin in Queensland remains relatively unknown. We look forward to new data from this area so as to have an understanding of the evolution of the margin as a whole. In many ways, the southern part of the East Australian margin has generated more questions than it has answered, and the many mismatches between the field data from southeast Australia and the theory and modelling of high-elevation passive continental margins are very striking. It is likely that further data from the northern (Queensland) part of the margin will complicate this picture even further.

Acknowledgements

The support of the Australian Research Council and Monash University is gratefully acknowledged. This chapter was written while PB held a Leverhulme Trust Visiting Fellowship in the Department of Geography at the University of Edinburgh. Gratitude is extended to the Leverhulme Trust for enabling this visit, and to the Edinburgh Department for its generous hospitality.

References

Arvidson, R., Becker, R., Shanabrook, A., Luo, W., Sturchio, N., Sultan, M., Lofty, Z., Mahmood, A.M. and Alfy, Z.E. 1994. Climatic, eustatic, and tectonic controls on Quaternary deposits and landforms, Red Sea coast, Egypt. *J. Geophys. Res.*, **99**, 12 175–12 190.

Bird, M.I. and Chivas, A.R. 1993. Geomorphic and palaeoclimatic implications of an oxygen-isotope chronology for Australian deeply weathered profiles. *Austr. J. Earth Sci.*, **40**, 345–358.

Bishop, P. 1982. Stability or change: a review of ideas on ancient drainage in eastern New South Wales. *Austr. Geogr.*, **15**, 219–230.

Bishop, P. 1985. Southeast Australian late Mesozoic and Cenozoic denudation rates: A test for late Tertiary increases in continental denudation. *Geology*, **13**, 479–482.

Bishop, P. 1986. Horizontal stability of the Australian continental drainage divide in south central New South Wales during the Cainozoic. *Austr. J. Earth Sci.*, **33**, 295–307.

Bishop, P. 1988. The Eastern Highlands of Australia: The evolution of an intra-plate highland belt. *Prog. Phys. Geog.*, **12**, 159–182.

Bishop, P. 1995. Drainage rearrangement by river capture, beheading and diversion. *Prog. Phys. Geog.*, **19**, 449–473.

Bishop, P. 1996. Discussion: Landscape evolution and tectonics in southeastern Australia (Ollier & Pain 1994). *AGSO J. Austr. Geol. Geophys.*, **16**, 315–317.

Bishop, P. and Brown, R. 1992. Denudational isostatic rebound of intraplate highlands: The Lachlan River valley, Australia. *Earth Surf. Processes Ldfms.*, **17**, 345–360.

Bishop, P. and Brown, R. 1993. Reply to Discussion: Denudational isostatic rebound of intraplate highlands: The Lachlan River valley, Australia. *Earth Surf. Processes Ldfms.*, **18**, 753–755.

Bishop, P. and Cowell, P. 1997. Lithological and drainage network determinants of the character of drowned, embayed coastlines. *J. Geol.*, **105**, 685–699.

Bishop, P. and Hunter, T. 1990. Pebble fabrics in the Rickabys Creek Gravel and their implications for the relationships between the Lapstone Monocline and the Rickabys Creek Gravel. Geological Survey of NSW Unpublished Report, 1990/281.

Bishop, P., Hunt, P. and Schmidt, P. 1982. Limits to the age of the Lapstone Monocline, N.S.W. – a palaeomagnetic study. *J. Geol. Soc. Austr.*, **29**, 319–326.

Bishop, P., Young, R.W. and McDougall, I. 1985. Stream profile change and long-term landscape evolution – Early Miocene and modern rivers of the east Australian highland crest, central New South Wales, Australia. *J. Geol.*, **93**, 455–474.

Branagan, D.F. 1983. The Sydney Basin and its vanished sequence. *J. Geol. Soc. Austr.*, **30**, 75–84.

Brown, C.M. 1983. Discussion: A Cainozoic history of origin of Australia's southeast highlands. *J. Geol. Soc. Austr.*, **30**, 483–486.

Brown, M.C. 1983. Discussion: Origin of coastal lowlands near Ulladulla, N.S.W. *J. Geol. Soc. Austr.*, **30**, 247–248.

Brown, M.C. 1996. Discussion: Geomorphic and tectonic significance of Early Cretaceous lavas on the coastal plain, southern New South Wales. *Austr. J. Earth Sci.*, **43**, 688–689.

Brown, R.W., Rust, D.J., Summerfield, M.A., Gleadow, A.J.W. and De Wit, M.C.J. 1990. An early Cretaceous phase of accelerated erosion on the south-western margin of Africa: Evidence from apatite fission track analysis and the offshore sedimentary record. *Nucl. Tracks Rad. Measur.*, **17**, 339–350.

Brown, R.W., Summerfield, M.A. and Gleadow, A.J.W. 1994. Apatite fission track analysis: Its potential for the estimation of denudation rates and implications for models of long-term landscape evolution. In: M.J. Kirkby (ed.) *Process Models and Theoretical Geomorphology*, Wiley, Chichester, pp. 23–53.

Colwell, J.B., Coffin, M.F. and Spencer, R.A. 1993. Structure of the southern New South Wales continental margin, southeastern Australia. *BMR J. Austr. Geol. Geophys.*, **13**, 333–343.

Conaghan, P.J., Jones, J.G., McDonnell, K.L. and Royce, K. 1982. A dynamic fluvial model for the Sydney Basin. *J. Geol. Soc. Austr.*, **29**, 55–70.

Craft, F.A. 1931. The physiography of the Shoalhaven River

Valley. i. Tallong-Bungonia. *Proc. Linnean. Soc. N.S.W.*, **56**, 99–132.

Davies, P.J. 1975. Shallow seismic structure of the continental shelf, southeast Australia. *J. Geol. Soc. Austr.*, **22**, 345–359.

Department of Mineral Resources. 1983. *1:100 000 Geological Series Sydney Sheet. Sheet 9130*. Geological Survey of NSW, Department of Mineral Resources, Sydney.

Department of Mineral Resources. 1985. *1:100 000 Geological Series Wollongong – Port Hacking Sheet. Sheet 9029–9129*. Geological Survey of NSW, Department of Mineral Resources, Sydney.

Department of Mineral Resources. 1997. *1:100 000 Geological Series Gosford – Port Macquarie Sheet. Sheet 9131–9231*. Geological Survey of NSW, Department of Mineral Resources, Sydney.

Dewey, J.F. and Bird, J.M. 1970. Mountain belts and the new global tectonics. *J. Geophys. Res.*, **75**, 2625–2647.

Dietrich, W.E., Wilson, C.J. and Reneau, S.L. 1986. Hollows, colluvium, and landslides in soil-mantled landscapes. In: A.D. Abrahams (ed.) *Hillslope Processes*, Allen and Unwin, Boston, pp. 31–53.

Dingle, R.V. and Hendey, Q.B. 1984. Late Mesozoic and Tertiary sediment supply to the eastern Cape Basin (SE Atlantic) and palaeo-drainage systems in southwestern Africa. *Mar. Geol.*, **56**, 13–26.

Driscoll, N.W. and Karner, G.D. 1994. Flexural deformation due to Amazon fan loading: a feedback mechanism affecting sediment delivery to margins. *Geology*, **22**, 1015–1018.

Dulhunty, J.A. 1973. Potassium–argon ages and their significance in the Macquarie Valley, New South Wales. *J. Proc. R. Soc. N.S.W.*, **106**, 104–110.

Dumitru, T.A., Hill, K.C., Coyle, D.A., Duddy, I.R., Foster, D.A., Gleadow, A.J.W., Green, P.F., Kohn, B.P., Laslett, G.M. and O'Sullivan, A.J. 1991. Fission track thermochronology: application to continental rifting of south-eastern Australia. *APEA J.*, **31**, 131–142.

Duncan, R.A. and McDougall, I. 1989. Volcanic time–space relationships. In: R.W. Johnson (ed.) *Intraplate Volcanism in Eastern Australia and New Zealand*, Cambridge University Press, Cambridge, pp. 43–54.

Ewart, A.E. 1989. Fractionation, assimilation, and source melting: a petrogenetic overview. In: R.W. Johnson (ed.) *Intraplate Volcanism in Eastern Australia and New Zealand*, Cambridge University Press, Cambridge, pp. 324–333.

Fabel, D. and Finlayson, B.L. 1992. Constraining variability in south-east Australian long-term denudation rates using a combined geomorphological and thermochronological approach. *Z. Geomorph.*, **36**, 293–305.

Falvey, D.A. 1974. The development of continental margins in plate tectonic theory. *APEA J.*, **14**, 95–106.

Falvey, D.A. and Mutter, J.C. 1981. Regional plate tectonics and the evolution of Australia's passive continental margins. *BMR J. Austr. Geol. Geophys.*, **6**, 1–29.

Fried, A.W. and Smith, N. 1992. Landscape history in the Glen Innes – Inverell area of the eastern highlands of Australia: the influence of rock control, lava extrusion and tectonism in longterm geomorphic history. *Earth Surf. Processes Ldfms.*, **17**, 375–385.

Gale, S.J. 1992. Long-term landscape evolution in Australia. *Earth Surf. Processes Ldfms.*, **17**, 323–343.

Gallagher, K., Hawkesworth, C.J. and Mantovani, M.J.M. 1994. The denudation history of the onshore continental margin of SE Brazil inferred from fission track data. *J. Geophys. Res.*, **99**, 18 117–18 145.

Gardner, T.W. 1983. Experimental study of knickpoint and longitudinal profile evolution in cohesive, homogeneous material. *Geol. Soc. Amer. Bull.*, **94**, 664–672.

Gardner, T.W. and Sevon, W.D. (eds) 1989. Appalachian geomorphology. *Geomorphology*, **2**, 1–318.

Gilchrist, A.R. and Summerfield, M.A. 1990. Differential denudation and flexural isostasy in formation of rifted-margin upwarps. *Nature*, **346**, 739–742.

Gilchrist, A.R. and Summerfield, M.A. 1994. Tectonic models of passive margin evolution and their implications for theories of long-term landscape development. In: M.J. Kirkby (ed.) *Process Models and Theoretical Geomorphology*, Wiley, Chichester, pp. 55–84.

Gilchrist, A.R., Kooi, H. and Beaumont, C. 1994. Post-Gondwana geomorphic evolution of southwestern Africa: Implications for the controls on landscape development from observations and numerical experiments. *J. Geophys. Res.*, **99**, 12 211–12 228.

Goldrick, G. 1994. A one-person, one-instrument method for precision barometric altimetry. *Earth Surf. Processes Ldfms.*, **19**, 801–808.

Goldrick, G. and Bishop, P. 1995. Distinguishing the roles of lithology and relative uplift in the steepening of the long profiles of bedrock rivers. *J. Geol.*, **103**, 227–231.

Grimes, K.G. 1980. The Tertiary geology of north Queensland. In: R.A. Henderson and P.J. Stephenson (eds) *The Geology and Geophysics of Northeastern Australia*, Queensland Division, Geological Society of Australia, Brisbane, pp. 329–347.

Hack, J.T. 1957. Studies of longitudinal stream profiles in Virginia and Maryland. *U.S. Geol. Surv. Prof. Pap.*, **294-B**, 45–97.

Hack, J.T. 1975. Dynamic equilibrium and landscape development. In: W.N. Melhorn and R.C. Flemal (eds) *Theories of Landform Development*, State University of New York Press, Binghamton, pp. 87–102.

Haworth, R.J. and Ollier, C.D. 1992. Continental rifting and drainage reversal: The Clarence River of Eastern Australia. *Earth Surf. Processes Ldfms.*, **17**, 387–397.

Hill, S.M. 1999. Mesozoic regolith and palaeo-landscape features in southeastern Australia: significance for interpretations of the evolution of the eastern highlands. *Austr. J. Earth Sci.* (in press).

Horton, R.E. 1945. Erosional development of streams and their drainage basins: hydrophysical approach to quantitative geomorphology. *Geol. Soc. Amer. Bull.*, **56**, 275–370.

Howard, A.D., Dietrich, W.E. and Seidl, M.A. 1994. Modeling fluvial erosion on regional to continental scales. *J. Geophys. Res.*, **99**, 13 971–13 986.

Howard, A.D. and Kerby, G. 1983. Channel changes in badlands. *Geol. Soc. Amer. Bull.*, **94**, 739–752.

Jennings, J.N. 1972. The age of Canberra landforms *J. Geol. Soc. Austr.*, **19**, 371–378.

Johnson, R.W. (ed.) 1989. *Intraplate Volcanism in Eastern Australia and New Zealand*, Cambridge University Press, Cambridge.

Jones, O.T. 1924. The longitudinal profiles of the Upper Towy drainage system. *Quart. J. Geol. Soc. Lond.*, **80**, 568–609.

Jongsma, D. and Mutter, J.C. 1978. Non-axial breaching of a rift valley: evidence from the Lord Howe Rise and the southeastern Australian margin. *Earth Planet. Sci. Lett.*, **39**, 226–234.

Keller, E.A. and Rockwell, T.K. 1984. Tectonic geomorphology, Quaternary chronology and paleoseismicity. In: J. E. Costa and P.J. Fleisher (eds) *Development and Applications of Geomorphology*, Springer-Verlag, Berlin, pp. 203–239.

Kooi, H. and Beaumont, C. 1994. Escarpment evolution on high-elevation rifted margins: Insights derived from a surface process model that combines diffusion, advection, and reaction. *J. Geophys. Res.*, **99**, 12 191–12 209.

Lambeck, K. and Stephenson, R. 1986. The post-Palaeozoic uplift history of south-eastern Australia. *Austr. J. Earth Sci.*, **33**, 253–270.

Leopold, L.B., Wolman, M.G. and Miller, J.P. 1964. *Fluvial Processes in Geomorphology*, Freeman, San Francisco.

Li, Shu, 1995. Long-term landscape evolution: a case study from the lower Snowy river, Australia. Unpublished PhD thesis, University of Melbourne.

Li, Shu, Webb, J.A. and Finlayson, B.L. 1996. Discussion: Landscape evolution and tectonics in southeastern Australia (Ollier & Pain 1994). *AGSO J. Austr. Geol. Geophys.*, **16**, 323–324.

Lister, G.S. and Etheridge, M.A. 1989. Detachment models for uplift and volcanism in the Eastern Highlands, and their application to the origin of passive margin mountains. In: R.W. Johnson (ed.) *Intraplate Volcanism in Eastern Australia and New Zealand*, Cambridge University Press, Cambridge, pp. 297–313.

Lister, G.S., Etheridge, M.A. and Symonds, P.A. 1986. Detachment faulting and the evolution of passive margins. *Geology*, **14**, 246–250.

McDougall, I. and Duncan, R.A. 1988. Age progressive volcanism in the Tasmantid seamounts. *Earth Planet. Sci. Lett.*, **89**, 207–220.

McKeown, F.A., Jones-Cecil, M., Askew, B.L. and McGrath, M.B. 1988. Analysis of stream-profile data and inferred tectonic activity, eastern Ozark Mountains region. *U.S. Geol. Surv. Bull.*, **1807**.

Merritts, D.J., Vincent, K.R. and Wohl, E.E. 1994. Long river profiles, tectonism, and eustasy: A guide to interpreting fluvial terraces. *J. Geophys. Res.*, **99**, 14 031–14 050.

Middleton, M.D. and Schmidt, P.W. 1982. Palaeothermometry of the Sydney Basin. *J. Geophys. Res.*, **87**, 5351–5359.

Moore, M.W., Gleadow, A.J.W. and Lovering, J.F. 1986. Thermal evolution of rifted continental margins: new evidence from fission tracks in basement apatites from southeastern Australia. *Earth Planet. Sci. Lett.*, **78**, 255–270.

Montgomery, D.R. and Dietrich, W.E. 1992. Channel initiation and the problem of landscape scale. *Science*, **255**, 826–830.

Nott, J.F. 1992. Long-term drainage evolution in the Shoalhaven catchment, southeast highlands, Australia. *Earth Surf. Processes Ldfms.*, **17**, 361–374.

Nott, J.F. 1996. Discussion: Landscape evolution and tectonics in southeastern Australia (Ollier & Pain 1994). *AGSO J. Austr. Geol. Geophys.*, **16**, 319–321.

Nott, J.F., Idnurm, M. and Young, R.W. 1991. Sedimentology, weathering, age and geomorphological significance of Tertiary sediments on the far south coast of New South Wales. *Austr. J. Earth Sci.*, **38**, 357–373.

Nott, J.F. and Owen, J.A. 1992. An Oligocene palynoflora from the middle Shoalhaven catchment, NSW and the Tertiary evolution of flora and climate in the southeast Australian highlands. *Palaeogeog. climat. ecol.*, **95**, 135–151.

Nott, J. and Purvis, A.C. 1995. Geomorphic and tectonic significance of Early Cretaceous lavas on the coastal plain, southern New South Wales. *Austr. J. Earth Sci.*, **42**, 145–149.

Nott, J. and Purvis, A.C. 1996. Reply: Geomorphic and tectonic significance of Early Cretaceous lavas on the coastal plain, southern New South Wales. *Austr. J. Earth Sci.*, **43**, 689–692.

Nott, Y., Young, R. and McDougall, I. 1996. Wearing down, wearing back, and gorge extension in the long-term denudation of a highland mass: Quantitative evidence from the Shoalhaven catchment, southeast Australia. *J. Geol.*, **104**, 224–232.

Ollier, C.D. 1978. Tectonics and geomorphology of the eastern highlands. In: J.L. Davies and M.A.J. Williams (eds) *Landform Evolution of Australasia*, ANU Press, Canberra, pp. 5–47.

Ollier, C.D. 1982. The Great Escarpment of eastern Australia: tectonic and geomorphic significance. *J. Geol. Soc. Austr.*, **29**, 13–23.

Ollier, C.D. 1985. Morphotectonics of continental margins with great escarpments. In: M. Morisawa and J.T. Hack (eds) *Tectonic Geomorphology*, Allen and Unwin, Boston, pp. 3–25.

Ollier, C.D. and Pain, C.F. 1994. Landscape evolution and tectonics in southeastern Australia. *AGSO J. Austr. Geol. Geophys.*, **15**, 335–345.

Ollier, C.D. and Pain, C.F. 1996. Reply: Landscape evolution and tectonics in southeastern Australia (Ollier & Pain 1994). *AGSO J. Austr. Geol. Geophys.*, **16**, 325–331.

O'Reilly, S.Y. 1989. Xenolith types, distribution, and transport. In: R.W. Johnson (ed.) *Intraplate Volcanism in Eastern Australia and New Zealand*, Cambridge University Press, Cambridge, 249–253.

Orr, M. 1996. Discussion. Geomorphic and tectonic

significance of early Cretaceous lavas on the coastal plain, southern New South Wales. *Austr. J. Earth Sci.*, **43**, 687–688.

O'Sullivan, P.B., Kohn, B.P., Foster, D.A. and Gleadow, A.J.W. 1995. Fission track data from the Bathurst Batholith: evidence for rapid mid-Cretaceous uplift and erosion within the eastern highlands of Australia. *Austr. J. Earth Sci.*, **42**, 597–607.

O'Sullivan, P.B., Foster, D.A., Kohn, B.P. and Gleadow, A.J.W. 1996. Tectonic implications of early Triassic and middle Cretaceous denudation in the eastern Lachlan Fold Belt, NSW, Australia. *Geology*, **24**, 563–566.

Pain, C.F. and Ollier, C.D. 1986. The Comboyne and Bulga Plateaus and the evolution of the Great Escarpment in New South Wales. *J. Proc. R. Soc. N.S.W.*, **119**, 123–130.

Paton, T.R., Mitchell, P.B. and Hunt, P.A. 1989. Isotope dating of the Australian regolith. *Nature*, **337**, 22–23.

Pazzaglia, F.J. and Gardner, T.W. 1994. Late Cenozoic flexural deformation of the middle U.S. Atlantic passive margin. *J. Geophys. Res.*, **99**, 12 143–12 157.

Pickett, J. and Bishop, P. 1992. Aspects of landscape evolution in the Lapstone Monocline area, N.S.W. *Austr. J. Earth Sci.*, **39**, 21–28.

Pickett, J.W., Macphail, M.K., Partridge, A.D. and Pole, M.S. 1997. Middle Miocene palaeotopography at Little Bay, near Maroubra, N.S.W. *Austr. J. Earth Sci.*, **44**, 509–518.

Reed, J.C. 1981. Disequilibrium profile of the Potomac River near Washington, D.C. – A result of lowered base level or Quaternary tectonics along the Fall Line? *Geology*, **9**, 445–450.

Roach, I.C., McQueen, K.G. and Taylor, G. 1996. Discussion: Landscape evolution and tectonics in south-eastern Australia (Ollier & Pain 1994). *AGSO J. Austr. Geol. Geophys.*, **16**, 309–313.

Roy, P.S. and Thom, B.G. 1991. Cainozoic shelf sedimentation model for the Tasman Sea margin of southeast Australia. In: M.A.J. Williams, P. de Deckker and A.P. Kershaw (eds) *The Cainozoic of Australia: A Re-appraisal of the Evidence. Geol. Soc. Austr. Spec. Publ.*, **18**, 119–136.

Rust, D.J. and Summerfield, M.A. 1990. Isopach and borehole data as indicators of rifted margin evolution in southwestern Africa. *Mar. Petrol. Geol.*, **7**, 277–287.

Ruxton, B.P. and Taylor, G. 1982. The Cainozoic geology of the Middle Shoalhaven Plain. *J. Geol. Soc. Austr.*, **29**, 239–246.

Schmidt, P.W. and Embleton, B.J.J. 1981. Magnetic overprinting in southeastern Australia and the thermal history of its rifted margin. *J. Geophys. Res.*, **86**, 3998–4008.

Schmidt, P.W., Lackie, M.A. and Anderson, J.C. 1995. Palaeomagnetic evidence for the age of the Lapstone Monocline, NSW. *Austr. Coal Geol.*, **10**, 14–19.

Seeber, L. and Gornitz, V. 1983. River long profiles along the Himalayan arc as indicators of active tectonics. *Tectonophysics*, **92**, 335–367.

Seidl, M.A., Dietrich, W.E. and Kirchner, J.W. 1994. Longitudinal profile development into bedrock: an analysis of Hawaiian channels. *J. Geol.*, **102**, 457–474.

Seidl, M.A., Weissel, J.K. and Pratson, L.F. 1996. The kinematics and pattern of escarpment retreat across the rifted continental margin of SE Australia. *Basin Res.*, **12**, 301–316.

Smith, V. 1979. The Cainozoic geology and construction material resources of the Penrith–Windsor area, Sydney basin, New South Wales. Geological Survey of New South Wales, Unpublished Report, 1979/074.

Steckler, M.S. and Omar, G.I. 1994. Controls on erosional retreat of the uplifted rift flanks at the Gulf of Suez and northern Red Sea. *J. Geophys. Res.*, **99**, 12 159–12 173.

Steckler, M.S., Omar, G.I., Karner, G.D. and Kohn, B.P. 1993. Pattern of hydrothermal circulation within the Newark basin from fission-track analysis. *Geology*, **21**, 735–738.

Stephenson, R. and Lambeck, K. 1985. Erosion-isostatic rebound models for uplift: an application to south-eastern Australia. *Geophys. J. R. Astr. Soc.*, **82**, 31–55.

Summerfield, M.A. 1991. *Global Geomorphology*, Longman, London.

Sun, S.-S., McDonough, W.F. and Ewart, A. 1989. Four component model for Australian basalts. In: R.W. Johnson (ed.) *Intraplate Volcanism in Eastern Australia and New Zealand*, Cambridge University Press, Cambridge, pp. 333–347.

Taylor, G., Taylor, G.R., Bink, M., Foudoulis, C., Gordon, I., Hedstrom, J., Minello, J. and Whippy, F. 1985. Pre-basaltic topography of the Northern Monaro and its implications. *Austr. J. Earth Sci.*, **32**, 65–71.

Taylor, G., Truswell, E.M., McQueen, K.G. and Brown, M.C. 1990. Early Tertiary palaeogeography, landform evolution, and palaeoclimates of the Southern Monaro, NSW, Australia. *Palaeogeog. climat. ecol.*, **78**, 109–134.

Taylor, T.G. 1911. A discussion of the salient features in the physiography of eastern Australia. *Commwlth Bur. Met. Bull.*, **8**.

Tucker, G.E. and Slingerland, R.L. 1994. Erosional dynamics, flexural isostasy, and long-lived escarpments: A numerical modeling study. *J. Geophys. Res.*, **99**, 12 229–12 243.

Twidale, C.R. 1991. A model of landscape evolution involving increased and increasing relief amplitude. *Z. Geomorph.*, **35**, 85–109.

Twidale, C.R. 1994. Gondwanan (Late Jurassic and Cretaceous) palaeosurfaces of the Australian craton. *Palaeogeog. climat. ecol.*, **112**, 157–186.

Twidale, C.R. and Campbell, E.M. 1995. Pre-Quaternary landforms in the low latitude context: the example of Australia. *Geomorphology*, **12**, 17–35.

van der Beek, P., Andriessen, P. and Cloetingh, S. 1995. Morphotectonic evolution of continental rifted margins: Inferences from a coupled tectonic–surface processes model and fission track thermochronology. *Tectonics*, **14**, 406–421.

van der Beek, P. and Braun, J. 1999. Controls on post-mid-

Cretaceous landscape evolution in the Southeastern Highlands of Australia: Insights from numerical surface process models. *J. Geophys. Res.*, in press.

Veevers, J.J. (ed.) 1984. *Phanerozoic Earth History of Australia*. Clarendon Press, Oxford.

Weissel, J.K. and Hayes, D.E. 1977. Evolution of the Tasman Sea reappraised. *Earth Planet. Sci. Lett.*, **36**, 77–84.

Weissel, J.K. and Seidl, M.A. 1997. Influence of rock strength properties on escarpment retreat across passive continental margins. *Geology*, **25**, 631–634.

Wellman, P. 1974. Potassium–argon ages of the Cainozoic rocks of eastern Victoria. *J. Geol. Soc. Austr.*, **21**, 359–376.

Wellman, P. 1979a. On the Cainozoic uplift of the south-eastern Australian highlands *J. Geol. Soc. Austr.*, **26**, 1–9.

Wellman, P. 1979b. On the isostatic compensation of Australian topography, *BMR J. Austr. Geol. Geophys.*, **4**, 373–382.

Wellman, P. 1987. Eastern Highlands of Australia: their uplift and erosion. *BMR J. Austr. Geol. Geophys.*, **10**, 277–286.

Wellman, P. 1996. The effective elastic thickness of the Australian crust: a major control on the wavelength of topography and basins. *AGSO Res. Newslett.*, **24**, 16.

Wellman, P. and McDougall, I. 1974. Potassium–argon ages on the Cainozoic volcanic rocks of New South Wales, Australia. *J. Geol. Soc. Austr.*, **21**, 247–272.

Young, R.W. 1977. Landscape development in the Shoalhaven River catchment of southern New South Wales. *Z. Geomorph.*, **21**, 262–283.

Young, R.W. 1981. Denudational history of the south-central uplands of New South Wales. *Austr. Geogr.*, **15**, 77–88.

Young, R.W. 1983. The tempo of geomorphological change: evidence from southeastern Australia. *J. Geol.*, **91**, 221–230.

Young, R.W. 1989. Crustal constraints on the evolution of the continental divide of eastern Australia. *Geology*, **17**, 528–530.

Young, R.W. and McDougall, I. 1982. Basalts and silcretes on the coast near Ulladulla, southern New South Wales. *J. Geol. Soc. Austr.*, **29**, 425–430.

Young, R.W. and McDougall, I. 1983. Reply: Origin of coastal lowlands near Ulladulla, N.S.W. *J. Geol. Soc. Austr.*, **30**, 248–249.

Young, R.W. and McDougall, I. 1985. The age, extent and geomorphological significance of the Sassafras basalt, south-eastern New South Wales. *Austr J. Earth Sci.*, **32**, 323–331.

Young, R.W. and McDougall, I. 1993. Long-term landscape evolution: Early Miocene and modern rivers in southern New South Wales, Australia. *J. Geol.*, **101**, 35–49.

Young, R.W., Cope, S., Price, D.M., Chivas, A.R. and Chenall, B.E. 1996. Character and age of lateritic weathering at Jervis Bay, New South Wales. *Austr. Geog. Studs.*, **34**, 237–246.

Zuber, M.T., Bechtel, T.D. and Forsyth, D.W. 1989. Effective elastic thicknesses of the lithosphere and mechanisms of isostatic compensation in Australia. *J. Geophys. Res.*, **94**, 9353–9367.

Appendix: Theoretical derivation of a general form of the equilibrium long profile

We introduce here in summary form a new derivation of the form of the equilibrium long profile.

The simplest case for investigating stream long profiles and grade is that of a stream which flows over lithologies of uniform erosional resistance throughout its length, which is experiencing uniform rates of denudation throughout its catchment (the whole catchment including the divide and the base level are being lowered at the same rate), and which exhibits a systematic increase in discharge (Q) downstream such that Q is some power function of downstream distance (L):

$$Q = lL^{\lambda} \tag{A1}$$

where l and λ are constants. The rate of stream incision (I) along any reach is a function of stream power which is proportional to the product of Q and stream slope (S) (Leopold *et al.*, 1964). Therefore the rate of incision is given by:

$$I = iSQ \tag{A2}$$

where i describes the proportion of stream power that is expended in incision.

In order to maintain a stable form for the long profile over time all points along the stream must incise at the same rate. If any reach incises less quickly, then the slope at the downstream extremity of that reach will increase, resulting in an increased rate of incision and a tendency to restore the original form of the long profile (and vice versa).

Equation A2 can be modified to describe the form of a graded stream:

$$I_{\text{grade}} = iSQ \tag{A3}$$

where I_{grade} is a constant rate of incision equal to the rate of denudation throughout the catchment. Substitution of equation A1 and rearrangement yields:

$$S = \frac{I_{\text{grade}}}{il}L^{-\lambda} \quad \text{or} \quad S = kL^{-\lambda}$$

$$\text{or} \quad \ln S = \gamma_1 - \lambda \ln L \tag{A4}$$

where $k = I_{\text{grade}}/il$ and $\gamma_1 = \ln k$ is a constant. Therefore, it is a condition of grade in this form that a plot of the logarithm of downstream distance against the logarithm of slope, which we name a DS plot, should describe a straight line. The slope of the DS plot, λ, depends on the form of the relationship between discharge and downstream distance; that is, whether or not the relationship between discharge and distance is linear. λ is thus a function of catchment geometry and rainfall/runoff characteristics. The intercept of the DS plot,

γ_1, reflects the proportion of total stream energy used in incision (a function of stream geometry), the denudation rate and the scale of the relationship between discharge and downstream distance.

The foregoing discussion has assumed that both the local base level and the divide are being lowered at the same rate, but given that these rates are determined by separate, independent processes, this is unlikely to be the case. As long as it is assumed that the rate differential does not exceed the stream's capacity to adjust, then differing rates of base-level and divide lowering should not complicate the model except to note that the overall slope of the profile, as expressed in γ_1, will change through time. For example, in a stream where the rate of base-level lowering exceeds the rate of divide lowering the value of γ will increase over time. This leads to the very important consequence that changes in the slope of the long profile between ancient and modern rivers need not necessarily constitute evidence of tilting or climatic change.

This allows us to consider uplift resulting in a knickpoint as an extreme example of the case where the rate of lowering of the base level relative to the divide exceeds, for a period of time, the ability of the stream to accommodate it. As a result, the stream is displaced from equilibrium. The stream reattains a graded profile by the headward propagation of the knick-point. Unlike Hack's (1975) model of the equilibrium long profile which cannot accommodate Gardner's (1983) observation that knickpoints decline as they propagate upstream, this model *demands* such a decline and that the post-uplift graded profile be steeper than the pre-uplift profile because the difference in elevation between the divide and the local base level has increased, at least until the divide itself is lowered by drainage head processes (Dietrich *et al.*, 1986).

Equation A3 can be expanded to accommodate the effects of lithological variation.

$$I_{grade} = \frac{iSQ}{R} \tag{A5}$$

where R is some measure of lithological resistance to erosion. After appropriate substitutions and simplification this equation yields

$$S = \frac{RI_{grade}}{il} L^{-\lambda} \quad \text{or} \quad \ln S = \gamma - \lambda \ln L \tag{A6}$$

γ is equal to $\ln(RI_{grade}/il)$ and is similar to γ_1 in equation A4 except that it also incorporates lithological resistance. In the absence of any corresponding change in stream geometry or in the relationship between discharge and stream length, the effect of a lithological change on the DS plot is a parallel shift (i.e. a change in the intercept with no change on the gradient of the line). In producing such a result, a change in lithological resistance is indistinguishable from a change in the factors affecting the scale of the scale of the length–discharge relationship.

The DS model presented here can be shown to incorporate Hack's SL model as a special case where $\lambda = 1$. For the more general case of $\lambda \neq 1$, the graded long profile has the form:

$$H = H_0 - k\frac{L^{1-\lambda}}{1-\lambda} \tag{A7}$$

The interpretation of H_0 requires some elaboration. As the divide is approached, stream power tends to zero so that at some point slope or drainage head processes (Dietrich *et al.*, 1986) rather than fluvial processes determine the local rate of lowering. H_0 must therefore be seen not as an estimate of the true elevation of the divide but as an estimate of the theoretical elevation of the divide if the slope processes were not active.

The form of the long profile derived here can be shown to be identical to the three broad types of process-based models of bedrock channel erosion (the shear-stress model of Howard and Kerby (1983); the stream power model of Seidl *et al.* (1994); and the sediment scour model of Howard *et al.* (1994)). Howard *et al.* (1994) have argued that all three can be expressed in the form:

$$\frac{dy}{dt} = -kA^mS^n \tag{A8}$$

Hack (1957), Montgomery and Dietrich (1992) and Seidl *et al.* (1994) have identified a relationship between catchment area (A) and stream length (L) of the form:

$$A = k_2L^a \tag{A9}$$

After appropriate substitution and rearrangement equation A8 can be simplified to:

$$S = -kL^{-\frac{am}{n}} \tag{A10}$$

Equation A10 has an identical form to equation A4, with λ equal to am/n. It follows that, although the DS model was developed on the basis of form, it could have been developed equally validly on the basis of process.

MODELLING LONG PROFILES

Equation A7 can be used to derive mathematical descriptions of long profiles. Rearranging equation A7 yields:

$$\ln(H_0 - H) = \ln\left(\frac{k}{1-\lambda}\right) + (1-\lambda)\ln L \tag{A11}$$

For any value of H_0 the value of λ and k can be determined from the slope of a least-squares linear regression of the log of downstream distance versus the log of the fall ($H_0 - H$). These can be substituted into equation A7 to give a mathematical description of the long profile. The goodness of fit of each description so generated can be evaluated by comparison with the observed profile using the criterion of the standard error of the estimate of elevation. This procedure is repeated and the best estimate of H_0 determined by converging on that

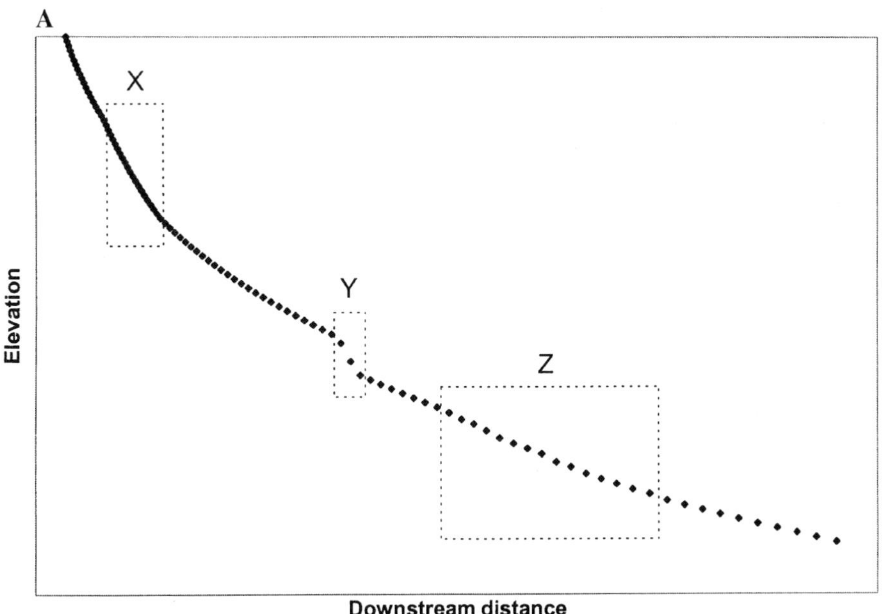

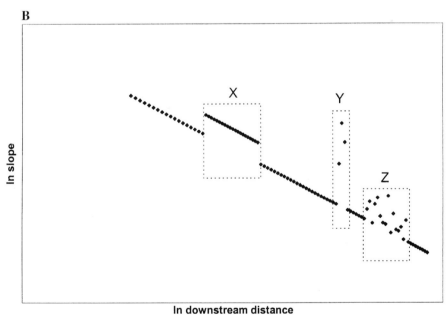

Figure A11.1 *The (A) long profile and (B) DS plot of a hypothetical stream showing a change in lithological resistance (X), a well-defined knickpoint (Y) and a broad zone of disequilibrium (steepening) (Z). Note that the DS plot is much more sensitive to these changes than the long profile*

value which yields the best fit between the modelled and observed long profile.

In summary, a DS plot for a graded river should consist of a number of straight line segments (each of which corresponds to a different lithology) having the same slope (λ) but different intercepts (γ) (Figure A11.1). Well-defined knickpoints are easily identifiable as outliers on the DS plot. DS plots can also be used to identify less well-defined under- or oversteepened reaches. Such reaches are represented on the DS plot by segments of different slope (λ), but great care needs to be exercised to rule out other possible causes of a change in the value of λ such as changes in catchment characteristics.

12

Morphotectonic evolution of the South Atlantic margins of Africa and South America

Roderick W. Brown, Kerry Gallagher, Andrew J.W. Gleadow and Michael A. Summerfield

12.1 Introduction

Denudation was recognized as a significant process in some of the earliest geodynamic models which sought to explain the evolution of 'Atlantic-type' (Dewey and Bird, 1970) rifted continental margins (Gilluly, 1964; Hsü, 1965; Sleep, 1971; Falvey, 1974). The impetus for explicitly addressing onshore denudation as an integral and important process in the tectonic development of these margins was the discovery of extensive sedimentary sequences, up to 12–15 km thick, in offshore marginal basins. Extensive drilling and modern high-resolution seismic reflection data from many passive margins have confirmed that the offshore accumulation of large volumes of clastic sediments is a general characteristic of rifted continental margins (Watkins *et al.*, 1979, 1992).

To understand how the topography, drainage patterns and sediment source areas of the sub-aerial parts of continental margins have changed over geological time scales we clearly need information on variations in rates of denudation over time spans of 10^6–10^8 a. In principle, sediment volume data have great potential for deciphering the history of onshore denudation in the adjacent hinterland (Poag and Sevon, 1989; Rust and Summerfield, 1990; Pazzaglia and Brandon, 1996). In practice, however, reconstructing onshore landscape evolution from offshore sedimentary sequences alone is inhibited by the inherent lack of control on the temporal and spatial variations of the source area of the sediment. Historically, quantifying long-term rates of post-break-up denudation and the associated topographic history of continental margins has proved to be a particularly difficult task. An obvious hindrance is the significant amount of post-break-up denudation that typically accompanies and follows margin formation and which effectively removes any stratigraphic record for this phase of the margin's history. It is here that the application of low-temperature thermochronologic techniques, such as apatite fission-track analysis (AFTA), can make a major contribution to an understanding of the morphotectonic history of continental margins by providing quantitative point estimates of the depth of rock removal over time.

Apatite fission-track analysis is a well-established thermochronological technique that is sensitive to temperatures of <~130°C on time scales of 1–100 Ma (Gleadow *et al.*, 1986; Green *et al.*, 1989; Hurford, 1991) (see Chapter 4). As the temperature in the Earth's crust generally increases with depth, denudation in response to both erosional and tectonic processes causes rocks to cool as they approach the surface. For <5 km of denudation the temperatures of interest will typically be <200°C, and so this technique is particularly well suited to the study of long-term denudation histories (Summerfield, 1991; Brown *et al.*, 1994a). In this chapter we discuss the implications of fission-track data for the interpretation of the morphotectonic history of the originally conjugate passive continental margins of South America (south-east Brazilian sector) and Africa (Namibian and South African sectors).

Geomorphology and Global Tectonics. Edited by Michael A. Summerfield.

12.2 Morphology of the Margins

The present-day maximum elevations along the margins of both southwestern Africa and southeastern Brazil are relatively high (1500–2000 m). The hinterland elevations, however, differ significantly, the average elevation in southern Africa being about 1200 m (Gilchrist and Summerfield, 1990) whereas it is <200 m over extensive areas in southern Brazil and northeastern Argentina (Figures 12.1–12.3). The morphological asymmetry between the two margins is most striking between 29°S and 42°S in South America and the pre-break-up conjugate sector between 20°S and 34°S in Africa (Figure 12.2). This part of the South American margin is underlain by a complex series of elongate, fault-bounded sedimentary basins oriented across the present continental margin (Figure 12.4), and it is clearly not a simple Atlantic-type rifted margin as it has experienced significant syn- and post-rift intracontinental deformation (Urien and Zambrano, 1973; Conceicao *et al.*, 1988; Nürnberg and Müller, 1991). Further north the mean elevation of the South American margin increases abruptly (from ~200 m to ~1200 m) and is matched on the African margin by the continuation of the elevated escarpment zone of northern Namibia and Angola (Figures 12.2 and 12.3).

The present-day margin of southeast Brazil is characterized by elevated (500–1500 m) regions around, and inland of, Rio de Janeiro. The Serra de Mantequira/Mar is composed of Precambrian crystalline rocks, and the Serra Geral escarpment south of Florianopolis has a capping of several hundred metres of Paraná flood basalt, directly overlying the Mesozoic–late Palaeozoic sediments of the Paraná Basin (Figure 12.4). The low elevation coastal plain in this region is underlain by Precambrian rocks of the Brasilia mobile belt (Trompette, 1994). Between these two regions lies the low-lying and dissected Ponta Grossa structural arch (running ~NW from ~25°S), within which Precambrian basement is also exposed. North of Rio de Janeiro, the geology is dominated by the São Françisco craton which was originally adjacent to the northwestern region of the Congo craton in West Africa prior to rifting. These two cratonic regions also effectively form the nuclei for the late Precambrian Ribeira–Mantiqueira and West Congo fold belt. The elevation in the São Françisco craton region is typically 500–1000 m, although there is no well-developed margin escarpment equivalent to that seen further south.

The topography in the southwest Cape region of South Africa is dominated by the high-relief mountains of the Cape Fold Belt which reach elevations of over 2000 m and are underlain by the folded and resistant quartzite-dominated lithologies of the Siluro-Ordovician Table Mountain Group rocks (Tankard *et al.*, 1982; Söhnge and Hälbich, 1983). Further inland, a well-defined escarpment with a mean summit elevation of ~1500m and relief of ~800 m is underlain by the flat-lying sedimentary strata of the Permo-Triassic Karoo sequence (Karoo Supergroup). The location of the escarpment in the southwest Cape coincides closely with the southern limit of exposed mid-Jurassic dolerite sills, which form a significant component (up to 70% in places) of the stratigraphic section in this area (Winter and Venter, 1970; Rowsell and De Swardt, 1976). Further north the escarpment is less well defined and is replaced by a highland region underlain by the crystalline metamorphic rocks of the Namaqua metamorphic belt which separate the high-relief coastal region from a low-relief interior plateau at ~1000 m elevation. This highland region is truncated at around 28°S by the Orange River valley, north of which the margin topography is again dominated by a well-defined escarpment region underlain by the gently east-dipping (<5°) sedimentary strata of the latest Precambrian Nama Group (Tankard *et al.*, 1982). In this region the escarpment is generally defined by the western boundary of the Nama Group sediments which form a near-horizontal capping to the highly deformed crystalline basement beneath. Between about 23°S and 28°S the region immediately inland of the escarpment has been dissected to varying degrees by the Fish River, a major tributary of the Orange River.

In Namibia the escarpment region extends northward to approximately 22°S where it merges with the NE-trending highland region of the Khomas Hochland which is underlain by the metamorphic and igneous rocks of the Pan-African (~500 Ma BP) Damara metamorphic belt (Figure 12.4). North of the Khomas Hochland the escarpment is replaced in northern Namibia by a gradual rise from the coast to the elevated interior. Superimposed on this regional morphology are the isolated and spectacular mountains underlain by early Cretaceous igneous intrusive complexes such as Brandberg (2559 m), Okenyenya (1902 m) and Erongo (2200 m) (Milner *et al.*, 1993, 1995a; Watkins *et al.*, 1994). In southern Angola (north of 18°S) the margin is once again characterized by a well-defined escarpment region which locally reaches elevations of over 2500 m.

A significant feature of the bathymetry of the South Atlantic are the two anomalously shallow aseismic ridges, the Walvis Ridge and the Rio Grande Rise, which extend, respectively, east and west of the present-day

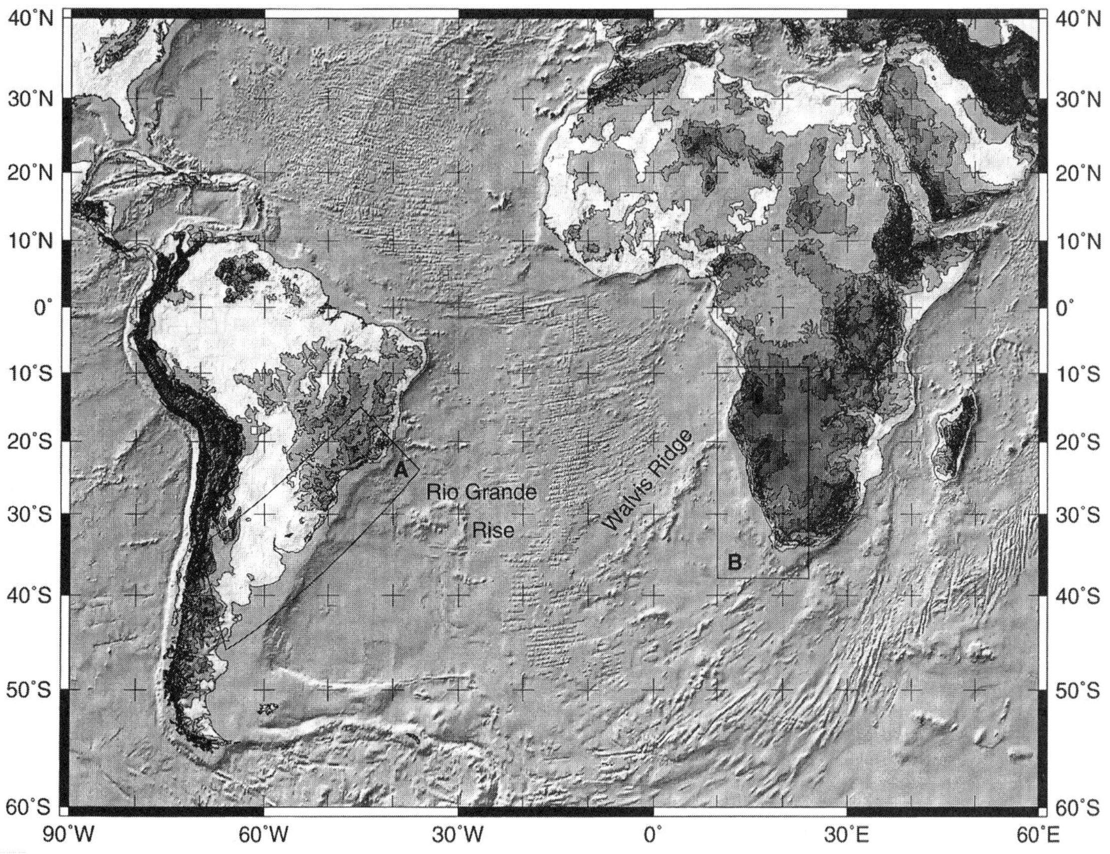

Figure 12.1 *Shaded relief image of South Atlantic bathymetry and the topography of South America and Africa. The image was created using the 5-minute ETOPO 5 digital data set using GMT software (Wessel and Smith, 1991). The contour interval for the sub-aerial topography is 300 m. The polygons labelled A and B indicate the regions shown in Figures 12.2 and 12.4. The Walvis Ridge and Rio Grande Rise are clearly visible offshore, and in both cases terminate at the respective continental margins at a regional 'low' along the margin topography. The most striking morphological difference between the two South Atlantic margins is the high topography of southern Africa relative to South America (Summerfield, 1996)*

mid-ocean ridge (Figure 12.1). These two ridges project onshore into two regions of relatively low elevation, the Ponta Grossa Arch in Brazil and the Etosha Basin in northern Namibia, which may represent once conjugate depressions in the pre-break-up regional topography (Gallagher and Hawkesworth, 1994).

12.3 Regional Geology and Palaeogeography

12.3.1 PRE-BREAK-UP PARANÁ AND KAROO BASINS

The pre-break-up geological evolution of southern Africa and Brazil was dominated by the development of the Palaeozoic–Mesozoic Karoo and Paraná Basins, respectively. The Paraná Basin, located primarily in southern Brazil, also extends into Paraguay, Uruguay and Argentina and covers a total area of about 1.4×10^6 km^2. Isopach data imply a broadly concentric depocentre with a maximum sediment thickness of ~5 km (Zalan *et al.*, 1987).

The Karoo Basin in South Africa formed as an extensive foreland basin (~0.85×10^6 km^2) adjacent to the Cape Fold Belt during the early Permian (Tankard *et al.*, 1982; Söhnge and Hälbich, 1983). Thick sedimentary sequences (Karoo Supergroup) accumulated along its southern margin, but the Karoo sedimentary sequence as a whole thins rapidly northwards. In the western Cape region of South Africa the Karoo

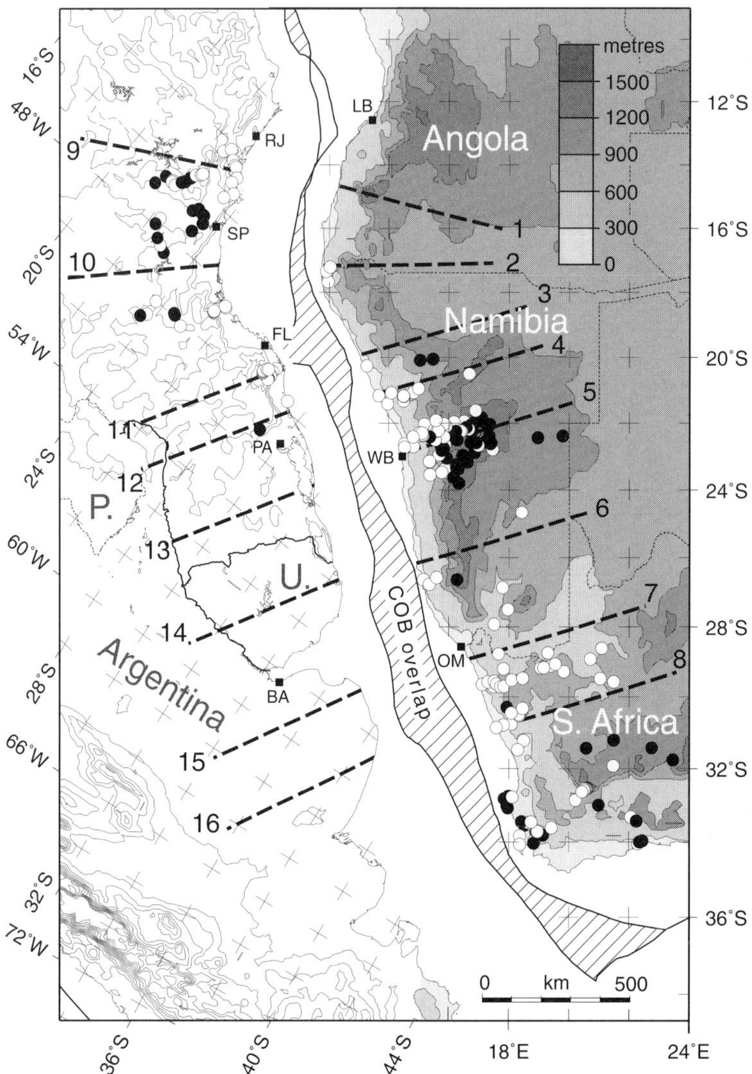

Figure 12.2 Schematic pre-break-up reconstruction of the regions indicated in Figure 12.1. The topography is from the digital ETOPO 5 data set and is shown contoured at 300 m intervals. The locations of fission-track samples are indicated by circles with the white and black representing, respectively, fission-track ages <134 Ma and >134 Ma. The locations of the cross-sections shown in Figure 12.3 are indicated by the numbered dashed black lines. Letters indicate the following: P – Paraguay, U – Uruguay, RJ–Rio de Janeiro, SP–São Paulo, FL–Florianopolis, PA–Pôrto Alegre, BA–Buenos Aires, LB–Lobito, WB–Walvis Bay, OM–Oranjemund, COB–Continent–Ocean Boundary

Figure 12.3 Topographic cross-sections across the South Atlantic margins of Africa and South America derived from the GLOBE 30' Digital Chart of the World data sets. The thin solid black line indicates the topography along the cross-section while the heavy solid line indicates the mean topography along a 20 km wide swath along the same transect. Mean elevations were calculated over 20 km sections of the cross-section (20×20 km blocks). The thin broken lines indicate the maximum and minimum elevations from within the 20×20 km blocks along the cross-section and the difference between them provides an indication of local relief. Apatite fission-track ages within 50 km of each cross-section are shown in their projected position relative to the adjacent cross-section and the age is indicated by the vertical scale on the right. As in Figure 12.2, white circles indicate samples with ages <134 Ma and black circles samples with ages >134 Ma

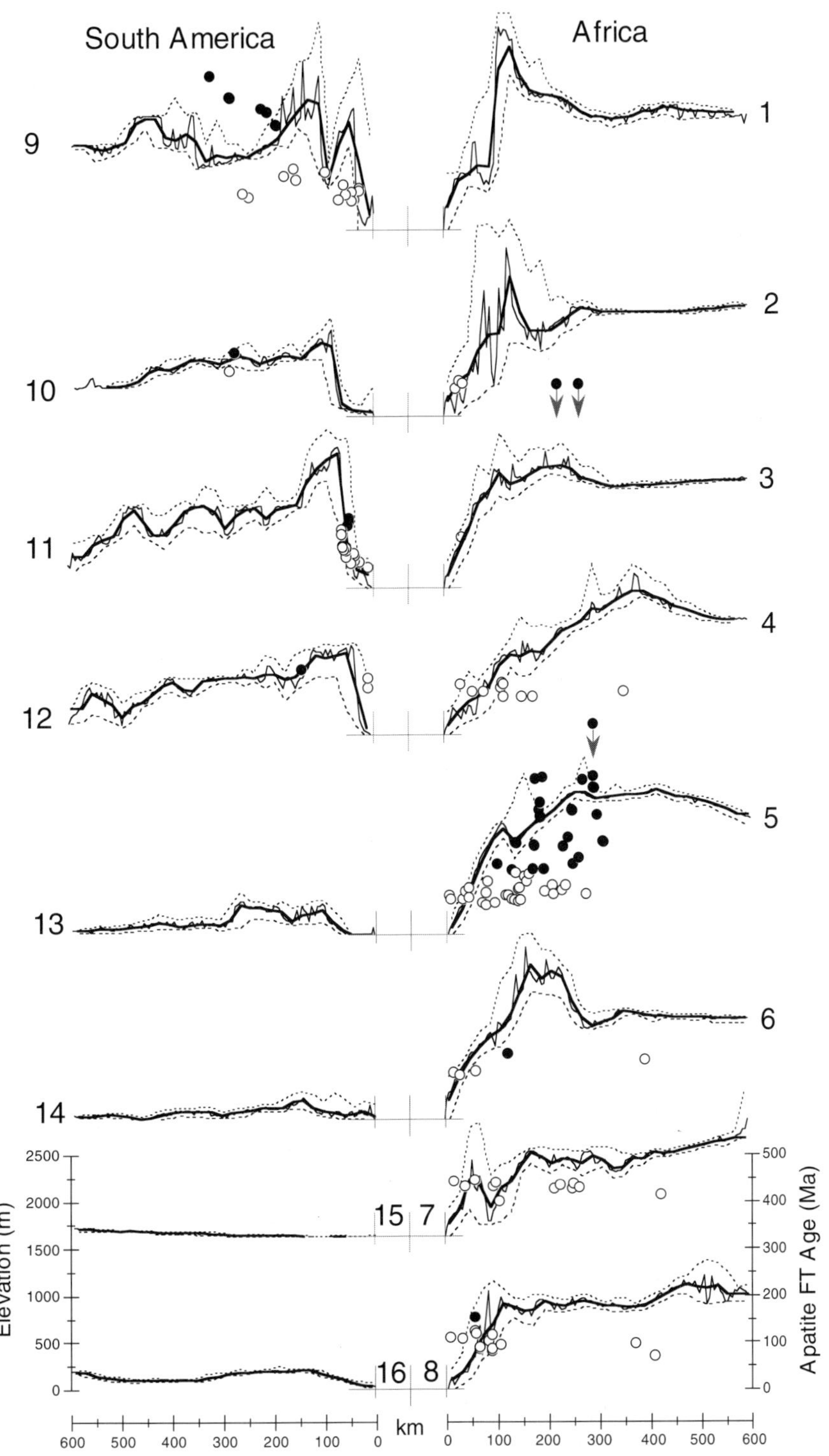
South America
Africa
9
1
10
2
11
3
12
4
13
5
14
6
15
7
16
8
Elevation (m)
2500
2000
1500
1000
500
0
Apatite FT Age (Ma)
500
400
300
200
100
0
km
600
500
400
300
200
100
0
0
100
200
300
400
500
600

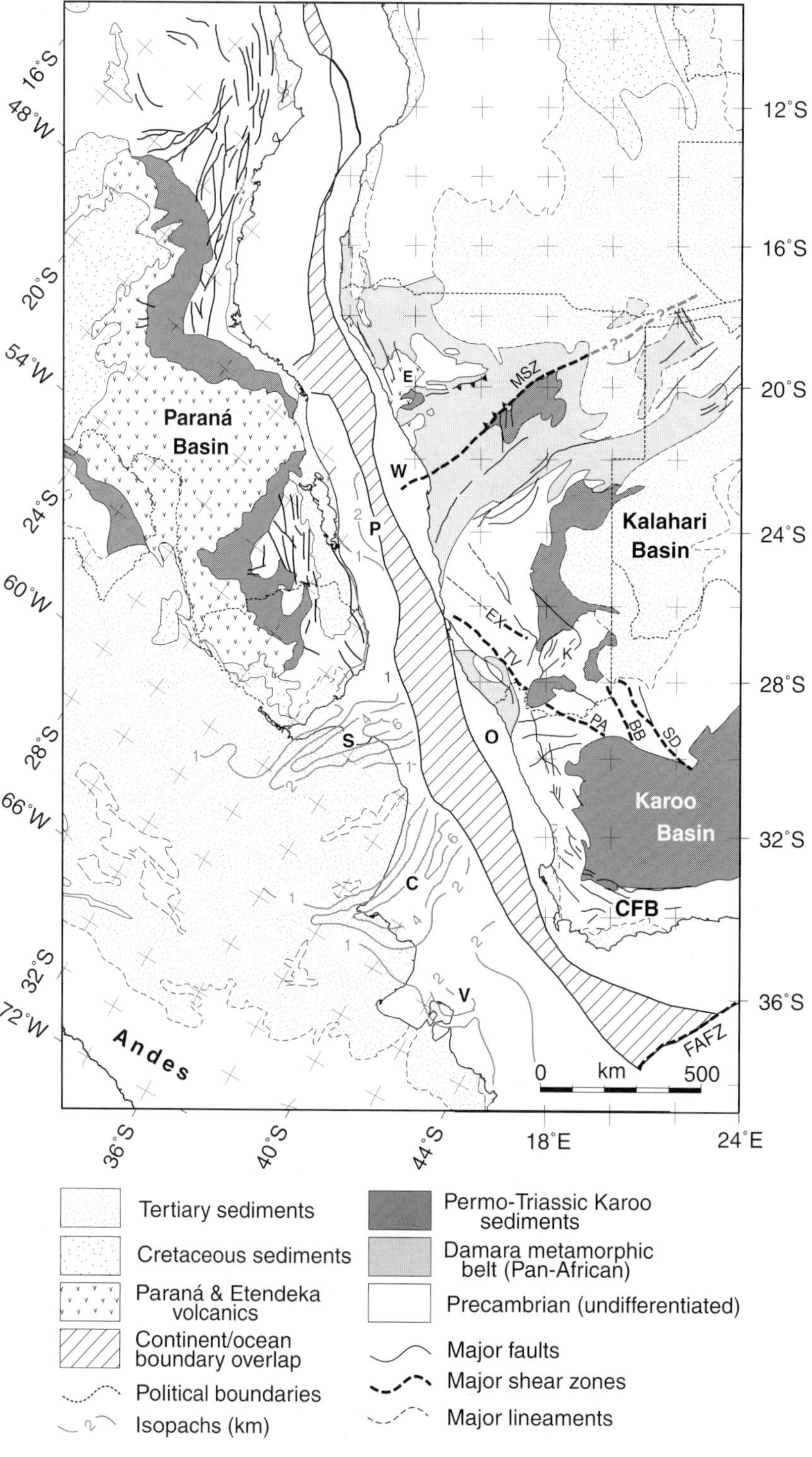

16°S
48°W
20°S
54°W
24°S
60°W
28°S
66°W
32°S
72°W
36°S
40°S
44°S
18°E
24°E
12°S
16°S
20°S
24°S
28°S
32°S
36°S
Paraná Basin
Kalahari Basin
Karoo Basin
Andes
E
W
P
S
C
V
O
K
MSZ
EX
TV
PA
BB
SD
CFB
FAFZ
0 km 500
Tertiary sediments
Cretaceous sediments
Paraná & Etendeka volcanics
Continent/ocean boundary overlap
Political boundaries
Isopachs (km)
Permo-Triassic Karoo sediments
Damara metamorphic belt (Pan-African)
Precambrian (undifferentiated)
Major faults
Major shear zones
Major lineaments

sequence reaches a maximum preserved thickness of ~4.5 km, but thins to less than a few hundred metres towards the northern (~29°S) and western margins of the basin (Cloetingh *et al.*, 1992; Visser, 1995). In Namibia the occurrence of Karoo sequence rocks is restricted to smaller basins, mainly in the southeast and northwest of the country. In the southeast the sequence has a maximum thickness of ~1.5 km, while in the northwest (Huab, Omingonde and Goboboseb Basins), where the spatial pattern of deposition was strongly controlled by synsedimentary faulting, maximum thicknesses vary between 250 m and ~700 m (Dingle *et al.*, 1983; Hegenberger, 1988; Horsthemke *et al.*, 1990). Except in the Etendeka region of Namibia, Karoo rocks do not occur along the coastal part of the margin, which is primarily underlain by a variety of Precambrian crystalline rocks. The present extent of the Karoo cover is therefore restricted to the interior regions, with the present westernmost exposures located between 100 and 300 km inland from the coastline.

In northern Namibia, remnants of the basal Karoo Dwyka Formation (Permian) occur in exhumed palaeovalleys where they underlie elevated river terraces (Martin, 1953, 1973). These glacial deposits, together with associated striated bedrock surfaces, such as those exposed at Nooitgedagt (near Kimberley, South Africa), indicate that the present land surface in southwest Africa locally represents an exhumed Permo-Carboniferous landscape (Visser, 1987, 1995). The Itararé Formation is considered to be the Brazilian equivalent of the Dwyka Formation (Eyles and Eyles, 1993; Visser 1989). The Dwyka/Itararé formations are mostly glacio-marine in southwestern South Africa, southeastern Brazil and southern Namibia, but terrestrial in northern Namibia (Visser, 1989; Horsthemke *et al.*, 1990).

Marine–terrestrial transitions provide conclusive information on absolute elevation (at least relative to contemporary sea level). Inferences regarding the stratigraphic position of the last major marine–terrestrial transition in the main Karoo Basin in South Africa are based on the interpretation of the Upper Permian Ecca Formation as shallow marine, and the successive early Triassic Beaufort Formation as predominantly fluvial (Visser, 1989; Horsthemke *et al.*, 1990). This dates a regionally extensive marine–terrestrial transition between 260 and 245 Ma across most of southern Africa (Visser, 1989). Similarly, in southern Brazil, Mesozoic deposition in the Paraná Basin was exclusively terrestrial following the development of a minor unconformity in the late Permian, (Zalan *et al.*, 1987). Apart from minor marine transgressions during the late Cretaceous and Cenozoic, which affected only the shelf zone and a narrow coastal strip (≤40 km wide), southwestern Africa has remained a terrestrial environment since the Triassic (Martin, 1973; Dingle *et al.*, 1983). In contrast, thick sequences of marine sediments (up to 2.5 km) within the Colorado, San Jorge and Salado Basins indicate that the southern sector of the South American margin was flooded during the late Cretaceous (Maastrichtian) and most of the Tertiary (Urien and Zambrano, 1973).

In Namibia, the terrestrial Etjo Formation, comprising aeolian sandstones and associated playa lake deposits, is interbedded with the lower Etendeka lavas, indicating a minimum age for these deposits of early Cretaceous (~135 Ma) (Horsthemke *et al.*, 1990; Swart, 1992; Milner *et al.*, 1995a, b). The massively dune-bedded Botacatu sandstones in southern Brazil also appear to be interbedded with the Paraná volcanics, suggesting a minimum age of early Cretaceous there as well (Bigarella, 1970; Zalan *et al.*, 1987, Horsthemke *et al.*, 1990). These two formations appear to be lithologically equivalent to the largely aeolian Clarens Formation in South Africa (Dingle *et al.*, 1983). This latter formation has a maximum thickness of 300 m in the upper Orange River valley, but is generally about 100 m thick and therefore of similar thicknesses to the Botacatu Sandstone in Brazil (Zalan *et al.*, 1987). These units collectively indicate terrestrial, arid to semi-arid conditions across southwestern Gondwana from the early Jurassic to early Cretaceous.

12.3.2 SYN- AND POST-BREAK-UP MAGMATISM

The early Cretaceous Paraná–Etendeka flood basalts were erupted sub-aerially and directly on to the aeolian dune-bedded Botacatu Sandstone in Brazil and Etjo Sandstone in Namibia. Intercalated sand dunes tens of metres thick between some flows show that the continental conditions prevailed during the eruptions (Rocha-Campos *et al.*, 1988; Milner *et al.*, 1995a, b).

Figure 12.4 *Simplified geological map of the areas shown in Figure 12.2. Shear zones: MSZ – Mwembeshi, EX – Excelsior, TV – Tantalite Valley, PA – Pofadder, BB – Brakbos, SD – Staussberg–Doornberg. Offshore basins: V – Valdes, C – Colorado, S – Salado, P – Pelotas, W – Walvis, O – Orange. Other features: CFB – Cape Fold Belt, K – Karas Mountains, E – Etendeka Volcanic Province, FAFZ – Falkland–Agulhas Fracture Zone*

The Paraná–Etendeka volcanics represent one of the largest continental flood basalt provinces on Earth, although there is a distinct asymmetry in the present-day distribution of volcanics in Brazil and Namibia. In Brazil the Paraná flood basalts are areally extensive, covering some 1.2×10^6 km^2, with a maximum preserved thickness of about 1700 m, whereas the Etendeka remnants in Namibia cover about 8×10^4 km^2, with a maximum preserved thickness of only ~900 m (Erlank *et al.*, 1984; Hawkesworth *et al.*, 1992; Milner *et al.*, 1995b). There is also a pronounced contrast in the petrological/geochemical nature of the volcanics, with more than 90% of the Paraná province consisting of tholeiites and the remainder being more silica-rich rhyolites/quartz latites. In contrast, rhyolites constitute ~50% of the exposed Etendeka volcanic sequence. The more silica-rich volcanics in the Paraná tend to occur only in the southeast of the region, close to the present-day coast, and occur as the uppermost components of the sequence (Hawkesworth *et al.*, 1992). Nevertheless, it is generally accepted that these two regions represent a single magmatic province, with some of the individual eruptive units extending up to a distance of 340 km and being identifiable on both sides of the Atlantic Ocean (Milner *et al.*, 1995b).

The age and duration of the magmatic episode has recently been estimated by isochron analysis of $^{40}Ar/^{39}Ar$ data acquired from laser-spot total-fusion measurements on surface samples and suites collected from boreholes through the lava pile in Brazil. The results suggest that magmatism commenced around 136–138 Ma BP, 500–1000 km from the present-day coastline of southeast Brazil, and the eruption age appears to decrease to about 127 Ma BP at the coast of Brazil (Turner *et al.*, 1994; Stewart *et al.*, 1996). The majority of the sequence appears to have been erupted between 133 and 127 Ma BP (Stewart *et al.*, 1996). Rather than being situated beneath the rift margin, it is inferred that the plume was initially located inland of the present Brazilian margin, and that South America moved northwestwards relative to it. Additionally, $^{40}Ar/^{39}Ar$ and palaeomagnetic analyses on intrusive and volcanic rocks from the Etendeka region of Namibia point to a more restricted period of magmatism of about 1 Ma duration occurring at 132 ± 0.7 Ma BP (Renne *et al.*, 1996).

The origin of this flood basalt province is ultimately attributable to the influence of a mantle plume, although there is considerable debate concerning the relative role of the plume and the overlying continental lithosphere as magmatic source regions (e.g. Gallagher and Hawkesworth, 1992, 1994; Hawkesworth *et al.*, 1992; White and McKenzie, 1995; Turner *et al.*, 1996). Offshore, the Rio Grande Rise and Walvis Ridge formed as a result of the passage of the spreading oceanic lithosphere over the mantle plume, which is now located close to the island of Tristan da Cunha. The morphology of the offshore ridges is also asymmetric, with approximately twice the excess volume of oceanic crust in the Rio Grande Rise relative to the Walvis Ridge (Gallagher and Hawkesworth, 1994).

In general, we would expect some dynamic uplift associated with a buoyant mantle plume, of the order of 500–1000 m, depending on the plume temperature and mantle viscosity. There is no direct stratigraphic evidence for such uplift in Brazil, as the prevailing sedimentary environment in the Paraná Basin was already continental, as indeed it was over a large region of southern Africa (Visser, 1995). Cox (1989) has suggested that present-day drainage patterns in Brazil and Namibia reflect long-lived uplift associated with plume-induced magmatic underplating in the late Jurassic–early Cretaceous. However, this argument is based primarily on the existence of drainage away from the present-day elevated margin, although this may be a consequence, at least in part, of erosional rebound and flexure of the margin following break-up (Gilchrist and Summerfield, 1990; Summerfield, 1990).

Early Cretaceous mafic dykes are ubiquitous along the southwest African margin. These dykes, which are related to final break-up (Reid *et al.*, 1990, 1991), have been dated in the southwest Cape (~130–135 Ma BP), northern Richtersveld in the northwest Cape (~134 Ma BP), and at Horingbaai in northern Namibia (~125 Ma BP). Most of the observed Ponta Grossa dykes in Brazil are oriented approximately perpendicular to the present-day margin. Geochemically, these dykes are similar to the high Ti/Y lavas (129–138 Ma BP) and are inferred to be their feeder dykes. A younger suite of coast-parallel dykes with more geochemical affinity to low Ti/Y lavas (127–133 Ma BP) appears to represent late-stage magmas associated with continental break-up and ultimately the Tristan da Cunha plume (Hawkesworth *et al.*, 1992; Stewart *et al.*, 1996). The depth of intrusion of these dykes is difficult to constrain, but depths less than 1–2 km seem unlikely. Their present exposure therefore indicates minimum post break-up denudation of this order.

Major early Cretaceous alkaline intrusive complexes (135–125 Ma BP) are common in Namibia (Watkins *et al.*, 1994; Milner *et al.*, 1995a) and Brazil (Gomes *et al.*, 1990). The Namibian intrusions also provide some indirect constraints on post-intrusive denudation. For example, the Brandberg complex in northwest Namibia has a peak elevation of ~2600 m, although the regional landsurface is ~2000 m lower, suggesting at least ~2000 m of local, post-intrusion denudation. Similarly, the Okenyenya intrusion located ~70 km further east has a relief of ~1000 m and therefore indicates that at least this much denudation has occurred locally. The Cape Cross and Messum centres nearer the continental margin are the same age but are deeply eroded, suggesting a minimum of 1–2 km for post-intrusive denudation by analogy with the two previous locations.

12.3.3 POST-BREAK-UP SEDIMENTATION AND MAGMATISM

After the eruption of the Paraná–Etendeka flood basalt in southern Brazil and Namibia, a thin (<250 m) veneer of continental sediments was deposited in Brazil (Bauru Group). Of probable late Cretaceous age, these contain conglomerates with basaltic fragments and fluvial sediments with crocodile/dinosaur fossils. Minor alluvial Cenozoic sediments are also distributed in the northwestern, eastern and southern regions of the Paraná Basin (Zalan *et al.*, 1987). In southern Africa the post-Etendeka geology is dominated by the extensive but thin (generally less than 200 m) Kalahari Basin sediments which cover much of central southern Africa (Thomas and Shaw, 1990). The age of the base of the Kalahari Basin sequence is thought to be late Cretaceous to earliest Cenozoic (Thomas and Shaw, 1990; Partridge, 1993). In the northern Etendeka province of Namibia the lava sequence is preserved within narrow, coast-parallel, fault-bounded half-graben (downthrown to the west). In the same area a conglomeratic deposit, consisting entirely of basaltic clasts derived from the west, was deposited within an active half-graben structure (Ward and Martin, 1987). These structures clearly post-date the volcanism, and indicate significant tectonism and erosion of the lava sequence at some time after ~124 Ma BP.

Numerous intrusive kimberlite and associated alkaline pipes of late Cretaceous age occur within South Africa and southern Namibia. At some of these sites crater facies sedimentary sequences are still preserved indicating minimal net denudation (probably less than 100 m) in these areas since the late Cretaceous–early Cenozoic (Smith, 1986; de Wit *et al.*, 1992). In Namibia, terrestrial sedimentation during the Cenozoic was largely restricted to a 150-km-wide zone along the southern sector of the margin – the region now occupied by the Namib Sand Sea (Ward, 1987). The sequence here is thin (generally less than 300 m) and the oldest unit, the Tsondab Sandstone Formation which overlies the basement unconformity surface, is probably diachronous with a possible maximum age of early Palaeocene (Ward, 1987, 1988; Ward and Corbett, 1990). The chronology of the upper part of this sequence has been clarified through the correlation of successive struthious egg types with associated micromammals and indicated an age range of Miocene to Quaternary (Senut *et al.*, 1994; Senut and Pickford, 1995).

12.4 Continental Break-Up and Rifting in the South Atlantic

12.4.1 BREAK-UP AND RIFTING CHRONOLOGY

Continental rifting between South America and Africa began during the late Jurassic (~150 Ma BP) (Urien and Zambrano, 1973; Zambrano and Urien 1974; Urien *et al.*, 1976; Nürnberg and Müller, 1991). (It should be noted that in this chapter we have used the chronological scale of Harland *et al.* (1990) which differs from the time scale of Haq *et al.* (1987, 1988) used provisionally by Brown *et al.* (1995) to calibrate the seismic sequence stratigraphy of South African offshore basins. The stage boundaries of Harland *et al.* (1990) are typically 5 to 15 Ma older than those used by Brown *et al.* (1995).) This rifting followed a prolonged period of complex extensional and strike-slip motion between several microplates which now constitute the Patagonian massifs and the Falkland Plateau of South America (Marshall, 1994a, b; Platt and Philip, 1995). Considering primarily the Namibian margin, Maslanyj *et al.* (1992) initially suggested that rifting began as early as the Permo-Triassic, but they have subsequently presented a revised picture (Light *et al.*, 1992, 1993) indicating rift onset during mid–late Jurassic (Rift Stage I), with a later Hauterivian–Barremian (130–120 Ma BP) rift stage (Rift Stage II) coinciding with the initiation of sea-floor spreading in the south, and slightly post-dating the beginning of Paraná–Etendeka volcanism (~132 Ma BP) in the north (Turner *et al.*, 1994; Milner *et al.*, 1995b; Renne *et al.*, 1996; Stewart *et al.*, 1996). A similar chronology has been described by Chang *et al.* (1992) who considered the stratigraphy and structural evolution of the east Brazilian margin basins. In both eastern Brazil and southwestern Africa, post-late-Jurassic sediment fill is predominantly terrestrial, and sub-aerially erupted basalts occur in most of the South American marginal basins (Salado, Colorado, Campos, Pelotas, Santos) and in the Walvis and Orange Basins on the African margin (Urien and Zambrano, 1973; Dingle *et al.*, 1983; Chang *et al.*, 1992).

The rifting seems to have propagated from south to north towards the Walvis Ridge–Rio Grande Rise. North of the Falklands–Agulhas fracture zone the oldest magnetic anomaly clearly identifiable on oceanic crust on both the African and South American plates is M4 (130 ± 1 Ma BP), whereas further north, near the Walvis Ridge, the oldest identified anomaly is M0 (~125 Ma BP). Older anomalies (back to M12, 138 ± 1 Ma BP) identified by Rabinowitz and LaBrecque (1979), are inferred to be due to the widespread sub-aerial eruption and intrusion of basaltic magma during the initial rift stage (Wickens and McLachlan, 1990; Nürnberg and Müller, 1991). North of the Walvis Ridge–Rio Grande Rise, rifting began during the latest Jurassic–earliest Cretaceous (Castro, 1987; Chang *et al.*, 1992) (Tithonian–Barremian, 152–125 Ma BP). Full ocean spreading and opening of the equatorial Atlantic began in post-M0 times, with the oldest identified sea floor being late Aptian (~112 Ma BP) (Austin and Uchupi, 1982; Cande *et al.*, 1988; Nürnburg and Müller, 1991; Chang *et al.*, 1992).

12.4.2 REGIONAL TECTONIC SETTING AND STRUCTURAL CONTROL

Pre-rift plate reconstructions across the South Atlantic cannot be accurately achieved if the present continents of Africa and South America are assumed to have behaved as ridged continental lithosphere (Martin *et al.*, 1981; Unternehr *et al.*, 1988; Nürnberg and Müller, 1991). Current plate tectonic models for the opening of the South Atlantic have therefore attempted to overcome these problems by invoking complex zones of intracontinental deformation (second-order plate boundaries) within both the South American and African plates. These zones are inferred to have accommodated the required intracontinental deformation, during and subsequent to break-up, along a series of intracontinental rifts (e.g. the Recôncavo–Tucano-Jatobá and Rio do Peixe basins in Brazil, and the Benue Trough and southern Sudanese rift basins in Africa) and sub-continental strike-slip faults and shear zones (Pindell and Dewey, 1982; Fairhead, 1988; Unterhner *et al.*, 1988; Nürnberg and Müller, 1991; Sénant and Popoff, 1991; de Matos, 1992; Magnavita *et al.*, 1994).

A complex mosaic of interconnecting rift and horst structures, with two dominant trends of NNW–SSE and E–W, developed between southwestern Africa and southern South America during the initial rift phase (Dingle *et al.*, 1983; Uliana and Biddle, 1988). Final break-up occurred along the locus of the easternmost NNW–SSE-trending rifts (North Malvinas, Orange, Luderitz, Pelotas and Walvis Basins). Consequently, there is a marked structural asymmetry between the two margins along this sector as all the major E–W graben (the San Jorge, Colorado and Salado Basins) remained attached to the South American plate. Most of the early intra-plate rift basins were subsequently abandoned after the onset of the main break-up phase during the Barremian–earliest Aptian (Chang *et al.*, 1992; Magnavita *et al.*, 1994; Castro, 1987). However, many of these basins have undergone significant post-break-up tectonic inversion and erosion. Strata of the Recôncavo–Tucano-Jatobá Basin in eastern Brazil, for instance, have anomalously high levels of maturation considering their present burial depth, and Albian marine sediments occur 400 km inland at a present elevation of 800 m above sea level (Magnavita *et al.*, 1994).

Most of the syn- and post-break-up structural control represents reactivation of pre-existing basement features. For example, the Pernambuco/Patos shear zones in northeastern Brazil (Sénant and Popoff, 1991) and the Central African and Mwembeshi shear zones were active during the Pan-African/Braziliano and Kibaran orogenic phases (Coward and Daly, 1984). However, others, such as the Gastre fault system in Patagonia (Rapela and Pankhurst, 1992) and the Falklands–Agulhas fracture zone (Ben-Avraham *et al.*, 1993; Marshall, 1994a, b), truncate older basement structures and appear to have formed during the earliest stages of Gondwana fragmentation during the Triassic–early Jurassic. Many of the pre-existing structures were clearly also active during the Permo-Triassic and controlled the location and structural style of the continental 'Karoo' basins (Rosendahl, 1987; Hegenberger, 1988; Daly *et al.*, 1989, 1991; Eyles and Eyles, 1993). In northern Namibia the regional basement structure is controlled by the NE–SW strike of the intracontinental branch of the Pan-African Damara metamorphic belt (Tankard *et al.*, 1982), although this alignment changes to a coast-parallel trend near the coast in, for example, the Kaoko and Gariep belts (Light *et al.*, 1993; Trompette, 1994). The coastal branch of the Damara belt is matched on the Brazilian margin by the Dom Feliciao belt, which itself abuts the eastern margin of the Rio de la Plata craton in Uruguay and the extreme south of Brazil.

In addition to their role in intracontinental tectonics, reactivation of these structures during and after break-up has also had a strong influence on the pattern and location of offshore sedimentation (Fuller, 1971; Le Pichon and Hayes, 1971; Dingle and Scrutton, 1974; Scrutton and Dingle, 1974; Chang *et al.*, 1992; Bruhn and Walker, 1995) and syn-rift volcanism (Marsh, 1973; Turner *et al.*, 1994; Watkins *et al.*, 1994; Milner *et al.*, 1995b). Deposition along both margins has occurred within discrete basins/depocentres separated along the strike of the margin by basement structural 'highs'. Seismic reflection studies along the Namibian margin have documented substantial cross-margin strike-slip structures associated with some of these ridges which were clearly active until the late Cretaceous (Light *et al.*, 1992). Some of these basement highs have also been correlated with major oceanic transform faults/fracture zones (Fuller, 1971; Le Pichon and Hayes, 1971).

12.4.3 OFFSHORE BASINS

Sedimentation within the developing marginal rift basins offshore of southwest Africa and southern South America was dominated by oxidized, terrestrial, volcaniclastic sandstones, evaporites and sub-aerial volcanics. Within the central rift the transition to marine depositional environments occurred shortly after break-up and is marked everywhere by a well-developed 'drift-onset' unconformity (6At1 Sequence Boundary, Brown *et al.*, 1995) at the end of the Hauterivian (~130 Ma BP). However, there is no record of marine incursions into the intracontinental basins of South America until the late Cretaceous (Urien and Zambrano, 1973; Dingle *et al.*, 1983).

The offshore part of the eastern Brazilian margin consists of six major sedimentary basins, from the Uruguayan border in the south to Recifé in the north – the Pelotas, Santos, Campos, Espirito Santo, Bahia Sul and Sergipe–Alagoas Basins. Maximum sediment thicknesses range from 2.5 to 10 km and the stratigraphic sequence consists of late Jurassic to recent sediments and volcanics (Asmus and Ponte, 1973; Chang *et al.*, 1992). The subsidence histories of these offshore basins are broadly consistent with conventional models of lithospheric extension and thermal subsidence, with average extension factors beneath the continental shelf of between 2 and 3 (Chang *et al.*, 1992). Further south, the Argentine sector of the South American margin is also underlain by thick (>7 km) sedimentary sequences deposited within several discrete depocentres (Urien and Zambrano, 1973). A rough estimate of the total volume of sediment preserved within all Atlantic marginal basins of South America (between ~32°S and ~54°S) can be obtained from the data presented by Zambrano and Urien (1974). The total volume of 1.8×10^6 km^3 is equivalent to an average depth of denudation of ~2 km from a strip 240 km wide along the entire 3800 km length of the margin.

The offshore margin of Namibia and South Africa is dominated by four main basins; from south to north these are the Orange, Luderitz, Walvis and Namibe Basins, the latter occurring to the north of the Walvis Ridge and having more in common with the Angolan shelf (Light *et al.*, 1993). The main depocentres of these basins contain sequences of clastic sediments and volcanics which are generally less than 6 km thick, but reach >12 km in thickness in the northern Walvis Basin (Gerrard and Smith, 1982; Rust and Summerfield, 1990; Maslanyj *et al.*, 1992). Several regional unconformities have been identified and mapped seismically within the African basins. The most significant of these include the 'drift-onset' unconformity (6At1) dated as late Hauterivian (~132 Ma BP), a major composite, and locally variable, unconformity (15At1–16At1) spanning the Cenomanian–Turonian (97–88 Ma BP), and an extensive upper Maastrichtian (~68 Ma BP) unconformity.

Although rifting began in the middle Jurassic, and break-up finally occurred during the early Cretaceous, the major volume of sediment within the Orange and Walvis basins was deposited during the late Cretaceous–early Tertiary (Rust and Summerfield, 1990; Brown *et al.*, 1995). Rust and Summerfield (1990) determined this volume to be ~2.8×10^6 km^3 (adjusted to equivalent rock volume). This equates to a depth of denudation of 1.8 km averaged over the present Orange River basin and the other Atlantic-draining catchments south of the latitude of the Walvis Ridge (an area of 1.55×10^6 km^2), a figure which is comparable to the rough estimate determined for the South American margin. These calculations suggest that both margins have experienced significant amounts of denudation since the middle Jurassic–early Cretaceous. Clearly some sectors of the margin, such as northeastern Argentina, have undergone subsidence and burial since this time. Therefore, other areas must have experienced much greater depths of denudation than the overall mean value of ~2 km.

The Walvis Ridge–Rio Grande Rise structure also acted as a significant divide in terms of ocean circulation up to about 90–60 Ma BP (Hu *et al.*, 1988). North of the Rio Grande Rise (Santos Basin and beyond), evaporites were deposited during the Aptian (124–112 Ma BP), while deeper water marine conditions existed further south. During the Turonian–Coniacian (90–87 Ma BP) the Santos Basin, adjacent to the Ponta Grossa Arch, experienced a period of detrital influx, although this sediment pulse is considerably less pronounced in the Campos Basin to the north of Rio de Janeiro (Kumar and Gambôa, 1979; Viviers and de Azevedo, 1988). The base of the Miocene (23 Ma BP) is represented by a widespread unconformity in many of the Brazilian offshore basins, although not in the Santos Basin. There is local evidence in, for instance, the Oligocene Enchova palaeocanyon in the Campos Basin described by Antunes *et al.* (1988) of even earlier erosional events affecting some of the marginal basins.

The sedimentary record in the offshore basins clearly records a variable history of clastic sediment input and local erosion since the formation of the South Atlantic continental margins. However, the distribution and total volumes of sediment clearly indicate that both the South American and African margins have experienced very significant amounts of denudation (an average of ~2 km at least) since their formation in the middle Jurassic–early Cretaceous. However, the chronology and spatial distribution of onshore denudation are likely to have been highly variable, depending on the post-break-up tectonics, the pattern of drainage development, styles of landscape evolution, lithological heterogeneity and long-term climatic variations.

12.4.4 POST-BREAK-UP PLATE MOTION CHANGES

Geological and geophysical studies have documented a series of early Cretaceous rift basins formed during the initial break-up between South America and Africa. The structural style of many of these basins indicates that they formed within major, continental-scale sinistral and dextral strike-slip zones. Basins in West and Central Africa were subsequently deformed during a compressional-shear episode during the latest Cretaceous which caused the folding of the sedimentary sequence in the Benue Trough and renewed extension within the southern Sudanese basins (Fairhead, 1988; Unternehr *et al.*, 1988; Fairhead and Binks, 1991; Binks and Fairhead, 1992). This later period of intracontinental deformation, which included the reactivation of the Central African Shear Zone (CASZ), has been ascribed to shear stresses related to major changes in the geometry and relative motions of the plates involved in the opening of the Central and South Atlantic Ocean basins (Fairhead and Binks, 1991; Janssen *et al.*, 1995).

Similar tectonic events are recorded within the Cretaceous basins in northeastern Brazil (Sénant and Popoff, 1991; Magnavita *et al.*, 1994) and were associated with the reactivation of the Pernambuco/Patos shear zone system. The lack of Cretaceous-age rocks within the interior of southern Africa has prevented any detailed investigation of the effects and extent of this period of tectonism in this region. Nevertheless, the existence of significant post-break-up differential motion between southern and northwestern Africa is supported by recent palaeomagnetic data from the Karoo basalts in Lesotho which show that the middle Jurassic (180 Ma BP) poles for southern and northwestern Africa are consistently 12–15° of arc apart for all current models of plate movement for the South Atlantic (Kosterov and Perrin, 1996).

The primary evidence for these plate tectonic changes is the dramatic reorientation of the geometry, and significant increase in the number of fracture zones within the Atlantic and western Indian Ocean basins (Cande *et al.*, 1988; Royer *et al.*, 1988; Fairhead and Binks, 1991; Nürnberg and Müller, 1991). This change occurred between magnetic anomalies C34 (83 Ma BP) and C31 (67 Ma BP) and was accompanied by a significant decrease in the spreading rate within the South Atlantic (34 to 16 mm a^{-1} (half rate)). Several other major changes in plate geometry and motion occurred at this time; south of the Agulhas fracture zone, for instance, a fragment of the present Falkland Plateau (the Malvinas microplate) underwent a clockwise rotation of ~30° (La Breque and Hayes, 1979), while the spreading ridge south of the fracture zone was abandoned reducing the offset by ~800 km during anomaly C32 (70–72 Ma BP) (Martin and Hartnady, 1986). A second prominent ridge-segment jump occurred south of the Rio Grande Rise (Cande *et al.*, 1988). These changes were partly accommodated (or driven?) by the onset of asymmetry in spreading rates in the South Atlantic. These were 20% faster on the American plate north of the Tristan da Cunha fracture zone between anomalies C34 (83 Ma BP) and C22 (50 Ma BP), but 20% slower south of the Gough fracture zone over a similar period (Cande *et al.*, 1988).

12.5 Neotectonics

Tectonic reactivation of a variety of pre-existing structures occurred on a range of scales in eastern Brazil during the Tertiary, and many of the larger features represent reactivated Precambrian shear zones (Ferreia *et al.*, 1987; Riccomini *et al.*, 1989; Saadi, 1993). Much of this neotectonic activity was associated with strike-slip motion, as is apparent not only in localized sedimentary basins, such as the onshore Taubaté basin (late Quaternary) between São Paulo and Rio de Janeiro (Riccomini *et al.*, 1989), but also from earthquake focal mechanisms (Ferreia *et al.*, 1987). The general inference from earthquake fault plane solutions and epicentre distributions is that there is a regional E–W compression. The plate-scale present-day stress state of Brazil has been simulated by Coblentz and Richardson (1996), and their results are broadly consistent with observations indicating that much of eastern Brazil is in a stress regime compatible with strike-slip deformation. This is attributable to the balance of large-scale tectonic factors, such as ridge push and the regional topography of the continental margin, but local deformation seems to be influenced by other factors, such as crustal density variations and bending stresses related to the geometry of the margin (McQueen, 1986).

Evidence for neotectonic activity along large-scale structures in southern Africa has been well documented by studies of recent seismicity (Fairhead and Girdler, 1969; Maasha and Molnar, 1972; Reeves, 1972; Reeves and Hutchins, 1975; Scholz *et al.*, 1976; Hartnady, 1985, 1990). These studies have mapped a well-defined seismic zone, interpreted as a zone of incipient rifting (Reeves, 1972; Scholz *et al.*, 1976). It is broadly coincident with the Precambrian Mwembeshi shear zone which extends in a southwesterly direction across southern Africa from the Luangwa valley in Zambia through northern Botswana and Namibia (Coward and Daly, 1984; Daly *et al.*, 1989) (Figure 12.4). Fault plane solutions, shallow seismic refraction data and interpretations of satellite imagery from northern Botswana indicate that the seismicity is associated with active SW-trending normal faults, dipping 60° NW with throws of 100–200 m, which displace Cenozoic sediments of the Kalahari Basin (Scholz *et al.*, 1976; Mallick *et al.*, 1981). Isopachs of these sediments reveal an elongated, SW-trending depocentre located immediately northwest of this seismic zone in northwestern Botswana (Thomas and Shaw, 1990), which suggests that the fault activity dates back to at least the early Cenozoic.

Evidence of surface uplift as a result of neotectonic crustal deformation is limited to elevated coastal marine sediments. For instance, fossil invertebrates and mammals provide biostratigraphic age constraints on a series of littoral deposits in the Hondeklip Bay area in the northwestern Cape which reach up to 70–90 m above present sea level (Pether, 1986). This highest terrace can be correlated with the Pliocene Varswater Formation in the Saldanha Bay area in the southwest Cape, and, allowing for a decrease in global sea level of 25–35 m since a possible mid-Pliocene maximum, it indicates a surface uplift of up to 55–65 m over the past 3 Ma. However, any extrapolation of this rate of surface uplift back in time, or to areas inland, is entirely speculative.

Daly *et al.* (1989, 1991) have argued that intracontinental extensional and strike-slip deformation, driven by collisional processes along the southern margin of Gondwana during the Permo-Triassic, were accommodated by reactivation of steeply dipping, SW-trending Precambrian shear zones (Coward and Daly, 1984). Tectonic studies of the Cenozoic Malawi and Tanganyika rifts provide further support for the view that continental-scale, pre-existing basement structures within southern Africa have been reactivated during multiple periods of intracontinental deformation (Ring, 1994; Wheeler and Rosendahl, 1994). In contrast to South America, present-day stress indicators in southern Africa indicate large E–W extensional stresses (Coblentz and Sandiford, 1994). This is compatible with sinistral and dextral extensional strike-slip motion occurring, respectively, on major SW- and SE-trending structures.

12.6 Denudation Chronologies from Fission-Track Data

Fission-track analysis provides a means of establishing thermal histories over a temperature range of ~60–110°C (Carlson, 1990; Corrigan, 1991; Gallagher, 1995) (see Chapter 4). This range is characteristic of the upper 4–5 km of the continental crust and so apatite fission-track data are ideally suited to inferring long-term denudation chronologies (Summerfield, 1991; Brown *et al.*, 1994a). In order to convert the thermal history information extracted from fission-track data into estimates of denudation some constraint on the palaeogeothermal gradient is required. This information can be obtained from a series of selected samples, such as vertical profiles of samples over a range of topographic elevations, or a series of samples from deep boreholes (such as those typically drilled during hydrocarbon exploration). The temperature information derived from the fission-track data can then be combined with spatial information (the vertical offset between samples) to obtain an estimate of the palaeogeothermal gradient (see Chapter 4). The dependence of the fission-track age on the underlying track-length distribution means that a meaningful interpretation of a measured age, in terms of timing an event such as continental break-up or a discrete cooling event, can *only* be achieved through analysing the distribution of track lengths. It is therefore crucial to consider the track-length data in detail as this is where most thermal history information is contained (Corrigan, 1991; Gallagher, 1995) (see Chapter 4).

12.7 Fission-Track Data from the South Atlantic Margins

A substantial fission-track data set (over 240 samples in total) is available from the southeast Brazilian and southwest African margins (Figure 12.2). Although the coverage is still sparse over large areas, the available data do provide some important insights into the timing and distribution of long-term denudation across these passive continental margins. The implications of these data are discussed in the following sections with specific emphasis on their utility for estimating the chronology and spatial pattern of denudation.

12.7.1 THE SOUTHEAST BRAZILIAN MARGIN

Gallagher *et al.* (1994) have discussed fission-track data for a suite of samples collected from the southeastern margin of Brazil, between Rio de Janeiro and Porto Alegre, seaward of the present-day outcrop of the Paraná continental flood basalt province. These data show features typical of other passive margins around the world (Gallagher and Brown, 1997). There is a regional trend of young fission-track ages close to the coast, with age increasing with distance inland. The youngest ages found on the coastal plain in Brazil are ~60 Ma, although ages back to 90–100 Ma BP are also found in this region (Figures 12.5 and 12.6) (Plate 12.1). The rate of increase in age varies spatially (but with no obvious trend), and the age variation is complicated by the fact that the lithologies sampled represent rocks of various stratigraphic ages.

Two samples were collected from the Paraná flood basalt province, which is the highest volcanic unit forming the southern Serra Geral escarpment. One of the samples was collected from the top of the escarpment (~1750 m), and has an apatite fission-track age (132 ± 5 Ma) coincident with the eruption age of the basalt (~132 ± 1 Ma) (Turner *et al.*, 1994; Stewart *et al.*, 1996). The distribution of track lengths within this sample is unimodal, with a mean length of 15.1 ± 0.3 μm, indicating that the sample has not experienced temperatures in excess of ~50°C since being quenched at the time of eruption. The other basalt sample was collected from 100 m lower in the volcanic sequence and yielded a slightly younger fission-track age of 127 ± 6 Ma, although this is also statistically indistinguishable from the eruption age. The mean track length of 14.14 ± 0.19 μm obtained for this sample is slightly shorter than the stratigraphically higher sample, but it still indicates only minor post-eruption heating.

The Paraná basalt has a relatively low thermal conductivity (~1.86 W m^{-1} K^{-1}), and so for the present-day heat flow of ~70 mW m^{-2} (Hurter and Pollack, 1996) a temperature of ~50°C equates to a depth of

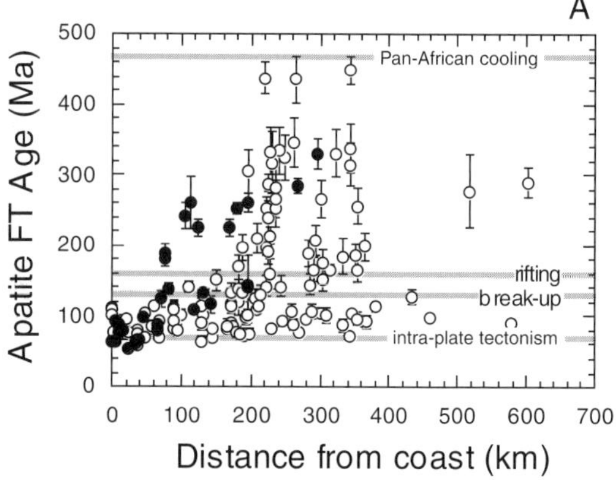

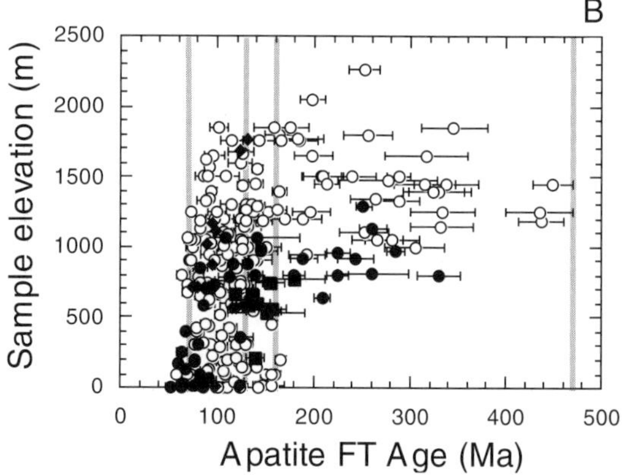

Figure 12.5 *Apatite fission-track age in relation to distance from the coast (A) and sample elevation (B). Filled circles represent Brazilian samples, open circles southwest African samples. The thick grey lines represent the timing of significant tectonic events*

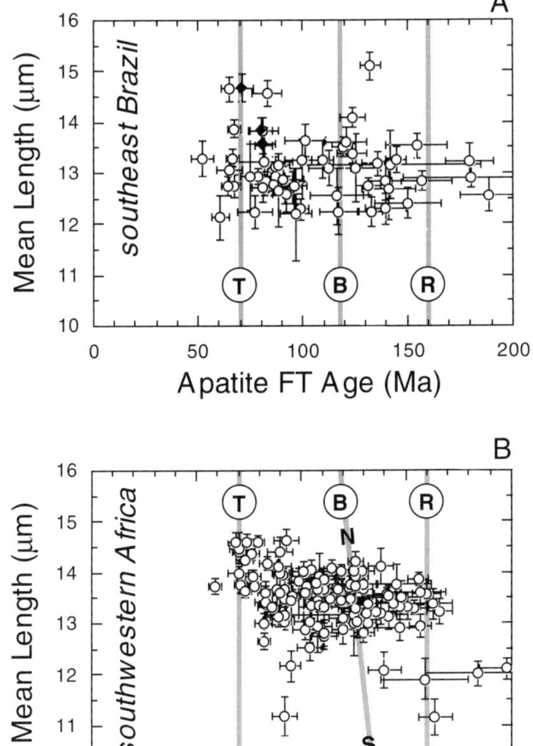

Figure 12.6 *Boomerang plots (Gallagher and Brown, 1997) of apatitie fission-track age plotted as a function of mean fission-track length for all data from southeastern Brazil and southwestern Africa. The thick grey lines indicate the initiation of rifting (R), the time of break-up (B) and the time of a major plate motion reorganization within the South Atlantic and Indian Ocean basins (T). For the time of break-up along the southwest African margin, the slanting thick grey line indicates the decreasing age of this event along the margin from south (S) (~134 Ma BP just north of the Falkland–Agulhas fracture zone) to north (N) (~118 Ma BP at the latitude of the Walvis Ridge)*

~1300 m (for a surface temperature of 20°C). This provides a maximum estimate for the amount of denudation from the present escarpment summit. This estimate of 1300 m of eroded section, combined with the preserved thickness of basalt of ~500–600 m, is consistent with the maximum thickness of ~1800–1900 m for the Paraná flood basalts observed further east (~1700 m, Zalan *et al.*, 1987) which coincides with the general depocentre of the Paraná Basin. We consider 1300 m to be a reasonable maximum estimate for the net amount of section eroded from the hinterland immediately behind the present-day escarpment over the past ~130 Ma, giving a mean denudation rate over this period of ~10 m Ma^{-1}.

Samples from the Paraná Basin sediments below the basalt all yield fission-track ages younger than their stratigraphic ages, although most pre-date the time of break-up. This indicates that these samples have experienced maximum temperatures in excess of 50–60°C since deposition, but the older samples were probably never heated to greater than 110–120°C. Apatite fission-track ages between 75 and 130 Ma, with a mean of ~90 Ma, were obtained for several sandstone samples collected a few metres below the lowest basalt unit. Gallagher *et al.* (1994) have shown that these samples would have been totally reset at the time of the basalt eruption, although the direct thermal influence of the hot lava is restricted to a length scale similar to the thickness of the flows (which are typically only several metres thick). A more important, regional-scale thermal effect arises from the burial of these sediments under 1–2 km of low thermal conductivity basalt at ~132 Ma BP, equivalent to a temperature increase of ~40°C for every 1000 m of basalt cover. For a reasonable surface temperature estimate of ~20°C, 1500 m of basalt would increase the temperature of the samples immediately below the basalt to ~80°C. The thermal conductivity of the sediments typically varies between 2.0 and 3.0 W m^{-1} K^{-1}, with the more quartz-rich formations generally having higher values (Hurter and Pollack, 1996). The temperature gradients for these conditions are 23–35°C km^{-1}, so a rock 1000 m below the basalt–sediment contact (approximately equivalent to the present-day sea level) would have been at temperatures between 100 and 120°C – sufficient to induce significant to total annealing of fission tracks in apatite. Thus, ~2500–3000 m is a reasonable estimate of the minimum amount of total post-rift denudation from the coastal plain in southeastern Brazil, with the amount of denudation decreasing inland towards the escarpment.

As indicated above, fission-track ages from Pre-cambrian basement rocks exposed in the coastal plain in southeastern Brazil range from ~60 to ~90 Ma; that is, considerably younger than the time of rifting and the eruption of the Paraná flood basalts. The mean track lengths for these samples are reduced (12–13.5 μm) and the track-length distributions are typically negatively skewed. Samples from further inland have older fission-track ages, with a maximum of ~330 Ma (Plate 12.1). All of these ages are significantly younger than the stratigraphic age of the host rocks, but only those samples proximal to the present-day continental margin have ages younger than the age of rifting (Plate 12.1 and Figure 12.2). There is also a progressive change in the form of track-length distributions moving inland; as the age increases, so the mean tracklength falls (the minimum observed is ~11 μm), and the track-length distributions become distinctly bimodal. However, the oldest measured fission-track age (330 ± 22 Ma) from a sample ~350 km inland from the present coastline has a relatively long mean track length of 13.06 ± 0.11 μm, with a narrow, unimodal track-length distribution. The distinctive geographical pattern in fission-track ages and form of the track-length distributions suggest that each sample has experienced different maximum pre-rift palaeotemperatures, and this trend indicates different amounts of post-rift denudation. The modelling results presented by Gallagher *et al.* (1994, 1995) show that these data are consistent with variable amounts of denudation across the margin, with in excess of 4 km removed from what is now the low-elevation coastal plain, and <1.5 km from the hinterland 200–300 km inland.

Fission-track data are also available for four samples from the vicinity of the late Cretaceous Poços de Caldos intrusion, just to the north of the Paraná Basin. Whole rock K–Ar dating has yielded an age for this intrusion of between 63 and 80 Ma (Amaral *et al.*, 1967). The fission-track ages for these samples range from 71 ± 5 to 81 ± 6 Ma, with mean track lengths between 13.54 ± 0.25 and 14.67 ± 0.4 μm. These data indicate that only a moderate amount of post-intrusion denudation (~2 km, or maximum post-intrusion temperatures of ~60–70°C) has occurred at these sites. This depth of denudation is similar to that inferred above from the occurrences of late Cretaceous alkaline igneous complexes in Namibia (see Section 12.3.2).

Preliminary data from Archean rocks in north-eastern Brazil (eastern São Françisco craton) indicate relatively young apatite fission-track ages close to the present-day coastline (83–104 Ma); these are approximately 20–30 Ma younger than the inferred age of break-up (Harman *et al.*, 1998). The mean track lengths for these samples are relatively long (13.4–14.2 μm) indicating that the rocks were at depths equivalent to palaeotemperatures greater than ~120°C until about 100 Ma ago. Further inland the fission-track ages are older – generally around 180 Ma – indicating maximum pre-break-up palaeotemperatures of less than ~100°C. However, two samples from ~600 km inland, and within 10 km of the Pan-African Pernambuco shear zone, yielded apatite ages of 76 ± 5 and 85 ± 7 Ma. The Pernambuco shear zone is a major structure running ~E–W at ~8°S in northern Brazil (Sénant and Popoff, 1991) which correlates directly with the Foumban shear zone in Cameroon, West Africa (Martin *et al.*, 1981; de Wit *et al.*, 1988) representing the eastern extension of the Central African shear zone system. These new data suggest that the Pernambuco shear zone was also reactivated during the late Cretaceous. The timing of this reactivation, and the relationship with an ancient, continental-scale structural zone, is remarkably similar to the Cretaceous history inferred from fission-track data for the Mwembeshi shear zone in southern Africa.

12.7.2 THE SOUTHWEST AFRICAN MARGIN

12.7.2.1 South Africa

The stratigraphic ages of 66 samples analysed from the Atlantic margin of South Africa range from Precambrian (Namaqua metamorphic belt) through to late Triassic (Stormberg Group, upper Karoo sequence). Despite this wide range of stratigraphic ages, virtually all of the samples analysed yielded apatite fission-track ages ranging between 166 ± 6 Ma and 70 ± 5 Ma, with a conspicuous lack of ages younger than ~70 Ma (Figures 12.5 and 12.6). Apatite ages pre-dating break-up at ~134 Ma BP were only obtained from samples from the interior regions of the continent and at elevations in excess of ~1500 m. Significantly, however, this is not a ubiquitous characteristic of the continental interior, as some of the youngest ages (~70 Ma) occur up to 600 km inland (Figures 12.2 and 12.5). This regional pattern is in marked contrast to that exhibited by the Brazilian data for which there is a relatively simple relationship between apatite age and distance from the continental margin (Figure 12.5). In general, ages post-dating break-up (~134 Ma) occur over the full sampled elevation range, and although they are typical of the coastal region, they also occur within the interior. Apatite age generally increases systematically with increasing elevation for specific localities.

The fact that all of the apatite fission-track ages are significantly younger than the stratigraphic age of the host rocks indicates that all the sampled rocks have been subjected to substantially higher temperatures since deposition or formation (mostly in excess of ~110°C). Almost all the samples with Cretaceous apatite ages have mean confined track lengths greater than ~12.5 to ~13 μm. The track-length distributions for these samples are generally unimodal, with the mode being between ~13 and ~14 μm, and they are also generally negatively skewed with 'tails' of shorter tracks <~10 μm in length. This demonstrates that most of the tracks have experienced only a moderate degree of thermal annealing (shortening) at temperatures <~70°C. The majority of the samples must therefore have cooled from maximum palaeotemperatures close to, or greater than, ~110°C during the Cretaceous.

More information on the chronology of cooling and the range of palaeotemperatures can be derived from an examination of the variation of mean track length with apatite age. This relationship is best illustrated by a 'boomerang' plot (Figure 12.6). The ages at which maxima occur in the track-length data suggest that discrete cooling episodes occurred at approximately those times (see Brown *et al.* (1994a) and Gallagher and Brown (1997) for a more detailed explanation). The plot indicates that some of the samples (those with mean track lengths >~14 μm) cooled from palaeotemperatures in excess of ~110°C to near surface temperatures during the early Cretaceous (140–120 Ma BP), whereas others have cooled from similar temperatures during the mid-Cretaceous (100–80 Ma BP). This later phase of cooling is also recorded by a vertical sequence of five samples, collected over an elevation range of 600 m, from Paarlberg in the southwest Cape. All these samples yielded similar ages with a mean of 90 ± 3 Ma and a mean track length of 14.15 ± 0.12 μm. The significantly shortened mean track lengths for most of the remaining samples indicate that they spent a protracted period at temperatures between 110 and 60°C prior to cooling to surface temperatures during the late Cretaceous.

12.7.2.2 Namibia

All of the 115 samples analysed from Namibia were collected from Precambrian to Cambrian metamorphic, igneous and sedimentary rocks. The apatite fission-track ages range from 449 ± 20 Ma to 59 ± 3 Ma. The youngest ages (60–75 Ma) are associated with narrow, unimodal track-length distributions (standard deviations of 0.95 to 1.59 μm) with long mean lengths (14.61 ± 0.1 to 13.64 ± 0.12 μm). Older ages are associated with broader, more complex, and sometimes distinctly bimodal, track-length distributions, with short mean lengths (~11–12.5 μm). The regional distribution of ages is similar to that seen in South Africa, with the older ages being limited to samples from inland and elevations greater than ~1000 m, and the youngest ages concentrated along the coast. A vertical sample profile from the escarpment region near Gamsberg (23° 20'S; 16°15'E) and an E–W transect across the margin from Okahandja to Swakopmund indicate that apatite age and mean track length generally vary systematically with sample elevation, with apatite age increasing, and mean track length decreasing, with increasing sample elevation.

The most significant difference between the data from Namibia and South Africa is the maximum age and the proportion of older apatite ages. All samples yielding apatite ages older than ~150 Ma occur in northern Namibia, and the age distribution defines a distinct NE-trending corridor (at about 22°S) of ages younger than ~100 Ma separating regions characterized by substantially older ages (≥~200 Ma) (Plate 12.1). The orientation of this corridor is similar to the regional structural trend of the intracontinental branch of the Pan-African Damara metamorphic belt, and appears to be coincident with the northern central zone as defined by the structural and metamorphic architecture of the belt (Miller, 1983). The sinistral ductile shear zones within the Damara metamorphic belt define the southwestern extension of the continental-scale Mwembeshi shear zone (Figure 12.2) (Coward and Daly, 1984).

Although there are insufficient data from southern Namibia to define clearly a regional pattern, it nevertheless seems significant that apatite ages of ~70 Ma were obtained from samples at Karasburg (18°45'E, 28°02'S) south of the Groot Karasberge. These mountains are bounded by steeply dipping NNE-trending reverse faults associated with a series of major NW-trending shear zones and transverse faults (Münch, 1974; Genis and Schalk, 1984). The northwest orientation of these major structures is similar to the Brakbos–Dagbreek–Strausberg shear zone system (Hartnady *et al.*, 1985) and associated Doornberg lineament (Coward and Potgieter, 1983) within the Gordonia sub-province of southern Africa. These are major transcurrent structures adjacent to the western margin of the Kaapvaal craton, and samples from this region in South Africa have yielded apatite ages of 70 ± 5 Ma and 73 ± 4 Ma with relatively long mean track lengths of 13.98 ± 0.13 μm and 14.23 ± 0.09 μm, respectively (de Wit, 1988).

Overall the data from Namibia indicate substantial cooling during the Cretaceous, with a discrete phase of accelerated cooling identified in the latest Cretaceous and earliest Cenozoic (~80–60 Ma BP) (Brown *et al.*, 1990). The geographical extent and magnitude of this latter episode was clearly highly variable. The cooling within the interior appears to be closely related to the geometry of pre-existing crustal structures. Consequently, the pattern of denudation implied by these data cannot be simply related to the formation of the passive margin during the late Jurassic–early Cretaceous. An appropriate regional model must account for an early Cretaceous phase of enhanced denudation linked to the early rift stage of margin development, as well as a later, more variable, phase of accelerated denudation apparently associated with the tectonic reactivation of these regional crustal structures. Other factors that may have played an important role in controlling denudation rates along the margin include the evolution of the Orange drainage basin and other Atlantic-draining systems (Dingle and Hendey, 1984; Rust and Summerfield, 1990; de Wit, 1993), regional lithological effects, including the distribution, both vertically and horizontally, of Karoo basalts, dolerite sills and sediments as well as basement lithologies, and a change from a temperate, seasonally-wet climate to a more arid regime during the latest Cretaceous–early Cenozoic (Ward, 1987; Ward and Corbett, 1990; Rayner *et al.*, 1991; Partridge, 1993).

12.8 Mapping Denudation Over Time

Combining large fission-track data sets with recent advances in deriving quantitative thermal history information from fission-track data (Gallagher, 1995) provides a powerful new means of constructing maps of long-term denudation. Palaeotemperature estimates for any given time can be derived directly from the detailed thermal history information for each sample site. An estimate of the amount of denudation that has occurred since that time can then be obtained by dividing the palaeotemperature values by the thermal gradient at each site after subtracting the appropriate palaeosurface temperature.

Here we have assumed that the eroded rock had an average thermal conductivity of 2.2 W m^{-1} K^{-1} and we have used surface heat flow data from Brazil and southern Africa (Pollack *et al.*, 1993) to derive estimates of the near-surface temperature gradient. This approach allows for spatially varying thermal gradients. Temporally varying heat flow and conductivity values, as derived from an independent thermal model for continental rifting for example, could easily be incorporated, but we have not used this strategy here. Modelling studies indicate that the thermal effects of rifting are unlikely to be significant within the shallow crust (<~10 km depth) of the onshore regions of passive margins (e.g. Buck *et al.*, 1988; Brown *et al.*, 1994b; Gallagher *et al.*, 1994). Furthermore, available global heat flow data indicate that the variation in temperature gradients in continental interiors is relatively restricted, with a mean between 20 and 30°C km^{-1} (Pollack *et al.*, 1993). Errors arising from anomalous transient thermal gradients could be significant, but even in extreme cases are unlikely to be greater than a factor of two (see Chapter 4).

In Plates 12.2 and 12.3, we present maps showing the estimated palaeotemperature at 158 Ma and 118 Ma of rocks presently cropping out at the surface. These times broadly represent the time of rift-onset and drift-onset. For regions covered by the late Cretaceous–Cenozoic Kalahari Basin sediments within southern Africa (Thomas and Shaw, 1990) and Bauru Group sediments in Brazil (Zalan *et al.*, 1987; de Wit *et al.*, 1988) palaeotemperatures were set to the surface temperature during periods of deposition. The palaeotemperature estimates were converted to equivalent depth as described above, and maps representing the amount of syn-rift (158–118 Ma BP) and post-rift (118–0 Ma BP) denudation are shown in Plates 12.4 and 12.5, respectively.

Both margins are characterized by moderate to low (~1 km) amounts of syn-rift denudation. In contrast, the post-rift maps show that substantial amounts (~3–5 km) of denudation occurred following break-up. The depth of denudation generally decreases inland from maximum values at the coast. However, the spatial distribution is highly variable, both along the margin and within the continental interior. For example, over 5 km of post-rift denudation is indicated for an isolated region ~500 km inland, within the NW-trending Gordonia sub-province (Hartnady *et al.*, 1985) adjacent to the Kaapvaal craton of southern Africa (29°S, 21°30'E), whereas an average of ~3–4 km is typically indicated for the coastal region (Plate 12.5B). Similar variations are evident for the Brazilian margin where 3–4 km of denudation occurred within the region overlying the NE-trending Ponta Grossa Arch (25°S, 50°W), compared with estimates of 1–2 km for the coastal region immediately to the north (Plate 12.5A).

12.9 Post-Break-Up Landscape Evolution

The starting point for discussions of the relationship between tectonics and long-term landscape development across the fragments of the Gondwana supercontinent continues to be the broad synthesis constructed by L.C. King and based on his pediplanation model (King,

1967), although a number of modifications have subsequently been proposed, particularly in relation to the timing of the proposed denudational cycles. King's scheme was founded upon a number of assumptions as to how landscapes behave through geological time. Elevated landscape elements of low relief were held to represent erosion surfaces originally 'graded' to sea level, but which had been subsequently uplifted. Under pediplanation, downwearing is regarded as minimal in comparison with the backwearing of scarps, and new denudational cycles initiated by 'semisynchronous' tectonic uplift events of global extent were thought to propagate inland as waves of denudation through knickpoint retreat along river channels and the parallel retreat of slopes (King, 1983). Variations in lithology and changes in climate were treated as minor irritants which did not distort the general story, while local tectonic activity was thought to produce only minor variations on the general theme. This approach to reconstructing landscape history and relating it to tectonic events rested very largely on the correlation of usually widely scattered remnants of erosion surfaces (landscape elements of low relief), originally thought to have been continuous, and was provided with some chronological control by occasional overlying deposits or through extrapolation to unconformities in marine sequences along adjacent continental margins.

In specifically commenting on the continental margins of the South Atlantic, King (1983) argued that vertical tectonic displacements had operated "in concert" since break-up, and that Africa and Brazil had "experienced the same interludes of cyclic denudation throughout the long period spanning from the Jurassic to the present day" (King, 1983, p. 184). In southern Africa King (1976, 1983) proposed six denudational cycles and five intervening tectonic episodes. The (first) 'Gondwana planation' extended to the end of the Jurassic and was terminated by the fragmentation of Gondwana. The subsequent post-Gondwana ('Kratacic planation'), initiated in the early Cretaceous, was terminated by continent-wide surface uplift in the mid–late Cretaceous. There followed the 'African' or 'Moorland' planation from the late Cretaceous to the end of the Oligocene which was terminated by widespread epeirogenic uplift. During the Miocene broad valley-floor pediplains developed in the African surface forming the 'Rolling Landsurface'. This denudational cycle was interrupted by moderate uplift at the end of the Miocene, but valley planation was renewed in the Pliocene giving rise to the 'Widespread Landscape' in the Pliocene. Strong uplift and flexure of continental margins was inferred for the late Pliocene/early Quaternary which precipitated the 'Youngest Cycle' episode of river incision and gorge cutting.

A parallel chronology for landscape development in Brazil was proposed (King, 1956, 1967). The dominant erosion surface of purported early Cenozoic age was attributed to the 'Sul-Americana planation', a denudational cycle considered to be "analogous in extent, importance and age with the 'African' denudation cycle" (King, 1967, p. 322). Remnants of earlier 'Gondwana' and 'post-Gondwana' cycles were identified in the high terrain south of Belo Horizonte, while a subsequent late Cenozoic 'Velhas' planation was thought to have penetrated along all the major drainage systems, including the São Françisco and Paraná, and destroyed large areas of the Sul-Americana surface. As with the margins of the African continent, King (1956) proposed significant Plio-Pleistocene surface uplift of up to several hundreds of metres along the coastal flank of Brazil, with a gradual rise in elevation of the resulting deformed landsurfaces towards a maximum elevation of around 1000 m along the rifted crest of the São Françisco valley, and a steeper decline southeastwards across a coastal monocline towards the Atlantic (King, 1967). As with the African margins, this tectonic event is considered to have precipitated a phase of rapid fluvial incision along the coastal flanks.

Various attempts have been made to modify King's general scheme whilst retaining the essential elements of his approach and the assumptions inherent to it. For instance, Partridge and Maud (1987) have presented a revised landscape chronology for southern Africa. This argues for the initiation of the Great Escarpment at break-up owing to the pre-existing high elevation of the southern African portion of Gondwana, with a subsequent 'African' cycle of advanced planation from the late Jurassic/early Cretaceous until the early Miocene (~18 Ma BP) operating at two levels, one above and one below the Great Escarpment. They suggest that this prolonged phase of denudation was terminated by a slight westward tilting of the African surface, which precipitated a 'Post-African I cycle' which persisted until the late Pliocene when asymmetrical uplift across southern Africa caused westward tilting of previously formed landsurfaces. This, in turn, is considered to have promoted a 'Post-African II cycle' of valley incision since the late Pliocene with the development of coastal gorges. Partridge and Maud (1987) argue that their approach to denudation chronology in southern Africa improves on previous attempts in that it links coastal deposits to erosion-surface remnants inland, and uses deep weathering profiles and duricrust cappings purportedly "diagnostic" of the oldest (African) surface.

In spite of detailed field observations of weathering profiles, it is not apparent how particular weathering deposits can be used as a chronological marker given the numerous factors, other than time, that can influence their physical, chemical and mineralogical properties. Weathering profiles and duricrusts with similar characteristics are more likely to be indicative of a palimpsest of weathering and denudational regimes than diagnostic of the 'age' of the landsurface with which they are associated. Moreover, fundamental difficulties remain with the entire strategy of landscape history reconstruction using the kind of approach pursued by L.C. King. Some of these were highlighted at the time these ideas were being presented (e.g. Wellington, 1955; De Swardt and Bennet, 1974), but were largely ignored. Helgren (1979) has identified three axioms on which King's strategy is dependent – continents are subject to widespread episodic uplift, all slopes experience parallel retreat for long distances, and river knickpoints retreat inland over long distances – and argued that none of them are supported by either current geomorphological theory or empirical evidence. To these we could add the assumption that extensive, low-relief surfaces can only be formed in relation to a base level represented by sea level. This is fundamental to the entire approach because the persistence of local or regional base levels over long periods of time due, for instance, to particularly resistant lithologies, would undermine the link between particular surfaces and the datum provided by sea level and make it impossible to quantify the tectonic displacement of erosion surfaces. This is a critical problem given the need among geophysicists for empirical data on spatial variations in surface uplift.

Our assessment is that the approach advocated by King and adopted by other researchers is founded on unverified assumptions and therefore does not provide a viable basis for assessing the relationships between tectonics and macroscale landscape development. Our assessment of the apatite fission-track data available for southwest Africa and southeast Brazil, coupled with offshore sediment volume data, is that it is possible to reconstruct the post-break-up *denudational* histories of passive continental margins in broad outline. The temperature sensitivity of the apatite fission-track system inevitably means that the resolution is poor for the more recent denudational history, although the application of He-apatite thermochronology and cosmogenic isotope analysis should help to alleviate this problem in the future. The fundamental difficulty is not with establishing a denudational record, but with determining changes in landsurface *elevation* through time. It is this information that is critical to linking tectonics to landscape development, yet we simply do not have sound empirical data on such changes, notwithstanding 'estimates' based on the displacement of landsurfaces of supposed known age assumed to have originally been 'graded' to sea level. The lack of Cretaceous or younger marine deposits across southern Africa means that there is no sea-level palaeodatum from which post-Jurassic tectonic displacements can be calculated (Sahagian, 1988), and this is the reason for continuing discrepancies between proposed 'uplift' histories; for instance, compare the surface uplift chronology of Partridge and Maud (1987) outlined above with the scenario presented by Burke (1996). The latter argues for a low-lying African continent prior to 30 Ma BP after which time there was widespread surface uplift leading to the basin and swell topography and escarpments which constitute the present-day landscape.

Given the difficulties with the conventional approach to denudation chronology, we favour a conservative assessment of the links between tectonics and landscape development for the South Atlantic margins of Africa and South America. The significant pulse of denudation recorded along both margins, during and following rifting and break-up, strongly implies the existence of significant local relief at this time. This is compatible with the presence of high topography along the newly created continental margins and the establishment of a new, lower base level for denudation. High elevation could have been inherited from the pre-break-up topography (perhaps caused in part by underplating associated with the Paraná–Etendeka magmatic event) or have been associated with the dynamic and thermal effects of rifting. There appear to be no compelling arguments against the persistence of high elevation along both margins since the time of break-up, since claims for significant Pliocene–Quaternary surface uplift rest on hypothetical correlations between moderately uplifted coastal deposits and supposedly contemporaneous inland erosion-surface remnants now at much higher elevations. The greater depths of denudation generally recorded close to the margin compared with the continental interior imply a flexural isostatic response that would tend to maintain an upwarped topography parallel to the margin (Gilchrist and Summerfield, 1990). Regional and local-scale variations in amounts of post-break-up denudation recorded by AFTA data suggest the role of post-break-up tectonics, probably associated with the re-activation of ancient crustal structures and intraplate deformation, but may also reflect lithological and climatic factors, as well as the possible shifting of drainage divides through time.

12.10 Conclusions and Outstanding Research Questions

The fission-track data summarized here indicate that there has been substantial denudation across the South Atlantic margins of Africa and South America since the break-up of Gondwana. This observation argues strongly against previous ideas concerning the very long-term survival of erosion surfaces in these areas, and also severely undermines a strategy of reconstructing landscape histories that rely on a record of tectonic uplift and base-level change being reflected in the erosion surfaces that remain in the modern landscape. The key message here is that, in general, attention is more beneficially paid to establishing variations in rates of denudation across landscapes, rather than attempting to diagnose the amount and timing of tectonic uplift on the basis of correlating erosion-surface remnants.

On a regional scale, the geomorphic response to continental rifting and break-up, indicated by the chronology and magnitude of denudation, has varied significantly along, and between, the originally conjugate sectors of the South Atlantic margins of Africa and South America. For instance, the substantial amount of post-rift denudation indicated for the southwest African margin contrasts markedly with the history of sediment accumulation on the originally conjugate Argentinean sector of South American margin. These differences are mirrored in continental margin morphology and primarily reflect the contrasting tectonic histories of the respective margins, with the geometry, timing and style of post-rift tectonic reactivation of major intracontinental structures probably being particularly important. The morphological and tectonic asymmetry between the southern sectors contrasts with the more symmetrical margin morphology and broadly similar denudational histories indicated for the northern Namibian and southeast Brazilian margins.

Overall the fission-track data are broadly consistent with models of landscape development which predict a major phase of denudation following continental rifting (e.g. Gilchrist and Summerfield, 1994; Gilchrist *et al.*, 1994; Kooi and Beaumont, 1994). The record of denudation inferred from the apatite fission-track analysis is also broadly similar to estimates derived from the offshore sedimentary record (Brown *et al.*, 1990; Rust and Summerfield, 1990). However, the timing and distribution of denudation are not compatible with simple escarpment retreat models which predict only moderate amounts (~1 km) of post-rift denudation inland of the scarp. This is particularly true for southwestern Africa, and may be partly a consequence of the post-break-up tectonic history of the continental interior. A possible explanation for at least a part of this discrepancy is that discrete tectonic episodes, inferred to have occurred during the late Cretaceous, and which included reactivation of major intracontinental structures, led to locally accelerated phases of denudation being superimposed on the secular regional pattern. Refinement of numerical long-term landscape evolution models to allow for such episodes will require a more detailed and explicit treatment of post-break-up tectonic events. Other factors, such as patterns of continental drainage development, the effects of regional variations in lithology, and climatic change, may also have played a role.

Further detailed studies of key structural boundaries would clearly help to document and quantify the geomorphic effects of post-rift tectonic episodes more completely. Further work is also needed in assessing the temporal variation of temperature gradients as a consequence of real variations in heat flow, thermal conductivity and heat transfer mechanisms. This will allow more robust denudation estimates to be derived from fission-track data, and should provide meaningful resolution bounds on these estimates. Such variations could be readily constrained in southern Africa by analysing detailed fission-track profiles from the numerous and widely distributed deep boreholes, both onshore and offshore, that have been drilled since the 1970s as part of extensive mineral and hydrocarbon exploration programmes. In addition, explicitly incorporating reliable stratigraphic information which provides bounds on the timing and depth of denudation, such as terrestrial deposits of known age overlying older basement (e.g. Smith, 1986; Ward and Corbett, 1990; de Wit *et al.*, 1992; Partridge, 1993), into the thermal history modelling procedures would improve the calibration of the thermal histories and reliability of the model interpolations. It would also be useful to apply AFTA studies in conjunction with cosmogenic isotope analysis since the latter can potentially provide critical data on site-specific rates of landscape change for key landscape components (such as escarpments, river channels and interfluves) over time scales (typically 10^4–10^6 a) intermediate between those addressed by fission-track thermochronology and those accessed by modern denudation rate data (Bierman, 1994). The finer resolution of spatial variations in denudation rates provided by cosmogenic isotope analysis compared with AFTA would also enable questions such as the mode and rate of escarpment evolution to be tackled more directly.

Apatite fission-track analysis clearly has immense potential in providing quantitative estimates of long-term denudation on a continental scale. In regions such as southern Africa, which have experienced a protracted and complex history of denudation, and consequently lack any extensive stratigraphy which records this history directly, the method is particularly appropriate. The challenge for future research is to build a quantitative model of this complex morphotectonic history by improving and exploiting the technique's potential in this area while taking full account of complementary offshore and onshore stratigraphic information.

Acknowledgements

Fission-track research at La Trobe University is supported by grants from the Australian Research Council and the Australian Institute of Nuclear Science and Engineering. Part of this research was supported by Natural Environment Research Council Grants GR9/1573 to KG and RWB, and GR3/6693 to MAS. We also acknowledge contributions from the University of London Central Research Fund, the Nuffield Foundation, the British Council, the Royal Society, London, the Royal Society of Edinburgh, and the Carnegie Trust for the Universities of Scotland. Chris Hawkesworth and Marta Mantovani were instrumental in setting up the initial work in Brazil. We thank Anglo American Research Laboratories Pty Ltd and De Beers Consolidated Mines Ltd for logistical and financial support. Logistical support from the Geological Survey of the Ministry of Mines and Energy, Namibia, is also gratefully acknowledged. RWB was supported by an Australian Research Council Postdoctoral Research Fellowship.

References

Amaral, G., Bushee, J., Cordani, U.G., Kawashita, K. and Reynolds, J.H. 1967. Potassium–argon ages of alkaline rocks from southern Brazil. *Geochim. Cosmochim Acta*, **31**, 117–142.

Antunes, R. L., Sonoki, N.T. and Carminatti, M. 1988. The Enchova paleocanyon (Campos Basin – Brazil): Its Oligocene–Miocene history based on calcareous nannoplankton stratigraphy and seismostratigraphy. *Rev. Bras. Geosci.*, **18**, 283–290.

Asmus, H.E. and Ponte, F.C. 1973. The Brazilian marginal basins. In: A.E.M. Nairn and F.G. Stehli (eds) *The Ocean Basins and Margins, Vol. 1, The South Atlantic*, Plenum Press, New York, pp. 87–133.

Austin, J.A. and Uchupi, E. 1982. Continental–oceanic crustal transition off southwest Africa. *Amer. Ass. Petrol. Geol. Bull.*, **66**, 1328–1347.

Ben-Avraham, Z., Hartnady, C.J.H. and Malan, J.A. 1993. Early tectonic extension between the Agulhas Bank and the Falkland Plateau due to the rotation of the Lafonia microplate. *Earth Planet. Sci. Lett.*, **117**, 43–58.

Bierman, P.R. 1994. Using in situ cosmogenic isotopes to estimate rates of landscape evolution: A review from the geomorphic perspective. *J. Geophys. Res.*, **99**, 13 885–13 896.

Bigarella, J.J. 1970. Continental drift and palaeocurrent analysis. In: *Second Gondwana Symposium: Proccedings and Papers*, Council for Scientific and Industrial Research, Pretoria, 73–97.

Binks, R.M. and Fairhead, J.D. 1992. A plate tectonic setting for the Mesozoic rifts of West and Central Africa. *Tectonophysics*, **213**, 141–151.

Brown, L.F.Jr, Benson, J.M., Brink, G.J., Doherty, S., Jollands, A., Jungslager, E.H.A., Keenen, J.H.G., Muntingh, A. and van Wyk, N.J.S. 1995. *Sequence Stratigraphy in Offshore South African Divergent Basins, An Atlas on Exploration for Cretaceous Lowstand Traps by Soekor (Pty) Ltd., Amer. Ass. Petrol. Geol. Studs. Geol.*, **41**, 184 pp.

Brown, R.W., Rust, D.J., Summerfield, M.A., Gleadow, A.J.W. and De Wit, M.C.J. 1990. An accelerated phase of denudation on the south-western margin of Africa: evidence from apatite fission track analysis and the offshore sedimentary record. *Nucl. Tracks Rad. Measur.*, **17**, 339–350.

Brown, R.W., Summerfield, M.A. and Gleadow, A.J.W. 1994a. Apatite fission track analysis: Its potential for the estimation of denudation rates and implications for models of long-term landscape development. In: M.J. Kirkby (ed.) *Process Models and Theoretical Geomorphology*, Wiley, Chichester, pp. 23–53.

Brown, R., Gallagher, K. and Duane, M. 1994b. A quantitative assessment of the effects of magmatism on the thermal history of the Karoo sedimentary sequence. *J. Afr. Earth Sci.*, **18**, 227–243.

Bruhn, C.H.L. and Walker, R.G. 1995. High-resolution stratigraphy and sedimentary evolution of coarse-grained canyon-filling turbidites from the upper Cretaceous transgressive megasequence, Campos Basin, offshore Brazil. *J. Sed. Res.*, **65**, 426–442.

Buck, W.R., Matinez, F., Steckler, M.S. and Cochran, J.R. 1988. Thermal consequences of lithospheric extension: pure and simple. *Tectonics*, **7**, 213–234.

Burke, K. 1996. The African Plate. *S. Afr. J. Geol.*, **99**, 341–409.

Cande, S.C., LaBreque, J.L. and Haxby, W.F. 1988. Plate kinematics of the South Atlantic: Chron 34 to present. *J. Geophys. Res.*, **93**, 13 479–13 492.

Carlson, W.D. 1990. Mechanisms and kinetics of apatite fission-track annealing. *Amer. Mineral.*, **75**, 1120–1139.

Castro A.C.M. Jr. 1987. The northeastern Brazil and Gabon basins; a double rifting system associated with multiple crustal detachment surfaces. *Tectonics*, **6**, 727–738.

Chang, H.K., Kowsmann, R.O., Figueiredo, A.M.F. and Bender, A.A. 1992. Tectonics and stratigraphy of the East Brazil Rift system: an overview. *Tectonophysics*, **213**, 97–138.

Cloetingh, S., Lankreijer, A., De Wit, M. and Martinez, I. 1992. Subsidence history analysis and forward modelling of the Cape and Karoo Supergroups. In: M.J. De Wit and I.G.D. Ransome (eds) *Inversion Tectonics of the Cape Fold Belt, Karoo and Cretaceous Basins of Southern Africa*, Balkema, Rotterdam, pp. 239–248.

Coblentz, D.D. and Richardson, R.M. 1996. Analysis of the South American intraplate stress field. *J. Geophys. Res.*, **101**, 8643–8657.

Coblentz, D.D. and Sandiford, M. 1994. Tectonic stresses in the African plate: constraints on the ambient lithospheric stress state. *Geology*, **22**, 831–834.

Conceicao, de Jesus, J.C., Zalan, P.V. and Wolf, S. 1988. Mechanisms, evolution and chronology of South Atlantic rifting. *Bol. Geosci. Petrobras*, **2**, 255–265.

Corrigan, J. 1991. Inversion of apatite fission track data for thermal history information. *J. Geophys. Res.*, **96**, 10 347–10 360.

Coward, M.P. and Daly, M.C. 1984. Crustal lineaments and shear zones in Africa: their relationship to plate movements. *Precambrian Res.*, **24**, 27–45.

Coward, M.P. and Potgieter, R. 1983. Thrust zones and shear zones of the margin of the Namaqua and Kheis mobile belts, Southern Africa. *Precambrian Res.*, **21**, 39–54.

Cox, K.G. 1989. The role of mantle plumes in the development of continental drainage patterns. *Nature*, **342**, 873–877.

Crowley, K.D., Cameron, M. and Schaefer, R.L. 1991. Experimental studies of annealing of etched fission tracks in apatite. *Geochim. Cosmochim. Acta*, **55**, 1449–1465.

Daly, M.C., Chorowicz, J. and Fairhead, J.D. 1989. Rift basin evolution in Africa: the influence of reactivated steep basement shear zones. In: M.A. Cooper and G.D. Williams (eds) *Inversion Tectonics, Geol. Soc. Spec. Publ.*, **44**, 309–334.

Daly, M.C., Lawrence, S.R., Kimun'a, D. and Binga, M. 1991. Late Palaeozoic deformation in central Africa: a result of distant collision? *Nature*, **350**, 605–607.

de Matos, R.M D. 1992. The northeast Brazilian rift system. *Tectonics*, **11**, 766–791.

De Swardt, A.M.J. and Bennet, G. 1974. Structural and physiographic development of Natal since the late Jurassic. *Trans. Geol. Soc. S. Afr.*, **77**, 309–322.

De Wit, M.C.J. 1988. Aspects of the geomorphology of the north-western Cape, South Africa. In: G.F. Dardis and B.P. Moon (eds) *Geomorphological Studies in Southern Africa*, Balkema, Rotterdam, pp. 57–69.

De Wit, M.C.J. 1993. Cainozoic evolution of drainage systems in the north-west Cape. Unpublished PhD thesis, University of Cape Town.

De Wit, M., Jeffery, M., Bergh, H. and Nicolaysen, L. 1988. *Geological Map of Sectors of Gondwana Reconstructed to their Disposition ~150 Ma*. American Association of Petroleum Geologists and University of the Witwatersrand, South Boulder, Tulsa, Oklahoma.

De Wit, M.C.J., Ward, J.D. and Spaggiari, R. 1992. A reappraisal of the Kangnas dinosaur site, Bushmanland, South Africa. *S. Afr. J. Sci.*, **88**, 504–507.

Dewey, J.F. and Bird, J.M. 1970. Mountain belts and the new global tectonics. *J. Geophys. Res.*, **75**, 2625–2647.

Dingle, R.V. and Hendey, Q.B. 1984. Late Mesozoic and Tertiary sediment supply to the eastern Cape basin (S. E. Atlantic) and palaeo-drainage systems in southwestern Africa. *Mar. Geol.*, **56**, 13–26.

Dingle, R.V. and Scrutton, R.W. 1974. Continental breakup and the development of the post-Paleozoic sedimentary basins around southern Africa. *Geol. Soc. Amer. Bull.*, **85**, 1467–1474.

Dingle, R.V., Siesser, W.G. and Newton, A.R. 1983. *Mesozoic and Tertiary Geology of Southern Africa*, Balkema, Rotterdam.

Erlank, A.J., Marsh, J.S., Duncan, A.R., Miller, R.McG., Hawkesworth, C.J., Betton, P.J. and Rex, D.C. 1984. Geochemistry and petrogenesis of the Etendeka volcanic rocks from SWA/Namibia. *Spec. Publ. Geol. Soc. S. Afr.* **13**, 195–245.

Eyles, N. and Eyles, C.H. 1993. Glacial geologic confirmation of an intraplate boundary in the Paraná basin of Brazil. *Geology*, **21**, 459–462.

Fairhead, J.D. 1988 Mesozoic plate reconstructions of the Central–South Atlantic Ocean: the role of the West and Central African Rift System. *Tectonophysics*, **155**, 181–191.

Fairhead, J.D. and Binks, R.M. 1991. Differential opening of the Central and South Atlantic Oceans and the opening of the West African rift system. *Tectonophysics*, **187**, 191–203.

Fairhead, J.D. and Girdler, R.W. 1969. How far does the rift system extend through Africa? *Nature*, **221**, 1018–1020.

Falvey, D.A. 1974. The development of continental margins in plate tectonic theory. *Aust. Petrol. Explor. Assoc. J.*, **14**, 95–106.

Ferreia, J.M., Takeva, M., Costa, J.M. and Moreira, J.A. 1987. A continuing intraplate earthquake sequence near Joao Camara, Northeastern Brazil – preliminary results. *Geophys. Res. Lett.*, **14**, 1042–1045.

Fuller, A.O. 1971. South Atlantic fracture zones and lines of old weakness in Southern Africa. *Nature*, **231**, 84–85.

Gallagher, K. 1995 Evolving thermal histories from fission track data. *Earth Planet. Sci. Lett.*, **136**, 421–435.

Gallagher, K. and Brown, R.W. 1997 The onshore record of

continental break-up inferred from apatite fission track analysis. *J. Geol. Soc. Lond.*, **154**, 451–457.

Gallagher, K. and Hawkesworth, C.J. 1992 Dehydration melting and the generation of continental flood basalts. *Nature*, **358**, 57–59.

Gallagher, K. and Hawkesworth, C. 1994. Mantle plumes, continental magmatism and asymmetry in the South Atlantic. *Earth Planet. Sci. Lett.*, **123**, 105–117.

Gallagher, K., Hawkesworth, C. and Mantovani, M. 1994. The denudation history of the onshore continental margin of S.E. Brazil inferred from fission track data. *J. Geophys. Res.*, **99**, 18 117–18 145.

Gallagher, K., Hawkesworth, C. and Mantovani, M. 1995. Denudation, fission track analysis and the long-term evolution of passive margin topography: application to the S.E. Brazilian margin. *J. S. Amer. Earth Sci.*, **8**, 65–77.

Genis, G. and Schalk, K.E.L. 1984. *The Geology of Area 2618: Keetmanshoop. Explanation of Sheet 2618, Scale 1:250000*, Geological Survey of Namibia, Windhoek.

Gerrard, I. and Smith, G.C. 1982. Post Paleozoic succession and structure of the southwestern African continental margin. *Amer. Ass. Petrol. Geol. Mem.*, **34**, 49–74.

Gilchrist, A.R., Kooi, H. and Beaumont, C. 1994. Post-Gondwana geomorphic evolution of southwestern Africa: Implications for the controls on landscape development from observations and numerical experiments. *J. Geophys. Res.*, **99**, 12 211–12 288.

Gilchrist, A.R. and Summerfield, M.A. 1990. Differential denudation and flexural isostasy in formation of rifted margin upwarps. *Nature*, **346**, 739–742.

Gilchrist, A.R. and Summerfield, M.A. 1994 Tectonic models of passive margin evolution and their implications for theories of long-term landscape development. In: M.J. Kirkby (ed.) *Process Models and Theoretical Geomorphology*, Wiley, Chichester, pp. 55–84.

Gilluly, J. 1964. Atlantic sediments, erosion rates and the evolution of the continental shelf: some speculations. *Bull. Geol. Soc. Amer.*, **75**, 483–492.

Gleadow, A.J.W. 1981. Fission track dating methods: what are the real alternatives? *Nucl. Tracks. Rad. Measur.*, **5**, 3–14.

Gleadow, A.J.W., Duddy, I.R., Green, P.F. and Lovering, J.F. 1986. Confined fission track lengths in apatite: A diagnostic tool for thermal history analysis. *Contrib. Min. Petrol.*, **94**, 405–415.

Gomes, C.B., Ruberti, E. and Morbidelli, L. 1990. Carbonatite complexes from Brazil: a review. *J. S. Amer. Earth Sci.*, **3**, 51–63.

Green, P.F. 1988. The relationship between track shortening and fission track age reduction in apatite: Combined influences of inherent instability, annealing anisotropy, length bias and system calibration. *Earth Planet. Sci. Lett.*, **89**, 335–352.

Green, P.F., Duddy, I.R., Gleadow, A.J.W. and Lovering, J.F. 1989. Apatite fission track analysis as a palaeotemperature indicator for hydrocarbon exploration. In: N.D. Naeser and T.H. McCulloh (eds) *Thermal Histories of Sedimentary Basins: Methods and Case Histories*, Springer-Verlag, New York, pp. 181–195.

Green, P.F., Duddy, I.R., Gleadow, A.J.W., Tingate, P.R. and Laslett, G.M. 1986. Thermal annealing of fission tracks in apatite, 1. A qualitative description. *Chem. Geol.*, **59**, 237–253.

Haq, B.U., Hardenbol, J. and Vail, P.R. 1987. Chronology of fluctuating sea level since the Triassic. *Science*, **235**, 1156–1166.

Haq, B., Hardenbol, J. and Vail, P.R. 1988. Mesozoic and Cenozoic chronostratigraphy and eustatic cycles of sea-level change. In: C.K. Wilgus, B.S. Hastings, C.G.St.C. Kendall, H.W. Posamentier, C.A. Ross, and J.C. Van Wagoner (eds) *Sea-Level Changes: An Integrated Approach, Soc. Econ. Paleont. Mineral. Spec. Publ.*, **42**, 71–108.

Harland, W.B., Armstrong, R.L., Cox, A.V., Craig, L.E., Smith, A.G. and Smith, P.G. 1990. *A Geologic Time Scale 1989*, Cambridge University Press, Cambridge.

Harman, R., Gallagher, K., Brown, R., Raza, A. and Bizzi, L. 1998. Accelerated denudation and tectonic/geomorphic reactivation of the cratons of north-east Brazil during the mid late Cretaceous. *J. Geophys. Res.*, **103**, 27 091–27 105

Hartnady, C.J.H. 1985. Uplift, faulting, seismicity, thermal spring and possible incipient volcanic activity in the Lesotho–Natal region, SE Africa: the Quathlamba hotspot hypothesis. *Tectonics*, **4**, 371–377.

Hartnady, C.J.H. 1990. Seismicity and plate boundary evolution in southeastern Africa. *S. Afr. J. Geol.*, **93**, 473–484.

Hartnady, C., Joubert, P. and Stowe, C. 1985. Proterozoic crustal evolution in southwestern Africa. *Episodes*, **8**, 236–244.

Hawkesworth, C.J., Gallagher, K., Kelley, S., Mantovani, M., Peate, D.W., Regelous, M. and Rogers, N. 1992. Paraná magmatism and the opening of the South Atlantic. In: B.C. Storey, T. Alabaster and R.J. Pankhurst (eds) *Magmatism and the Causes of Continental Break-Up, Geol. Soc. Spec. Publ.*, **68**, 221–240.

Hegenberger, W. 1988. Karoo sediments of the Erongo mountains, their environmental setting and correlation. *Comm. Geol. Surv. S.W. Africa/Namibia*, **4**, 51–57.

Helgren, D.M. 1979. Rivers of diamonds: An alluvial history of the Lower Vaal Basin, South Africa. *University of Chicago, Department of Geography, Research Paper*, **185**.

Horsthemke, E., Ledendecker, S. and Porada, H. 1990. Depositional environments and stratigraphic correlation of the Karoo Sequence in northwestern Namibia. *Comms. Geol. Surv. Namibia*, **6**, 63–73.

Hsu, K.J. 1965. Isostasy, crustal thinning, mantle changes and the disappearance of ancient land masses. *Amer. J. Sci,*, **263**, 97–109.

Hu, X., Wang, Y.L. and Schmitt, R.A. 1988. Geochemistry of sediments on the Rio Grande Rise and the redox evolution of the South Atlantic Ocean. *Geochim. Cosmochim. Acta*, **52**, 201–207.

Hurford, A.J. 1991. Uplift and cooling pathways derived from fission track analysis and mica dating: a review. *Geol. Rund.*, **80**, 349–368.

Hurter, S.J. and Pollack, H.N. 1996. Terrestrial heat flow in the Paraná basin, southern Brazil. *J. Geophys. Res.*, **101**, 8659–8671.

Janssen, M.E., Stephenson, R.A. and Cloetingh, S. 1995. Temporal and spatial correlations between changes in plate motions and the evolution of rifted basins in Africa. *Geol. Soc. Amer. Bull.*, **107**, 1317–1332.

King, L.C. 1956. A geomorfologia do Brasil oriental. *Rev. Geog. Bras.*, **18**, 147–265.

King, L.C. 1967. *The Morphology of the Earth* (2nd edn), Oliver and Boyd, Edinburgh.

King, L.C. 1976. Planation remnants upon high lands. *Z. Geomorph.*, **20**, 133–148.

King, L.C. 1983. *Wandering Continents and Spreading Sea Floors on an Expanding Earth*, Wiley, Chichester.

Kooi, H. and Beaumont, C. 1994. Escarpment evolution on high-elevation rifted margins: Insights derived from a surface processes model that combines diffusion, advection and reaction. *J. Geophys. Res.*, **99**, 12 191–12 209.

Kosterov, A.A. and Perrin, M. 1996. Palaeomagnetism of the Lesotho basalt, southern Africa. *Earth Planet. Sci. Lett.*, **139**, 63–78.

Kumar, N. and Gambôa, L.A.P. 1979. Evolution of the Sao Paulo Plateau (southeastern Brazilian margin) and implications for the early history of the South Atlantic. *Geol. Soc. Amer. Bull.*, **90**, 281–293.

LaBrecque, J.L. and Hayes, D.E. 1979. Seafloor spreading history of the Agulhas basin. *Earth Planet. Sci. Lett.*, **45**, 411–428.

Laslett, G.M., Gleadow, A.J.W. and Duddy, I.R. 1984. The relationship between fission track length and track density in apatite. *Nucl. Tracks Rad. Measur.*, **9**, 29–38.

Laslett, G.M., Green, P.F., Duddy, I.R. and Gleadow, A.J.W. 1987. Thermal annealing of fission tracks in apatite, 2. A quantitative analysis. *Chem. Geol.*, **65**, 1–13.

Le Pichon, X. and Hayes, D.E. 1971. Marginal offsets, fracture zones and early opening of the South Atlantic. *J. Geophys. Res.*, **76**, 6283–6293.

Light, M.P.R., Maslanyj, M.P. and Banks, N.L. 1992. New geophysical evidence for extensional tectonics on the divergent margin offshore Namibia. In: B.C. Storey, T. Alabaster and R.J. Pankhurst (eds) *Magmatism and the Causes of Continental Break-Up, Geol. Soc. Spec. Publ.*, **68**, 257–270.

Light, M.P.R., Maslanyj, M.P., Greenwood, R.J. and Banks, N.L. 1993. Seismic sequence stratigraphy and tectonics offshore Namibia. In: G.D. Williams and A. Dobb (eds) *Tectonics and Seismic Sequence Stratigraphy, Geol. Soc. Spec. Publ.*, **71**, 163–191.

Maasha, N. and Molnar, P. 1972. Earthquake fault parameters and tectonics in Africa. *J. Geophys. Res.*, **77**, 5731–5743.

Magnavita, L.P., Davison, I. and Kusznir, N.J. 1994. Rifting, erosion and uplift history of the Reconcavo–Tucano–Jatoba Rift, northeast Brazil. *Tectonics*, **13**, 367–388.

Mallick, D.I.J., Habgood, F. and Skinner, A.C. 1981. *A Geological Interpretation of Landsat Imagery and Air Photography of Botswana*, Overseas Geology and Mineral Resources, **56**, HMSO, London.

Marsh, J.S. 1973. Relationships between transform directions and alkaline igneous rock lineaments in Africa and South America. *Earth Planet. Sci. Lett.*, **18**, 317–323.

Marshall, J.E.A. 1994a. The Falklands Islands: A key element in Gondwana paleogeography. *Tectonics*, **13**, 499–514.

Marshall, J.E.A. 1994b. The Falkland Islands and early fragmentation of Gondwana: implications for hydrocarbon exploration in the Falkland Plateau. *Mar. Petrol. Geol.*, **11**, 631–636.

Martin, A.K. and Hartnady, C.J.H. 1986. Plate tectonic development of the south east Indian Ocean: A revised reconstruction of East Antarctica and Africa. *J. Geophys. Res.*, **91**, 4767–4786.

Martin, A.K., Hartnady, C.J.H. and Goodlad, S. 1981. A revised fit of South America and South Central Africa. *Earth Planet. Sci. Lett.*, **54**, 293–305.

Martin, H. 1953. Notes on the Dwyka succession and on some pre-Dwyka valleys in South West Africa. *Trans. Geol. Soc. S. Afr.*, **56**, 37–43.

Martin, H. 1973. The Atlantic margin of southern Africa between latitude 17° south and The Cape of Good Hope. In: A.E.M. Nairn and F.G. Stehli (eds) *The Ocean Basins and Margins, Vol.1, The South Atlantic*, Plenum Press, New York, pp. 277–300.

Maslanyj, M.P., Light, M.P.R., Greenwood, R.J. and Banks. N.L. 1992. Extension tectonics offshore Namibia and evidence for passive rifting in the South Atlantic. *Mar. Petrol. Geol.*, **9**, 590–601.

McQueen, H.W.S. 1986. The inference of inplane stress in the lithosphere using numerical models. PhD thesis, Australian National University, Canberra.

Miller, R. McG. 1983. Evolution of the Damara Orogen of South West Africa/Namibia. *Geol. Soc. S. Afr. Spec. Publ.*, **11**.

Milner, S.C., Le Roex, A.P. and O'Connor, J.M. 1995a. Age of Mesozoic igneous rocks in northwestern Namibia, and their relationship to continental breakup. *J. Geol. Soc. Lond.*, **152**, 97–104.

Milner, S.C., Duncan, A.R., Whittingham, A.M. and Ewart, A. 1995b. Trans-Atlantic correlation of eruptive sequences and individual silicic volcanic units within the Paraná-Etendeka igneous province. *J. Vol. Geotherm. Res.*, **69**, 137–157.

Milner, S.C., Le Roex, A.P. and Watkins, R.T. 1993. Rb–Sr age determinations of rocks from the Okenyenya igneous complex, northwestern Namibia. *Geol. Mag.*, **130**, 335–343.

Münch, H.-G. 1974. The tectonics of the northern part of the Klein Karas and Groot Karas Mountains, South West Africa. *Ann. Geol. Surv. S. Afr.*, **9**, 107–109.

Naeser, C.W. 1979. Fission track dating and geologic annealing of fission tracks In: E. Jaeger and J.C. Hunziker (eds) *Lectures in Isotope Geology*, Springer-Verlag, Heidelberg, pp. 154–169.

Nürnburg, D. and Müller, R.D. 1991. The tectonic evolution

of the South Atlantic from late Jurassic to present. *Tectonophysics*, **191**, 27–53.

Partridge, T.C. 1993. The evidence for Cainozoic aridification in southern Africa. *Quat. Internat.*, **17**, 105–110.

Partridge, T.C. and Maud, R.R. 1987. Geomorphic evolution of southern Africa since the Mesozoic. *S. Afr. J. Geol.*, **90**, 179–208.

Pazzaglia, F.J. and Brandon, M.T. 1996. Macrogeomorphic evolution of the post-Triassic Appalachian mountains determined by deconvolution of the offshore basin sedimentary record. *Basin Res.*, **8**, 255–278.

Pether, J. 1986. Late Tertiary and early Quaternary marine deposits of the Namaqualand coast, Cape Province: New perspectives. *S. Afr. J. Sci.*, **82**, 464–470.

Pindell, J.L. and Dewey, J.F. 1982. Permo-Triassic reconstructions of western Pangea and the evolution of the Gulf of Mexico/Caribbean region. *Tectonics*, **1**, 179–211.

Platt, N.H. and Philip, P.R. 1995. Structure of the southern Falklands Islands continental shelf: initial results from new seismic data. *Mar. Petrol. Geol.*, **12**, 759–771.

Poag, C.W. and Sevon, W.D. 1989. A record of Appalachian denudation in postrift Mesozoic and Cenozoic sedimentary deposits of the U.S. middle Atlantic margin. *Geomorphology*, **2**, 119–157.

Pollack, H.N., Hurter, S.J. and Johnson, J.R. 1993. Heat flow from the Earth's interior: analysis of the global data set. *Rev. Geophys.*, **31**, 267–280.

Rabinowitz, P.D. and LaBrecque, J. 1979. The Mesozoic South Atlantic Ocean and evolution of its continental margin. *J. Geophys. Res.*, **84**, 5973–6002.

Rapela, C.W. and Pankhurst, R.J. 1992. The granites of northern Patagonia and the Gastre Fault System in relation to the break-up of Gondwana. In: B.C. Storey, T. Alabaster and R.J. Pankhurst (eds) *Magmatism and the Causes of Continental Break-Up, Geol. Soc. Spec. Publ.*, **68**, 209–220.

Rayner, R.J., Waters, S.B., McKay, I.J., Dobbs, P.N. and Shaw, A.L. 1991. The mid-Cretaceous palaeoenvironment of central southern Africa (Orapa, Botswana). *Palaeogeog. climat. ecol.*, **88**, 147-156.

Reeves, C.V. 1972. Rifting in the Kalahari? *Nature*, **237**, 95–96.

Reeves, C.V. and Hutchins, D.G. 1975. New data on crustal structures in central southern Africa. *Nature*, **254**, 408–410.

Reid, D.L., Cooper, A.F., Rex, D.C. and Harmer, R.E. 1990. Timing of post-Karoo alkaline volcanism in southern Namibia. *Geol. Mag.*, **127**, 427–433.

Reid, D.L., Erlank, A.J. and Rex, D.C. 1991. Age and correlation of the False Bay dyke swarm, south-western Cape, Cape Province. *S. Afr. J. Geol.*, **94**, 155–158.

Renne, P.R., Glen, J.M., Milner, S.C. and Duncan, A.R. 1996. Age of Etendeka flood volcanism and associated intrusions in southwestern Africa. *Geology*, **24**, 659–662.

Riccomini, C., Peloggia, A.U.G., Saloni, J.C.L., Kohnke, M.W. and Figueira, R.M. 1989. Neotectonic activity in the Serra do Mar rift system (southeastern Brazil). *J. S. Amer. Earth Sci.*, **2**, 191–197.

Ring, U. 1994. The influence of preexisting structure on the evolution of the Cenozoic Malawi rift (East African rift system). *Tectonics*, **13**, 313–326.

Rocha-Campos, A.C., Cordani, U.G., Kawashite, K., Sonoki, H.M. and Sonoki, I.K. 1988. Age of the Paraná volcanism. In: E.M. Piccirillo and A.J. Melfi (eds) *The Mesozoic Flood Volcanism of the Paraná Basin: Petrogenetic and Geophysicsl Aspects*, IAG-USP Press, pp. 25-47.

Rosendhal, B.R. 1987. Architecture of continental rifts with special reference to East Africa, *Ann. Rev. Earth Planet. Sci.*, **15**, 445–503.

Rowsell, D.M. and De Swardt, A.M.J. 1976. Diagenesis in Cape and Karoo sediments and its bearing on their hydrocarbon potential. *Trans. Geol. Soc. S. Afr.*, **79**, 81–129.

Royer, J.-Y., Patriat, P., Bergh, H.W. and Scotese, C.R. 1988. Evolution of the Southwest Indian Ridge from the Late Cretaceous (anomaly C34) to the Middle Eocene (anomaly 20). *Tectonophysics*, **155**, 235–260.

Rust, D.J. and Summerfield, M.A. 1990. Isopach and borehole data as indicators of rifted margin evolution in southwestern Africa. *Mar. Petrol. Geol.*, **7**, 277–287.

Saadi, A. 1993. Neotectonica da plataforma Brasileira: esboco e interpretacai prelimares. *Geonomos*, **1**, 1–15.

Sahagian, D. 1988. Epeirogenic motions of Africa as inferred from Cretaceous shoreline deposits. *Tectonics*, **7**, 125–138.

Scholz, C.H., Koczynski, T.A. and Hutchins, D.G. 1976. Evidence for incipient rifting in southern Africa. *Geophys. J. R. Astr. Soc.*, **44**, 135–144.

Scrutton, R.A. and Dingle, R.V. 1974. Basement control over sedimentation on the continental margin west of southern Africa. *Trans. Geol. Soc. S. Afr.*, **77**, 253–260.

Sénant, J. and Popoff, M. 1991. Early Cretaceous extension in northeast Brazil related to the South Atlantic opening. *Tectonophysics*, **198**, 35–46.

Senut, B. and Pickford, M. 1995. Fossil eggs and Cenozoic continental biostratigraphy of Namibia. *Palaeont. Afr.*, **32**, 33–37.

Senut, B., Pickford, M. and Ward, J.D. 1994 Biostratigraphie de éolianites néogènes du Sud de la Sperrgebiet (Désert de Namib, Namibie). *C. R. Acad. Sci. Paris*, **318**, 1001–1007.

Sleep, N.H. 1971. Thermal effects of formation of Atlantic continental margins by continental breakup. *Geophys. J. R. Astr. Soc.*, **24**, 325–351.

Smith, R.M.H. 1986. Sedimentation and palaeoenvironments of Late Cretaceous crater-lake deposits in Bushmanland, South Africa. *Sedimentology*, **33**, 369–386.

Söhnge, A.P.G. and Hälbich, I.W. 1983. *Geodynamics of the Cape Fold Belt, Geol. Soc. S. Afr. Spec. Pub.*, **12**.

Stewart, K., Turner, S., Kelley, S., Hawkesworth, C., Kirstein, L. and Mantovani, M. 1996. 3-D ^{40}Ar/^{39}Ar geochronology in the Paraná continental flood basalt province. *Earth Planet. Sci. Lett.*, **143**, 95–109.

Summerfield, M.A. 1990. Geomorphology and mantle plumes. *Nature*, **344**, 387–388.

Summerfield, M.A. 1991. *Global Geomorphology*, Longman, London.

Summerfield, M.A. 1996. Tectonics, geology, and long-term landscape development. In: W.H. Adams, A.S. Goudie and A.R. Orme (eds) *The Physical Geography of Africa*, Oxford University Press, Oxford, pp. 1–17.

Swart, R. 1992. Note: Cretaceous synvolcanic conglomerates on the coastal margin of Namibia related to the break-up of West Gondwana. *Comm. Geol. Surv. Namibia*, **8**, 137–141.

Tankard, A.J., Jackson, M.P., Eriksson, K.A., Hobday, D.K., Hunter, D.R. and Minter, W.E.L. 1982. *Crustal Evolution of Southern Africa: 3.8 Billion Years of Earth History*, Springer-Verlag, New York.

Thomas, D.S.G. and Shaw, P.A. 1990. The deposition and development of the Kalahari Group sediments, Central Southern Africa. *J. Afr. Earth Sci.*, **10**, 187–197.

Trompette, R. 1994 Geology of Western Gondwana (200–500 Ma), Balkema, Rotterdam.

Turner, S., Regelous, M., Kelley, S., Hawkesworth, C. and Mantovani, M. 1994. Magmatism and continental break-up in the South Atlantic: High precision $^{40}Ar/^{39}Ar$ geochronology. *Earth Planet. Sci. Lett.*, **121**, 333–348.

Turner, S.P., Hawkesworth, C.J., Gallagher, K., Stewart, K., Peate, D. and Mantovani, M. 1996. Mantle plumes, flood basalts and thermal models for melt generation beneath continents: assessment of a conductive heating model. *J. Geophys. Res.*, **101**, 11 503–11 518.

Uliana, M.A. and Biddle, K.T. 1988. Mesozoic–Cenozoic paleogeographic and geodynamic evolution of southern South America. *Rev. Bras. Geosci.*, **18**, 172–190.

Unternehr, P., Curie, D., Olivet, J.L., Goslin, J. and Beuzart, P. 1988. South Atlantic fits and intraplate boundaries in Africa and South America. *Tectonophysics*, **155**, 169–179.

Urien, C.M. and Zambrano, J.J. 1973. The geology of the basins of the Argentine continental margin and Malvinas Plateau. In: A.E.M. Nairn and F.G. Stehli (eds) *The Ocean Basins and Margins, Vol. 1, The South Atlantic*, Plenum Press, New York, pp. 135–169.

Urien, C.M., Martins, L.R. and Zambrano, J.J. 1976. The geology and tectonic framework of southern Brazil, Uruguay and northern Argentina continental margin: their behaviour during South Atlantic opening. *Annales Braz. Acad. Sci.*, **48,** 365–376.

Visser, J.N.J. 1987. The palaeogeography of part of southwestern Gondwana during the Permo-Carboniferous glaciation. *Palaeogeog. climat. ecol.*, **61**, 205–219.

Visser, J.N.J. 1989. The Permo-Carboniferous Dwyka Formation of southern Africa: Deposition by a predominantly sub-polar marine ice-sheet. *Palaeogeog. climat. ecol.*, **70**, 377–391.

Visser, J.N.J. 1995. Post-glacial Permian stratigraphy and geography of southern and central Africa: boundary conditions for climatic modelling. *Palaeogeog. climat. ecol.*, **118**, 213–243.

Viviers, M. C. and de Azevedo, R.L.M. 1988. The south-eastern area of the Brazilian continental margin: its evolution during the middle and late Cretaceous as indicated by paleoecological data. *Rev. Bras. Geosci.*, **18**, 291–298.

Wagner, G.A. 1968. Fission track dating of apatites. *Earth Planet. Sci. Lett.*, **4**, 411–415.

Ward, J.D. 1987. The Cenozoic succession in the Kuiseb Valley, Central Namib Desert. *Geol. Surv. S. W. Afr./Nam. Mem.*, **9**.

Ward, J.D. 1988. Geology of the Tsondab Sandstone Formation. *J. Sed. Geol.*, **55**, 143–162.

Ward, J.D. and Corbett, I. 1990. Towards an age for the Namib. In: M.K. Seely (ed.) *Namib Ecology: 25 Years of Namib Research, Transvaal Museum Mono.*, **7**, 17–26.

Ward, J.D. and Martin, H. 1987. A terrestrial conglomerate of Cretaceous age – A new record from the Skeleton Coast, Namib Desert. *Comm. Geol. Surv. S. W. Afr./Nam.*, **3**, 57–58.

Watkins, J.S., Montadert, L. and Wood Dickerson, P. (eds) 1979. *Geological and Geophysical Investigations of Continental Margins, Amer. Assoc. Petrol. Geol. Mem.*, **29**.

Watkins, J.S., Zhiqiang, F. and McMillen, K.J. (eds) 1992. *Geology and Geophysics of Continental Margins, Amer. Assoc. Petrol. Geol. Mem.*, **53**.

Watkins, R.T., McDougall, I. and Le Roex, A.P. 1994. K–Ar ages of the Brandberg and Okenyenya igneous complexes, north-western Namibia. *Geol. Rund.*, **83**, 348–356.

Wellington, J.H. 1955. *Southern Africa: A Geographical Study, Vol. 1. Physical Geography*, Cambridge University Press, Cambridge.

Wessel, P. and Smith, W.H.F. 1991. Free software helps to map and display data. *EOS, Trans. Amer. Geophys. Un.*, **72**, 441–446.

Wheeler, W.H. and Rosendahl, B.R. 1994. Geometry of the Livingstone Mountains border fault, Nyasa (Malawi) rift, East Africa. *Tectonics*, **13**, 303–312.

White, R.S. and McKenzie, D.P. 1995. Mantle plumes and flood basalts. *J. Geophys. Res.*, **100**, 17 543–17 585.

Wickens, H.deV. and McLachlan, I.R. 1990. The stratigraphy and sedimentology of the reservoir interval of the Kudu 9A-2 and 9A-3 boreholes. *Comm. Geol. Surv. Namibia*, **6**, 9–22.

Winter, H. de la R. and Venter, J.J. 1970. Lithostratigraphic correlation of recent deep boreholes in the Karroo–Cape Sequence. In: *Second Gondwana Symposium: Proccedings and Papers*, Council for Scientific and Industrial Research, Pretoria, pp. 395–408.

Zalan, P.V., Wolff, S., Conceicao, C.J., Astolfi, M.A.M., Vieira, I.S., Appi, V.T., Zanotto O.A. and Marques, A. 1987. Tectonics and sedimentation of the Paraná Basin. *Ann Simp. Gondwana VII*, 83–117.

Zambrano, J.J. and Urien, C.M. 1974. Pre-Cretaceous basins in the Argentine continental shelf. In: C.A. Burk and C.L. Drake (eds) *The Geology of Continental Margins*, Springer-Verlag, Berlin, pp. 463–469.

13

Late Cenozoic landscape evolution of the US Atlantic passive margin: insights into a North American Great Escarpment

Frank J. Pazzaglia and Thomas W. Gardner

13.1 Introduction

> No one now regards a river and its valley as ready-made features of the Earth's surface. All are convinced that rivers have come to be what they are by slow processes of natural development, in which every peculiarity of river course and valley form has its appropriate cause. Being fully persuaded of the gradual and systematic evolution of topographical forms, it is now desired, in studying the rivers and valleys of Pennsylvania, to seek the causes of streams in their present courses; to go back, if possible to that early date when central Pennsylvania was first raised above the sea, and trace the development of the several river systems then implanted upon it from their ancient beginning to the present time.
>
> (Davis, 1889)

The preceding paragraph is from William Morris Davis's landmark paper on the development of Appalachian drainage. Five generations later we still ponder and debate the ideas and concepts espoused in that paper 'The rivers and valleys of Pennsylvania'. A succeeding paper which formulated the cycle of erosion (Davis, 1899) provided a conceptual model for general large-scale landscape evolution in a passive margin setting. Davis's model gained a large and enthusiastic following, culminating with Douglas Johnson's *Stream Sculpture on the Atlantic Slope* (Johnson, 1931). Nevertheless, criticisms of Davisian theory appeared almost as soon as it was proposed (e.g. Tarr, 1898). Alternative landscape evolution models (e.g. Meyerhoff and Olmstead, 1936) would ultimately be integrated to form the foundation of dynamic equilibrium (Hack, 1960), the other most popular conceptual model for long-term landscape evolution.

Davisian and Hackian views are end-member concepts that cannot capture all of the complexities of long-term landscape evolution (Young and McDougall, 1993; Kooi and Beaumont, 1996). In this regard, Davisian theory holds that relief in a landscape should decay exponentially as interfluves are rounded and lowered, and valley bottoms widen. Hackian dynamic equilibrium holds that a characteristic relief will be maintained as the landscape becomes well adjusted to rates of denudation, structure, rock type and climate. Geomorphologists still tend to explain long-term passive margin landscape evolution in Davisian (Ollier, 1985a, b) or Hackian terms (Summerfield, 1985; Gilchrist and Summerfield, 1990, 1991), and proponents of both agree on one critical point: that the erosional and isostatic response of a landscape following a landsurface uplift event is both complex and significantly longer than the original orogenic event (Schumm and Rea, 1995). In this sense, the eastern US Atlantic passive margin embodies several intriguing geomorphological paradoxes. The volume of sediment in Atlantic margin basins indicates that no less than 7 km of crustal section has been removed from the Appalachians in the past 180 Ma, yet some portions of the Appalachian mountains still have a high mean elevation and local relief. In contrast, other portions of the orogen are low-standing and display regional erosion surfaces cut across rock types of variable resistance. And unlike the southern continents, the US Atlantic passive margin lacks a conspicuous Great Escarpment (Ollier, 1985a).

A Great Escarpment, in the topographic sense, is the steep, seaward-facing cliff etched into a broad, landward tilted asymmetric bulge ('Randschwellen') on passive margins (Ollier, 1985a). Conceptually, this topographic feature could develop from two very different tectonic settings (see Chapter 3). A Type-1 Great Escarpment, like those found on the southern continents, develops on initial high-elevation margins where the rift-flank escarpment is maintained by minimal post-rift offshore sediment loading and flexural subsidence. Such escarpments reflect the landward retreat of the initial rift-flank uplift (Gilchrist *et al.*, 1994; Kooi and Beaumont, 1994; Tucker and Slingerland, 1994). In contrast, a Type-2 Great Escarpment develops on the seaward side of a post-rift flexural bulge along a margin dominated by post-rift offshore sediment loading (Keen and Beaumont, 1990). Such escarpments are localized by initial topography and crustal structure, drainage divides and rock types.

Eastern North America may not have a margin-long conspicuous Great Escarpment like the southern continents, but at least along the southern Atlantic margin, the Blue Ridge stands as an imposing geological and landscape feature. Davis (1903, p. 214) alluded to a topographic Great Escarpment when he wrote: "The divide between the eastern and western streams is known as the Blue Ridge; but in southern Virginia and North Carolina it is not a ridge, with a crestline and well-defined slopes on either side; it is an escarpment, descending from the hilly mountainous upland of the western drainage area to the rolling and hilly lower land of the eastern streams." Davis (1903, p. 222) quotes Marius Campbell, a strong proponent for uplift and arching of the Piedmont province (e.g. Hayes and Campbell, 1894; Campbell, 1929), as stating "exceptional conditions have prevailed in this region which permitted the formation of the scarp; the exceptional conditions are those of unsymmetrical uplift or warping, the unsymmetrical uplift having its axis east of and presumably near the present front of the Blue Ridge with its longer and more gentle slope to the west and its steeper slope to the east". More recently, Hack (1982, p. 1) favoured a tectonic origin for the Blue Ridge escarpment stating that: "The origin of the escarpment is probably a monoclinal flexure or movement along a series of faults as much as 30 km southeast of the present position of the escarpment".

Ollier (1985b) reviews these previous studies of the Blue Ridge and concludes that it is the most likely candidate for an eastern North American Great Escarpment.

Following from Davis (1903), Hack, (1982) and Ollier, (1985b), we explore the possibility that the Blue Ridge is eastern North America's Great Escarpment. In doing so, we will focus on some of the complexities of post-orogenic landscape evolution, attempt to reconstruct the margin's late Cenozoic morphotectonic history, and resolve whether seaward-facing escarpments in eastern North America have a Type-1 or Type-2 origin.

13.2 Morphological and Geological Setting

> This broad lowland is a lowland no longer. It has been raised from over the greater part of its area into a highland, with an elevation from one to three thousand feet, sloping gently eastward and descending under the Atlantic level near the present margin of the Cretaceous formation. The elevation seems to have taken place in Tertiary time, and we will refer to it as of that date. Opportunity was then given for the revival of the previously exhausted forces of denudation, and as a consequence we now see the formerly even surface of the plain greatly roughened by the incision of deep valleys and the opening of broad lowlands on its softer rocks. Only the harder rocks retain indications of the even surface which once stretched continuously across the whole area.
>
> (Davis, 1889)

The landscape Davis is describing is the middle and southern US Atlantic passive margin which lies roughly between lat. 30–44°N, long. 84–68°W (Figure 13.1 and Plate 13.1). This region encompasses the Appalachian highlands, the Coastal Plain, and offshore depositional basins beneath the continental shelf and rise. The transition between the erosional setting of the Appalachian highlands and depositional setting of the Coastal Plain and offshore basins is the Fall Zone, a region with up to 150 m of relief that separates the resistant metamorphic rocks of the Appalachian Piedmont from the soft, unconsolidated sediments of the Coastal Plain (Figure 13.1 and Plate 13.1).

Elevation and local relief vary considerably across the different morphological provinces of the Appa-

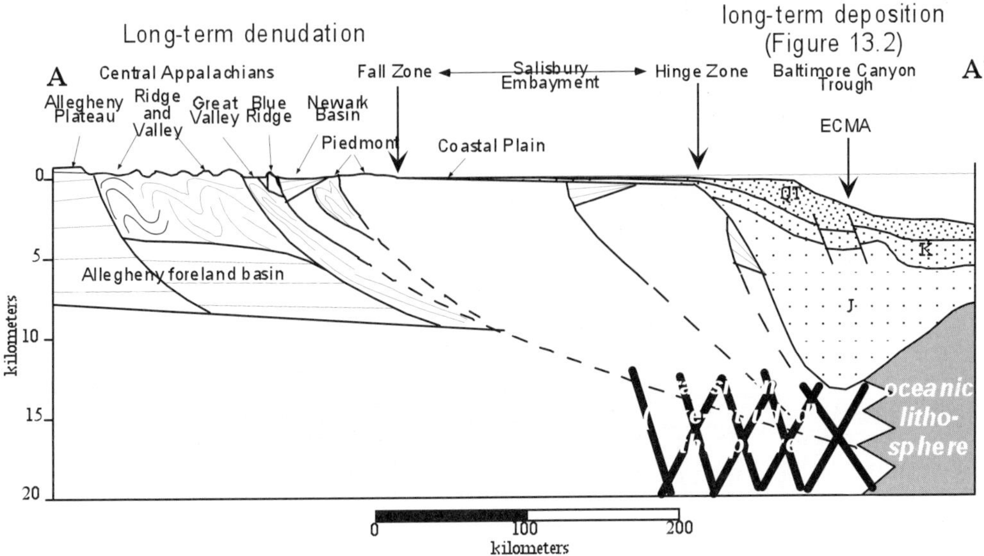

Figure 13.1 *Cross section A–A' from Plate 13.1 (modified from Pazzaglia and Gardner, 1994). J = Jurassic age sediment, K = Cretaceous age sediment, QT = Cenozoic age sediment, ECMA = East Coast Magnetic Anomaly*

lachians (Figure 13.1 and Plate 13.1). The Piedmont upland lies about 250 m above sea level. It is underlain by high-grade metamorphic rocks and exhibits an upland surface of low relief (<20 m) punctuated by river gorges where local relief does not exceed 180 m. Similarly, the Ridge and Valley of Pennsylvania, underlain by folded Palaeozoic sedimentary rocks, rises to only 650 m above sea level with about 300 m of local relief. Up to 9000 m of Palaeozoic foreland basin sedimentary rocks remain beneath the Ridge and Valley and Allegheny Plateau (Figure 13.1). In contrast, the Adirondack, White and Green Mountain ranges of the New England Appalachians, and the Blue Ridge of Virginia are considerably higher and steeper than the Ridge and Valley. High peaks in the Adirondack and White Mountains rise to over 1500 m above sea level. Summits of the Blue ridge reach greater than 1000 m above sea level and loom 800 m above the Piedmont along the Blue Ridge escarpment. Particularly in New England, erosion has exhumed structurally deep parts of the Appalachian orogen exposing resistant mid-crustal Proterozoic and lower Palaeozoic metamorphic and igneous rocks.

The Appalachian Mountains were built by several Palaeozoic orogenic events culminating with the Permian Alleghenian orogeny. Size and relief of the Permian Appalachians may have been similar to the modern central Andes, which have a mean elevation of about 3500–4500 m (Slingerland and Furlong, 1989) (see Chapter 9). Erosion during the Permian and early Triassic presumably removed most of the Alleghenian topography with virtually all of the detritus being shed west into, and beyond, the Appalachian foreland basin. An increase in elevation and relief was reintroduced into the Appalachian landscape in the late Triassic and early Jurassic in association with continental rifting which ultimately led to the opening of the Atlantic Ocean (Judson, 1975). East-flowing Atlantic drainage commenced with the formation of late Triassic and Jurassic rift basins (Meyerhoff and Olmstead, 1936; Mackin, 1938; Judson, 1975). The modern offshore sedimentary basins formed during, and subsequent to, rifting, and have served as an effective trap for detritus shed eastward from the post-rift margin.

The modern drainage divide (Plate 13.1) separating Atlantic from Gulf of Mexico drainage lies far to the west in New England and on the Allegheny Plateau in New York and Pennsylvania. In southern Virginia, the divide moves eastward across the Ridge and Valley to the Blue Ridge where it remains all the way into northern Georgia (Judson, 1975). The eastward jump of the drainage divide in southern Virginia is coincident with a deepening Bouguer gravity low centred beneath the Blue Ridge (Plate 13.1) (Judson, 1975).

The Blue Bidge mountains (reviewed in Hack, 1989), the most likely candidate for an eastern North American Great Escarpment, are an important focus of this study. The northern Blue Ridge (BRn of Plate 13.1) is a narrow (~15 km) anticlinorium whose high ridges are underlain, and held up by, Proterozoic and early Palaeozoic quartzites, phyllites, a massive basalt plateau, and a coarse-grained hypersthene granodiorite. Drainage in this portion of the Blue Ridge is to the east and the Atlantic Ocean. In contrast, the southern Blue Ridge (BRs of Plate 13.1) is much broader (~120 km) and underlain by a more diverse suite of Proterozoic and early Palaeozoic sandstones, quartzites, granites and gneisses. In most instances, the elevation of the southern Blue Ridge is coincident with rock erodibility, with the more resistant rocks standing higher; however, there are locations where topographic features are not sensitive to distinct structural and rock-type trends. The most spectacular example of the latter is the Blue Ridge escarpment itself which in many places represents nothing more than an abrupt topographic rise across identical high-grade metamorphic rocks of the low-standing Piedmont and Blue Ridge crest (Mills *et al.*, 1987). Most of the southern Blue Ridge is drained to the west and into the Gulf of Mexico. The divide between steep, east-flowing streams on the escarpment and the larger, west-flowing drainages follows the crest of the range. Numerous examples of past and imminent stream capture attest to the westward march of both the divide (Davis, 1903; Thompson, 1939; Hack, 1982) and escarpment. The high elevations and steep relief of the Blue Ridge have been attributed to rock-type erodibility (Hack, 1982, 1989), Cenozoic uplift along a fault or fault system (White, 1950; Hack, 1982), monoclinal flexures (Hayes and Campbell, 1894; Hack, 1982), or some type of dynamic topography produced by deep crustal discontinuities (Battiau-Queney, 1988, 1989) or asthenospheric flow (Vogt, 1991).

Conventional wisdom holds that eastern North America has been tectonically quiescent since it entered the drift stage in the late Jurassic; however, post-rift flexural deformation of the margin (Keen and Beaumont, 1990; Pazzaglia and Gardner, 1994) has been documented. Smaller tectonic features, such as high-angle reverse faults, are superimposed on the flexural deformation (Mixon and Newell, 1977; Mixon and Powars, 1984; Newell, 1985; Prowell, 1988; Gardner, 1989). It is important to point out that these faults are small, with the known maximum total displacement on the order of tens to possibly 100 m. Despite the paucity of structural and stratigraphic evidence in the Appalachian highlands suggesting major post-orogenic uplift (Pazzaglia and Brandon, 1996), studies of offshore sedimentary basins have clearly demonstrated several dramatic increases in sediment accumulation rates (Poag, 1985, 1992; Poag and Ward, 1993) (Figure 13.2). The implication is that these events are related to increases in mechanical denudation rates driven by rock uplift or climatically modulated changes in rock erodability.

13.3 Previous Work

Our present understanding of Atlantic passive margin long-term landscape evolution has been fundamentally shaped by early studies related to the genesis and preservation of peneplains, and more recent studies including detailed stratigraphic investigations of offshore sedimentary basins, fission-track thermochronology, and large-scale geodynamic and landscape evolution models. Here we borrow heavily from excellent, detailed reviews on the history of Appalachian geomorphological thought presented in Mills *et al.* (1987) and Morisawa (1989).

Folio mapping by the US Geological Survey on the Piedmont and Coastal Plain in, and around, Washington, DC, revealed a complex history of Appalachian uplift and denudation as recorded in the landscape and in the wedge of detrital material beneath the Coastal Plain (McGee, 1888a, b; Bascom *et al.*, 1902, 1909). Davis (1889, 1899) and Johnson (1931) recognized the importance of this early mapping and used it to argue for the age of peneplains and the drainages superimposed on them (Figure 13.3). Other studies recognized the limited, but intriguing, stratigraphic data preserved in the Appalachian landscape (Bashore, 1894, 1896; Shattuck, 1901, 1902; Shattuck *et al.*, 1906; Salisbury and Knapp, 1917; Barrell, 1920; Darton, 1939) and began to demonstrate that a simple evolution of the drainage divide and Appalachian drainages as features inherited from a Cretaceous sedimentary cover (Johnson, 1931; Mackin, 1938) was not supported by the geological evidence (Meyerhoff and Olmstead, 1936; Rich, 1938; Thompson, 1939). The alternative models of landscape evolution favoured an establishment of consequent drainages, adjusted to rock type and structure, and the westward march of the drainage divide by stream capture mechanisms after Alleghenian orogenesis in the Permian.

The age and origin of peneplains (Johnson, 1931) was challenged by studies which focused on upland gravels in the Piedmont and Ridge and Valley and the

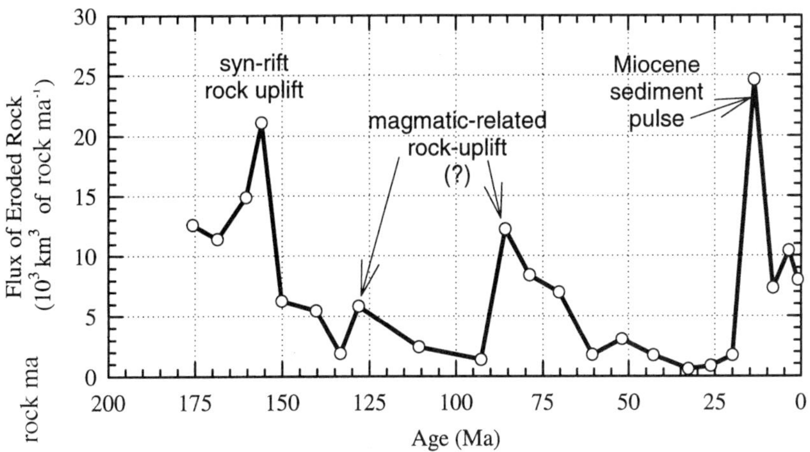

Figure 13.2 Sediment flux history of siliciclastic detrius from the New England and central Appalachians to the offshore basins of the middle Atlantic margin. Data compiled and modified from Poag and Sevon (1989) and Poag (1992). (Modified from Pazzaglia and Brandon, 1996)

correlation of these gravels across the Fall Zone to the Coastal Plain sediments (Bascom, 1921; Knopf, 1924; Stose, 1928; Campbell, 1929; Ashley, 1930, 1933). Based on the limited knowledge of the volume and extent of Coastal Plain sediments, Ashley (1930) argued that peneplains, if they were real, could not be older than Miocene. Following from Jamison (1882), Ashley (1933) suggested that the origin of deeply incised Piedmont drainages may have resulted from an upwardly flexed peripheral bulge in front of the Pleistocene ice sheets. Related studies recognized the possibility of Appalachian uplift driven by isostatic adjustments to regional denudation (Knopf and Jonas, 1929). Knopf and Jonas (1929) and Ashley (1930, 1933) had, at this early stage, identified the most critical geological observations against the Davisian model that would be used later by Hack (1960), Schumm (1963), Judson (1975), Matthews (1975) and Poag and Sevon (1989).

More recent studies (e.g. Poag, 1985, 1992; Grow *et al.*, 1988; Klitgord *et al.*, 1988; Poag and Sevon, 1989), which have documented the structure and stratigraphy of Atlantic margin basins, support a youthful modern Appalachian landscape. The offshore sediment basins contain the detritus of no less than 7 km of crustal section, presumably eroded from the Appalachians in the past 180 Ma (Poag and Sevon, 1989; Poag, 1992; Poag and Ward, 1993). Detritus from the central and New England Appalachians lies in the 400-km-long, 18-km-deep Baltimore Canyon Trough, the largest and deepest continental basin along the Atlantic margin (Figure 13.1 and Plate 13.1). Along the southern Atlantic margin, detritus is preserved in the somewhat smaller Carolina Trough (Popenoe, 1985). The sedimentation history of both depocentres could be summarized as rapid siliciclastic sediment fluxes in the Jurassic, several distinct pulses of siliciclastic sedimentation in the Cretaceous separated by periods of carbonate deposition, slow siliciclastic sediment fluxes throughout the Palaeogene, and a large pulse in siliciclastic sediment flux beginning in the Miocene and continuing to the present (Figure 13.2).

The driving mechanisms for changes in the siliciclastic sediment flux to the Atlantic margin basins are poorly understood. These mechanisms may be related to tectonic (Hack, 1982), thermal (magmatic) (de Boer *et al.*, 1988), and climatic (Barron, 1989; Pazzaglia *et al.*, 1997) processes (Poag and Sevon, 1989; Poag, 1992). Mass-balance modelling approaches assuming an elevation-dependent erosion law (Pazzaglia and Brandon, 1996) strongly favour thickening of the crust, driven by magmatic (de Boer *et al.*, 1988) and asthenospheric flow (Vogt, 1991; Gurnis, 1992) processes, leading to rock uplift and erosion. In any case, no less than a depth of 1.1 km of rock, spread evenly across the modern Atlantic drainage basins, is needed to account for the Miocene to present volume of sediment in these offshore basins (Braun, 1989).

Fission-track thermochronology data for the Appalachian basin support a late Cenozoic increase in denudation rates (Roden and Miller, 1989; Roden, 1991; Kohn *et al.*, 1993; Boettcher and Milliken, 1994).

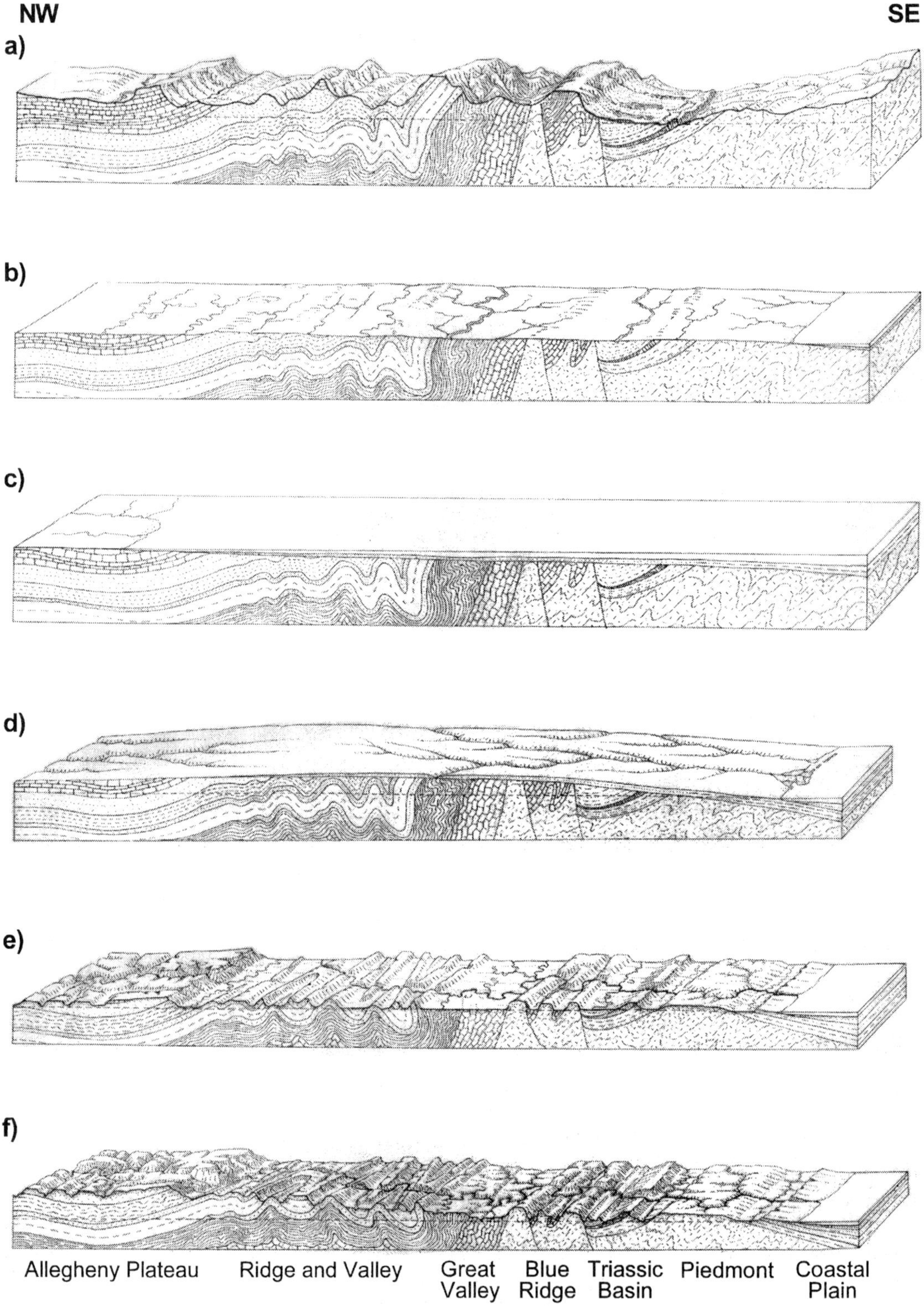

NW
SE
a)
b)
c)
d)
e)
f)
Allegheny Plateau
Ridge and Valley
Great Valley
Blue Ridge
Triassic Basin
Piedmont
Coastal Plain

The fission-track data indicate the removal of 1.5 ± 0.5 km of crustal section in the last 20 Ma (Boettcher and Milliken, 1994) which is consistent with the offshore sediment data. Such a slab of eroded rock was not derived uniformly from the Appalachian drainage basins; rather, higher standing areas should have been eroded more deeply and over a longer period of time than the low standing regions near the coast. The fission-track data do, in fact, support a non-uniform pattern of erosion. Denudation of the New England Appalachians was concentrated in the Cretaceous (Zimmerman *et al.*, 1975; Miller and Duddy, 1989), which corresponds reasonably well with intrusive and volcanic activity of the White Mountains at that time. In contrast, most of the Piedmont and Ridge and Valley, which now stand relatively low, experienced most erosion from the Permian through the Jurassic (Zimmerman, 1979; Roden and Miller, 1989; Boettcher and Milliken, 1994). These data place severe limitations on the importance of rock type in controlling long-term rates of erosion (Hack, 1980) because they demonstrate along-strike variations in erosional histories for areas that are otherwise underlain by similar rock types.

The US Atlantic margin is fundamentally different to the passive margins of the southern continents in that the thick accumulations of post-rift offshore sediments flexurally depress the margin (Keen and Beaumont, 1990). Geodynamic modelling of the Atlantic margin (Keen and Beaumont, 1990) indicates that the modern seaward-facing escarpments would be of Type-2 origin, etched into the flexural peripheral bulge. Any initial Type-1 escarpment is interpreted to have been destroyed by flexural subsidence of the rift-flank uplift.

13.4 Late Cenozoic Flexural Deformation of the Middle and Southern Atlantic Margin

The middle US Atlantic passive margin, encompassing the Appalachian Piedmont, Coastal Plain (Salisbury Embayment) and Baltimore Canyon Trough (Plate 13.1), contains a wide array of geomorphic and stratigraphic data that can be used to investigate the late Cenozoic morphotectonic evolution of the passive margin. Our goal here is to present the results of two simple geodynamic models that have been constructed around the hypothesis that the late Cenozoic Atlantic margin is affected by flexural isostatic processes which play an important role in the genesis and maintenance of distinct margin topographic features such as the Fall Zone, the drainage divide and the Blue Ridge escarpment. We demonstrate our geodynamic modelling approach by reviewing previous work which involved a simple one-dimensional line load model for the middle Atlantic margin (Pazzaglia and Gardner, 1994). We then build upon the results of that earlier model with a more realistic two-dimensional flexural model which simulates late Cenozoic flexural deformation along the entire Atlantic margin. We wish to point out that these geodynamic models are highly simplified and not intended to capture geological complexities that are commonly built into more rigorous, physically based numerical models. It is not a goal of this chapter to develop and test robust geodynamic models and we will not present a detailed description of the mathematical basis for the models. Consequently, we caution against using the results of these simple models to draw unequivocal conclusions about the actual processes or rates of late Cenozoic Atlantic margin deformation.

The Atlantic margin is particularly well suited for flexural isostatic geodynamic modelling given that the offshore basins are among the most thoroughly studied and best understood passive margin sedimentary basins (e.g. Grow *et al.*, 1988). These basins have a well-known depositional history and lithostratigraphic and chronostratigraphic reference sections (Poag, 1985, 1992; Popenoe, 1985; Poag and Ward, 1993). In addition, the stratigraphy and age of Coastal Plain marine (Owens and Gohn, 1985; Ward and Strickland, 1985) and fluvial deposits (Owens and Minard, 1979; Pazzaglia, 1993) are well known, and fluvial terraces of

Figure 13.3 *Classic interpretation of the evolution of Appalachian drainage for the middle Atlantic margin (modified from Johnson, 1931): (a) syn-rift uplift of eroded, post-Alleghenian Appalachians; (b) erosional bevelling in the Jurassic to produce the Fall Zone peneplain; (c) transgression of the Cretaceous seas across the entire orogen, and the deposition of a Cretaceous marine cover; (d) uplift and arching of the Fall Zone peneplain and its Cretaceous marine cover in the late Cretaceous or early Tertiary, with regional superimposition of Atlantic drainages destroying the Fall Zone peneplain and creating the Schooley peneplain; (e) uplift and incision of the Schooley peneplain and creation of the Harrisburg peneplain on erodable rock types; and (f) continued uplift in the late Tertiary forces incision of the Harrisburg peneplain and further dissection of the Schooley peneplain, with erosion of the Somerville peneplain on just the weakest rock types, and subsequently Quaternary uplift and incision of the Somerville peneplain to produce the modern landscape. (Used by permission of Columbia University Press)*

the largest Atlantic margin stream, the Susquehanna River, have been mapped and dated by longitudinal and petrographic correlation to the Coastal Plain stratigraphy (Pazzaglia and Gardner, 1993; Engel and Gardner, 1996).

13.4.1 OFFSHORE BASINS

Sediment derived from east-flowing Atlantic margin drainages has been deposited in sedimentary basins underlying the Coastal Plain, continental shelf and continental rise (Plate 13.1). Poag (1985, 1992) and Poag and Sevon (1989) subdivided the offshore basin stratigraphy into 23 informal time-stratigraphic units. They subsequently collapsed these into 12 formal allostratigraphic formations (Poag and Ward, 1993). Poag's inventory accounts for all significant sedimentary accumulations including deep-sea sediments of the continental rise. The siliciclastic volumes have been recalculated (Pazzaglia and Brandon, 1996) using the time scales of Harland *et al.* (1990) and Cande and Kent (1992) and the depth–porosity curve for the COST-B2 well (Scholle, 1977). The fluxes that produced these volumes are plotted against geological age in Figure 13.2 and are given in terms of the solid-rock volume delivered into the basins per million years (Pazzaglia and Brandon, 1996). Sediment loads are calculated as the product of sediment thickness represented on the isopach maps (Popenoe, 1985; Poag and Sevon, 1989) (Plate 13.2) and sediment density (Scholle, 1977).

13.4.2 CONTINENTAL DENUDATION

Rock removed from the Atlantic margin by both mechanical and chemical processes defines a negative load that is the product of rock density and the vertical thickness of denudation. Geochemical mass balance studies of saprolite production rates suggest average total denudation rates in the Appalachian Piedmont ranging from about 5 to 50 m Ma^{-1} (Cleaves *et al.*, 1970, 1974; Pavich, 1985, 1989; Cleaves, 1989, 1993; Pavich *et al.*, 1989). Similar, drainage-basin-wide rates for erosion based on modern sediment yield data have been reported for major Atlantic drainages such as the Susquehanna, Juniata, Delaware and Potomac Rivers (Judson and Ritter, 1964; Sevon, 1989; Milliman and Syvitski, 1992). Considering that modern sediment yield data producing the higher end of the observed denudation range probably reflect anthropogenic effects and may not be representative of actual long-term rates of denudation, we adopt a uniform, mean total (sum of chemical and mechanical) denudation rate of 10 m Ma^{-1}.

13.4.3 COASTAL PLAIN

The middle Atlantic Coastal Plain is the sub-aerially exposed portion of the Salisbury Embayment, a large arcuate-shaped basin approximately 300 km in diameter flanked on the north by the South Jersey Arch, on the south by the Norfolk Arch, and on the west by the Fall Zone (Owens, 1970) (Plate 13.1). Detailed stratigraphic reconstructions obtained from field mapping and borehole analyses have demonstrated that the Salisbury Embayment has a complex Cenozoic depositional history attributable to isostatic, tectonic and eustatic processes (e.g. Brown *et al.*, 1972; Newell and Rader, 1982; Newell, 1985; Owens and Gohn, 1985; Ward and Strickland, 1985).

Appalachian Piedmont rocks continue as Salisbury Embayment basement seaward of the Fall Zone, dipping seaward more steeply than anywhere else along the Atlantic Coastal Plain (Hack, 1982). Cretaceous to Eocene, predominantly marine, deposits unconformably overlie the basement, filling the basin in a seaward-thickening wedge (Owens and Gohn, 1985; Ward and Strickland, 1985). A basin-wide unconformity of middle Oligocene age (Ward and Strickland, 1985; Miller *et al.*, 1996) represents a sustained period of sub-aerial erosion which separates the early Cenozoic and Cretaceous deposits from the overlying Chesapeake Group.

Several upper Coastal Plain and Fall Zone fluvial deposits, like those present at the mouth of the Susquehanna River, represent the proximal, updip equivalents to the marine Chesapeake Group facies (Figure 13.4a). Petrography-based lithostratigraphic correlations along the Fall Zone and downdip into the Salisbury Embayment have established a regional chronostratigraphic framework for these fluvial deposits (Pazzaglia, 1993).

13.4.4 PIEDMONT TERRACES

Fluvial terraces, mapped and correlated on the basis of petrography and elevation, flank the lower Susquehanna River (Pazzaglia and Gardner, 1993) (Figure 13.4b). These terraces (Tg1, Tg2, Tg3 and QTg) are degraded strath terraces cut into Piedmont bedrock and mantled with a thin colluvium of fluvially rounded pebbles and cobbles. The petrography of these deposits begins almost exclusively as massive vein quartz for the highest and oldest terrace Tg1, becomes progressively more heterolithic with the introduction of quartzite clasts in Tg2 and more labile sandstones and siltstones in Tg3, and very heterolithic with extrabasinal granites and gneisses in QTg. These petrographic trends mirror those of the upland gravels and Coastal Plain deposits and allow for downstream correlation to dated deposits (Figure 13.4a, b).

Our model for terrace genesis along the lower Susquehanna River requires a graded river (Knox, 1975; Leopold and Bull, 1979) with a fixed base level to attain and maintain a characteristic longitudinal profile on an isostatically dominated margin (Pazzaglia and Gardner, 1993). Terrace genesis along major streams, such as the Susquehanna, in the Appalachian landscape spans a time range commensurate with a period of Coastal Plain deposition. The minimum age of a fluvial terrace would be approximately equal in age to the unconformity at the top of a Coastal Plain unit deposited during the coincident period of terrace genesis. Thus, correlation of terraces, Fall Zone upland gravels, Coastal Plain deposits, and offshore basin deposits can be used to define stratigraphic horizons of nearly synchronous age (Figure 13.4c).

13.4.5 TIME LINES

We have correlated four stratigraphic horizons along cross section B–B' (Plate 13.1) and used them as time lines (Figure 13.4c) to constrain the progressive flexural deformation of the margin in response to offshore sediment loading and continental denudation. The present elevation of a stratigraphic time line with respect to modern sea level is the sum of the original land surface or coastal plain depositional contact, the change in eustatic sea level, and the change in elevation attributable to isostatic deformation. Depositional contacts and change in eustatic sea level can be obtained by regional geomorphic and stratigraphic relationships and from published sources (Haq *et al.*, 1987; Greenlee *et al.*, 1988; Pazzaglia and Gardner, 1994). Values for flexural isostatic deformation are generated by the geodynamic model.

13.5 Geodynamic Models

13.5.1 ONE-DIMENSIONAL MODEL

The geodynamic models simulate flexural deformation assuming that the US Atlantic passive margin is in isostatic equilibrium (Karner and Watts, 1982). Time lines can be reconstructed from geological and geomorphic data, and the passive margin lithosphere can be simulated as a uniformly thick, perfectly elastic plate, without horizontal stresses, that will respond flexurally to strike-averaged, vertically applied loads.

The one-dimensional model treats the passive margin lithosphere as an infinite, unbroken, elastic plate of uniform thickness. The one-dimension approximation of flexure for a thin, unbroken elastic plate under a line load is given in Turcotte and Schubert (1982, pp. 125–132) and outlined in Pazzaglia and Gardner (1994); the interested reader is referred to these sources for a complete mathematical representation and parameterization of the model. The one-dimensional model is composed of 17 equally spaced, 50-km-wide cells, aligned parallel to cross-section B–B'; (Pazzaglia and Gardner, 1994) (Plate 13.1). The unbroken, elastic plate is allowed to flex under vertical stresses applied as positive sediment loads offshore, and negative denudational loads for the continent. We have produced four models for the upper Oligocene–lower Miocene, middle Miocene, upper Miocene, and Plio-Pleistocene. Finally, flexural deformation of the plate under these loads is compared to stratigraphic horizons corresponding to each of the four time lines (Figure 13.4c).

13.5.2 TWO-DIMENSIONAL MODEL

The two-dimensional model also treats the Atlantic margin lithosphere as an unbroken, infinite, perfectly elastic plate of finite thickness. The two-dimensional approximation for the bending of an elastic plate under the vertical stress of a distribution of point loads is given in Turcotte (1979). Several recent geodynamic studies have shown how this solution could be used to estimate flexure of plates constrained by stratigraphic (Kruse and Royden, 1994) and topographic and denudational data (Stephenson, 1984). The interested reader is referred to these studies for an excellent treatment of the assumptions and mathematical details of two-dimensional flexure.

We have ultilized the freely distributed UNIX computer program Generic Mapping Tools (GMT) to perform the two-dimensional flexure calculations (Wessel and Smith, 1991). With GMT, we have created a gridded representation of the offshore sediment thickness (Plate 13.2) for the four stratigraphic time lines (Figure 13.4b). We have also prepared a gridded representation of the total upper Oligocene–Plio-Pleistocene offshore sediment thickness and thickness of section eroded from the continent over the past

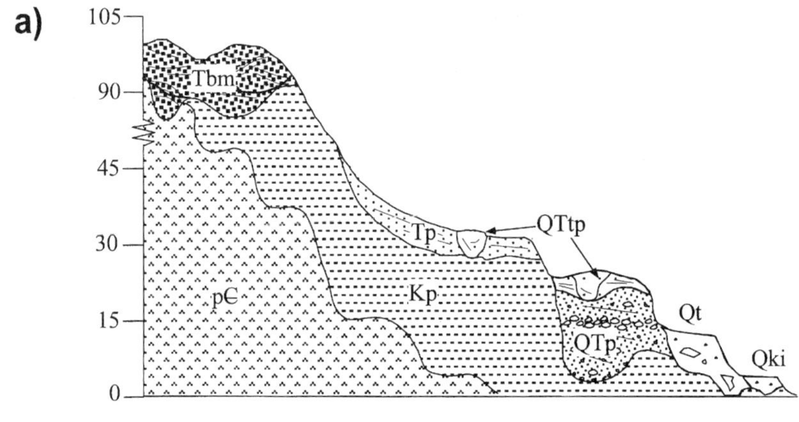
a)
105
90
45
30
15
0
Tbm
Tp
QTtp
pЄ
Kp
Qt
QTp
Qki

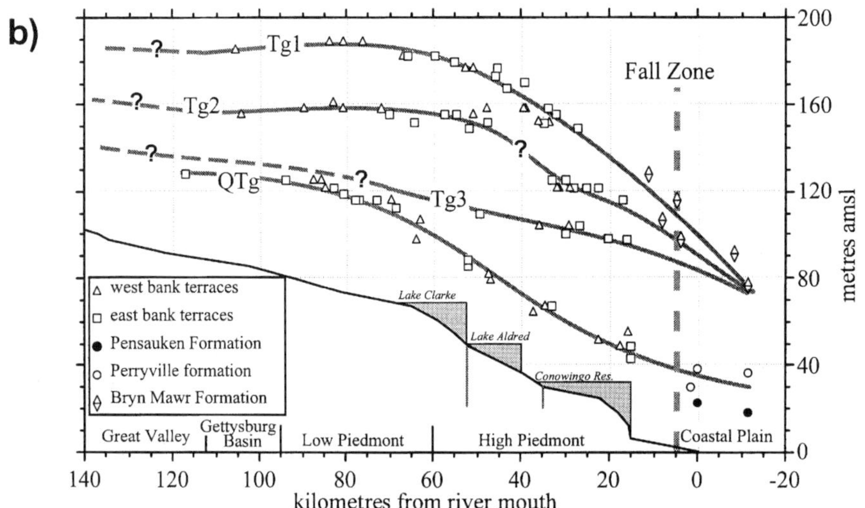
b)
Tg1
Tg2
QTg
Tg3
Fall Zone
west bank terraces
east bank terraces
Pensauken Formation
Perryville formation
Bryn Mawr Formation
Lake Clarke
Lake Aldred
Conowingo Res.
Great Valley
Gettysburg Basin
Low Piedmont
High Piedmont
Coastal Plain
metres amsl
kilometres from river mouth

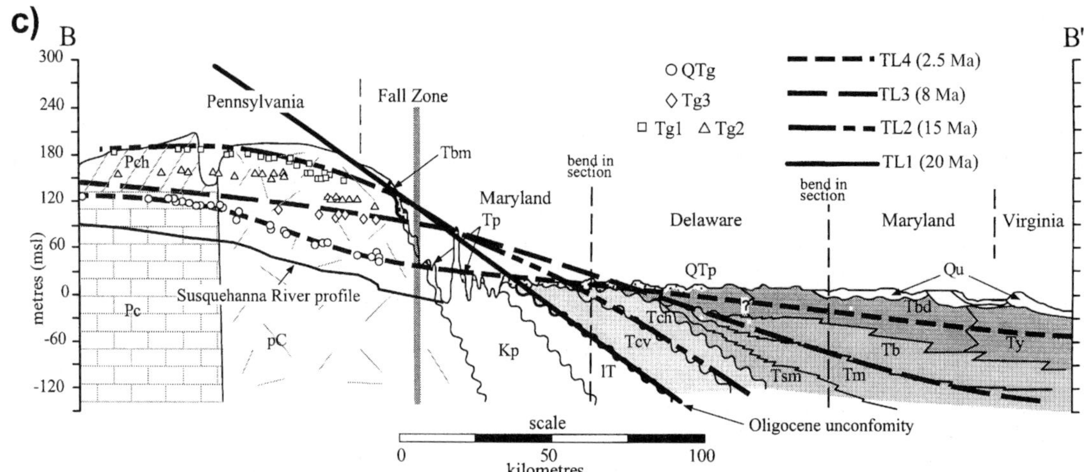
c)
B
B'
Pennsylvania
Fall Zone
QTg
Tg3
Tg1
Tg2
TL4 (2.5 Ma)
TL3 (8 Ma)
TL2 (15 Ma)
TL1 (20 Ma)
Tbm
bend in section
Maryland
Tp
Delaware
Maryland
Virginia
Pch
Pc
Susquehanna River profile
pC
Kp
QTp
Qu
Tbd
Tch
Tcv
Tb
Ty
IT
Tsm
Tm
metres (msl)
scale
0
50
100
kilometres
Oligocene unconfomity

20 Ma (200 m for our erosion rate of 10 m Ma^{-1}). We have used GMT to solve for the isostatic compensation of these gridded sediment loads with a two-dimensional FFT routine that computes the isostatic compensation from a topographic load in the frequency domain before transforming back to the space domain. GMT allows us to specify the thickness of the elastic plate, density of sediment load, density of underlying mantle, and density of material (water) infilling the flexural depression. Our area of interest is surrounded by a wide margin of grid cells of null value to simulate a plate of 'infinite' dimension and minimize model boundary effects. The result is a flexed surface that we can compare to our stratigraphic time lines as well as important topographic features such as the Fall Zone and Blue Ridge. We have constructed and run five separate two-dimensional models. Four of the models consider the cumulative, flexural effect of sediment loading only for the late Oligocene–early Miocene, middle Miocene, late Miocene and Plio-Pleistocene, and no concomitant continental denudation. The fifth model considers the cumulative effect of sediment loading since the late Oligocene as well as continental denudation at a rate of 10 m Ma^{-1} for that same time period.

13.5.3 METHODOLOGICAL AND DATA ERRORS

The observed data used for our modelling approach have large and relatively obvious sources of systematic error including age, elevation, sediment volume and eustasy. We wish to estimate error for our stratigraphic time lines as a qualitative way of tempering our comparison of those time lines to the geodynamic models. The 95% confidence interval reported in Table 4 of Cande and Kent (1992) provides a rough measure of the uncertainties in Δt, the time durations assigned to the stratigraphic intervals used to construct Figure 13.2 and the time lines of Figure 13.4c. Averaging and conversion of their confidence intervals to units of standard error suggests a relative standard error, RSE(Δt), of ~5%. The total vertical elevation range of Coastal Plain deposits and Piedmont fluvial terraces is not greater than 400 and 200 m, respectively (Figure 13.4c). Reasonable estimates for the error in calculating the depositional gradient for the stratigraphic time lines are about ±10 m for the terrace data (RSE(timelines) ~0.25%) and about ±25 m for the Coastal Plain subsurface data (RSE(Coastal Plain) ~3%). We estimate a sea level standard error SE(ΔSL) of ~30 m based on the residuals of the Haq *et al.* (1987) eustatic curve relative to a best-fit long-term curve (Pazzaglia and Brandon, 1996) which show about 60 m of total eustatic variation in the past 20 Ma. This variation contributes a relative standard error to the stratigraphic times lines of ~25% (RSE(ΔSL) ~25%). Propagation of all errors indicates a RSE(total) of about 30%.

Errors in the magnitude of flexural deformation are dominated by the uncertainties of the actual offshore sediment volume. Differences between Poag and Sevon (1989) and Poag (1992) suggest that estimates of sediment volume are probably within ~20% of their true value. Conversion of seismic two-way travel time isochron maps to sediment thickness in metres probably carries an uncertainty of ~200 m or a RSE(Δs) ~10%. Thus, we adopt a total relative standard error for sediment thickness of about 20% (RSE(Δs) ~20%). Model sensitivity analyses (Pazzaglia and Gardner, 1994) have demonstrated that these uncertainties in the sediment volume do not significantly affect the flexural results.

13.6 Results

The results of the one-dimensional line load model have already been presented and discussed in detail in Pazzaglia and Gardner (1994). Here we provide a short

Figure 13.4 *(a) Inner Coastal Plain and Fall Zone fluvial deposits at the head of Chesapeake Bay (modified from Pazzaglia, 1993). p€ = Precambrian bedrock, Kp = Cretaceous Potomac Group, Tbm = Bryn Mawr Formation, Tp = Perryville Formation, QTp = Pensauken Formation, Qt = Talbot Formation, Qki = Kent Island Formation. (b) Terrace longitudinal profiles of the lower Susquehanna River through the Pennsylvania and Maryland Piedmont (modified from Pazzaglia and Gardner, 1993). (c) Cross-section B–B' of Plate 13.1 showing the four stratigraphic time lines constructed by correlation of Susquehanna River terraces (Figure 13.4b) through Inner Coastal Plain and Fall Zone fluvial deposits (Figure 13.4a) and into Coastal Plain deposits of known age in the Salisbury Embayment (modified from Pazzaglia and Gardner, 1994). p€ = Precambrian bedrock, Pc = Conestoga Group, Pch = Chilhowee Group, Kp = Cretaceous Potomac Group, Tbm = Bryn Mawr Formation, Tp = Perryville Formation, QTp = Pensauken Formation, lT = lower Tertiary deposits, Tcv = Calvert Formation, Tch = Choptank Formation, Tsm = St Marys Formation, Tm = Manokin Formation, Tb = Bethany Formation, Tbd = Beaverdam Formation, Ty = Yorktown Formation, Qu = Quaternary deposits undivided*

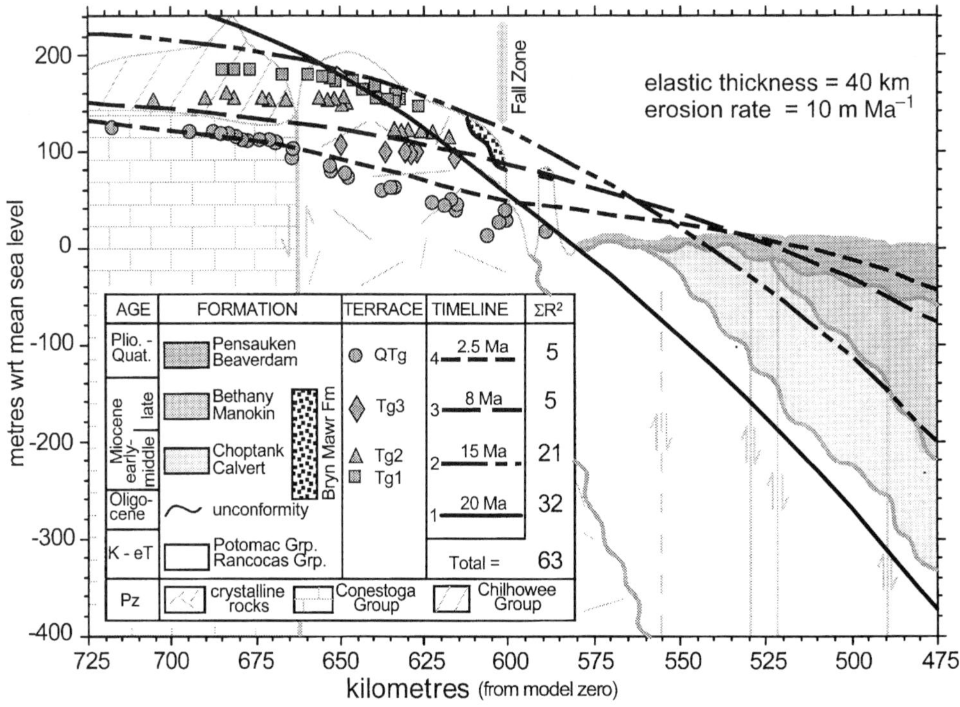

Figure 13.5 *Results of the one-dimensional geodynamic model. Model shown is parameterized with a flexural rigidity of* 4×10^{23} *Nm and a uniform total denudation rate of 10 m* Ma^{-1}. *(modified from Pazzaglia and Gardner, 1994)*

synopsis of these results that serves as a basis for the two-dimensional model results.

The one-dimensional results show the best fit to the stratigraphic time lines occurs when the model is parameterized by a flexural rigidity of 4×10^{23} Nm (effective elastic thickness, h, of 40 km) and an erosion rate of 10 m Ma^{-1} (Figure 13.5). The best fit for our one-dimension models was calculated as the sum of the squares of the residuals (ΣR^2) between the model-generated flexural profiles (thick, dark lines in Figure 13.5) and the stratigraphic time lines (shaded background and terrace data in Figure 13.5). When the one-dimensional model was parameterized with higher flexural rigidities (effectively thicker elastic plates), the flexural profiles tended to flatten. Conversely, with lower flexural rigidities (effectively thinner elastic plates), the flexural profiles become steeper. In either case, these profiles produced rather poor fits to the Coastal Plain portions of the stratigraphic time lines. In contrast, erosion rates strongly influence the fit of the flexural profiles to the terrace portion of the stratigraphic time lines. Variance in the erosion rates of just 5 m Ma^{-1} produced flexural profiles either tens of metres above or tens of metres below the mapped terrace elevations. The sensitivity of this simple one-dimensional model is such that an effective elastic thickness of 40 km and an erosion rate of 10 m Ma^{-1} represents geologically reasonable parameterization for flexure of the middle Atlantic margin. The total amount of rock uplift through the Pennsylvania Piedmont in the past 15 Ma ranges from 35 m at the Fall Zone to 130 m at the Great Valley.

The one-dimensional model captures, to a first-order, flexural deformation of the Atlantic margin as long as the flexural profiles are aligned orthogonal to the coast and in, or near, the centre of Coastal Plain depositional basins such as the Salisbury Embayment (the alignment of cross-section A–A' of Plate 13.1). Application of the one-dimensional model on a cross-section coincident with a known Coastal Plain arch (such as the Norfolk, Arch of Plate 13.1) could not produce flexural profiles that closely matched stratigraphic time lines. The obvious implications of this observation are that: (1) the arches are not the product of margin flexural deformation, but rather reflect tectonic or dynamic topographic processes, or (2) the arches are in fact related to flexural deformation that simply cannot be captured by a simple, one-dimensional line load model.

Shorter wavelength structural features, such as the Norfolk Arch, are better addressed with the two-dimensional model which considers a much larger portion of the margin and the combined effects of offshore sediment loading for both the middle and southern Atlantic margin.

Results of the two-dimensional model, parameterized with a flexural rigidity of 4×10^{23} Nm and no continental erosion, show a progressive flexural subsidence of the offshore depocentres since the late Oligocene culminating with up to 400 m of subsidence by the Pleistocene (Plate 13.3). It is important to reiterate that for the models shown in Plate 13.3, only offshore sediment loading is considered; there is no erosion of the sub-aerially exposed portion of the margin. The onshore peripheral bulge from the offshore loading coincides with the Appalachian Piedmont and does not show more than 35 m of rock uplift. Both the position and magnitude of the peripheral bulge are nearly coincident to that predicted by the one-dimensional model (Pazzaglia and Gardner, 1994). The most striking feature of Plate 13.3 is the close coincidence of the flexural hinge or null point, that is, the geographic position of neither flexural uplift or subsidence, with the Fall Zone (Figure 13.5). Even rather slight changes in the location of the Fall Zone, such as the abrupt, westward step in North Carolina (see arrow, Plate 13.3d) are well explained by flexural warping of the margin in response to the offshore load. Shorter wavelength features of the Atlantic margin such as the Norfolk Arch are not resolved by the simple two-dimensional models shown in Plate 13.3. Because we do not consider the flexural effects of the offshore basins north and south of the Baltimore Canyon Trough and Carolina Trough, the model generated feature coincident with the Cape Fear Arch (Plate 13.4d) is not taken as a true representation of that feature.

The two-dimensional flexural model that incorporates continental erosion (Plate 13.4) supports the results of the other four two-dimensional models and provides a first-order estimate for the amount of post-late Oligocene rock uplift that could be attributed to flexural processes alone. The effects of continental denudation are important in that up to 80 m of rock uplift is predicted for a broad region stretching across the Piedmont and into the Blue Ridge for the southern portion of the margin, or into the Ridge and Valley for the more central portion of the margin. The peripheral bulge of this model (Plate 13.4) is much broader and further inland than the models which consider offshore loading only (Plate 13.3). The flexural hinge coincident with the Fall Zone is exacerbated by an overall steeper profile. Again, the flexural hinge is seen as a very abrupt and steep feature for the Salisbury Embayment, and a broader, more gentle feature for the southern portion of the margin. In this model, the Cape Fear Arch does emerge as a short wavelength feature of the margin, but the Norfolk Arch is still not resolved. At the only location for which we have well-constrained, stratigraphic time lines (cross-section A–A') (Plate 13.1 and Fig. 13.4c), the two-dimensional model comes within a factor of two of predicting the observed subsidence of Coastal Plain sediments and uplift of Piedmont fluvial terraces.

13.7 Discussion and Conclusions

13.7.1 THE FALL ZONE, PIEDMONT AND COASTAL PLAIN

The boundary between hard, resistant rocks of the Piedmont, and soft, non-resistant rocks of the Coastal Plain is expressed as a seaward-facing escarpment with 50 to 150 m of relief called the Fall Zone. The location of the Fall Zone is coincident with the seaward-facing portion of the flexural peripheral bulge in both our one- and two-dimensional models (Figure 13.5, Plates 13.3 and 13.4). For the models with no continental erosion (Plate 13.3) location of the flexural peripheral bulge is a function only of the offshore load and model flexural rigidity. Because we carefully characterized both of these parameters with the one-dimensional model (Pazzaglia and Gardner, 1994), the green colours of Plate 13.3 mark the location landward of the offshore load where a 40-km-thick elastic plate will begin bending downward to accommodate that load, a location coincident with the Fall Zone.

Incorporation of erosion into the two-dimensional model (Plate 13.4) further enhances the flexural bending at the Fall Zone but does not change its location, an observation consistent with the stratigraphic record. Along the westward margin of the Salisbury Embayment (Plates 13.1 and 13.3) where the Fall Zone is a prominent topographic feature with upward of 150 m of relief, marine sediments of major Cretaceous and Cenozoic eustatic transgressions onlap, but do not extend westward across the Piedmont (Owens, 1970; Owens and Gohn, 1985). Further south, the Fall Zone has less relief and does not mark such an abrupt transition between the Piedmont and the Coastal Plain (Hack, 1982). Here, Cretaceous and early Cenozoic marine transgressions locally extended westward over the Piedmont (Christopher *et al.*, 1980). Clearly, a Piedmont of low relief must have existed along the southern Atlantic margin between the base of the Blue Ridge and the Fall Zone to accommodate and preserve these deposits. Late Cenozoic flexure, greater along the middle than southern Atlantic margin, exacerbates fluvial and marine etching into the flexural peripheral bulge, accentuating the topographic expression of the Fall Zone. The two-dimensional model captures the distinction between the Fall Zone of the Salisbury Embayment and that of the southern Atlantic margin with a steep flexural gradient in the north, and a much more gentle gradient in the south (Plates 13.3 and 13.4).

Given that the flexural deformation of the Atlantic margin is determined by the thick accumulations of offshore sediment, margin topographic features like the Fall Zone are ultimately influenced by the conditions that controlled the formation of the offshore depocentres. The post-rift structural setting of the Atlantic margin has been inherited from pre-rift orogenic events (Battiau-Queney, 1988; Hatcher *et al.*, 1989). For example, the Baltimore Canyon Trough is a lower plate basin developed above a low-angle detachment of the Triassic rift margin (Klitgord *et al.*, 1988). The proximity of the Baltimore Canyon Trough to the modern shoreline and the strong arcuate shape of the Salisbury Embayment is a reflection of a major salient of Alleghenian deformation situated between the centre of Taconic deformation to the north, and Acadian deformation to the south. Ultimately, the Palaeozoic orogenies which built the Appalachians are superimposed, and in part controlled, from even earlier structures associated with the late Proterozoic Grenville orogeny and subsequent rift opening of the Iapetus Sea.

Differential uplift and warping in the Appalachian Piedmont (Hayes and Campbell, 1894; Campbell, 1929; Reed, 1981; Hack, 1982) and Coastal Plain (Owens, 1970; Owens and Gohn, 1985) also reflect a complex interaction between pre-rift and post-rift structures. Streams that traverse the Piedmont of the middle Atlantic margin, such as the Susquehanna and Potomac Rivers, tend to be more deeply incised than streams of the southern Piedmont. The high Piedmont uplands dissected by these streams are underlain by rock types of variable resistance, yet on average they stand higher than the Piedmont uplands of the southern Appalachians. And locally, the regions of deep incision are not related to resistant rock type, as is the case for the Talladega belt of northern Georgia which is underlain largely by phyllite (Hack, 1989). At the Fall Zone, all streams of the middle Atlantic margin make a sharp bend to the southwest before turning eastward across the Coastal Plain (see the Salisbury Embayment in Plate 13.1). Locally, these bends may be structurally controlled (Higgins *et al.*, 1974; Newell, 1985; McCartan, 1989; Pazzaglia, 1993); however, the consistent southwest deflection of every major channel is also reflected in the southwestern migration of the locus of Chesapeake Group marine deposition (Ward and Strickland, 1985; Pazzaglia, 1993) in the Salisbury Embayment. The two-dimensional model shows greater late Cenozoic flexural uplift in the northern portion of the Salisbury Embayment (Figure 13.5, Plates 13.3 and 13.4) which is consistent with these geological observations. Overall, the shorter wavelength structural features underlying the Coastal Plain, such as the basins and arches (Plate 13.1), are not well represented in our geodynamic models. We conclude that the model is either not parameterized properly to capture these features, or the basins and arches are the result of non-flexural (tectonic) processes.

13.7.2 THE BLUE RIDGE AND DRAINAGE DIVIDE

Consideration of continental erosion in the geodynamic modelling produces somewhat disparate results for the landscape west of the Piedmont (Plate 13.3 vs. Plate 13.4). For the models with no continental erosion (Plate 13.3), the late Cenozoic peripheral bulge is centred on the Piedmont and the Blue Ridge actually lies on the landward side of the asymmetric peripheral bulge. In contrast, the model with continental erosion (Plate 13.4) produces a wider peripheral bulge further to the west that is nearly coincident with the Blue Ridge along the southern margin and significantly landward west of the Allegheny Front along the northern margin. The location of the peripheral bulge in this model is strongly controlled by the distribution of erosion in the model grid, which for our case is assumed to be uniform. Our present body of knowledge on the rates of erosion in the Appalachian landscape (reviewed in Sevon, 1989) do not allow us to assume non-uniform erosion rates and subsequent variations in the rate of isostatically driven rock uplift.

Late Cenozoic uplift of the Blue Ridge and location of the drainage divide are intricately related. The two-dimensional model with erosion (Plate 13.4) shows that the Blue Ridge should have experienced approximately 70 to 80 m of post-Oligocene flexurally driven rock uplift. This amount of rock uplift, in and of itself, cannot be responsible for the high mean elevation of the Blue Ridge with respect to the adjacent Piedmont, but it just might be enough to maintain the position of the drainage divide and Blue Ridge escarpment, an idea first proposed by Hack (1982) and demonstrated by the geodynamic modelling results of Keen and Beaumont (1990). These results offer an alternative explanation for the location of the drainage divide as not being primarily controlled by poorly understood deep crustal structures as proposed by Battiau-Queney (1988).

In contrast, the drainage divide of the middle Atlantic margin lies on the Allegheny Front, 200 km to the west of the narrow, broken Blue Ridge outcrop belt (Plate 13.1). Here the drainage divide is significantly seaward of the maximum peripheral bulge (Plate 13.4). Between the southern Blue Ridge and Allegheny Front, the drainage divide steps abruptly to the west (Plate 13.1), a step not mimicked by the crest of the flexural peripheral bulge (Plate 13.4). Steep flexural gradients imposed on the middle Atlantic margin throughout its post-rift history have given central Appalachian streams such as the palaeo-Delaware, Susquehanna, Potomac and James Rivers a short, steep path to base level that favours westward extension of headwaters. Further south, where the offshore flexural depression is less and maximum flexural uplift is centred beneath the Blue Ridge, the flexural profiles are much less steep. Here, the Piedmont streams have a longer distance to base level, and they are not as aggressive in advancing their headwaters. We recognize the importance and complications of upstream controls on stream power and a stream's ultimate ability to cut headward; however, we cannot dismiss the role that base level plays in the long-term development of a stream's gradient (Pazzaglia and Gardner, 1993) and its concomitant ability to extend its headwaters. We conclude that the combination of the ability for a stream to extend westward, structural controls on the width of the Piedmont and Blue Ridge outcrop belt, and nearly 80 m of post-20 Ma BP flexural uplift are the primary factors in determining the modern drainage divide location.

13.7.3 A NORTH AMERICAN GREAT ESCARPMENT

The results of our geodynamic modelling, taken in the context of previous geodynamic models (Keen and Beaumont, 1990) and conceptual models for long-term landscape evolution, allow us to evaluate the tectonic origin of the margin's more spectacular topographic features including the Fall Zone, Allegheny Front and Blue Ridge escarpment. Our results are consistent with a complex landscape response following rifting, likely to have been shaped by the pre-rift geological, structural and topographic setting, and strongly influenced by late Cenozoic flexural deformation. The first-order, post-rift response of the margin has been subsidence (Keen and Beaumont, 1990; Pazzaglia and Gardner, 1994), and that is why the Atlantic margin has a well-developed Coastal Plain. Continued offshore subsidence supports a landward flexural bulge that has been documented by deformed late Cenozoic morphological and stratigraphic horizons (Pazzaglia and Gardner, 1994).

Along both the middle and southern Atlantic margin, the Fall Zone stands as a distinct escarpment etched into the seaward portion of the flexural peripheral bulge. The Fall Zone escarpment is a classic Type-2 escarpment and not a candidate for the Great Escarpment of eastern North America. Similarly, the Allegheny Front along the middle Atlantic margin (Plate 13.4) is also primarily a Type-2 escarpment. In general the drainage divide of the middle Atlantic margin, largely defined by the Allegheny Front, lies significantly seaward of the flexural peripheral bulge and is well adjusted to rock type and structure. These observations are consistent with a Type-2 escarpment. If rifting along the middle Atlantic margin generated a Type-1 escarpment in the Blue Ridge or Piedmont provinces, the large degree of post-rift flexural subsidence, and rapid elongation and growth of major drainage systems such as the Susquehanna River, destroyed that feature and set the stage for the evolution of the current Type-2 escarpment.

A vestige of a true Type-1 escarpment may remain on the southern Atlantic margin represented by the Blue Ridge escarpment. For the southern Atlantic margin, post-rift flexural subsidence was not as great as that along the middle or northern margin (Keen and Beaumont, 1990). Here, the headwaters of east-flowing streams have collectively extended the drainage divide to the Blue Ridge, nearly coincident with the location of the maximum, late Cenozoic peripheral bulge, but not beyond. A relatively gentle seaward-sloping late Cenozoic flexural gradient has not favoured the elongation and subsequent growth of Atlantic slope streams capable of breaching Blue Ridge rocks. Retreat of the escarpment has left the Piedmont uplands as the dissected remnant of a seaward-sloping pediment. Given that the southern Piedmont uplands locally preserve Cretaceous and early Cenozoic deposits (Christopher *et al.*, 1980; Hack, 1982; Pavich, 1989; Markewitz *et al.*, 1990; Howard *et al.*, 1993) we favour rather rapid post-rift retreat of the Blue Ridge escarpment to, or near, its present location.

The near coincidence of the Blue Ridge, the drainage divide, and maximum flexural bulge along the southern Atlantic margin are features consistent with the Type-1 escarpments of the southern continents.

We conclude that the southern Blue Ridge escarpment is now clearly localized by post-rift flexural effects, but most probably evolved directly from an initial rift-flank escarpment not completely destroyed by early post-rift flexural subsidence. In this sense, the Blue Ridge escarpment is a hybrid between a true Type-1 and Type-2 escarpment and contains some components of a Type-1 Great Escarpment. Its relatively subdued topographic expression with respect to the more well-known great escarpments may be related to differences in margin elevation during rifting, the generally relatively large degree of post-rift flexural downwarping of the US Atlantic margin, or the fact that the Atlantic margin inherited numerous geological structures from the complex Appalachian orogen. In any case, our results support the earlier claims of a southern Blue Ridge Great Escarpment for eastern North America by Davis (1903) and Ollier (1985b).

Acknowledgements

Acknowledgement is made to the Donors of The Petroleum Research Fund, administered by the American Chemical Society, for the partial support of this research. Additional support to FP was furnished through a Shell Fellowship by the Penn State Department of Geosciences, an Earth System Science Fellowship through EOS NASA grant NAS 5-30556, and a Student Research Grant of the Geological Society of America. Part of this work was completed and supported by a NSF Post-Doctoral Research Fellowship (EAR-9302661) at Yale University. The authors wish to express thanks to Mark Brandon for early guidance on the conceptualization of the two-dimensional model. We thank Merri Lisa Formento-Trigilio for a helpful review of an early version of this manuscript. Formal reviews by Chris Beaumont and Mike Summerfield have improved this chapter; however, we assume full responsibility for the data and interpretations advanced within.

References

Ashley, G.H. 1930. Age of the Appalachian peneplains. *Bull. Geol. Soc. Amer.*, **41**, 695–700.

Ashley, G.H. 1933. The scenery of Pennsylvania. *Pennsyl. Topo. Geol. Surv. Bull.*, **G-6**, 91 pp.

Barrell, J. 1920. Piedmont terraces of the northern Appalachians. *Amer. J. Sci.*, **49**, 227–258, 327–362, 407–428.

Barron, E.J. 1989. Climate variations and the Appalachians from the late Paleozoic to the present: Results from model simulations. *Geomorphology*, **2**, 99–118.

Bascom, F. 1921. Cycles of erosion in the Piedmont province of Pennsylvania. *J. Geol.*, **29**, 540–559.

Bascom, F. *et al.* 1902. Philadelphia Folio. *U.S. Geol. Surv. Atlas U.S.*, **162**, 12 pp.

Bascom, F. *et al.* 1909. Trenton Folio, New Jersey–Pennsylvania. *U.S. Geol. Surv. Atlas U.S.*, **167**, 24 pp.

Bashore, H.B. 1894. The Harrisburg terraces (Pa). *Amer. J. Sci., 3rd Series*, **47**, 98–99.

Bashore, H.B. 1896. Notes on the glacial gravels in the lower Susquehanna valley. *Amer. J. Sci., 4th Series*, **1**, 281–282.

Battiau-Queney, Y. 1988. Long term landform development of the Appalachian Piedmont (USA), a continental margin without 'Randschwellen'. *Geogr. Ann.*, **70**, 369–374.

Battiau-Queney, Y. 1989. Constraints from deep crustal structure on long-term landform development of the British Isles and eastern United States. *Geomorphology*, **2**, 53–70.

Boettcher, S.S. and Milliken, K.L. 1994. Mesozoic–Cenozoic unroofing of the southern Appalachian basin: Apatite fission track evidence from middle Pennsylvanian sandstones. *J. Geol.*, **102**, 655–663.

Braun, D.D. 1989. The depth of the middle Miocene erosion and the age of the present landscape. In: W.D. Sevon (ed.) *The Rivers and Valleys of Pennsylvania Then and Now*, Harrisburg Geological Society Field Trip Guidebook, Harrisburg, PA, pp. 16–20.

Brown, P.M., Miller, J.A. and Swain, F.M. 1972. Structural and stratigraphic framework, and spatial distribution of permeability of the Atlantic Coastal Plain, North Carolina to New York. *U.S. Geol. Surv. Prof. Pap.*, **796**, 79 pp.

Campbell, M.R. 1929. Late geologic deformation of the Appalachian Piedmont as determined by river gravels. *Nat. Acad. Sci. Proc.*, **15**, 156–161.

Cande, S.C. and Kent, D.V. 1992. A new geomagnetic polarity time scale for the late Cretaceous and Cenozoic. *J. Geophys. Res.*, **97**, 13 917–13 951.

Christopher, R.A., Prowell, D.C., Reinhardt, J. and Markewich, H.W. 1980. The stratigraphic and structural significance of Paleocene pollen from Warm Springs, Georgia. *Palynology*, **4**, 105–124.

Cleaves, E.T. 1989. Appalachian Piedmont landscape from the Permian to the Holocene. *Geomorphology*, **2**, 159–179.

Cleaves, E.T. 1993. Climate impact on isovolumetric weathering of a coarse-grained schist in the northern Piedmont Province of the central Atlantic States. *Geomorphology*, **8**, 191–198.

Cleaves, E.T., Godfrey, A.E. and Bricker, O.P. 1970. Geochemical balance of a small watershed and its geomorphic implications. *Geol. Soc. Amer. Bull.*, **81**, 3015–3032.

Cleaves, E.T., Godfrey, A.E. and Bricker, O.P. 1974. Chemical weathering of serpentinite in the eastern Piedmont of Maryland. *Geol. Soc. Amer. Bull.*, **85**, 437–444.

Darton, N.H. 1939. Gravel and sand deposits of eastern Maryland. *U.S. Geol. Surv. Bull.*, **906A**, 39 pp.

Davis, W.M. 1889. The rivers and valleys of Pennsylvania. *Nat. Geog. Mag.*, **1**, 183–253.

Davis, W.M. 1899. The geographical cycle. *Geogr. J.*, **14**, 481–504.

Davis, W.M. 1903. The stream contest along the Blue Ridge. *Bull. Geog. Soc. Philadelphia*, **3**, 213–244.

de Boer, J.Z., McHone, J.G., Puffer, J.H., Ragland, P.C. and Whittington, D. 1988. Mesozoic and Cenozoic magmatism. In: R.E. Sheridan and J.A. Grow (eds) *The Atlantic Continental Margin, U.S.*, The Geology of North America, **I-2**, Geological Society of America, Boulder, CO, pp. 217–241.

Engel, S.A. and Gardner, T.W. 1996. Quaternary soil chronosequences on terraces of the Susquehanna river, Pennsylvania. *Geomorphology*, **17**, 273–294.

Gardner, T.W. 1989. Neotectonism along the Atlantic passive margin. *Geomorphology*, **2**, 71–97.

Gilchrist, A.R. and Summerfield, M.A. 1990. Differential denudation and flexural isostasy in formation of rifted-margin upwarps. *Nature*, **346**, 739–742.

Gilchrist, A.R. and Summerfield, M.A. 1991. Denudation, isostasy, and landscape evolution. *Earth Surf. Processes Ldfms.*, **16**, 555–562.

Gilchrist, A.R., Kooi, H. and Beaumont, C. 1994. The post-Gondwana geomorphic evolution of south-western Africa: Qualitative comparison of observations with a numerical model. *J. Geophys. Res.*, **99**, 12 211–12 228.

Greenlee, S.M., Schroeder, F.W. and Vail, P.R. 1988. Seismic stratigraphy and geohistory analysis of Tertiary strata from the continental shelf off New Jersey – calculations of eustatic fluctuations. In: R.E. Sheridan and J.A. Grow (eds) *The Atlantic Continental Margin, U.S.*, The Geology of North America, **I-2**, Geological Society of America, Boulder, CO, pp. 437–444.

Grow, J.A., Klitgord, K.D. and Schlee, J.S. 1988. Structure and evolution of the Baltimore Canyon Trough. In: R.E. Sheridan and J.A. Grow (eds) *The Atlantic Continental Margin, U.S.*, The Geology of North America, **I-2**, Geological Society of America, Boulder, CO, pp. 269–290.

Gurnis, M. 1992. Long-term controls on eustatic and epeirogenic motions by mantle convection. *GSA Today*, **2**, 141, 144–145, 156–157.

Hack, J.T. 1960. Interpretation of erosional topography in humid temperate regions. *Amer. J. Sci.*, **258A**, 80–97.

Hack, J.T. 1980. Rock control and tectonism – Their importance in shaping the Appalachian Highlands. *U.S. Geol. Surv. Prof. Pap.*, **1126-B**, B1–B17.

Hack, J.T. 1982. Physiographic divisions and differential uplift in the Piedmont and Blue Ridge. *U.S. Geol. Surv. Prof. Pap.*, **1265**, 49 pp.

Hack, J.T. 1989. Geomorphology of the Appalachian highlands. In: R.D. Hatcher, Jr, W.A. Thomas and G.W. Viele (eds) *The Appalachian–Ouachita Orogen in the United States*, The Geology of North America, **F-2**, Geological Society of America, Boulder, CO, pp. 459–470.

Haq, B.V., Hardenpol, J. and Vail, P.R. 1987. Chronology of fluctuating sea levels since the Triassic. *Science*, **235**, 1156–1167.

Harland, W.B., Armstrong, R.L., Cox, A.V., Craig, L.E., Smith, A.G. and Smith, D.G. 1990. *A Geologic Time Scale 1989*, Cambridge University Press, Cambridge.

Hatcher, R.D., Jr, Thomas, W.A. and Viele, G.W. (eds) 1989. *The Appalachian–Ouachita Orogen in the United States*, The Geology of North America, **F-2**, Geological Society of America, Boulder, CO, 767 pp.

Hayes, C.W. and Campbell, M.R. 1894. Geomorphology of the southern Appalachians. *Nat. Geog. Mag.*, **6**, 63–126.

Higgins, M.W., Zietz, I. and Fisher, G.W. 1974. Interpretation of aeromagnetic anomalies bearing on the origin of upper Chesapeake Bay and river course changes in the central Atlantic seaboard region: speculations. *Geology*, **2**, 73–76.

Howard, J.L., Amos, D.F. and Daniels, W.L. 1993. Alluvial soil chronosequence in the inner Coastal Plain, central Virginia. *Quat. Res.*, **39**, 201–213.

Jamison, T.F. 1982. On the cause of the depression and re-elevation of the land during the glacial period. *Geol. Mag.*, **9**, 461.

Johnson, D. 1931. *Stream Sculpture on the Atlantic Slope*, Columbia University Press, New York, 142 pp.

Judson, S. 1975. Evolution of Appalachian topography. In: W.N. Melhorn and R.C. Flemal (eds) *Theories of Landform Development, Proceedings of the 6th Geomorphology Symposium*, Publications in Geomorphology, State University of New York, Binghamton, New York, pp. 29–44.

Judson, S. and Ritter, D.F. 1964. Rates of regional denudation in the United States. *J. Geophys. Res.*, **69**, 3395–3401.

Karner, G.D. and Watts, A.B. 1982. On Atlantic-type continental margins. *J. Geophys. Res.*, **87**, 2923–2948.

Keen, C.E. and Beaumont, C. 1990. Geodynamics of rifted continental margins. In: M.J. Keen and G.L. Williams (eds) *Geology of the Continental Margin of Eastern Canada*. The Geology of North America, **I-1**, Geological Society of America, Boulder, CO, pp. 391–472.

Klitgord, K.D., Hutchinson, D.R. and Shouten, H. 1988. U.S. Atlantic continental margin, structure and tectonic framework. In: R.E. Sheridan and J.A. Grow (eds) *The Atlantic Continental Margin U.S.*, The Geology of North America, **I-2**, Geological Society of America, Boulder, CO, pp. 19–55.

Knopf, E.B. 1924. Correlation of residual erosion surfaces in the eastern Appalachian highlands. *Bull. Geol. Soc. Amer.*, **35**, 633–668.

Knopf, E.B. and Jonas, A.I. 1929. Geology of the McCalls Ferry – Quarryville district, Pennsylvania. *U.S. Geol. Surv. Bull.*, **799**, 156 pp.

Knox, J.C. 1975. Concept of the graded stream. In: W.N. Melhorn and R.C. Flemal (eds) *Theories of Landform Development: Proceedings from the 6th Annual Geomor-*

phology Symposia, State University of New York, Binghamton, New York, pp. 169–198.

Kohn, B.P., Wagner, M.E., Lutz, T.M. and Organist, G. 1993. Anomalous Mesozoic thermal regime, central Appalachian Piedmont: Evidence from sphene and zircon fission track dating. *J. Geol.*, **101**, 779–794.

Kooi, H. and Beaumont, C. 1994. Escarpment evolution on high-elevation rifted margins: Insights derived from a surface processes model that combines diffusion, advection, and reaction. *J. Geophys. Res.*, **99**, 12 191–12 209.

Kooi, H. and Beaumont, C. 1996. Large-scale geomorphology: classical concepts reconciled and integrated with contemporary ideas via a surface processes model. *J. Geophys. Res.*, **101**, 3361–3386.

Kruse, S.E. and Royden, L.H. 1994. Bending and unbending of an elastic lithosphere: The Cenozoic history of the Apennine and Dinaride foredeep basins. *Tectonics*, **13**, 278–302.

Leopold, L.B. and Bull, W.B. 1979. Base level, aggradation and grade. *Proc. Amer. Phil. Soc.*, **123**, 168–202.

Mackin, J.H. 1938. The origin of Appalachian drainage – a reply. *Amer. J. Sci.*, 5th Series, **36**, 27–53.

Markewitz, H.W., Pavich, M.J. and Buell, G.R. 1990. Contrasting soils and landscapes of the Piedmont and Coastal Plain, eastern United States. *Geomorphology*, **3**, 417–447.

Matthews, W.H. 1975. Cenozoic erosion and erosion surfaces of eastern North America. *Amer. J. Sci.*, **275**, 818–824.

McCartan, L. 1989. Atlantic Coastal Plain sedimentation and basement tectonics southeast of Washington, D.C. *28th International Geological Congress, Field Trip Guidebook T214*, American Geophysical Union, Washington, DC, 25 pp.

McGee, W.J. 1888a. The geology of the head of Chesapeake Bay. *U.S. Geological Survey, 7th Annual Report (1885–86)*, Washington, DC, 537–646.

McGee, W.J. 1888b. Three formations of the middle Atlantic slope. *Amer. J. Sci., 3rd Series*, **35**, 120–143, 328–330, 367–388, 448–466.

Meyerhoff, H.A. and Olmstead, E.W. 1936. The origins of Appalachian drainage. *Amer. J. Sci.*, **232**, 21–42.

Miller, D.S. and Duddy, I.R. 1989. Early Cretaceous uplift and erosion of the northern Appalachian Basin, New York, based on apatite fission track analysis. *Earth Planet. Sci. Lett.*, **33**, 35–49.

Miller, K.G., Mountain, G.S. *et al.* 1996. The Leg 150 shipboard party and members of the New Jersey Coastal Plain drilling project, Drilling and dating New Jersey Oligocene-Miocene sequences: Ice volumes, global sea level and Exxon records. *Science*, **271**, 1092–1095.

Milliman, J.D. and Syvitski, J.P.M. 1992. Geomorphic tectonic control of sediment discharge to the ocean: The importance of small mountainous rivers. *J. Geol.*, **100**, 525–544.

Mills, H.H., Brackenridge, G.R., Jacobson, R.B., Newell, W.L., Pavich, M.J. and Pomeroy, J.S. 1987. Appalachian mountains and plateaus. In: W.L. Graf (ed.) *Geomorphic Systems of North America*, The Geology of North America, Special Centennial Volume, **2**, Geological Society of America, Boulder, CO, pp. 5–50.

Mixon, R.B. and Newell, W.L. 1977. Stafford fault system – structures documenting Cretaceous and Tertiary deformation along the Fall Line in northeastern Virginia. *Geology*, **5**, 437–441.

Mixon, R.B. and Powars, D.S. 1984. Folds and faults in the inner Coastal Plain of Virginia and Maryland: Their effect on the distribution and thickness of Tertiary rock units and local geomorphic history. In: N.O. Frederiksen and K. Kraft (eds) *Cretaceous and Tertiary Stratigraphy, Paleontology, and Structure, Southwestern Maryland and Northeastern Virginia, Field Trip Volume and Guidebook*, American Association of Palynologists, Washington, DC, pp. 112–122.

Morisawa, M. 1989. Rivers and valleys of Pennsylvania, revisited. *Geomorphology*, **2**, 1–22.

Newell, W.L. 1985. Architecture of the Rappahannock estuary – neotectonics in Virginia. In: M. Morisawa and J.T. Hack (eds) *Tectonic Geomorphology*, Allen and Unwin, Boston, pp. 322–342.

Newell, W.L. and Rader, E.K. 1982. Tectonic control of cyclic sedimentation in the Chesapeake Group of Virginia and Maryland. In: P.T. Lyttle (ed.) *Central Appalachian Geology, Northeast–Southeast Geological Society of America Meeting, Field Trip Guidebooks*, American Geological Institute, Falls Church, VA, pp. 1–26.

Ollier, C.D. (ed.) 1985a. Morphotectonics of passive continental margins. *Z. Geomorph. Suppl.*, **54**, 117 pp.

Ollier, C.D. 1985b. Morphotectonics of continental margins with great escarpments. In: M. Morisawa and J.T. Hack (eds) *Tectonic Geomorphology*, Allen and Unwin, Boston, pp. 3–25.

Owens, J.P. 1970. Post-Triassic tectonic movements in the central and southern Appalachians as recorded by sediments of the Atlantic Coastal Plain. In: G.W. Fisher, F.J. Pettijohn and J.C. Reed (eds) *Studies of Appalachian Geology: Central and Southern*, Wiley Interscience, pp. 417–427.

Owens, J.P. and Gohn, G.S. 1985. Depositional history of the Cretaceous series in the U.S. Coastal Plain: Stratigraphy, paleoenvironments, and tectonic controls on sedimentation. In: C.W. Poag (ed.) *Geologic Evolution of the Atlantic Margin*, Van Nostrand Reinhold, New York, pp. 25–86.

Owens, J.P. and Minard, J.P. 1979. Upper Cenozoic sediments of the lower Delaware Valley and the northern Delmarva Peninsula, New Jersey, Pennsylvania, Delaware, and Maryland. *U.S. Geol. Surv. Prof. Pap.*, **1067-D**, 47 pp.

Pavich, M.J. 1985. Appalachian Piedmont morphogenesis: Weathering, erosion and Cenozoic uplift. In: M. Morisawa and J.T. Hack (eds) *Tectonic Geomorphology*, Allen and Unwin, Boston, pp. 299–319.

Pavich, M.J. 1989. Regolith residence time and the concept of surface age of the Piedmont 'Peneplain'. *Geomorphology*, **2**, 181–196.

Pavich, M.J., Leo, G.W., Obermeier, S.F. and Estabrook, J.R. 1989. Investigations of the characteristics, origin, and residence time of the upland mantle of the Piedmont of Fairfax County, Virginia. *U.S. Geol. Surv. Prof. Pap.*, **1352**, 114 pp.

Pazzaglia, F.J. 1993. Stratigraphy, petrography, and correlation of late Cenozoic middle Atlantic Coastal Plain deposits: Implications for late-stage passive margin geologic evolution. *Geol. Soc. Amer. Bull.*, **106**, 1617–1634.

Pazzaglia, F.J. and Brandon, M.T. 1996. Macrogeomorphic evolution of the post-Triassic Appalachian mountains determined by deconvolution of the offshore basin sedimentary record. *Basin Res.*, **8**, 255–278.

Pazzaglia, F.J. and Gardner, T.W. 1993. Fluvial terraces of the lower Susquehanna River. *Geomorphology*, **8**, 83–113.

Pazzaglia, F.J. and Gardner, T.W. 1994. Late Cenozoic flexural deformation of the middle U.S. Atlantic passive margin. *J. Geophys. Res.*, **99**, 12 143–12 157.

Pazzaglia, F.J., Robinson, R. and Traverse, A. 1997. Palynology of the Bryn Mawr Formation (Miocene): insights on the age and genesis of Middle Atlantic margin fluvial deposits. *Sed. Geol.*, **108**, 19–44.

Poag, C.W. (ed.) 1985. *Geologic Evolution of the United States Atlantic Margin*, Van Nostrand Reinhold, New York.

Poag, C.W. 1992. U.S. middle Atlantic continental rise: Provenance, dispersal, and deposition of Jurassic to Quaternary sediments. In: C.W. Poag and P.C. de Graciansky (eds) *Geologic Evolution of Atlantic Continental Rises*, Van Nostrand Reinhold, New York, pp. 217–263.

Poag, C.W. and Sevon, W.D. 1989. A record of Appalachian denudation in postrift Mesozoic and Cenozoic sedimentary deposits of the U.S. middle Atlantic continental margin. *Geomorphology*, **2**, 119–157.

Poag, C.W. and Ward, L.W. 1993. Allostratigraphy of the U.S. middle Atlantic continental margin – characteristics, distribution, and depositional history of principal unconformity-bounded upper Cretaceous and Cenozoic sedimentary units. *U.S. Geol. Surv. Prof. Pap.*, **1542**, 81 pp.

Popenoe, P. 1985. Cenozoic depositional and structural history of the North Carolina Margin from seismic stratigraphic analyses. In: C.W. Poag (ed.) *Geologic Evolution of the United States Atlantic Margin*, Van Nostrand Reinhold, New York, pp. 125–187.

Prowell, D.C. 1988. Cretaceous and Cenozoic tectonism on the Atlantic coastal margin. In: R.E. Sheridan and J.A. Grow (eds) *The Atlantic Continental Margin*. The Geology of North America, **I-2**, Geological Society of America, Boulder, CO, pp. 557–564.

Reed, J.C. 1981. Disequilibrium profile of the Potomac River near Washington, D.C. – A result of lowered base level or Quaternary tectonics along the Fall Line. *Geology*, **9**, 445–450.

Rich, J.L. 1938. Recognition and significance of multiple erosion surfaces. *Bull. Geol. Soc. Amer.*, **49**, 1695–1722.

Roden, M.K. 1991. Apatite fission-track thermochronology of the southern Appalachian basin: Maryland, West Virginia, and Virginia. *J. Geol.*, **99**, 41–53.

Roden, M.K. and Miller, D.S. 1989. Apatite fission-track thermochronology of the Pennsylvania Appalachian basin. *Geomorphology*, **2**, 39–51.

Salisbury, R.P. and Knapp, G.N. 1917. The Quaternary formations of New Jersey. *New Jersey Geological Survey Final Report*, **8**, 218 pp.

Scholle, P.A. (ed.) 1977. Geological studies on the COST No. B-2 well, U.S. mid-Atlantic outer continental shelf area. *U.S. Geol. Surv. Circ.*, **750**, 71 pp.

Schumm, S.A. 1963. Disparity between rates of denudation and orogeny. *U.S. Geol. Surv. Prof. Pap.*, **454-H**, H1–H13.

Schumm, S.A. and Rea, D.K. 1995. Sediment yield from disturbed earth systems. *Geology*, **23**, 391–394.

Sevon, W.D. 1989. Erosion in the Juniata River drainage basin, Pennsylvania. *Geomorphology*, **2**, 303–318.

Shattuck, G.B. 1901. The Pleistocene problem of the north Atlantic Coastal Plain. *Johns Hopkins University Circular*, Johns Hopkins University, Baltimore, Maryland, pp. 74–75.

Shattuck, G.B. 1902. *Cecil County*, Maryland Geological Survey County Reports, 315 pp.

Shattuck, G.B. *et al.* 1906. The Pliocene and Pleistocene deposits of Maryland. In: G.B. Shattuck (ed.) *The Pliocene and Pleistocene*, Maryland Geological Survey Bulletin, pp. 21–137.

Slingerland, R. and Furlong, K.P. 1989. Geodynamic and geomorphic evolution of the Permo-Triassic Appalachian mountains. *Geomorphology*, **2**, 23–37.

Stephenson, R. 1984. Flexural models of continental lithosphere based on the long-term erosional decay of topography. *Geophys. J. R. Astr. Soc.*, **77**, 385–413.

Stose, G.W. 1928. High gravels of the Susquehanna River above Columbia, Pennsylvania. *Bull. Geol. Soc. Amer.*, **39**, 1073–1086.

Summerfield, M.A. 1985. Plate tectonics and landscape development on the African continent. In: M. Morisawa and J.T. Hack (eds) *Tectonic Geomorphology*, Allen and Unwin, Boston, pp. 27–52.

Tarr, R.S. 1898. The peneplain. *Amer. Geol.*, **21**, 351–370.

Thompson, H.D. 1939. Drainage evolution in the southern Appalachians. *Bull. Geol. Soc. Amer.*, **50**, 1323–1356.

Tucker, G.E. and Slingerland, R.L. 1994. Erosional dynamics, flexural isostasy, and long-lived escarpments: A numerical modeling study. *J. Geophys. Res.*, **99**, 12 229–12 243.

Turcotte, D.L. 1979. Flexure. *Adv. Geophys.*, **21**, 51–86.

Turcotte, D.L. and Schubert, G. 1982. *Geodynamics*, Wiley, New York, 450 pp.

Vogt, P.R. 1991. Bermuda and Appalachian–Labrador rises: common non-hotspot processes? *Geology*, **19**, 41–44.

Ward, L.W. and Strickland, G.L. 1985. Outline of Tertiary stratigraphy and depositional history of the U.S. Atlantic Coastal Plain. In: C.W. Poag (ed.) *Geologic Evolution of the Atlantic Margin*. Van Nostrand Reinhold, New York, pp. 87–123.

Wessel, P. and Smith, W.H.F. 1991. Free software helps map and display data. *EOS Trans. Amer. Geophys. Un.*, **72**, 441.

White, W.A. 1950. Blue Ridge front – a fault scarp. *Bull. Geol. Soc. Amer.*, **61**, 1309–1346.

Young, R. and McDougall, I. 1993. Long-term landscape evolution: Early Miocene and modern rivers in Southern New South Wales, Australia. *J. Geol.*, **101**, 35–49.

Zimmerman, R.A. 1979. Apatite fission track evidence of post-Triassic uplift in the central and southern Appalachians. *Geol. Soc. Amer. Absts with Progs.*, **11**, 219.

Zimmerman, R.A., Reimer, G.M., Foland, K.A. and Faul, H. 1975. Cretaceous fission track ages of apatites from northern New England. *Earth Planet. Sci. Lett.*, **28**, 181–188.

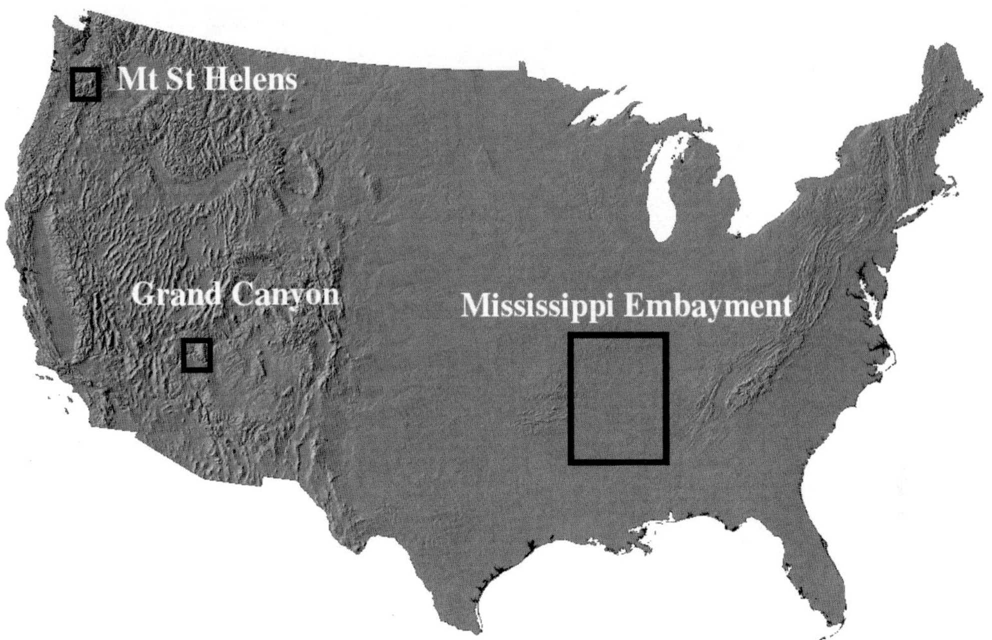

Plate 2.1 *Location map of US DEMs referred to in this chapter. The three areas are: Mount St Helens, in the Cascade volcanic arc; Grand Canyon on the Colorado Plateau; and the New Madrid seismic zone, in the lower Mississippi River valley. The base map is the digital shaded relief map of the United States, MAP I-2206, by Gail P. Thelin and Richard J. Pike, US Department of the Interior, US Geological Survey, 1991*

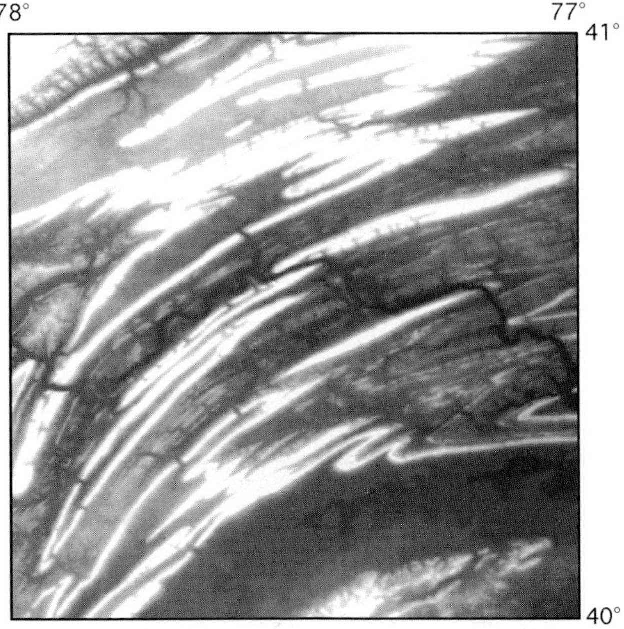

Plate 2.2 *Harrisburg West, Pennsylvania, 1-degree quadrangle DEM. The DEM allows for the calculation of elevation histograms to determine whether there are quantifiable data that may represent surfaces. With the addition of geological information and GIS, one could correlate elevation and rock type*

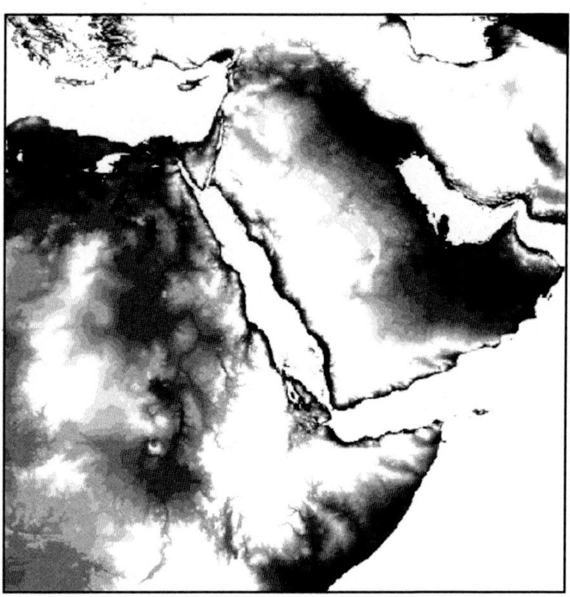

Plate 2.3 *Thresholded DEM for Africa showing high elevations in saturated white, and lower elevations (above sea level) in progressively darker greys. Regional topographic data permits the rapid measurement of topographic features for possible geodynamic modelling, and quantitiative description. The spectacular escarpments on the east side of the Red Sea give way to long downward topographic ramps. The East African rifts show topographic segmentation. Notice the almost circular shapes of the low areas which follow the Nile to the Mediterranean Sea. The data come from GRID at EROS Data Center, the Digital Chart of the World (see Appendix)*

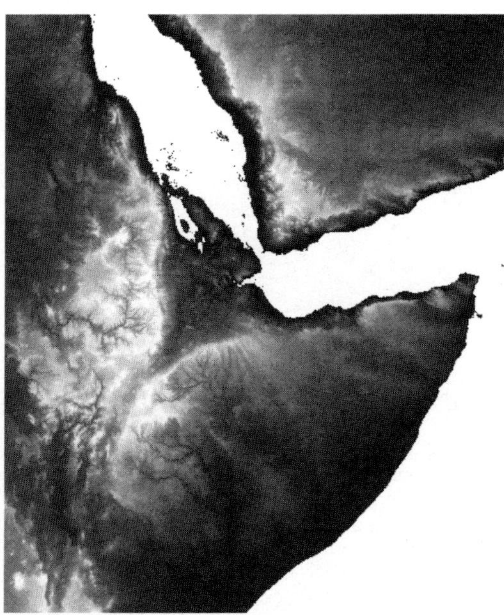

Plate 2.4 *DEM of East Africa centred on Afar, Ethiopia, displaying the topographic complexity of the rift and the importance it plays in river pattern, water-sediment distribution, rainfall, and possibly even evolution of early hominids. The rift valley lies essentially at the highest part of the topography. The Arabian plate sheds little sediment into the gulf due to the position of the escarpment and the general slope of the landscape*

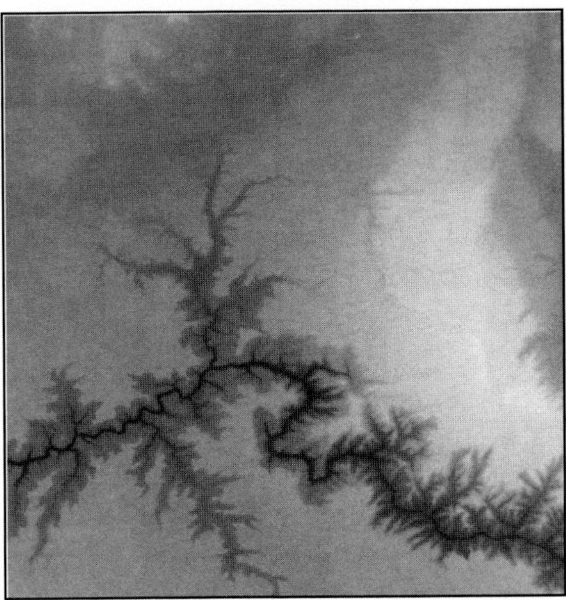

Plate 2.5 *Grand Canyon East, 1-degree quadrangle DEM, showing the Colorado River cutting into the Colorado Plateau. The pronounced northward erosion into the Colorado plateau is Kanab Creek cutting into the Kaibab limestone. The Kaibab plateau is bounded on the west by several down-to-the-west faults. At the intersection of these structures with the Colorado River in the Grand Canyon, one clearly sees that the pattern of plateau dissection is different, distinctively separating a more dissected western Grand Canyon from a less dissected eastern Grand Canyon. Note how the Colorado River wraps around the Kaibab uplift. DEMs also permit calculation of the amount of material removed by the Colorado River, which is quite useful for related sedimentological investigations*

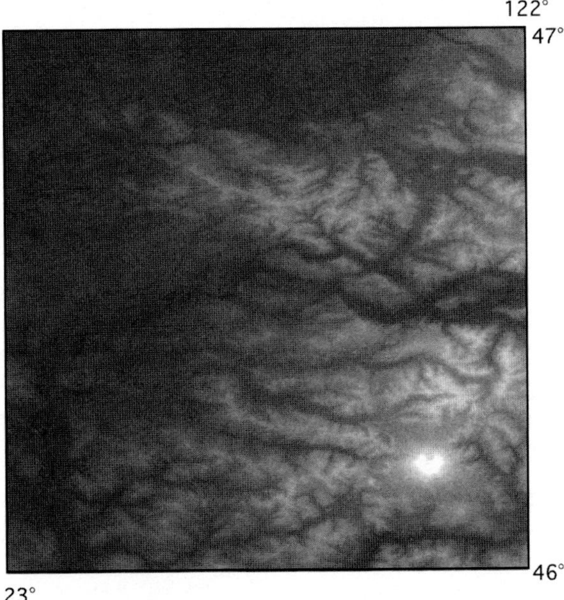

Plate 2.6 *A section of the Hoqium East, Washington, 1-degree quadrangle DEM showing the topography around the Mount St Helens volcano after the 18 May, 1980 eruption. Note the distinctive tectonic structures which cut across the topographic grain of the landscape. The DEM has been enhanced using hill shading*

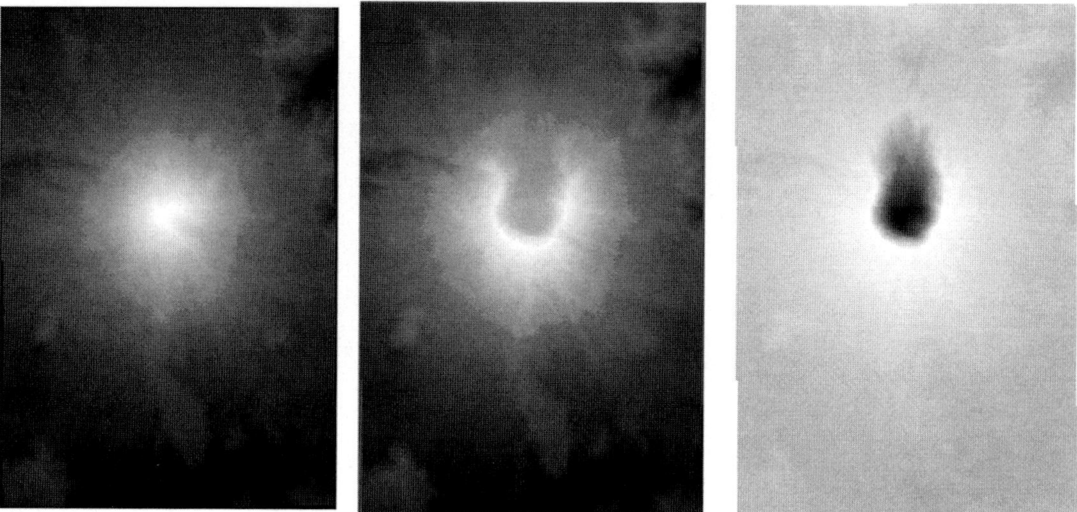

Plate 2.7 *Series of 7.5 minute DEMs for Mount St Helens volcano in the Cascade volcanic range, southern Washington State, northwestern United States. The image on the left represents a pre-eruption topography, the centre image a post-eruption topography, and the right image is the arithmetic difference of the images. The right image is therefore a quantitative estimate of the volumetric change caused by the eruption*

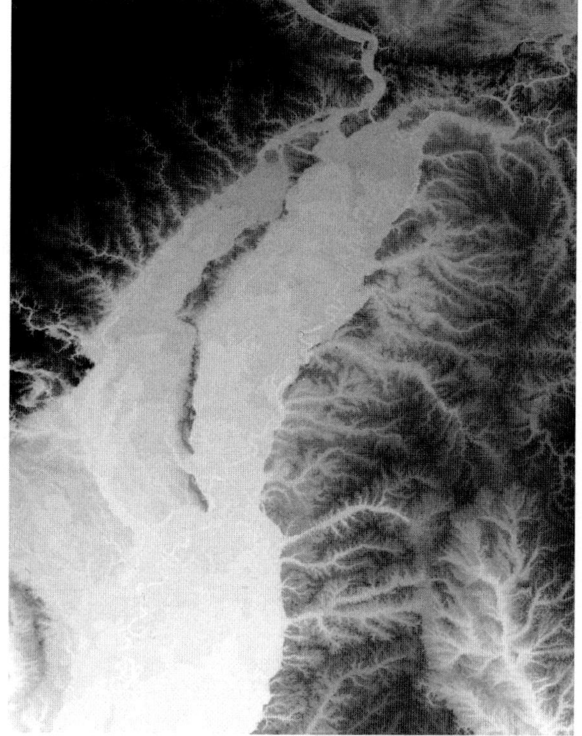

Plate 2.8 *Digital topography of the midcontinental USA centred on the lower Mississippi valley. New Madrid is located near the distinctive meander loop of the Mississippi River and was the approximate location of several great earthquakes in 1811-1812*

Plate 2.9 *Filtered topography showing interfluves, ridges and other textures in the grey pattern. Colour-coded elevations have been superimposed on the image. Note that the distinctive Mississippi River meander pattern is recorded in the filtered topography along the east side of the river valley and that textural patterns are visible in the meander belt*

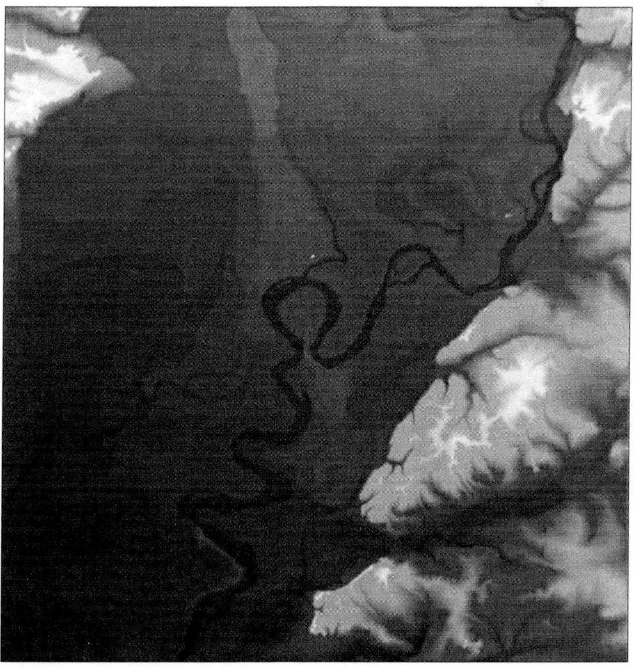

Plate 2.10 *Digital topography of the Dyersburg West, 1-degree quadrangle. In the approximate centre of the image is the New Madrid loop of the Mississippi River. The river flows from top to bottom in the image, and intersects a positive feature at the location of the loop. This area also coincides in part with the Lake County uplift, a tectonic dome*

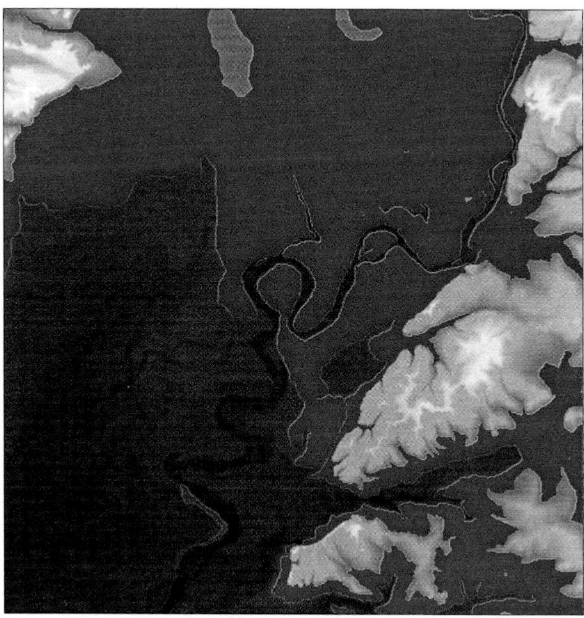

Plate 2.11 *Digital topography of the Dyersburg quadrangle. The red band represent a density slice through the topography. The goal of highlighting a particular altitude band is to identify areas of anomalous topographic pattern or areas which are positive or negative relative to a simple mode of valley formation. In this image several anomalous features exist. Note the high area adjacent to the river in the lower part of the image. The low area just south of the loop is Reelfoot Lake, suggested to be a tectonic depression*

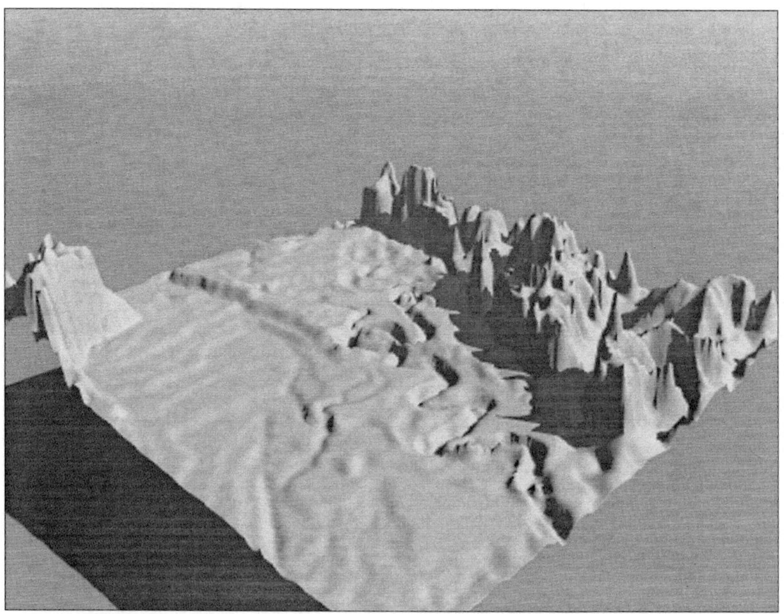

Plate 2.12 *Three-dimensional rendered view of the region around the New Madrid loop of the Mississippi River, looking north with the sun in the east. On the right are loess cliffs which border the river at this location. On the left is part of Crowleys ridge, an enigmatic topographic feature which may in part be tectonic and in part erosional in origin. Image processing such as this allows visualisation of geomorphic hypotheses on the origin of the topography. See text for discussion*

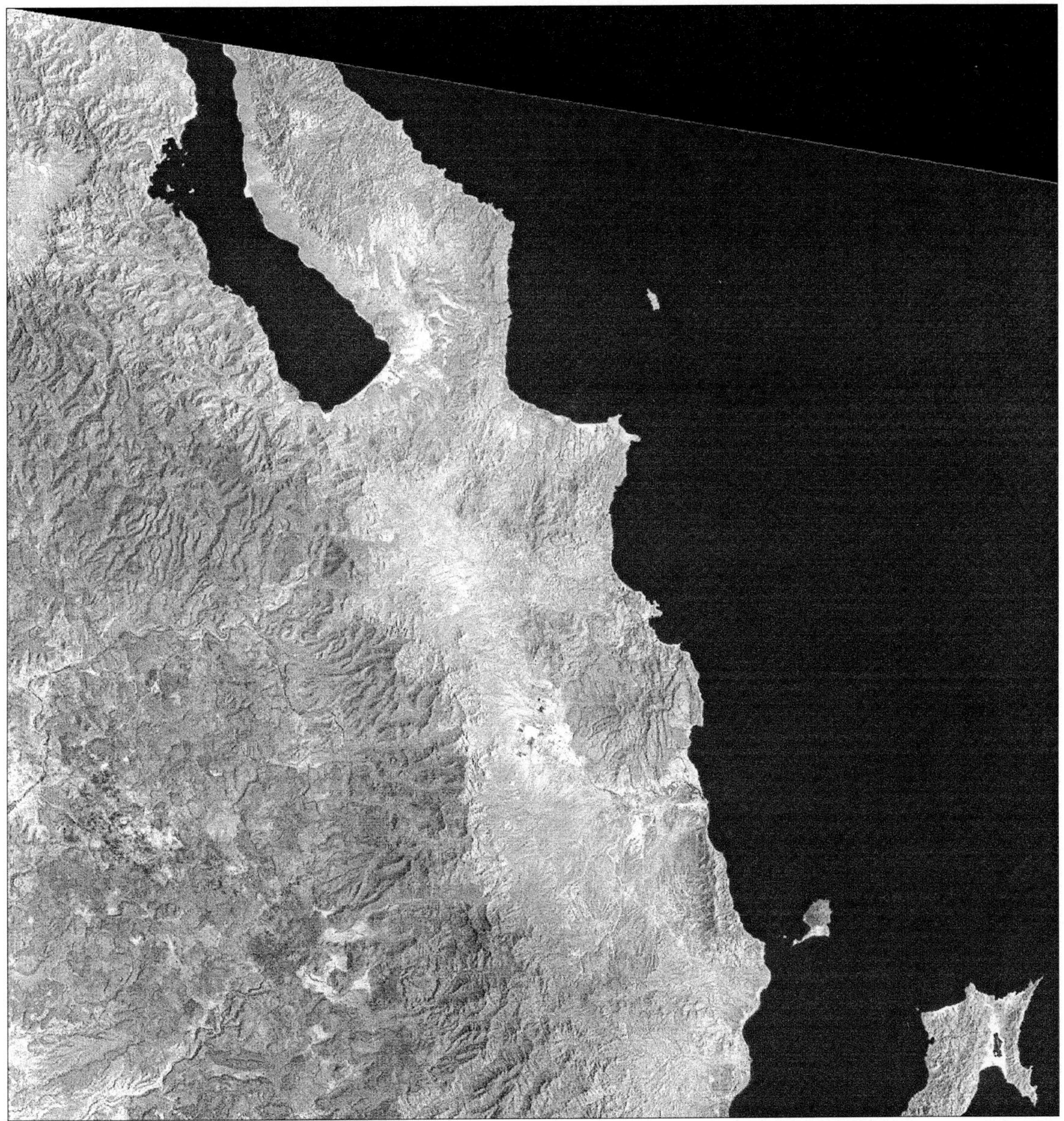

Plate 2.13 *Landsat MSS image of Baja Sur centred on the Mencenares volcanic complex Baja California, Mexico. The image is of Row 35 Path 42, Landsat WRS2, part of the Pathfinder data. North–south and northwest–southeast-trending faults cut the Mencenares volcanic complex. Many other structures can be mapped on this image; however the topographic variation is difficult to extract*

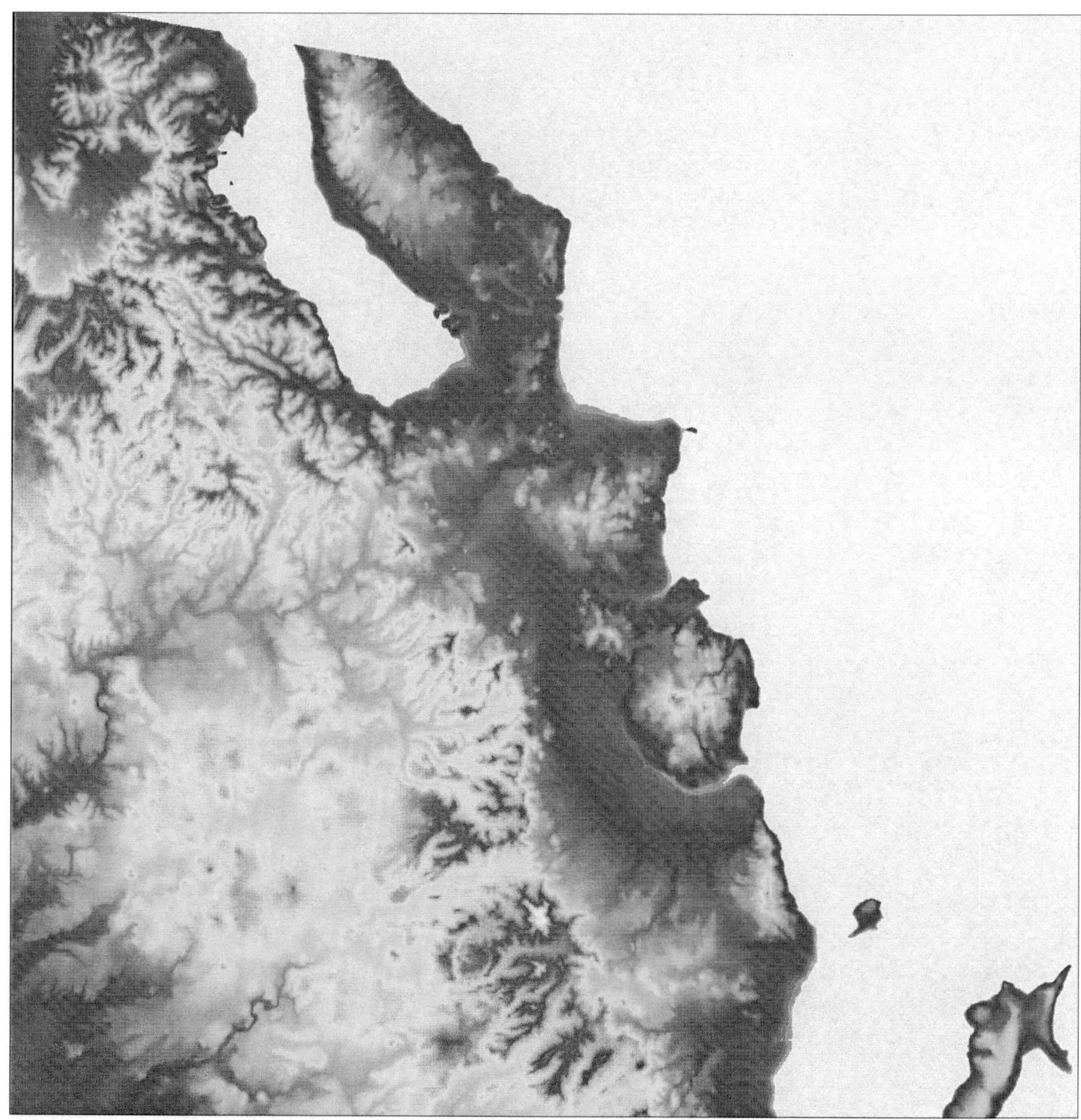

Plate 2.14 *DEM of the portion of Baja Sur centred on the Mencenares volcanic complex, Row 35 Path 42, Landsat WRS2, part of the Pathfinder data (from INEGI). The pronounced escarpment is the Main Gulf Escarpment (MGE) which has almost continuous expression along Baja. The escarpment is in places fault controlled, but in other places is believed to be dominantly erosional. Topographic barriers which extend from the MGE to the coastline may indicate segmentation boundaries. Note how the escarpment topography in the southern part of the image appears to have a continuous expression across the Loreto basin. The Loreto basin was a rapidly subsiding basin during the Pliocene*

Plate 2.15 *Rendered digital topography looking north from a position just north of the proposed accommodation zone. In the centre of the image is the Mencenares volcano. The San Juanico embayment is just north of the Mencenares. The top of the image is the approximate location of another accommodation zone, separating the Loreto segment from the Mulege segment. The higher topography and change in MGE may indicate the accommodation structure*

Plate 2.16 *Rendered digital topography looking south from a position just north of the proposed accommodation zone separating the Loreto segment from the segments to the south. The large island is that part of Isla Carmen which is in view, the smaller one nearer to the viewer being Isla Coronado. Note the line of east–west-trending hills in the centre of the image which form the current drainage divide*

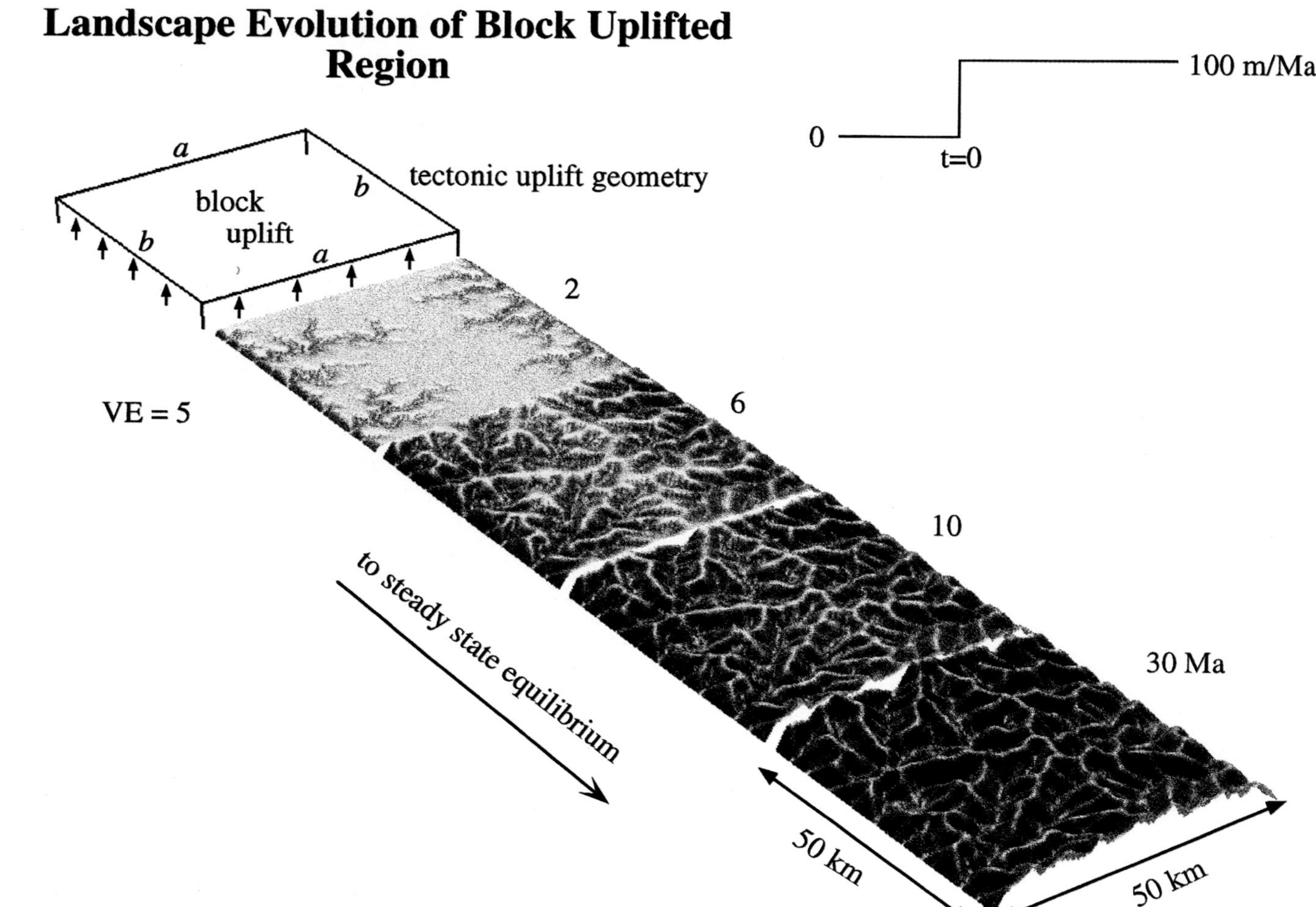

Plate 3.1 *Landscape evolution of a block uplifted at a velocity of 100 m Ma^{-1} and subject to denudation by the SPM described in Section 3.3.1. Initial topography is flat with small-amplitude white noise. Uplift starts at* t = 0 *and panels show evolution at 2, 6, 10 and 30 Ma toward a steady state. Boundary conditions are constant elevation and flux transmission on* b, *and increasing elevation (increasing at uplift rate), and flux reflection on* a. *The model therefore represents part of a long, linear region in which model properties repeat in the longitudinal direction at a 50 km interval. Model properties are spatially and temporally uniform:* $K_s = 0.2\ m^2\ a^{-1}$, $K_f = 0.01$, $l_f = 100$ km, *and the precipitation is* $1\ m\ a^{-1}$

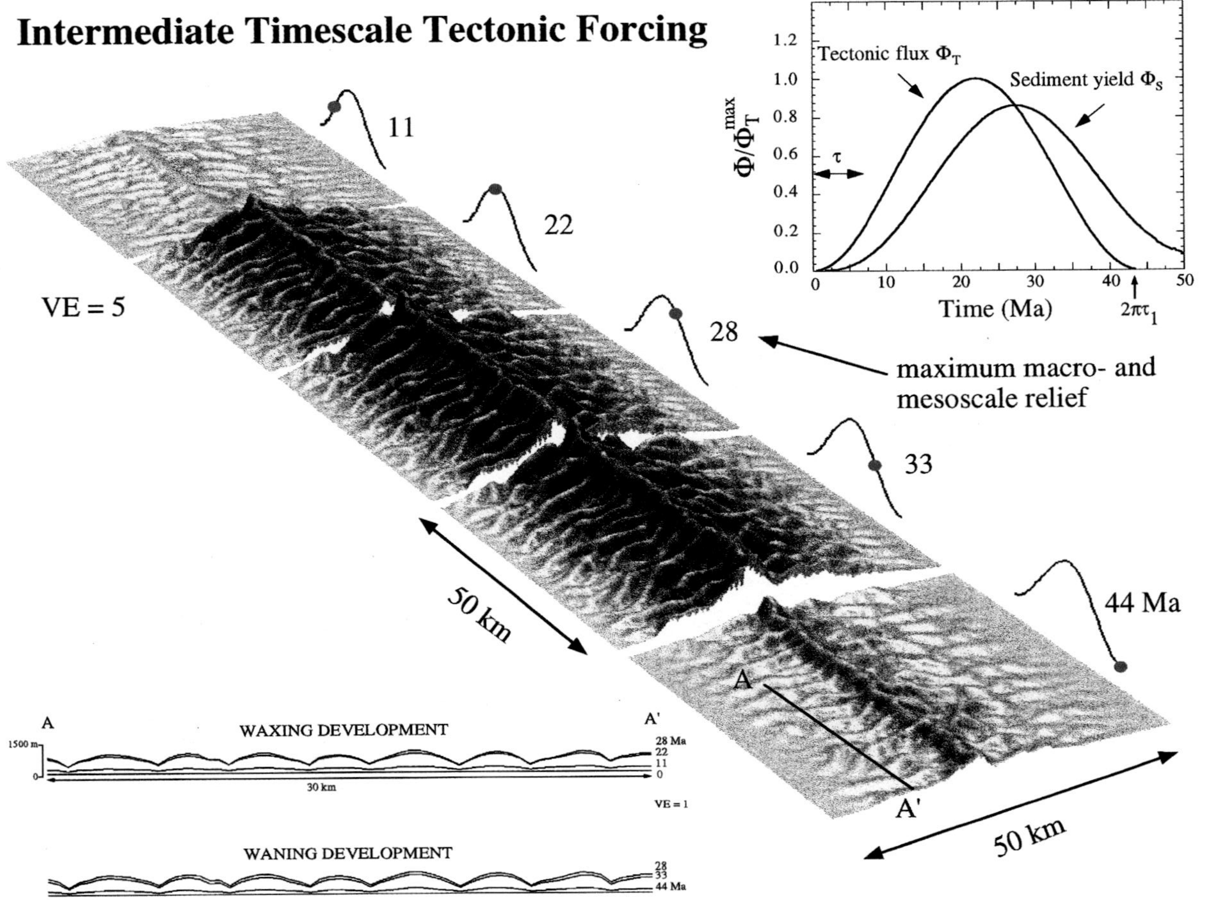

Plate 3.2 *Example of landscape evolution (panels 11–44 Ma) in which a region is uplifted with a triangular or wedge-shaped geometry and is acted upon by the SPM described in Section 3.3.1. Tectonic forcing is shown by tectonic volume flux, Φ_T (upper right panel) and the response is shown by the total sediment yield (sediment volume flux, Φ_S) at the model boundaries. The position of each model panel in the evolution of the tectonic flux is shown by the dot on the projected tectonic flux curve next to the panels. The period of the tectonic forcing τ_1 is approximately equal to the landscape response time. The upper right inset shows that the response (sediment yield) lags the forcing (tectonic flux). The boundary conditions are the same as those for the model of Plate 3.1. Topographic evolution across the interfluves along A–A′ is shown at the bottom left. Note how the relief characteristics also lag the tectonic forcing (see Kooi and Beaumont (1996) for more details). Model properties are spatially and temporally uniform:* $K_s = 0.2\ m^2\ a^{-1}$, $K_f = 0.01$, $l_f = 100$ *km and the precipitation is* $1\ m\ a^{-1}$

Escarpment Evolution with Complex Upland Drainage

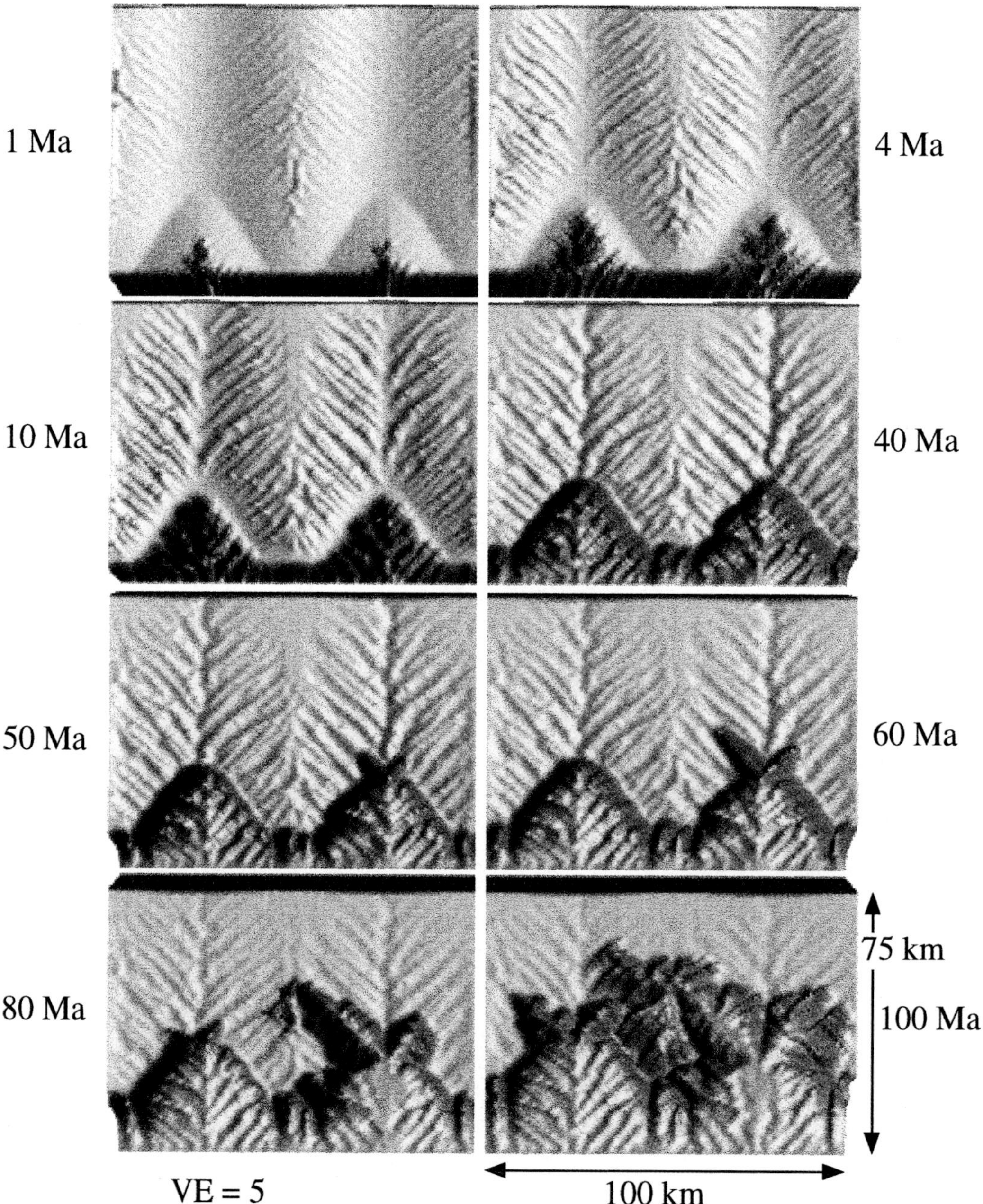

Plate 3.3 *Model landscape evolution of a section of escarpment (bottom of 1Ma panel) in which the upland above the escarpment has a complex drainage, partly draining towards, and partly draining away from, the escarpment. The model evolution demonstrates that the inherited drainage divides on the upland exert a major control on the escarpment evolution in this SPM experiment. Model properties are spatially and temporally uniform:* $K_s = 1.0\ m^2\ a^{-1}$, $K_f = 0.01$, $l_f = 100$ *km and the precipitation is* $0.3\ m\ a^{-1}$

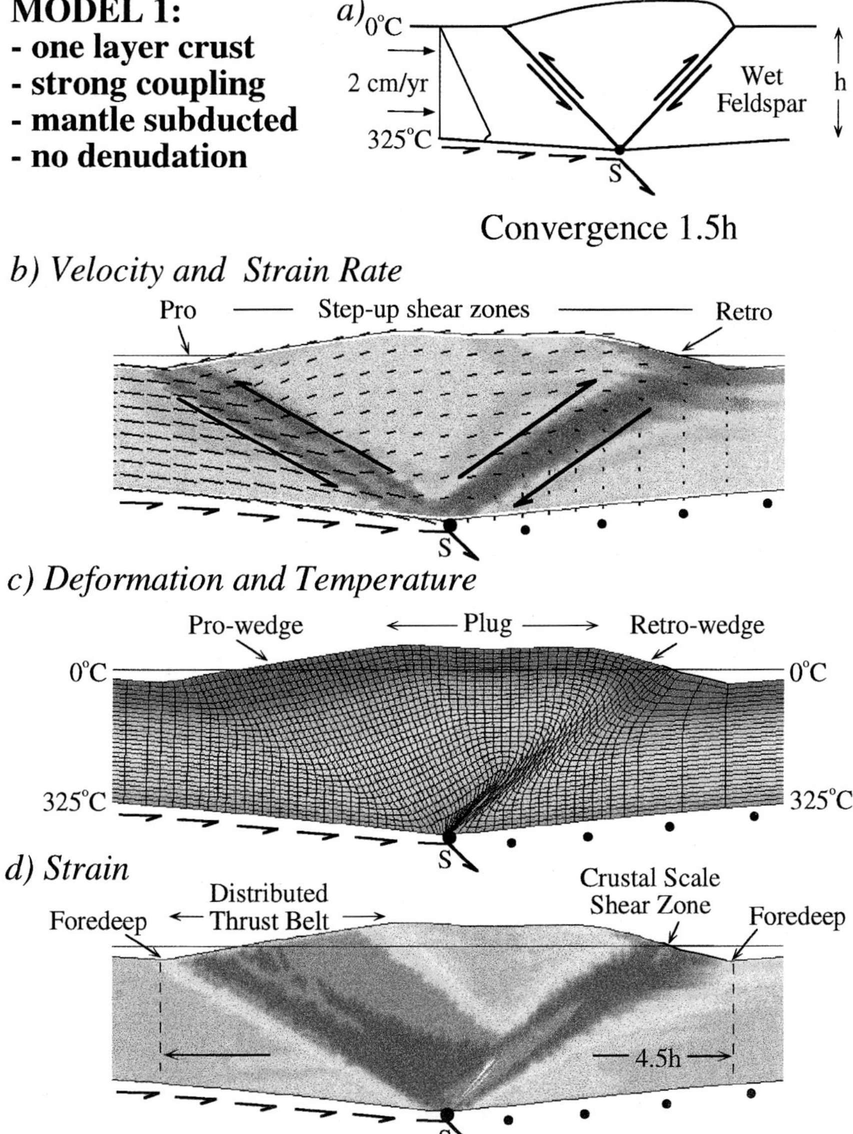

Plate 3.4a *Plane-strain finite element model results showing deformation of a crustal layer (thickness,* h*) that is strongly coupled to the mantle (not shown) and in which the mantle detaches and subducts at S. The general problem is shown in Figure 3.6 and in (a). The results are shown for convergence,* $\Delta x = 1.5$ h*. In both Plate 3.4a and 3.4b (b) shows velocity by lines indicating left to right motion with a value of 2 cm* a^{-1} *at left. Strain rate (second invariant of deviatoric strain rate)* ~10^{-14} s^{-1} *in red, and pink step-up shears* $<10^{-15}$ s^{-1} *in blue areas. Diagrams (c) show deformation of the Lagrangian tracking grid and advection of temperature field (coloured). Diagrams (d) show total strain (natural logarithm of second invariant of deviatoric strain) >1.0 in red and pink deformed regions, <0.1 in largely undeformed blue regions. Model 1 (Plate 3.4a) has no surface denudation. Crustal thickening causes flexural isostatic compensation and formation of foredeeps and the model orogen is 4.5* h *wide. Model 2 (Plate 3.4b) has the same properties as Model 1 but is totally denuded showing a narrow orogen, 2.5* h *wide, with pro-crust exposed in hangingwall of the retro-step-up shear*

MODEL 2:
- one layer crust
- strong coupling
- mantle subducted
- total denudation

a)

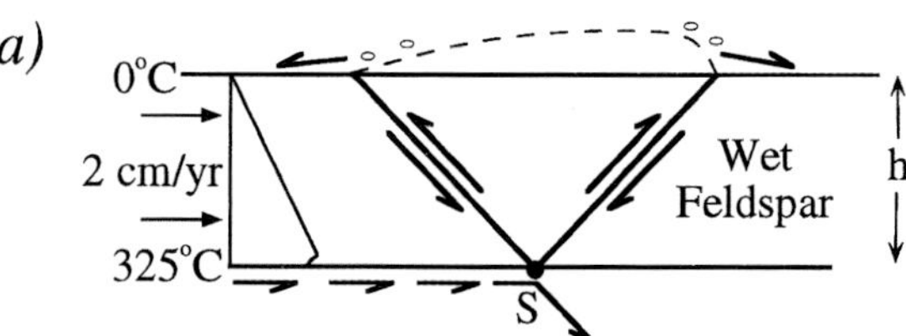

Convergence 1.5h

b) Velocity and Strain Rate

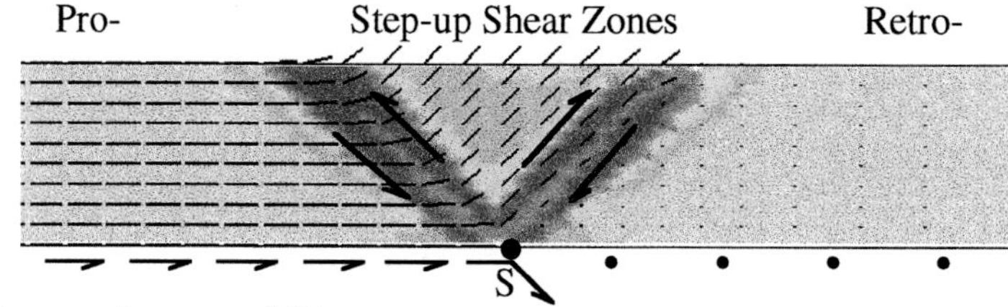

c) Deformation and Temperature

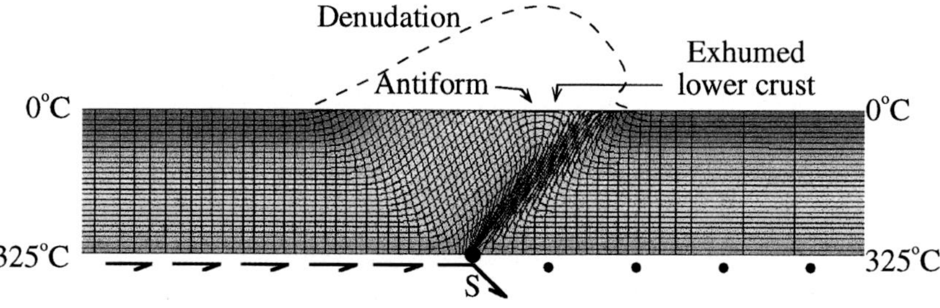

d) Strain

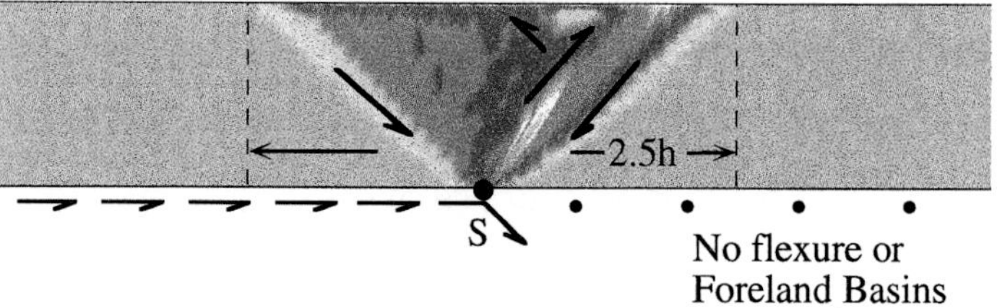

Plate 3.4b

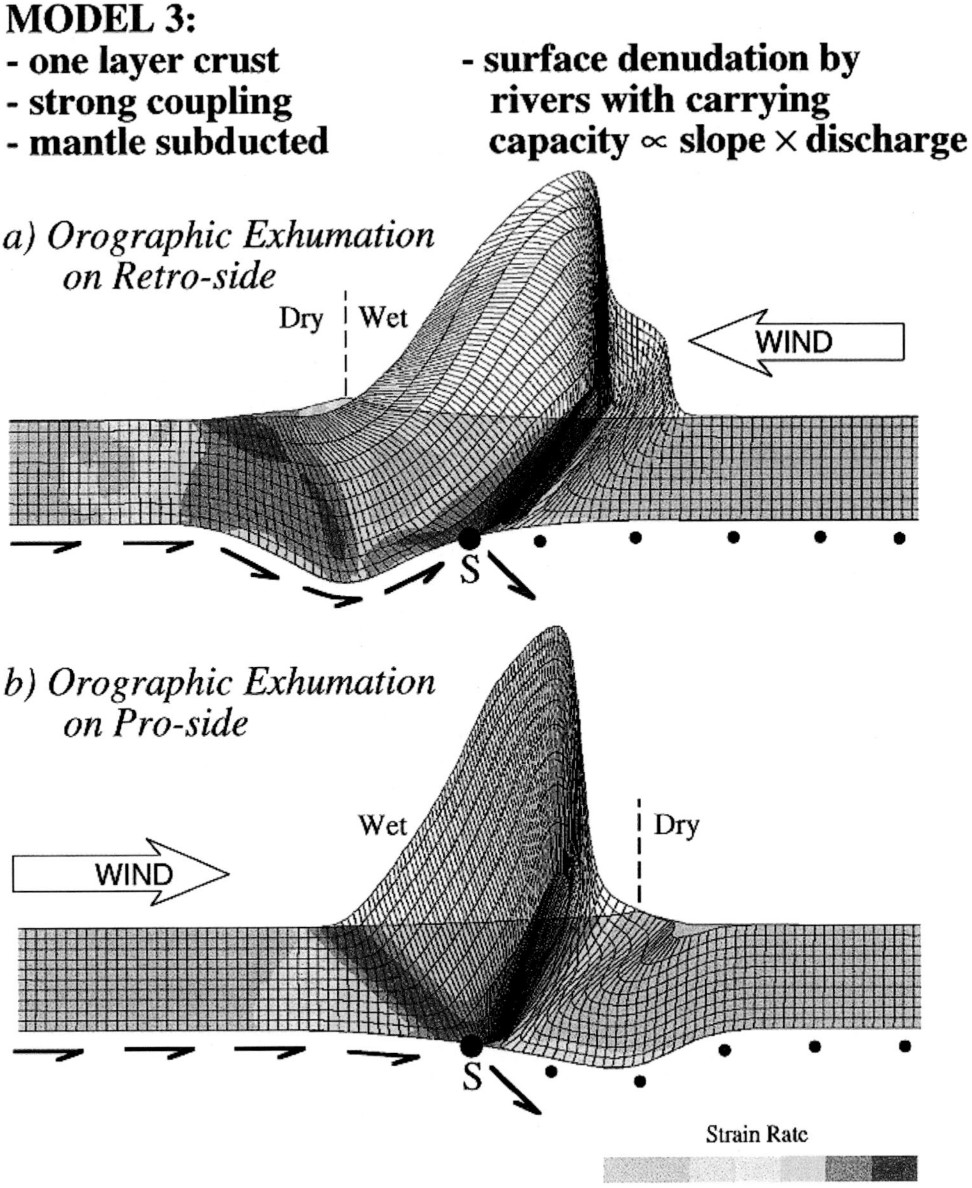

MODEL 3:

- **one layer crust**
- **strong coupling**
- **mantle subducted**
- **surface denudation by rivers with carrying capacity $\propto$ slope $\times$ discharge**

a) Orographic Exhumation on Retro-side

Dry | Wet

WIND

S

b) Orographic Exhumation on Pro-side

Wet | Dry

WIND

S

Strain Rate

Plate 3.5 *Plane-strain finite element results for models like those shown in Plate 3.4 but in which the denudation is by one-dimensional rivers charged by orographically enhanced precipitation. The models have the same tectonic polarity but opposite climatic polarities (shown by wind arrows). The Lagrangian grid tracks the motion and shows deformation principally associated with the step-up shear zones. The grid shown above the surface provides a measure of crust removed by denudation. Colours show strain rates; high rates are red. In a) denudation focused on the retro-side causes maximum rock exposure in the hangingwall of the retro-step-up shear (high-grade rocks exposed at low elevation). In Plate 3.5b b) denudation focused on the pro-side causes maximum rock exposure on 'short-circuit' shear (high-grade rocks exposed at high elevation). Note also the difference in position of the highest point in the model orogen with respect to S and, therefore, contrasting drainage distributions of the two models*

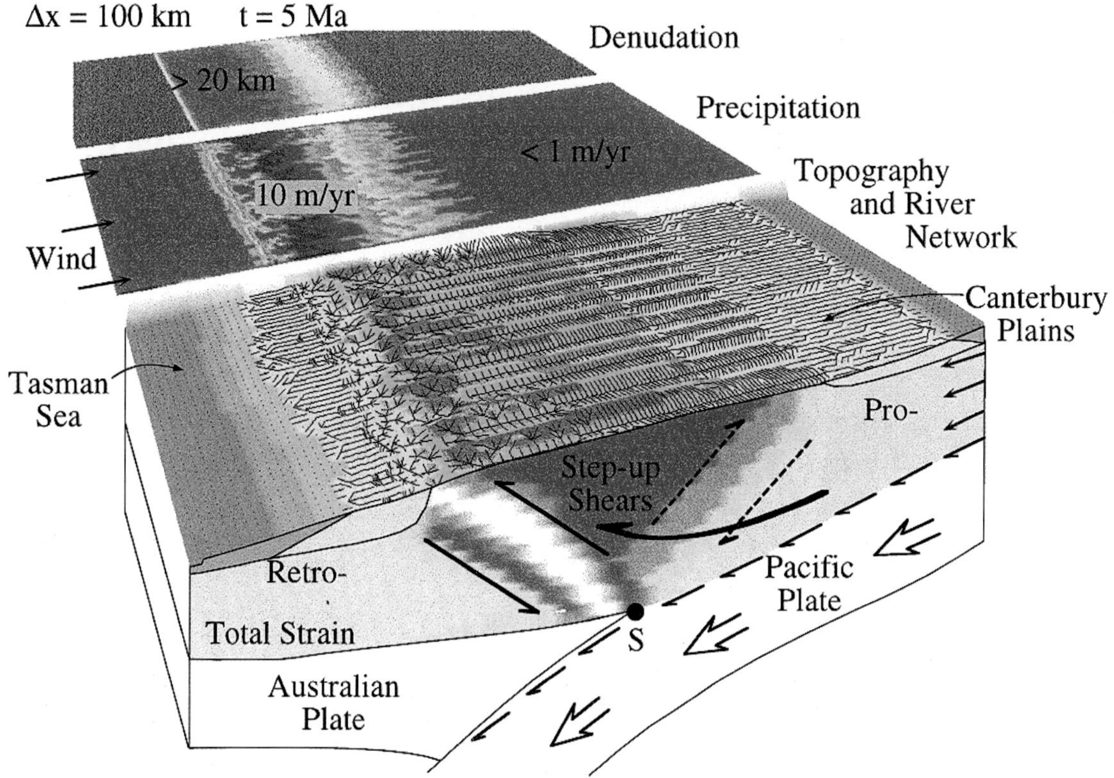

Plate 3.6 *Plane-strain finite element model like that shown in Plate 3.4 (but opposite polarity) coupled to a planform surface processes model in which fluvial denudation is coupled to orographically enhanced precipitation. The model is shown after 100 km of convergence at 2 cm* a^{-1}. *The tectonic and denudational setting are similar to the Southern Alps in the continental collison zone of central South Island, New Zealand, except that convergence is not oblique in the model. The model shows how an orogen develops with a steep windward flank sited above the retro-step-up shear (similar to Plate 3.5a). It also shows that precipitation is strongly focused on the windward flank and that correspondingly focused denudation is at a maximum at the retro-step-up shear. The coupled tectonics and climate-driven denudation have brought the base of the crust up to the surface in the hangingwall of the step-up shear zone. The numerical model combines the finite element techniques with the SPM described in Section 3.3.1. Geographic names are for reference. The model does not represent a complete description of processes in South Island, New Zealand*

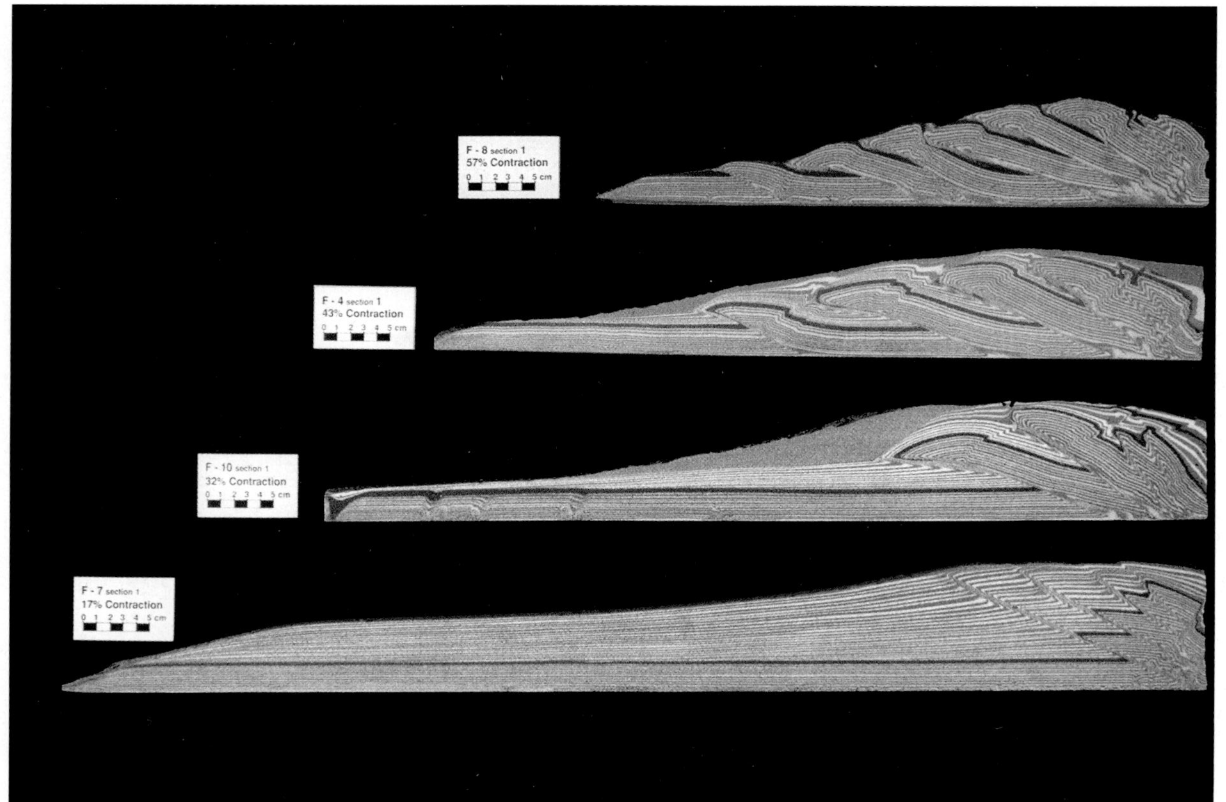

Plate 3.7 *Results of sandbox models demonstrating effect of sedimentation over an actively forming critical wedge, analogous to an accretionary wedge or fold-and-thrust belt of a convergent orogen. Model materials are sand/mica layers. Initial conditions for each of the four models were the same. The results show the influence of increasing amounts of sand added to the surface of the model during deformation. The top panel illustrates no sediment added, the bottom panel most sediment added. (From Storti and McClay, 1995)*

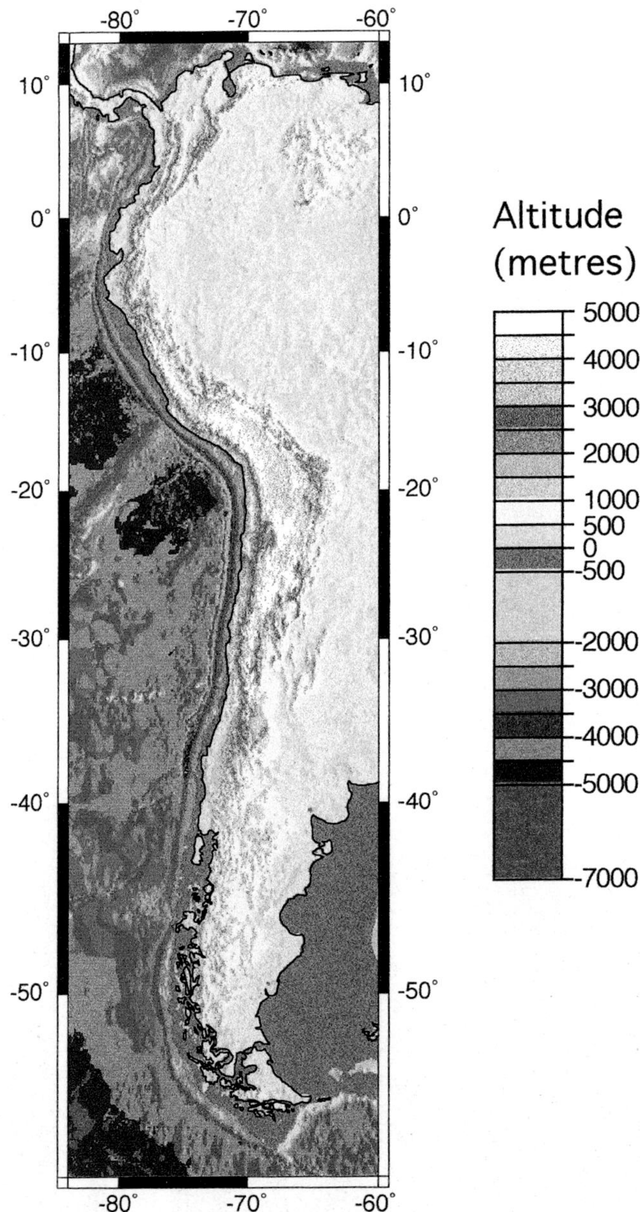

Plate 9.1 *Colour-shaded relief map of western South America, illuminated from the east, generated from 3-minute topographic data. The pronounced topographic low off the western coast is the Peru-Chile trench, where Pacific oceanic crust is subducting beneath the continental margin. The most prominent feature of the Andes is the 'Central Andean Plateau' between 10°S and 30°S, which changes trend abruptly at ~ 19°S. Outlying uplifts are clear to the east of the main Andean range as the plateau tapers to north and south. The quite distinct character of the northern Andes, where high cordilleras are separated by broad, low–altitude valleys, and the generally low relief of the southern Andes are also apparent*

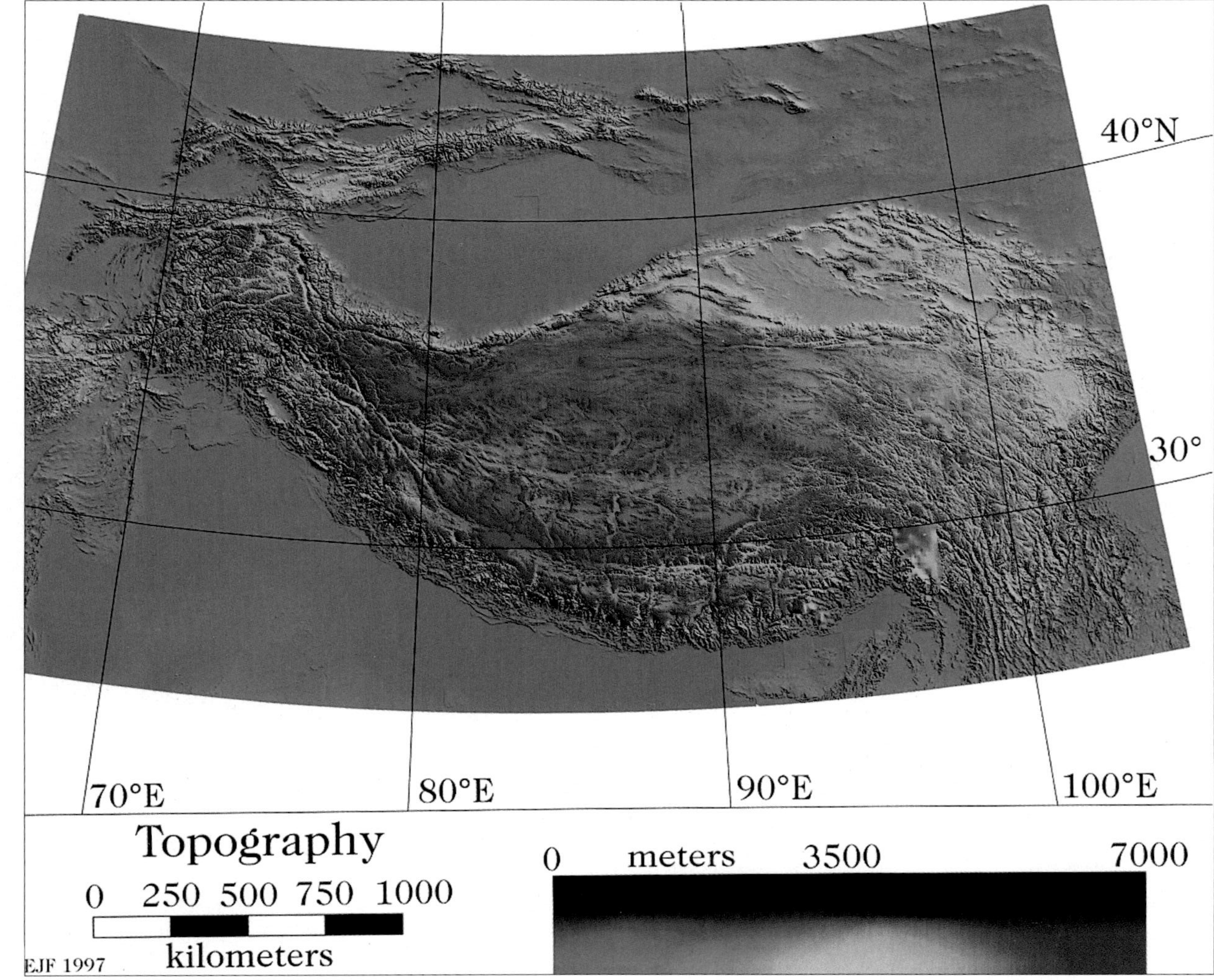

Plate 10.1 *Shaded relief image of the topography of central Asia at ~500 m resolution. Illumination is from the northeast. Hues are set by elevation as shown on the colour scale. Modified from Fielding* et al. *(1994)*

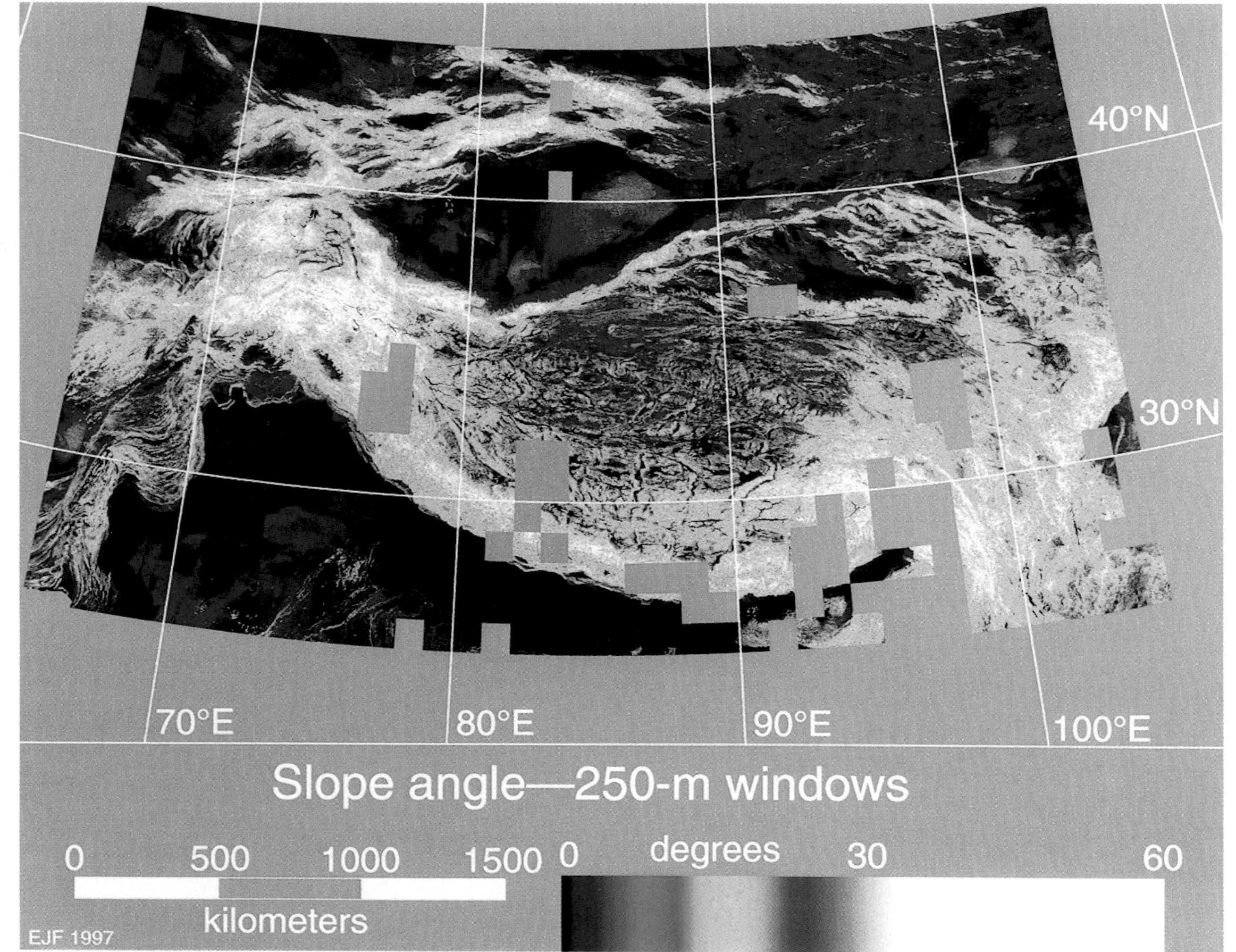

Plate 10.2 *Slope map for central Asia calculated from 3-arcsecond topography. Slopes are measured within ~250 m windows and sampled at ~1 km resolution. Grey areas are missing data. Colour scale bar shows slope values. Modified from Fielding* et al. *(1994)*

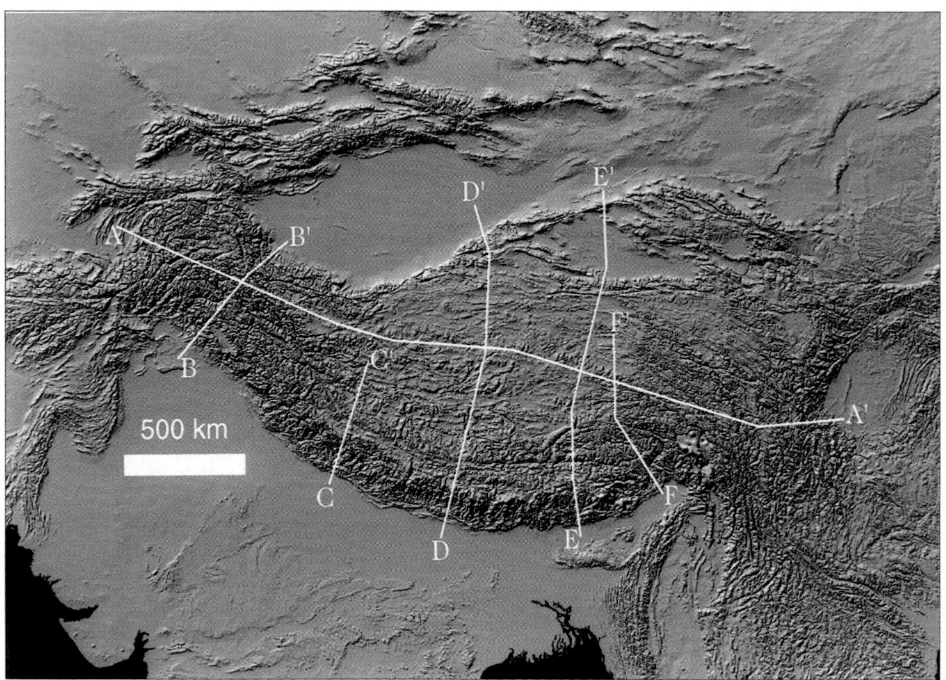

Plate 10.3 *Shaded relief of the topography of central Asia at ~1 km resolution. Topography on a 30-arcsecond grid from Eros Data Center (1996) was projected into a Lambert conformal conic projection and illuminated from the northeast. White lines show locations of topographic profiles in Figure 10.7*

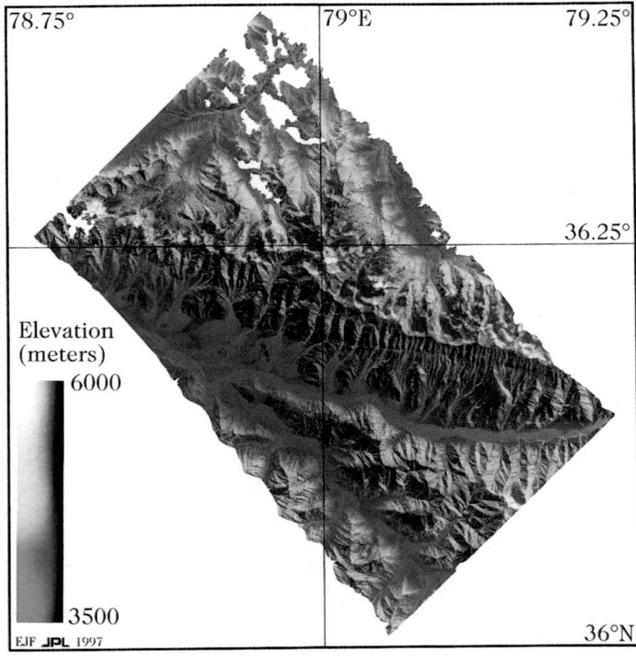

Plate 10.4 *DEM of Karakax valley and Kun Lun mountains at the northwest margin of the Tibetan Plateau, derived from SIR-C interferometry (see Figure 10.9 for location). Elevation is colour-coded according to the colour scale shown, and the image intensity is proportional to the L-band (24 cm wavelength) radar image amplitude*

Plate 10.5 *Three–dimensional perspective view of the Karakax valley DEM. Elevation is shown without vertical exaggeration with the SIR-C radar images overlain. The L-band amplitude is assigned to red, L- and C-band (24 and 6 cm wavelengths) average to green, and C-band to blue. The active strand of the Altyn Tagh fault is visible as a sharp break in slope running diagonally up the valley side*

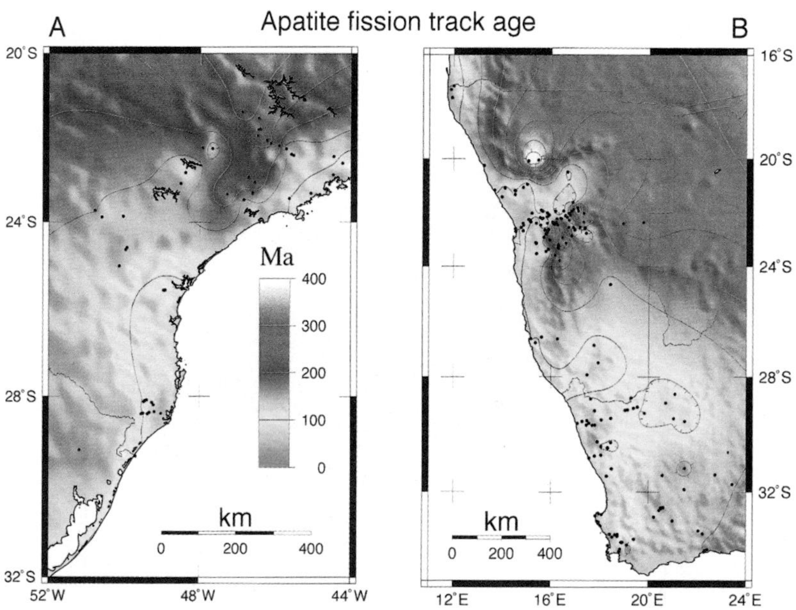

Plate 12.1 *Apatite fission-track age for southeast Brazil (A) and southwest Africa (B) shown as a colour drape image over the present topography. The topography is represented by shading under artificial illumination at an azimuth of 60°. Dots in this plate and Plates 12.2–12.5 are sample locations. Fission-track ages tend to be young near the present-day coast, but young ages also occur inland. Note the NE-trending corridor of younger apatite ages (<100 Ma) in northern Namibia (B), and the presence of young ages in the interior of South Africa*

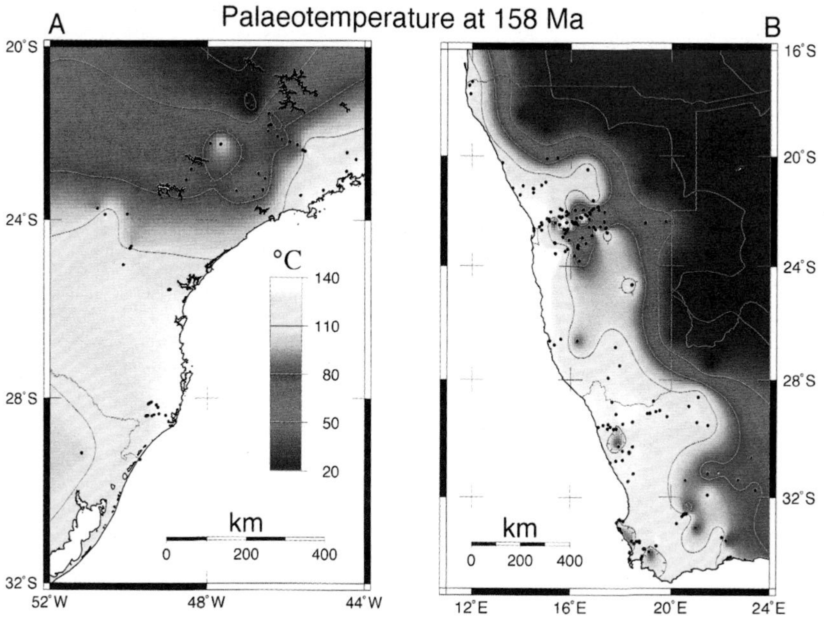

Plate 12.2 *Maps of estimated pre-rift palaeotemperature of the present surface at 158 Ma BP for southeast Brazil (A) and southwest Africa (B)*

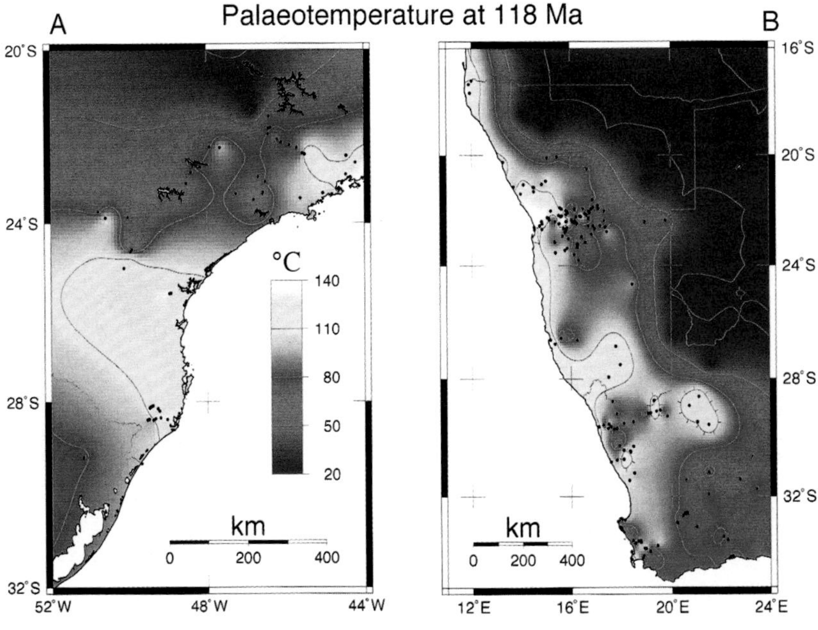

Plate 12.3 *Maps of estimated palaeotemperature of the present surface shortly after the time of break-up (118 Ma BP) for southeast Brazil (A) and southwest Africa (B)*

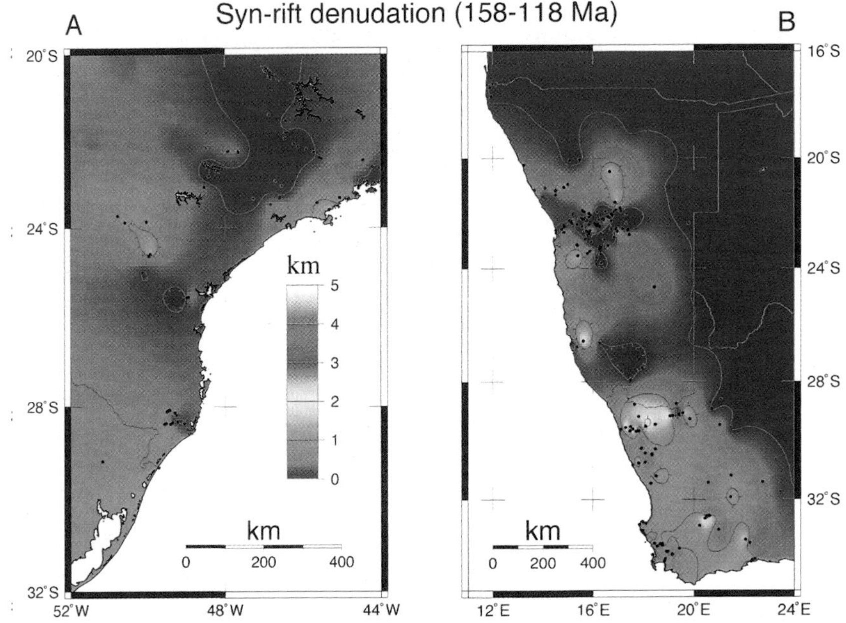

Plate 12.4 *Maps of estimated syn-rift denudation (158–118 Ma BP) for southeast Brazil (A) and southwest Africa (B). Denudation estimates were derived from the palaeotemperature differences between maps shown in Plates 12.2 and 12.3 (see text for details)*

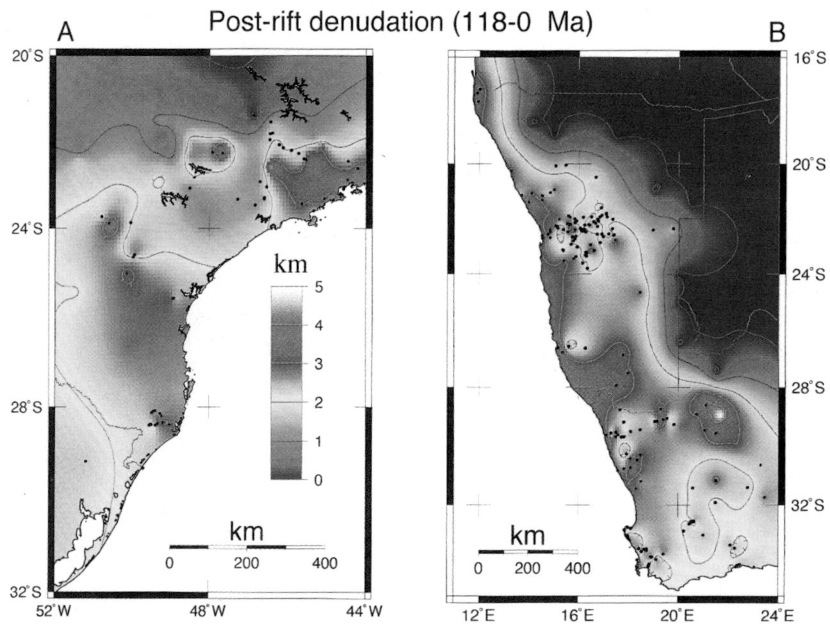

Plate 12.5 *Maps of estimated post-rift denudation (118–0 Ma BP) for southeast Brazil (A) and southwest Africa (B). Denudation estimates were derived from the palaeotemperature map shown in Plate 12.3 (see text for details)*

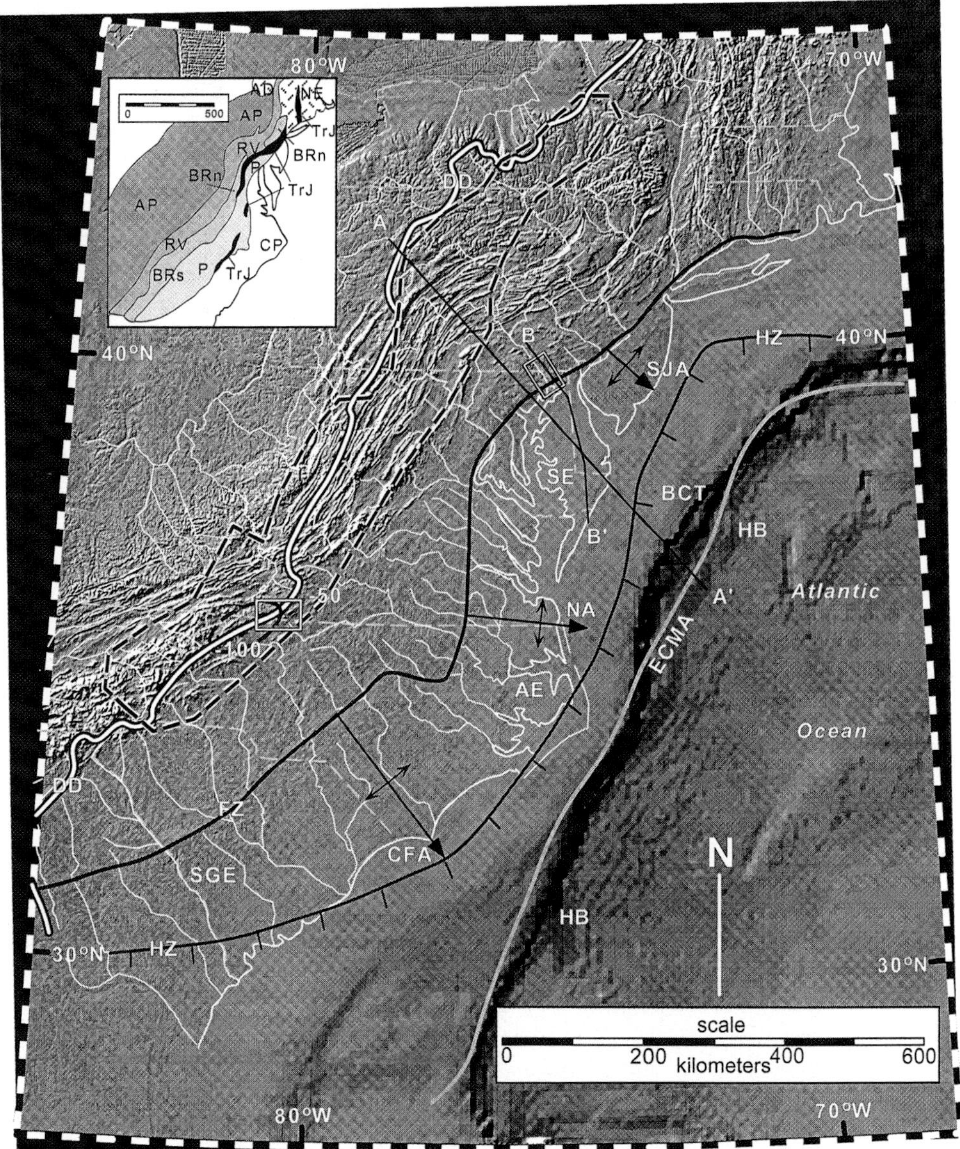

Plate 13.1 *Digital shaded topography of the Atlantic passive margin constructed from 30-second (USGS TOPO30) and 5-minute (NOAA ETOPO5) resolution digital elevation data. Major rivers are shown in a thin, solid white line. State boundaries are shown in a thin, dashed white line. Cross-section A–A′ is shown in Figure 13.1 and cross-section B–B′ is shown in Figure 13.4c. The boxed area along cross-section B–B′ is depicted in Figures 13.4a, b. Major morphological provinces in the inset map are labelled as: AP = Allegheny Plateau, RV = Ridge and Valley, BRn = northern Blue Ridge, BRs = southern Blue Ridge, NE = New England uplands, P = Piedmont, TrJ = Triassic–Jurassic fault basins, CP = Coastal Plain, AD = Adirondack dome. Major Coastal Plain basin and arch features shown are: SJA = South Jersey Arch, SE = Salisbury Embayment, AE = Albermerle Embayment, NA = Norfolk Arch, CFA = Cape Fear Arch, SGE = Southeast Georgia Embayment. Major offshore features are: BCT = Baltimore Canyon Trough, ECMA = East Coast Magnetic Anomaly, HZ = hinge zone, HB = Hatteras Basin. The Fall Zone (FZ) and drainage divide (DD) are shown by solid, thick black and white lines, respectively. Dashed black lines show the Bouguer gravity anomaly in milligals (from Judson, 1975)*

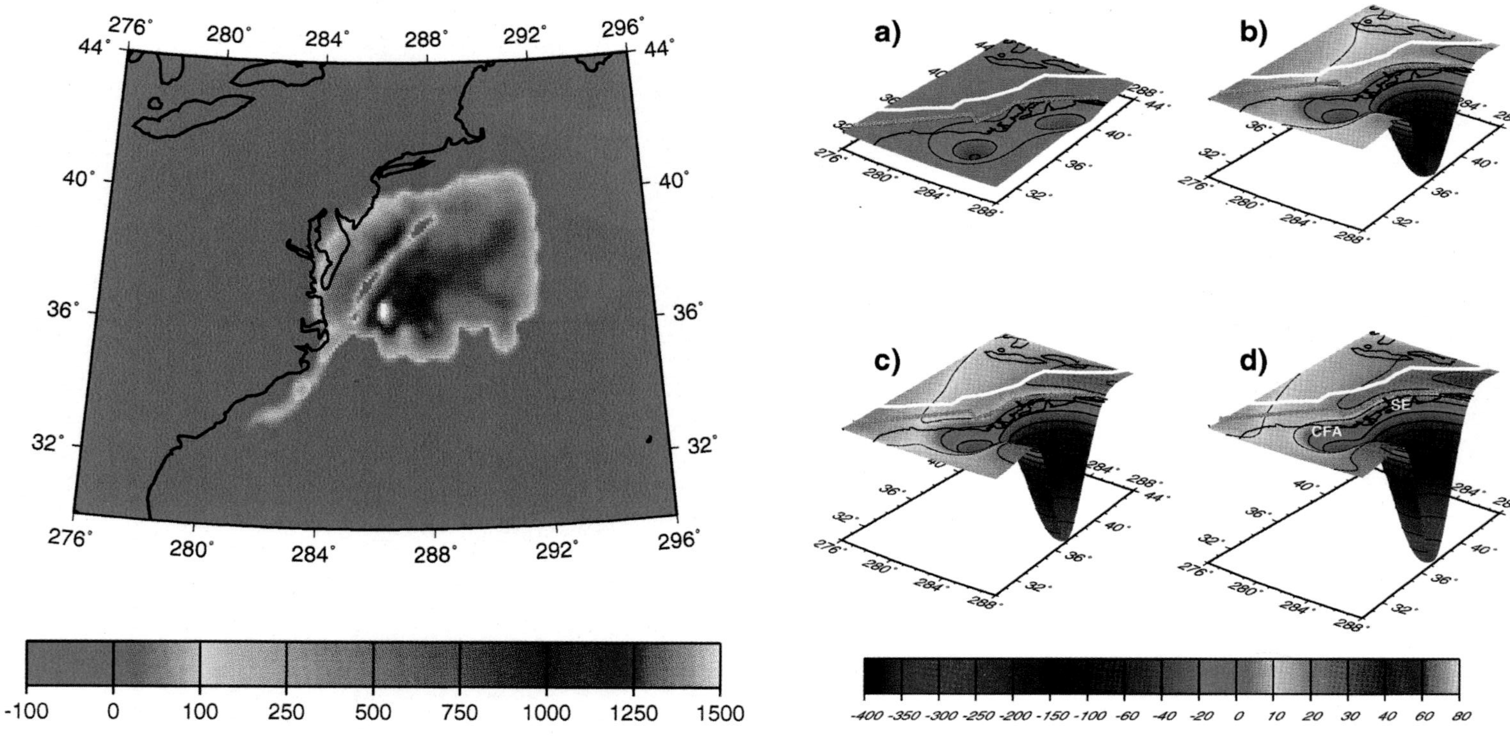

Plate 13.2 *Example of a sediment isopach map showing the volume of sediment in the offshore basins for a given stratigraphic time line. This figure shows the combined offshore sediment load for the lower Oligocene–upper Miocene and middle Miocene. The isopach data have been recalculated and recontoured from Popenoe (1985) and Poag (1992). Sediment thickness, in metres, is estimated to be 1000 times the two-way travel times shown on the isochron maps (Poag, 1992). Scale in metres*

Plate 13.3 *Results of the two-dimensional model. All models shown in this figure are parameterized by a flexural rigidity of 4 x 10^{23} Nm. These models do not incorporate continental erosion. The thin black line is the modern US Atlantic coastline, the thick grey line is the location of the Fall Zone, and the thick white line is the drainage divide. Cumulative, late Cenozoic flexure of the margin is shown for the (a) late Oligocene–early Miocene, (b) middle Miocene, (c) late Miocene, and (d) Plio-Pleistocene. CFA = Cape Fear Arch, SE = Salisbury Embayment. Scale is in metres*

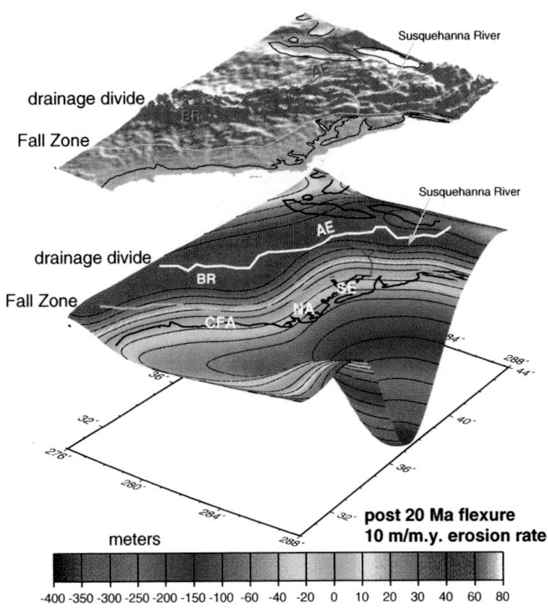

Plate 13.4 *Results of the two-dimensional model illustrating the flexural effects of cumulative late Oligocene to Plio-Pleistocene offshore loading and concomitant erosion of the continent at a rate of 10 m Ma^{-1} (200 m section removed in 20 Ma). Note location of the Fall Zone (thick grey line), drainage divide (thick white line), the Blue Ridge (BR), Abermale Embayment (AE), Cape Fear Arch (CFA), Salisbury Embayment (SE) and Norfolk Arch (NA). Scale is in metres*

Plate 14.1 *A Miocene landscape in Arena valley, Quatermain Mountains, showing an avalanche tongue containing volcanic ash 11.3 Ma old overlain by a sequence of blocky moraines deposited by an expanded Taylor Glacier. The glacier has overridden the tongue without modifying it*

Plate 14.2 *View eastward down Wright Valley (Dry Valleys area) towards the Ross Sea coast, showing the channelled form of the 'Labyrinth' leading into the deepened trough in the middle distance. The higher valley benches and dissection of the mountains on the side of Wright Valley have been attributed at various times to either glacial or pre-glacial fluvial action. US Navy photograph TMA 1564*

Plate 14.4 *View of the main escarpment and (in the background) Shapeless Mountain, Dry Valleys area. In the foreground dissected mountains of the Olympus Range have been isolated as a result of escarpment retreat. Photograph by G.H. Denton*

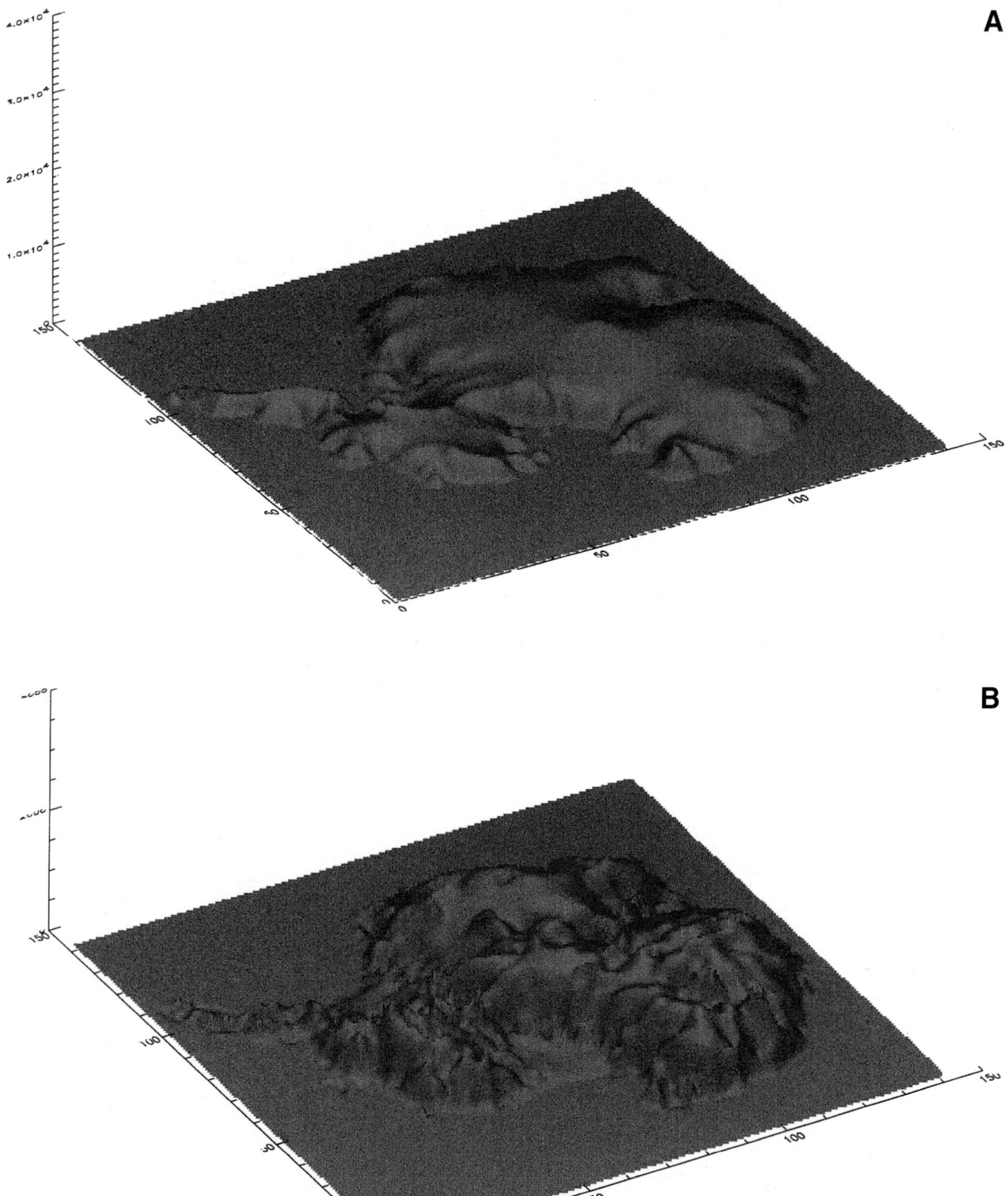

Plate 14.3 *(a) The Antarctic ice sheet surface. West Antarctica and the Antarctic Peninsula lie in the foreground and East Antarctica in the backgound. Ice thicknesses in the centre reach 4 km. (b) The magnitude of bed rebound following removal of ice assuming only local (Airy) isostatic compensation* (T_e *=0). Note the high-frequency components, which reflect the undulating bedrock topography. In both (a) and (b), vertical axis in metres, horizontal axes in grid units of 70km*

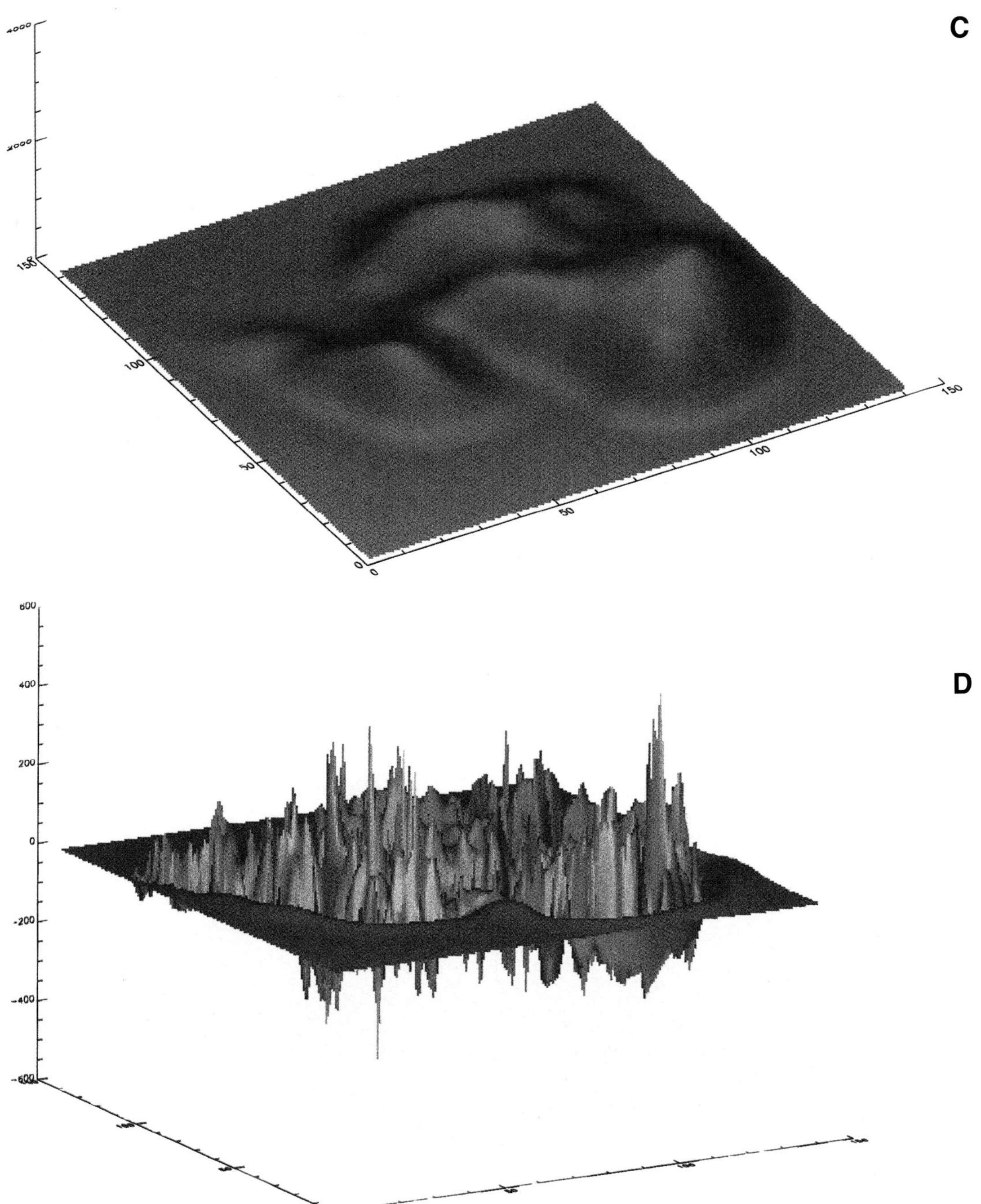

Plate 14.3 *(c) The magnitude of bed rebound following removal of ice assuming a very high flexural rigidity (T_e = 120 km). Note that the high-frequency undulations in ice thickness have been filtered out. (d) The difference between the magnitudes of rebound of Antarctic bedrock assuming no flexural rigidty (T_e = 0) and very high flexural rigidity (T_e = 120 km). Of note are the high-frequency components, but over the whole of Antarctica the difference in bed rebound is less than 100–200 m. This reflects the large wavelength of the ice sheet, and thus the fact that bed rebound essentially reflects the density contrast between ice and asthenosphere. In both (c) and (d), vertical axis in metres, horizontal axes in grid units of 40 km*

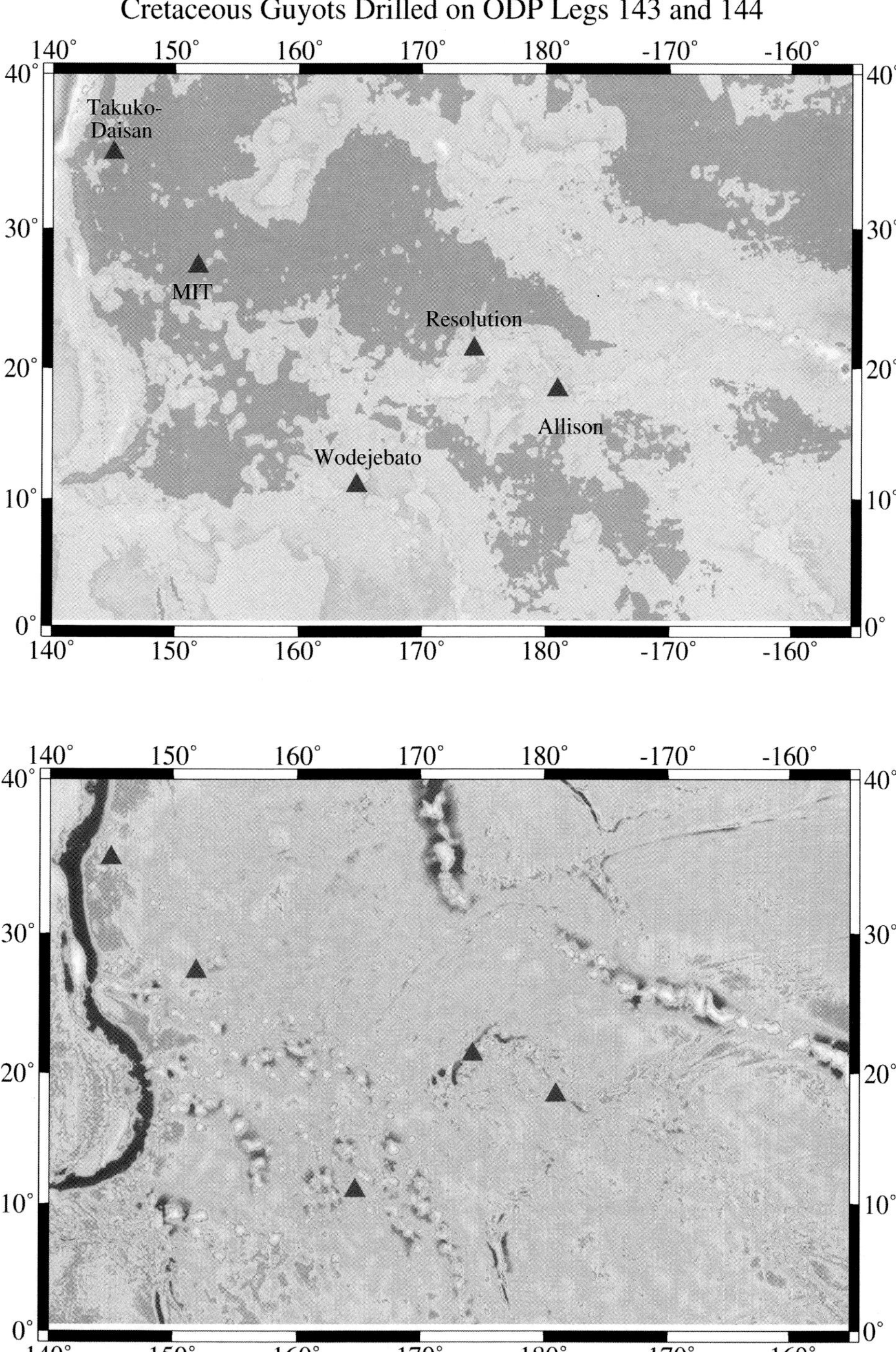

Plate 16.1 *Topography and satellite-derived gravity anomaly map of the western Pacific Ocean showing the location (red triangles) of the five drilled Cretaceous guyots analysed in Figure 16.9. (a) Topography map based on the GEODAS data set. (b) Satellite-derived gravity anomaly map based on Smith and Sandwell (1997)*

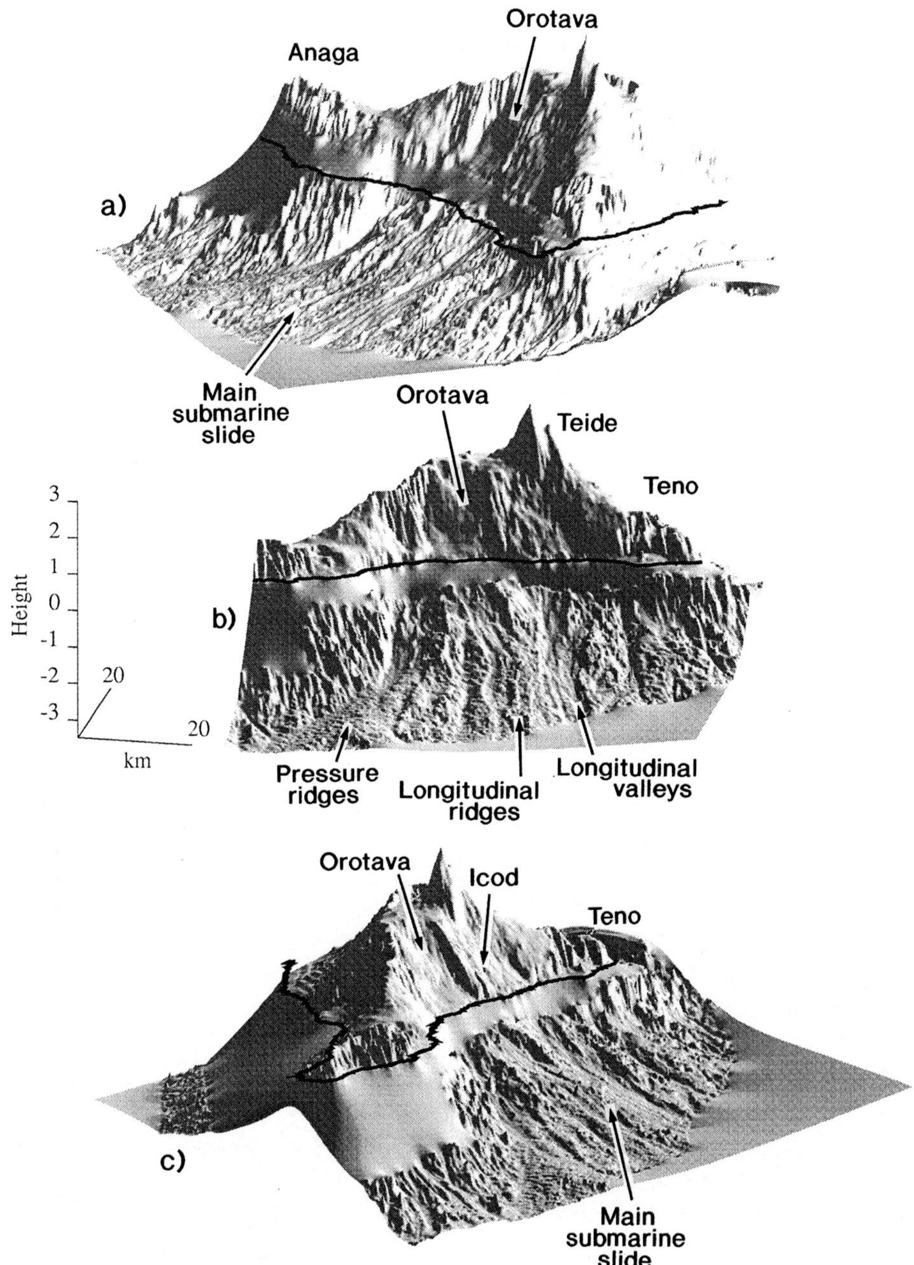

Plate 16.2 *Perspective images of the north flank of Tenerife (Canary Islands) as viewed from (a) WNW (285°), (b) NNW (350°) and (c) ENE (41°). The images were constructed from swath bathymetry data offshore and Spanish military topograph maps onshore using GMT software (Wessel and Smith, 1991). The surface of the images represents actual topographic heights above and below sea level. The grey shades reflect the intensity of the horizontal gradients of the topography in the direction of an artificial sun. The elevation of the sun was fixed at 15° and the sun angle was varied from 220° for (a), 230° for (b) and 270° for (c). White is high reflectivity (due to a steep gradient in the sun direction) and black is low reflectivity*

14

Linking tectonics and landscape development in a passive margin setting: the Transantarctic Mountains

Andrew Kerr, David E. Sugden and Michael A. Summerfield

14.1 Introduction

Two contrasting approaches have generally been applied to the problem of long-term landscape development in passive continental margin settings. One involves the modelling of tectonic mechanisms, based on geophysical data on endogenic processes and lithospheric properties, to yield coarse-scale predictions of changes in topography through time. The other approach uses geomorphological data to reconstruct histories of landscape development, with little explicit attempt to explain the tectonic events that have contributed to the sequence of landscape change. The former strategy involves forward modelling in which the aim is to identify the particular combination of lithospheric properties and tectonic processes responsible for creating the present large-scale topography. The latter starts with the present landscape and aims to reconstruct its history by identifying age relationships between different landscape elements.

Over the past decade it has been increasingly appreciated that the internally driven, tectonic mechanisms operating on passive margins do not act independently, but rather interact with surface processes which redistribute material through denudation and deposition (Gilchrist and Summerfield, 1990, 1994) (see Chapter 3). It has also become evident that the effective analysis of such interactions between the Earth's internal and external systems requires inputs from both the tectonic and geomorphological modelling strategies. In other words, geomorphological evidence of landscape change as well as geophysical data and tectonic models are necessary components of any comprehensive explanation of large-scale, long-term landscape evolution on passive margins.

In this chapter we assess the utility of these tectonic and geomorphological modelling approaches to landscape analysis in the context of the high-elevation passive margin represented by Transantarctic Mountains, and specifically in relation to the Dry Valleys area. Rising to elevations in excess of 4000 m and extending for more than 3000 km, the Transantarctic Mountains provide a valuable case study for the assessment of the relative role of tectonic and surface processes in landscape development. This is because the persistence of a frigid polar environment throughout a significant part of the late Cenozoic has led to the detailed preservation of a range of ancient landscape features and associated terrestrial deposits to an extent that has not occurred elsewhere. It is likely that ice has been present as a geomorphic agent in Antarctica since at least the Oligocene (Robert and Chamley, 1992), and the present-day dry, polar environment of the Transantarctic Mountains appears to have existed with little change since the mid-Miocene (Denton *et al.*, 1993). In contrast to virtually all other terrestrial environments, the role of running water as a geomorphic agent has been largely suppressed for millions of years and this provides a setting within which to investigate the relationships between tectonics and an unusually limited suite of

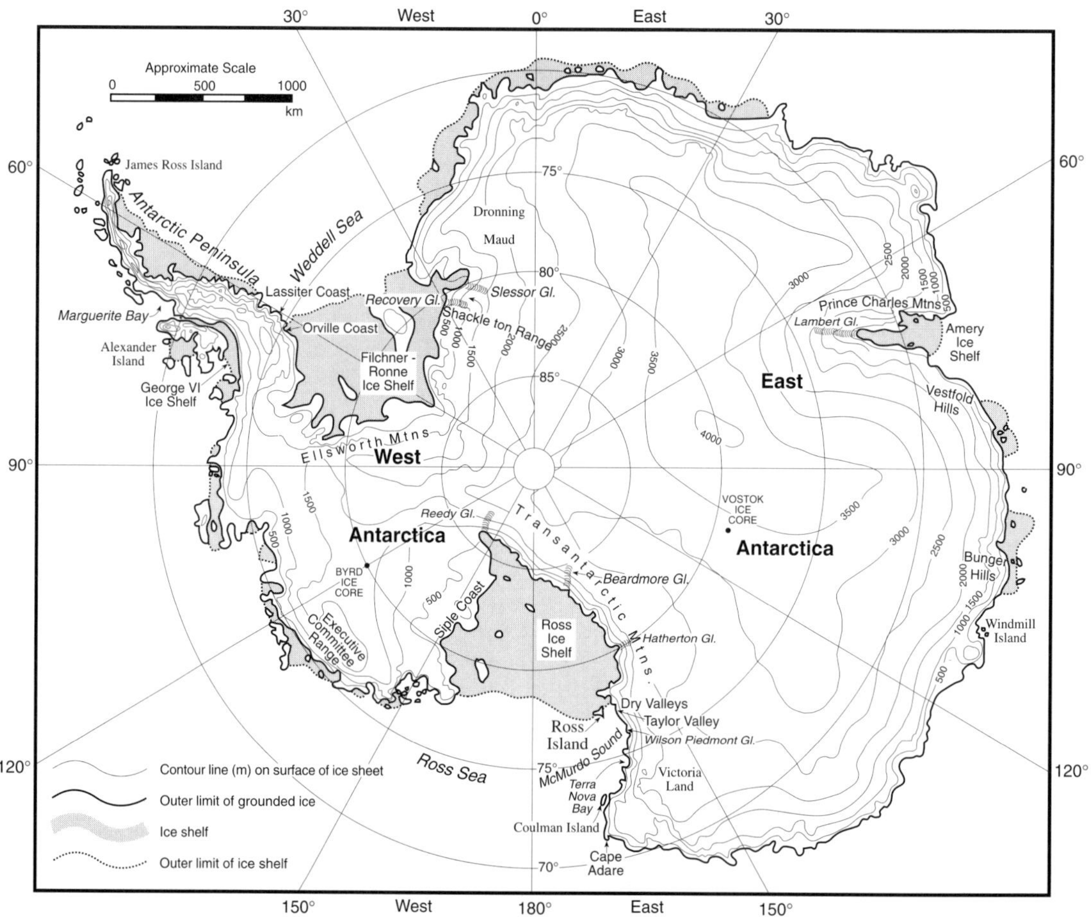

Figure 14.1 *The Transantarctic Mountains in the context of the East and West Antarctic Ice Sheets. The locations of the Dry Valleys area, Taylor Valley and Beardmore Glacier, features specifically discussed in the text, are shown*

surface geomorphic processes. The Transantarctic Mountains therefore represent a valuable benchmark against which to compare the morphotectonic evolution of other glaciated passive margins, notably those fringing the North Atlantic Ocean.

Out specific aim is to assess the role of the tectonic and surface geomorphic processes that have shaped the topographic evolution of the Transantarctic Mountains. First, we describe their present morphology and geology. We then outline the available geophysical, geochemical and chronological evidence for lithospheric processes, and the geomorphological evidence for surface processes. Finally, we consider existing tectonic and geomorphological models, focusing in particular on how the two sets of models can best be integrated.

14.2 The Transantarctic Mountains

14.2.1 MORPHOLOGY

The Transantarctic Mountains sweep in a broad arc across Antarctica from northern Victoria Land on the Pacific coast to the Thiel Mountains at the junction of the East and West Antarctic Ice Sheets, and continue as a coherent topographic feature in a series of ranges as far as the Theron Mountains which terminate near the South Atlantic Ocean (Figure 14.1). Rather than representing a mountain range in the generally accepted sense of the term, the Transantarctic Mountains in fact consist of a large amplitude (~2000–4000 m), short wavelength (50–200 km) upwarp that forms the high rim of an extensive plateau that rises gradually

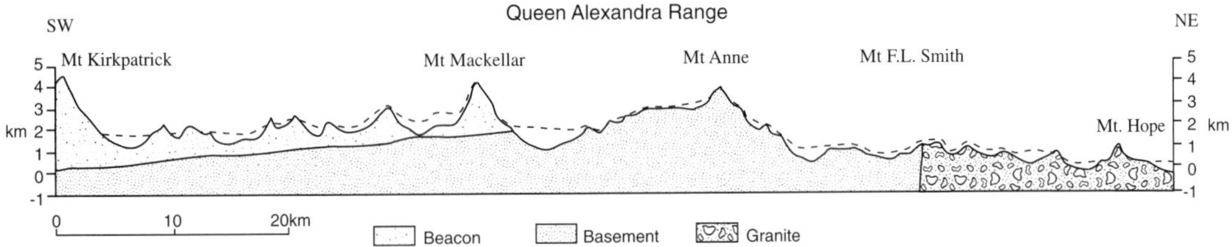

Figure 14.2 *Cross-section of the Transantarctic Mountains in the Queen Alexandra Range, parallel to the northern flank of the Beardmore Glacier (see Figure 14.1), showing the nature of the dissected mountains lying on the coastal side of the main escarpment. The gentle dip of the basement and the overlying Beacon Supergroup sediments is towards the East Antarctic Ice Sheet (from Fitzgerald, 1994)*

from the interior of East Antarctica where it is fully submerged beneath the East Antarctic Ice Sheet (Figure 14.2). The outer flank of this plateau, which lies immediately adjacent to the highest, axial section of the upwarp, is marked by a dramatic, major escarpment, or locally a broader, stepped topographic discontinuity (the Transantarctic Mountains escarpment). This is most prominent where it flanks the structural basins occupied by the Ross Sea and Ross Ice Shelf in the Pacific sector, and the Filchner Ice Shelf in the Atlantic sector, and is least evident in the vicinity of the South Pole where it is buried by the East Antarctic Ice Sheet. The highest section of the plateau rim, flanking the Ross Sea embayment, is in the Queen Elizabeth, Queen Alexandra and Queen Maud Mountains (between 83° and 86°S) and in the Royal Society Range (78°S) where individual peaks rise above the general elevation to reach altitudes of 4000–4500 m. Summit levels are lower at each end of the Transantarctic Mountains, averaging around 2000 m in the Admiralty Mountains in northern Victoria Land and 1200–2000 m in the Theron Mountains and the Shackleton Range near the Atlantic coast.

The plateau summit is typically flat-topped along its inside flank while the outer edge of the plateau, where it protrudes above the ice surface, is dissected to varying degrees by a network of deep valleys separated by sharp-crested ridges and peaks. Radio-echo sounding of the sub-glacial bed has revealed that the high plateau is some 50 km wide in the vicinity of the McMurdo Dry Valleys, but up to 200 km across in the Queen Alexandra Mountains area (Calkin, 1974a). In some locations, for instance in the Royal Society Range, the flanking escarpment forms a more or less continuous topographic feature with a combined vertical relief of 2000–2800 m, but in other areas, such as northern Victoria Land, it is fragmented by intervening valleys and its form is less distinct.

Throughout its length the plateau forming the Transantarctic Mountains is divided into discrete blocks by outlet glaciers that drain from the East Antarctic Ice Sheet and flow into adjacent ice shelves or ocean. In the Ross Sea sector there are 32 such glaciers between latitudes 74 and 87°S with a mean spacing of 47 km over a distance of 1500 km (Webb, 1994). There is a tendency for there to be fewer, but larger, glaciers crossing the Transantarctic Mountains plateau where its elevation is highest. For example, the large Beardmore Glacier, which crosses the plateau between the Alexandra and Queen Maud Ranges in one of its highest sectors, has a width which varies from 15 to 45 km. The ice thickness is around 2000 m, giving a bedrock relief between the glacier bed and the flanking plateau summits of some 6000 m (Webb, 1994).

14.2.2 GEOLOGY

The central Transantarctic Mountains comprise gently tilted blocks of sedimentary Beacon Supergroup rocks (Devonian–Triassic) overlying Precambrian–Devonian basement (Figure 14.3). This sequence was subsequently intruded by dolerite sills and overlain by basalts of the Ferrar Group. The extrusive rocks (Kirkpatrick Basalt and associated tuffaceous and pyroclastic sequences), which represent remnants of an originally much more extensive continental flood basalt province, have been shown by $^{40}Ar/^{39}Ar$ dating to have been erupted over an interval of less than about 1 Ma at 176.6 ± 1.8 Ma BP up to a distance of more than 1200 km along the Transantarctic Mountains (Heimann *et al.*, 1994). Although they have not been shown to be physically connected to these basalts, the geochemically similar Ferrar Group dolerite dykes and sills are probably genetically and temporally related. Most of the higher summits in the Ross Sea sector are formed in the

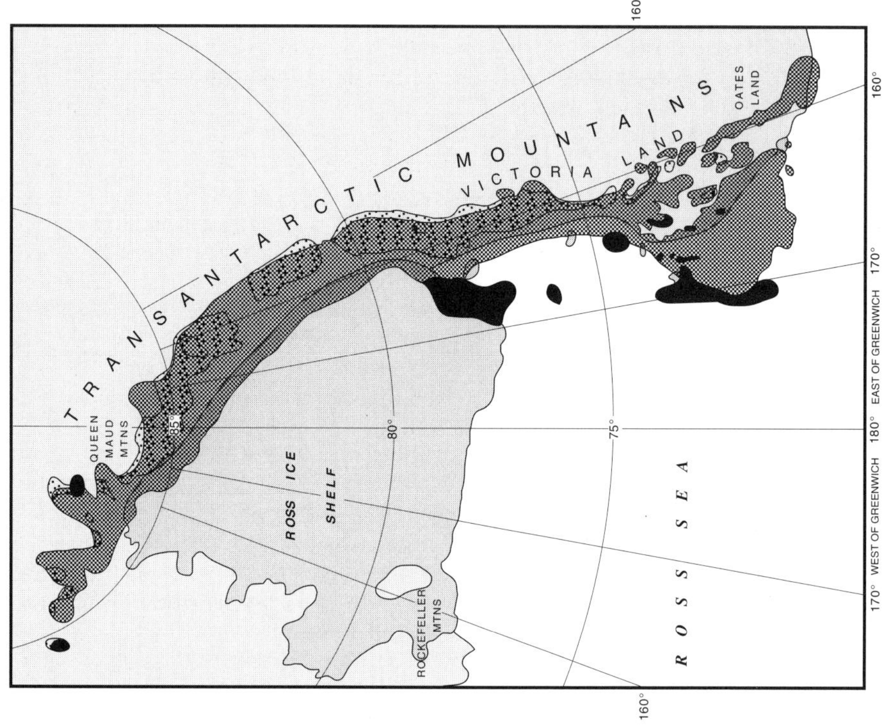

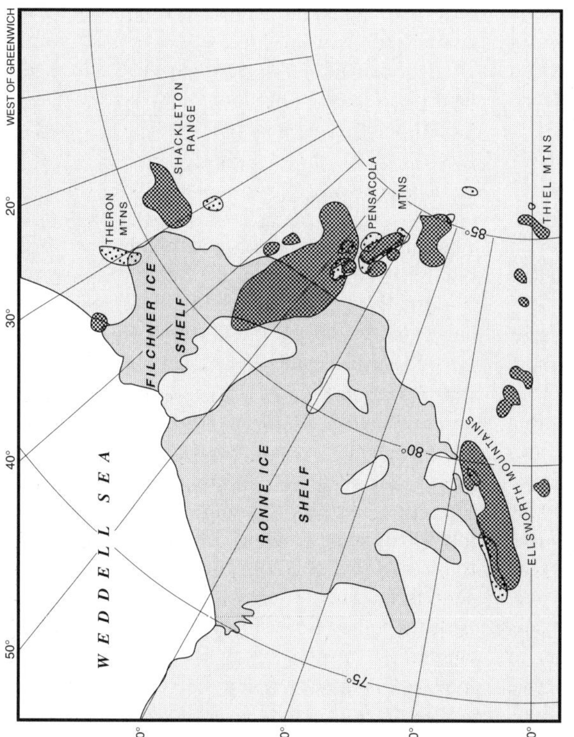

Figure 14.3 *Simplified geology of the Transantarctic Mountains*

sandstones of the Beacon Supergroup, or are capped by Ferrar Group dolerites or basalts, but in much of northern Victoria Land and in the Shackleton Range basement rocks outcrop on the summits. Nevertheless, in these latter areas there is also considerable lithological variation; for example, the two massifs on either side of the Shackleton Range (the Whichaway Nunataks and the Theron Mountains) are capped by Beacon Supergroup sediments.

14.3 Geophysical, Geochemical and Geochronological Data

The geophysical, geochemical and geochronological data on the Transantarctic Mountains region comprise magnetic and gravity anomaly maps, estimates of the cooling history of rocks derived from apatite fission-track thermochronology, deep seismic profiles, radar measurements of ice thickness, and geochemical signatures of source areas for volcanic rocks. Each of these data sources differs in the temporal information that it can provide. Clearly, magnetic and gravity anomaly maps represent snapshots of crustal structure that provide us with no direct indication of change through time. Similarly, radar measurements and deep seismic profiles define the present state of the East Antarctic Ice Sheet and the lithosphere, but do not yield any rates of change. Apatite fission-track thermochronology, by contrast, can provide information on the timing of fault movements and on denudational history, although the temporal resolution of events is relatively poor. Similarly, structural geology provides us with temporal data on relative motions along fault planes, while valuable geochronological information can be gleaned from volcanic events.

14.3.1 GRAVITY AND MAGNETIC DATA

Gravity and magnetic measurements across the Transantarctic Mountains, although sparse in places, have provided ample, though non-unique, data for the interpretation of their underlying crustal structure. The most notable features are an extensive negative free air gravity anomaly over the sub-glacial Wilkes Basin lying parallel to, and inland from, the Transantarctic Mountains escarpment (Bentley, 1983), and a similarly elongated positive anomaly lying offshore in the Ross Sea. The gravity low has been explained both as the result of locally compensated sediments within the Wilkes Basin (Drewry, 1976; Steed, 1983), and, more recently, as a result of lithospheric flexure along the East Antarctic margin (Stern and ten Brink, 1989). Since it is always possible to devise models which satisfy observed gravity anomalies, as Robinson and Splettstoesser (1984) have demonstrated for the Transantarctic Mountains escarpment, the quality of any particular model depends on the overall coherence of the predictions with other geophysical phenomena. In this case, the Stern and ten Brink (1989) model provides a link with observed gravity anomalies across the Transantarctic Mountain escarpment from the Wilkes Basin to the Ross Sea, and is thus more robust than models which consider only Airy-type compensation at one locality. It fails, however, to replicate seismic profiles of the crustal structure underlying the Transantarctic Mountains escarpment (ten Brink *et al.*, 1993). Bouguer gravity anomalies indicate a substantial regional gradient across the Transantarctic Mountains escarpment from the inland, sub-glacial Wilkes Basin to the Ross Sea (Bentley, 1983). This is most readily explained as a rapid change in the thickness of the crust, of the order of 10–20 km, across the escarpment (Behrendt *et al.*, 1991).

Magnetic anomaly maps provide evidence of numerous high-amplitude, short wavelength anomalies along the escarpment flank of the Transantarctic Mountains. The rocks inferred to cause these anomalies are intrusive or extrusive bodies ranging in age from Precambrian to the Cenozoic (Behrendt *et al.*, 1991). Although it is not possible to differentiate ages from magnetic data alone, Behrendt *et al.* (1991) have argued that rift-related, late Cenozoic volcanics are the most likely source of many of the anomalies. Other sources include Jurassic bodies such as the Dufek intrusion or the Kirkpatrick basalts.

14.3.2 FISSION-TRACK DATA

One of the most powerful tools for elucidating shallow lithospheric processes is the thermochronologic technique of apatite fission-track analysis which can provide information on the cooling history of rocks in the upper few kilometres of the crust (see Chapter 4). Although cooling occurring as a result of the relaxation of a thermal event or tectonic denudation can account for some thermal histories recorded by fission-track data, in most tectonic settings cooling is dominated by the effects of crustal stripping by sub-aerial denudation (Brown *et al.*, 1994). If palaeogeothermal gradients can be constrained or reasonably assumed, cooling histories can be converted into records of denudation. In the Transantarctic Mountains the

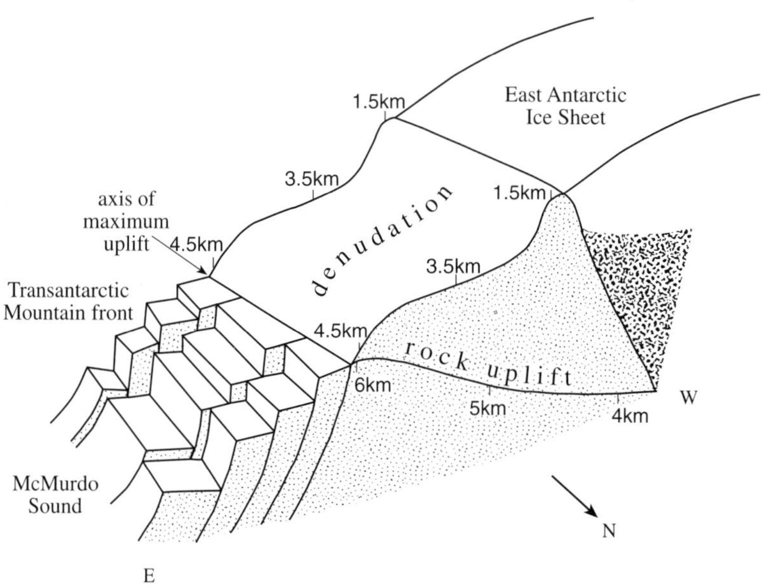

Figure 14.4 *Idealized cross-section of the pattern of denudation and associated faulting in the Dry Valleys area (from Fitzgerald, 1992)*

dominance of crustal cooling through sub-aerial denudation is indicated by apatite fission-track ages that post-date the Jurassic thermal event marked by the Ferrar Group dolerites and basalts. In the Dry Valleys region of southern Victoria Land an apatite fission-track age of about 55 Ma at the 'break-in-slope' of the age–elevation profile demonstrates that a phase of accelerated denudation began at this time (Gleadow and Fitzgerald, 1987; Fitzgerald, 1992). The denudation tells us little about surface uplift unless an initial topography is known or assumed. Since this is a notoriously difficult value to specify it is important to note the limiting assumptions inherent in obtaining these values (Gleadow and Fitzgerald, 1987). Constraints on the palaeogeothermal gradient indicate 4.8–5.3 km of denudation at a mean rate of about 100 ± 5 m Ma^{-1} in the Dry Valleys area. In addition to information on denudation, vertical displacement of neighbouring apatite fission-track age profiles constrains relative movement of large tilt blocks along normal faults trending parallel to sub-parallel to the present coastline, an interpretation supported by the off-setting of Jurassic dolerite sills (Figure 14.4). Data from the central Transantarctic Mountains in the Queen Alexandra–Queen Elizabeth Range area reveal a more complex thermal history with evidence for a phase of accelerated denudation (totalling < 2 km) beginning in the early Cretaceous on the inland flank of the upwarp, and a later episode of rapid denudation of possibly up to 10 km near the coast (Fitzgerald, 1994).

14.3.3 SEISMIC AND STRUCTURAL DATA

Studies complementary to those of the onshore Cenozoic denudational history have been undertaken using offshore seismic reflection profiles. Although difficulties exist in determining the source area for much of the sediment, detailed studies of the profiles provide a record of ice sheet grounding and scouring episodes (Anderson and Bartek, 1992). This work provides a chronology of depositional and erosional events across the near-coast margin which can be used to elucidate onshore events. Offshore boreholes constrain interpretations of seismic profiles and generate scenarios of sediment transport and deposition across the Transantarctic Mountains escarpment (Barrett, 1989). Of particular interest structurally are the very deep basins which lie parallel to the Transantarctic Mountains escarpment. For example, the Victoria Land Basin, which has a half-graben structure, is believed to contain up to 14 km of stratified sediments, the upper 4–6 km of which are inferred to be Cenozoic marine deposits (Cooper *et al.*, 1991). It was initially thought that the existence of the Victoria Land Basin was indicative of a broken plate boundary between the

East Antarctic craton and the stretched lithosphere beneath the Ross Sea (Stern and ten Brink, 1989), but a deep seismic profile across this boundary (ten Brink *et al.*, 1993) has shown a smooth change in lithospheric thickness, indicating that structurally the Victoria Land Basin is a localized phenomenon.

Recently, Wilson (1995) mapped fault arrays along the Transantarctic Mountains escarpment and concluded that they run obliquely to the axis of the mountains. This implies that the regional rift boundary is not controlled by continuous rift border faults as was previously assumed. It was inferred that a dextral transtensional regime existed through the Cenozoic and that dominantly transcurrent motion occurred during the most recent faulting episode. The implications of this structural interpretation are profound, since this tectonic model is incompatible with large-magnitude crustal stretching and associated isostatic uplift of the Transantarctic Mountains–West Antarctica rift system during the Cenozoic. Wilson postulated that it is more likely that crustal thinning across the rift system took place during the Mesozoic, when major West Antarctic crustal block motions took place.

14.3.4 GEOCHEMICAL AND GEOCHRONOLOGICAL DATA FROM VOLCANIC ROCKS

The geochemical composition of volcanic rocks provides information on the magma source and can assist in the identification of lithospheric processes. Drill cores, such as that from the CIROS-1 hole in the Ross Sea Basin, indicate volcanic activity of Oligocene age and older (Barrett, 1989), while exposed alkaline rocks along the Transantarctic Mountains escarpment have ages of up to 25 Ma. Jones and Fitzgerald (1984) have noted the existence of extensive volcanism associated with the separation of Australia and Antarctica in the late Eocene. In general, magmatic activity in the region involved the generation of alkaline basaltic magma from mantle sources, largely within the aesthenosphere and without major modification, and differentiation of other basaltic magma in crustal reservoirs by a variety of fractional crystallization processes (Behrendt *et al.*, 1991). Volcanic rocks have also been invaluable in constraining amounts of surface uplift along the oceanward flank of the Transantarctic Mountains escarpment. For instance, the present elevation of sub-aerially erupted volcanic rocks in Taylor Valley in the Dry Valleys block which range in age from 3.89 to 1.50 Ma constrains surface uplift in this area, in one specific case to a maximum of around 300 m since about 2.5 Ma BP (Wilch *et al.*, 1993).

14.4 Evidence for Surface Processes

There have been three periods of intense debate about the surface processes responsible for shaping the Transantarctic Mountains. The first occurred early in the 20th century and arose from the work of geologists and geographers on the expeditions of Scott and Shackleton. The second followed the research carried out in association with the International Geophysical Year in 1957, while the most recent debate has focused on the question of the stability of the East Antarctic Ice Sheet. In spite of these more or less discrete episodes of research activity, there has been an underlying continuity to discussions about the nature of landscape evolution in the Transantarctic Mountains. Points of dispute current today have their roots in differences in interpretation that can be traced back to the beginning of the century, with each debate being concerned with three key questions: Does the present landscape reflect the actions of current glacial processes? If it does, at what rate is the landscape being changed? If it does not, then how and when was it formed?

14.4.1 ROLE OF RECENT GLACIAL EROSION

Of these questions, the first is now couched in terms of the protective role of cold-based, and the erosive power of warm-based, ice. In some areas there is startling evidence of the preservation of fragile landforms beneath ice which has presumably been cold-based (Plate 14.1). For instance, an unconsolidated volcanic ash deposit in Arena Valley in the Quartermain Mountains of the Dry Valleys region consists of material so loose it can be excavated by hand, and yet it was covered by an enlarged Taylor Glacier which draped it with large boulders without modifying its form (Marchant *et al.*, 1993a). Similarly, observations on the lack of debris in Antarctic glaciers and the absence of rock flour in meltwater streams led Ferrar (1907) and Taylor (1922) to conclude that ice is more conservative than erosive. Elsewhere, there is limited evidence of the processes of basal regelation and glacial erosion. The debris exposed in basal ice at the snout of Taylor Glacier, for example, has been derived from the bed up-glacier (Robinson, 1984), while observations take from submersibles of basal debris and sub-glacial meltwater at the snout of the Mackay Glacier in South Victoria Land demonstrate that some outlet glaciers are warm-based and capable of erosion (Powell *et al.*, 1996). This debate is important in the context of landscape evolution since it places constraints on the timing and transformation of the landscape by glacial action.

14.4.2 MODELS OF LANDSCAPE EVOLUTION

If the present glaciation is having a limited effect on the landscape, what of earlier glaciation? Taylor (1922) argued that the landscape of Victoria Land was essentially glacial in origin, while David and Priestley (1914) envisaged an ice sheet superimposed on a pre-glacial water-eroded landscape, and a similar view was held by Priestley (1909), who noted the existence of glacial facets in northern Victoria Land which he thought represented the glacial straightening of formerly sinuous river valleys.

This early debate continues today with the coexistence of two widely differing interpretations of landscape development. The view that the present landscape is the result of glacial dissection of an original low-relief planation surface has been developed for northern Victoria Land by van der Wateren and colleagues (van der Wateren and Verbers, 1990; Verbers and van der Wateren, 1992). They have recognized a uniform pattern of topography regardless of rock type, involving flat summits, glacially eroded rock terraces 100–400 m below the summits, and glacial troughs with a drift limit above the present ice. This landscape is explained in terms of successive downcutting by glaciers associated with conditions similar to those of the present, and dissection of the original plateau surface is ascribed to tectonic uplift, with the present topography representing differences in the rate and timing of uplift. At locations such as the western shoulder of the Rennick graben they consider that an initial radial pattern of ice flow was subsequently replaced by an incised linear trough pattern related to outlet glaciers. The view that outlet glaciers and cirque glaciers created the main erosional features of the Transantarctic Mountains is also implicit in the early work in the Dry Valleys area where valley benches were attributed to earlier phases of trough excavation and high-level valleys were interpreted as cirques (Bull *et al.*, 1962; Denton *et al.*, 1971; Nichols, 1971; Selby and Wilson, 1971; Calkin, 1974b) (Plate 14.2).

A contrasting view of landscape development argues for a primary fluvial origin for the main valleys with more limited subsequent modification by ice. For instance, Webb (1994) considers that the valleys now occupied by glaciers such as the Beardmore were eroded down to near sea level by rivers as well as being overdeepened by glaciers, while Sugden *et al.* (1995a) have concluded that the landscape in the Dry Valleys area is essentially fluvial in origin, with diagnostic features such as sinuous valleys, major long profiles graded to near sea level, fluvial valley benches and dendritic valley patterns all still apparent. They argue that glacial modification, even over the long term, has been selective, being limited to the straightening of certain troughs and the partial scouring of pre-existing surfaces. Glacial modification of the landscape under present-day climatic conditions is thought to be highly selective and probably limited to erosion beneath major outlet glaciers.

14.4.3 CLIMATIC INTERPRETATION OF LANDSCAPE DEVELOPMENT

Views as to the age of landscape dissection by glaciers have also been influenced by the morphogenetic impact that the present climatic regime is thought to have had. Taylor (1922), for example, argued that the cutting of typical glacial features such as cirques must have occurred at an early stage of glaciation when the climate was much warmer than at the present. In a similar vein, evidence of sub-glacial 'alpine' topography on the continental flank of the Transantarctic Mountains has more recently been interpreted as reflecting earlier conditions when the glaciers were warm-based (Drewry, 1982). Using the same reasoning, the deposition of the warm-based tills associated with the Sirius Group deposits, which occur at high altitudes at over 30 locations throughout the Transantarctic Mountains, have long been attributed to conditions before the creation of the present cold-based ice sheet (Mercer, 1968; Barrett and Powell, 1982; Webb *et al.*, 1984; McKelvey *et al.*, 1991; Barrett *et al.*, 1992). The glacial modification of areas such as the Dry Valleys region, involving areal scouring and the formation of striations and meltwater channels, has been related to overriding by a higher warm-based ice sheet (Denton *et al.*, 1984). These views reflect the difficulty of explaining the production of glacially eroded landforms under the conditions of cold-based ice prevalent today, the main issue being the time when warm conditions last occurred. According to these interpretations of the morphological and sedimentary evidence the most recent warm episode is now thought to date back to the mid-Miocene, or even the Oligocene (Denton *et al.*, 1993). Other researchers, by contrast, have argued that such warm conditions may have prevailed as recently as 2–3 Ma ago on the basis of a proposed Pliocene age for at least some Sirius Group deposits (Webb *et al.*, 1984; Barrett *et al.*, 1992).

If it is believed that present glaciers are capable of carving the entire suite of glacial forms under the present climatic regime, then there is no need to relate the landscape development to earlier, warmer conditions. In the case of northern Victoria Land, van der Wateren and Verbers (1992) have attempted to constrain the timing of glacial incision by cosmogenic surface exposure dating. This indicates ages of 1–2 Ma on the upper surface, with progressively younger ages at lower elevations. They interpreted these surface exposure age relationships as indicating that dissection by glaciers of this sector of the Transantarctic Mountains occurred during the past 2 Ma. However, this interpretation does not adequately take into account the possible effects on exposure ages of different rates of weathering and denudation as a function of rates of biological activity and salt deposition, which are in turn related to elevation and distance from the sea. Their dates should therefore be regarded as minima for the most recent phase of glacial incision.

If glacial landforms are very old then this has implications for periglacial processes and their effect on the landscape. Priestley (1909) and Taylor (1922) recognized the insignificance of water action in slope evolution in the Transantarctic Mountains and pointed to the probable importance of wind and chemical action. These early ideas evolved into a powerful argument by Selby (1971, 1974) about the significance of salt weathering in landscape evolution. He maintained that over long periods of time salt crystal growth causes significant rock disintegration, and that in combination with mass movement and wind action this process has been able to transform an original landscape of glacial troughs into one characterized by sharp valley divides flanked by slopes with angles controlled by rock strength and rate of weathering. He identified four typical slope elements; an upper free face, whose angle reflects the rock mass strength, a rectilinear, weathering-limited slope at 31–37°, a flat basal pediment eroded across bedrock, and a concave basal slope with debris accumulation. A similar model of slow slope retreat under periglacial conditions is also argued to account for much of the landscape of the Dry Valleys area as reflected by evidence of the great age of soils (Ugolini and Bull, 1965; Campbell and Claridge, 1978).

In summary, there is no clear consensus about the answers to the three questions outlined above. Different views coexist about evidence on the nature and efficacy of present processes in the Transantarctic Mountains, their rate of change under present conditions, and indeed to what extent the landscape is an inheritance from the past. Bearing in mind the uncertainties in the data concerning the tectonic and surface processes responsible for shaping the Transantarctic Mountains, we now look at how different models have been used to interpret the data. There are two sets of models and again these reflect their different origins, namely, tectonic and geomorphological.

14.5 Tectonic Models of Landscape Development

Numerous models of the tectonic structure of the Transantarctic Mountains have been developed since early workers described the 'Great Antarctic Horst' (David and Priestley, 1914), but there is limited understanding of the causes of surface uplift that created the major topographic feature observed today. One suite of morphotectonic models has cited glaciation as the primary factor. Voronov (1964), for instance, considered the continental margin represented by the Transantarctic Mountains to be upwarped due to ice loading inland, while Grindley (1967) argued that ice loading promoted a mantle bulge and hence surface uplift at the margins of the East Antarctic Ice Sheet. Taking a similar approach, McGinnis *et al.* (1985) postulated that crustal bending due to the glacial load together with consequent reheating was one factor that had driven surface uplift. A more quantitative analysis was undertaken by Drewry (1983) who used flexural modelling to quantify the isostatic effect of the ice sheet on the Transantarctic Mountain margin of Antarctica. He concluded that the Transantarctic Mountains are depressed by approximately 500 m. In a similar vein, the role of glaciers as geomorphic agents leading to isostatic unloading was considered by Wellman and Tingey (1981) and Tingey (1985), who assessed the contribution of glacial erosion in denuding and transporting large volumes of sediment through glacial troughs. Using isostatic considerations they showed that mountains flanking deep glacial troughs can be uplifted by hundreds of metres. However, all such analyses are limited since a full quantitative study of all relevant parameters, such as the effective elastic thickness of the lithosphere, the capacity for glacial erosion, and the isostatic effect of glacial loading, has yet to be undertaken. Bed rebound for Antarctica as a whole, assuming both Airy isostasy (effective elastic thickness ($T_e = 0$), and flexural isostasy ($T_e = 120$ km) is illustrated in Plate 14.3.

A second suite of models ignores the impact of glaciation, and instead postulates an aesthenospheric cause for the Transantarctic upwarp. For example, Smith and Drewry (1984) argued that geochemical phase changes in the deep crust could contribute to surface uplift. These are considered to result from the existence of hot asthenosphere close to the surface resulting from crustal stretching during the rifting along the Ross Sea margin. The Smith and Drewry model is attractive because it links the thermal event associated with the break-up of Australia from Antarctica at 90 Ma BP with the initiation of uplift of the Transantarctic Mountains at approximately 60 Ma BP; however, the necessary time delays between crustal stretching and surface uplift implicit in the model do not equate with known events. In addition, there are physical problems with the necessary distribution of the slab of eclogite at the mantle–crust interface required to cause the uplift.

Recently, a more quantitative approach to the morphotectonic evolution of the Transantarctic Mountains has been made possible through the application of thermochronological techniques and the refinement of thermal and mechanical models of the lithosphere. Fitzgerald *et al.* (1986) suggested that the depth of denudation across the Transantarctic Mountains implied by apatite fission-track analysis data, together with the resulting topography, is consistent with crustal thinning due to asymmetric extension of the crust of the kind originally suggested by Wernicke (1985). Utilizing these new data, Stern and ten Brink (1989) provided the first in a series of analyses differentiating the tectonic processes responsible for the marginal upwarp that has created the Transantarctic Mountains. Using gravitational data and the morphology of the margin, they modelled the tectonic evolution of the Transantarctic Mountains by means of the elastic flexure of two cantilevered lithospheric plates with a stress-free edge between East Antarctica and the rifted Ross Embayment. Three principal uplift forces were identified: thermal uplift, the isostatic response to denudational unloading, and the Vening Meinesz effect resulting from footwall unloading after elastic failure of the lithosphere. Their model was fitted to data using an effective elastic thickness for the East Antarctic craton of 115 km; however, since the observed topographic deflection was modelled solely as the result of a line load acting at the coastline, this value is probably too high (van der Beek *et al.*, 1994). This work was extended through a comparison of the escarpment flanking the Transantarctic Mountains with the Great Escarpment of southern Africa, and it was again concluded that it was not possible to explain the surface uplift required to generate the observed topography from a single mechanism (ten Brink and Stern, 1992). Taking a different approach Bott and Stern (1992) used a viscoelastic finite element model and interpreted the present state of uplift and subsidence across the margin as a response to lithospheric tension and basal upthrust produced by low-density mantle underlying the Transantarctic Mountains. Their work suggests that the main plate interior stress field is less important than locally derived loading stresses in the support of the present topography.

Subsequent field work designed to test these modelling studies has revealed no evidence of deep crustal faulting through the thinned crust below the Ross Embayment, and the broken plate model therefore appears to be inapplicable in this region (ten Brink *et al.*, 1993). The field data in fact show a smooth transition from rifted to unrifted crust, with no sedimentary trough in front of the Transantarctic Mountains escarpment as predicted by the broken plate model, no indication of crustal underplating, and little indication of tectonic activity. These findings were interpreted by ten Brink *et al.* (1993) to mean that the main phase of stretching of the Ross Embayment crust must have preceded uplift of the adjacent Transantarctic Mountains by 20 Ma or more. Coupled with the current aseismicity of Antarctica, this interpretation implies a lack of recent significant tectonic uplift or subsidence and casts doubt on the claim that major surface uplift has taken place in the last 1–2 Ma (Behrendt and Cooper, 1991; Webb *et al.*, 1994). Although ten Brink *et al.* (1993) argue that the uplift of the Transantarctic Mountains has occurred largely as a result of thermal effects, this interpretation appears to have arisen through the elimination of other possible mechanisms rather than because the data supporting thermal uplift are conclusive. Indeed, the detailed parameter study by van der Beek *et al.* (1994) has indicated that flexural uplift as a result of lithospheric necking is more likely to be a dominant surface uplift mechanism than thermal buoyancy since the amount of uplift and narrow wavelength of the topography are inconsistent with uplift generated by thermal effects alone. They have further maintained that differential denudation across the margin has not been a primary surface uplift mechanism since the effective elastic thicknesses for the lithosphere required in such models

(Gilchrist and Summerfield, 1990) are not representative of cratonic values. Kerr and Gilchrist (1996), however, used a parameter study of erosion across the Transantarctic Mountains escarpment linked to flexural isostasy to show that this need not be the case as long as certain other conditions are met, such as a retreating escarpment and a significant difference in rates of denudation across the escarpment. Such conditions are likely across the Transantarctic Mountains escarpment where the removal of a coastal wedge of material has been documented (Gleadow and Fitzgerald, 1987). Van der Beek *et al.* (1994) concluded that the amount of uplift of the Transantarctic margin requires a simple shear mechanism with lithospheric stretching concentrated under the uplifted escarpment. The kinematics of extension are determined to a large extent by the rheological coupling of strong upper/middle crust and weak lower crust.

This discussion of the tectonic processes responsible for the uplift of the Transantarctic Mountains points to activity early in the evolution of the mountains and offshore basins. It is interesting that the tenor of such arguments has been overshadowed by the recent debate on the stability of the East Antarctic Ice Sheet during the late Cenozoic (Barrett, 1991; Sugden, 1992). At the heart of this controversy is the question of whether the Transantarctic Mountains have experienced substantial surface uplift of the order of 1–2 km since the mid-Pliocene (Behrendt and Cooper, 1991; Webb *et al.*, 1994), or whether the mountains have remained at essentially their present altitude since the Miocene. Clearly, these different viewpoints have major implications for the tectonic and surface processes involved. At this stage it is worth noting that the likely processes of tectonic uplift, and indeed the alignment of existing faults (Wilson, 1995), do not support the idea of significant uplift during the Pliocene.

In summary, numerous mechanisms have been cited that can contribute to the formation of the upwarp comprising the Transantarctic margin. Recent opinion favours a combination of flexural uplift due to lithospheric necking (van der Beek *et al.*, 1994), thermal buoyancy (ten Brink *et al.*, 1993), and differential denudation across the margin (Kerr and Gilchrist, 1996). The available geophysical data are less clear with regard to the relative importance of tectonic mechanisms and the timing of the events which led to the tectonic uplift. Nonetheless, the implication is that many of the lithospheric processes assumed to have contributed to the tectonic uplift of the rifted margin occurred during the Mesozoic.

14.6 Towards a Model of Landscape Evolution in the Dry Valleys Region

Current models of landscape evolution in the Dry Valleys area reflect new evidence about the antiquity of the Transantarctic Mountains landscape and the efficacy of present-day surface processes. The advent of new dating techniques has brought the realization that some landform features date back to the mid-Miocene and has led to a more detailed understanding of the relationships between landscape evolution and surface processes.

14.6.1 LANDSCAPE COMPONENTS

The landscape of the Dry Valleys comprises three main elements (Sugden *et al.*, 1995a). The first consists of a series of erosion surfaces bounded by escarpments forming three giant steps rising from the coast (Plate 14.4). Residuals rising above the uppermost surface exceed an altitude of 3000 m and form the highest summits in the area. The upper escarpment, which has a relief of 1200 m, is broken up into a network of valleys and residual mountains, such as the Olympus and Asgard ranges. The landscape is characterized by rectilinear slopes which rise at angles of 26–36° from low-gradient surfaces and which are often topped by steeper free faces; most are cut in Beacon Supergroup sediments and many are capped by Ferrar dolerites. The second landscape component consists of the main 'Dry Valleys' that cut across the escarpment front of the Transantarctic Mountains to the sea. These major valleys, which are sinuous in places and incised to near, or below, present sea level, form a dendritic network linked to high-level tributary valleys. The third landscape element comprises glacial landforms. These include straight, steep-walled troughs with truncated spurs, cirques, surfaces of areal scouring (marked by rock hollows, streamlining and roches moutonnées) and integrated sub-glacial meltwater channel systems (Sugden *et al*, 1991), such as the area known as the Labyrinth in upper Wright Valley (Plate 14.2).

14.6.2 DENUDATION CHRONOLOGY

Significant parts of the Dry Valleys area lack features characteristic of glacial erosion and therefore appear to represent landform features relict from a much earlier fluvially dominated erosional regime. Support for this interpretation is provided by the retention of sinuous valley forms as a result of a lack of truncation of spurs. Glacial action subsequently led to selective modification of this essentially fluvial landscape as is illustrated by the way in which areal scouring by ice affects many erosion surfaces and valley slopes, the straightening and lowering of central Wright Valley to form a cliffed trough, the dissection of the bench of upper Wright Valley by meltwater channels, and the pattern of cliffs and truncated spurs in that part of central Taylor Valley which still retains its sinuous fluvial form. This relationship is also seen in the case of cirques in the Asgard Mountains which are incised into the rectilinear slopes of the arcuate heads of the upper valleys. The glacial cirques are often very much smaller than the valley heads.

There is evidence of differential denudation in the relationship between the main landforms and the underlying basement rocks. In the Dry Valleys region Beacon Supergroup sediments overlie the Ordovician–Devonian Kukri unconformity (Gunn and Warren, 1962). At the heads of Taylor and Wright Valleys the palaeo-erosion surface represented by this unconformity, which dips inland at 2–3°, is at an altitude of 700 m and 900 m, respectively. It rises in discrete blocks towards the coast, reaching an altitude of 1800 m at the main escarpment front. Basement rocks are fully exposed on the coastal piedmont, the lowest step of the three main erosion surfaces. Dislocation of the Kukri unconformity and other stratigraphic boundaries shows that there has been faulting parallel to the coast (Ferrar, 1907; Wright and Priestley, 1922), or obliquely with respect to it (Tessonsohn and Worner, 1991; Fitzgerald, 1992).

In interpreting the significance of the present altitude of the Kukri unconformity it is possible to develop an early argument used by Wright and Priestley (1922) who realized that Beacon Supergroup sediments must have been stripped from the coast. If it is assumed that the Kukri surface was once approximately horizontal, then these relationships suggest that a wedge of rock exceeding the thickness of the Beacon Supergroup sediments and dolerites in the area (~2.5 km according to Barrett (1981)) has been removed from the coast. Reconstructing the dips of the surface and allowing for the relief in the basement rocks indicates the removal of over 4 km of section at the coast (Sugden *et al.*, 1995a).

There is clear evidence of the great age of even small-scale landform features in the Dry Valleys block. It has long been realized that the presence of glacial and marine sediments of Miocene age in the mouths of the Dry Valleys themselves show that they had already been cut by this time (McGinnis, 1991; Barrett and Hambrey, 1992; Ishman and Rieck, 1992; Prentice *et al.*, 1993). The presence of small volcanic scoria cones of Pliocene age on the slopes of Taylor Valley confirms this interpretation (Wilch *et al.*, 1993b). Even more remarkable is the suite of volcanic ash deposits trapped in surficial deposits at 15 sites on the slopes of the Quartermain and Asgard ranges (Marchant *et al.*, 1993a, b). These reveal a range of ages from 3.9 to 15.25 Ma. The ashes occur in avalanche deposits and frost-wedge casts on valley floors, rectilinear slopes and surfaces of areal scouring. In one location they even overlie a relict of Miocene glacier ice over 8 Ma old (Sugden *et al.*, 1995b). They demonstrate that even minor elements of the landscape, at least at higher altitudes, are relicts from the mid-Miocene (Plate 14.1). Moreover, they demonstrate that areal scouring by overriding ice and the cutting of the meltwater channels, which post-date the formation of the pre-existing fluvial landscape, had also taken place by the mid-Miocene. They also imply remarkably low rates of denudation, a conclusion confirmed by recent work using cosmogenic isotope analysis (Ivy-Ochs *et al.*, 1995; Brook *et al.*, 1996).

14.6.3 MODEL OF LANDSCAPE EVOLUTION

The relationships described above allow us to create a model of landscape evolution for the Dry Valleys area. This is covered in more detail elsewhere (Sugden *et al.*, 1995a), but Figure 14.5 reveals the main stages. An original flat-lying basement, covered in sedimentary rocks, is planed from the coast to expose the basement at the coast. Rock uplift is demonstrated by the upwarping of the basement at the mountain front, and may well have coincided with faulting. By the Miocene the lower reaches of the main valleys were graded to near sea level and, as indicated by the presence of Miocene marine deposits, were subsequently inundated by the sea. Subsequent uplift of a few hundred metres near the coast has raised the valley thresholds and created basins with internal drainage; it is worth noting that this latter feature is not characteristic of neighbouring blocks of the Transantarctic Mountains, such as that of the Royal Society Range.

It is not clear how widely this Dry Valley's reconstruction applies to the Transantarctic Mountains as a whole. But there are hints that other landscapes are equally old. For example, dating of lava in a valley in the Southern Cross area of northern Victoria Land shows that the valley had been cut by 12.5 Ma ago. The Sirius Group deposits, which pre-date dissection of the mountains, indicate temperate glacial conditions on the highest of the Transantarctic Mountains and

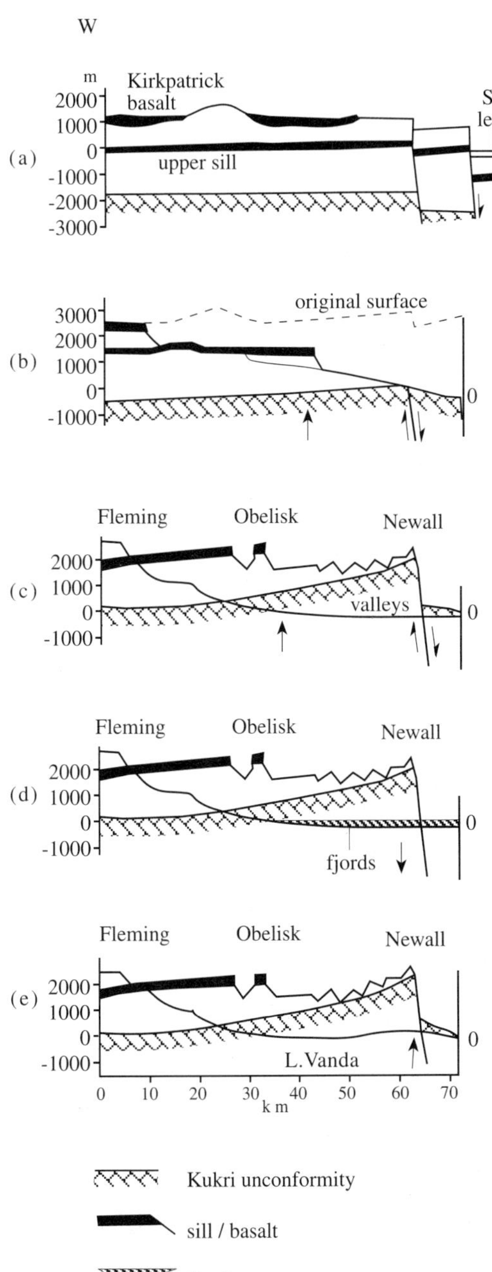

Figure 14.5 *Schematic model of post-rifting landscape evolution in the Dry Valleys area of the Transantarctic Mountains: (a) initial lithospheric extension and rifting; (b) planation with respect to a new, lower base level; (c) valley down-cutting; (d) subsequent phase of subsidence; and (e) subsequent uplift of ~300 m at the escarpment front. The arrows indicate inferred vertical crustal movements (after Sugden* et al., *1995a)*

occur at over 30 sites throughout the mountains (Denton *et al.*, 1993). If the great age of the deposits in the Dry Valleys area is confirmed elsewhere, then it implies that essentially unmodified Miocene surfaces occur widely. If so, then the Dry Valleys reconstruction may indeed be representative of a wider area. Thus it is worth summarizing the main conclusions raised by the reconstruction of landscape evolution. They are:

1. A wedge of rock over 4 km thick at the coast and thinning inland has been removed from the coastal margin of the mountains.
2. Most of this erosion was accomplished under fluvial conditions before a full Antarctic ice sheet built up.
3. Dissection of the mountains was largely accomplished before the mid-Miocene.
4. Glacial modification of the mountains has been selective and, with the main exception of the major outlet glaciers and valley glaciers of northern Victoria Land, often modest.
5. Except at low altitudes where there has been significant summer melting, rates of denudation under existing frigid conditions are extremely low and have remained so for over 15 Ma.

14.7 Integration of Geomorphic and Tectonic Approaches: Broader Research Issues

In this chapter we have assessed both tectonic and geomorphological approaches to the evolution of the Transantarctic Mountains. It has been possible to identify tectonic processes which would be expected to leave their mark on the landscape, and also to identify ways in which an understanding of geomorphology should help constrain the tectonic processes. Both approaches to landscape evolution have their parallels in other mountain areas of the world, and in the Transantarctic Mountains as elsewhere it is difficult to achieve an integration between the two. The fundamental problem is the mismatch of time scales. Tectonic processes typically evolve over tens of millions of years. Surface processes usually create landscapes on time scales of hundreds of thousands of years; the temporal constraints are good but restricted to the more recent period.

The Transantarctic Mountains are probably unique in permitting significant overlap between the two sets of processes and their different time scales. The surface deposits give a remarkable window into a mid-Miocene landscape where even minor details of the landscape are apparently perfectly preserved. If, as seems possible, the Sirius Group deposits mark the initial glaciation of Antarctica before the build up of the present full ice sheet, then in scattered locations the window may extend back as far as the Oligocene. In such a case we can see surfaces which have survived sub-aerially with minimal modification for tens of millions of years. Moreover, such surfaces are distributed throughout large areas of the mountains, offering scope for a broader understanding of mountain development. Presumably, this remarkable landscape preservation is related to the lack of water which prevents weathering by physical, chemical and biological activity.

By way of conclusion, it is useful to highlight the potential of two-way links between geomorphology and tectonics in the Transantarctic Mountains. Geomorphology constrains the understanding of tectonic processes in several ways. First, there is the issue of the timing of uplift, and especially whether there was a possibility of surface uplift of 1–2 km in the Pliocene. The geomorphological evidence seems to rule this out, at least in the Dry Valleys area of the Transantarctic Mountains where the idea first took hold. Second, geomorphological relationships are also able to help constrain the timing of any faulting which may have been associated with uplift. For example, uniform rectilinear slopes of presumed Miocene age truncate the faulting at the mountain front on the side of Wright Valley and demonstrate that the valley side was cut after the faulting. Third, knowledge of the location and depth of denudation can be used to constrain predictions of lithospheric flexure. In the case of the Dry Valleys it is possible to map out the geometry of the coastal wedge of eroded rock and thus refine estimates of the role of flexure in uplift. It should also be possible to integrate the effects of the erosion of glacial troughs on the uplift of adjacent parts of the mountains.

Finally, it is possible that the study of landforms may be able to contribute to the solution of a major unknown in any reconstruction of the tectonic and surface processes affecting uplifted passive margins, namely changes in the elevation of the land surface over time. This profoundly affects both tectonic and geomorphological models. One approach in the Dry Valleys has been to use cosmogenic surface exposure dating on high-altitude surfaces. Brook *et al.* (1996) were able to use the data they obtained in the Dry Valleys to rule out the possibility of Pliocene uplift of the area. With further precision it should be possible to place constraints on the minimum elevation of a surface over varying lengths of time. Another more speculative approach is to look at the gas content in buried ice of Miocene age. At the time of writing, the ice from Beacon Valley in the Quartermain Mountains is being analysed to see how successful such an approach may be.

14.8 Conclusions

The purpose of this chapter has been to explore how tectonic and geomorphological approaches, if combined, contribute to a better understanding of the evolution of the Transantarctic Mountains. In most parts of the world the mismatch between time scales makes it difficult to relate landscape development to the underlying tectonic processes with confidence. In the Transantarctic Mountains, however, the great span of ages represented by the present landscape makes it possible to use geomorphological observations to evaluate competing tectonic models directly. The landscape is old and its window into a Miocene world holds great potential for further refinement of models of rifted margin evolution.

Acknowledgements

The research on which this chapter is based was supported by the UK Natural Environment Research Council through grant no. GR3/9128 to DES and MAS. We also wish to thank George Denton and acknowledge the support of the Division of Polar Programs of the National Science Foundation.

References

Anderson, J.B. and Bartek, L.R. 1992. Cenozoic glacial history of the Ross Sea revealed by intermediate resolution seismic reflection data combined with drill site information. In: J.P. Kennet and D.A. Warnke (eds) *The Antarctic Environment: A Perspective on Global Change, Ant. Res, Ser.*, **56**, American Geophysical Union, Washington, pp. 231–263.

Barrett, P.J. 1981. History of the Ross Sea region during the deposition of the Beacon Supergroup 400–180 million years ago. *J. R. Soc. N.Z.*, **11**, 447–458.

Barrett, P.J. (ed.) 1989. Antarctic Cenozoic history from the CIROS-1 drillhole, McMurdo Sound. *Bulletin in the Miscellaneous Series of the New Zealand Department of Scientific and Industrial Research*, **245**, 254.

Barrett, P.J. 1991. Antarctica and global climate change: a

geological perspective. In: C.M. Harris and B. Stonehouse (eds) *Antarctica and Global Climate Change*, Belhaven Press, London, and SPRI, Cambridge, pp. 35–50.

Barrett, P.J., Adams, C.J., McIntosh, W.C., Swisher, C.C. and Wilson, G.S. 1992. Geochronological evidence supporting Antarctic deglaciation three million years ago. *Nature*, **359**, 816–818.

Barrett, P.J. and Hambrey, M.J. 1992. Plio-Pleistocene sedimentation in Ferrar Fiord, Antarctica. *Sedimentology*, **39**, 109–123.

Barrett, P.J. and Powell, R.D. 1982. Middle Cenozoic glacial beds at Table Mountain, Southern Victoria Land. In: C. Craddock (ed.) *Antarctic Geoscience Symposium on Antarctic Geology and Geophysics*, University of Wisconsin Press, Madison, pp. 1059–1067.

Behrendt, J.C. and Cooper, A. 1991. Evidence of rapid Cenozoic uplift of the shoulder escarpment of the Cenozoic West Antarctic rift system and a speculation on possible climate forcing. *Geology*, **19**, 315–319.

Behrendt, J.C., Lemasurier, W.E., Cooper, A.K., Tessonsohn, F., Trehu, A. and Damaske, D. 1991. Geophysical studies of the West Antarctic rift system. *Tectonics*, **10**, 1257–1273.

Bentley, C.R. 1983. Crustal structure of Antarctica from geophysical evidence – a review. In: R.L. Oliver, P.R. James and J.B. Jago (eds) *Antarctic Earth Science*, Australian Academy of Science, Canberra, pp. 491–497.

Bott, M.H.P. and Stern, T.A. 1992. Finite element analysis of Transantarctic Mountain uplift and coeval subsidence in the Ross Embayment. *Tectonophysics*, **201**, 341–356.

Brook, E.J., Brown, E.T., Kurz, M.D., Ackert, R.P., Raisbeck, G.M. and Yiou, F. 1996. Constraints on age, erosion, and uplift of Neogene glacial deposits in the Transantarctic Mountains determined from in situ cosmogenic ^{10}Be and ^{26}Al. *Geology*, **23**, 1063–1066.

Brown, R.W., Summerfield, M.A. and Gleadow, A.J. 1994. Apatite fission track analysis: its potential for the estimation of denudation rates and implications for models of long-term landscape development. In: M.J. Kirkby (ed.) *Process Models and Theoretical Geomorphology*, Wiley, Chichester, pp. 23–53.

Bull, C.B.B., McKelvey, B.C. and Webb, P.-N. 1962. Quaternary glaciations in Southern Victoria Land, Antarctica. *J. Glaciol.*, **4**, 63–78.

Calkin, P.E. 1974a. Subglacial geomorphology surrounding the ice-free valleys of southern Victoria Land, Antarctica. *J. Glaciol.*, **13**, 415–429.

Calkin, P.E. 1974b. Processes in the ice-free valleys of southern Victoria Land, Antarctica. In: B.D. Fahey and R.D. Thompson (eds) *Research in Polar and Alpine Geomorphology*, Department of Geography, University of Guelph, Ontario, pp. 167–186.

Campbell, I.B. and Claridge, G.G.C. 1978. Soils and Late Cenozoic history of the Upper Wright Valley area, Antarctica. *N.Z. J. Geol. Geophys.*, **21**, 635–643.

Cooper, A.K., Davey, F.J. and Behrendt, J.C. 1991. Structural and depositional controls on Cenozoic and (?) Mesozoic strata beneath the western Ross Sea. In: M.R.A. Thomson, J.A. Crame and J.W. Thomson (eds) *Geological Evolution of Antarctica*, Cambridge University Press, Cambridge, pp. 279–283.

David, T.W.E. and Priestley, R.E. 1914. Glaciology, physiography, stratigraphy and tectonic geology of South Victoria Land. In: *British Antarctic Expedition 1907–09, Reports on the Scientific Investigations*, Vol. 1, Geology, Heinemann, London.

Denton, G.H., Armstrong, R.L. and Stuiverm, M. 1971. The Late Cenozoic glacial history of Antarctica. In: K.K. Turekian (ed.) *The Late Cenozoic Glacial Ages*, Yale University Press, New Haven, pp. 267–306.

Denton, G.H., Prentice, M.L., Kellogg, D.E. and Kellogg, T.B. 1984. Late Tertiary history of the Antarctic ice sheet: Evidence from the Dry Valleys. *Geology*, **12**, 263–267.

Denton, G.H., Sugden, D.E., Marchant, D.R., Hall, B.L. and Wilch, T.I. 1993. East Antarctic ice sheet sensitivity to Pliocene climatic change from a Dry Valleys perspective. *Geogr. Ann.*, **75A**, 155–204.

Drewry, D.J. 1976. Sedimentary basins of the East Antarctic craton from geophysical evidence. *Tectonophysics*, **36**, 301–314.

Drewry, D.J. 1982. Ice flow, bedrock and geothermal studies from radio-echo sounding inland of McMurdo Sound, Antarctica. In: C. Craddock (ed.) *Antarctic Geoscience*, University of Wisconsin Press, Madison, pp. 977–983.

Drewry, D.J. (ed.) 1983. *Antarctica: Glaciological and Geophysical Folio*, Cambridge University Press, Cambridge.

Ferrar, H.T. 1907. *Report on the Field Geology of the Region Explored During the 'Discovery' Antarctic Expedition, 1901–4, National Antarctic Expedition 1901–4, Natural History*, vol. 1, *Geology*, British Museum, London.

Fitzgerald, P.G. 1992. The Transantarctic Mountains of Southern Victoria Land: the application of fission track analysis to a rift shoulder uplift. *Tectonics*, **11**, 634–662.

Fitzgerald, P.G. 1994. Thermochronologic constraints on post-Paleozoic tectonic evolution of the central Transantarctic Mountains, Antarctica. *Tectonics*, **13**, 818–836.

Fitzgerald, P.G., Sandiford, M., Barrett, P.J. and Gleadow, A.J.W. 1986. Asymmetric extension associated with uplift and subsidence in the Transantarctic Mountains and Ross Embayment. *Earth Planet. Sci. Lett.*, **81**, 67–78.

Gilchrist, A.R. and Summerfield, M.A. 1990. Differential denudation and flexural isostasy in formation of rifted-margin upwarps. *Nature*, **346**, 739–742.

Gilchrist, A.R. and Summerfield, M.A. 1994. Tectonic models of passive margin evolution and their implications for theories of long-term landscape development. In: M.J. Kirkby (ed.) *Process Models and Theoretical Geomorphology*, Wiley, New York, pp. 55–84.

Gleadow, A.J.W. and Fitzgerald, P.G. 1987. Uplift history and structure of the Transantarctic Mountains and new evidence from fission track dating of basement apatites in the Dry Valleys area, Southern Victoria Land. *Earth Planet. Sci. Lett.*, **82**, 1–14.

Grindley, G.W. 1967. The geomorphology of the Miller

Range, Transantarctic Mountains with notes on the glacial history and neotectonics of East Antarctica. *N.Z. J. Geol. Geophys.*, **10**, 557–598.

Gunn, B.M. and Warren, G. 1962. Geology of Victoria Land between the Mawson and Mulock Glaciers, Antarctica. *Bull. N.Z. Geol. Surv.*, **71**.

Heimann, A., Fleming, T.H., Elliot, D.H. and Foland, K.A. 1994. A short interval of Jurassic continental flood basalt volcanism in Antarctica as demonstrated by ^{40}Ar/^{39}Ar geochronology. *Earth Planet. Sci. Lett.*, **121**, 19–41.

Ishman, S.E. and Rieck, H.J. 1992. A Late Neogene Antarctic glacio-eustatic record, Victoria Land Basin margin, Antarctica. In: J.P. Kennett and D.A. Warnke (eds) *The Antarctic Paleoenvironment: A Perspective on Global Change, 1, Ant. Res. Ser.*, **56**, 327–347.

Ivy-Ochs, S., Schlüchter, C., Kubik, P.W., Dittrich-Hannen, B. and Beer J. 1995. Minimum ^{10}Be exposure ages of early Pliocene for the Table Mountain plateau and the Sirius Group at Mount Fleming, Dry Valleys, Antarctica. *Geology*, **23**, 1007–1010.

Jones, J.B. and Fitzgerald, M.J. 1984. Extensive volcanism associated with the separation of Australia and Antarctica. *Science*, **226**, 346–348.

Kerr, A.R. and Gilchrist, A.R. 1996. Glaciation, erosion and the evolution of the Transantarctic Mountains, Antarctica. *Ann. Glaciol.*, **23**, 303–308.

Marchant, D.R., Swisher, III, C.C., Lux, D.R., West, Jr, D.P. and Denton, G.H. 1993a. Pliocene paleoclimate and East Antarctic ice-sheet history from surficial ash deposits. *Science*, **260**, 667–670.

Marchant, D.R., Denton, G.H. and Swisher III, C.C. 1993b. Miocene–Pliocene–Pleistocene glacial history of Arena Valley, Quartermain Mountains, Antarctica. *Geogr. Ann.*, **75**A, 269–302.

McGinnis, L.D. (ed.) 1991. Dry Valleys Drilling Project. *Ant. Res. Ser.*, **33**, AGU, Washington, DC.

McGinnis, L.D., Bowen, R.H., Erickson, J.M., Allred, B.J. and Kreamer, J.L. 1985. East–West Antarctic boundary in McMurdo Sound. *Tectonophysics*, **114**, 341–356.

McKelvey, B.C., Webb, P.-N., Harwood, D.M. and Mabin, M.C.G. 1991. The Dominion Range Sirius Group: A record of the late Pliocene–early Pleistocene Beardmore Glacier. In: M.R.A. Thomson, J.A. Crane and J.W. Thomson (eds) *Geological Evolution of Antarctica*, Cambridge University Press, New York, pp. 675–682.

Mercer, J.H. 1968. Glacial geology of the Reedy Glacier area, Antarctica. *Geol. Soc. Amer. Bull.*, **79**, 471–486.

Nichols, R.L. 1971. Glacial geology of the Wright Valley, McMurdo Sound. In: L.O. Quam (ed.) *Research in the Antarctic*, American Association for the Advancement of Science, Washington, DC, pp. 293–340.

Powell, R.D., Dawber, M. and McInnes, J.N. 1996. Observations of the grounding-line area at a floating glacier terminus. *Ann. Glaciol.*, **22**, 217–223.

Prentice, M.L., Bockheim, J.G., Wilson, S.C., Burckle, L.H., Hodell, D.A., Schlüchter, C. and Kellogg, D.E. 1993. Late Neogene Antarctic glacial history: Evidence from central Wright Valley. In: J.P. Kennett and D.A. Warnke (eds) *The Antarctic Paleoenvironment: A Perspective on Global Change, 2, Ant. Res. Ser.*, **60**, 207–250.

Priestley, R.E. 1909. Scientific results of the western journey. In: E.H. Shackleton (ed.) *The Heart of the Antarctic*, Vol. 2, Heinemann, London, pp. 315–333.

Robert, C. and Chamley, H. 1992. Late Eocene–early Oligocene evolution of climate and marine circulation: deep-sea clay mineral evidence. In: J.P. Kennett and D.A. Warnke (eds) *The Antarctic Paleoenvironment: A Perspective on Global Change, 1, Ant. Res. Ser.*, **56**, 97–117.

Robinson, E.S. and Splettstoesser, J.F. 1984. Structure of the Transantarctic Mountains determined from geophysical surveys. In: M.D. Turner and J.F. Splettstoesser (eds) *Geology of the Central Transantarctic Mountains, Ant. Res. Ser.*, **36**, 119–162.

Robinson, P.H. 1984. Ice dynamics and thermal regime of Taylor Glacier, south Victoria Land, Antarctica. *J. Glaciol.*, **30**, 153–160.

Selby, M.J. 1971. Slopes and their development in an ice-free, arid area of Antarctica. *Geogr. Ann.*, **53**A, 235–245.

Selby, M.J. 1974. Slope evolution in an Antarctic oasis. *N.Z. Geogr.*, **30**, 18–34.

Selby, M.J. and Wilson, A.T. 1971. Possible Tertiary age for some Antarctic cirques. *Nature*, **229**, 623–624.

Smith, A.G. and Drewry, D.J. 1984. Delayed phase change due to hot aesthenosphere causes Transantarctic uplift. *Nature*, **309**, 536–538.

Steed, R.H.N. 1983. Structural interpretations of Wilkes Land. In: R.L. Oliver, P.R. James and J.B. Jago (eds) *Antarctic Earth Science*, Cambridge University Press, Cambridge, pp. 567–572.

Stern, T.A. and ten Brink, U.S. 1989. Flexural uplift of the Transantarctic Mountains. *J. Geophys. Res.*, **94**, 10 315–10 330.

Sugden, D.E. 1992. Antarctic ice sheets at risk? *Nature*, **359**, 775–776.

Sugden, D.E., Denton, G.H. and Marchant, D.R. 1991. Subglacial meltwater channel systems and ice sheet overriding, Asgard Range, Antarctica. *Geogr. Ann.*, **73**A, 109–121.

Sugden, D.E., Denton, G.H. and Marchant, D.R. (eds) 1993. The case for a stable East Antarctic ice sheet. *Geogr. Ann.*, **75**A, 351 pp.

Sugden, D.E., Denton, G.H. and Marchant, D.R. 1995a. Landscape evolution of the Dry Valleys, Transantarctic Mountains: tectonic implications. *J. Geophys. Res.*, **100**, 9949–9967.

Sugden, D.E., Marchant, D.R., Potter, N., Souchez, R., Denton, G.H., Swisher III, C.C. and Tison, J.-L. 1995b. Miocene glacier ice in Beacon Valley, Antarctica. *Nature*, **376**, 412–414.

Taylor, G. 1922. *The Physiography of McMurdo Sound and Granite Harbour Region*, British Antarctic (Terra Nova) Expedition, 1910–13, Harrison, London.

ten Brink, U.S. and Stern, T.A. 1992. Rift flank uplifts and hinterland basins: Comparison of the Transantarctic

Mountains with the great escarpment of southern Africa. *J. Geophys. Res.*, **97**, 569–585.

ten Brink, U.S., Bannister, S., Beaudoin, B.C. and Stern, T.A. 1993. Geophysical investigations of the tectonic boundary between East and West Antarctica. *Science*, **261**, 45–50.

Tessensohn, F. and Worner, G. 1991. The Ross Sea rift system. In: M.R.A. Thomson, J.A. Crane and J.W. Thomson (eds), *Geological Evolution of Antarctica*, Cambridge University Press, New York, pp. 273–277.

Tingey, R.J. 1985. Uplift in Antarctica. *Z. Geomorph., Suppl.*, **54**, 85–99.

Ugolini, F.C. and Bull, C. 1965. Soil development and glacial events in Antarctica. *Quaternaria*, **7**, 251–269.

van der Beek, P.A., Cloetingh, S. and Andriessen, P. 1994. Mechanisms of extensional basin formation and vertical motions at rift flanks: Constraints from tectonic modelling and fission track thermochronology. *Earth Planet. Sci. Lett.*, **121**, 317–330.

van der Wateren, F.M. and Verbers, A.L.L.M. 1990. Cenozoic glaciation of the Rennick Glacier area, the Everest Range and Yule Bay area, North Victoria Land, Antarctica. *Polarforschung*, **60**, 73–77.

van der Wateren, F.M. and Verbers, A.L.L.M. 1992. Cenozoic glacial geology and mountain uplift in northern Victoria Land, Antarctica. In: Y. Yoshida, K. Kaminuma and K. Shiraishi (eds) *Recent Progress in Antarctic Earth Science – Proceedings of the Sixth International Symposium on Antarctic Earth Science*, Terrapub, Tokyo, pp. 707–714.

Verbers, A.L.L.M. and van der Wateren, F.M. 1992. A glacio-geological reconnaissance of the southern Prince Albert Mountains, Victoria Land, Antarctica. In: Y. Yoshida, K. Kaminuma and K. Shiraishi (eds) *Recent Progress in Antarctic Earth Science – Proceedings of the Sixth International Symposium on Antarctic Earth Science*, Terrapub, Tokyo, pp. 715–719.

Voronov, P.S. 1964. Tectonics and neo-tectonics of Antarctica. In: R.J. Adie (ed.) *Antarctic Geology*, North-Holland, New York, pp. 692–700.

Webb, P.-N. 1994. Paleo-drainage systems of East Antarctica and sediment supply to West Antarctic rift system basins. *Terra Antarctica*, **1**, 457–461.

Webb, P.-N., Harwood, D.M. and Mabin, M.G.C. and McKelvey, B.C. 1994. Late Neogene uplift of the Transantarctic Mountains in the Beardmore Glacier region. *Terra Antarctica*, **1**, 463–467.

Webb, P.-N., Harwood, D.M., McKelvey, B.C., Mercer, J.H. and Stott, L.D. 1984. Cenozoic marine sedimentation and ice volume variation on the East Antarctic craton. *Geology*, **12**, 287–291.

Wellman, P. and Tingey, R.J. 1981. Glaciation, erosion and uplift over part of East Antarctica. *Nature*, **291**, 142–144.

Wernicke, B. 1985. Uniform-sense normal simple shear of the continental lithosphere. *Can. J. Earth Sci.*, **22**, 108–125.

Wilch, T.L., Lux, D.R., Denton, G.H. and McIntosh, W.C. 1993a. Minimal Plio-Pleistocene surface uplift in the Dry Valleys sector of the Transantarctic Mountains. *Geology*, **21**, 841–844.

Wilch, T.L., Denton, G.H., Lux, D.R. and McIntosh, W.C. 1993b. Limited Pliocene glacier extent and surface uplift in middle Taylor Valley, Antarctica. *Geogr. Ann.*, **75**A, 331–351.

Wilson, T.J. 1995. Cenozoic transtension along the Transantarctic Mountains – West Antarctic rift boundary, southern Victoria Land, Antarctica. *Tectonics*, **14**, 531–545.

Wright, C.S. and Priestley, R.E. 1922. Glaciology, *British Antarctic (Terra Nova) Expedition 1910–13*, vol. 2, Harrison, London.

15

Morphotectonic evolution of the Western Ghats, India

Y. Gunnell and L. Fleitout

15.1 Introduction

High-elevation passive margins have attracted the attention of geomorphologists since the pioneering observations of Jessen (1943) on 'marginal swells', and tentative classifications of these first-order landforms have so far relied on a variety of criteria: the age of rifting and 'maturity' (Birot, 1958; Vanney, 1982), gravity anomaly patterns (Birot, 1982), the relative importance of faulting versus monoclinal warping (Birot, 1958), the mode of piedmont development (Thomas, 1995), or the mechanical style of rifting (Gilchrist and Summerfield, 1994). Clearly, the chief objective has been to constrain three major parameters: the initial rifting process, the landform geometry and the life expectancy of these shoulder uplifts. Progress in understanding plate rifting mechanisms on the one hand, and cooling histories of rocks by fission-track thermochronology on the other, has gradually provided a unifying framework of analysis to reassess existing geomorphological passive margin typologies with the tools and concepts of plate tectonics theory. The following study of the Western Ghats makes a contribution to the ongoing debate by introducing new data on a so far patchily documented passive margin.

15.2 The Western Ghats Passive Margin Shoulder: Overview

The Western Ghats passive margin mountains fringe the west coast of peninsular India in the form of a Great Escarpment bordering the Deccan plateau and overlooking the Konkan and Malabar strips of coastal lowlands. Along-strike variation in relative relief and detailed sinuosities are due to contrasting lithological resistance to long-term erosion of broad geological units comprising the late Cretaceous flood basalts of Maharashtra, the Archean Dharwar craton in Karnataka and the Palaeozoic granulite fold belt in southern Kerala. In the region of interest (Figure 15.1), escarpment ridge crest elevations vary from 500 to 1900 m, but overall the remarkable continuity and length (~1500 km) of the Western Ghats in spite of structural variations remains the key issue, suggesting a single, post-Cretaceous process underlying both fairly uniform scarp recession and shoulder uplift.

15.3 Theoretical Models

Among the possible first-order mechanisms responsible for rift-flank uplift at passive margins, a variety of speculative models have been proposed in the literature (see reviews in Keen and Beaumont (1990) and Summerfield (1991a)), although offshore basin subsidence has tended to be a more important focus of research to date (see Chapter 3). Thermally driven models, for instance, such as active rifting triggered by hot spot swells or secondary mantle convection, predict a degree of uplift of the rift margins, but the uplift is transient due to expected thermal and convective decay. Pure shear kinematic models do not normally emphasize uplift of the onshore region, while simple shear rifting mechanisms are difficult to prove along mature passive margins. Even in the case of the North Sea, one of the best documented rifted basins in

Geomorphology and Global Tectonics. Edited by Michael A. Summerfield.

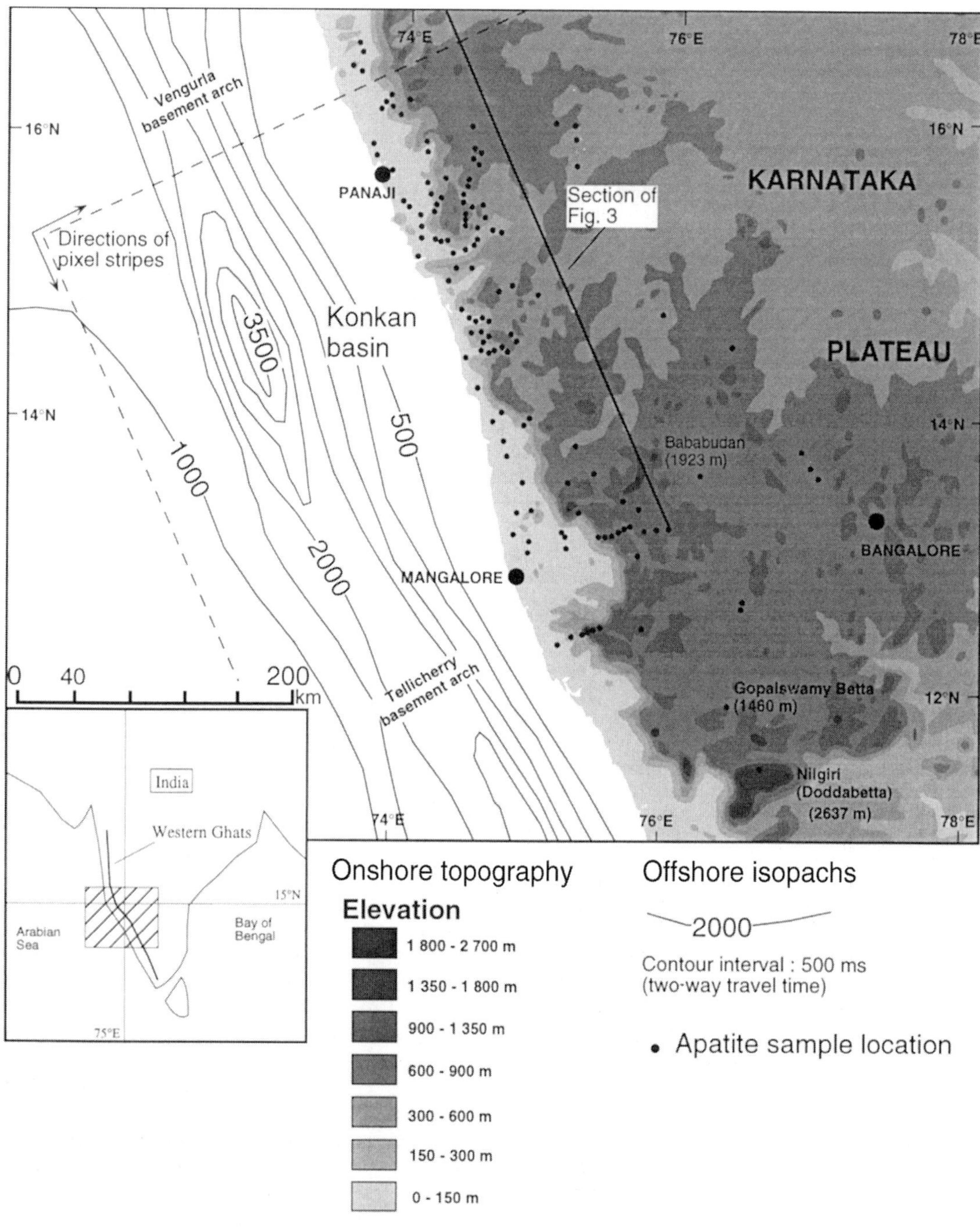

Figure 15.1 *The Indian margin: relief, apatite sample locations and offshore isopachs for Tertiary sediment (contours after Prasada Rao and Srivastava, 1984). the volume of sediment used for computing the offshore load was taken in the Konkan basin from the coastline to the 250 km offshore limit, which corresponds to the continental slope and/or the Laccadive ridge. The northern and southern limits are given by the 16°N and 12°N basement arches, where sediment thins out. The two-way-time values were converted to thickness values by considering that 1 s represents 1 km of decompacted sediment thickness*

the world, evidence for either pure or simple shear is equivocal.

Dynamic stretching models predict permanent uplift of the rift flanks and rely on the concept of lithospheric necking. The amplitude, wavelength and location of the eroding shoulder with regard to the rift axis depend upon the mechanical properties and thickness of the lithosphere (Chéry *et al.*, 1992; van der Beek *et al.*, 1994). However, the level of necking in the crust or upper mantle and the physical existence and nature of this key mechanical layer depend on the geothermal gradient and the age of the lithosphere, which need to be verified by geophysical data. It is, for instance, debatable whether the South Indian Archean lithosphere (Dharwar craton), where current geothermal gradients are 10°C km^{-1} (Verma, 1991), has a ductile lower crustal layer. Furthermore, if large uplifts of several kilometres such as those flanking the Red Sea or forming the Transantarctic Mountains are predicted to occur preferentially by rifting of thick cratonic lithosphere, it is surprising that the Western Ghats are not considerably higher than they actually are.

Lithospheric delamination models could also explain rift-flank uplift of continental margins. Delamination involves uplift caused by wasting away of a subcrustal lithospheric keel of cold mantle, which increases the buoyancy of the lithospheric and generates plateau uplift. The 'eroded' mantle is colder than the ambient oceanic mantle found at the same depth on either side of the continental mass and denser than the overlying crust. However, a recent study (Doin *et al.*, 1996) has suggested that the density of the cold mantle at the base of the thermal lithosphere is not significantly different from that of the surrounding, hotter mantle. This implies, contrary to customary assumptions made in delamination models, that the delamination process involves replacing the colder mantle by hotter mantle, which is different in terms of mineralogy, temperature and rheology, but not in overall terms of density. Hence, the process of replacing a mass of material at depth by a mass of different material which has equivalent density would not, in fact, be expected to induce surface readjustments leading to relief generation.

Finally, magmatic underplating of the crust also stands as one of the more likely candidates to explain durable surface uplift at passive margins. Underplating was, for instance, proposed to account for the high elevation of the Natal Drakensberg (Cox, 1992a, b). However, it was considered by the same author to be a very minor contributor to surface uplift (no more than 100 m, M. Widdowson, pers, comm. 1995) in the case of the Western Ghats volcanic province. Generally, the presence of underplated material has proved difficult to detect, whether by seismic procedures (Lister and Etheridge, 1989) or by gravity anomaly studies. As a cursory examination of free-air anomalies along the Indian margin confirms, gravity patterns are ambiguous in the context of thick, regionally compensated lithosphere where short wavelength gravity anomalies can be supported for long geological time spans.

In summary, many existing models may be invoked to explain the present-day Western Ghats shoulder amplitude and wavelength, the length and continuity of the escarpment, as well as its persistence in the landscape since the onset of ocean-floor spreading at 60 Ma BP. However, the western Indian margin has recorded strong erosion since the time of rifting, both in the coastal plain and on the Ghats themselves, and the need was felt to explore how such prolonged denudation could affect landscape evolution and contribute to producing the present-day topography.

15.4 Modelling Assumptions and Geomorphological Constraints

Following the idea that a thin elastic sheet resting on an inviscid asthenosphere approximates the mechanical behaviour of a cold slab of cratonic lithosphere, it has been hypothesized that higher denudational unloading rates controlled by the Arabian Sea base level, compared to the slower downwearing of the Karnataka plateau region, could generate an isostatic imbalance resulting in a flexurally supported lithospheric upwarp (Gilchrist and Summerfield, 1990). Assuming flexural isostasy, the highly asymmetrical geographical distribution pattern of long-term denudation on either side of the escarpment promotes parallel scarp retreat, thereby organizing drainage patterns and governing general landscape evolution in a self-regulating manner driven by elastic rebound (Figure 15.2). This mechanism is not only theoretically sufficient to explain the persistence of the Western Ghats escarpment, but it even predicts a degree of net surface uplift of the backslope region promoted by the massive erosion of the foreland. Implementation of the denudational model requires access to four sets of measurable parameters.

The first is the shape of the present-day observed topography. In order to avoid the fairly high variation in elevation and small-scale morphology along the strike of the Ghats (Figure 15.1) due to lithological heterogeneity and dissection by rivers, it was decided

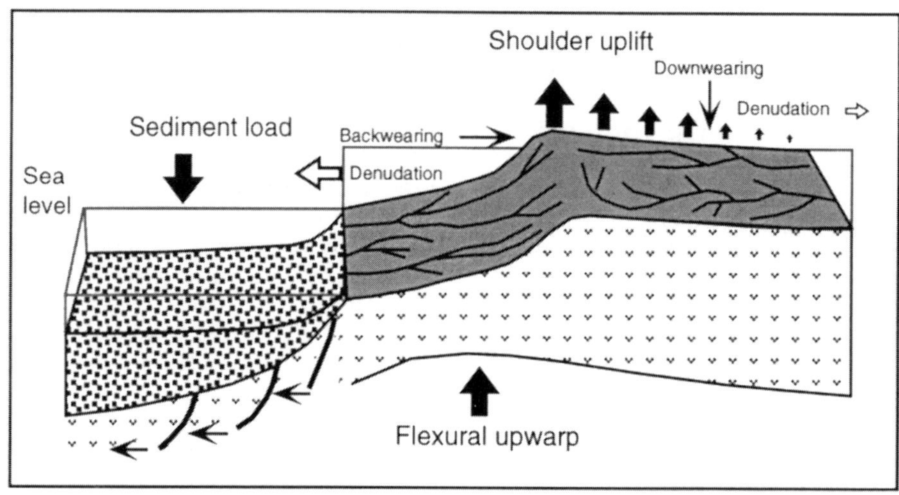

Figure 15.2 *Conceptual model of passive margin denudation and shoulder uplift by flexural rebound*

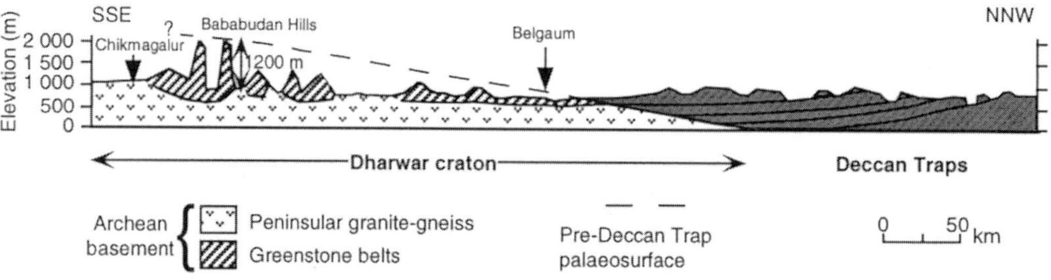

Figure 15.3 *Proposed Cretaceous palaeosurface. Sub-volcanic slope is derived from seismic data giving approximate depth to basement (from Gunnell, 1997). Minimum lost section in the Bababudan Hills region is found to be of the order of 1200 m*

that for the tentative modelling purpose of this study a mean profile would be more appropriate than an arbitrarily chosen individual profile. This was achieved by compounding a sequence of parallel topographic profiles, digitally extracted from an ETOPO 5 data base, and drawn as contiguous pixel stripes *perpendicularly* to the escarpment across the Karnataka plateau. The final curve represents the arithmetic mean, computed for each pixel stripe running *parallel* to the escarpment within the grid rectangle, of the elevation profiles. The selected grid spans four degrees of latitude (Figure 15.1). The quality of the data is mediocre in that the pixel grid misses most of the known topographic peaks and lows which, to the geomorphologist, are important markers of the true landscape evolution, but the general profile (see Figures 15.6 to 15.11) is believed to give a fair reflection of the regional elevation pattern.

Second, it is necessary to reconstitute the initial, pre-warping palaeosurface at the time of rifting (t_0 = 60 Ma in the model), in order to follow its subsequent evolution through time. It has been suggested that the few summits at 1800–1900 m on the Dharwar craton represent derived residuals (short of strictly being physical remains) of the deformed Mesozoic basement surface which was partly sealed at the K/T boundary by the Deccan flood basalts (Gunnell, 1997) (Figure 15.3). Modal elevations of the Karnataka plateau now range between 600 and 900 m, which implies that a minimal crustal section of 900–1200 m was removed during the course of the Cenozoic and Quaternary.

Third, a seismic thickness map of post-Cretaceous shelf deposits (Figure 15.1) was used to compute the volume of sediment deposited as a direct consequence of the Western Ghats scarp recession. Overall mass balance was established in the way illustrated in Figure

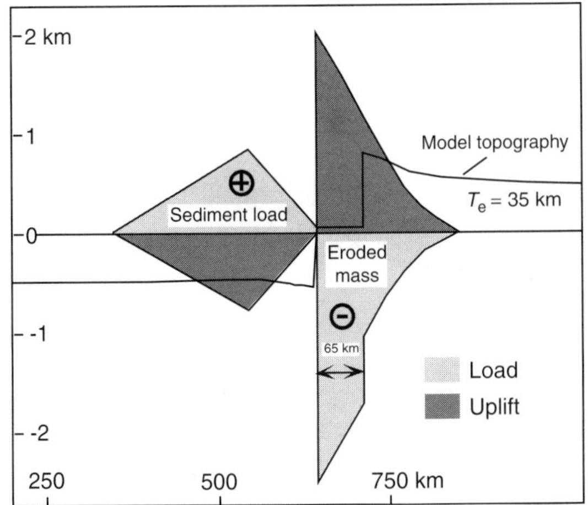

Figure 15.4 *Modelled offshore loading/onshore unloading and onshore uplift/offshore subsidence as weighted by measured fission-track and sediment thickness values*

15.4. Conveniently, only insignificant volumes of sediment are found in the Konkan onshore region, thereby facilitating the onshore/offshore dichotomy implied in the procedure. In every model run, the final sediment mass deposited offshore was set to match the observed values provided by Figure 15.1, and was therefore treated as a constant. Erosion in the source area (i.e. the Western Ghats foreland) was, however, treated as a variable. Therefore, depending on the accompanying assumptions used in the model (flexural rigidity values, continuous or 'broken' plate situation; see below), eroded mass in the source area (the light-shaded polygon forward of the escarpment on Figure 15.4) and sediment load offshore (the light-shaded triangle, offshore, on Figure 15.4) may differ.

Fourth, the load on the Karnataka plateau region, which is controlled by the Bay of Bengal base level, was obtained in the form of lost section values by converting cooling rates derived from apatite fission-track analysis data into denudation rates. This low-temperature thermochronological method is now becoming increasingly used in passive margin environments (see Chapter 4). A methodological overview is provided in Gunnell and Fleitout (1998) and full analytical results of the fission-track study are detailed in Gunnell *et al.* (1999). A total set of 75 samples was used (Figure 15.1) and treated according to standard methods presented in Hurford and Green (1982), Gleadow *et al.* (1986), Laslett *et al.* (1987), Green (1988), Green *et al.* (1989) and Hurford (1990). Cooling paths (Figure 15.5) were determined from the model age and track-length data using a genetic algorithm forward modelling technique discussed in Gallagher (1995).

The 75 apatite samples yielded two distinct cooling patterns which required separate interpretation. A first set of highly consistent results involves the samples collected on the backslope of the Western Ghats, where slow, monotonous cooling (Figure 15.5a), attributed to downwearing processes, predominate. The measured geothermal gradients for the Dharwar craton of between 8 and 11°C km^{-1} (Verma, 1991) provide mean Mesozoic denudation rates over the crystalline basement of 15–20 m Ma^{-1}. Remarkably, if extrapolated into the Tertiary (as suggested by several model cooling paths in Figure 15.5a), such rates amount to a loss of section of 900 to 1200 m in the past 60 Ma. This matches extremely well the lost slab estimated by the palaeosurface mapping method.

The second set of samples, collected along the coastal foreland, gives a less consistent picture as far as cooling paths are concerned but, in a context dominated by backwearing processes, it appears in any case ambiguous to express the cooling results in terms of surface lowering and therefore of lost section by denudation. Furthermore, no data on geothermal gradients in the coastal region are available. For these reasons, the use of offshore sediment thicknesses, known to have been eroded away from the Western Ghats foreland, was preferred (Figure 15.4). However, what the fission-track cooling paths for the majority of coastal and escarpment samples *do* reveal is a belated, mid-Tertiary acceleration in the rate of cooling (Figure 15.5b) which is not observed on the backslope of the Ghats. This acceleration signal, 35 to 40 Ma after the recorded rifting and flood basalt event at ~65 Ma BP, comes very late after the peak of tectonic activity. It is therefore doubtful whether it should at all be interpreted as a direct response of the denudation system to the new Arabian Sea base level. Indeed, in other rifted margin regions studied by fission-track analysis, such as southern Africa and the Red Sea, the erosional signal subsequent to rifting is recorded within 5 to 20 Ma. In the present case, there is evidence to suggest (Subrahmanyam *et al.*, 1995) that the rift hinge lies 250–300 km to the east of the present coastline, between Madagascar and the Laccadive micro-continents. Although this could explain that no clear-cut signature of the rifting event was registered in the fission-track record of mainland India (Kalaswad *et al.*, 1993), it remains difficult to conceive any genetic relationship between the rift hinge and the present-day escarpment.

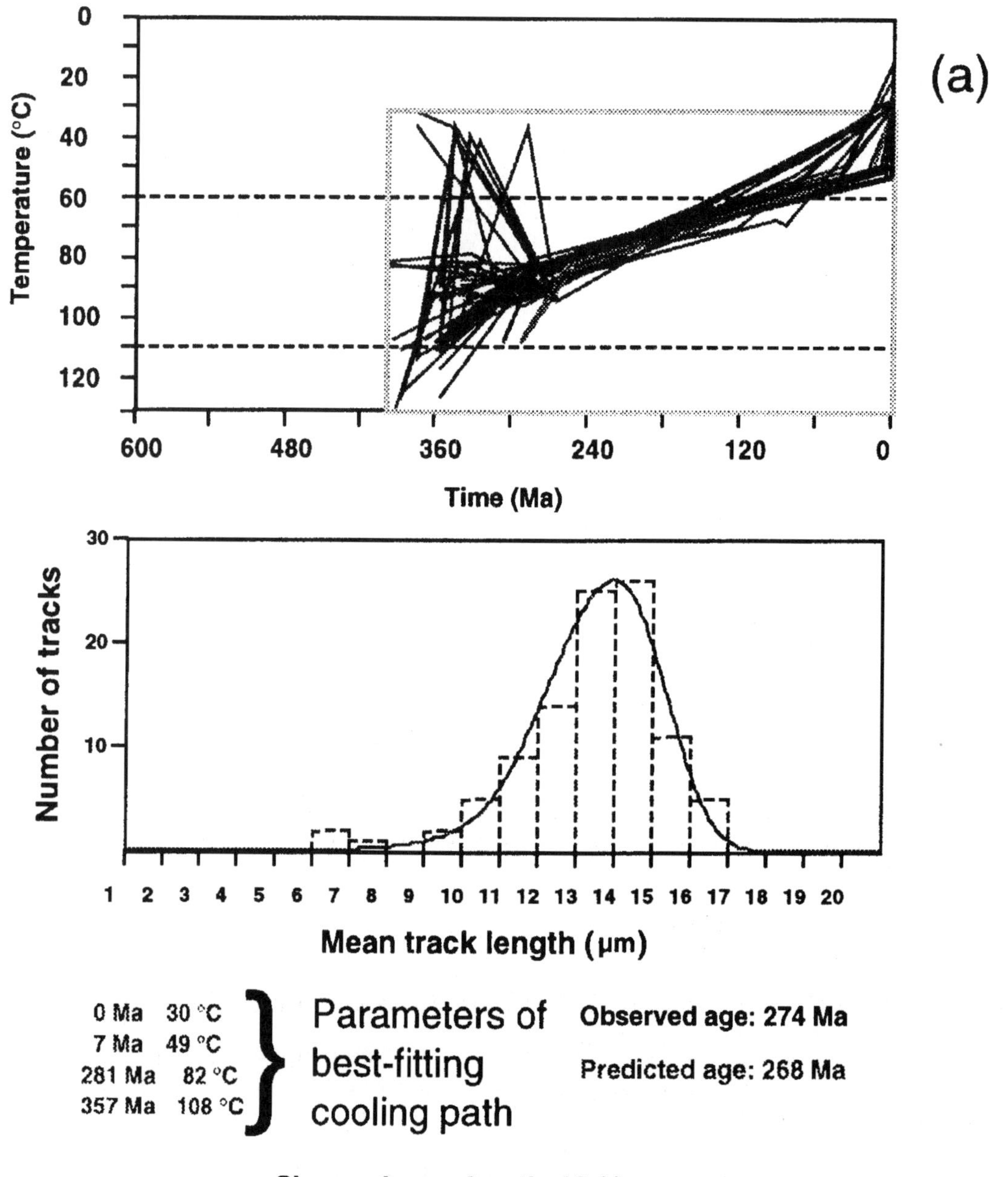

Figure 15.5 Typical cooling curves for (a) backslope and (b) coastal samples. The time–temperature curves show the trend of the 100 best-fitting genetic algorithm iterations (out of a total of 2000) as constrained by the track-length histogram and its best-fitting ('predicted') curve, shown below. Plot (b) shows the marked acceleration in cooling after 31 ± 11 Ma BP

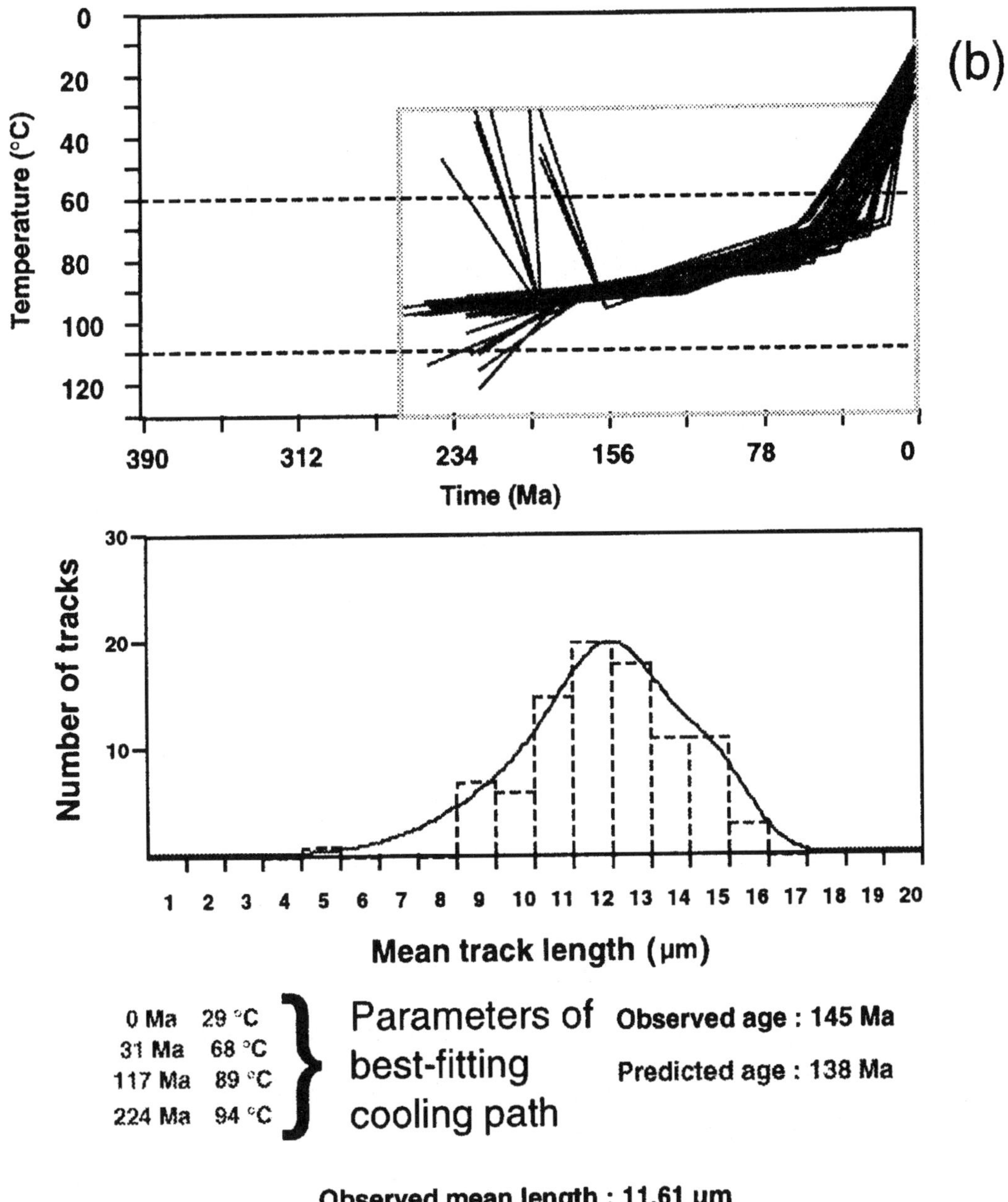

Figure 15.5 (*continued*)

Alternative evidence suggests that the boundary between stretched and unstretched crust lies somewhere quite close to the present-day coastline. A 'West Coast' rift-bounding fault, though contested by some (see review in Chandrasekharam, 1985), has long been held by geologists to mark the initial position of the escarpment, and is usually mapped quite close to the current coastline. In view of this evidence, the key factor appears to reside in the notion that the original Indian rifted margin displayed fairly low relief: the reason why cooling/denudation rates in the coastal region do not significantly deviate from the values found on the backslope of the Western Ghats until the Oligocene (Figure 15.5) can thus be explained by the fact that, below a certain threshold of continental freeboard, the resolution of the fission-track method is too coarse to record any related peak of denudation. This observation derives from the notion that the movement of rocks in fission-track analysis occurs relative to a thermal reference frame, but is also relief-dependent. Unlike the more usual 'mineral pair' method used in geochronology, the 'relief method' in apatite fission-track analysis uses the difference in cooling age for two samples of the same mineral collected at different elevations. When two samples separated by less than 500 m of relative elevation yield fission-track ages that are similar *within analytical error* (i.e. $\pm 2\sigma$), it cannot be asserted with confidence that the higher sample passed through the critical isotherm significantly earlier than the lower one. It therefore has to be inferred that the two adjacent crustal columns did not experience significantly different denudation rates, and that insufficient relief contrast between the two was the major control on the rate of denudational cooling.

The acceleration in denudation in the Oligocene must then be taken as evidence that a threshold of coastal unloading and escarpment relief amplification was exceeded around that time, leading to a subsequent increase in rates of denudation. This timing constraint may reflect the fact that the lateral distance of headward erosion at ~30 Ma BP had reached a threshold whereby cumulative unloading since rifting had generated the first major signal of flexural response in the history of the margin, very much in the way suggested by King (1955). Only from then on did escarpment amplification and retreat create sufficient and self-regulated surface uplift to generate a more aggressive phase of denudation, expressed thereafter in the fission-track record. This notion is directly supported by the observation that large volumes of prograding Neogene terrigenous sediment are present in the offshore basins (Biswas, 1987; Whiting *et al.*, 1994; Subrahmanyam *et al.*, 1995).

A final timing constraint is set by the steady aridification of the interior Deccan since the late Miocene (Gunnell, 1996; Gunnell and Bourgeon, 1998) despite a regional backdrop of southwest monsoon intensification (Prell and Niitsuma, 1989) attributed to Himalayan uplift. This apparent contradiction requires that the progressive uplift of the Ghats attained a critical threshold of surface elevation by the late Tertiary, allowing the rift shoulder to act as an effective climatic barrier across the path of the monsoonal airflow.

15.5 Flexural Model and Erosional History

The view that the late Cretaceous, pre-Deccan trap peneplain of Greater India had an average elevation no greater than 400–500 m, for the reasons mentioned above, permits us to set the initial K/T boundary topography of the modelled margin at ~500 m. In the context of the currently observed mean elevation of 720 m on the shoulder itself (see ETOPO profiles on Figures 15.6 to 15.11), such an elevation is already significant. The initial Dharwar craton topography was further arbitrarily modelled as a flat surface, setting optimal conditions for subsequent uplift and generation of the observed shoulder effect. Clearly, this assumption remains unproven, and young rift flanks usually exhibit a form of thermally induced warping. However, constraints on the rift geometry of a mature passive margin are inevitably speculative and may ultimately not be of critical importance (see discussion).

Using these assumed initial conditions, a finite difference computer model was written with a view to simulating, for different input values of lithospheric rigidity, the form of the flexural shoulder as it evolved during the course of the Tertiary. Variability was introduced into the model for the other parameters such as width of the coastal plain, location of the flexural node and initial elevation of the rift flank. On the backslope, the total lost section of 1–1.2 km was removed as a load along the crest of the Ghats decreasing to a minimum in the Deccan interior (light-shaded area eastward of the escarpment in Figure 15.4) at the longitude where the upland surface falls to its lowest mean elevation (~500 m) before rising again slightly over the Eastern Ghats hills.

The model simulation behaves in the following way: from an initial position at t_0 = 60 Ma, with a relief of

500 m and a flat-topped Deccan palaeosurface, the escarpment retreats by 65 km (assumed distance to the West Coast Fault) as a succession of ten 6.5 km increments, while progressively discharging the mass of crustal material to the continental shelf. The offshore basin simultaneously subsides under the load of incoming sediment. Meanwhile, the backslope is also progressively stripped by increments, with the eroded material feeding eastwards to the Bay of Bengal. The erosion law governing the denudation processes both in the uplands and on the foreland was chosen to be proportional, at each successive step of the simulation, to the increasing local relief between the surface elevation h reached at the iteration of interest and the initial reference elevation h_0 at t_0. This initial reference level, to the west of the retreating escarpment, is 0 m (mean sea level, supposed unchanged), and to the east, on the uplifting plateau surface, 500 m. This simulates the local plateau base level created by downcutting valley floors. The denudation rate Δh (in m Ma^{-1}) can thus be expressed in the form:

$$\Delta h = \alpha\,(h - h_0) \quad \text{with } \alpha \text{ in Ma}^{-1} \qquad (1)$$

The α parameter is a coefficient controlling the denudation rate. For a given flexural rigidity D, if α is too small, then erosion remains too low and produces insufficient rebound; if α is too large, erosion consumes too much of the rebound and the outcome yields insufficient relief. There is, therefore, an optimal, intermediate value of α for which maximum surface uplift is expected.

Linear diffusion, although widely used for denudational models in the literature, is clearly inoperative in a passive margin context where a large escarpment is not only maintained by parallel retreat but effectively grows with time. The overriding influence of local relief, a substitute for mean slope value, on denudation is also in keeping with empirical correlations established for the region (Gunnell, 1996a) as well as previous work in other regions (Ahnert, 1970; Summerfield, 1991b). Since the local relief on the seaward side at t_0 is 500 m while being zero in the hinterland, the rate of erosion in the coastal region remains from the start proportionately higher than on the backslope, an asymmetry which contributes to the maintenance of elevations in the coastal plain realistically close to sea level while, at the same time, providing a means of explaining the persistence of the escarpment relief.

Effective elastic thickness (T_e) of the lithosphere was obtained by trial and error. Flexural rigidity (D) is given by the relation:

$$T_e = \sqrt[3]{\frac{12(1-\nu^2)D}{E}} \qquad (2)$$

where E is Young's modulus and ν is the Poisson coefficient.

In addition to finding a value for flexural rigidity, it was also desirable to test the relative merits of an infinite elastic plate model, where the plate is modelled as a continuum from the cratonic interior to the oceanic lithosphere, and a semi-infinite plate model, where a form of decoupling is introduced between the onshore lithosphere and the margin in the form of a narrow segment where the value for D is very low. This simulates a faulted hinge zone with minimal flexural strength and is hereafter referred to, for simplicity, as the 'broken' plate scenario. In the context of a rifted margin, where a degree of crustal thinning (and therefore mechanical weakening) occurs below the continental shelf, the second alternative was deemed the more realistic of the two. Recent research (Subrahmanyam *et al.*, 1995) has confirmed the existence of a large number of discrete basins and ridges between the coast and the Laccadives, supporting the assumption of a very weak lithosphere. In both the infinite and semi-infinite plate scenarios, a bimodal elastic thickness was assumed (Watts, 1992), with a high value for the continent and a low value for the shelf and its oceanic continuation. Numerical results are illustrated in Figures 15.6 to 15.11.

In the case of a continuous elastic plate with a reduced elastic thickness in the offshore region ($T_e \leq 5$ km) surface uplift after 60 Ma is not predicted to exceed 100 m due to limited rebound. As illustrated by the flexural upwarp in Figure 15.6, where an input elastic thickness of 35 km was used, the erosional rebound can only match the actual shoulder uplift geometry if it is supplemented by an internal factor of buoyancy, such as underplating. In this case, erosion would therefore appear to explain only a small fraction of the overall uplift of the Western Ghats, indicating the partial inadequacy of a purely denudational model.

It is in the 'broken' elastic plate form (Figure 15.6) that the model attains its best performance in simulating the observed shoulder uplift evolution. For an elastic thickness value of 35 km the rebound reaches the desired maximum 300 m of mean surface uplift although the observed topography wavelength is slightly underestimated.

It appears from Figure 15.6 that the best-fitting flexural rigidity for the Dharwar craton is somewhere in the vicinity of 2.7×10^{23} N m. The 35 km value is

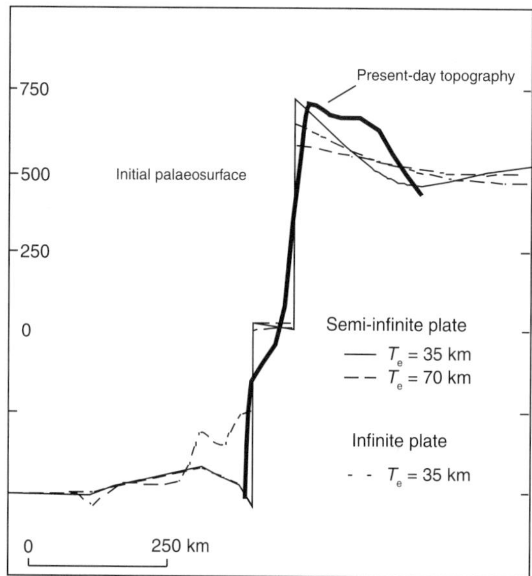

Figure 15.6 *Model topography for three different lithospheric configurations. The infinite plate scenario reveals insufficient rebound compared to the semi-infinite plate alternative. The latter appears to provide a better overall match to the mean onshore topographic profile. Note that vertical scale, as for all following figures, is highly exaggerated for clarity (elevation/horizontal distance ratio ≈ 1/500)*

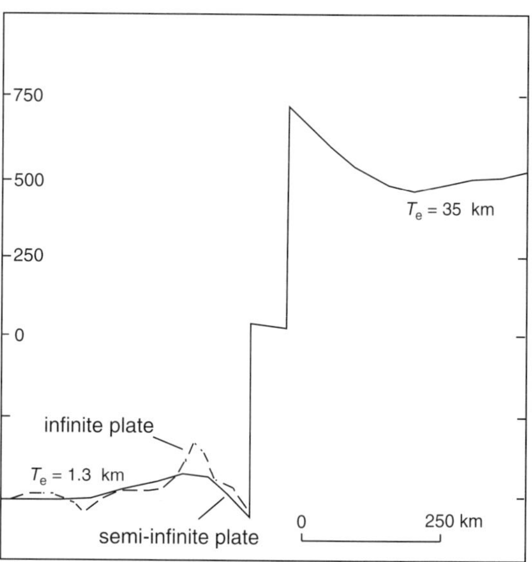

Figure 15.7 *Illustration of how onshore topography remains unaffected by choice of a 'broken' or an infinite elastic sheet model providing near-Airy isostasy*

already high, but not surprisingly so for an Archaean craton. Average values of T_e reaching 80 km were formerly postulated by Lyon-Caen and Molnar (1983) for the North Indian collisional margin and Watts and Cox (1989) for the Deccan trap region. Manglik and Singh (1992) inferred similar values from the present-day thermal structure of the Indian lithosphere. However, these imply a flexural rigidity one order of magnitude higher than the value proposed here. D varies as a cubic function of T_e, which explains the great sensitivity of this parameter to small variations in D, and raises questions as to which might be the most robust method of estimating the parameters of interest. However, on running the model with a 'broken' lithosphere plate, it was found that the wavelength and total uplift of the Dharwar craton is only moderately sensitive to variations in flexural rigidity, as illustrated in Figure 15.6. Therefore, as revealed by the goodness of fit in Figure 15.6, the exceptionally high T_e values proposed by the above-mentioned authors could, in retrospect, find some credibility if the appropriate modelling precautions are taken.

If, in turn, the rigidity of the shelf and oceanic lithospheres are held low while keeping the continental T_e constant at 35 km, Figure 15.7 reveals that the presence or absence of a fault makes no difference to the onshore topographic profile. In other words, as long as near-Airy isostasy is assumed offshore, the model onshore topography proves to be insensitive to the precise status (infinite or 'broken') of the plate. In the particular case of a 'broken' plate model, it was found that the onshore topography was fairly insensitive to T_e values offshore, which were set as high as 17.5 km without any significant overall change.

Finally, Figures 15.8 to 15.10 show successive iterations of scarp retreat and mean surface uplift for different input T_e values. Time labels mark successive locations of the 'great escarpment' in the Cenozoic in the case of an arbitrary steady rate of scarp recession (1.08 km Ma^{-1}). The effect is portrayed better still when maximum elevation is considered (Figure 15.11), reminding us that residual massifs reaching 1900–2000 m (Figures 15.1 and 15.3) survive as short wavelength, flexurally supported loads resting on top of the mean topography. The fission-track, sedimentary and palaeoclimatic constraints in fact suggest a tendency for rebound to accelerate with increasing escarpment relief (see above), implying that the rate of scarp recession actually increased with time; this interpretation can be satisfied by simply changing the time labels on Figures 15.8 to 15.11 and 15.13 since the iterations in the model, although reflecting regularly spaced time

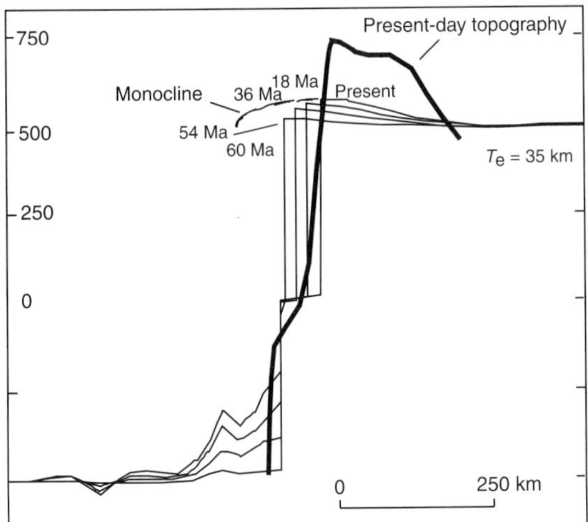

Figure 15.8 *Cenozoic scarp retreat and mean surface uplift by denudational rebound in the case of an infinite elastic plate with continental T_e = 35 km. Note low shoulder and monoclinal flexure*

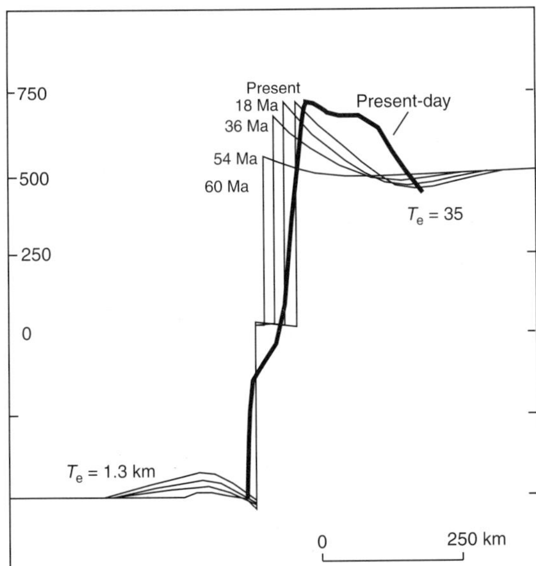

Figure 15.9 *Cenozoic scarp retreat and mean surface uplift by denudational rebound in the case of a semi-infinite elastic plate with continental T_e = 35 km. Note high shoulder, 'concave-up' profile and fairly good fit to observed topography*

intervals, are independent of actual rates of scarp retreat (and therefore of geological time). The role of the Western Ghats as a climatic barrier to the monsoon, as is the case today, may therefore have taken effect at some time in the Neogene.

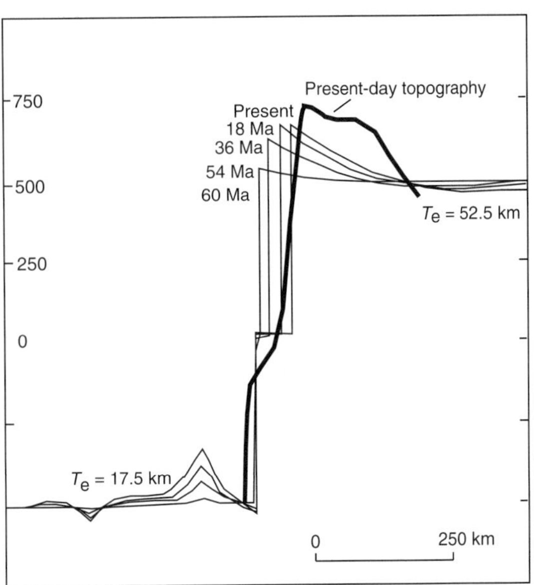

Figure 15.10 *Cenozoic scarp retreat and mean surface uplift by denudational rebound in the case of a semi-infinite elastic plate with continental T_e = 52.5 km. Note change in profile shape but relative insensitivity to maximum rebound value when compared to Figure 15.9*

15.6 Discussion

15.6.1 THE LITHOSPHERIC MODEL: STRENGTHS AND LIMITATIONS

Our proposed denudational model succeeds in fitting realistic geophysical parameters to field and fission-track data. As in all models, simplifying assumptions and arbitrary boundary conditions are introduced. Among the less contentious of these is the assumption of flexural rather than local isostasy in the context of a Precambrian craton. Some of the more contentious ones in the following list may serve as a basis for future discussion in similar contexts.

1. The thin elastic sheet assumption is often accepted for oceanic lithospheres but sometimes challenged for continental environments on grounds that viscoelastic (e.g. Beaumont, 1981) or viscoplastic models are more realistic – although arbitrary assumptions on the rheology of the lithosphere cannot be avoided. To others, the notion of effective elastic thickness has a concrete physical meaning in continental as well as oceanic lithosphere (Burov and Diament, 1995). In the present

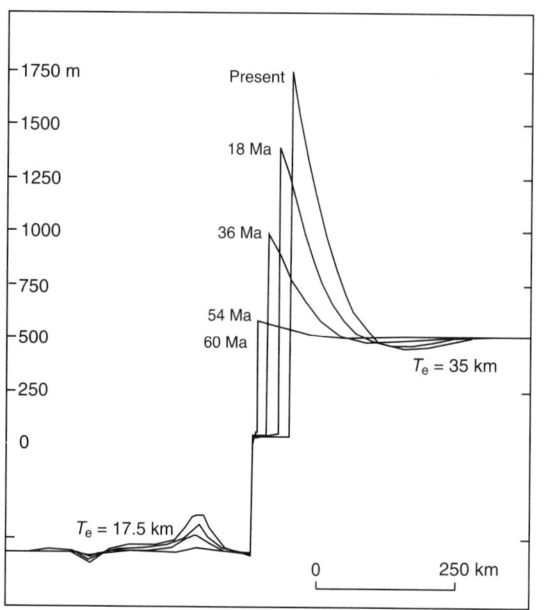

Figure 15.11 *Cenozoic scarp retreat and expected surface uplift of points where denudation is minimal (due, for instance, to more resistant lithologies). Parameters are equivalent to Figure 15.9 (a semi-infinite elastic plate with continental* T_e = 35 km*). The high summits reflect the present-day position of the Kudremukh and Bababudan massifs (Figure 15.3) and illustrate the growing importance of the Western Ghats as a climatic barrier to the monsoon during the Cenozoic*

case of an Archaean craton, this issue is not particularly problematic since most authors agree that elastic thickness is correlated with the age of the last thermotectonic event and fission-track analysis reveals that this event is at least Palaeozoic, if not older. Assumptions usually made when loading an oceanic plate may therefore be applied.

2. The model assumes, partly on the basis of trends observed in the fission-track results for a small number of control samples located in interior Karnataka, that denudation on the Deccan plateau was lower in the central peninsular region (flexural node) than on the immediate backslope of the Ghats. This would require to be substantiated by apatite sampling across the entire South Indian region. Nevertheless, additional simulation runs were performed in which a uniform erosion rate on the entire Deccan plateau was introduced: unlike in Figure 15.4, a uniform slab with a thickness of 1 km is removed over the 60 Ma period (Figure 15.12). The erosion law takes the form:

$$\Delta h = \alpha\,(h - h_0) + \beta \quad \text{with } \beta \text{, a constant.} \qquad (3)$$

The results (Figure 15.13) reveal that this hypothesis, which amounts to bulk plateau uplift, does not significantly alter the final outcome. Indeed, it appears that greater onshore elastic thicknesses (e.g. 70 km) yield better fits than lower values, thus reinforcing the case made by other authors that lithospheric rigidities for Archaean cratons are high (see above) and that denudation values in cratonic interiors may be higher than sometimes proposed (Brown *et al.*, 1994). Overall, the outcome illustrated in Figure 15.13 achieves the best-fitting rift shoulder profile in this study.

3. It also remains uncertain what exact volume of offshore sediment it is acceptable to reload onto the onshore coastal region. In the absence of sufficiently detailed well logs, it is impossible to be selective about lithologies. Knowing that the lower Tertiary was dominated by aggrading carbonate platform sedimentation (e.g. Whiting *et al.*, 1994), it is debatable whether thicknesses of non-clastic material should be reloaded onto the continent. In a tropical environment, it can however be argued that solute load in streams is far from negligible and that its delivery to the sea significantly contributes to the growth and proliferation of marine organisms. The question therefore remains open as how to quantify correctly over time the input from continental sources to shelf carbonate biomass accumulation.

4. Although the infinite elastic plate model was rejected for not yielding a sufficiently good topographic fit and/or a sufficiently realistic continental elastic thickness, the flexural camber obtained in this version of the model (Figure 15.8) raises the question of the tectonic *style* of deformation induced by flexure: indeed, the favoured version of the model (Figures 15.9 and 15.13) imposes an overall 'bending up' of the craton, which stands in apparent contradiction to recent work by Mitchell and Widdowson (1991) and Widdowson and Cox (1996). Following a technique of geochemical fingerprinting of Deccan lava units, these authors concluded that the lava pile displays a slight monoclinal camber over the Ghats escarpment towards the Arabian Sea. This suggests a 'bending down' of the craton. Dessai and Bertrand (1995) take the view that geochemical tracers should not be allowed to obscure the existence of faulted and rotated blocks in the Bombay region, which could explain the apparent monoclinal structure of the

Panvel flexure observed in that region (a feature superficially reminiscent of the Lebombo monocline in South Africa). The broader issue of lava flow dips, both in relation to the volcanic load at the time of eruption (Watts and Cox, 1989) and to later shoulder upwarp, is evidently complex and has not been satisfactorily resolved. Although the favoured outcome of our model apparently conflicts with the aforementioned evidence from the Deccan traps, the 'downwarp' alternative (Figure 15.8) leaves room none the less for possible agreement, pending further investigation. It imposes, however, a more piecemeal approach: for instance, Whiting *et al.* (1994) speculated that the Indus fan load contributed to the uplift of the Western Ghats, thereby also suggesting a degree of coupling between the offshore and onshore regions. However, the contribution to uplift was not predicted to exceed 60–80 m, and, moreover, the crystalline section of the Ghats of concern here was mapped as being located outside the range of the Indus fan forebulge effect. Considering that the Western Ghats extend as far as Cape Comorin and increase in elevation in that direction, another explanation must be found. The possibility of underplating is currently being explored by one of us (LF) through an integrated analysis of gravity patterns.

5. Ultimately, the success of the model hinges on the key issue of mechanical decoupling of the continent from the margin. In the case of a continuous plate, allowance in flexural models is rarely made for the consequences of post-rift passive cooling of the stretched offshore lithosphere. Cooling, however, implies the generation of excess density on the continental shelf, with an entrainment effect on the rest of the margin which is likely to reduce even more the (already modest) onshore uplift. Necking models provide a partial solution to this by predicting substantially elevated rift shoulders at the outset, and the belief that the initial topography at break-up consisted of a flat, regional planation surface is admittedly tenuous. However, in any model applied to a mature passive margin (as opposed, for instance, to the Red Sea rift), initial topography is probably the most difficult parameter to constrain. That much of Gondwana had reached a stage of advanced planation towards the end of the Mesozoic is a commonly held belief and is, on the balance of global geomorphological evidence, more difficult to refute than to accept. Nevertheless, it is virtually impossible to ascertain that the Mesozoic surface was not, for instance, warped by pre-rift updoming due to the Reunion mantle plume intumescence (Cox, 1989), therefore weakening the idea that the currently observed easterly tilt and drainage pattern of peninsular India was acquired solely in the Cenozoic as a consequence of denudational rebound along the western continental margin.

The present study, however, suggests that the aforementioned caveats are less of a problem if a form of mechanical decoupling is introduced: indeed, the 'broken' plate model relies so heavily on the denudational driving force that the passive cooling of the shelf region and the exact shape of the initial topography, unlike the situation with continuous plate models, are of little consequence to the final margin morphology, and are ultimately irrelevant.

15.6.2 FISSION-TRACK DATA IN GLOBAL GEOMORPHOLOGY: ADVANTAGES AND DISADVANTAGES

In addition to the contentious points discussed above, and to which additional data and more refined modelling may bring greater precision, the Western Ghats shoulder uplift model also provides insights into the potential of apatite fission-track analysis in shedding light both on global tectonic events and on landscape evolution (see Chapter 4). In terms of fission-track analysis, it has been argued that the technique can bracket palaeoelevations inasmuch as rapid cooling of a rifted margin in the past reflects the fact that sufficient relative relief was available to induce strong erosion: since strong erosion was not recorded until ~30 Ma after the onset of sea-floor spreading along the Indian margin (Figure 15.5b), it was inferred that the continental freeboard of India at 60 Ma BP was not in excess of 400–500 m. This finding can in turn shed light on the palaeoelevation of Gondwana and links up with the controversial question raised by Partridge and Maud (1987) on the pre-break-up elevation of southern Africa. These authors held the view that the region lay at the heart of the Gondwana landmass, which took the shape of a vast dome sloping away towards the periphery. If India is considered to be located in such a peripheral region, it could explain why the estimated freeboard is considerably lower than the 2000 m proposed by Partridge and Maud for southern Africa. The exact value of South Africa's Cretaceous palaeoelevation is, however, conjectural. For instance, ten Brink and Stern (1992) were compelled by their modelling assumptions to set the altitude of South Africa at

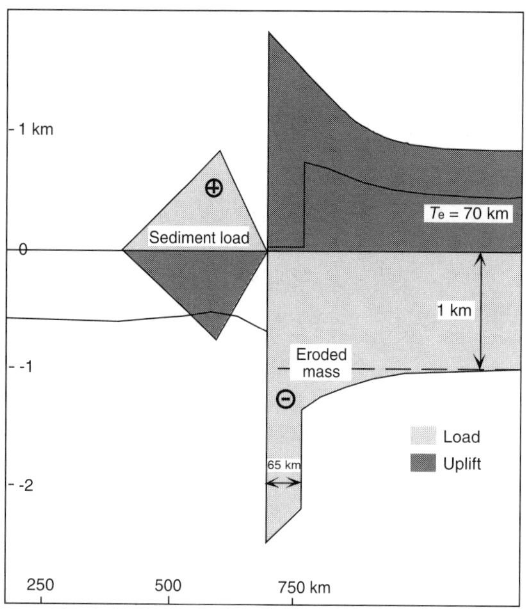

Figure 15.12 Modelled offshore loading/onshore unloading and onshore uplift/offshore subsidence for a uniformly distributed denudation thickness of 1 km on the Karnataka plateau, while maintaining the same erosion law as in the preceding examples

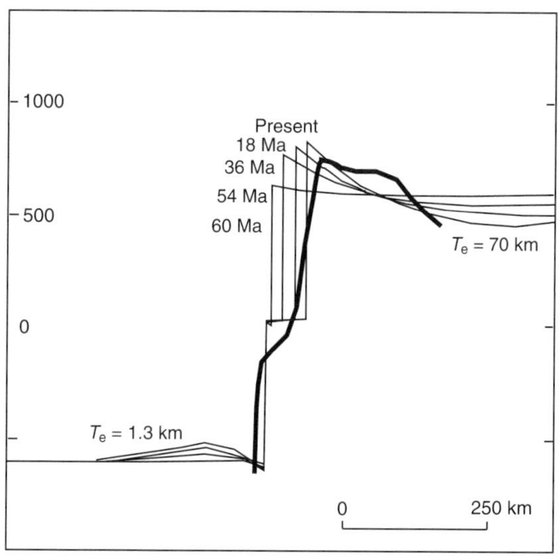

Figure 15.13 Illustration of scarp retreat and surface uplift by denudational rebound for a broken plate with high elastic thickness on the continent (70 km) and 1 km of lost section on the Karnataka plateau. Note similar outcome to previous cases (e.g. Figures 15.6, 15.9 and 15.10) and tendency to vindicate high T_e values

2500 m at the time of rifting, while Gilchrist and Summerfield (1994) opted for half that amount (1200 m). Whatever the correctness of the answers reached so far, it is the case that transgressive Mesozoic and Cenozoic marine sediments are scarce on most continental fragments of Gondwana (with the exception of Australia) by comparison with their Eurasian and North American counterparts. It therefore follows that the freeboard of many Gondwana fragments remained largely above the altitude of the highest marine transgressions, and that if the basement region of southern India was no higher than 500 m at the K/T boundary, it was presumably no lower than ~300 m (the value often taken from global eustatic curves as the maximum altitude of Cretaceous sea level). In terms of global tectonics, this implies that the behaviour of the Gondwana lithosphere differs from that of Laurasia in ways in which fission-track analysis could possibly help to unravel.

The information obtained on available relief following rifting events is, in a sense, a positive way of exploiting the resolution limits of the fission-track technique. These limits, however, also have an effective drawback: for low denudation rates and geothermal gradients characteristic of cratonic regions, fission tracks cannot record erosional signals which involve the removal of crustal sections less than ~500 m. It is one of the long-standing tenets of geomorphology that landscapes are generated in the form of denudational cycles, punctuated by pulses of uplift and rejuvenation and phases of tectonic quiescence during which lowering of divides largely ignores lithological contrast and structural diversity. Other theories have emphasized continuous and either slow or accelerating epeirogeny, predicting a variety of morphological outcomes. Clearly, tectonic regime is a major key to understanding landscape development but geomorphological theory leaves unanswered the whole question of equifinality in landscape development. For this reason, existing theories need further testing with the help of modern geochronological techniques. Fission-track analysis, which provides information on rock cooling trends, is expected to monitor accelerations or decelerations in denudation rates; if so, such variations may then be related to tectonic pulses, thereby contributing to establish the effective controls which presided over the evolution of a particular landscape. Tentative Cenozoic denudation rates on the Karnataka plateau, for instance, support the hypothesis that the observed residual summits (Figure 15.3) represent a relict Mesozoic surface preserved on the resistant Banded

Iron Formation. In support of this idea, cooling curves for Karnataka plateau samples in Figure 15.5a resolve slow cooling for the entire Mesozoic: this either vindicates a landscape development pattern driven by the slow uplift of a plateau having already low relief, accompanied by slow downwearing, or a long period of tectonic quiescence. Which of the two agrees best with the true landscape history of Gondwana for that period remains unresolved by the data, but both scenarios suggest that conditions conductive to achieving advanced Mesozoic planation were fulfilled. Using fission-track analysis in Kenya, Foster and Gleadow (1996) similarly attempted to bracket a palaeosurface level which had been previously inferred from bevelled and lateritized summits.

Discrepancies, however, begin to emerge between apatite fission-track information and field evidence for pulses of change presumed to have taken place in the Cenozoic, mainly because, by the Tertiary, upper crustal temperatures of the Deccan were too low to be confidently resolved by fission-track models (Figure 15.5a). For instance, if extrapolated to the Cenozoic (Gunnell *et al.*, 1999), the assumption of either slow plateau uplift or tectonic quiescence leaves unexplained the existence of at least one mid-Tertiary palaeosurface in the southern Deccan (Gunnell, 1997). Furthermore, partial planation later in the Neogene yielded several local levels and basins which can be identified with the help of fossil soils and drainage patterns (e.g. Gunnell and Bourgeon 1998). It seems that fission tracks in shield areas characterized by slow downwearing cannot unequivocally resolve *rates of change* in denudation rate below a rather coarse threshold. Shield regions may therefore not yet have found in fission-track analysis a fully satisfactory quantitative tool to either challenge or validate the tectonic regimes which underlie competing geomorphological theories. Fission tracks are thus equally unlikely to resolve low-magnitude denudation events (epicycles) caused, for instance, by climatic change. It may therefore also be the case that, even against the background of a continuous flexural response to progressive denudational unloading (Gilchrist and Summerfield, 1991), increments of rejuvenation below a certain threshold of magnitude can overprint cyclical manifestations of erosion which leave a mark on the landscape, but are not registered in the lithospheric response if flexural rigidity is sufficiently high. The impetus given to erosion, in the case of the Karnataka plateau, may indeed have been contributed by remote factors related to changing boundary conditions such as the Himalayan collision or Bay of Bengal sea levels, and not only or necessarily, as believed by King, to the rift-flank rebound itself.

5.7 Conclusion

The present study has explored several aspects of shoulder uplift and landscape development in southern India and raised the following major points.

(i) While the modelling in this study does not entirely reject the possibility that the shoulder uplift of the Western Ghats could have been achieved through tectonic uplift caused by Earth interior processes, including underplating, it suggests that the rim bulge could also have been obtained during the Cenozoic by passive denudational rebound alone if there is a weak shelf zone between the craton and the ocean. In the case of an infinite elastic plate, only a minor contribution from erosional rebound is obtained, so that a supplementary, uplift-inducing geophysical process needs to be called upon. Conversely, a 'broken' plate model reveals the combined advantage of being fairly insensitive to flexural rigidities both offshore and onshore, to denudation patterns in the highland region, to sediment volumes deposited offshore and to the initial topography at break-up. Two upland denudation models for the 'broken' plate scenario have been implemented. In the first, upland denudation is restricted to the immediate backslope of the Deccan plateau, and erosion is assumed to be insignificant in the cratonic interior (Figure 15.4); this yields best-fitting rift shoulder morphologies for only moderate values of elastic lithospheric thickness (~35 km). The second model more realistically assumes non-zero denudation in the Deccan interior (Figure 15.12) and this performs equally well in terms of surface uplift, but better still in terms of shoulder wavelength simulation (Figure 15.13). Furthermore, this model supports higher elastic thicknesses (e.g. 70 km), an outcome which concurs with independent estimates found in the literature. Overall, the infinite plate model reveals far less flexibility in terms of parametric constraints and boundary conditions, although this shortcoming does not, at this stage, completely undermine its validity.

(ii) The preference given to a high flexural rigidity for the Dharwar craton has major geomorphological consequences. Indeed, what may be background noise to the fission-track record, such as minor tectonic pulses or eustatic changes controlled from the Bay of Bengal, might nevertheless translate as conspicuous second-order geomorphological features supported by the elastic bending resistances of the craton. The worldwide tendency since the Mesozoic has been for the areal extent of successive planation events to diminish (see brief overview in Gunnell, 1997), suggesting a growing mismatch between large wavelength regional isostasy and the increasingly localized manifestations of landscape stripping and planation benches. This could be attributed to the growing role, during the Cenozoic, of lithological controls over erosion due to an increasingly discriminatory influence of climate on weathering patterns, and to a higher frequency of endogenic impulses of change which repeatedly arrested the completion of incipient erosion cycles.

(iii) Although global tectonics and mega-geomorphology are currently benefiting from advances in thermochronological techniques (e.g. fission-track analysis, $^{40}Ar/^{39}Ar$), and lithospheric simulation models, the resolution window of these scale-specific approaches has its limitations and needs auxiliary constraints which can provide additional fine tuning. Landscape development histories cannot ignore other strands of evidence such as drainage patterns, lithology, geological structure, soils, palaeosols or other datable material (e.g. Fabel and Finlayson, 1992; Gunnell, 1996a) and need to be treated as a nested hierarchy of clues and controls.

Acknowledgements

Processing of the apatite samples was funded by the CNRS-URA 1562 and NERC Grant GR9/963. The authors are grateful to Tony Hurford and Mike Widdowson for sharing their own sample data with us, and to Andy Carter at University College, London, for carrying out most of the laboratory analyses. Two anonymous reviewers provided useful criticisms and Mike Summerfield, who took the initiative of turning to us for this contribution on the Western Ghats, also suggested many improvements to the manuscript.

References

Ahnert, F. 1970. Functional relationships between denudation, relief and uplift in large mid-latitude drainage basins. *Amer. J. Sci.*, **268**, 243–263.

Beaumont, C. 1981. Foreland basins. *Geophys. J. R. Astr. Soc.*, **65**, 291–329.

Birot, P. 1958. *Morphologie structurale*, Presses Universitaires de France, Paris.

Birot, P. 1982. Quelques réflexions sur l'origine des bourrelets montagneux des marges passives. *Homm. Terr. Nord*, 1–8.

Biswas, S.K. 1987. Regional tectonic framework, structure and evolution of the western marginal basins of India. *Tectonophysics*, **135**, 307–327.

Brown, R.W., Summerfield, M.A. and Gleadow, A.J.W. 1994. Apatite fission track analysis: its potential for the estimation of denudation rates and implications for models of long-term landscape development. In: M.J. Kirkby (ed.) *Process Models and Theoretical Geomorphology*, Wiley, Chichester, pp. 23–53.

Burov, E.B. and Diament, M. 1995. The effective elastic thickness (T_e) of continental lithosphere: what does it really mean? (Constraints from rheology, topography and gravity). *J. Geophys. Res.*, **100**, 3908–3925.

Chandrasekharam, D. 1985. Structure and evolution of the western continental margin of India deduced from gravity, seismic, geomagnetic and geochronological studies. *Phys. Earth Planet. Interiors*, **41**, 186–198.

Chéry, J., Lucazeau, F., Daignières, M. and Villotte, J.P. 1992. Large uplift of rift flanks: a genetic link with lithospheric rigidity? *Earth Planet. Sci. Lett.*, **112**, 195–211.

Cox, K.G. 1989. The role of mantle plumes in the development of continental drainage patterns. *Nature*, **342**, 873–877.

Cox, K.G. 1992a. Continental magmatic underplating. *Phil. Trans. R. Soc. Lond.*, **342A**, 155–166.

Cox, K.G. 1992b. Karoo igneous activity, and the early stages of the break-up of Gondwanaland. In: B.C. Storey, T. Alabaster, and R.J. Pankhurst (eds) *Magmatism and the Causes of Continental Break-up*, *Geol. Soc. Spec. Publ.*, **68**, 137–148.

Dessai, A.G. and Bertrand, H. 1995. The 'Panvel flexure' along the Western Indian continental margin: an extensional fault structure related to Deccan magmatism. *Tectonophysics*, **241**, 165–178.

Doin, M.-P., Fleitout, L. and McKenzie, D. 1996. Geoid anomalies and the structure of continental and oceanic lithospheres. *J. Geophys. Res.*, **101**, 16 119–16 135.

Fabel, D. and Finlayson, B.L. 1992. Constraining variability in south-east Australian long-term denudation rates using a combined geomorphological and thermochronological approach. *Z. Geomorph.*, **36**, 293–305.

Foster, D.A. and Gleadow, A.J.W. 1996. Structural framework and denudation history of the flanks of the Kenya and Anza Rifts, East Africa. *Tectonics*, **15**, 258–271.

Gallagher, K. 1995. Evolution temperature histories from apatite fission-track data. *Earth Planet. Sci. Lett.*, **136**, 421–435.

Gilchrist, A.R. and Summerfield, M.A. 1990, Differential denudation and flexural isostasy in formation of rifted-margin upwarps. *Nature*, **346**, 739–742.

Gilchrist, A.R. and Summerfield, M.A. 1991. Denudation, isostasy and landscape evolution. *Earth Surf. Processes Ldfms.*, **16**, 555–562.

Gilchrist, A.R. and Summerfield, M.A. 1994. Tectonic models of passive margin evolution and their implications for theories of long-term landscape development. In: M.J. Kirkby (ed.) *Process Models and Theoretical Geomorphology*, Wiley, Chichester, pp. 55–84.

Gleadow, A.J.W., Duddy, I.R., Green, P.F. and Lovering, J.F. 1986. Confined fission-track lengths in apatite: a diagnostic tool for thermal analysis. *Contrib. Min. Petrol.*, **94**, 405–415.

Green, P.F. 1988. The relationship between track shortening and fission-track age reduction in apatite: Combined influences of inherent stability, annealing anisotropy, length bias and system calibration. *Earth Planet. Sci. Lett.*, **89**, 335–352.

Green, P.F., Duddy, I.R., Laslett, G.M., Hegarty, K.A., Gleadow, A.J.W. and Lovering, J.F. 1989. Thermal annealing of fission-tracks in apatite: quantitative modelling techniques and extension to geological timescales. *Chem. Geol.*, **75**, 155–182.

Gunnell, Y. 1996a. Géodynamique d'une moyenne montagne tropicale. La genèse des paysages dans le Ghat occidental du Deccan, sur son revers continental et son piémont maritime. Thèse de doctorat, Université Blaise-Pascal, Clermont-Ferrand.

Gunnell, Y. 1996b. Géographie comparative des héritages cuirassés sur les terres cristallines de l'Inde du Sud et d'Afrique de l'Quest. Leur signification dans l'évolution du milieu physique. *Ann. Géog.*, **591**, 451–479.

Gunnell, Y. 1997. Topography, palaeosurfaces and denudation over the Karnataka uplands, southern India. In: M. Widdowson (ed.) *Palaeosurfaces: Recognition, Reconstruction and Palaeoenvironmental Interpretation, Geol. Soc. Lond. Spec. Publ.*, **120**, 249–267.

Gunnell, Y. and Bourgeon, G. 1998. Soils and climatic geomorphology on the Karnataka Plateau, peninsular India. *Catena* (in press).

Gunnell, Y. and Fleitout, L. 1998. Shoulder uplift of the Western Ghats passive margin, India: A denudational model. *Earth Surf. Processes Ldfms.*, **23**, 391–404.

Gunnell, Y., Widdowson, M. and Hurford, A.J. 1998. Plume heads and rift shoulders. A fission-track study of the Western Ghats passive margin (in press).

Hurford, A.J. 1990. Standardization of fission-track dating calibration: recommendation by the Fission Track Working Group of the I.U.G.S. Subcommission on Geochronology. *Chem. Geol.*, **80**, 171–178.

Hurford, A.J. and Green, P.F. 1982. A user's guide to fission-track dating calibration. *Earth Planet. Sci. Lett.*, **59**, 343–354.

Jessen, O. 1943. Die Randschwellen der Kontinente. *Petermanns Geog. Mitt.*, **241**, 1–205.

Kalaswad, S., Roden, M.K., Miller, D.S. and Morisawa, M. 1993. Evolution of the continental margin of western India: new evidence from apatite fission-track dating. *J. Geol.*, **101**, 667–673.

Keen, C.E. and Beaumont, C. 1990. Géodynamique des marges continentales de divergence. In: M.J. Keen and G.L. Williams (eds) *Géologie de la Marge Continentale de l'Est du Canada*, Commission Géologique du Canada, pp. 421–508.

King, L.C. 1955. Pediplanation and isostasy: an example from South Africa. *Quart. J. Geol. Soc.*, **111**, 353–359.

Laslett, G.M., Green, P.F. Duddy, I.R. and Gleadow, A.J.W. 1987. Thermal annealing of fission-track in apatite, 2, A quantitative analysis. *Chem. Geol.*, **65**, 1–13.

Lister, G.S. and Etheridge, M.A. 1989. Detachment model for the uplift and volcanism of the Eastern Highlands. In: R.W. Johnson (ed.) *Intraplate Volcanism in Eastern Australia and New Zealand.* Cambridge University Press, Cambridge, pp. 297–313.

Lyon-Caen, H. and Molnar, P. 1983. Constraints on the structure of the Himalaya from an analysis of gravity anomalies and a flexural model of the lithosphere. *J. Geophys. Res.*, **88**, 8171–8191.

Manglik, A. and Singh, R.N. 1992. Rheological thickness and strength of the Indian continental lithosphere. *Proceedings of the Indian Academy of Sciences, Earth Planet. Sci.*, **101**, 339–345.

Mitchell, C. and Widdowson, M. 1991. A geological map of the southern Deccan Traps, India and its structural implications. *J. Geol. Soc. Lond.*, **148**, 495–505.

Partridge, T.C. and Maud, R.R. 1987. Geomorphic evolution of southern Africa since the Mesozoic. *S. Afr. J. Geol.*, **90**, 179–208.

Prasada Rao, R. and Srivastava, D.C. 1984. Regional seismic facies analysis of western offshore, India. *Bulletin of the ONGC*, **21**, 83–95.

Prell, W.L. and Niitsuma, L. 1989. Introduction, background, and major objectives for ODP Leg 117 Western Arabian Sea in search of ancient monsoons. In: *Proc. Ocean Drilling Prog. Initial Rep.*, College Station, Texas, pp. 5–9.

Subrahmanyam, V., Gopala Rao, D., Ramana, M.V., Krishna, K.S., Murty, G.P.S. and Gangadhara Rao, M. 1995. Structure and tectonics of the southwestern continental margin of India. *Tectonophysics*, **249**, 267–282.

Summerfield, M.A. 1991a. *Global Geomorphology*, Longman, London.

Summerfield, M.A. 1991b. Sub-aerial denudation of passive margins: regional elevation versus local relief models. *Earth Planet. Sci. Lett.*, **102**, 460–469.

ten Brink, U. and Stern, T. 1992. Rift flank uplifts and hinterland basins: comparison of the Transantarctic mountains with the Great Escarpment of southern Africa. *J. Geophys. Res.*, **97**, 569–585.

Thomas, M.F. 1995. Models for landform development on passive margins. Some implications for relief development in glaciated areas. *Geomorphology*, **12**, 3–15.

van der Beek, P.A., Cloetingh, S. and Andriessen, P. 1994. Mechanisms of extensional basin formation and vertical motions at rift flanks: constraints from tectonic modelling and fission-track thermochronology. *Earth Planet. Sci. Lett.*, **121**, 417–433.

Vanney, J.-R. 1982. Les bourrelets des boucliers anciens et leurs marges continentales. *Bull. Assoc. Géog. Franç.*, **489**, 231–238.

Verma, R.K. 1991. *Geodynamics of the Indian Peninsula and the Indian Plate Margin*, Oxford and IBH Publishing Co., New Delhi.

Watts, A.B. 1992. The effective elastic thickness of the lithosphere and the evolution of foreland basins. *Basin Res.*, **4**, 169–178.

Watts, A.B. and Cox, K.G. 1989. The Deccan Traps: an interpretation in terms of progressive lithospheric flexure in response to a migrating load. *Earth Planet. Sci. Lett.*, **93**, 85–97.

Whiting, B.M., Karner, G.D. and Driscoll, N.W. 1994. Flexural and stratigraphic development of the West Indian continental margin. *J. Geophys. Res.*, **99**, 13 791–13 811.

Widdowson, M. and Cox, K.G. 1996. Uplift and erosional history of the Deccan traps, India: evidence from laterites and drainage patterns of the Western Ghats and Konkan coast. *Earth Planet. Sci. Lett.*, **137**, 57–65.

16

The growth and decay of oceanic islands

A.B. Watts

16.1 Introduction

Scientists have been fascinated by the flora and fauna of oceanic islands ever since their discovery by the early European explorers. One of the first to recognize their geological significance was Charles Darwin. He showed (Darwin, 1842) that while most high islands were of volcanic origin, some low ones were capped by coral reefs in the form of atolls. He argued that the coral grew on the flanks of once high volcanic islands that had subsided below sea level. The notion that oceanic islands were not permanent features of the Earth's crust but were subject to vertical movements was initially disputed since it was not immediately clear why coral reefs would grow up to form an island.

Darwin's hypothesis was confirmed about 100 years later when Hess (1946) discovered a number of submerged, flat-top features on the sea floor which he named guyots (after the American geographer, Arthur Guyot). Hess believed that the flat tops represented wave-truncation surfaces and that, like atolls, guyots were once high volcanic islands which had subsided below sea level. Subsequent seismic reflection profile and drilling data have shown that many guyots are capped by coral which either had stopped growing or had been killed off prior to their submergence.

Atolls and guyots differ from high islands by their greater abundance, their concentration in a single ocean (the Pacific) and their progressive increase in number with age (Batiza, 1982). The present day hot-spot population is unable, however, to explain their distribution, suggesting a much greater density of hot spots in the past. There is evidence that magmatic activity in the Pacific was particularly intense during the late Cretaceous (Larson and Schlanger, 1981) when hot spots – often at a mid-ocean ridge crest – generated a large number of high islands which have since subsided to become oceanic plateaus (e.g. Manihiki), rises (e.g. Hess, Shatsky) and guyots (e.g. mid-Pacific mountains).

Because of their size, rapidity of formation and distribution, oceanic islands have played a significant role in the development of plate tectonics. They have provided key information on the absolute motions of the plates (Gripp and Gordon, 1990). In addition, they have featured prominently in field studies; for example, magnetic anomalies (e.g. Harrison *et al.*, 1975) have provided information on palaeomagnetic pole positions and polar wander curves, while gravity and geoid anomalies (e.g. Watts and Cochran, 1974) have been used to estimate the long-term (>10^6 a) thermal and mechanical properties of the lithosphere.

Oceanic islands are not passive features, however, that simply record the motions and physical properties of the plates. They are dynamically evolving, inherently unstable structures (McGuire and Saunders, 1993) that are prone to failure on a wide range of spatial and temporal scales. Islands display a wide variety of morphological features (Nunn, 1994). These include radial drainage patterns, planezes and amphitheatre-headed valleys. Landslides are one of the principal variables in the formation of amphitheatre-headed valleys. It has been known for some time (Holcomb and Searle, 1991) that landslides – and the debris avalanches that they initiate – have greatly modified the landscape of oceanic islands and, in some cases, have scalloped their shorelines.

Because of the narrowness of their insular shelves (Bloom, 1967), oceanic islands are one of the best recorders that we have of global sea-level changes in the geological past. Ancient shorelines may be expressed as a single elevated surface, a raised coral

Geomorphology and Global Tectonics. Edited by Michael A. Summerfield.

reef, or a series of vertically stacked notches (Nunn, 1994). In New Guinea, the ancient shorelines form a distinctive 'staircase' pattern which has been attributed (Chappell, 1974) to the interaction of relatively short-term, glacially driven sea-level changes with long-term local tectonic uplift.

Although many intraplate oceanic islands have experienced uplift (Wilson, 1963), most are dominated by long-term tectonic subsidence. Two mechanisms that have been invoked to explain this subsidence are thermal contraction of the underlying oceanic crust as it ages and cools (Detrick *et al.*, 1977) and stress relaxation due to volcanic loading (Bodine *et al.*, 1981). The purpose of this chapter is to show: (a) how the thermal and mechanical properties of the Earth's lithospheric plates may contribute to the uplift and subsidence history of oceanic islands; and (b) how these movements may, in some cases, influence their geomorphic development.

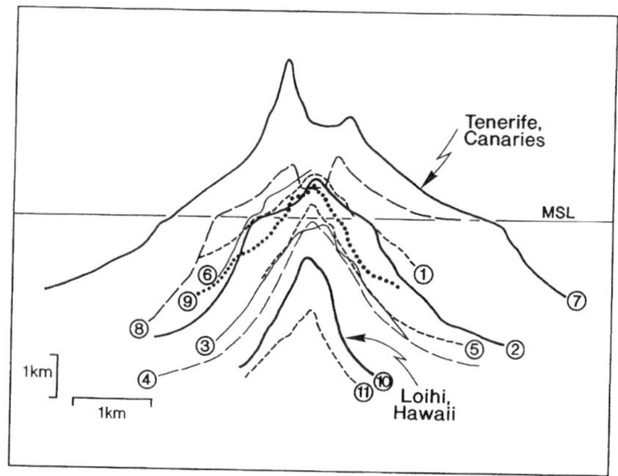

Figure 16.1 *Topographic profiles of active oceanic volcanoes based on Menard (1984): (1) Faial, Azores; (2) Socorro, Revillagegados; (3) Moua Pikaa, Society Islands; (4) Macdonald, Australs; (5) San Benedicto, Revillagegados; (6) Marion Island; (7) Tenerife, Canary Islands; (8) Fernandina, Galapagos; (9) Pinto, Galapagos; (10) Loihi, Hawaii; (11) Rocard, Society Islands*

16.2 Growth

According to Menard (1986), there are in excess of 3×10^5 volcanoes (active and inactive) with a relief of more than 500 m on the ocean floor. Some of these volcanoes represent ancient oceanic islands which have long since subsided below sea level. A majority, however, represent volcanoes that are in various stages of growth.

Unfortunately, we have only fragmentary knowledge of the distribution of young seamounts on the sea floor. They are known from various localities in the plate interiors (e.g. Loihi (Fornari *et al.*, 1988)), but, the best mapped occurrences are from young oceanic crust in the region of mid-oceanic ridges. At the East Pacific Rise crest, for example, Fornari *et al.* (1984), Barone and Ryan (1990) and Scheirer and Macdonald (1995) have identified numerous steep-sided, circular seamounts. The majority of these are not found at the ridge axis, but on 0.1–0.3-Ma-old oceanic crust where they reach heights of up to 200–1600 m above the mean depth of the surrounding sea floor.

At the East Pacific Rise crest, most seamounts have flat tops and some have well-defined summit depressions (Fornari *et al.*, 1984). Scheirer and Macdonald (1995) showed that small seamounts on the rise crest generally have steeper slopes (15–30°) than large ones. As seamounts increase in height, their slopes decrease. This is most probably due to the development of volcaniclastic aprons on their flanks. Despite the presence of such aprons, however, large seamounts retain their pointed form as they grow. Smith (1988) has studied more than 85 well-surveyed seamounts and concluded that away from the ridge crest small seamounts generally have steeper slopes than large ones. She also showed that the prevalent shape of large-sized seamounts is a sharp crest with outward dipping slopes of about 15°. Elements of this pointed form are retained in the seamounts that eventually break the sea surface to form high islands (Figure 16.1).

The existence of high seamounts on young oceanic crust suggests that they build up at quite rapid rates on the sea floor. According to Menard (1986), a 5-km-high intraplate volcano takes about 400 years to become 1 km high, 0.37 Ma to reach 4 km and about 1 Ma to achieve its final height.

Most large oceanic islands and seamounts are flanked by broad depressions in the depth of the surrounding sea floor (e.g. Figure 16.2). These depressions are infilled, at least in part, by substantial thicknesses of well- to poorly stratified material (e.g. ten Brink and Watts, 1985). Some of this material has been derived by mass wasting processes on the island and therefore contains an 'inverted' record of its growth history. The remainder is derived from pelagic sedimentation and, in the cases of islands that are located near a continental margin, from terrestrial sources. Unfortunately, few of the sediments in these flanking depressions have been drilled to any great depth.

The available drill data have been used together with field observations to derive generalized models for the growth of oceanic islands. Staudigel and Schmincke (1984), for example, recognize two main stages – each of which is characterized by distinct volcanic deposits (Figure 16.3). During the first deep-water stage, volcanic lavas intrude the pre-existing sedimentary layer of the oceanic crust. Once a solid volcanic edifice is established, pillow lavas will dominate in an extrusive phase. Typically, the deep water extrusive facies will include pillow lavas, clastic rocks and massive pillows. Clastic rocks probably dominate in the aprons. With time the seamount reaches a critical depth for explosive volcanism. During this second phase, clastic rocks become more important. Within the core of the seamount the volcanic rocks are dominated by scoria pillow breccias, scoria lapilli breccias and minor amounts of vesicular lavas. On the flanks, pillow fragments dominate the volcanic sequence.

The final stage is the construction of a sub-aerial edifice. Usually this process begins with the eruption in shallow water of pyroclastic cones which have little resistance to wave action. Many new islands therefore disappear soon after they break the surface. Some, however, persist and grow to form high islands. An important question (Nunn, 1994) is what processes allow a largely fragmentary edifice just below the sea surface to extend and eventually persist. Although tectonic uplift may certainly be a contributor, there is evidence that shifts in the eruption sequence are sufficient for an island to persist. On Surtsey, a new oceanic island which began to form in 1963 a few kilometres south of Iceland, shifts in the eruption centre have caused part of the pyroclastic cone to be draped by a lava layer which is more resistant to wave action (Fridriksson, 1975). Once an island acquires such a protective armour, and assuming that it still has access to a sufficient magma supply, then it can grow upwards and outwards to form a high island.

16.3 Flexure of the Lithosphere

The volcanic rocks that make up oceanic islands displace lower density material (i.e. seawater and air) and therefore represent a load on the surface of the crust which should bend or flex under their weight. The loading induces vertical movements of the crust and upper mantle which will have a profound effect on the geomorphic development of islands.

Most early studies (e.g. Walcott, 1970) inferred the flexure at oceanic islands from gravity anomaly data. More recently, the marine seismic reflection and refraction profile technique has been used (Watts *et al.*, 1985, 1997; Caress *et al.*, 1995; Ito *et al.*, 1995) to image the surfaces of flexure directly. Reflection data in the Canary Islands (Watts *et al.*, 1997) region, for example, show that volcanic loads have depressed the oceanic crust by as much as 2 s of two-way travel time (TWTT) – which corresponds to depths of 2–3 km. The region of subsidence extends for over 250 km from the island centre, and there is evidence of uplift in flanking areas which has raised the sea floor by up to several tens of metres.

The pattern of flexure that is observed in the immediate vicinity of oceanic islands suggests that the lithosphere responds to volcanic loads in a similar manner as would a thin elastic plate that overlies a weak fluid (Figure 16.4a). According to this model, the flexure due to a volcanic load is resisted by two main factors: the intrinsic strength of the lithosphere and the buoyancy of the substratum. The relative importance of these factors depends on the shape of the applied load: narrow loads are supported mainly by the strength of the lithosphere and wide ones by the buoyancy of the substratum. By comparing the observed flexure to calculations that take into account the amplitude and wavelength of the load it is possible to estimate the elastic thickness, T_e, of the lithosphere, which is a measure of its long-term ($>10^6$ a) mechanical strength.

Flexure studies of a wide range of geological features (e.g. oceanic islands and seamounts, deep sea trench outer rises and river deltas) suggest (Watts, 1978) that T_e depends on the age of the lithosphere *at the time of loading* (Figure 16.4b). Islands or seamounts that form on young sea floor (e.g. Emperor Seamounts) have a low T_e, whereas those that form on old sea floor (e.g. Hawaii) have high values. This result can be explained when it is considered that as the lithosphere ages and cools, it becomes more rigid in the way that it responds to geological loads. As Figure 16.4b shows, T_e is quite well described by the depth to the 450°C oceanic isotherm, based on cooling plate models. The figure also indicates that T_e is significantly less than the seismic thickness determined by surface wave studies (e.g. Leeds, 1975) which suggests that the oceanic lithosphere thickens to depths of >100 km with age. Although surface waves exhibit both azimuthal and polarization anisotropy (Regan and Anderson, 1985), these data suggest (e.g. Bodine *et al.*, 1981) that islands initially load a lithosphere which is relatively thick and rigid; some sort of stress relaxation then sets in as the lithosphere changes from its short-term (?seismic) to its long-term elastic thickness.

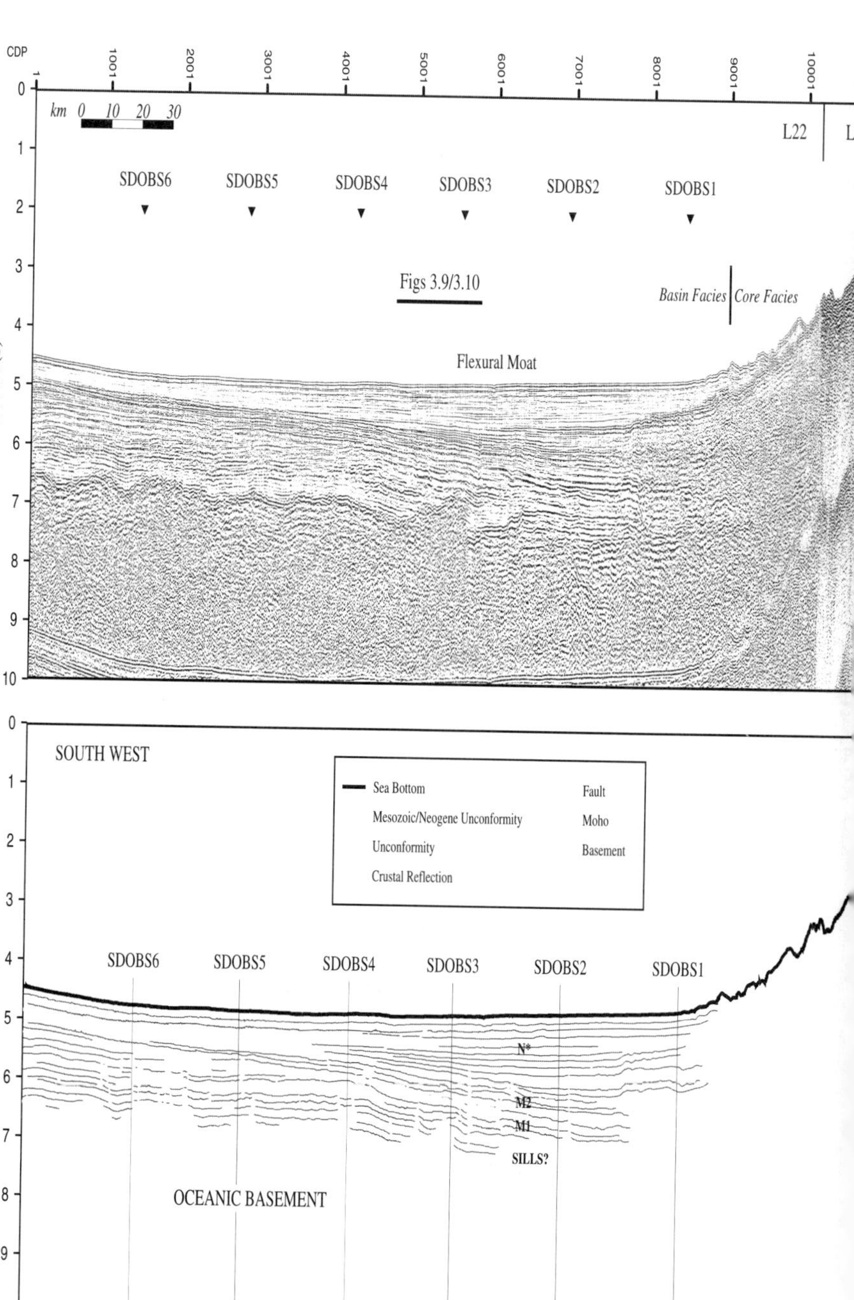

Figure 16.2 *Seismic reflection profile of the Canary Islands archipelago between La Gomera and Tenerife. The profile shows that the Canary Islands volcanic loads have flexed pre-existing oceanic crust and overlying Mesozoic sediments downwards. The flexure is partly infilled by Tertiary sediments*

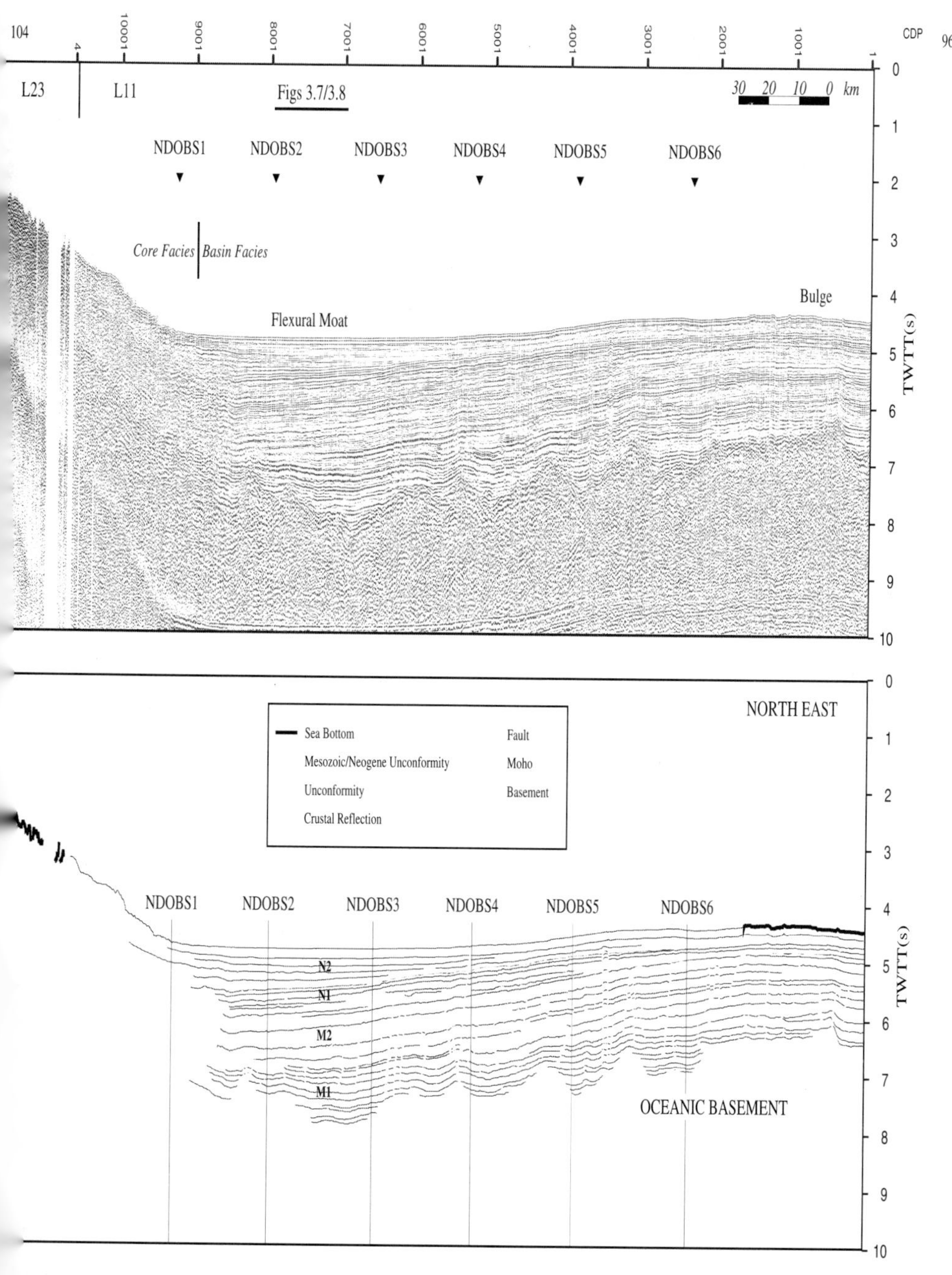

which progressively onlap the underlying Mesozoic sediments: (a) original seismic reflection profile; (b) stratigraphic interpretation

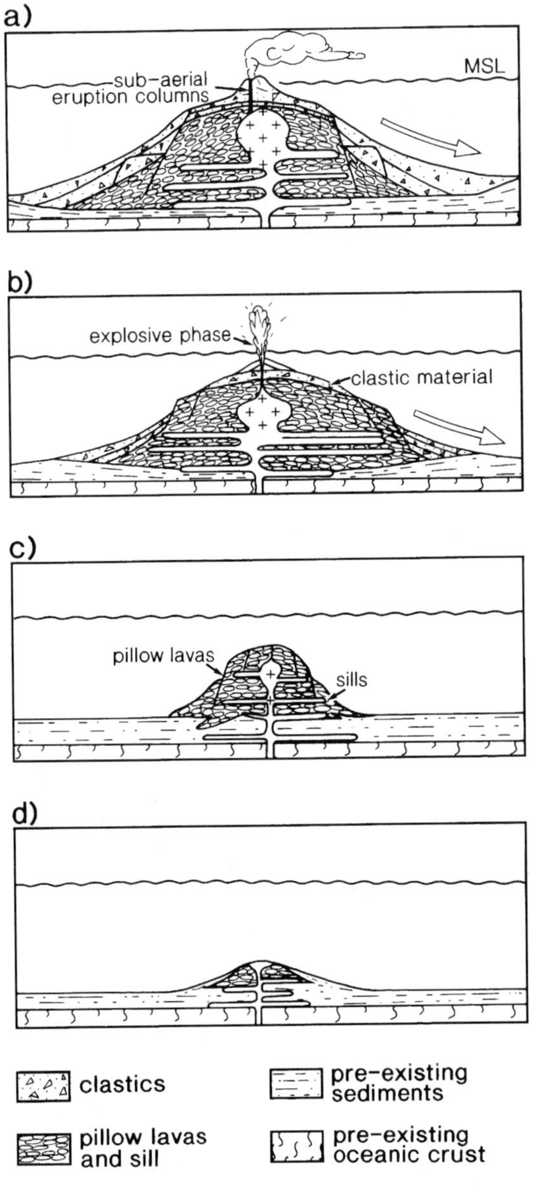

Figure 16.3 *Schematic diagram for the evolution of an oceanic island based on Staudigel and Schmincke (1984): (a) final phase; (b) explosive phase; (c) constructional phase; and (d) initial phase*

Simple calculations (Figure 16.5) for a typical size of island show that the effects of stress relaxation may be quite large. Consider, for example, a volcano that grows from a depth of sea floor of 3.5 km to a height of 2.6 km above sea level. Initially, the volcano will be supported by its short-term (?seismic) thickness and there will be little or no flexural deformation. As the lithosphere relaxes, however, from its short-term to its long-term elastic thickness, its strength reduces and the volcano will 'fall in' to the depression that has been created. In the example considered here, the strength reduction results in a decrease in the elevation of the volcano from 2.6 to 0.5 km above sea level. The depression extends to the flanking moat which is likely to be infilled, at least in part, by sediments derived by mass wasting from the island, the load of which might further decrease the height of the volcano. If the sediments were deposited during the relaxation then they should be characterized by offlap as the depocentre is gradually reduced in width. This seems to disagree with observations (e.g. Figure 16.2) which suggest onlap – indicating that the relaxation is completed in a relatively short time compared to the time that it takes mass wasting or other denudational processes to supply material to the depressions.

Unfortunately, because of the lack of deep drilling data, little is known about the time scale of the inferred relaxation processes; neither is anything known about the mechanisms by which stress relaxation is achieved in the lithosphere. The best evidence, perhaps, is the 100 year record of tide-gauge data in the Hawaiian Islands (Moore, 1970). These data show that the youngest island in the chain, Hawaii, is subsiding at an average rate of about 2 mm a^{-1}, which is significantly faster than other islands in the chain. The rates decrease towards Molokai and Maui, and Oahu appears to be stable.

Despite questions concerning the relaxation time, it is clear that flexure has contributed significantly to the vertical motion history of oceanic islands. These influences are probably best seen in the hot-spot-generated island chains. The addition of Hawaii to the south-eastern end of the Hawaiian–Emperor seamount chain, for example, has caused subsidence in the pre-existing islands of Molokai and Maui (Watts and ten Brink,

Figure 16.4 *Flexure of the oceanic lithosphere caused by volcanic loads. (a) Flexure of the oceanic crust caused by a 5-km-high, 120-km-wide volcanic load. The flexure has been computed for a flexural rigidity of the plate of 10^{23} Nm (equivalent to an elastic thickness of 25 km) and uniform densities of 2700, 2700 and 3300 kg m^{-3} for the load, infill and mantle, respectively. (b) Plot of elastic thickness against age of the oceanic lithosphere at the time of loading for seamounts and oceanic islands (light filled circles–French Polynesia; heavy filled circles–other oceanic islands/seamounts), deep sea trench island arc systems (filled triangles), river delta system (filled diamond) fracture zones (light horizontal bar) and mid-oceanic ridges (filled square)*

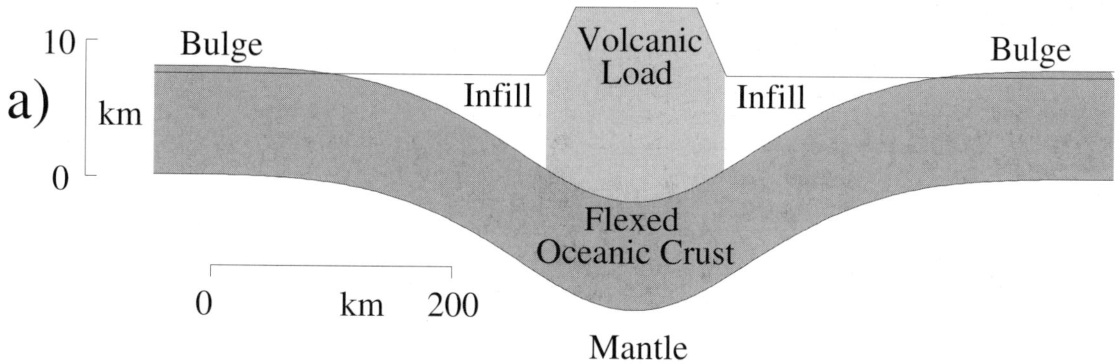

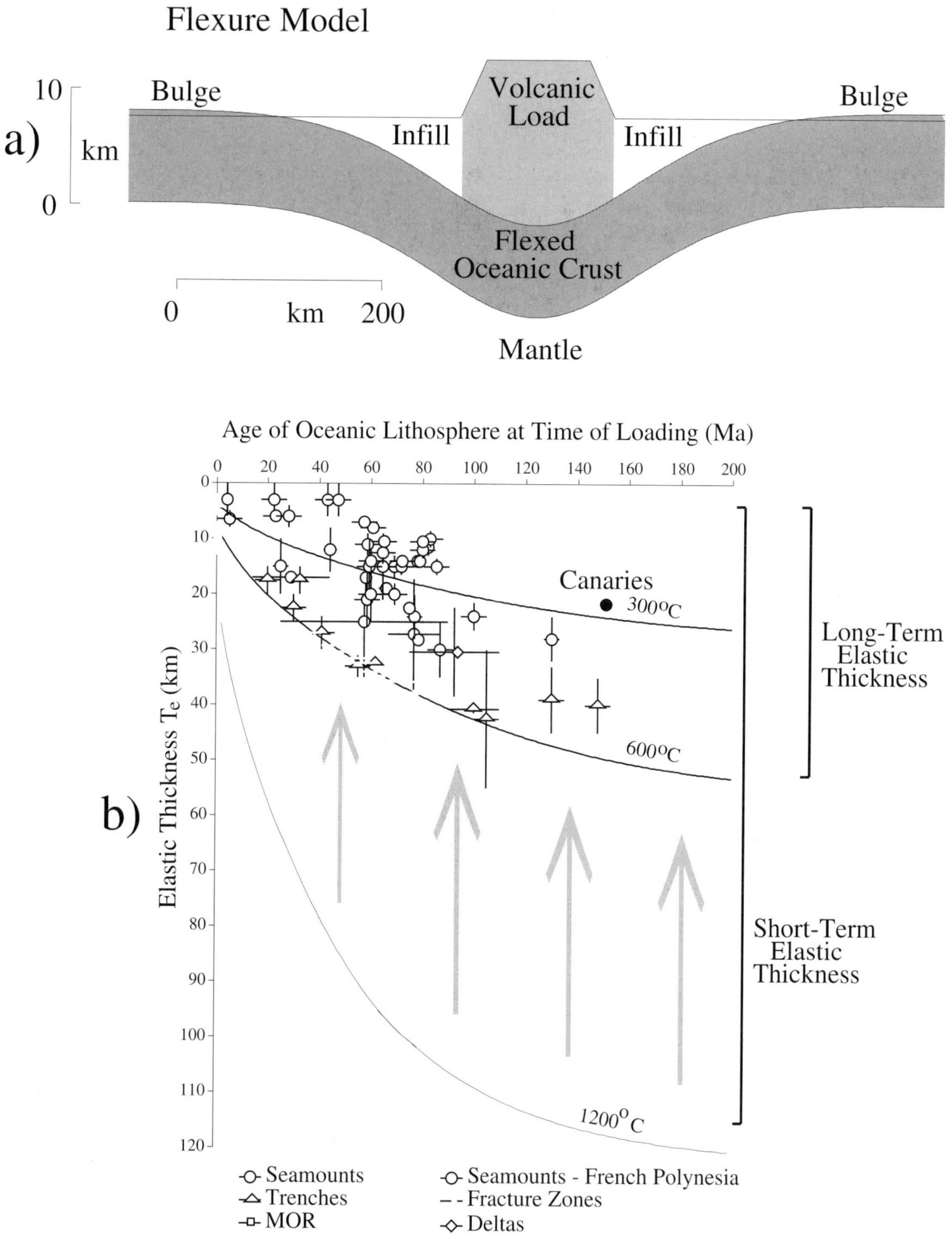

Flexure Model
a)
10
km
0
Bulge
Volcanic Load
Infill
Infill
Bulge
Flexed Oceanic Crust
0 km 200
Mantle
b)
Age of Oceanic Lithosphere at Time of Loading (Ma)
0 20 40 60 80 100 120 140 160 180 200
Elastic Thickness Te (km)
0 10 20 30 40 50 60 70 80 90 100 110 120
Canaries
300°C
600°C
1200°C
Long-Term Elastic Thickness
Short-Term Elastic Thickness
Seamounts
Trenches
MOR
Seamounts - French Polynesia
Fracture Zones
Deltas

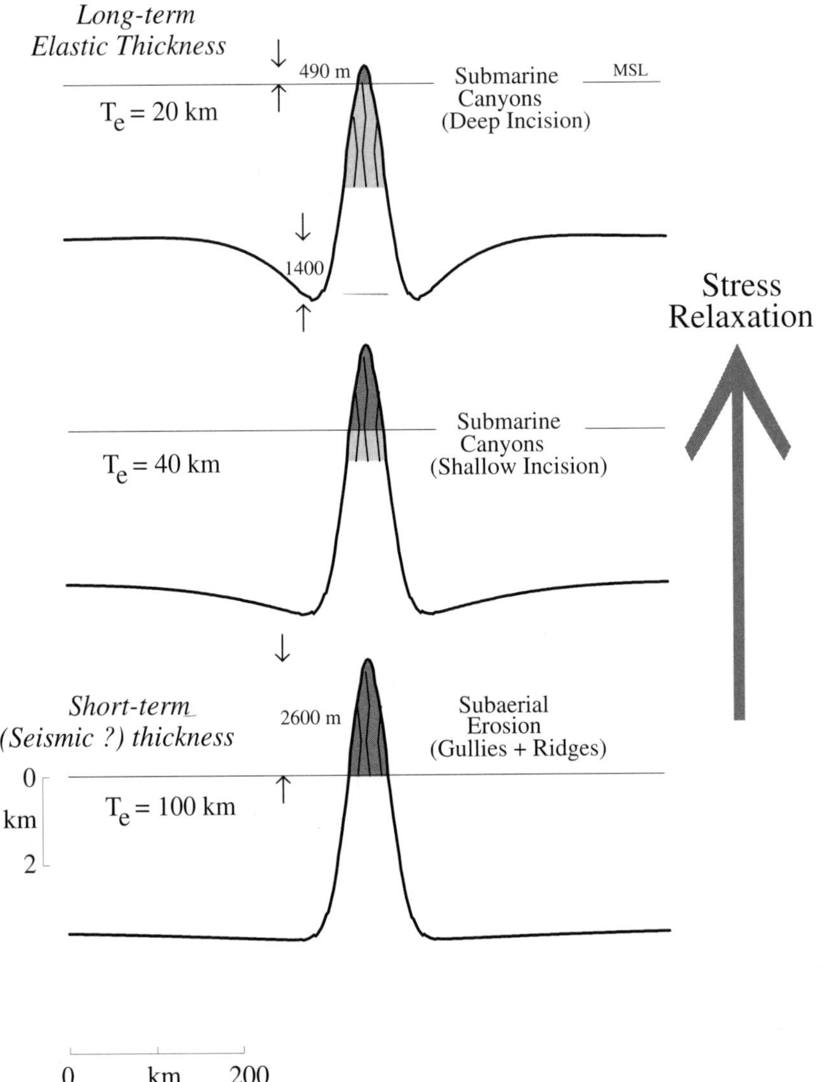

Figure 16.5 *Simple model to illustrate the consequences of stress relaxation on the morphological development of an oceanic island. (a) A volcano builds up on thick, rigid lithosphere to sea level, breaks the sea surface and forms a 2.6-km-high oceanic island. The high island would be subject to fluvial erosion and valleys/gullies may form. (b) Stress relaxation sets in causing subsidence beneath the island and nearby regions. The gullies and ridges formed during state (a) subside below sea level. (c) Stress relaxation is complete, the lithosphere reaches its long-term elastic thickness, and subsidence ceases*

1989), which find themselves in Hawaii's flexural depression (Figure 16.6). The subsidence accounts for the depth of several submarine terraces around Molokai and Maui (Coulbourn *et al.*, 1974) and may explain, at least in part, the reduction in elevation of these islands away from Hawaii. In addition, there is evidence of uplift on Oahu and Kauai which find themselves in Hawaii's flexural bulge. Both islands have extensive coral reef deposits that are located above sea level (Moore, 1987; Jones, 1992). Flexure also appears to have influenced the subsidence and uplift history of some more widely spaced, isolated, islands. For

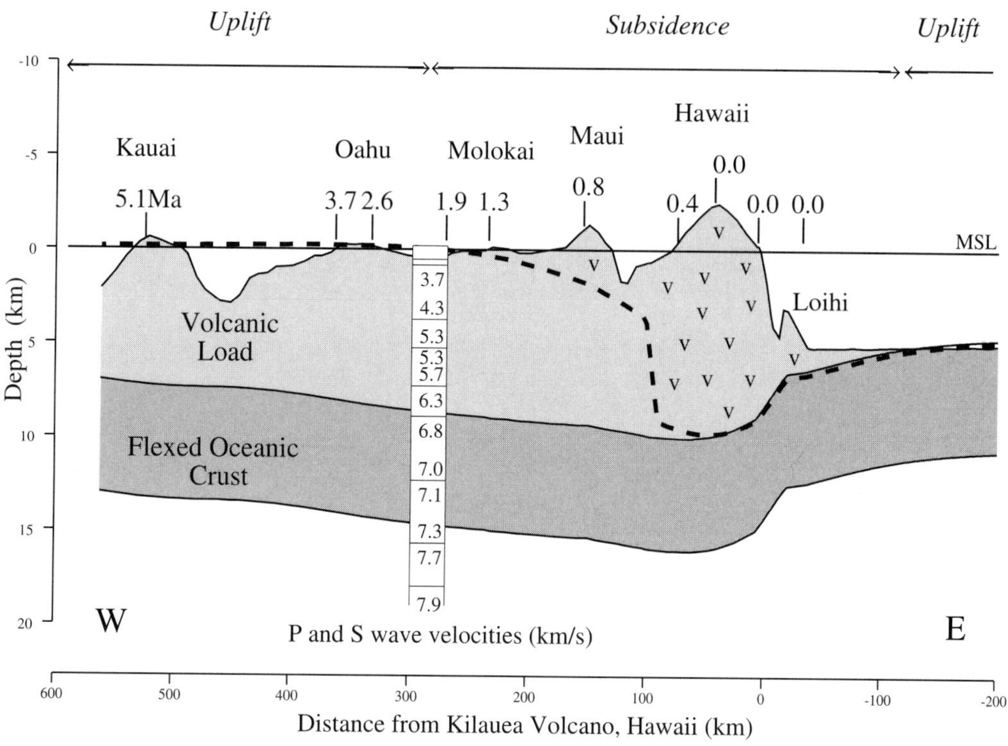

Figure 16.6 *Schematic cross-section of the crust and upper mantle structure along the crest of the Hawaiian Ridge between the islands of Hawaii and Kauai based on seismic, flexure and gravity modelling (Watts and ten Brink, 1989). The region with 'v' shading shows the load and flexure that is associated with the relatively young load of the island of Hawaii. The flexure curve (heavy dashed line) shows that Hawaii contributes to the uplift and subsidence history of pre-existing islands on the Hawaiian Ridge*

example, Rarotonga in the Cook Islands is a relatively young isolated volcano which has flexed the lithosphere over a broad area, causing subsidence in some pre-existing islands and uplift to others (McNutt and Menard, 1978).

There is evidence that loads other than new volcanoes may modify the elevation of oceanic islands. Island arc-deep sea trench systems, for example, are associated with vertical and horizontal loads which are capable of depressing the oceanic crust by up to 4 km below the depth that it should be on the basis of its thermal age. At some trenches the depression is flanked by a flexural bulge (Hanks, 1971) which can rise up to 700–900 m above the mean depth of the surrounding sea floor. An oceanic island that is carried toward a deep sea trench by plate motions may therefore experience uplift as the underlying oceanic crust 'rides' the bulge before being consumed at a trench. Examples of such islands include Niue and the Loyalty Islands (DuBois *et al.*, 1973) in the southwestern Pacific, and the Cocos (Keeling) Islands and Christmas Island (Woodroffe *et al.*, 1990) in the northeastern Indian Ocean.

Seismic reflection and refraction profile data have confirmed the existence of flexure at oceanic islands (Figure 16.7). They have also revealed several additional details of the mechanical behaviour of the lithosphere at oceanic islands. These include the possibility of inelastic yielding (ten Brink and Brocher, 1987), pre-existing weak zones in the lithosphere and, the addition, beneath some islands, of lower crustal bodies which 'underplate' the flexed oceanic crust (Watts *et al.*, 1985; Caress *et al.*, 1995).

16.4 Drowned Active Volcanoes

Ever since their discovery, the flat surfaces of guyots (Figure 16.8) have been taken as *prima facie* evidence that once high volcanic islands are likely to subside

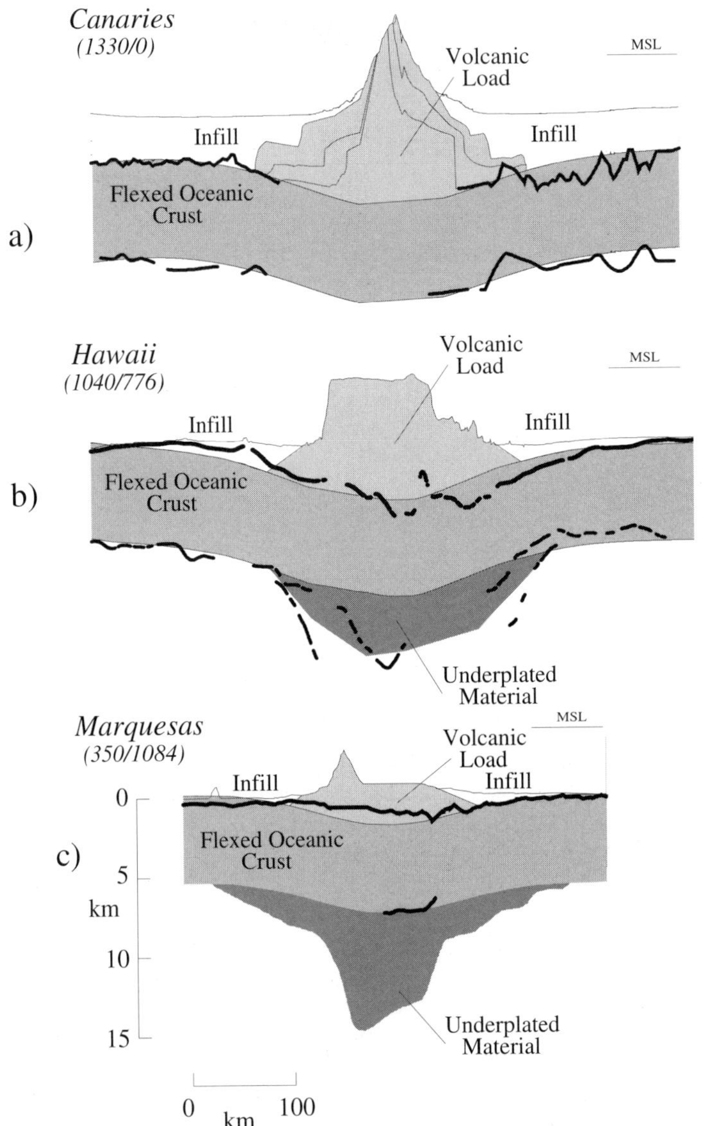

Figure 16.7 *Comparison of the crustal and upper mantle structure along transects of Hawaii, Marquesas and the Canary Islands (Watts* et al., *1985, 1997; Caress* et al., *1995). Heavy lines show prominent reflectors at the top and base of the oceanic crust, except on the Canary Islands where they correspond to the Moho depth based on refraction modelling. The two numbers in parentheses show the area (in km^2) of magmatic material that has been added, respectively, to the surface and base of the flexed crust. Note that the flexed crust at the Marquesas and Hawaiian Islands are underplated whereas the Canary Islands do not appear to be so*

below sea level. The general interpretation is that the flat surfaces are truncation surfaces which reduce the height of the island to sea level by processes that involve wave action, shelf widening and sub-aerial and submarine erosion.

It is well known that wave action is highly effective in eroding the coastlines of oceanic islands. Huge cliffs are a common feature of many islands. The width of the insular shelf that surrounds islands progressively increases with time (Menard, 1986); young islands

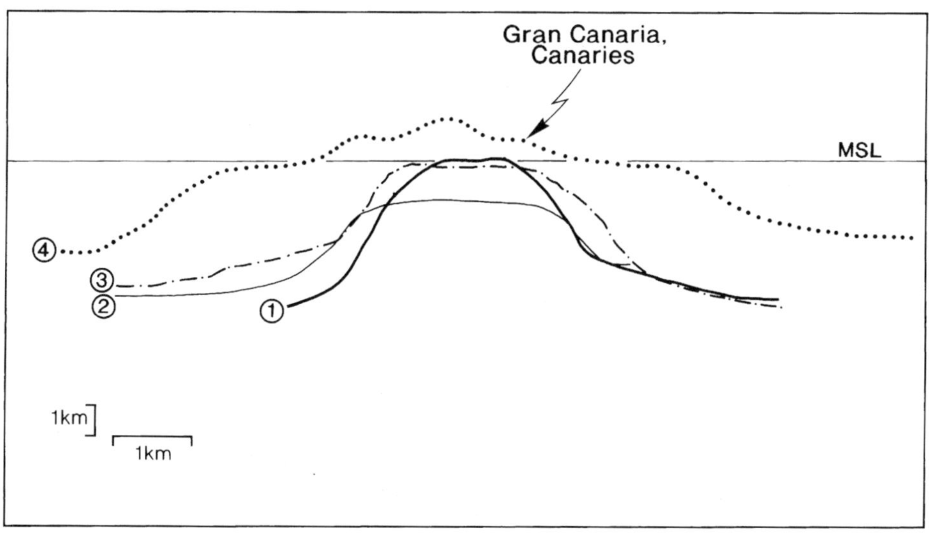

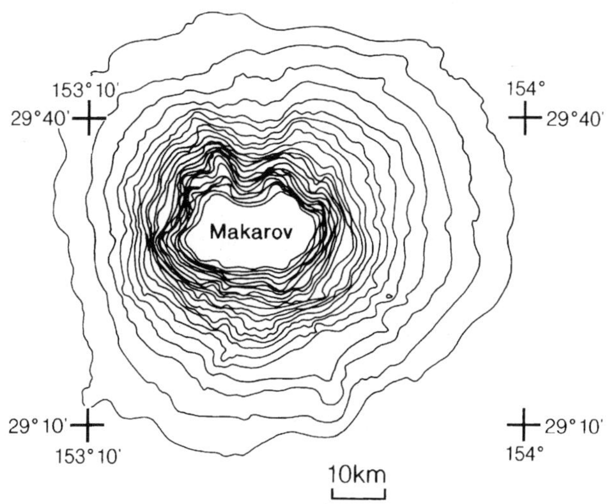

Figure 16.8 *Topography of atolls and guyots: (a) Profiles of (1) Laysan Island, Hawaiian Islands; (2) Wodejebato; (3) Bikini Atoll based on Menard (1986); and (4) Gran Canaria, Canary Islands, based on the swath bathymetry survey of Funck (1995). (b) Map of Makarov guyot based on multibeam bathymetry (Vogt and Smoot, 1984). Contour interval is 100 fathoms (= 183 m)*

have narrow shelves whereas old islands have wide ones. There is even evidence of differences in the widths of shelves on the windward and lee of islands suggesting that wind direction is an important factor. Interestingly, no relation exists between the depth of the shelf break and island age, which indicates that despite differences in volcano age some sort of balance is maintained between erosional and depositional processes.

Atolls are special cases of drowned oceanic islands which, despite subsidence, have remained at, or close to, sea level for some time. These low islands form in warm seas and are built up mainly of coral reefs. Darwin thought that the coral grew up on the rough surface of an extinct volcano which had subsided below sea level. If this is correct, then the basement beneath an atoll should be a rough and irregular surface, and magnetic anomaly and seismic reflection profile data show that many of them are. Some atolls (e.g. Maldives), however, appear to be underlain by a smooth and flat volcanic basement (Aubert and

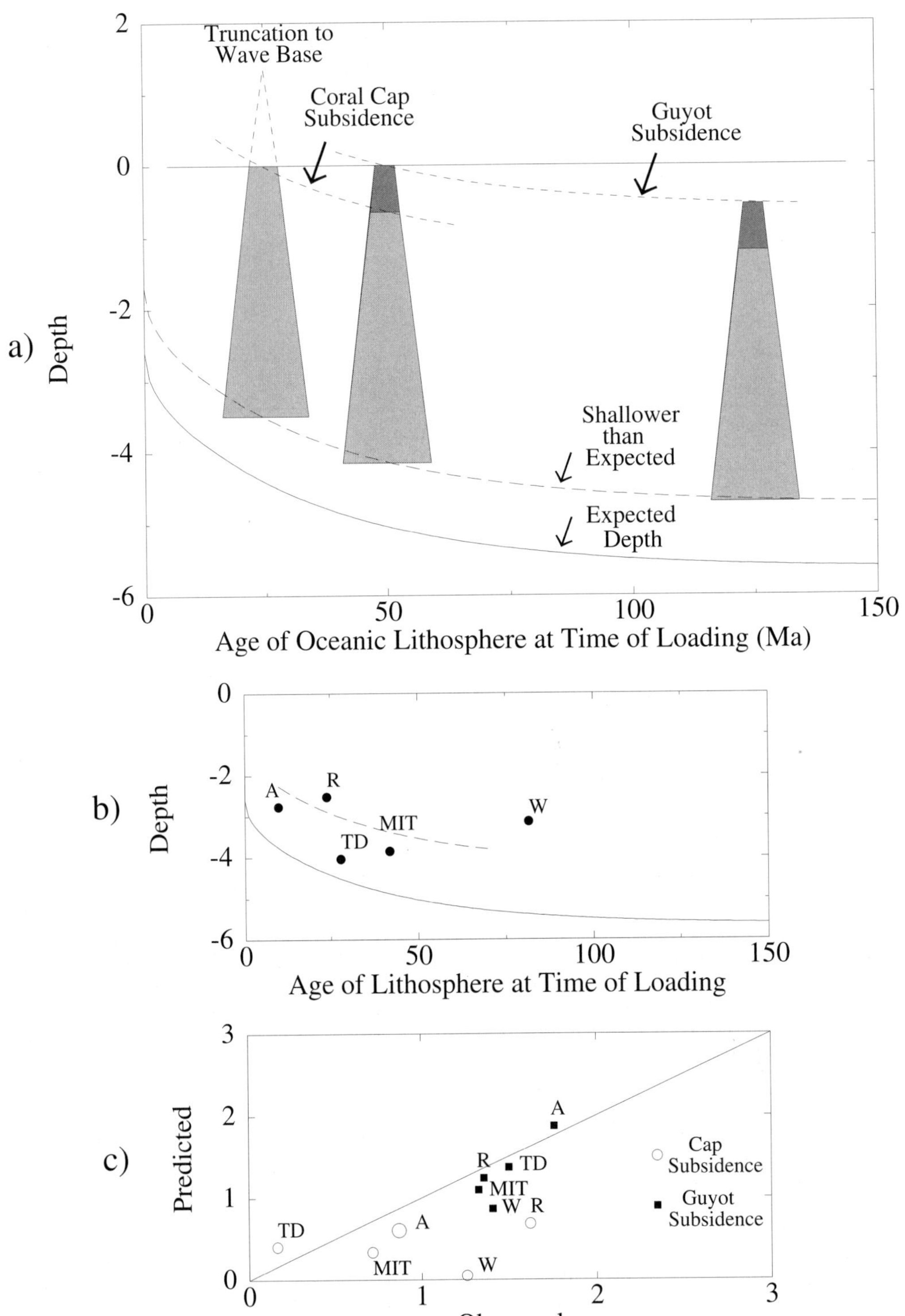
a)
Depth
Truncation to Wave Base
Coral Cap Subsidence
Guyot Subsidence
Shallower than Expected
Expected Depth
Age of Oceanic Lithosphere at Time of Loading (Ma)
b)
Depth
A
R
MIT
TD
W
Age of Lithosphere at Time of Loading
c)
Predicted
Observed
Cap Subsidence
Guyot Subsidence
TD
A
MIT
W
R

Droxler, 1996). The mechanism that could cause such a flat surface is not clear, but it has been speculated that the basements of some atolls have been elevated and truncated to sea level prior to accumulating their carbonate platforms. Another possibility is that plate motions cause some guyots, which already have a flat surface, to drift into warmer seas and, as a result, acquire a carbonate platform.

Eventually, both guyots and atolls subside and the question arises of the origin of the vertical movements associated with the drowning of these once high active volcanoes. Some of the movements can be attributed to loading subsidence, but the movements that provide the accommodation space for coral reefs to accumulate persist for much longer periods of time (up to 30 Ma) than can be explained by this mechanism. It is well known that once created at a mid-ocean ridge, oceanic crust subsides as it cools and increases in density with age. Islands that are emplaced on the oceanic crust should therefore subside with it, as the sea floor ages. This is true irrespective of whether individual islands are emplaced on thinner or thicker than normal oceanic crust. A surprising result then is that, despite large differences in the age of the underlying oceanic crust, the depth to the top of many western Pacific guyots is so similar (~1.5 km). Menard (1964) constructed a contour map of the difference in depth between the surrounding sea floor and the guyot top (i.e. guyot relief) and showed that a number of the guyots must have formed on unusually shallow sea floor. Menard did not take into account the thickness of any coral cap on these guyots. However, if he had then the area of the shallow region would have been even larger. As it was, he was able to show that the shallow region extended over a vast area of the western Pacific which he termed the Darwin Rise.

Figure 16.9a shows a simple model of guyot formation. The model assumes that a once high island is emplaced on unusually shallow sea floor and is then truncated to sea level. After some time, the island subsides along with the underlying oceanic crust and, because it is located in warm waters, a carbonate cap grows. This phase of subsidence is referred to as the cap subsidence. Initially, carbonate accumulation keeps pace with subsidence and an atoll forms. Eventually, however, carbonate growth is terminated and the once high island subsides below sea level. This phase of subsidence is referred to as the guyot subsidence. Why so many drowned guyots with carbonate caps exist in the coral seas is difficult to explain. It may be because the coral cap was unable to keep pace with subsidence, or a global sea-level rise killed off the carbonate, or some combination of these factors may have applied. Alternatively, plate motions caused the guyot to drift into colder seas where coral was unable to accumulate.

A few of the western Pacific guyots (i.e. Resolution, Allison, MIT, Wodejebato (formerly Sylvania) and Takuko-Daisan) that were studied by Menard have now been drilled by the Ocean Drilling Project (ODP) (Plate 16.1) (Premoli Silva *et al.*, 1995; Winterer and Sager, 1995). The cap and guyot subsidence at these guyots (Table 16.1) can therefore be compared to predictions based on simple thermal cooling models. Two main results emerge from the comparison: (a) the guyots formed on sea floor that was shallower by about 1.5 km than predicted, and (b) the form of the subsidence during cap and guyot formation is generally similar to that expected for the age of the underlying oceanic crust. There is a suggestion at Wodejebato guyot that the observed subsidence during cap and guyot formation is *greater* than expected for the age of the underlying sea floor.

A greater than expected subsidence has already been suggested by Detrick and Crough (1978) for Bikini and Enewetak, a pair of atolls some 250 km to the west of Wodejebato guyot. They argue that the oceanic lithosphere that underlies these atolls was thermally rejuvenated (i.e. thinned and uplifted) at the time of volcano emplacement such that its subsidence was 're-set' to that of much younger lithosphere. Subsequently, Lincoln and Schlanger (1991) showed that a thermal

Figure 16.9 *Simple model for the formation of a guyot with a coral cap. (a) A once high island formed on shallower than expected oceanic crust was truncated to sea level and then subsided like normal oceanic crust. (b) Comparison of observed and predicted subsidence during the coral cap (open circles) and subsequent guyot (filled squares) phases of subsidence. Note that the coral cap and guyot subsidence for Allison (A), MIT and Takuyo Daisan (TD) guyots is similar to that of a mid-oceanic ridge. The cap and guyot subsidence for Wodejebato (W) guyot differs, however, by being generally greater than expected. Resolution (R) guyot appears to have a greater than expected coral cap subsidence but a similar to expected guyot subsidence. (c) Comparison of 'observed' pedestal depths (dashed line) to calculated depths based on thermal modelling (solid line). All the guyots show shallower than expected depths, with Wodejebato showing the greatest discrepancy (2 km) and Allison, MIT and Takuyo Daisan the least (0.8 km)*

Table 16.1 *Summary of guyot data used to construct Figure 16.9*

Guyot name	Age of seafloor (Ma)	Age of volcano (Ma)	Age of top of coral cap (Ma)	Guyot depth g_d (km)	Cap thickness c_t (km)	Regional depth r_d (km)
Resolution	150	126	98	1.36	1.62	5.50
Allison	122	112	98	1.76	0.87	5.50
MIT	160	118	97	1.33	0.72	5.90
T. Daisan	141	113	97	1.50	0.17	5.70
Wodejebato	156	74.5	66	1.41	1.26	5.80

Pedestal height = $r_d - g_d - c_t$.

re-set age of 25 Ma accounts quite well for the cap subsidence at Bikini and Enewetak. Thermal rejuvenation requires that a large amount of heat is rapidly added to the base of the lithosphere and Detrick and Crough (1978) proposed that the most likely source of such heat was a hot spot in the underlying mantle. Crough (1978) showed that hot-spot-generated islands such as Hawaii, Cape Verdes and Bermuda – despite being located on different oceanic crust – were associated with topographic swells, each of which has a minimum depth of about 4.2 km. This is the approximate depth of 25 Ma oceanic crust which is the same age as deduced from the subsidence studies.

Despite these observations, questions remain concerning the extent of thermal rejuvenation at oceanic islands. Not all hot-spot-generated islands, for example, have well-developed swells. Some (e.g. Canary Islands) have small swells while others (e.g. French Polynesia) have unusually large ones. More importantly, most of the recently drilled guyots in the western Pacific subside in a similar way to the adjacent oceanic crust, yet these guyots appear to be associated with shallower than expected depths at the time they were formed. These observations suggest a mechanism which can cause uplift that persists for long periods of time. Such mechanisms include: (a) magmatic underplating (McKenzie, 1984) which would re-thicken the oceanic crust and produce permanent uplift, and (b) some form of dynamically maintained flow (Gurnis, 1990) which is capable of producing long-term subsidence and uplift of the deformable lithospheric 'lid'.

16.5 Decay

There is accumulating evidence that oceanic islands are associated with gravity driven processes that move material downslope. Landslides have been recognized as a major contributor to these downslope movements, with single events capable of transporting material hundreds and sometimes thousands of kilometres. Evidence of landsliding on volcanoes has been known for some time, but it was not until the 1980s, when sonar surveys were first carried out on the submerged flanks of oceanic islands, that the full extent of the phenomenon became clear. Some of the best examples of submarine slides are those described by Moore *et al.* (1989) from the southeastern end of the Hawaiian Ridge. Side-scan sonar data has revealed two types of slide which Moore *et al.* (1989) termed 'slumps' and 'debris avalanches'. The slumps are wide, with transverse ridges and steep toes, and were inferred to be slow moving. The debris avalanches, in contrast, are long compared to their width, comparatively thin, and thought to be fast moving. The volumes of individual Hawaiian slides are huge, ranging up to 1000 km^3, which is much greater than the 2.5 km^3 associated with volcanoes on continental crust such as Mount St Helens.

Figure 16.10 *Topography of Tenerife, Canary Islands. (a) Location map; the heavy box shows the location of the detailed topography maps in (b) and (c); GC, Gran Canaria; H, Hierro; LP, La Palma; L, Lanzaroti. Bathymetric contours (dotted lines) are at 0.5 km intervals. The shaded region to the north of Tenerife shows the extent of the Orotava/Icod debris avalanche. (b) Topography of Tenerife based on military maps onshore and swath bathymetry data offshore. The labels show the locations of the Orotava, Icod and Guimar landslide valleys onshore, and the submarine slide offshore. The thick grey dashed lines show the three-element 'Mercedes' pattern of recent rifts on Tenerife according to Carracedo (1994). The arrows indicate the directions of the three perspective plots shown in Plate 16.2. Contour interval is 0.25 km; annotation interval is 1 km. (c) Reconstruction showing the topography of Tenerife prior to sliding*

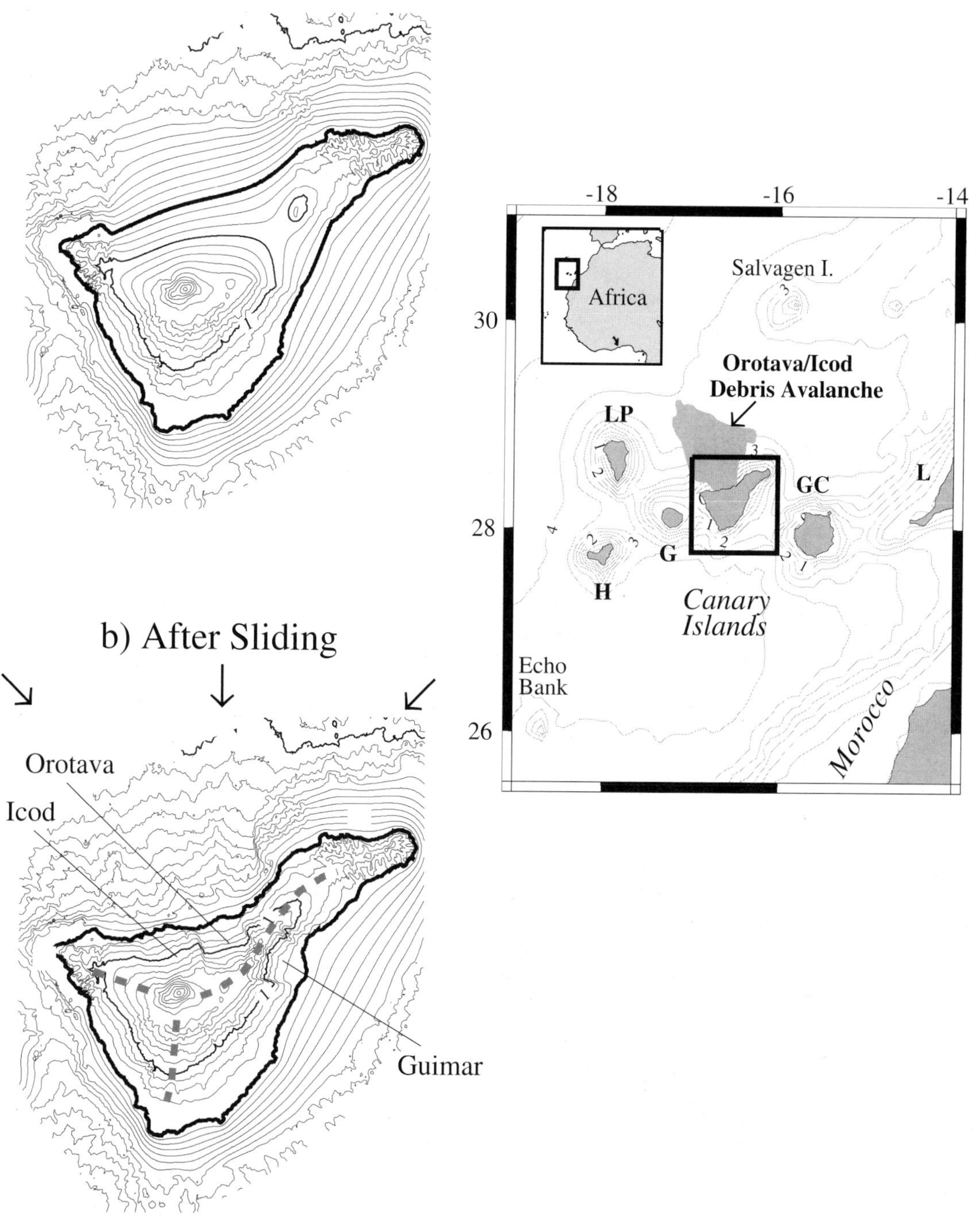
a) Before Sliding
b) After Sliding
Orotava
Icod
Guimar
-18
-16
-14
30
28
26
Africa
Salvagen I.
Orotava/Icod
Debris Avalanche
LP
GC
L
G
H
Canary
Islands
Echo
Bank
Morocco

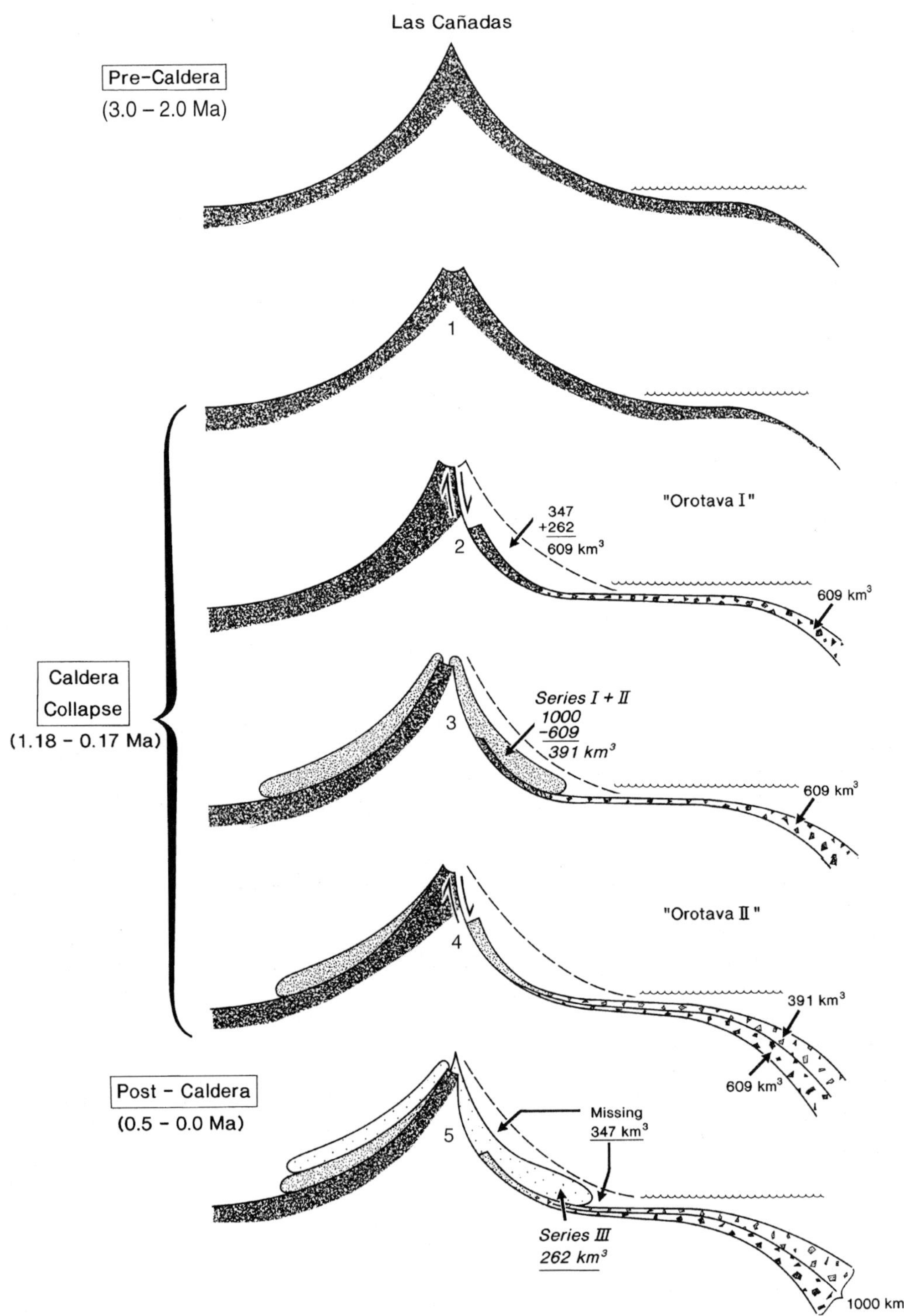
Las Cañadas
Pre-Caldera
(3.0 – 2.0 Ma)
1
Caldera
Collapse
(1.18 - 0.17 Ma)
"Orotava I"
347
+262
609 km³
2
609 km³
Series I + II
1000
-609
391 km³
3
609 km³
"Orotava II"
4
391 km³
609 km³
Post - Caldera
(0.5 - 0.0 Ma)
Missing
347 km³
5
Series III
262 km³
1000 km

Following the Hawaiian discoveries, giant slides have been identified on the submarine flanks of Réunion in the Indian Ocean (Lenat *et al.*, 1990) and the Canary Islands (Watts and Masson, 1995; Masson, 1996) in the Atlantic. An example from Tenerife, Canary Islands, is shown in Plate 16.2. Swath bathymetry data show a relatively deep region of more subdued topography than the surrounding, often dissected, slope regions which Watts and Masson (1995) have interpreted as a region of submarine sliding. As it is traced downslope, the region of subdued topography becomes broader, shallower and more sinuous. The subdued topography has a distinct flow fabric as is evidenced by longitudinal ridges and valleys and Watts and Masson 1995) have interpreted the region as a debris avalanche. Side-scan sonar data generally support these interpretations, and show, in addition, a characteristic speckled back-scatter pattern in the eastern part of the debris avalanche region which reflects the blocky nature of the sea floor. The lack of a speckled pattern in the data in the western part suggests that the blocks here may have been covered by a pelagic drape. Therefore, the sonar data suggest that the debris avalanche may have been caused by more than one individual event.

The debris avalanches on the north flank of Tenerife are located downslope of the Orotava and Icod valleys (Plate 16.2 and Figure 16.10) which Spanish geologists have long considered to be of landslide origin. The sub-aerial valleys are partly bounded by a steep topographic scarp, and it seems likely that these scarps are the source of one or more of the slides which initiated the debris avalanches mapped offshore. K–Ar dating of basaltic lavas (Ancochea *et al.*, 1990) from scarps in the Icod valley suggest an age as young as 0.17 Ma for the sliding; the Orotava valley, however, appears to be older (~0.37 Ma BP). These ages for the slides are similar to the ages of volcanigenic turbidites recovered from ODP drilling in the Madeira Abyssal Plain (Masson and Watts, 1995). These turbidites, which have been geochemically sourced to the Canary Islands, attest to the ability of oceanic island slides to transport material over vast distances across the sea floor.

As Watts and Masson (1995) have pointed out, the volume of material that is now in Tenerife's debris avalanche is much greater than the amount that is missing in the sub-aerial and submarine slide region. The volume of material in the avalanche is estimated to be 1000 km^3, compared to a combined total of 350 km^3 for the volume of rock missing from the sub-aerial (70 km^3) and submarine slide (280 km^3) region. These differences are reduced if account is taken of the fact that the landslide valleys are partially infilled by basalts that represent flank eruptions of Las Cañadas, the most recent volcano on Tenerife (Figure 16.11). It is estimated, based on the cross-sections of Coello (1973), that the total volume of these basalts is about 260 km^3. This increases the amount of missing rock involved in the slide from 350 km^3 to 610 km^3, but this is still only two thirds of the total rock volume in the debris avalanche. The most likely cause of this discrepancy is that the landslide valleys were filled by an earlier series of basalts (the Series I and II basalts) which have since slid out. At present, the older basalts are only preserved in the Tigaiga massif, which separates the Orotava and Icod landslide valleys. The massif therefore appears to be a 'remnant' which escaped the main collapse of the north flank of Tenerife and so, in the process, reveals a more or less complete stratigraphic sequence.

The replenishment of landslide valleys by magmatic material, and the transport of large volumes of material from oceanic islands into the deep sea by mass wasting processes, has widespread environmental implications. Moore *et al.* (1989), for example, has discussed the possibility that sub-aerial and submarine sliding could generate giant waves that could have impacts on shorelines some distance away. They cite as an example a gravel deposit on Lanai in the Hawaiian Islands which is presently 375 m above sea level and may have been deposited during a giant wave that was generated by a landslide on a neighbouring island in the chain.

Other consequences of landsliding are their effects on human settlements. Unfortunately, any attempt to assess the potential for sliding on an oceanic island is

Figure 16.11 *Schematic diagram showing the interplay between magma supply and landsliding on the north flank of Tenerife. The diagram shows five main stages. Stage 1: Initial construction of the Las Cañadas volcano and caldera. Stage 2: Development of the initial landslide scar which removed the north wall of the caldera. Stage 3: Partial infilling of the landslide valleys by an early basaltic flow (Series I + II basalts?). Stage 4: Sliding out of the basaltic flow. Stage 5: Partial infilling of the landslide scar by later basalt flows (Series III basalts?). The volume of the initial scar has been estimated from the combined volume of the missing topography and the volume of the Series III basalts. The volume of the early basaltic flow has been estimated from the difference between the volume of the initial scar and the total volume of the submarine flow*

complicated because of the lack of geochronological data on the occurrence and frequency of previous slides. However, the data that are available suggest that giant landslides are relatively rare events, with islands such as Tenerife (Masson and Watts, 1995) experiencing only a few such events during the past 1 Ma. What is perhaps clearer is the region of a volcano flank where the next slide may occur. The crests of many islands (and some seamounts) are characterized by a three-component 'Mercedes-type' pattern of rift-type structures along which there is evidence of recent volcanism. Landslides appear to develop within those segments of the volcano flank which are defined by rift zones. Within these 'segments' some parts of the volcano flank have failed in the past, but others have not and so may do so in the future.

16.6 Discussion

It should be clear from the discussion thus far that a wide range of geomorphic processes occur on oceanic islands which include wave cutting, landslides, debris avalanches and fluvial erosion. These processes act in the presence of certain tectonic movements that arise from, for example, thermal contraction and uplift of the underlying oceanic crust and volcanic loading. Oceanic islands are therefore a 'natural laboratory' to study the interactions between geomorphic and plate tectonic processes. The problem is in separating the role of tectonics from the other factors that may influence island landforms such as climate, lithology, magma supply and, especially, global sea-level change.

It has already been pointed out that it is relatively easy to separate tectonic effects from global sea-level change at some islands. For example, in the Huon Peninsula, New Guinea, the tectonic signal can be relatively easily separated from global sea-level change by the staircase pattern of coral terraces along the coast. Unlike New Guinea, Hawaii is located in the interior of a plate. However, absolute plate motions are sufficiently fast (>100 mm a^{-1}) that the flexural effects of a new load (e.g. Hawaii) will be spread out along an island chain, as is illustrated in Figure 16.6. Plate motions here, therefore, give rise to a simple pattern in the uplift and subsidence history of islands 'upstream' of Hawaii which can be relatively easily separated into their global sea level and tectonic components.

But what of other islands, such as Tenerife, that are located on the slow-moving African plate? One advantage of such a setting is that the flexural effects of a new load will be focused on nearby pre-existing volcanoes rather than being spread out along an island chain as they are at Hawaii. Repeated volcanism could therefore lead to excessive amounts of subsidence of nearby pre-existing volcanoes because they remain longer in the flexural moat of the new load. Tenerife comprises three relatively old (~3–8 Ma BP) volcanoes, which originally may have formed distinct islands, and the relatively young (<2 Ma BP) Las Cañadas volcano. This long history of volcanic loading is probably the main reason that today Tenerife displays such a wide variety of landforms.

Several observations need to be made, however, before there can be any attempt to quantify the relationship between landforms and tectonics on Tenerife. First, the young part of the island stands some 2–4 km above sea level, in contrast to the older volcanoes on the island, Anaga and Teno, which are now only about 0.5 km high. Second, there is evidence that only the high, young, part of the island is associated with landslides. Third, the old part of the island is deeply 'gullied' by fluvial erosion. Finally, seismic data suggest that the volcanic load of Tenerife has flexed the underlying oceanic crust downwards by about 2–3 km below the expected depth of >140 Ma old sea floor.

A schematic model that attempts to relate the plate tectonic and geomorphic development of Tenerife is illustrated in Figure 16.12. For simplicity, only four stages are shown. The first stage corresponds to the shield building phase when two of the oldest volcanoes on Tenerife (Teno and Anaga) stand high because they have not been on the lithosphere long enough for any stress relaxation to have set in. At this time, giant landslides develop on the unbuttressed flanks of the volcanoes, initiating debris avalanches which transport large amounts of material downslope and into the adjacent deep ocean. The removal of this material unloads the volcano and may result in uplift. In the second stage fluvial erosion sculptures the volcano into a complex pattern of radiating ridges and troughs. A few million years later, during the third stage, stress relaxation sets in as the elastic thickness of the lithosphere decreases from its short-term to its long-term value. The resulting decrease in the strength of the plate causes the volcano to sink into its own depression and the sub-aerially eroded surface of the volcano to subside. Since part of the subsidence occurs below sea level, the gullies will be flooded and submarine canyons will be formed. Once the relaxation is completed there should be no further reduction in the strength of the plate. Wave action may therefore start to widen the shelf and oversteepen the remainder of the sub-aerially

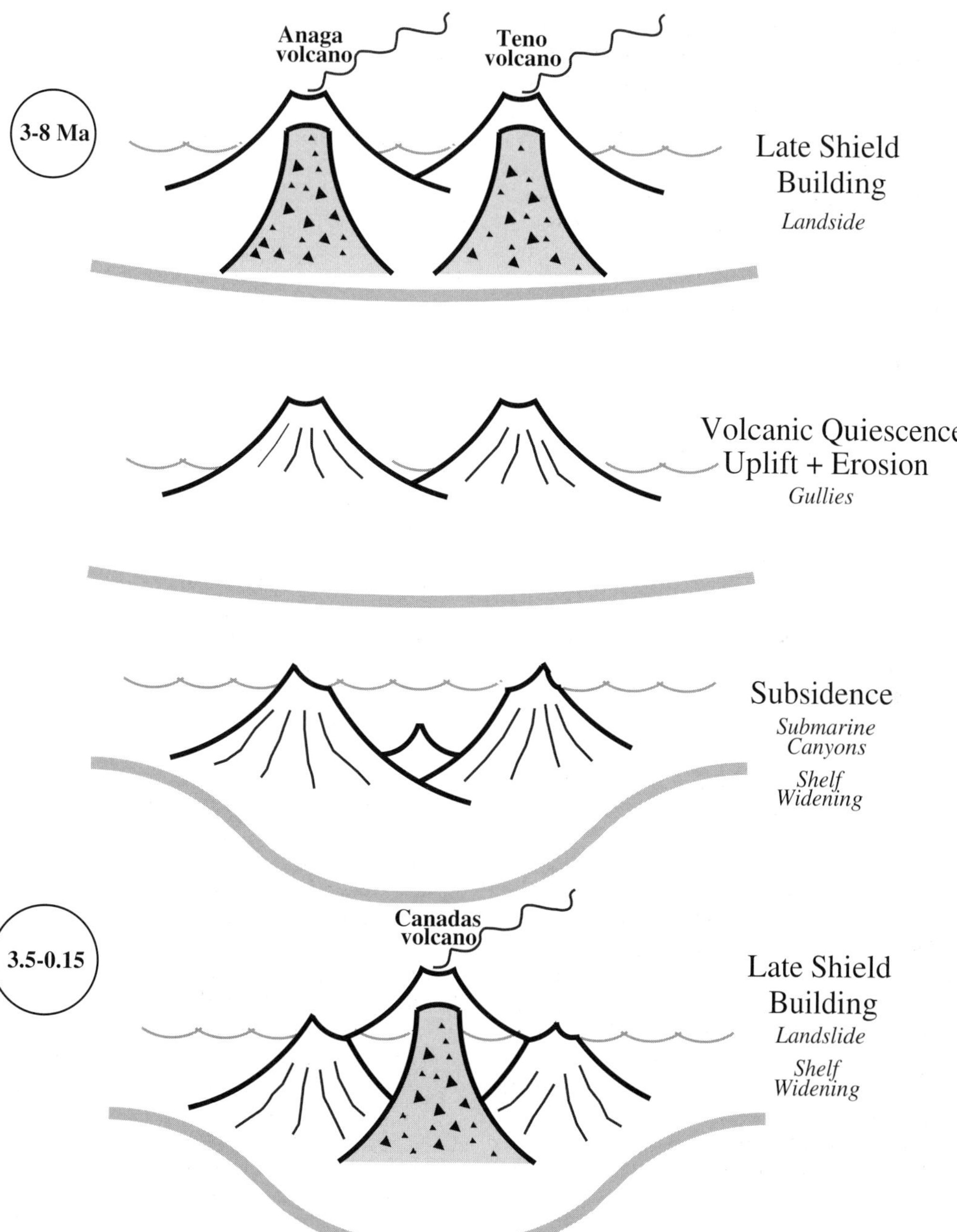

Figure 16.12 *Simple model showing the relationship between the morphological development of Tenerife and tectonics based on Watts and Masson (1995). The model shows the development of the island at four stages in the past: a late shield building phase when the Anaga and Teno volcanoes were constructed, an uplift and erosion phase when the flanks of these volcanoes were carved into 'gullies', a subsidence phase when the gullies were transformed into submarine canyons and, a late shield building phase when Las Cañadas volcano was constructed. Ages are only approximate. Debris avalanches, it is believed, are associated with the late shield building phase of oceanic islands. The thick grey line schematically indicates the response of the lithosphere to volcano loading at each stage*

eroded ridge and valley topography. The fourth stage corresponds to the present day and hence the late stages in the development of Las Cañadas, the most recent volcano on Tenerife. A giant landslide has developed on the unbuttressed northern flank of the volcano. The landslide extends downslope from the regions of slope failure to the deep sea and in the process infills the submarine canyons of the sub-aerially eroded flanks of the older volcanoes.

Simple thermal and mechanical models suggest that a volcano the size of Tenerife which has an initial short-term thickness of 100 km and a long-term thickness of 20 km would experience a *maximum* loading subsidence of about 2.5 km. This estimate is in excellent agreement with observations. For example, sonar data suggest that the submarine canyons extend to water depths of as much as 2.5 km. Also, seismic data suggest that about 2–3 km of flexure has occurred in the vicinity of the Canary Islands.

It should be emphasized that the model in Figure 16.12 is probably oversimplified. Apart from seismic and sonar data, we know of no other evidence that the older volcanoes on Tenerife have undergone such large amounts of loading subsidence. We also know of no direct evidence that Las Cañadas will experience large amounts of loading subsidence in the future. The evidence that is available is conflicting. Other islands in the Canary Islands (e.g. Lanzarote) (Driscoll *et al.*, 1965) appear, for example, to be characterized by uplift rather than subsidence.

By considering Tenerife, an oceanic island located in the centre of a tectonically stable slow-moving plate, there may have been bias in the assessment of the importance of tectonics in controlling landforms. Nevertheless, the model presented here, it is believed, is a useful first step to understanding the interplay between geomorphic and plate tectonic processes in an oceanic island setting. Additional sonar data on the submarine flanks of oceanic islands, together with the development of more refined thermal and mechanical models, offer the most promise of addressing this problem in the future.

Acknowledgements

I would like to thank Marc Audet, Jenny Collier, the late Keith Cox, Rupert Dalwood, Tim Henstock, Niels Hovius, Hugh Jenkyns, Doug Masson and Christine Peirce for helpful discussions on different aspects of ocean island structure and evolution. I also thank the US Office of Naval Research who provided much of the initial support for the study of Pacific seamounts and the US National Science Foundation (OCE-07655) and the UK Natural Environment Research Council (GR3/8554) for their support of the Hawaiian and Canary islands seismic experiments, respectively. Finally, I would like to acknowledge the late Bill Menard whose book on *Islands* was a major stimulus for this chapter.

References

Ancochea, E., Fúster, A.M., Ibarrola, E., Cendrero, A., Coello, J., Hernan, F., Cantagrel, J.M. and Jamond, C. 1990. Volcanic evolution of the island of Tenerife (Canary Islands) in the light of new K–Ar data. *J. Volcan. Geotherm. Res.*, **44**, 231–248.

Aubert, O. and Droxler, A.W. 1996. Seismic stratigraphy and depositional signatures of the Maldive Carbonate system (Indian Ocean). *Mar. Petrol. Geol.*, **13**, 503–536.

Barone, A. and Ryan, W.B.F. 1990. Single plume model for asynchronous formation of the Lamont seamounts and adjacent East Pacific Rise terrain. *J. Geophys. Res.*, **95**, 10 801–10 827.

Batiza, R. 1982. Abundances, distribution and sizes of volcanoes in the Pacific Ocean and implications for the origin of non-hotspot volcanoes. *Earth Planet. Sci. Lett.*, **60**, 195–206.

Bloom, A. 1967. Pleistocene shorelines: A new test of isostasy. *Geol. Soc. Amer. Bull.*, **78**, 1477–1494.

Bodine, J.H., Steckler, M.S. and Watts, A.B. 1981. Observations of flexure and the rheology of the oceanic lithosphere. *J. Geophys. Res.*, **86**, 3695–3707.

Caress, D.W., McNutt, M.K., Detrick, R.S. and Mutter, J.C. 1995. Seismic imaging of hotspot-related crustal underplating beneath the Marquesas Islands. *Nature*, **373**, 600–603.

Carracedo, J.C. 1994. The Canary Islands: an example of structural control on the growth of large oceanic-island volcanoes. *J. Volcano. Geotherm. Res.*, **60**, 225–241.

Chappell, J. 1974. Geology of coral terraces, Huon Peninsula, New Guinea: a study of Quaternary tectonic movements and sea level changes. *Geol. Soc. Amer. Bull.*, **85**, 553–570.

Coello, J. 1973. Las series volcánicas en subsuelos de Tenerife. *Estud. Geol.*, **29**, 491–512.

Coulbourn, W.T., Campbell, J.F. and Moberly, R. 1974. Hawaiian submarine terraces, canyons, and Quaternary history evaluated by seismic reflection profiling. *Mar. Geol.*, **17**, 215–234.

Crough, S.T. 1978. Thermal origin of mid-plate hot-spot swells. *Geophys. J. R. Astr. Soc.*, **55**, 451–469.

Darwin, C. 1842. *The Structure and Distribution of Coral Reefs*, Smith Elder, London.

Detrick, R.S. and Crough, S.T. 1978. Island subsidence, hot spots, and lithospheric thinning. *J. Geophys. Res.*, **83**, 1236–1244.

Detrick, R.S., Sclater, J.G. and Thiede, J. 1977. The subsidence of aseismic ridges. *Earth Planet. Sci. Lett.*, **34**, 185–196.

Driscoll, E.M., Hendry, G.L. and Tinkler, K.J. 1965. The geology and geomorphology of Los Ajaches, Lanzarote. *Geol. J.*, **4**, 321–334.

DuBois, J., Launay, J. and Recy, J. 1973. Uplift movements in New Caledonia – Loyalty Islands area and their plate tectonics interpretation. *Tectonophysics*, **24**, 133–150.

Fornari, D.J., Ryan, W.B.F. and Fox, P.J. 1984. The evolution of craters and calderas on young seamounts: Insights from Sea MARC I and SeaBeam sonar surveys of a small seamount group near the axis of the East Pacific Rise at ~10°N. *J. Geophys. Res.*, **89**, 11 069–11 084.

Fornari, D.J., Garcia, M.O., Tyce, R.C. and Gallo, D.G. 1988. Morphology and structure of Loihi seamount based on SeaBeam sonar mapping. *J. Geophys. Res.*, **93**, 15 227–15 238.

Fridriksson, S. 1975. *Surtsey: Evolution of Life on a Volcanic Island.* Butterworths, London.

Funck, T. 1995. Structure of the volcanic apron north of Gran Canaria deduced from reflection, seismic, bathymetric and borehole data. PhD thesis, University of Kiel.

Gripp, A.E. and Gordon, R.G. 1990. Current plate velocities relative to the hotspots incorporating the NUVEL-1 global plate motion model. *Geophys. Res. Lett.*, **17**, 1109–1112.

Gurnis, M. 1990. Ridge spreading, subduction, and sea level fluctuations. *Science*, **250**, 970–972.

Hanks, T.C. 1971. The Kuril trench–Hokkaido rise system: Large shallow earthquakes and simple models of deformation. *Geophys. J. R. Astr. Soc.*, **23**, 173–189.

Harrison, C.G.A., Jarrard, R.O., Vacquier, V. and Larson, R.L. 1975. Palaeomagnetism of Cretaceous Pacific seamounts. *Geophys. J. R. Astr. Soc.*, **42**, 859–882.

Hess, H.H. 1946. Drowned ancient islands of the Pacific basin. *Amer. J. Sci.*, **244**, 772–791.

Holcomb, R.T. and Searle, R.C. 1991. Large landslides from oceanic volcanoes. *Mar. Geotech.*, **10**, 19–32.

Ito, G., McNutt, M.K. and Gibson, R.L. 1995. Crustal structure of the Tuamotu Plateau, 15°S, and implications for its origin. *J. Geophys. Res.*, **100**, 8097–8114.

Jones, A.T. 1992. Holocene coral reef on Kauai, Hawaii: evidence for a sea-level highstand in the central Pacific. *Soc. Sediment. Geol. Spec. Publ.*, **48**, 267–271.

Larson, R.L. and Schlanger, S.O. 1981. Cretaceous volcanism and Jurassic magnetic anomalies in the Nauru Basin, western Pacific Ocean. *Geology*, **9**, 480–484.

Leeds, A.R. 1975. Lithospheric thickness in the western Pacific. *Phys. Earth Planet. Interiors*, **11**, 61–64.

Lenat, J.-F., Bachelery, P., Bonneville, A., Galdeano, A., Labazuy, P., Rousset, D. and Vincent, P. 1990. Structure and morphology of the submarine flank of an active basaltic volcano: Piton de la Fournaise (Réunion island, Indian ocean). *Oceanologica Acta, Special*, **10**, 211–224.

Lincoln, J.M. and Schlanger, S.O. 1991. Atoll stratigraphy as a record of sea level change: problems and prospects. *J. Geophys. Res.*, **96**, 6727–6752.

Masson, D.G. 1996. Catastrophic collapse of the volcano island of Hierro 15 ka ago and the history of landslides in the Canary Islands. *Geology*, **24**, 231–234.

Masson, D.G. and Watts, A.B. 1995. Slope failures and debris avalanches on the flanks of volcanic oceanic islands – the Canary Islands, off NW Africa. *Landslide News*, **9**, 21–24.

McGuire, B. and Saunders, S. 1993. Recent earth movements at active volcanoes: A review. *Quat. Proc.*, **3**, 33–46.

McKenzie, D.P. 1984. A possible mechanism for epeirogenic uplift. *Nature*, **307**, 616–618.

McNutt, M. and Menard, H.W. 1978. Lithospheric flexure and uplifted atolls. *J. Geophys. Res.*, **83**, 1206–1212.

Menard, H.W. 1964. *Marine Geology of the Pacific*, McGraw-Hill, New York.

Menard, H.W. 1984. Origin of guyots: The *Beagle* to *Seabeam. J. Geophys. Res.*, **89**, 11 117–11 123.

Menard, H.W. 1986. *Islands*, W.H. Freeman, Oxford.

Moore, J.G. 1970. Relationship between subsidence and volcanic load, Hawaii. *Bull. Volcanol.*, **34**, 562–576.

Moore, J.G. 1987. Subsidence of the Hawaiian Ridge. *U.S. Geol. Surv. Prof. Pap.*, **1350**, 85–100.

Moore, J.G., Clague, D.A., Holcomb, R.T., Lipman, P.W., Normark, W.R. and Torresan, M.E. 1989. Prodigious submarine landslides on the Hawaiian Ridge. *J. Geophys. Res.*, **94**, 17 465–17 484.

Nunn, P.D. 1994. *Oceanic Islands*, Blackwell, Oxford.

Premoli Silva, L., Haggerty, J., Rack, F. *et al.* 1995. *Proc. Ocean Drilling Prog. Initial Rep.*, **144**, College Station, Texas.

Regan, J. and Anderson, D.L. 1985. Anisotropic models of the upper mantle. *Phys. Earth Planet. Interiors*, **35**, 227–263.

Sandwell, D.T. and Smith, W.H.F. 1997. Marine gravity anomaly from Geosat and ERS 1 satellite altimetry. *J. Geophys. Res.*, **102**, 10 039–10 054.

Scheirer, D.S. and Macdonald, K.C. 1995. Near-axis seamounts on the flanks of the East Pacific Rise, 8°N to 17°N. *J. Geophys. Res.*, **100**, 2239–2259.

Smith, D.K. 1988. Shape analysis of Pacific seamounts. *Earth Planet. Sci. Lett.*, **90**, 457–466.

Staudigel, H. and Schmincke, H.-U. 1984. The Pliocene seamount series of La Palma/Canary islands. *J. Geophys. Res.*, **89**, 11 195–11 215.

ten Brink, U.S. and Brocher, T. 1987. Multichannel seismic evidence for a subcrustal intrusive complex under Oahu and a model for Hawaiian volcanism. *J. Geophys. Res.*, **92**, 13 687–13 707.

ten Brink, U.S. and Watts, A.B. 1985. Seismic stratigraphy of the flexural moat flanking the Hawaiian Islands. *Nature*, **317**, 421–424.

Vogt, P.R. and Smoot, N.C. 1984. The Geisha Guyots: multibeam bathymetry and morphometric interpretation. *J. Geophys. Res.*, **89**, 10 085–11 107.

Walcott, R.I. 1970. Flexure of the lithosphere at Hawaii. *Tectonophysics*, **9**, 435–446.

Watts, A.B. 1978. An analysis of isostasy in the world's oceans: 1. Hawaiian–Emperor Seamount Chain. *J. Geophys. Res.*, **83**, 5989–6004.

Watts, A.B. and Cochran, J.R. 1974. Gravity anomalies and flexure of the lithosphere along the Hawaiian–Emperor seamount chain. *Geophys. J. R. Astr. Soc.*, **38**, 119–141.

Watts, A.B. and Masson, D.G. 1995. A giant submarine slide on the north flank of Tenerife, Canary Islands. *J. Geophys. Res.*, **100**, 24 487–24 498.

Watts, A.B., ten Brink, U., Buhl, P. and Brocher, T. 1985. A multichannel seismic study of lithospheric flexure across the Hawaiian–Emperor seamount chain. *Nature*, **315**, 105–111.

Watts, A.B. and ten Brink, U.S. 1989. Crustal structure, flexure and subsidence history of the Hawaiian Islands. *J. Geophys. Res.*, **94**, 10 473–10 500.

Watts, A.B., Peirce, C., Collier, J., Dalwood, R., Canales, J.P. and Henstock, T.J. 1997. A seismic study of crustal structure in the vicinity of Tenerife, Canary Islands. *Earth Planet. Sci. Lett.*, **146**, 431–447.

Wilson, J.T. 1963. Pattern of uplifted atolls in the main ocean basins. *Science*, **139**, 592–594.

Winterer, E.L. and Sager, W.W. 1995. Synthesis of drilling results from the mid-Pacific mountains: regional context and implications. In: E.L. Winterer, W.W. Sager, J.V. Firth and J.M. Sinton (eds) *Proc. Ocean Drilling Prog. Sci. Results*, **143**, College Station, Texas, pp. 497–535.

Woodroffe, C., McLean, R., Polach, H. and Wallensky, E. 1990. Sea level and coral atolls: late Holocene emergence in the Indian Ocean. *Geology*, **18**, 62–66.

Index